Alphonso Wood

Descriptive Botany

Being a Succinct Analytical Flora

Alphonso Wood

Descriptive Botany
Being a Succinct Analytical Flora

ISBN/EAN: 9783337270018

Printed in Europe, USA, Canada, Australia, Japan

Cover: Foto ©berggeist007 / pixelio.de

More available books at **www.hansebooks.com**

FLORA ATLANTICA.

DESCRIPTIVE BOTANY;

BEING

A SUCCINCT ANALYTICAL FLORA,

INCLUDING ALL THE PLANTS GROWING IN THE UNITED
STATES FROM THE ATLANTIC COAST TO
THE MISSISSIPPI RIVER.

FROM THE AMERICAN BOTANIST AND FLORIST.

By ALPHONSO WOOD, A.M.,
AUTHOR OF THE CLASS-BOOK OF BOTANY, ETC.

A. S. BARNES & COMPANY,
NEW YORK, CHICAGO, AND NEW ORLEANS.
1879.

WOOD'S BOTANY.

OBJECT LESSONS IN BOTANY, pp. 340, 12mo. An introduction to the Science, full of lively description and truthful illustrations; with a limited Flora, but a *complete* System of Analysis. Price (postpaid) $1.25.

THE BOTANIST AND FLORIST, pp. 620, 12mo. A thorough text-book, comprehensive and practical; with a Flora, and System of Analysis equally *complete*. "I have been deeply impressed, almost astonished, (writes Prof. A. Winchell, of University of Michigan) at the evidence which this work bears of skillful and experienced authorship—nice and constant adaptation to the wants and conveniences of students in Botany," etc. Price (postpaid) $2.25.

THE CLASS-BOOK OF BOTANY, pp. 850, 8vo. The principles of the Science more fully announced and illustrated—the Flora and Analysis complete, with all our plants portrayed in language both scientific and popular. "The whole science (writes Prof. G. H. Perkins, of Vermont University), so far as it can be taught in a college course, is well presented, and rendered unusually easy of comprehension. I regard the work as most admirable." Price (postpaid) $3.00.

THE PLANT RECORD—a beautiful book, for classes and amateurs, showing, in a few pages, how to analyze a plant—any plant, and furnishing tablets for the systematic record of the analysis. Price (postpaid) 70 cents.

FLORA ATLANTICA, or WOOD'S DESCRIPTIVE FLORA, pp. 448, 12mo. This work is equivalent to the Part IV of the Botanist and Florist, being a succinct account of all the plants growing East of the Mississippi River, both native and cultivated, with a system of analytical tables well-nigh perfect. Price (postpaid) $1.50.

WOOD'S BOTANICAL APPARATUS—a complete outfit, for the field and the herbarium. It consists of a portable trunk, a Wire Drying Press, a Knife-trowel, a Microscope, and Forceps. Price $8.00.

Copyright, 1870, 1879, by A. S. Barnes & Co.

"FOURTEEN WEEKS" IN EACH SCIENCE,

By J. Dorman Steele, Ph. D., F. G. S., Etc.

Now Ready:

| PHILOSOPHY. | PHYSIOLOGY. | ZOOLOGY. |
| CHEMISTRY. | GEOLOGY. | ASTRONOMY. |

A KEY to Practical Questions in Steele's Works,
Seven volumes, each, postpaid $1.25.

PREFACE.

THIS Flora will be found a phenomenon in brevity. Within the space of 426 duodecimo pages, in fair leaded type, we have recorded and defined nearly 4,500 species — all the known Flowering and Fern-like plants, both native and cultivated (not excepting the Sedges and Grasses), growing in the Atlantic half of the country. This conciseness has been attained, not by the omission of anything necessary to the complete definition and prompt recognition of every species, but simply by *avoiding repetitions*. In the *final* definition of the species (see, for example, *R. bulbosus*, the Bulbous Buttercup, p. 20) we give but one, two, three, rarely four lines. This cannot, of course, include its full portraiture. It includes only those few features which have not already been given elsewhere, and which *here* serve to distinguish the *R. bulbosus* from the two preceding species with which it stands grouped in the table. But the full description of *R. bulbosus* (and of every species) will nevertheless be found in the Flora. Some of its features are given under its genus, Ranunculus; some under its Order; some under its Cohort; others under its Class, its Province, and its Sub-kingdom. Moreover, all along the path of its analysis through the tables its characters are announced and recognized; so that if all the statements descriptive of *R. bulbosus* were collected, we should have nearly a half-page of text, and no important character left unnoticed.

Between the cultivated exotics and the wild native or naturalized species constituting our own flora, a distinction is made *in the type.* The names of the latter are expressed in full-face, **Roman** for the species, and *Italic* for the varieties. The names of the exotics are in SMALL CAPITALS.

The geographical limits of the present flora are the same as those adopted in the Class-Book; viz., all the States of the American Union lying east of the Mississippi River. This will necessarily include so many of the plants of the States bordering on the western shore of the Mississippi, that the book may be regarded as well adapted to those States also.

It gives me great pleasure to acknowledge my obligations to the friends whose names occur below and in many other parts of our work, for their contributions of new and rare plants, and for valuable information concerning them:—first, and especially, to Prof. THOS. C. PORTER, of Lafayette College; to E. L. HANKENSON, Newark, N. Y.; to JOHN WOLF, Canton, Ill.; to CHAS. H. PECK, Albany, N. Y.; to WM. R. GIRARD, Esq., Poughkeepsie, N. Y.; to N. COLMAN, Iowa; to Rev. J. H. CARRUTH, Kansas; to Dr. W. MATTHEWS, Dakota; to H. MAPES, Michigan, &c., &c.

And as a just tribute to the memory of my LAMENTED WIFE, I would add that whatever is new and peculiar in the plan of the present Flora, that on which its definite conciseness depends, *is due to her alone.* She first indicated the method, and for years assiduously advocated its adoption.

CUJUS NOMINI AC MEMORIÆ CARISSIMÆ,
HOC OPUS, IN MEDIO DOLORE AC DESIDERIO CONFECTUM,
DEDICAT CONJUX.

HINTS FOR COLLECTING, DRYING, Etc.

SPECIMENS for analysis and for the herbarium should have leaves, flowers, and fruit. Care should also be taken to represent the varying forms and kinds of leaf and flower. In collecting, a strong knife, or knife-trowel, is requisite for digging and cutting, and a close tin box, or trunk, fifteen inches in length and of a portable form. Enclosed in such a box, with a little moisture, specimens will remain fresh for many days.

In drying for the herbarium, much care and effort is needed in order to retain the natural aspect, form, and colors. The true secret of the art consists in extracting all the moisture before decomposition takes place.

The Drying Press, invented by us, consists of a dozen quires of unsized paper, folded 10 × 14 inches, inclosed between two stout sheets of woven wire of the same size, with folded edges, secured by several leathern straps with buckles. When in use, suspend this press in the wind and sunshine, or in rainy weather by the fire. In such circumstances specimens dry well without once changing; but if boards be used instead of the wires, the papers will require to be changed and dried daily. Succulent plants may be immersed in boiling water before pressing, to hasten their desiccation, and thick or fleshy stems and roots may be divided lengthwise for the same reason.

The Lens, either single, double, or triple, is almost indispensable in analysis. In viewing minute flowers, or parts of flowers, its use cannot be too highly appreciated. Together with the lens, a needle inserted in a handle, a penknife, and a pair of delicate forceps are required in dissection.

ABBREVIATIONS AND SIGNS.

§ BOTANICAL TERMS OFTEN RECURRING IN DESCRIPTIONS.

ach. achenia.
æst. æstivation.
alter. alternate.
amplex. amplexicaul.
anth. anther.
axill. axillary.
cal. calyx.
caps. capsule.
cor. corolla.
cyp. cypsela.
decid. deciduous.
diam. diameter.
ellip. elliptical.
emarg. emarginate.
epig. epigynous.
f. or *ft.* feet.
fil. filaments.
fl. flower; *fls.* flowers.

fr. fruit.
gl. glume; *gls.* glumes.
hd. head; *hds.* heads.
hyp. hypogynous.
imbr. imbricate.
inf. inferior.
invol. involucre.
irreg. irregular.
leg. legume.
lf. leaf; *lvs.* leaves.
lfts. leaflets.
lom. loment.
opp. opposite.
ova. ovary.
pap. pappus.
ped. peduncle.
pet. petals.
perig. perigynous.

perig. perigynium.
pls. pales.
pn. pinnæ.
pnl. pinnulæ.
recep. receptacle.
reg. regular.
rhiz. rhizoma.
rt. root.
sc. scale, scales.
sds. seeds.
seg. segment.
sep. sepals.
st. stem.
sta. or *stam.* stamens.
stig. stigmas.
sty. styles.
var. variety

§ TIMES OF FLOWERING, AND LOCALITIES.

1. Names of the Months and Seasons are abbreviated in the usual manner, as, *Jan.* January; *Apr.* April; *Spr.* Spring; *Aut.* Autumn; *Sum.* Summer; &c.
2. The names of States and Territories of the U. S. are abbreviated precisely as in other works, thus:—*Ala.* Alabama; *Ark.* Arkansas; *Conn.* Connecticut, &c.
3. Sections of States are thus designated:—*N. N. Y.* Northern New York; *W. Pa.* Western Pennsylvania; *E. Fla.* East Florida; *S. Ill.* Southern Illinois, &c.
4. Names of foreign Countries:—*Eur.* Europe; *Afr.* Africa; *S. Afr.* South Africa; *Aust.* Australia; *Can.* Canada; *Mex.* Mexico; *S. Am.* South America, &c.
5. *E.* East, Eastward, indicates the States of the Atlantic seaboard from Maine to Virginia inclusive; *N-E.* or *N. Eng.* denotes the New England States.
6. *M.* is used to denote the Middle States; viz., N. Y., Penn., N. J., and Del.
7. *N.* North, Northward, indicates generally the territory north of 42° N. latitude.
8. *N-W.* Northwest, indicates Wis., Minn., and parts of Ill. and Mich.
9. *S.* South, Southward, is used to indicate the Southern States in general,—all lying south of Virginia and Kentucky.
10. *S-W.* Southwest, viz., Miss., La., Ark., and perhaps Tennessee and Texas.
11. *W.* West, denotes the States lying due north of Tennessee and Arkansas.

ABBREVIATIONS AND SIGNS.

§ SIGNS.

① An annual Herb.
② A biennial Herb.
♃ A perennial Herb.
♄ An undershrub, deciduous.
♄ An undershrub, evergreen.
♄ A Shrub, deciduous.
♄ A Shrub, evergreen.
♄ A Tree, deciduous.
♄ A Tree, evergreen.
♄ An herbaceous Vine, ① or ②.
♄ A perennial Vine, ♃.
♄ Woody Vine, deciduous.

♄ Woody Vine, evergreen.
⌐ Trailing Herb, ① or ②.
⌐ Trailing Herb, ♃.
≈ An aquatic Plant.
☿ Flowers perfect.
♂ Flowers staminate.
♀ Flowers pistillate.
☿ Monœcious.
♂ ♀ Diœcious.
♂ ☿ ♀ Polygamous.
0 Wanting, or none.
∞ Numerous, or indefinite.

§ A Plant introduced and naturalized;
† Plant cultivated for ornament; } at the end of the description.
‡ Plant cultivated for use;

o— Cotyledons accumbent;
o| Cotyledons incumbent; } used only in the Cruciferæ. (Page 34.)
o)) Cotyledons conduplicate;

! (Note of exclamation), used technically, denotes certainty.
? (Note of interrogation), implies doubt or uncertainty.
f (with or without a period), a foot..........
′ (a single acute accent), an inch............ } after a number.
″ (a double accent), a line =1-12 of an inch..

---

§ AUTHORS' NAMES CITED IN THIS WORK.

Adans.	Adanson.	*Dill.*	Dillenius.
A. DC.	Alphonse De Candolle.	*Desv.*	Desvaux.
Ait.	Aiton.	*Dougl.*	Douglas.
All.	Allione.	*Ehrh.*	Ehrhart.
Anders.	Andersson.	*Ell.*	Elliott.
Arn.	Arnott.	*Endl.*	Endlicher.
Aub.	Aublet.	*Engel.*	Engelmann.
Bart.	Barton.	*Fisch.*	Fischer.
Bartl.	Bartling.	*F. & M.*	Fischer & Meyer.
Beauv.	Beauvois.	*Frœl.*	Frœlich.
Benth.	Bentham.	*Gært.*	Gærtner.
Bernh.	Bernhardt.	*Gmel.*	Gmelin.
Berl.	Berlandier.	*Good.*	Goodenough.
Bois.	Boissier.	*Gr.*	A. Gray.
Bong.	Bongard.	*Grev.*	Greville.
Bork.	Borkhausen.	*Griseb.*	Grisebach.
Br.	Brown.	*Gron.*	Gronovius.
Bw.	Bigelow.	*Hedw.*	Hedwig.
Cass.	Cassini.	*Hoffm.*	Hoffman.
Cav.	Cavanilles.	*Hook.*	Hooker (W. J.)
Cham.	Chamisso.	*Hook. f. (filius)*	Hooker (J. D.)
Darl.	Darlington.	*Hornem.*	Hornemann.
DC	De Candolle.	*Huds.*	Hudson. [Kunth.
Desf.	Desfontaines.	*H. B. K.*	Humboldt, Bonpland &
Dew.	Dewey.	*Jacq.*	Jacquin.

ABBREVIATIONS AND SIGNS.

AUTHORS' NAMES—(Continued).

Juss.	Jussieu.	*Richn.*	Richardson.
A. Juss.	Adrien Jussieu.	*Rœm.*	Rœmer.
L. or *Linn.*	Linnæus.	*Salisb.*	Salisbury.
Lag.	Lagasca.	*Schk.*	Schkuhr.
Lam.	Lamarck.	*Schrad.*	Schrader.
Lamb.	Lambert.	*Schreb.*	Schreber.
Ledeb.	Ledebour.	*Schult.*	Schultes.
Lehm.	Lehmann.	*Schw.*	Schweinitz.
Lesq.	Lesquereux.	*Scop.*	Scopoli.
Lestib.	Lestibudois.	*Ser.*	Seringe.
L'Hér.	L'Heritier.	*Soland.*	Solander.
Lindl.	Lindley.	*Spreng.*	Sprengel.
Mart.	Martins.	*Steud.*	Steudel.
Mich.	Micheli.	*Sulliv.*	Sullivant.
Michx. or *Mx.*	Michaux.	*Thunb.*	Thunberg.
Mx. f.	Michaux (the younger).	*Torr.*	Torrey.
Mill.	Miller.	*T. & G.*	Torrey & Gray.
Mitch.	Mitchell.	*Tourn.*	Tournefort.
Muhl.	Muhlenberg.	*Trautv.*	Trautvetter.
Nees.	Nees von Esenbeck.	*Trin.*	Trinius.
Nutt. or *N.*	Nuttall.	*Tuckm.*	Tuckerman.
Pal.	Pallas.	*Vaill.*	Vaillant.
Pav.	Pavon.	*Vent.*	Ventenat.
Pers.	Persoon.	*Vill.*	Villars.
Ph.	Pursh.	*Wahl.*	Wahlenberg.
Pluk.	Plukenet.	*Walp.*	Walpers.
Plum.	Plumier.	*Walt.*	Walter.
Poir.	Poiret.	*Wangh.*	Wangenheim.
R. Br.	Robert Brown.	*Willd.*	Willdenow.
Raf.	Rafinesque.	*With.*	Withering.
Reichenb.	Reichenbach.	*Wulf.*	Wulfen.
Rich.	Richard.		

ANALYSIS OF THE NATURAL ORDERS,

Founded on the most obvious or artificial characters: designed as a key for the determination of the Order of any plant, native, or naturalized, or cultivated, growing within the limits of this Flora.

PROVINCES, CLASSES, AND COHORTS.

Sub-kingdom I. The Flowering Plants..(See, next, Provinces 1, 2)...**PHÆNOGAMIA**.
Sub-kingdom II. The Flowerless Plants..(See the Provinces 3, 4)....**CRYPTOGAMIA**.
  Province 1. Leaves net-veined. Flowers never completely 3-parted (mostly ♂ and ♀). Embryo with 2 or more cotyledons. Wood (if any) in annual circles..(See Classes 1, 2)..................*EXOGENS*.
  Province 2. Leaves parallel-veined (rarely netted). Flowers 3-parted. Bark, wood, and pith commingled. Embryo with but one cotyledon..(See Classes 3, 4)......................................*ENDOGENS*.
  Province 3. Stem and leaves distinguishable..(**H**)...................*ACROGENS*.
  Province 4. Stem and leaves undistinguishable..(**K**).............*THALLOGENS*.
    Class 1. Stigmas present. Seeds enclosed in vessels..(*).......ANGIOSPERMS.
    Class 2. Stigmas 0. Seeds naked (Pines, Firs, Cedars,&c.) (**)..GYMNOSPERMS.
    Class 3. Flowers without glumes. Perianth colored or green..(†)..PETALIFERÆ.
    Class 4. Flowers with green alternate glumes. No perianth..(††)..GLUMIFERÆ.
      * Cohort 1. Corolla with the petals distinct..(**A**)............**Polypetalæ**.
      * Cohort 2. Corolla with the petals united..(**B**)............**Gamopetalæ**.
      * Cohort 3. Corolla none. Calyx often none..(**C**)..................**Apetalæ**.
      ** Cohort 4. The cone-bearing plants (same as Class 2)..(**D**).......**Conoids**.
      † Cohort 5. Fls. on a spadix, apetalous or incomplete..(**E**)...**Spadicifloræ**.
      † Cohort 6. Flowers complete, with a true perianth..(**F**).......**Florideæ**.
      †† Cohort 7. The Grasses, Grains, &c. (same as class 4)..(**G**)..**Graminoids**.

A. Cohort 1. POLYPETALOUS EXOGENS.

* Herbs with the leaves alternate or all radical..(12)
* Herbs with the leaves opposite on the stem..(9)
* Shrubs, trees, or undershrubs..(2)
  2 Flowers regular or nearly so..(3)
  2 Flowers irregular (or the fruit a legume) (§ 165)..(*r*)
    3 Polyandrous,—stamens 3–10 times as many as the petals..(4)
    3 Oligandrous, stamens 1–2 times as many as the petals or fewer..(6)

ANALYSIS OF THE NATURAL ORDERS.

 4 Leaves opposite..(*s*)
 4 Leaves alternate..(5)
  5 Stamens on the torus or the hypogynous corolla..(*t*)
  5 Stamens and petals on the calyx tube..(*v*)
 6 Ovaries simple, distinct, or one only. Vines or erect shrubs..(*w*)
 6 Ovary compound, and wholly adherent to the calyx..(*x*)
 6 Ovary compound and free from the calyx or nearly so..(7)
  7 Stamens opposite to the petals and of the same number..(*y*)
  7 Stamens alternate with the petals or of a different number (8)
   8 Leaves opposite on the stems..(*z*)
   8 Leaves alternate, and compound..(*yy*)
   8 Leaves alternate and simple..(*zz*)
 9 Polyandrous—stamens 3—10 times as many as the petals..(*m*)
 9 Oligandrous,—stamens 1—2 times as many as the petals or fewer..(10)
  10 Pistils separate and distinct, few or solitary, simple..(*n*)
  10 Pistils united into a compound ovary free from the calyx..(11)
  10 Pistils united into a compound ovary adherent to the calyx..(*o*)
   11 Stamens opposite to the petals and of the same number..(*p*)
   11 Stamens alternate with the petals or of a greater number..(*q*)
 12 Flowers regular or nearly so. Fruit never a legume..(14)
 12 Flowers irregular (rarely regular and the fruit a legume)..(13)
  13 Stamens numerous, 3 or more times as many as the petals..(*k*)
  13 Stamens few and definite, 4—12..(*l*)
   14 Stamens (or anthers) 3—10 times as many as the petals..(15)
   14 Stamens few and definite. Ovary free from the calyx..(17)
   14 Stamens few and definite. Ovary adherent to the calyx..(*j*)
 15 Stamens hypogynous—inserted on the torus..(16)
 15 Stamens perigynous—inserted on the corolla at the base..(*c*)
 15 Stamens perigynous—inserted on the calyx at the base..(*d*)
  16 Pistils few or many, distinct (at least as to the styles)..(*a*)
  16 Pistils (and styles if any) completely united..(*b*)
 17 Pistils *one*, or indefinite and distinct, simple..(*e*)
 17 Pistils definitely—* 2 united, the short styles combined into *one*..(*f*)
   —* 2, 3 or 4 united, styles or stigmas, 2, 3, 4 or 6..(*g*)
   —* 5, distinct or united, with 5 distinct styles..(*h*)
   —* 5, united and the styles also combined into *one*..(*i*)

a Petals 5 or more, deciduous. Leaves never peltate..................RANUNCULACEÆ.
a Petals 3 or numerous. Water plants with peltate leaves.... }
 b Sepals 4—6, equal. Petals ∞, imbricated in the bud... } ....NYMPHÆACEÆ.
 b Sepals 5, equal. Petals 5, imbricate. Leaves tubular..........SARRACENIACEÆ. 8
 b Sepals 5, unequal. Petals 5, convolute. Flowers of 2 sorts..........CISTACEÆ. 15
 b Sepals 2, with—*bb* 5 petals imbricated in the bud.............PORTULACCACEÆ. 20
   —*bb* 4 or 8 petals usually crumpled in bud...........PAPAVERACEÆ. 9
 c Filaments united into a tube. Anthers 1-celled.................MALVACEÆ. 23
 d Sepals 2, persistent, capping the lid of the pyxis.............PORTULACCACEÆ. 20
 d Sepals 3—5, valvate in the bud. Pod long, 2-carpelled...............TILIACEÆ. 25
 d Sepals 3—5.—*dd* Petals imbricate in bud. Fruits simple.............ROSACEÆ. 44
   —*dd* Petals convolute in bud. Fruit compound.........LOASACEÆ. 55
e Stamens opposite to the petals and of the same number. Pistil 1 only..BERBERIDACEÆ. 6
e Stamens alternate with the petals or more numerous................RANUNCULACEÆ. 1
 f Stamens 6, tetradynamous. Pod 2-ceiled. Flowers cruciform......CRUCIFERÆ. 11
 f Stamens 4—32, not tetradynamous. Pod 1-celled................CAPPARIDACEÆ. 12
 g Sepals 5, unequal. Flowers perfect, numerous, minute ....CISTACEÆ. 15
 g Sepals 5, equal. Flowers monœcious. Herbs woolly or scurfy .....ORDER 113

ANALYSIS OF THE NATURAL ORDERS. 7

g Sepals 5, or 3, equal, and the stamens twice as many.........GERANIACEÆ. 30
g Sepals 5, and the stamens (anthers) of the same number..(*gg*)
    gg Sterile filam. numerous, in several whorls. Climbing..PASSIFLORACEÆ. 57
    gg Sterile filaments numerous, in 5 clusters. Herb erect..SAXIFRAGACEÆ. 45
    gg Sterile filaments 0..(*)
        * Flowers white, racemed. Climbing.......................ORDER 106
        * Flowers yellow. Plants erect.......................TURNERACEÆ. 56
        * Flowers cyanic. Herbs stemless...................DROSERACEÆ. 17
h Stamens 5, alternate with the 5 petals. Styles 5 or 3. Seeds ∞......LINACEÆ. 28
h Stamens 5, opposite to the 5 petals. Styles 5, but the seed 1............ORDER 83
h Stamens twice as many as the petals..(*hh*)
    hh Stamens 6. Leaves peltate...........................NYMPHÆACEÆ. 7
    hh Stamens 6—24, distinct.............................CRASSULACEÆ. 46
    hh Stamens 10, united at base..........................GERANIACEÆ. 30
i Ovary 1-celled. Leaves all radical, spinescent, irritable.......DROSERACEÆ. 17
i Ovary 3-5-celled. Leaves mostly radical, not dotted...............ORDER 73
i Ovary 3-5-celled. Leaves cauline, pinnate, dotted............RUTACEÆ. 31
j Style 1, but the carpels as many as the petals (2—6)..........ONAGRACEÆ. 54
j Styles 3—5, ovary 3-5-celled, 3-5-seeded, wholly adherent......ARALIACEÆ. 64
j Styles 3—8, ovary 1-celled, half-adherent. Sepals 2..........PORTULACACEÆ. 20
j Styles 2, carpels 2, fewer than the (5) petals.—* Seeds several..SAXIFRAGACEÆ. 45
                        —* Seeds 2........UMBELLIFERÆ. 63
k Ovaries many, or few, rarely 1, always simple................RANUNCULACEÆ. 1
k Ovary compound, 3-carpelled, open before ripe..............RESEDACEÆ. 13
l Sepals (4 or 5) produced into 1 slender spur behind, petals 2 or 5....GERANIACEÆ. 30
l Sepals 2 (or vanished), petals 4 (2 pairs) with 1 or 2 blunt spurs....FUMARIACEÆ. 10
l Sepals 5, very unequal; petals 3. Stamens 6 or 8. No spur......POLYGALACEÆ. 42
l Sepals and petals each of the same number, viz...(*ll*)
    ll 4, the flowers slightly irregular. Stamens 6—32. No spur..CAPPARIDACEÆ. 12
    ll 4, the flowers moderately irregular. Stamens 8. A vine......SAPINDACEÆ. 37
    ll 5, with 5 stamens, and generally a blunt spur....................VIOLACEÆ. 14
    ll 5, with 10 or more stamens. No spur. Fruit a legume.......LEGUMINOSÆ. 43
m Pistils many, entirely distinct, simple......................RANUNCULACEÆ. 1
m Pistils 3—5, united more or less completely....................HYPERICACEÆ. 16
m Pistils 5—10, united, with sessile stigmas and many petals..........FICOIDEÆ. 61
n Pistil solitary, simple. Petals 6—9. Stamens 12—18........BERBERIDACEÆ. 6
n Pistils 3 or more, distinct, simple. Flowers all symmetrical..CRASSULACEÆ. 46
n Pistils 2, consolidated with the 5 stamens. Juice milky............ORDER 100
    o Carpels as many as the sepals..(*nn*)
    o Carpels fewer in number than the sepals..(*oo*)
        nn Anthers opening at the top. Flowers 4-parted....MELASTOMACEÆ. 52
        nn Anthers opening laterally. Styles united into 1......ONAGRACEÆ. 54
        nn Anthers opening laterally. Styles or stigmas distinct..HALORAGEÆ. 48
            oo Each carpel ∞-seeded. Styles 2............SAXIFRAGACEÆ. 45
            oo Each carpel 1-seeded. Styles 2 or 3.................ARALIACEÆ. 64
            oo Each carpel 1-seeded. Style 1 (double)............CORNACEÆ. 65
    p Style 3-cleft at the summit. Flowers 5-parted......PORTULACACEÆ. 20
    p Style and stigma 1, undivided. Flowers 7-parted................ORDER 81
q Leaves pinnate, with interpetiolar stipules...................ZYGOPHYLLACEÆ. 29
q Leaves simple, toothed or lobed. Flowers cruciform. Stamens 6....CRUCIFERÆ. 11
q Leaves simple, toothed or lobed. Flowers 5-merous. Stamens 10. GERANIACEÆ. 30
q Leaves simple, entire..(*qq*)
    qq Petals and stamens on the throat of the calyx............LYTHRACEÆ. 58
    qq Petals on the torus..(*)

ANALYSIS OF THE NATURAL ORDERS.

* Flowers irregular, unsymmetrical........................POLYGALACEÆ. 41
* Flowers regular, 2-(or 2-)parted throughout..............ELATINACEÆ. 18
* Flowers regular, 5-parted. Leaves punctate.............HYPERICACEÆ. 16
* Flowers regular, 5-parted. Leaves dotless..........CARYOPHYLLACEÆ. 19
r Pistil a simple carpel, becoming a legume. Stamens 10—100......LEGUMINOSÆ. 43
r Pistil compound, viz..(rr)
  rr 3-carpelled. Flowers perfect. Leaves digitate............SAPINDACEÆ. 37
  rr 3-carpelled. Flowers monœcions. Cultivated.............BEGONIACEÆ. 59
  rr 5-carpelled.—* Stipules present. Cultivated.................GERANIACEÆ. 30
        —* Stipules none. Native................................ORDER 73
s Stamens on the receptacle, in several sets. Leaves dotted........HYPERICACEÆ. 16
s Stamens on the receptacle, in 1 set. Lvs. fleshy. (S. Fla)...*Clusia*. GUTTIFERÆ. (21)
  Stamens on the calyx..(ss)
    ss Sepals, petals, and ovaries indefinite........................CALYCANTHACEÆ. 3
    ss Sepals, &c., definite. Leaves dotted, entire....................MYRTACEÆ. 51
    ss Sepals, &c., definite. Leaves dotless, entire..................LYTHRACEÆ. 58
    ss Sepals, &c., definite. Leaves dotless, subdentate..........SAXIFRAGACEÆ. 45
t Filaments united into 1 set (monadelphous). Petals convolute..(u)
t Filaments united into 1 or several sets. Petals imbricate..(uu)
t Filaments distinct..(tt)
  tt Petals 6, valvate, lurid. Erect shrubs............................ANONACEÆ. 4
  tt Petals 3—9, imbricate. Trees or shrubs....................MAGNOLIACEÆ. 2
  tt Petals 4—8, imbricate. Climbing or trailing..............MENISPERMACEÆ. 5
  tt Petals 4, imbricate. Shrubs, S........................CAPPARIDACEÆ. 12
    u Anthers 1-celled. Sepals valvate in the bud..................MALVACEÆ. 23
    u Anthers 2-celled. Sepals valvate. Handsome tree......STERCULIACEÆ. 24
    u Anthers 2-celled. Sepals imbricate. A large tree in S. Fla..CANELLACEÆ. (22)
      uu Leaves punctate with pellucid dots, jointed to stalk..AURANTIACEÆ. 32
      uu Leaves opaque..(*)
        * Sepals valvate. Flowers small ....TILIACEÆ. 25
        * Sepals imbricate. Flowers large................CAMELLIACEÆ. 26
v Style 1, with many stigmas. Green fleshy shrubs..................CACTACEÆ. 60
v Styles several or 1, each with 1 stigma. Woody trees or shrubs...ROSACEÆ. 44
v Style 1, with 1 stigma. Stam. in 5 sets, long, red, very showy...MYRTACEÆ. 51
w Trailing vines, with crimson fls. Ovaries ∞, in a little spike...MAGNOLIACEÆ. 2
w Climbing vines, with white-greenish fls. Ova. 2—6, capitate...MENISPERMACEÆ. 5
w Erect shrubs, with yellow flowers, 6-parted. Pistil only 1.......BERBERIDACEÆ. 6
w Erect shrubs (S. Fla.) with yellow fls. Pistils 5, 2-ovuled, 1-sded..SURIANACEÆ. (62)
w Trees, with greenish fls.,—* and pinnate lvs. Pist. 3-5, 1-ovuled..SIMARUBACEÆ. 34
        —* and simple leaves. Follicles 3—5...STERCULIACEÆ. 24
  x Flowers 4-parted. Stamens 8. (Fls. red or roseate, drooping)..ONAGRACEÆ. 54
  x Flowers 4-parted. Sta. 8. Fls. light yellow. Coasts, S. Fla..RHIZOPORACEÆ. (49)
  x Flowers 4-parted. Stamens 4. Flowers whitish, in cymes......CORNACEÆ. 65
  x Flowers 5-parted..(xx)
    xx Ovary 5-carpelled, 5-styled, 5-seeded........................ARALIACEÆ. 64
    xx Ovary 5-carpelled, 1-styled, 1-seeded. S. Fla..........COMBRETACEÆ. 50
    xx Ovary 2-4-carpelled, ∞-seeded......................SAXIFRAGACEÆ. 45
      y Leaves opposite. Stem climbing with tendrils or radicles..VITACEÆ. 41
      y Lvs. alternate. St. erect, or climbing without tendrils..RHAMNACEÆ. 40
z Leaves simple. Stamens 5. Carpels 3—5, style 1, short..........CELASTRACEÆ. 39
z Leaves simple. Sta. 10. Carpels and sty. 3. S. Fla..*Byrsonima*. MALPIGHIACEÆ. (39)
z Leaves pinnate, or palmately lobed. Carpels and styles 2 or 3.:....SAPINDACEÆ. 37
z Leaves pinnate..(*)
  * Stamens 10. Small tree with blue flowers. S. Fla........ZYGOPHYLLACEÆ. 29
  * Stamens 2. Carpels 1 or 2. Style 1..................................ORDER 101

ANALYSIS OF THE NATURAL ORDERS. 9

 * Stamens 8. Carpel and style 1............................Burseraceæ. 35
 yy Filaments 10, united into a tube or cup. Flowers in panicles......Meliaceæ. 27
 yy Filaments 6—10, distinct. Flowers small, white, in racemes....Burseraceæ. 35
 yy Filaments 6—10, distinct. Fls. small, white or hoary, paniculate..Sapindaceæ. 37
 yy Filaments 5, distinct...(*)
  * Leaves pellucid-punctate ....Rutaceæ. 31
  * Leaves opaque. Ovary 1-celled, 1-seeded..............Anacardiaceæ. 36
 zz Petals 4, yellow, strap-shaped, appearing in late Autumn......Hamamelaceæ. 47
 zz Petals 4—7, cyanic (rarely yellow), rounded or short..(†)
  † Style 0, the stigmas 1, 4, or 5, sessile. Drupe 4–6-seeded...........Order 74
  † Styles (or stigmas) 3, but the drupe only 1-seeded ....Anacardiaceæ. 36
  † Styles 3, capsule many-sded. Lvs. minute and scale-form..Tamariscineæ. 24 *bis*
  † Style 1,..(‡)
   ‡ Capsule 3-seeded. Seeds with a scarlet aril.............Celastraceæ. 38
   ‡ Caps. ∞-seeded. Clusters fragrant. Lvs. evergreen. Cult...Pittosporaceæ.
   ‡ Capsule with few or many seeds. Native shrubs...............Order 73

B. Cohort 2. GAMOPETALOUS EXOGENS.

§ Stamens (6 — ∞) more numerous than the lobes of the corolla..(9)
§ Stamens (2—12) fewer than the corolla lobes or of the same number..(2)
 2 Ovary inferior,=adherent to the tube of the calyx..(3)
 2 Ovary superior,=free from the tube of the calyx..(4)
  3 Stamens cohering by their anthers..(c)
  3 Stamens entirely distinct..(d)
   4 Flowers regular and the stamens symmetrical..(5)
   4 Flowers regular and the stamens reduced to 2 or 4..(n)
   4 Flowers irregular. Stamens (except in 3 or 4 species) unsymmetrical..(e)
    5 Stamens opposite to the lobes of the corolla (and distinct)..(e)
    5 Stamens alternate with the corolla lobes (rarely connate)..(6)
     6 Shrubs, trees, with the carpels or stigmas 3—6..(f)
     6 Herbs 1-10 carpelled, or shrubs 2-carpelled..(7)
      7 Ovary 1, deeply 4-parted or 4-partible, forming 4 achenia..(g)
      7 Ovaries 2, distinct (often covered by the stamens)..(h)
      7 Ovary 1 compound,—* one-celled..(k)
         —* two-six-celled..(m)
 9 Flowers irregular (rarely regular and the fruit a legume)..(a)
 9 Flowers regular and the fruit never a legume (§ 165)..(b)
  a Flowers 1- or 2-sided, with 1 or 2 blunt spurs. Stamens 6, in 2 sets...Order 10
  a Flowers 1-sided, no spur..(*)
   * Leaves compound. Fruit a legume ....Order 43
   * Leaves simple. Fruit 2-celled, 2-seeded.....................Order 42
   * Leaves simple. Fruit 5-celled ....Ericaceæ. 73
  b Corolla lobes convolute in bud. Stamens ∞, united into 1 tube....Order 23
  b Corolla lobes imbricate in bud. Stamens ∞, in 1 or several sets....Order 26
  b Corolla lobes imbricate or valvate..(u)
   u Stamens 10–24. Styles 5—12 ....Order 46
   u Stamens 5—10. Style 1. Capsule 5-celled.................Ericaceæ. 73
   u Stamens 8 – ∞. Style 1. Nut 1-5-seeded................Styraceæ. 76
   u Stamens 8. Styles 4. Berry 8-seeded ....Ebenaceæ. 77
   u Stamens 8. Style 1. Drupe 1-seeded....... ....Olaceæ. 80 (p. 447)

ANALYSIS OF THE NATURAL ORDERS.

 c Flowers in a compact head surrounded by an involucre.............COMPOSITÆ. 70
 c Flowers separate, irregular, perfect. Plants erect or trailing........LOBELIACEÆ. 71
 e Flowers separate, regular, imperfect. Weak vines.... ....ORDER 58
  d Leaves alternate. Flowers 5-parted, regular, separate.....CAMPANULACEÆ. 72
  d Leaves alternate. Fls. irregular, 5-parted. S. Fla..*Scœvola.* GOODENIACEÆ. (71)
  d Leaves opposite, with stipules between, or verticillate.... ....RUBIACEÆ. 67
  d Leaves opposite. Stipules none..(*v*)
   v Stamens 5—4. Ovaries 2-5-celled.....................:...CAPRIFOLIACEÆ. 66
   v Stamens 2—3. Ovaries 1-celled..................VALERIANACEÆ. 68
   v Stamens 4. Flowers capitate ....DIPSACEÆ. 69
  e Herbs. Ovary with 5 styles and but 1 seed.................PLUMBAGINACEÆ. 83
  e Herbs. Ovary with 1 style and many seeds.....PRIMULACEÆ. 81
  e Trees or shrubs. Appendages between the stamens.........SAPOTACEÆ. 78
  e Trees or shrubs. No appendages between the stam. S. Fla..MYRSINACEÆ. (79)
  f Leaves opposite. Style 1. Drupe 4-seeded. Herbs, shrubs..VERBENACEÆ. 90
  f Leaves alternate..(*w*)
   w Drupe 4-6-seeded. Shrubs, trees.................AQUIFOLIACEÆ. 74
   w Drupe 1-seeded. Thorny. S. Fla...........*Ximenia.* OLACACEÆ. (80)
   w Capsule 2-5-celled, ∞-seeded ....ERICACEÆ. 73
  g Herbs, with alternate leaves, generally rough-hairy.....BORRAGINACEÆ. 92
   h Stigmas connate. Flower bud convolute ....APOCYNACEÆ. 99
   h Stigmas connate. Flower bud valvate ...........ASCLEPIADACEÆ. 100
   h Stigmas distinct. Flowers minute, yellow........CONVOLVULACEÆ. 95
  k Ovule solitary. Corolla limb entire.............................ORDER 103
  k Ovules several. Leaves cleft and lobed ............HYDROPHYLLACEÆ. 93
  k Ovules several. Leaves or leaflets entire..(*x*)
   x Flowers not spicate.............................GENTIANACEÆ. 97
   x Flowers spicate.................}
  m Leaves all radical. Flowers spiked ...} ..............PLANTAGINACEÆ. 82
  m Leaves opposite. Ovary 2-celled.......................LOGANIACEÆ. 98
  m Leaves alternate..(*y*)
   m Leaves opposite. Ovary 3-celled. Not twining..}
   y Ovary 3-celled. Not twining.............:...} ....POLEMONIACEÆ. 94
   y Ovary 2-4-celled. Twining......................CONVOLVULACEÆ. 95
   y Ovary 2-4-celled, 4-seeded. Erect..........BORRAGINACEÆ. 92
   y Ovary 2-celled, ∞-seeded. —*z* Styles 2.......HYDROPHYLLACEÆ. 93
        —*z* Style 1...............SOLANACEÆ. 96
  n Stamens 4. Ova. 4-(rarely 1- or 2-)celled, with as many sds..VERBENACEÆ. 90
  n Stamens 2. Ovary 2-celled, forming 1 or 2 seeds.............OLEACEÆ. 101
 s Ovary deeply 4-parted, forming 4 (or fewer) achenia..(*p*)
 o Ovary entire. 4-ovuled, 4- or fewer-seeded. Leaves opposite..VERBENACEÆ. 90
 o Ovary entire, ∞-ovuled, ∞- or several-seeded..(*s*)
  p Leaves opposite. Stems square. Stamens 2—4............ ..LABIATÆ. 91
  p Leaves alternate. Stems round. Stamens 5...........BORRAGINACEÆ. 92
 s Trees or climbing shrubs. Seeds winged.....................BIGNONIACEÆ. 86
 s Trees. Seeds not winged .....SCROPHUL. 88. Erect shrubs ....ERICACEÆ. 73
  Herbs.—*ss* Leafless parasites. Native. Ovary 1-celled.....OROBANCHACEÆ. 85
   —*ss* Leafy at base or in the water. Flowers spurred..LENTIBULACEÆ. 84
   —*ss* Leafy. Flowers large, spurless. Ovary 1-celled...GESNERIACEÆ. 87
   —*ss* Leafy. Spurless. Fruit 4- or 5-celled ....§ BIGNONIACEÆ. 86
   —*ss* Leafy. Fruit 2-celled..(*t*)
 t Seeds on hooks or cups. Corolla mostly convolute ....ACANTHACEÆ. 89
 t Seeds without hooks. Corolla imbricated in the bud.....SCROPHULARIACEÆ. 88
 t Seeds without hooks. Corolla mostly plicate............... ......SOLANACEÆ. 96

ANALYSIS OF THE NATURAL ORDERS. 11

C. Cohort 3. APETALOUS EXOGENS.

¶ Plants herbaceous, the flowers not in aments (except Humulus, 115)..(*i*)
¶ Plants woody,—shrubs or trees..(8)
  2 Flowers with a regular calyx (or a calyx-like involucre). (3)
  2 Flowers achlamydeous,—neither calyx nor corolla..(*k*)
    3 Calyx tube adherent to the ovary, limb lobed, toothed, or entire (9)
    3 Calyx free from the ovary, sometimes enclosing it..(4)
  4 Ovaries several, entirely distinct, each 1-styled, 1-ovuled..(*g*)
  4 Ovary 1 only, simple or compound..(5)
    5 Style or stigma 1 only..(6)
    5 Styles or stigmas 2—12..(7)
      6 Ovary 1-ovuled, bearing but 1 seed..(*c*)
      6 Ovary many-ovuled, bearing many seeds..(*d*)
      7 Ovary 1-3-ovuled, 1-3-seeded..(*e*)
      7 Ovary 4-∞-ovuled, 4-∞-seeded..(*h*)
    8 Flowers not in aments, with the leaves opposite..(*n*)
    8 Flowers not in aments, with the leaves alternate..(10)
    8 Flowers imperfect, the sterile only in aments..(*v*)
    8 Flowers imperfect, both the fertile and sterile in aments..(*x*)
      9 Stamens 1—12, as many or twice as many as the stigmas..(*a*)
      9 Stamens 2—10, not symmetrical with the 1 or 2 stigmas..(*b*)
    10 Style or stigma 1. Fruit 1-seeded..(11)
    10 Styles or stigmas 2..(*s*)
    10 Styles or stigmas 3—9..(*t*)
      11 Calyx free from the ovary..(*p*)
      11 Calyx adherent to the ovary..(*r*)
        a Stigmas and cells of the ovary 1—4. Stamens 1—8....... Orders 48, or 54
        a Stigmas and cells of the ovary 6. Stamens 6 or 12...Aristolochiaceæ. 102
          b Styles 2. Ovary many-seeded. Stamens 8—10..............Order 45
          b Style 1. Ovary 1- or 2-seeded. Stamens 5...........Santalaceæ. 110
        c Flowers perfect. Calyx 4-lobed. Stamens 1—4 ....Order 44
        c Flowers perfect. Calyx entire, funnel-shaped, colored..Nyctaginaceæ. 101
        c Flowers diclinous. Calyx 4-5-parted, green..............Urticaceæ. 115
          d Stamens 4, opposite to the 4 sepals. Leaves numerous .....Order 53
          d Stamens 4, opposite to the 4 sepals. Leaves about 6.......Order 143
          d Stamens 5, alternate with the 5 sepals......................Order 81
          d Stamens ∞. Leaves large and showy. Cultivated..........Order 9
        e Fruit 3-(rarely 6-)seeded, with 3 (often cleft) styles .....Euphorbiaceæ. 113
        e Fruit 1-seeded. Stipules sheathing the stems..........Polygonaceæ. 104
        e Fruit 1-celled, mostly 1-seeded. Stipules none..(*f*)
          f Calyx with scarious bractlets outside..................Amarantaceæ. 107
          f Calyx naked (double in 1 genus). Lvs. alternate..Chenopodiaceæ. 106
          f Calyx naked. Leaves opposite............................Order 19
        g Stamens hypogynous—on the torus ....Order 1
        g Stamens perigynous—on the calyx ....Order 41
          h Leaves opposite. Fruit circumscissile, a pyxis ....Order 61
          h Leaves opposite. Fruit 4-5-valved, a capsule..... ......Order 19
          h Leaves alternate ..(*i*)
            i Fruit 5-horned, 5-celled, a capsule......................Order 46
            i Fruit a fleshy 4-10-seeded berry............Phytolaccaceæ. 105
            i Fruit circumscissile, a utricle..................Amarantaceæ. 107
        k Flowers on a spadix with a spathe. Monocotyledons.......... Order 130
        k Flowers in a long naked spike. Stamens 6 or 7.........Saururaceæ. 15
        k Flowers solitary, axillary, minute. Aquatic plants (*m*)

ANALYSIS OF THE NATURAL ORDERS.

    m Stamen 1, styles 2. Leaves opposite.............CALLITRICHACEÆ. 116
    m Stamens 2, styles 2. Leaves alternate, dissected..PODOSTEMIACEÆ. 117
    m Sta. 12–24, style 1. Lvs. verticillate, dissected..CERATOPHYLLACEÆ. 118
n Fruit a double samara (2-winged)......................ORDER 37
n Fruit a single samara (1-winged), or a drupe. Stamens 2........ORDER 101
n Fruit not winged,—*o* 3-seeded. Stamens 4.............EUPHORBIACEÆ. 113
    —*o* 1-seeded. Stamens 4 or 8...........ELEAGNACEÆ. 112
    —*o* 1-seeded. Stamens 3. Parasites..LORANTHACEÆ. 109
p Anthers opening by valves. Calyx colored................LAURACEÆ. 108
p Anthers opening by slits.—*q* Calyx colored. Stam. 8....THYMELACEÆ. 111
    —*q* Calyx greenish; racemed..........ORDER 37
    —*q* Cal. green; spiked. S. Fla..COMBRETACEÆ. (50)
r Ovary and seed only 1, in the juicy drupe. Trees............ORDER 65
r Ovaries 2–4, seed 1. Fruit a drupe or nut. Shrubs...SANTALACEÆ. 110
s Stamens numerous...ORDER 47
s Stamens as many as the calyx lobes..................§ 1. URTICACEÆ. 114
    t Leaves pinnate. Pistils 5, scarcely united...................ORDER 31
    t Leaves simple, linear, evergreen. Shrubs heath-like..EMPETRACEÆ. 119
    t Leaves simple, expanded. Fls. 3-parted. Fruit dry..EUPHORBIACEÆ. 113
    t Leaves simple, expanded. Fls. 4- or 5-parted. Fruit fleshy....ORDER 40
v Nut drupaceous, naked. Leaves pinnate ....JUGLANDACEÆ. 121
v Nut or nuts in a cup or involucre. Leaves simple........CUPULIFERÆ. 122
x Fruit fleshy, aggregated (sorosis). Juice (or sap) milky...§ 2. URTICACEÆ. 114
x Fruit dry. Plants with a watery juice or sap..(*y*)
    y Aments globular, racemed. Nutlets 2-celled, woolly.............ORDER 65
    y Aments globular, solitary. Nutlets 1-celled, 1-seeded....PLATANACEÆ. 120
    y Aments cylindrical or oblong..(*z*)
        z Ovary 2-celled, 2-ovuled, 1-seeded. Fruit often winged..BETULACEÆ. 123
        z Ovary 1-celled, 1-seeded. Fruit often fleshy..........MYRICACEÆ. 124
        z Ovary many-ovuled, many-seeded. Seeds comous......SALICACEÆ. 125

D. Cohort 4. THE CONOIDS

* Leaves pinnate. Stem simple, palm-like. Sterile flowers in cones....CYCADACEÆ. 126
* Leaves simple. Stem branching. Fertile flowers in cones.............CONIFERÆ. 127
* Leaves simple. Stem branching. Fertile flowers solitary...............TAXACEÆ. 128

E. Cohort 5. THE SPADICEOUS ENDOGENS.

¶ Trees or shrubs with palmi-cleft leaves all from one terminal bud, } ...PALMACEÆ. 129
    and a branching "spadix" from a spathe......................
¶ Herbs with simple, rarely ternate leaves. Spadix simple..(2)
    2 Plants frond-like, minute, floating loose on the water..............LEMNACEÆ. 131
    2 Plants with stem and leaves, rooting and fixed..(3)
        3 Spadix evident, in a spathe or on a scape ....ARACEÆ. 130
        3 Spadix obscure or spike-like. Stems leafy..(4)
            4 Flowers with no perianth, densely spicate or capitate......TYPHACEÆ. 132
            4 Flowers with a perianth or not. Plants submersed.......NAIADACEÆ. 139

ANALYSIS OF THE NATURAL ORDERS. 13

F. Cohort 6. FLORIDEÆ, or FLOWERING ENDOGENS.

¶ Flowers (not on a spadix) in a small, dense, involucrate head..(o)
¶ Flowers (not on a spadix) solitary, racemed, spicate, &c..(2)
  2 Perianth tube adherent to the ovary wholly or partly..(4)
  2 Perianth free from the ovary. (3)
    3 Petals and sepals differently colored (except in Medeola, 147)..(e)
    3 Petals and sepals similarly colored..(5)
      4 Flowers imperfect (♂ ♀ or ♂ ♥ ♀)..(a)
      4 Flowers perfect..(b)
        5 Leaves net-veined, broad..(k)
        5 Leaves parallel-veined..(6)
          6 Styles and often the stigmas also united into one..(m)
          6 Styles and stigmas 3, distinct..(n)
a Low aquatic herbs..HYDROCHARIDACEÆ. 135
a Climbing shrubby vines..DIOSCORIACEÆ. 143
  b Anthers 1 or 2, on the pistil (gynandrous)....................ORCHIDACEÆ. 137
  b Anthers 1 or 5, free from the pistil. Leaves ample..........SCITAMINEÆ. 138
  b Anthers 3 or 6..(c)
    c Perianth woolly or mealy outside. Ovary half free....HÆMADORACEÆ. 141
    c Perianth glabrous outside..(d)
      d Anthers 3, opening crosswise, inward............BURMANNIACEÆ. 136
      d Anthers 3, opening lengthwise, outward...............IRIDACEÆ. 142
      d Anthers 6, opening inward ......................AMARYLLIDACEÆ. 139
e Pistils 3—∞, distinct, forming achenia in fruit...................ALISMACEÆ. 134
  Pistils 3 only, more or less united..(g)
    g Leaves verticillate, in 1 or 2 whorls. Stigmas 3..............TRILLIACEÆ. 146
    g Leaves alternate..(h)
      h Stigmas 3. Plants with dry leaves, often epiphytes.....BROMELIACEÆ. 140
      h Stigmas united into 1................................COMMELYNACEÆ. 151
    k Flowers perfect, 4-parted................................ROXBURGHIACEÆ. 145
    k Flowers diœcious, 6-parted.....................................SMILACEÆ. 144
      m Flowers colored, regular. Stamens 6 (4 in one species).....LILIACEÆ. 147
      m Flowers colored, irregular or else triandrous.........PONTEDERIACEÆ. 149
      m Flowers greenish, glume-like or scarious..................JUNCACEÆ. 150
    n Leaves rush-like. Ovary of 3 1-seeded carpels } .. MELANTHACEÆ. 148
    n Leaves linear, lanceolate, &c. Ovary 6-∞-seeded... }
      o Petals yellow, small but showy. Plant acaulescent........XYRIDACEÆ. 152
      o Petals white, minute, fringed. Plant acaulescent....ERIOCAULONACEÆ. 154

G. Cohort 7. GRAMINOIDEÆ, or GRASS-LIKE ENDOGENS.

¶ Flowers with 6 bracts in 2 whorls (sepals and petals). Culms solid ....ORDER 150
¶ Flower with a single bract (glume). Culm solid, sheaths entire.......CYPERACEÆ. 153
¶ Flower with several bracts (glumes and pales). Culm hollow, } .. ....GRAMINEÆ. 156
  Sheaths split on one side. Ovary 1-seeded. Styles 2...... }

II. PROVINCE, ACROGENS.

§ Plants with well-developed foliage..(¶)
  ¶ Leaves few, mostly ample and from subterranean rhizomes..(a)

  a Fruit borne on the leaves which are often more or less contracted...FILICES. 159
  a Fruit borne at the base of the radical, entire or lobed leaves..MARSILEACEÆ. 156
 ¶ Leaves numerous, small, mostly spirally imbricated on the stem..(*b*)
  b Fruit axillary, sessile, opening by a slit ....LYCOPODIACEÆ. 157
  b Fruit mostly terminal and usually stalked, opening by a lid...........MUSCI.*
 ¶ Leaves numerous, small, imbricated on the stem in 2 rows. ⎱
§ Plants with the leaves and stem confounded, thallus-like....... ⎰ ....HEPATICÆ.*
§ Plants with verticillate branches instead of leaves..(*c*)
  c Fruit in terminal spikes, and of one kind only...............EQUISETACEÆ. 158
  c Fruit lateral, scattered on the branches, and of two kinds.........CHARACEÆ.*

K. PROVINCE, THALLOGENS

Plants aquatic, with a colored thallus. Fruit immersed in the frond.............ALGÆ.*
Plants on dry rocks, logs, or bark of trees, thalloid or granular................LICHENS.*
Plants growing on decaying organisms. Thallus cotton-like, the fruit very ⎱ ....FUNGI.*
 different, all without chlorophyll or starch.......................... ⎰

---

  * Those Orders, the lower Cryptogams, are omitted in this work.

PART FOURTH.

DESCRIPTIVE BOTANY, OR PHYTOLOGY,

COMPRISING A TABULAR FLORA OF

THE UNITED STATES AND CANADA

(WITHIN THE LIMITS STATED IN THE PREFACE).

Sub-Kingdom, PHÆNOGAMIA, the Flowering Plants, having stamens and pistils, producing seeds with an embryo. (For sub-kingdom Cryptogamia, see page 412.)

Province, EXOGENÆ, the Dicotyledonous Plants. Stems composed of bark, wood, and pith, exogenous (§ 405) in growth. Leaves mostly net-veined. Flowers 5-parted or 4-parted, rarely in 3s. Embryo with 2 or more opposite cotyledons. (Province Endogenæ, p. 316.)

Class I, ANGIOSPERMÆ. Pistils complete, with stigma and ovary, the latter enclosing the ovules, and in fruit enclosing the seeds. Cotyledons only 2. (Class II, Gymnospermæ, p. 311.)

Cohort 1, DIALYPETALÆ, the Polypetalous Exogens. Flowers having a double perianth, both calyx and corolla, the latter composed of distinct petals. (Cohort 2, p. 144.)

ORDER I. RANUNCULACEÆ. CROWFOOTS.

Herbs (or woody climbers) with a colorless, acrid juice. *Leaves* mostly divided, exstipulate, with half-clasping petioles. *Sepals* 3-15, green or petaloid. *Petals* 3-15, distinct, sometimes irregular or none. *Stamens* hypogynous, indefinite. *Ovaries* many or few, distinct, 1—∞-ovuled. *Fruit* either dry achenia, or follicles, or baccate, 1 — ∞-seeded. *Seeds* anatropous, embryo straight in horny albumen.—Abounding in cool regions.

Illustrated in figs. 33, 39, 83, 84, 109, 127, 132, 159, 155, 156, 212, 234, etc.

TRIBES AND GENERA.

Sepals valvate in the bud. Achenia tailed. (Tribe I.)
Sepals imbricated in the bud.—*a* Ovaries 1-seeded, acheniate. (2)
    —*n* Ovaries 2—∞ -seeded (3)
  2 Corolla 0, or undistinguishable from the colored calyx. (Tribe II., *b*)
  2 Corolla and calyx distinct either in color or form. (Tribe III., *c*)
    3 Sepals as permanent as the stamens. Fruit follicular. (Tribe IV., *d*)
    3 Sepals caducous sooner than the stamens. (Tribe V., *g*)
    3 Sepals persistent with the follicular fruit. (Tribe VI.)

I. CLEMATIDEÆ.—Petals 0, or stamen-like. Leaves all opposite.	CLEMATIS.	1
II. ANEMONEÆ. *b* Sepals deciduous with the stamens. Stem-leaves opposite.	ANEMONE.	2
b Sepals deciduous with the stamens. Leaves all radical.	HEPATICA.	3
b Sepals caducous.—Leaves ternately compound.	THALICTRUM.	4
—Leaves palmate, simple. Flowers ⚥.	TRAUTVETTERIA.	5
III. RANUNCULEÆ. *c* Sepals not appendaged. Petals red or yellow, no scale.	ADONIS.	6
c Sepals not appendaged. Petals xanthic, a scale at base.	RANUNCULUS.	7
c Sepals appendaged. Plant small. Leaves radical.	MYOSURUS.	8
IV. HELLEBOREÆ.—*d* Perianth regular. (*e*)		
e Petals 0. Sepals white.	ISOPYRUM.	9
e Petals 0. Sepals 6–9, yellow.	CALTHA.	10
e Petals slender, tubular at apex. Roots yellow.	COPTIS.	11
e Petals minute, tubular at base, 1-lipped.	TROLLIUS.	12
e Petals small, tubular, 2-lipped. Sepals persistent.	HELLEBORUS	13
e Petals small, concave, 2-lobed. Fls. racemed. Rt. yel.	ZANTHORHIZA.	14
e Petals larger than the colored sepals, 3-lobed.	NIGELLA.	15
e Petals larger than the colored sepals, spur-like, equal.	AQUILEGIA.	16
—*d* Perianth irregular. (*f*)		
f Upper sepal spurred, containing two spurred petals.	DELPHINIUM.	17
f Upper sepal hooded, covering two deformed petals.	ACONITUM.	18
V. CIMICIFUGEÆ. *g* Flowers numerous, in long, spicate racemes.	CIMICIFUGA.	19
g Flowers many, in short racemes. Fruit baccate.	ACTÆA.	20
g Flower 1 only. Plant 2-leaved. Berry compound.	HYDRASTIS.	21
VI. PÆONIÆ.—Petals plane, large, showy. Disk sheathing the follicles.	PÆONIA.	22

1. CLEMATIS, L. VIRGIN'S BOWER. Calyx of 4 (4—9 in the exotics) colored sepals, in æstivation valvate-induplicate. Petals 0, or if present, more like sterile filaments. Stamens shorter than the sepals, the outer or all sometimes sterile. Ovaries ∞ in a head. Achenia caudate with the lengthened plumous or pubescent styles ♃. ♄ Somewhat woody, climbing by the clasping petioles. Leaves opposite. Fig. 359.

SUBGENERA AND SPECIES.

§ ATRÁGENE. Onter stamens petal-like. Lvs. verticillate. Fls. solitary. Vine...No. 1
§ CLEMATIS *proper*. Petals none. Leaves opposite...(*)
  * Erect herbs. Lvs. simple. Fls. solitary, large, terminal, nodding. May..Nos. 9-11
  * Climbing.—*a* Fls. panicled, white, often diclinous, sepals thin................Nos. 2–4
    —*a* Fls. solitary, nodding,—*b* bell-shaped, pale bluish purple...Nos. 5, 6
          —*b* ovoid, dark purple ....Nos. 7, 8
Exotic.—* Flowers in clusters, white. Leaves pinnate................... ...Nos. 12, 13
    * Flowers single, large.—*x* Leaves simple. Sepals 4 ....Nos. 14, 15
         —*x* Leaves compound. Sepals 4, open ......Nos. 16, 17
         —*x* Leaves compound. Sepals 6—9, open...Nos. 18, 19

Order 1.—RANUNCULACEÆ. 17

1 C. verticillàris DC. Lvs. in whorls of 4, each ternate, and 2 large purple fls. at each node. Highland woods, Me. to Ga., W. to Rky. Mts. 15f. May, June. Rare.
2 C. Virginiàna L. Glabrous; lvs. ternate, lfts. lobed and cut-dentate; achenia long, plumed, in feathery tufts. Thickets, Can. to Ga., W. to Mo. 15f. Aug. †
3 C. Catesbyàna Ph. Pubescent; lvs. biternate, lfts. ovate, mostly 3-lobed, lobes entire; ach. short-plumed; sep. small, linear-oblong. Coast, S. Car. to Fla. 12f. July.
4 C. holoserícea Ph. Silky-pubescent; lvs. ternate, lfts. lance-oblong, entire; fls. in small corymbous clusters; sep. linear; ach. long-plumed. Carolina. Diœcious.
5 C. crispa L. Lvs. ternate, pinnate, or decompound, lfts. varying from ovate to lanceolate, and linear, acute, thin, smooth; ach. tails short, pubescent. Va. to Ga. and La. Lfts. 3—15. Fls. elegant, 15″ long. (C. Walteri Ph., C. cylindrica Sims, &c.)
6 C. reticulàta Walt. Lvs. ternate or pinnate, lfts. 3—7, obtuse at each end, at length rigid and prominently veined, often lobed; tails silky. Fla. Sep. 12—15″ long.
7 C. Viórna L. *Leather-flower.* Lvs. pinnate, lfts. ovate, acute, smooth; sep. lance-ovate, the cuspidate points reflexed; ach. tails long, plumous. Woods, O. to Ga. 10—15f. Peduncles with a pair of simple leaves. Summer. Rare.
8 C. Pitcheri T. & G. Leaves pinnate, leaflets coriaceous, roughened with the netted veins; sepals lance-ovate; ach. tails short, glabrous. Ill., Iowa, to Ark.
9 C. ochroleùca Ait. Lvs. silky-pubescent beneath, ovate, entire; sep. silky, yellowish within; ach. plumes long, straw-color. ♃ Woods, L. I. to Ga. Rare. 1f.
10 C. ovàta Ph. Leaves glabrous, glaucous beneath, broad-ovate; flower on a short peduncle, purple; sepals ovate, pointed. ♃ N. Car. to Fla. 1—2f. Leaves entire.
11 C. Baldwínii T. & G. Lvs. oblong to lance-linear, the lower 3-lobed or cleft; flower on a long peduncle, purplish. ♃ Fla. 1—2f. Plumous tails 2′ long.
12 C. ERÉCTA. Stem 3f, weak, inclining; lfts. lance-ovate. ♃ Europe. August.
13 C. FLÁMMULA. Climbing 12—20f; leaflets oval to oblong-linear, often lobed, acute, smooth; clusters terminal, fragrant. From France. August, September.
14 C. INTEGRIFÒLIA. Upright; lvs. lance., entire, smooth; fls. nodding, blue. Eur. 2f.
15 C. CIRRHÒSA. Climbing; lvs. ovate, subcordate, toothed; fls. fragrant, white. Eur.
16 C. VITICÉLLA. Lfts. 3—15, ovate or oval, entire; sep. obovate, purp., 15″. Eur. Sum.
17 C. GRAVÈOLENS. Lfts. 3—5, lanceolate, acute; sep. oblanceolate, ylw., 9″. Thibet.
18 C. FLÓRIDA. Lvs. ternate and bitern.; sep. ovate, pointed, wh. or purplish. Japan
β. SIEBÓLDTII. Fls. 4′ broad, creamy-white and purple, double. Splendid.
19 C. CŒRÙLEA. Lvs. ternate, hairy; fls. very large; sep. lance-ovate, blue, &c. Japan
β. AZUREA-GRANDIFLORA. Flowers 5—7′ broad, azure, or lilac-blue. July.

2. ANEMÒNE, L. WIND-FLOWER. Involucre remote from the flower, of 3 divided leaves, calyx regular, of 3—15 colored sepals. Corolla 0. Ovaries ∞, free, collected into a roundish or oblong head. Achenia with a short, rarely a lengthened beak. Seeds suspended. ♃ Lvs. radical. Stem leaves 2 or 3, opposite, forming the involucre. Figs. 116, 176.

§ PULSATÍLLA. Carpels many (50—75), with long plumous tails. One large flower..No. 1
§ ANEMONÁNTHEA. Carpels hairy, but neither tailed nor grooved...(a)
  a Pistils many (50—70) in a head, densely matted with wool in fruit...(b)
  a Pistils fewer (15 -20) in a head, merely pubescent in fruit............Nos. 2, 3
    b Stem leaves (involucre) sessile, with a single flower.............Nos. 4 –6
    b Stem leaves (involucre) petiolate, with 2 or 3 flowers............Nos. 7—9
§ SYNDÉSMON. Carpels few, not caudate, glabrous and grooved..............No. 10
                      Exotic, cultivated species........Nos. 11—13

1 A. pàtens L. β. *Nuttalliàna. Pasque-flower.* Clothed with long silky hairs; lvs. many-cleft, with linear segments, developed after the large spreading pale-purple flower. Dry hills, Ill., Wisc. to Dak. (Matthews). 1′—1f. Sepals 5 or 6, 1′. April.
2 A. nemoròsa L. Smooth, 1-flowered; leaves of the invol. 3, petiolate, 2-5-par. est. segm. cleft and lobed. Copses, com., 6—9′. Fl. white, purple outside. April, May.

3 A. Pennsylvánica L. Hairy, 1-, finally 2- or 3-flowered; leaves of the invol. sessile, large. veiny, 3-parted, acuminate-lobed and toothed. Prairies, Can. to Penn., W. to the Miss. 12—20′. Flowers pure white. June—August.

4 A. Caroliniàna Walt. Lvs. 3-parted into cuneate-linear, twice trifid segm.; involucre similarly cleft half-way; sepals obtuse, 15—20; carpels in an oblong head. Car. to Ill., and Nebr. 6—10′. Flower white-purple, pretty, fragrant. April, May.

5 A. heterophýlla Nutt. Lvs. of roundish-oval, crenate segments, invol. linear-cleft to the base; sepals acute, 5—13; carpels in a cylindrical head. Ga. to La. and Ark. 8—16′. Flower white-green, scentless. March, April.—Varies toward No. 4.

6 A. parviflòra Mx. Leaves of involucre 2, 3-cleft, segments cuneiform, 3-cleft, crenate-lobed; sepals 5 or 6; carpels in a globular head. L. Sup., and N. 3—12′. White.

7 A. multífida DC. *Red Anémone.* Involucre short-petioled; lateral peduncles involucellate; head of carpels oval. N. Vt. to L. Sup. Rare. Red-white. 1f. June.

8 A. Virginiàna L. Invol. long-petioled; lateral ped. involucellate; head of carp. oblong. Can. to Car. 2—3f. Fls. white-green, on long stalks. Sepals 5. Ju.—Aug.

9 A. cylíndrica Gray. Invol. long-petioled; peduncles all naked, long; head of carpels cylindrical. N. H., Mass., to Iowa. Silky pubescent. 2f. White-green. May.

10 A. thalictroides L. *Rue Anémone.* Glabrous, slender; invol. of 2 sessile biternate (apparently of 6-petioled ternate) lvs., lfts. 3-lobed; fls. umbelled; sep. 5—10. Woods, Can. to Ga., W. to Iowa. 6—10′. Root tuberous. Fls. white-purp., 1′. Apr., May.

11 A. CORONÀRIA. Lvs. multifid, segm. linear; sep. 6, roundish, close. Levant. May.

12 A. HORTÉNSIS. Lvs. 3-parted, with cuneate cut-dentate lobes; invol. sessile; sep. 10—12, oblong. Italy. Varieties are double, semidouble, red, white, blue, &c. May.

13 A. JAPÓNICA. Lvs. of the involucre and involucels broadly 3-5-lobed; fls. many, 18″ broad, white and red; sepals in 2 rows, roundish, widely spreading. Autumn.

3. HEPÁTICA, Dill. LIVERLEAF. LIVERWORT.

Invol. of 3 entire, ovate, obtuse bracts, resembling a calyx, situated a little below the flower. Calyx of 5—9 petaloid sepals, disposed in 2 or 3 rows. Cor. 0. Achenia awnless. ♃ Lvs. all radical, cordate, 3-lobed, thick, evergreen. Flowers single, on hairy scapes, appearing in early Spring before the new leaves Figs. 332, 431. Cultivated as a border flower.

1 H. tríloba Chaix. *Round-lobed L.* Lvs. with 3 round-obtuse lobes; bracts of the invol. obtuse. Woods, N. Eng. Scapes and leaf-stalks 3—4′. Fls. blue, varying to white, neat and elegant, becoming double in cultivation.

2 H. acutíloba DC. *Acute-leaved L.* Lvs. with 3 acute lobes, bracts of the invol. acute. Borders of woods, Vt. to Wis. 4—5′. Flowers violet-blue to rose-purple.

4. THALÍCTRUM, Tourn. MEADOW RUE.

Calyx colored, of 4—5 concave, caducous sepals. Petals 0. Filam. dilated upward, longer than the sepals. Ov. 4—15. Ach. stiped or sessile, ribbed or inflated, short-beaked. ♃ Lvs. ternately compounded, with stalked leaflets. Lfts. 3-7-lobed. Flowers paniculate, often diclinous, of no beauty.

* Flowers dioecious, in loose panicles. Styles slender. Achenia sessile or nearly so, ovoid, conspicuously angled and grooved..................................Nos. 1—3

* Fls. perfect, few in the corymbed clusters. Sty. short. Ach. long-stipitate....No. 4

1 T. dioìcum L. Slender, glaucous, glabrous (1—2f); leaves all petiolate (with the general petiole); fls. in slender panicles, purplish or greenish; fil. capillary, drooping, achenia about 8. Hilly woods: common. Leaflets thin, 5-7-lobed. April, May.

2 T. cornùti L. Stouter, tall (3—4f), smoothish; stem leaves sessile (no common petiole); lfts. thickish, veiny, with acutish lobes; anthers on white erect filaments achenia about 12, substipitate. Meadows. Leaflets 3-lobed. July, August.

ORDER 1.—RANUNCULACEÆ. 19

3 T. purpuráscens L. Stem tall (3—6f), purple; stem leaves sessile, or nearly so, lfts. thick and firm, with rolled edges, pale and often glandular-downy beneath; anth. linear, drooping; achenia sessile, as long as their stigmas. Hilly woods. June, July.

4 T. clavàtum DC. Slender (1—2f); lvs. petiolate, biternate. lfts. obtusely lobed; ach. curved, 5—10, short-pointed, long-stipe. Mts., N. Car. to Ala. White. July.

5. TRAUTVETTÈRIA, Fisch. & Meyer. Sep. 4 or 5, colored, caducous. Pet. 0. Filam. petaloid. Ach. 15—20 in a head, membranous, inflated, angular, tipped with the short hooked style. ♃ Leaves palmately lobed, alternate. Flowers corymbous, white.

T. palmàta F. & M.—Prairies and woods. Can. to Va., W. to the Cascade Mts.! 3—5f. Radical lvs. large, 5-9-lobed; stem lvs. few; corymb terminal. July, August.

6. ADÒNIS, L. PHEASANT'S-EYE. Sepals 5. Petals 5—15, the claw naked (no scale). Achenia spiked on the torus, ovate, pointed with the persistent style. Herbs with dissected leaves, and bright, showy flowers.

1 A. vernàlis. Fls. cup-shaped, yellow, of 10—12 oblong petals. ♃ Eur. 6—10'. May.

2 A. autumnàlis. Fls. globular, red, of 5—8 concave petals. ① Eur. 1f. Aug., Sept.

7. RANÚNCULUS, L. CROWFOOT. BUTTERCUPS. Sepals 5, ovate. Pet. 5—10, roundish, shining, each with a honey-scale (Fig. 39) or pore at the base inside. Ach. flattened, pointed, crowded in a head. ♃ ① Leaves alternate. Flowers generally yellow. Figs. 39, 83, 84, 109, 118, 159, 212, 234, 415, 416.

§ BATRÁCHIUM. Petals white, with a yellow, naked honey-pore on the claw. Seeds (achenia) transversely wrinkled. Leaves multifid, in water............................ No. 1

§ RANÚNCULUS. Petals (yellow) with a honey-scale on the claw of each...(*)
 * Achenia rough with points or prickles. Leaves palmate-parted. ①...... Nos. 18, 19
 * Achenia smooth,—*x* numerous, in an oblong head. Wet places........... Nos. 7—9
  —*x* many, in a rounded head...(*a*)
  a Leaves many-cleft, in thread-like segments, under water........................ No. 2
  a Leaves all undivided, entire or toothed. In wet places.............. Nos. 3—6
  a Lvs., at least the lowest ones, undivided, merely lobed or crenate... Nos. 10—12
  a Leaves all deeply divided, the lower—*y* pinnately with stalked lfts.. Nos. 13—15
   —*y* palmately with sessile lfts... Nos. 16, 17
    Exotic, cultivated..... Nos. 20, 21

1 R. aquátilis L. β. *trichophýllus* Chaix. *White Water-C.* Leaves all filiformly dissected and submersed. ♃ In slow streams. July, Aug. (R. divaricatus Schrank.)
γ. *heterophýllus* DC. Upper leaves floating, 3-5-lobed. Near Boston (Bigelow, now lost). In Idaho (Walker). Submersed leaves as in β.

2 R. multifidus Ph. *Yellow Water-C.* Floating or creeping; some of the leaves emersed, reniform, 3-5-parted, and cleft. Sepals reflexed; carpels with a straight beak, heads globous. Ponds and muddy shores, 1—2—3f. Petals 5—8. May, June.

3 R. Flámmula L. *Spearwort.* Stem erect from an ascending base; lvs. all lance-shaped, on sheathing petioles; ach. roundish, twice longer than its beak. Can. to Car., W. to Oreg. 8—16'. Lvs. 3—6'. Fls. showy. Sum. (R. alismæfolius Geyer.)

4 R. reptans L. Stem creeping, geniculate, rooting, filiform; nodes 1-flowered; lvs. linear or oblong; pet. 5—10, bright. N. Eng. to Oreg. Delicate. Fls. 4''. Lvs. 1'. Jl.

5 R. pusíllus Poir. Erect; lvs. all petiolate, lower ovate, upper lance-linear; pet. 3 (1—5) short; stam. 8—10; carp. scarcely pointed. N. Y. to Ga., and La. 6—12'. May

6 R. oblongifòlius Ell. Erect, diffuse; lvs. lance-ovate and lanceolate, all stalked pet. 5, stam. 20; carp. pointless. Ill. to Tex. June 2f. (R. Texensis Eng.)

7 R. Cymbalària Ph. St. filiform, creeping, rooting; lvs. reniform-cordate, crenate-dentate above; scapes 1-5-flowered (2—6'); petals 5—8, oval; carpels striate, beak short, uncinate. Brackish shores, N. J. to Dak. (Matthews). June.

8 R. sceleràtus Ph. Erect, smooth: root lvs. 3-lobed, lower stem lvs. 3-parted and cut-crenate; fls. small; carp. point' ss. Wet. Can. to Ga. 1f. Head 3". Jn.—Aug.

9 R. Pennsylvánicus L. Very .rsute; leaves ternate, lfts. subpetiolate, deeply 3-lobed and cut; sep. reflexed, longer than the 5 pet.; carp. beaked. Wet. 2f. Ju.-Aug.

10 R. abortìvus L. Very smooth; root lvs. roundish cordate, crenate, petiolate; upper leaves in 3 linear segments; sepals reflexed, longer than the very short petals. Woods: common. 8—16'. Flowers very small. Pretty. May, June.

11 R. recurvàtus Poir. Hirsute with thin spreading hairs; leaves all similarly 3-parted, lobes incised; sepals recurved, longer than the petals; carpels with a hooked beak. Woods. 1f. Pale green. Flowers small. May—July.

12 R. rhomboìdeus Goldie. Hairy, much branched; root lvs. rhomboid-ovate, crenate-dentate, long-stalked; sep. spreading, shorter than the petals; achenia smooth, with a very short beak. Prairies, Ill., Mich., Wis., Can. 6—10'. May.

13 R. fasciculàris Muhl. *Early C.* Erect; root a fascicle of fleshy fibres; root leaves appearing pinnate; peduncles terete; carpels scarcely margined, beak slender. Rocky hills. 5—10'. Hairs silky. Flowers 1' broad. April, May.

14 R. repens L. Root fibrous; later stems creeping, long; root leaves ternate, with stalked leaflets; pedicels furrowed; carpels broadly margined and stout-beaked. Moist shades. 1—3f. Flowers showy. Hairy or smooth. Very variable.

15 R. bulbòsus L. Hairy: stem erect, bulbous at the base; root leaves ternate, segments petiolate, incised; ped. furrowed; sepals reflexed. Fields, N. Eng., to Pa. 1f. May, Jn. The cup-shaped flower, golden-yellow, is larger and handsomer than No. 17.

16 R. palmàtus Ell. Erect; leaves 3-5-cleft, with the sinus at the base closed, segments all sessile, cut-dentate, or lobed; carpels margined and straight-beaked. Pine woods, Car. to Fla. 1f—18'. Pubescent. Flowers small (7"). April, May.

17 R. acris L. *Buttercups.* Erect; leaves deeply trifid, the base segments divaricate, all laciniate and sessile; pedicels terete; carpels with a short recurved beak. Common in N. Eng. and Can. Hairy. 2f. Flowers large, 1' broad. June—Sept.

18 R. muricàtus L. Glabrous; carpels aculeate, strongly margined, ending in a stout recurved beak. Va. to La., also in Cal. 1f. Leaves lobed and toothed.

19 R. parvifloròrus L. Villous; carpels rounded, granulated, tipped with a very short beak. Va. to La. 6—12'. Flowers small. March, April.

20 R. Asiáticus. *Garden Ranunculus.* Erect; leaves ternate or biternate, segments incised or lobed; head of carpels cylindric. Levant. 1f. Flowers variegated endlessly, of every form and hue. Not hardy.

21 R. aconitifòlius. Branching and many-flowered; leaves palmately 3-7-parted and cut-toothed, the upper sessile, with lance-linear lobes; calyx appressed; petals pure white From Europe. A fine old border flower, deep green, the flowers often double.

8. MYOSÙRUS, Dill. MOUSE-TAIL.

Sep. 5, produced downward at base below their insertion. Petals 5, with slender, tubular claws. Stamens 5—20. Achenia spicate on the spindle-shaped torus. ① Leaves linear, entire, radical. Scapes 1-flowered. Fig. 132.

M. mìnimus L. Low grounds, Ill. to La., W. to Oreg.! A curious little plant, remarkable for its tall torus, covered with numerous blunt carpels. Pet. yellow. Apr.

9. ISOPỲRUM, L. FALSE RUE ANEMONE.

Sep. 4, petaloid, deciduous. Pet. 5, small, tubular, sometimes 0. Follicles 3 or more, subsessile, pointed with the style, with 2 or more seeds. Delicate herbs. Leaves ternately compound, lfts. 2-3-lobed. Flowers pedunculate, white. Fig. 33.

ORDER 1.—RANUNCULACEÆ.

I. biternàtum T. & G. Glabrous, erect; stems clustered; pet. 0; follicles 3—6, strongly veined, 2-seeded. ♃ Damp shades, O. to Ark. 4—10'. May. Very pretty.

10. CALTHA, L. COWSLIP. MARSH MARIGOLD. Sepals 5—9, petaloid. Petals 0. Follicles 5—10, oblong, pointless, spreading, ∞-seeded. ♃ Very glabrous, aquatic.

C. palústris L. Stem hollow, thick; leaves thickish, large, orbicular or reniform, crenate or entire; flowers yellow. Wet meadows. 1f. Flowers 18'' broad. May.

11. COPTIS, Salisb. GOLD-THREAD. Sepals 5—7, oblong, concave, colored, deciduous. Petals 5—7, clavate, tubular at apex. Follicles 5—10, stipitate, rostrate, divergent, 4-6-seeded. ♃ Low, smooth, with radical leaves and flowers on a scape.

C. trifòlia Salisb. Leaves 3-foliate, leaflets sessile; scapes 1-flowered; pet. small and stamen-like; rhizome thread-like, of a golden yellow. Penn. to Can. 3—4'. Flowers white, the small yellow petals inconspicuous. Root bitter, tonic.

12. TRÓLLIUS, L. GLOBE-FLOWER. Sep. 5—15, petaloid. Pet. 5—25, small and inconspicuous, linear, tubular at base. Stam. and pistils ∞; follicles ∞-seeded. ♃ Smooth, with palmately-parted leaves.

1 T. laxus Salisb. Sepals 5, rounded, spreading; petals shorter than the stamens, orange-colored. Swamps, Can. to Penn. and Del. Rare. 1f. Flowers 18'' broad; sepals yellow, greenish outside. Pods about 10. June.
2 T. EUROPÆUS. Sepals 15, incurved, concave; petals 5—10, as long as the stamens. From Europe. 2f. Yellow. June, July. Hardy, and very ornamental.
3 T. ASIÁTICUS. Sepals 10, partly open; petals 10, longer than the stamens. From Asia. 2f, with ample foliage and orange-red flowers, varying to yellow. June, July.

13. HELLÉBORUS, L. HELLEBORE. Sepals 5, mostly greenish, persistent. Petals 8—10, very short, tubular, 2-lipped. Stigmas 3—10, orbicular. Follicles ∞-seeded. ♃ Leaves coriaceous, palmately or pedately divided. Flowers large, nodding. Fig. 494.

1 H. víridis L. Glabrous; rt. lvs. pedate, cauline palmate, sessile; fls. often in pairs; sepals round-ovate, acute, pale yellowish-green, spreading 1'. From Eur. 1f. § Apr.
2 H. NIGER. *Christmas Rose.* Root lvs. pedate; scape naked, bracted, 1- or 2-flowered; fls. 2' broad, white, pink, and finally green. In England, it flowers about Christmas 1f. Leaves thick, evergreen, and shining. March, April.

14. ZANTHORHÌZA, L. YELLOW-ROOT. Sep. 5. Pet. 5, of 2 roundish lobes raised on a claw. Stam. and pistils 5—10. Ova. 2- or 3-ovuled, follicles mostly 1-seeded, seed suspended. ♄ Roots and bark yellow and bitter. Leaves pinnate. Racemes axillary. Flowers dark purple.

Z. apiifòlia L'Her.—River banks, N. Y. to Ga. Lvs. clustered at top of the short, thick stem; leaflets 5, sessile, incised; racemes compound. Fls. 3'' broad. Apr.

5. NIGÉLLA, L. FENNEL-FLOWER. Sep. 5, petaloid. Pet. 5, 3-cleft. Pistils 5, becoming as many follicles which are distinct or united. ① Lvs. 1-2-pinnately divided into linear-subulate segments. Fig. 343.

1 N. DAMASCÈNA. *Ragged Lady.* Flowers in a leafy involucre; carpels united into a roundish, tumid capsule. From Spain. 2f. Flowers light blue. June—Aug.
2 N. SATÌVA. *Nutmeg-flower.* Hairy; flowers not involucrate; carpels distinct. Egypt

16. AQUILÈGIA, L. COLUMBINE. Sepals 5, equal, ovate, spreading, colored. Petals 5, all alike, horn-shaped, attached by the margin of the dilated mouth, produced to a honey spur behind. Pistils 5, follicles 5, many-seeded. ♃ Leaves bi-triternate, leaflets lobed. Flowers large and handsome, nodding. April—June. Figs. 127, 155, 156.

* Flowers scarlet, red, and orange-colored. Spurs of the petals straight....Nos. 1—3
* Flowers blue and white. Spurs straight in No. 4,....incurved in........Nos. 5—7

1 **A. Canadénsis** L. Very smooth, 1—2f; lfts. 3—9, round-wedge-form; fls. nodding, yellow within; stamens and styles yellow, exserted. Rocky woods, and cultivated.
2 **A.** SKÍNNERI. Like No. 1, but with larger fls., the spurs and sep. greenish. Mexico.
3 **A.** FORMÒSA. Sepals and spurs much longer than the petals; sta. included. Kamt.
4 **A.** CŒRÙLEA. Like No. 3, but the fls. all larger, blue and white, 2½′ long. R. Mts.
5 **A.** VULGÀRIS. *Common C.* Spurs little longer than the limb; stam. scarcely exserted. Europe.—Varies to purple, and white; also with double flowers,—spur within spur.
6 **A.** SIBÍRICA. Stem smooth, nearly naked, few-flwd., 1½f; spur some longer than the white-tipped limb; sepals very obtuse, violet. Very fine and choice like the next.
7 **A.** GLANDULÒSA. Glandular-hairy above; stems bracted, 1-2-flwd., 1f; spurs half as long as the snow-white limb; sepals sky-blue, acute, 1′ long. From Siberia.

17. DELPHÍNIUM, L. LARKSPUR. Flowers irregular. Sepals 5, colored, the upper one spurred behind. Petals 4, very unequal, the two upper spurred and enclosed in the spurred sepal. Styles and follicles 1—5. Handsome herbs, with palmately-divided leaves. Flowers of the cyanic series, never yellow. Figs. 26, 87, 88, 126.

§ CONSÓLIDA. Petals united into one piece. Style and follicle 1. ①........Nos. 4, 5
§ DELPHINÁSTRUM. Pet. 4, distinct. Pistils and follicles 2—5, mostly 3. ♃...(*a*)
   a Species indigenous, Penn., South and West, often cultivated..........Nos. 1—3
   a Species exotic, cultivated, natives of Siberia and California..........Nos. 6—9

1 **D. tricórne** Mx. Low (6—12′); leaf-lobes linear; raceme few-flwd., loose; spur ascending, straight; pods recurved. Uplands. Fls. 6—12, blue, white. April, May.
2 **D. azùreum** Mx. Erect (1—2f); leaf-lobes all narrow-linear; raceme strict; spur ascending; pods erect. Wis. to Ark. Flowers ∞, azure, or light blue. May, June.
3 **D. exaltàtum** L. Tall (2—4f); leaf-lobes wedge-lanceolate; rac. strict, ∞-flowered; spur straight; pods erect. Mich. to Car. Rac. panicled; fls. purp.-blue. July.
4 **D. Consólida** L. *Field L.* Branching; lvs. finely cut; fls. loosely racemed, scattered; pod smooth. Fields, gardens. 3—4f. Fls. blue, variable. Aug., Sept. § Eur.
5 **D.** AJÀCIS. *Rocket L.* Subsimple; leaves finely cut; flowers many, in crowded racemes; pod pubescent. Alps. 1—2f. Flowers pink, rose, white, often double.
6 **D.** ELÀTUM. *Bee L.* Pubescent, tall (5—6f); leaf-segments 5, cuneate, cut-trifid; rac. long; spur curved downward; petals hairy, resembling a bee inside the flower. Blue.
7 **D.** GRANDIFLÒRUM. Lvs. 5-7-parted, segm. 3-cleft, linear, distant; petals shorter than the calyx. Stem 2f. Flowers large, dark or purplish blue, often double.
8 **D.** CHILÁNTHUM. Leaf-lobes 3 or 5, oblong, acuminate; pods pubescent; sep. shorter than the calyx; spur decurved. Siberia. 2f. Dark blue.—Var. FORMÒSUM is very beautiful, blooming from July to Nov., the large flowers light blue, white at centre.
9 **D.** CARDINÀLE. Glabrous; lvs. 3-parted, segm. cleft into long acute lobes; fls. scarlet, large; spur longer than the sepals. California. 1—2f. Splendid, but not hardy.

18. ACONÌTUM, Tourn. WOLFBANE. MONK'S-HOOD. Sep. 5, irregular, colored, upper one (helmet) vaulted. Petals 2 (the 3 lower minute or 0), spurred at apex, on long claws, concealed beneath the helmet. Sty. and pods 3—5. ♃ Lvs. palmate. Fls. racemed or panicled. Poisonous. Fig. 29.

ORDER 1.—RANUNCULACEÆ.

1 A. uncinàtum L. Erect, weak (2f); leaf-divisions rhomb-lanceolate, cut-dentate; helmet obtusely conical, erect, short-beaked in front; flowers blue. Mts., N.Y. to Ga. Leaves thick, 4—5' wide. Branches divergent. Panicle loose. June, July.
2 A. reclinàtum Gray. Trailing (2—7f); leaf-divisions wedge-shaped, cut or lobed; helmet elongated-conical, with a straight beak; flowers white. Mountains, Va.
3 A. NAPÉLLUS. *Common Monk's-hood*, or *Aconite*. Smooth and rigidly erect, 3f; lvs. 5-parted, and cut into broad-linear segm. channelled above; fls. densely racemed, dark blue (or white in β. ALBUM), the hood broader than high. From Europe. Summer.
4 A. ANTHÒRA. Erect (1—2f); lvs. multifid with narrowly linear segm.; fls. panicled, large (as in the others), purple with yellow; hood rather high-crowned. Europe.
5 A. JAPÓNICUM. Smoothish, veiny, 3—5f; fls. deep blue, in panicled spikes; hood or helmet very high-crowned and inflated, with a thickened inflexed spur. Japan.
6 A. VARIEGÀTUM. Erect (3—4f), very smooth; leaves with rhomb-ovate divisions; fls. loosely panicled, blue, edged with white; helmet crown high, curved forward. Jn.+

19. CIMICÍFUGA, L. BUGBANE. Sepals 4 or 5, caducous. Petals stamen-like, 1—8, clawed, 2-horned at apex; follicles 1—8, dry, dehiscent. Leaves ternately decompound. Flowers white, in long racemes.

§ MACRÒTIS. Pistil 1, with a broad stigma and seeds in two rows..............No. 1
§ CIMICÍFUGA. Pistils 3—8, with a minute stigma, seeds in one row.........Nos. 2, 3

1 C. racemòsa Ell. *Black Snakeroot.* Tall (5—8f); rac. very long (1—3f), plume-like with its innumerable white stamens. Woods, Can. to Ga. Fetid. July.
2 C. Americàna Mx. Leaves triternate, thin; racemes slender, panicled; ovaries mostly 5, pods obovate, stiped. Mountains, Penn. to N. Car. 3—4f. Aug., Sept.
3 C. cordifòlia Ph. Leaves biternate, thick; racemes panicled, slender; ovaries 2 or 3; pods oblong, sessile. Mountains, N. Car. 3—4f. Sept.

20. ACTÆA, L. BANEBERRY. Sep. 4 or 5, caducous. Pet. 4—8, spatulate, long-clawed. Fil. slender. Ov. 1, with a sessile, 2-lobed stigma. Berry globous, with a lateral furrow, 1-celled, ∞-seeded. ♃ Lvs. ternately divided. Lfts. ovate, cut-lobed and toothed. Fls. white, in a short raceme.

A. spicàta L. β. *rubra* Mx. Raceme hemispherical; petals acute; pedicels slender; berries red, ovoid-oblong. Woods, Can. to Penn., and W. 1½—2f. Lvs. ample. Raceme as broad as long. May. These plants are often described as species.
γ. *alba* Mx. Raceme oblong; petals truncate; berries white, on thick stalks. Can. to Ga. Common. White berries sometimes occur with slender pedicels, and *vice versa*. Foliage exactly as in β. *Var. a.* is European.

21. HYDRÁSTIS, L. TURMERIC-ROOT. Sepals 3, petaloid, caducous. Pet. 0. Ovaries 12 or more, becoming a baccate fruit, resembling a raspberry; acines 1- or 2-seeded. Roots yellow, a tangled mass, sending up a single radical leaf and a stem which is 2-leaved and 1-flowered. Fig. 101.

H. Canadénsis L.—In damp woods, Can. to Car. and Ky. 1f. Leaves palmately 3-5-lobed. Flower terminal, reddish-white. Fruit crimson. June.

22. PÆÒNIA, L. PÆONY. Sepals 5, unequal, leafy, persistent. Petals 5. Ovaries 3—5, surrounded by an annular disk. Follicles ∞-seeded. ♃ Root fasciculate. Leaves ternately or pinnately compound. Flowers large, terminal, solitary. Figs. 36, 241.

§ Stems shrubby, perennial. Ovaries and pods 5. China...............Nos. 1, 2
§ Stems herbaceous, annual.—*x* Leaflets entire or cut-lobed. Ovaries 2 or 3 . Nos. 3, 4
—*x* Leaflets many-cleft. Ovaries 5..............Nos. 5, 6

1 P. MOUTAN. *Tree Pæony.* Ovaries distinct, half enveloped in the disk. 3—4f, widely branching. Flowers large, double, purple varying to white. June.

2 P. papaverácea. Ovaries closely united into a globous capsule. 3f. Fls. white, with a purple centre, 8—10′ broad, single or double, varying to rose. May, June.

3 P. officinàlis. *Common Red P.* Lfts. lance-ovate, incised: carpels 2, pubescent, suberect. Alps. Fls. double, red, rose, pink, flesh-colored, and white. June.

4 P. albiflòra. *Chinese P.* Lfts. lance-elliptic, entire; carpels 2 or 3, recurved, smooth; calyx bracteate. Tartary. Fls. smaller, white, rose, carmine, &c.

5 P. anómala. Leaf-segments lance-linear; carp. depressed, smooth; cal. bracted. Siberia. Fls. concave, rose-colored, pink, &c. May, June.

6 P. tenuifòlia. *Fennel P.* Segments many linear lobes, very smooth; carpels downy, spreading. Siberia. 2—3f. Fls. red, concave, open the first of May.

Order II. MAGNOLIACEÆ. Magnoliads.

Trees or shrubs, often aromatic, with alternate, undivided leaves, and regular, polygynous, hypogynous, trimerous, imbricated flowers. *Sepals* and *petals* in several circles, often similar. *Anthers* adnate. *Ovaries* imbricated or verticillate on the enlarged torus, 1 or 2-ovuled. *Fruit* dry or baccate, distinct or coherent into a cone-like head (sorosis) *Embryo* minute, at the base of fleshy albumen. Illust. figs. 274, 278, 331.

```
WINTEREÆ.   Stipules 0. Fls. ☿. Carpels arranged in a circle............ ...ILLICIUM.     1
MAGNOLIEÆ.  Stipules caducous. Fls. ☿. Carpels imbricated. ∞-rowed. (a)
 a Anthers introrse. Leaves folded lengthwise in bud................... ........MAGNOLIA.  2
 a Anthers extrorse. Leaves folded crosswise in the bud....................LIRIODENDRON.   3
§ SCHIZANDREÆ.  Stip. 0.  Fls. ♂ ♀. Carpels in many rows, baccate..........SCHIZANDRA.    4
```

1. ILLÍCIUM, L. Star Anise. (Lat. *illicio*, to attract; alluding to its fragrance.) Sep. 3—6, colored. Pet. 6—30. Carpels capsular, dry, arranged circularly, each with 1 smooth, shining seed. ♄ The smooth lvs., when bruised, exhale the odor of Anise. In wet grounds. May.

1 I. Floridànum Ellis. Lvs. acuminate; petals 21—30, purple. Fla. to La. 4—8f.
2 I. parviflòrum Mx. Lvs. acute; petals 6—12, yellow. Ga. Fla. Fls. smaller.

2. MAGNÒLIA, L. (Named for *Prof. Magnol*, a French botanist of the 17th century.) Sep. 3. Pet. 6—9. Anth. longer than the filaments, introrse. Ov. imbricated, 1-celled, 2-ovuled, becoming in fruit a fleshy, cone-like *sorosis*. Seeds berry-like, suspended from the opening follicles by a slender funiculus. ♄ and ♄, with large fragrant flowers. Lvs. conduplicate in bud, with membranous deciduous stipules. Fig. 331.

```
* Leaves cordate or auriculate at the base. Trees............... ............Nos. 5, 6, 7
* Leaves acute at the base,—rusty or glaucous beneath, coriaceous...........Nos. 1, 2
                       —green (not shining) both sides, thin.............Nos. 3, 4
                                            Exotic species, cultivated....Nos. 8–10
```

1 M. grandiflòra L. *Big Laurel.* Trees; lvs. evergreen, rusty-downy beneath; pet. obovate, white. Swampy woods, S. States. 80f. Fls. 9′ broad, lvs. 7×4′. May.

2 M. glauca L. *White Bay.* Shrub or small tree; lvs. obtuse, glaucous-white beneath; pet. ovate-roundish, erect. Coast, Ms. to La. 5—20f. Fls. 2′, cup-shaped. strongly fragrant, with white concave petals. Lvs. nearly evergreen. South. May–July.

3 M. acumináta L. *Cucumber Tree.* Lvs. oval, acuminate, scattered; fls. small (3—4′ broad). petals obovate. S. States, rare in N. Y. 70f. The cones of fruit bear some resemblance to a small cucumber. May.

4 **M. umbrélla** Lam. *Umbrella Tree.* Lvs. cuneate-lanceolate, whorled at the ends of the branches (like an umbrella); sep. reflexed; pet. lanceolate, acute. S. States, rare in N. Y. and O. 25f. Lvs. and fls. very large. White. May.

5 **M. cordàta** Mx. Lvs. broadly ovate, subcordate, pubescent beneath; petals 6—9, oblong, yellow, with reddish lines. Ga. Car. 40f. Lvs. downy beneath.

6 **M. Fràseri** Walt. Lvs. obovate-spatulate, auricled at the narrow base; pet. 6, pure white. Va. Ky. to Fla. 30f. Fls. 6'. Lvs. 1f. A slender tree.

7 **M. macrophýlla** Mx. Lvs. obovate-spatulate, cordate; pet. 6, rhomb-ovate, white, with a purple base inside. S. States. 20—30f. A small tree, with immense lvs. (2—3f) and fls. (petals 8' long). June.

8 **M.** conspícua. *Yulan.* Sep. 0 or very small; pet. 6—9, erect, of a creamy white, appearing before the leaves in early Spring. Lvs. acuminate. 15f.

9 **M.** purpùrea. Sep. 3; pet. 6, erect, lilac-purple outside, preceding the obovate lvs., which are pointed at both ends. China. 10—15f.

3. LIRIODÉNDRON, L. Tulip Tree. Whitewood. ($\Lambda\epsilon\iota\rho\iota\text{o}\nu$, a Lily, $\delta\epsilon\nu\delta\rho\text{o}\nu$, a tree.)
Sep. 3. Pet. 6, in 2 rows, erect. Anth. opening outward. Carpels 1 or 2-seeded, imbricated into a cone, indehiscent, separating from each other at maturity. ♄ Large, with showy, bell-shaped, upright flowers. Lvs. 4-lobed, retuse-truncate at apex, induplicate in bud, with large, caducous stipules. Figs. 274, 278.

L. tulipífera L.—A noble tree, beautiful in foliage and flowers; trunk 5—8f diameter; 100f or more high; lvs. very smooth; fls. greenish-yellow, orange within, abounding in honey. May, June.

4. SCHIZÁNDRA, Mx. ($\Sigma\chi\iota\zeta\omega$, to cut, $\check{\alpha}\nu\delta\rho\alpha$, stamens.)
Sep. and pet. 9—12, gradually larger inward. ♂ Stam. 5—15, monadelphous, anth. cells distinct. ♀ Carp. ∞, at first imbricated in a head, in fruit baccate, and loosely spicate on the lengthened torus. ♄ Lvs. pellucid-punctate, deciduous. Fls. solitary.

S. coccínea Mx. Lvs. ovate or oval, pointed; fls. on slender peduncles, small, red; stam. 5, in the upper fls. chiefly. Berries and torus red. Vine 12f. South

Order III. CALYCANTHACEÆ. Calycanths.

Shrubs with opposite, simple, exstipulate leaves, and axillary, solitary, often aromatic flowers. *Sepals* and *petals* ∞-rowed, imbricated on a tubular torus, the outer bract-like. *Filaments* ∞, inserted on the top of the torus, short. *Anthers* adnate, extrorse. *Carpels* ∞, 1-seeded, distinct, included in the green fleshy torus. *Seed* erect, without albumen.

CALYCÁNTHUS, L. Sweet-scented Shrub. (Κάλυξ, calyx, ανθος, flower.) Sep. and pet. oblong, undistinguishable, the inner gradually shorter. Stam. apiculate, the outer longer, inner sterile. Fruit, the enlarged green torus loosely enclosing few or many achenia. ♄ Fls. lurid purple, with the fragrance of strawberries.

1 **C. flóridus** L. Lvs. oval or elliptical, acute or acuminate, scabrous, downy beneath; fls. on very short axillary branches; sep. and pet. about 20, near 1' in length. S. States: common in gardens. Lvs. 2—5'. Shrub 4—8f. Apr. May.

2 C. lævigatus Willd. Lvs. thin, oval, obtuse or merely acute, nearly glabrous both sides; fls. smaller, sometimes inodorous. Pa., & S. to Fla. Mar. Apr.
3 C. glaucus Willd. Lvs. ovate, acuminate, large (4—7'), glaucous beneath; sep. and pet. lance-oblong, 1' in length. Mt. woods, Ga. to N. Car. 6—8f. May, June. †

Order IV. ANONACEÆ. Anonads.

Trees or shrubs with naked buds, entire, alternate lvs. destitute of stipules. *Flowers* usually green or brown, axillary, hypogynous, valvate in æstivation. *Sepals* 3. *Petals* 6, in two circles, sometimes coherent *Stamens* ∞, with an enlarged connectile, short filament, on a large torus. *Ovaries* several or ∞, separate or coherent, fleshy or not, in fruit. *Embryo* minute in the end of the ruminated albumen. Illust. fig. 314.

ASIMINA, Adans. PAPAW. Sep. 3. Pet. 6, the outer row larger than the inner. Stam. densely packed in a spherical mass. Pistils several, distinct, ripening but few, which become large, oblong, pulpy fruits, with many flat seeds. Shrubs or small trees, with brownish, axillary, solitary, flowers.

* Flowers appearing before the leaves. Petals purple..................Nos. 1, 2
* Flowers appearing with the leaves. Outer petals yellowish..................Nos. 3, 4

1 A. triloba Dunal. Lvs. obovate-oblong, acuminate; pet. dark purple, the outer orbicular, 3 or 4 times as long as the sepals; fruit ovoid-oblong. N. Y., S. and W. 15—20f. Lvs. 10', smooth. Fls. 1', Mar. Apr. Fr. 3', eatable in Oct.
2 A. parviflòra Dunal. Lvs. obovate-oval; pet. oval, green-purple, twice longer than sep. Woods, coastward. Car. to Fla. 2—3f. Lvs. 5'. Fls. 6''. Fr. 1', roundish.
3 A. grandiflòra Dunal. Lvs. obov.-obl. obtuse, grayish-tomentous; outer pet. very large (2' long), yellowish white. Ga. Fla. 2—3f. Fr. small, obovate. Mar. Apr.
4 A. pygmæa Dunal. Lvs. coriaceous, evergreen, narrowly oblong or oblanceolate, smooth; pet. obov.-obl., yellowish and brownish. Ga. Fla. 6—12'. Carp. 1'. May.

Order V. MENISPERMACEÆ. Menispermads.

Shrubs twining or climbing, with alternate, palmate-veined, exstipulate leaves. *Flowers* diœcious, rarely ☿ or ♀ ☿ ♂, hypogynous, 3–6-gynous. *Sepals* and *petals* similar, in 3 or more circles, imbricated in the bud. *Stamens* equal in number to the petals, and opposite to them, or 3 or 4 times as many. *Fruit* a 1-seeded drupe, with a large or long curved embryo in scanty albumen. Illust. 347.

♂ Stamens 12–20. Sep. 4–8, nut moon-shaped. Lvs. peltate..................MENISPERMUM. 1
♂ Stamens 6. Sep. 6, nut moon-shaped. Lvs. sinuate, 3-lobed..................COCCULUS. 2
♂ Stamens 6. Sep. 6, nut cup-shaped. Lvs. deeply 5-lobed..................CALYCOCARPUM. 3

1. MENISPÉRMUM, L. MOON-SEED. ($M\acute{\eta}\nu\eta$, the moon, $\sigma\pi\acute{\epsilon}\rho\mu\alpha$, seed; from the crescent form of the seed.) Fls. ♀ ♂. Sep. 4—8. Pet. 4—8, minute, retuse. ♂ Anth. 12—20, 4-celled. ♀ Ovaries and styles 2—4. ♄ Drupes 1-3-seeded. Seeds lunate and compressed. Fls. white, in axillary clusters. Fig. 347

ORDER 6.—BERBERIDACEÆ.   27

M. Canadénse L. St. climbing; lvs. 5–7-angled or lobed, peltate, the petiole inserted near the base; rac. compound; petals 6—7, small. ♄ Thickets: common. 8—12f. Drupes black, resembling grapes, ripe in Sept. Fls. in July.

2. **CÓCCULUS**, DC. (Diminutive, from Lat. *coccum*, a berry.) Fls. ♀ ♂. Sep., pet., and stam. 6. Anth. 4-celled. ♀ Ov. 3 to 6. Drupe globular-compressed, nut curved as in Menispermum. ♄ Fls. in axillary panicles, small, greenish.

C. Caroliniànus DC.—S. Ill. to Fla. 10—15f. Lvs. ovate or cordate, entire or lobed. Drupes red, 1–3 together, as large as a pea. June, July.

3. **CALYCOCÁRPUM**, Nutt. CUP-SEED. ($Κάλυξ$, a cup, $καρπός$, fruit.) Sep. 6. Pet. 0. ♂ Stam. 12. Anth. 2-celled. ♀ Stam. 6, abortive. Ov. 3. Stig. fimbriate-radiate. Drupe oval, with the putamen deeply excavated in front and cup-shaped. ♄ Fls. greenish-white, in long axillary panicles.

C. Lyòni Nutt.—Ga. to Ky. Vine 20—30f. Lvs. 6—8′ diam., lobes acuminate; drupe 1″ oval, greenish. Fls. small, 2″ diameter. June.

ORDER VI. BERBERIDACEÆ. BERBERIDS.

Herbs or shrubs with alternate leaves and with perfect, hypogynous, regular flowers. *Sepals* and *petals* imbricated in bud, each in one or several rows. *Stamens* as many as the petals, and opposite to them, rarely more. *Anthers* opening mostly by valves, hinged at top. *Pistil* 1. *Style* short or none. *Fruit* a berry or capsule. *Seeds* several, albuminous. Illust. 49, 91, 92, 189, 364, 403, 426.

§ Shrubs, with bristly-serrate leaves, yellow flowers and acid berries..............BERBERIS.   1
§ Herbs.—* Anthers opening by 2 valves hinged at the top...(a)
  a Stamens 6. Fruit 2, drupe-like, soon-naked seeds................CAULOPHYLLUM.   2
  a Stamens 6. Berry 1-4-seeded. Petals white, larger than sep.....DIPHYLLEIA.   3
  a Stamens 8. Pod opening by a lid. Petals 8....................JEFFERSONIA.   4
 —* Anthers opening by slits. Stamens 9—18...................PODOPHYLLUM.   5

1. **BÉRBERIS**, L. BERBERRY. (Name from the Arabic.) Calyx of 6 obovate, spreading, colored sepals, with the 3 outer ones smaller. Corolla of 6 suborbicular petals, with 2 glands at the base of each. Fil. 6, flattened. Anth. opening by uplifted valves. Style 0. Berry oblong, 1-celled. Seeds 2 or 3. ♄ with yellow wood and yellow fls. Figs. 91, 92, 403.

1 **B. vulgàris** L. Spines (reduced lvs.) 3-forked; lvs. simple, serratures terminated by soft bristles; raceme pendulous, many-flowered; pet. entire; berries oblong. N. States. 6—9f. Rac. 12-flowered. Berries red, very tart. May, June.

2 **B. Canadénsis** Ph. Lvs. repandly-toothed, teeth with short, soft bristles; rac few (6–8)-flowered; pet. notched; berries oval. Mts. Va. to Ga. 2—3f. May, June.

3 **B. Aquifòlium** Ph. Lvs. pinnate; lfts. 7—11, coriaceous, polished, evergreen, spinulous-toothed; clusters erect, crowded. Oregon. 3—5f. Berries globular. April.

2. **CAULOPHÝLLUM**, Mx. Conosh. ($Καυλός$, stem, $φύλλον$, leaf; the stem appearing as the stalk of the compound leaf.) Cal. of 6 green

sepals, 3-bracted at base. Cor. of 6 short, gland-like thickened petals, opposite the sepals. Stam. 6. Ov. 2-ovuled, becoming a thin pericarp, which soon breaks away after flowering, and the 2 round drupe-like seeds ripen naked. ♃ Glabrous and glaucous, arising from a knotted rhizome. Lvs 2 only, 2 and 3-ternate.

C. thalictroides Mx. *Pappoose Root.*—Can. to Car. and Ky. 1—2½f. Lfts. lobed 2—3'. Fls. greenish, in a simple terminal panicle. Seeds on thick st¹pes, blue, as large as peas. May.

3. DIPHYLLEIA, Mx. UMBRELLA-LEAF. ($\delta i s$, twice, $\varphi \iota \lambda \lambda o \nu$, leaf.) Calyx of 5 sepals, caducous. Cor. of 6 oval petals larger than the sepals. Stam. 6. Ov. eccentric. Stigma subsessile. Berry few-seeded, seeds attached laterally below the middle. ♃ Glabrous, arising from a thick, horizontal root-stock. Lvs. simple, peltate, 1 or 2 only.

D. cymòsa Mx.—Mts. Va. to Ga. and Tenn. 1—2f. Leaf centrally peltate, or if 2, alternately reniform-peltate, ample, lobed. Fls. white. June. Berries blue.

4. JEFFERSONIA, Bart. TWIN-LEAF. (In honor of *President Jefferson*, a patron of science.) Sep. 4. Pet. 8, spreading. Anth. 8, linear. Stig. peltate. Caps. obliquely obovate, stiped, circumscissile, opening by a lid. ♃ Rhizome and matted fibres blackish. Scape bearing a single flower, as tall as the 2-parted or binate leaves. Figs. 49, 189, 364, 426.

J. diphýlla Bart.—N. Y., W. and S. 1f. Fl. handsome, white. April. A singular plant, called *Rheumatism Root.* The pod has a persistent lid.

5. PODOPHÝLLUM, L. MAY APPLE. ($\Pi o \tilde{v} s, \pi o \delta \acute{o} s$, foot, $\varphi \iota \lambda \lambda o \nu$, leaf.) Sep. 3, concave, caducous. Pet. 6—9, obovate, concave. Anth. 9—18, linear. Berry large, ovoid, 1-celled, crowned with the solitary stigma. ♃ Barren stems with 1 centrally peltate leaf, flowering stems with 2 equal, opposite broad cordate-peltate leaves, and a large white flower between.

P. peltàtum L.—In rich shady soils. 1f. Fl. nodding, 2'. May. Fruit the size of a plum, with flavor of strawberry. July. Lvs. and roots poisonous.

ORDER VII. NYMPHÆACEÆ. NYMPHIADS.

Herbs perennial, aquatic (in deep water), with rhizomes submersed, scapes one-flowered (rarely a leafy stem), and leaves peltate or deep-cordate. *Flowers* regular, showy, hypogynous (rarely epigynous), with imbricated petals and sepals. *Carpels* 3—∞, distinct or united. *Ovules* parietal, never on the ventral suture. *Seeds* with the embryo enclosed in a sac at the end of copious albumen, or (in Nelumbium) exalbuminous. Illust. 202, 407–414, 505, &c.

§ CABOMBEÆ. Sepals 3. Petals 3. Carpels distinct, few-ovuled. Flowers small. (α)
  α Stam. 6. Carpels 3. Submersed leaves dissected...................CABOMBA. 1
  α Stam. 6–18. Carpels 6– ∞. Leaves all peltate.....................BRASENIA. 2
§ NELUMBONEÆ. Sep. 4 or 5. Pet. and stam. ∞. Carp. immersed in the torus, distinct, exalbuminous. Fls. very large............................ NELUMBIUM. 3

ORDER 7.—NYMPHÆACEÆ.   29

§ NYMPHEÆ. Sep. 4—6. Pet. and stam. ∞. Carp. united. Fls. large, showy. (b)
  b Pet. (stamen-like) and stam. hypogynous. Fls. yellow........................NUPHAR.   4
  b Pet. petaloid. Stamens epigynous (on the torus raised into a disk). ....NYMPHEA.   5
  b Pet. (petaloid), sep. and stamous epigynous. Lvs. peltate..................VICTORIA.   6

1. BRASENIA, Schreb. WATER TARGET. Sep. 3 or 4, colored within, persistent. Stam. 12—24. Pet. 3 or 4. Carp. 6—18, oblong, 2 (or by abortion 1)-seeded. ♃ The stems and under surface of the leaves are covered with a viscid jelly. Lvs. all floating, entire, elliptical.

B. peltàta Ph. Pools and muddy shores. The slender ped. and petioles long as the depth of the water. Lvs. 2½ × 1'. Fls. purple, 6'' broad. July.

2. CABOMBA, Aublet. Sep. 3, petaloid. Pet. 3. Stam. 6. Pistils 3 (rarely 2 or 4), nearly the length of stamens, and half as long as the petals and sepals. Carp. few-seeded. ♃ Lvs. opposite, mostly submersed and filiformly dissected. Fls. in the axils of the floating lvs.

C. Caroliniàna Gray. Floating lvs. few and small (6'' × 1''), immersed lvs. many. Stems branched. Fls. white, 6'', strictly trimerous. July, Aug.

3. NELUMBIUM, Juss. (*Nelumbo* is the name of the species in Ceylon.) Pet. and stam. ∞, hypogynous, in many rows. Carp. ∞, separate, becoming 1-seeded nuts, imbedded in as many cavities on the large, obconic, fleshy torus. Seed with large cotyledons, very short radicle and no albumen. Rhizome horizontal. Lvs. peltate, emersed. Scape 1-flowered. There are only 2 species, N. speciosum of E. India, and

N. lùteum L. Petals yellowish; anth. lengthened beyond the cells to a clavate appendage. A magnificent aquatic, frequent S. and W. In Sodus Bay, N. Y. (Hankenson), Lyme, Ct., near Philadelphia (Parish). Lvs. erect, round, centrally peltate, 10—18'. Fls. several times larger than those of Nymphæa odorata, fragrant. Nuts as large as acorns. June—Aug.

4. NUPHAR, Smith. YELLOW POND-LILY. (*Neufar* is the Arabic name.) Sep. 5 or 6, concave. Pet. ∞, small, linear, inserted with the ∞ stamens on the torus. Stig. discoid, with prominent rays. Caps. ∞-celled, ∞-seeded. ♃ Lvs. sagittate-cordate at the base, entire at the margin, on stout stalks.

1 **N. ádvena** Ait. Lvs. floating or erect, oval; lobes rounded, petioles half terete; stig. 12-24-rayed; sep. 6, unequal. Slow streams and muddy pools. Lvs. thick and large. Fls. deep yellow (save the 3 outer sep.), 2' diam., globular. June, July.
2 **N. Kalmiàna** Ait. Lvs. floating and submersed, the latter membranous, reniform-cordate; stig. 8-14-rayed, crenate; sepals 5, equal. Plant small and delicate. Floating leaves oval, 1—3' long, the lobes nearly meeting. Flowers about 1' diam. Sum.
3 **N. sagittifòlia** Ph. Leaves oblong, sagittate-cordate, obtuse; sep. 6; pet. 0; anth. subsessile. Slow waters, N. Car. to Ga. Lvs. 10—15'. Fls. 2', globular. June, July.

5. NYMPHÆA, L. WATER-LILY. Sepals 4 or 5. Pet. ∞, gradually passing into stamens, adherent to the ovary. Stamens ∞, the outer with broad filaments. Stigma surrounded with rays. Seeds ∞, arillate. ♃ ☞ Flowers white, roseate, or blue, very lovely. Figs. 202, 407 414.

1 **N. cœrùlea.** Lvs. crenate, lobes partly united, becoming peltate; pet. sky blue Egypt

2 **N. odorāta** L. Lvs. orbicular, entire, cleft at base to the insertion of the petiole fls. very fragrant, open from 6 A. M. to 3 P. M. upon the water's surface, white, varying to rose-color; seeds oblong. June—Aug.

3 **N. tuberòsa** Paine. Lvs. reniform-orbicular, cordate-cleft, 1f wide; rhizome bearing tubers, which separate spontaneously; fls. nearly scentless; seeds globular. N. Y. (Oneida Lake; Sodus Bay (Hankenson), and westward. Aug.

6. VICTŎRIA, Lindl.

(Name in honor of *Queen Victoria*.) Carp. immersed in the cup-form torus, united. Sep. 4. Pet. ∞, graduated into stamens, as in Nymphæa. Lvs. spiny, floating, strongly veined.

V. règia is the only species, native of the rivers of Trop. Am.; rarely cultivated. The lvs. are several feet in diam. Fls. like immense Water Lilies.

ORDER VIII. SARRACENIACEÆ. WATER PITCHERS.

Herbs, aquatic, in bogs, with fibrous roots, perennial, and with the *leaves* all radical, urn-shaped, or trumpet-shaped, and large flowers on scapes. *Floral envelopes* 4—10, imbricated, the outer greenish, sepaloid. *Stamens* ∞, hypogynous. *Carpels* united into a several-celled capsule. A curious family, remarkable for its leaves, which are of that class called *ascidia* (§ 322), holding water. Figs. 392, 393, 394.

1. SARRACÈNIA, Tourn. PITCHER PLANT.

(In honor of *Dr. Sarrazen*, of Quebec.) Sep. 5, colored, persistent, subtended by 3 bractlets. Pet. 5, incurved, deciduous. Stig. 5, united into a large peltate, persistent membrane, covering the ovary and stamens. Caps. 5-celled, 5-valved. Seeds very numerous. ♃ Lvs. all radical, urn-shaped or trumpet-shaped, with a wing on the front side and a hood (the lamina) at top. Fl. large, nodding.

§ Lamina inflected over the throat of the tube........................Nos. 1, 2
§ Lamina erect or nearly so, the throat open. (*)
　　* Leaf-tube pitcher-shaped, with a broad wing............................No. 3
　　* Leaf-tube trumpet-shaped, with a narrow wing..........................No. 4

1 **S. psittacìna** Mx. Lvs. short, reclined, with a broad semi-ovate wing; fls. deep purple. Bogs, Fla. Ga. La. 1f. Tube nearly closed. The leaf resembles a *parrot* in form, hence the specific name. March.

2 **S. variolàris** Mx. Lvs. elongated, suberect, mottled with white on the back; fls. yellow. Bogs, S. Car. to Fla. Lvs. 12—18′, scape shorter.

3 **S. purpùrea** L. *Side-saddle Flower*. Lvs. short, recumbent, inflated most near the middle; lamina broad-cordate. Bogs: common. Scapes 14—20′, each bearing large handsome deep-purple flower, in June.

　β. *heterophýlla* Torr. Fls. greenish yellow. No purple veins in the lvs. Ms
　γ. *alata*. Fls. large, yellow. Lvs. slender, erect, wing but 6′′′ broad. La. 1—2f.

4 **S. Gronòvii** Wood. *Trumpet-leaf*. Lvs. tall, erect, tube gradually enlarged to the open throat, wing narrowly linear, lamina roundish, contracted at base. Swampy pine-woods, S. States. 2—3f. Fls. very large, 4—5′ broad.

　α. *flava*. Foliage yellowish green, fls. yellow. Plant large.
　β. *rubra*. Foliage with purple veins, fls. red-purple. Plant smaller.
　ε. *Drummondii*. Lvs. mottled above, with purple veins and white diaphanous interstices. Plant very large. Fla.

Order IX. PAPAVERACEÆ. Poppy-worts.

Herbs with alternate, exstipulate leaves, and generally a milky or colored juice. *Flowers* solitary, on long peduncles, never blue, hypogynous, regular, ☿ or ⚥. *Sepals* 2, rarely 3, caducous, and *petals* 4, rarely 6, all imbricated. *Stamens* indefinite, but some multiple of 4. *Anthers* 2-celled, innate. *Ovaries* compound. *Style* short or 0. *Stigmas* 2, or if more, stellate upon the flat apex of ovary. *Fruit* either pod-shaped, with 2 parietal placentæ, or capsular, with several. *Seeds* ∞, minute. *Embryo* minute, at the base of oily albumen. Illust. 148, 344, 404, 405, 406, 463, 493.

¶ Plants with a white juice. Petals 4, crumpled in bud.........................PAPAVER. 6
¶ Plants with a watery juice. Calyx a mitre, falling off whole................ESCHSCHOLTZIA. 7
¶ Plants with a red juice. Petals 8, plane in the bud..........................SANGUINARIA. 1
¶ Plants with a yellow juice. Petals crumpled in the bud. (*)
 * Stigmas and placentæ 2 only. Capsule long, pod-shaped. (*a*)
 * Stigmas and placentæ 3, 4, or 6. Capsule ovoid. (*b*)
  a Pod 1-celled, smooth. Lvs. pinnate.............................CHELIDONIUM. 2
  a Pod 2-celled, rough. Lvs. palmate..............................GLAUCIUM. 3
  b Style distinct, but short....................................MECONOPSIS. 4
  b Style none, stigma sessile...................................ARGEMONE. 5
 † No petals. Juice reddish......................................BOCCONIA. 6

1. SANGUINÀRIA, L. BLOOD-ROOT.
(Latin *sanguis*, blood; all its parts abound in a red juice.) Sep. 2, caducous. Pet. 8—12, in 2 or 3 rows, the outer longer. Stam. about 24. Stig. sessile, 1 or 2-lobed. Capsule silique-form, oblong, 1-celled, 2-valved, acute at each end, many-seeded. ♃ A low, acaulescent plant, with a white flower, and a glaucous, palmate-veined leaf. Fig. 463.

S. Canadénsis L. An interesting flower, appearing in early Spring; common in the woods. 6'. From each bud of the root-stalk there springs a single large, glaucous leaf, and a scape with a single flower. Leaf kidney-shaped, with roundish lobes separated by rounded sinuses. Fl. of a quadrangular outline, white, scentless, and of short duration. The juice is emetic and purgative.

β, Leaf not lobed, margin undulate. Bainbridge, Ga., and elsewhere.

2. CHELIDÒNIUM, L. CELANDINE.
($\chi\epsilon\lambda\iota\delta\grave{\omega}\nu$, the swallow, being supposed to flower with the arrival of that bird, and to perish with its departure.) Sep. 2. Pet. 4, roundish, contracted at base. Stam. 24—32, shorter than the petals. Stig. small, sessile, bifid. Capsule silique-form, linear, 2-valved, 1-celled. Seeds crested. ♃ Fragile, pale green, with saffron-yellow juice. Figs. 344, 493.

C. màjus L. Lvs. pinnate; lfts. lobed, segments rounded; fls. in umbels. By fences, roadsides, &c. 1—2f. Fls. in loose umbels, yellow, very fugacious. May—Oct.

3. GLAUCIUM, Tourn. HORN POPPY.
($\Gamma\lambda\alpha\upsilon\kappa\grave{\omega}\nu$, glaucous, the hue of the foliage.) Sep. 2. Pet. 4. Style none. Stig. 2-lobed. Pod 2-celled, linear, very long, rough. ① or ② sea-green herbs, with clasping leaves, yellow juice, and solitary, yellow flowers.

G. lùteum Scop. Sparingly naturalized near the coast, from the Potomac southward. 2f. Lvs. 5-7-lobed. Fls. 2', of short duration. Pods 6–9'. June, Aug.

Order 9.—PAPAVERACEÆ.

4. MECONÓPSIS, Viguier. YELLOW POPPY. (*Μήκων*, a poppy, *όψις*, resemblance.) Sep. 2, hirsute. Pet. 4. Style conspicuous. Stig. 4—6, radiating, convex, free. Capsule ovoid, 1-celled, opening by 4 valves. ♃ Herbs with a yellow juice, pinnately-divided leaves, and stems 2-leaved, bearing an umbel.

> **M. diphýlla** DC. Lvs. sinuately 5–7-lobed, the cauline but 2, opposite; fls. few. large (?), yellow; pod bristly, oval. Woods, W. States. 12—18′. Pet. orbicular; style surpassing the stamens; pod 3′. May.

5. ARGEMÒNE, L. PRICKLY POPPY. ("*Αργεμος*, a disease of the eye, which this plant was supposed to cure.) Sep. 2 or 3, caducous, smaller than the 4 or 6 roundish petals. Stig. sessile, capitate, 4 or 6-rayed. Capsule ovoid, prickly, opening at the top by valves. ①. Herbs with yellow juice, spinous-pinnatifid leaves, and showy flowers.

> **A. Mexicàna** L. Calyx prickly; caps. prickly, 6-valved; fls. axillary and terminal, 2—3′ diam., yellow, varying to white. Waste grounds, South.

6. PAPÀVER, L. POPPY. (Celtic, *papa*, pap, a soporific food for children, composed of poppy seeds, &c.) Sep. 2, caducous. Pet. 4. Caps. 1-celled, opening by pores under the broad, persistent 4—20-rayed stigma. Exotic herbs, with white juice, abounding in opium. Fl. buds nodding, erect in flower and fruit. Figs. 148, 404–6.

> 1 **P. somníferum** L. *Opium Poppy*. Glabrous and glaucous; lvs. clasping, cut-dentate; caps. globous. ① with large white or purplish flowers, often double. 1½—3f. Extensively cultivated for opium. June, July. §.
> 2 **P. dùbium** L. St. hispid with spreading hairs; lvs. pinnately-parted; segm. incised; sep. hairy; caps. club-shaped. ① Fields. 2f. Slender. Fls. light red or scarlet. June, July. §.
> 3 **P. Rhǽas** L. St. many-flowered, hairy; lvs. incisely pinnatifid; caps. globous. ① Fls. very large, deep scarlet, more or less double. June, July.
> 4 **P. orientàle** L. St. 1-flowered, rough; lvs. scabrous, pinnate, serrate; caps. smooth. ♃ Levant. 3f. Fls. very large, scarlet, too brilliant to be looked upon in the sun. June.

7. ESCHSCHÓLTZIA, Cham. (Named for *Eschscholtz*, a German botanist well known for his researches in California.) Sep. 2, cohering, caducous. Pet. 4. Stam. ∞, adhering to the claws of the petals. Stig. sessile. Caps. pod-shaped, cylindric, 10-striate, many-seeded. ① Lvs. finely pinnatifid, glaucous. The juice, which is colorless, exhales the odor of hydrochloric acid.

> 1 **E. Douglásii** Hook. St. branching, leafy; torus obconic; cal. ovoid, with a very short, abrupt acumination; pet. bright yellow, with an orange spot at base. Cal. Oreg. Foliage smooth, abundant, and rich. Fls. 2′—3′ broad.
> 2 **E. Califórnica** Hook. St. branching, leafy; torus funnel-form, with a much-dilated limb; cal. conic, with a long acumination; flowers orange-yellow. Cal.

8. BOCCÒNIA, Plum. Sep. 2, colored. Pet. 0. Sty. bifid. Caps. 2-valved, 1–3-seeded. ♃ Cult. for the handsome glaucous lvs. Fls. in panicles.

> 1 **B. cordàta**. Lvs. roundish, cordate, many-lobed, veiny; flowers white or yellowish, numerous in the ample pyramidal panicle, in Summer. From China. Hardy.
> 2 **B. frutéscens**. Lvs. oblong, large, sinuate-lobed, splendid; fls. in Spr. wh. W. Ind

ORDER 10.—FUMARIACEÆ.   33

ORDER X. FUMARIACEÆ. FUMEWORTS.

Herbs smooth and delicate, with a watery juice. *Leaves* exstipulate, alternate, many-cleft. *Flowers* irregular. *Sepals* 2, very small. *Petals* 4, parallel, one or both of the outer saccate, 2 inner cohering at apex. *Stamens* 6, diadelphous. *Anthers,* 2 outer 1-celled, middle 2-celled. *Ovaries* superior, 1-celled. *Fruit* a nut 1-2-seeded, or a capsule ∞-seeded. *Seeds* shining, arilled. *Albumen* fleshy. Illust. 61, 252–4.

* Corolla equally 2 spur red or 2-saccate at base. (*a*)
* Corolla unequal, only 1 of the petals spurred. (*b*)
   a Petals slightly united or distinct, mostly deciduous. Not climbing............DICENTRA. 1
   a Petals firmly united, persistent. Plants climbing....................ADLUMIA. 2
   b Ovary with several seeds, forming a slender pod.....................CORYDALIS. 3
   b Ovary with 1 seed, forming a globular nut........................FUMARIA. 4

1. DICÉNTRA, Borkh. EAR-DROP. Sep. 2, very small, sometimes disappearing. The 2 outer petals alike, saccate at base, with spreading tips; the 2 inner alike, spoon-shaped, crested, meeting face to face over the stam. and pistil. Fil. flat, in 2 sets, united at top. Stig. 2-crested. Pod many-seeded. ♃ Lvs. ternately divided or cleft. Fls. racemed, nodding. Delicate and beautiful plants. Figs. 61, 252–4.

* Herbs native, acaulescent, the sepals small but manifest....................Nos. 1, 2, 3
* Herbs exotic, caulescent, the sepals obsolete or wanting....................No. 4

1 **D. cucullària** DC. *White Ear-drop.* Root bulb-like; spurs of the fls. divergent, acute, straight; flower nearly as broad as long. Woods, Can. to Ky. 6—10′. Lvs. all radical of numerous oblong linear segm. The bulb consists of reddish, scale-like tubers. Apr. May.

2 **D. Canadénsis** DC. *Squirrel-corn.* Root bearing yellow tubers as large as peas; rac. simple; fls. white, cordate-ovate; spurs rounded, incurved. Rocky woods, Can. to Ky. 6—8′. Lvs. as in No. 1. Fls. fragrant. May, June.

3 **D. exímia** DC. *Purple E.* Rhizome scaly; rac. paniculate; fls. cordate-oblong, rose-purple, spurs blunt, incurved; sep. ovate, acute; lvs. triternate, segm. cut into oblong, acute lobes N. Y. to Oreg.! 10—15′. Fls. all summer. †

4 **D. spectábilis.** *Bleeding Heart.* Stems recurved, branched; lvs. biternate, segm 2 or 3-lobed; fls. in spreading racemes, bright purple; cor. broad, heart-shaped; sep. obsolete. China. Very fine and showy.

2. ADLÙMIA, Raf. MOUNTAIN FRINGE. Sepals 2, minute. Petals 4, united into a cellular, monopetalous corolla, persistent, bi-gibbous at base, 4-lobed at apex. Stam. united in 2 equal sets. Pod 2-valved, many-seeded. ② ♃ Delicate, with tripinnate leaves, and ample pendulous cymes.

A. **cirrhòsa** Raf.—Rocky hills, Can. to N. Car. 20f. The leaf-stalks serve for tendrils. Leaflets 3-lobed. Flowers pinkish white. June—Aug.

3. CORÝDALIS, DC. Sepals 2, small. Petals 4. Corolla with a single spur at base on the upper side. Capsule silique-form, many-seeded. Seeds crested or arilled. Herbs caulescent, with multifid leaves. Racemes bracted, with ebracteolate pedicels.

1 **C. glaúca** Ph. Glaucous, erect; fls. red, yellow at the tip; pods erect; lobes of the leaflets obtuse, bracts minute. ② Rocky woods, Can. to N. Car. 1—4f. Raceme terminal. Flowers horizontal, spur short, blunt. May, June.

2 C. aùrea Willd. Low, diffuse, finally ascending; leaf-lobes acute; rac. opposite the lvs. and terminal; fls. secund, bright yellow. spur deflected; pods pendulous, torulous; seeds turgid, polished. ① Rocky shades. 8—12′. Cor. 6″. Bracts lance-ovate. Apr.—July.

 β. *macrántha.* Fls. 10″, spur nearly as long as limb; bracts and leaf-lobes linear. Dakota; sent by Dr. W. Matthews.

 γ. *flávula.* Fls. 3—4″, pale yellow, spur very short, petals pointed. Common.

3 C. montàna Engelm.? Ascending; rac. terminal; leaf-lobes obtuse, bracts lanceolate; cor. yellow, spur ascending, nearly as long as limb, lower petal at length pendent; pods erect; seeds lenticular. La. Tex.!

4. FUMÀRIA, L. Fumitory. (Lat. *fumus,* smoke; from its disagreeable odor.) Sep. 2, caducous. Pet. 4, unequal, 1 of them spurred at the base. Nut ovoid or globose, 1-seeded, and indehiscent. Lvs. cauline, finely dissected.

F. officinàlis L. Diffusely branched, erect; lvs. bipinnate; rac. loose; fls. minute, purple at the tip; calyx serrated; ped. erect, twice longer than bract; nut round-retuse. ② Waste grounds, §. 1f. July, Aug.

Order XI. CRUCIFERÆ. Crucifers.

Herbs with a pungent, watery juice, and alternate, exstipulate leaves, with *flowers* cruciform, tetradynamous, generally in racemes, and bractless. *Sepals* 4, deciduous. *Petals* 4, hypogynous, with long claws and spreading limbs. *Stamens* 6, the 2 outer opposite ones shorter than the 4 interior. *Ovary* 2-carpeled, 2-celled by a false partition, with parietal placentæ. *Fruit* a silique, or silicle, usually 2-celled. *Stigmas* 2, sessile. *Seeds* 2-rowed in each cell, but often so intercalated as to form but 1 row. *Embryo* with the 2 cotyledons variously folded on the radicle. *Albumen* 0. Illust. 55, 104, 192, 193, 239, 336, 429, 506.

A large and important Order, difficult of analysis. The Genera cannot be well distinguished by their flowers, so nearly alike are they in all. Their characters are taken from the fruit and seeds. Hence it is indispensable that specimens for analysis should be *in fruit* as well as in flower. DeCandolle arranged the Genera into Tribes according to the folding of the cotyledons upon the radicle. This occurs in three different modes, as follows:

Cotyledons incumbent, when they are so bent or folded as to apply the back of one of them to the radicle, as in the seed of Capsella, fig. 1.

Cotyledons accumbent when they are so turned as to apply their edges to the radicle, as seen in the seed of Arabis Canadensis, fig. 2.

Cotyledons conduplicate, when they are not only *incumbent,* as in the first case, but also folded on and partly embracing the radicle, as in Mustard, fig. 3.

Order 11.—CRUCIFERÆ.

In the following table we endeavor to combine with the systematic arrangement of DeCandolle a more practical artificial method:

* Crucifers native, or cultivated for food. (§)
* Crucifers exotic, cultivated for ornament or art. (§ §)

§ Fruit a long pod, silique (§ 166), opening by 2 valves. (a)
§ Fruit a short pod, silicle (§ 166), opening by 2 valves. (e)
§ Fruit a jointed pod, loment, partitioned across................................Nos. 28, 29
   a Flowers cyanic.—b Seeds arranged in a double row in each cell..................Nos. 1, 2
                 —b Seeds in 1 row.—c Pods sessile on the torus..................Nos. 3, 4, 5
                             —c Pods on a slender stipe............................No. 12
   a Flowers yellow.—d Seeds flat, wing-margined..............................No. 6
               —d Seeds ovate or oblong.......................... Nos. 9, 10, 11
               —d Seeds globular..No. 15
   e Flowers bright yellow. Silicle turgid, or slightly flattened............... Nos. 1, 20, 21
   e Flowers cyanic.—f Silicle turgid, with a broad partition................Nos. 19, 22
               —f Silicle flattened parallel with a broad partition...............Nos. 16, 18
               —f Silicle flattened contrary to the narrow partition..............Nos. 24, 26
§ § Fruit a silique or long pod, opening by 2 valves......................Nos. 7, 8, 13, 14
§ § Fruit a silicle—g with 1 seed only, and indehiscent..........................No. 27
    —g with 2 or more seeds.—h Petals all equal......................Nos. 16, 17
               —h Petals unequal..No. 23

Tribe I. ARABIDEÆ.—Pods mostly elongated. Seed oval or orbicular, more or less flattened. Cotyledons accumbent (—o).
  1 Seeds small, turgid, in a turgid, oblong or oval pod................Nasturtium.
  2 Seeds flattened, in a long, linear pod. Plants very erect..................Turritis.
  3 Silique linear, seeds in 1 row, not bordered. Purple......................Iodanthus.
  4 Silique linear, each valve with 1 central vein, not opening elastically........Arabis.
  5 Silique linear or lanceolate, valves veinless, opening elastically..........Cardamine.
  6 Silique oblong, flattened, seeds wing-margined. Leaves radical..........Leavenworthia.
  7 Silique long, ∞-seeded. Stigmas distinct, 2-horned....................Matthiola.
  8 Silique long, ∞-seeded. Stigmas capitate. Leaves entire. Flowers yellow..Cheiranthus.
  9 Silique 4-angled, 2-edged, rigid. Leaves lyrate-pinnatifid...................Barbarea.

Tribe II. SISYMBRIEÆ.—Pod elongated. Seeds oblong. Cotyledons incumbent (∞), oblong.
 10 Calyx erect. Pods 4-sided, valves strongly 1-veined. Leaves lanceolate......Erysimum.
 11 Calyx half spreading. Pods subterete. Leaves dissected or incised..........Sisymbrium.
 12 Very smooth herbs, with the white flowers in corymbs. South ....Warea.
 13 Stigma of 2 converging lobes. Petals entire, oblique. Leaves lanceolate.....Hesperis.
 14 Stigma lobes connate. Petals pinnatifid, involute in æstivation..........Schizopetalon.

Tribe III. BRASSICEÆ.—Pods elongated. Seeds globular, ((o.
 15 Pod terete or 4-sided..Brassica.

Tribe IV. ALYSSINEÆ.—Fruit short, septum broad. Seeds in 2 rows. Cotyledons — o.
 16 Silicle mostly orbicular, flattened. Cells 1-4 seeded........................Alyssum.
 17 Silicle very large, orbicular-oval, very flat, stipitate. Cultivated............Lunaria.
 18 Silicle oblong or elliptical. Seeds ∞, not margined. Pet. entire or 2-cleft....Draba.
 19 Silicle globular or ellipsoid. Seeds few. Flowers white.....................Armoracia.
 20 Silicle globular, inflated, thin, veinless. Flowers yellow...................Vesicaria.

Tribe V. CAMELINEÆ.—Pods mostly short. Septum broad. Cotyledons | o.
 21 Silicle obovoid, with ventricous valves, many seeds. Flowers yellow........Camelina.
 22 Silicle oval, turgid, few-seeded. Leaves linear, radical. Flowers white......Subularia.

Tribe VI. THLASPIDEÆ.—Pods short, septum narrow. Cotyledons accumbent. (23)..Iberis.

Tribe VII. LEPIDINEÆ.—Pods short, septum narrow. Cotyledons incumbent.
 24 Silicle triangular, many-seeded. Flowers white..........................Capsella.
 25 Silicle oval-orbicular, 2-seeded. Flowers white, often incomplete.........Lepidium.
 26 Silicle didymous, each half 1-seeded. Flowers minute.....................Senebiera.

Tribe VIII. ISATIDEÆ.—Silicle short, 1-celled, 1-seeded, indehiscent. (27) Cult...Isatis.

Tribe IX. CAKALINEÆ.—Pod 2-jointed. Cotyls. — o. (28) Fleshy sea-side herbs..Cakile.

Tribe X. RAPHANEÆ.—Pod moniliform. Cotyledons ((o. (29) Leaves lyrate... Raphanus

1. **NASTÚRTIUM,** R. Br. WATER-CRESS. (Lat. *nasus tortus*, nose tortured; alluding to the pungent qualities.) Sep. spreading. Siliques subterete, turgid, generally curved upward, often shortened to a silicle, valves veinless. Seeds small, ∞, turgid, generally arranged in a double row in each cell (= ○). ※ with pinnate or pinnatifid leaves.

* Petals white. Siliques rather long (10—12″)..................................No. 1
* Petals yellow, minute. Siliques shortened (4—8″), but longer than the pedicels. (*a*)
* Petals yellow. Siliques or silicles (1—6″), shorter than the pedicels. (*b*)
     a Leaves pinnate or pinnatifid. Diffusely branched......................Nos. 2, 3
     a Leaves lyrate, or merely toothed. Stems erect........................Nos. 4, 5
       b Petals not longer than the calyx, obscure... ...Nos. 6, 7
       b Petals longer than the calyx, bright yellow, the flowers showy.....Nos. 8, 9

1 **N. officinàle** R. Br. *English W.* Lvs. pinnate, lfts. ovate, subcordate, repand; petals white, longer than the calyx. ⚷ Springs, &c. May, June. § ‡
2 **N. tanacetifòlium** Hook. Upper leaf-segm. confluent, lower distinct, oblong, or roundish, sinuate-toothed, teeth obtuse; pods 4—6′, ped. ½ as long. ② South.
 β. *obtúsum.* Lfts. mostly distinct, obtuse, oval. Pods shorter (3—5″). Miss. R.
3 **N. Wálteri** Wood. Segments of the leaves all distinct, narrow, with a few linear, acute lobes or teeth; pods linear (5″), ped. 2—3″. ⚷ South. 3—5′. March, April.
4 **N. limòsum** N. Lvs. lanceolate, toothed, the lower lyrate; pods elliptic-oblong, 3—4″, ped. much shorter. ② Rivers, La. 10—15′. Fls. minute. Too near the next.
5 **N. sessiliflòrum** N. Lvs. wedge-obovate, repandly-toothed or subentire; pods linear-oblong, 5—6″, subsessile. ② Miss. Riv. Stem erect. Fls. minute. Apr.—June.
6 **N. palústre** DC. *Marsh Cress.* Glabrous; lvs. pinnately lobed, amplexicaul, lobes confluent, dentate; rt. fusiform; pet. as long as the sepals; silicle spreading, turgid, twice longer than wide. ⚷ Wet places. 1—2f. Pod 3″. June—Aug.
7 **N. hispidum** DC. Villous; lvs. runcinate-pinnatifid, lobes obtusely dentate; silicles tumid, ovoid, or globular, the pedicels longer, ascending; pet. scarcely as long as the calyx. ② Streams, 1—3f. Pod 1″. Ped. 2—3″. June—Aug.
8 **N. sylvéstre** R. Br. *Wood Cress.* Lvs. pinnately divided, segm. serrate or incised; pods linear, style very short. ⚷ Meadows, Ms. to Pa. Rare. June, July. §
9 **N. sinuàtum** Nutt. Lvs. pinnatifid, segm. lance-oblong, nearly entire; pods oblong, acute, with a slender style. ⚷ Rivers, St. Louis to Oreg. June.

2. **TURRÍTIS,** Dill. TOWER MUSTARD. (Lat. *turris*, a tower; from the strict form of the plants.) Sep. erect, converging. Seeds flattened, minute, in 2 rows in each cell of the long, narrowly-linear 2-edged silique; valves plane, 1-veined. Embryo = ○. Glabrous and strictly erect, stem-leaves sagittate-clasping. (Runs into Arabis.)

1 **T. glàbra** L. Fls. cream-white, erect; silique long (3′), strictly erect; stem lvs. ovate-lanceolate. ① Can., to Pa.(Porter.) 2—3f. Glaucous. Lvs. entire. July.
2 **T. stricta** Graham. Fls. rose-white, erect; silique long (3′), erect, finally ascending or spreading; stem lvs. linear-lanceolate. ② Rocks, N. Y. (rare) to Oreg. 1—2f. May.
 β. *brachycárpa.* Fls. and siliques spreading, the latter shorter (1′). Westward.

3. **IODÁNTHUS,** T. & G. FALSE ROCKET. (Ιώδης, violet-colored, ἄνθος, flower.) Calyx closed, shorter than the claws of the petals. Silique linear, terete, veinless. Seeds arranged in a single row in each cell (= ○). ⚷ Glabrous, with violet-purple flowers in panicled racemes. Leaves lanceolate.

I. hesperioìdes Torr & Gr. Penn. to Ill. and Ark. 2—3f. Lvs. serrate or the lower pinnatifid-lyrate. Pods 15—20″, spreading. May, June. (Arabis, Gr.)

ORDER 11.—CRUCIFERÆ. 37

4. ÁRABIS, L. ROCK-CRESS. Sepals mostly erect; silique linear, compressed; valves plane, each with 1 or 3 longitudinal veins, seeds in a single row in each cell, mostly margined, cotyledons accumbent or oblique. Flowers white. Figs. 336, 506.

* Leaves (all or at least the radical) pinnatifid. Stems clustered............Nos. 1, 2, 3
* Leaves all undivided, toothed or entire, often clasping..(a) (*Exotic.* No. 10.)
    a Siliques short (6—12″) and straight. Sds. not winged. Stems clustered..Nos. 4, 5
    a Siliques longer (1—2′), straight or curved. Sds. not winged. St. simple..Nos. 6, 7
    a Siliques long (3′), curved, pendent. Seeds winged......................Nos. 8, 9

1 A. Ludoviciàna Meyer. All the leaves pinnatifid or pinnate, smoothish; stems branched at base; siliques ascending; seeds bordered. ① South. 6—10′. March.

2 A. lyràta L. Upper leaves smooth, linear, entire; radical leaves lyrately pinnatifid, often pilous; st. branched at base; pedicels spreading; siliques erect, seeds not bordered, obliquely =o. ② Hills, Can. to Va. 6—12′. Pods 1½—2′. Pet. 3″ long. Apr., May.

3 A. petræa Lam. Upper leaves linear, entire, minute, radical pinnatifid, very small; stems clustered; pods ascending (1—1½′); seeds bordered, =o. ♃ Rocks (Greenwich), Ct., Vt., O., Mich. 6—12′. Flowers white or roseate. June.

4 A. Thaliàna L. St. clustered, erect; lvs. pilous, oblong, nearly entire; pet. twice longer than calyx; pods erect, squarish (9″); seeds obliquely |o. ② Fields, Vt. to Ill. and Car. (Wayne Co., N. Y. Hankenson.) 4—12′. Fls. small. May. (Sisymbrium, Gay.) §

5 A. dentàta T. & G. Stems clustered, diffuse; lvs. oblong, sharply toothed; petals hardly longer than the calyx; pods spreading. ① N. Y. to Mo. 1f. Fls. small. May.

6 A. pàtens Sull. Erect, pubescent; cauline leaves coarsely toothed; siliques spreading and curved upward, beaked with a distinct style. ② O. to Tenn. 1—2f. May.

7 A. hirsùta Scop. Erect, hirsute; radical leaves oblong-ovate, cauline lanceolate, sagittate-clasping, entire or toothed; siliques straight, erect; style none. ② Can. to Va., and W. 1—2f. June.

8 A. lævigàta DC. Tall, glaucous, smooth; stem leaves linear-lanceolate and linear, sagittate-clasping, upper entire: siliques very long, linear, at length spreading and pendulous. ② Can. to Tenn., and W. 2f. Pod 3′. May.

   β. *minor* (Porter). Plant smaller, 10—15′, with the lvs. sessile—not clasping. Penn.

9 A. Canadénsis L. *Sickle-pod.* Tall, pubescent; stem leaves lanceolate, pointed both ways, sessile; silique subfalcate, veined, pendulous. ② Rocky hills. 2—3f. Petals small, but twice longer than sepals. Pods 3′. May, June.

10 A. ALPÌNA. Erect, 8—12′, hoary with stellate hairs; lvs. oblong, with slender teeth, clasping; fls. showy, pure wh., in many little long-stalked corymbs. Alps. Mar.—May.

5. CARDAMÌNE, L. BITTER CRESS. Calyx a little spreading. Silique linear or lanceolate, with flat, veinless valves narrower than the dissepiment, and often opening elastically from the base. Stigma entire. Seeds not margined, =o. Flowers white or purple.

§ DENTÀRIA. Pod lance-linear. Rhizome thickish, knotted. Stem with 2 or 3 palmated leaves near the middle. Flowers large, corymbed...(*)
  * Leaves of the stem subopposite or subverticillate...................Nos. 1, 2, 3
  * Leaves of the stem alternate...Nos. 4, 5
§ CARDAMÌNE. Pod linear. Root tuberous or fibrous. Leaves alternate...(†)
  † Leaves pinnate, with many leaflets.....................................Nos. 6, 7
  † Leaves simple or partly ternate...(a)
    a Siliques pointed with a slender style. In low, wet grounds........Nos. 8, 9
    a Siliques tipped with the sessile stigma. In high mountains.......Nos. 10, 11

1 C. diphýlla. Stem 2-leaved; leaflets subovate; rhizome continuous, toothed. ♃ Damp woods, Can. to Car. 1f. Leaves 3-parted, nearly opposite. Root-stock pungent, aromatic. May.

2 C. laciniàta. Cauline lvs. 3, 3-parted, the divisions lanceolate or linear-oblong obtuse, lobed, toothed or entire; rhizome moniliform. ♃ Woods. 1f. Apr. May.

3 C. multífida. Cauline lvs. mostly 3, and verticillate, rarely 2, multifid with numerous linear lobes; rhizome tuberous. ♃ Woods, N. Car. to Ala. Rare. 9′.

4 C. máxima. Stem about 3-leaved (2 to 7); lfts. 3, ovate, toothed or cleft; rhizome moniliform, the tubers toothed. ♃ N. Y. and Penn. Rare. 1—2f. May.

5 C. heterophýlla. Stem about 2-leaved (2 or 3), leaflets 3, lanceolate and nearly entire; root-lvs. of 3 ovate-oblong, toothed, and cut-lobed leaflets; rhizome moniliform, scarcely toothed. ♃ Penn. Va. Ky. 6′. Flowers purple. June.

6 C. hirsùta L. Stem (hirsute in Europe) glabrous, erect; leaves pinnately 5-11-foliate, terminal leaflet largest; flowers (white) small, silique erect, linear or filiform; stigma minute, sessile. ② Wet. Variable. Stem 3—12′, slender or thick. Leaflets obtuse. Pod 1′. March—June.

 β. *sylvática.* Slender and delicate; leaflets 1 or 2-toothed; pods filiform, incurved. Grows in dryer places. 6′. (C. Virginica Mx.)

7 C. praténsis L. *Cuckoo Flower.* Stem ascending, simple; leaves pinnately 7-15-foliate; leaflets petiolate, subentire, lower ones suborbicular, upper linear-lanceolate; style distinct. ♃ Swamps, N. Y. to Arc. Am. 10—16′. Flowers large. Apr. May.

8 C. rhomboìdea DC. Stems simple, erect or ascending, tuberiferous at base; siliques linear-lanceolate; rt. lvs. roundish, entire, st. lvs. rhomboidal. ♃ May. 8—14′.

 β. *purpurea.* Slender, erect, few-leaved and purple-flowered. N. Y., O., Wisc.

9 C. rotundifòlia Mx. Stems decumbent, branching, finally stoloniferous; leaves all petiolate; pod linear-subulate; rt. fibrous. ♃ Cool springs. Pa. to Car. 1—2f. May, Jn.

10 C. bellidifòlia L. Leaves smooth, orbicular-ovate, nearly entire, petiolate; cauline entire or 3-lobed; siliques erect. ♃ White Mts. &c. 1½—3′. July.

11 C. spatulàta Mx. Lvs. hirsute, the radical spatulate, petiolate; cauline sessile, siliques spreading. ① Mts. of Car. and Ga. Trailing. 6—8′. April.

6. LEAVENWÓRTHIA, Torr.

(Named for *Dr. Leavenworth,* the discoverer.) Petals cuneate, retuse, or truncate. Silique flat, oblong, valves indistinctly veined. Seeds in a single row, flattened, wing-margined. Embryo nearly straight, curving toward an accumbent form. ② Low, smooth herbs with lyrate-pinnatifid leaves. Pet. yellow at base.

L. Michaùxii (and aurea) Torr.—Rocks, Ky. to Tex. 2—6′. Lvs. mostly radical. Fls. 1- 4.

7. MATTHÌOLA, R. Br. STOCK.

(In honor of *P. A. Matthioli,* physician to Ferdinand of Austria, and botanic author.) Calyx closed, 2 of the sepals gibbous at base. Siliques terete; stigmas connivant, thickened or cornute at the back. Herbaceous or shrubby, oriental plants, clothed with a hoary, stellate pubescence.

1 M. incàna. *Common Stock. Brompton S. July-flower.* Erect, branching from the woody base; lvs. lanceolate, entire. ② ♃ Eur. 2f. Fls. often double, white, purple.

2 M. ánnua. *Ten-weeks Stock.* Erect, branched; lvs. lanceolate, obtuse, toothed. ① S. Eur. 2f. Flowers infinitely various, mostly double. June—Nov.

8. CHEIRÁNTHUS, L. WALL-FLOWER.

($X \varepsilon \iota \rho$, the hand, $\check{\alpha} \nu \vartheta o \varsigma$, flower.) Calyx closed, 2 of the sepals gibbous at base. Silique terete or compressed. Stigma 2-lobed or capitate. Seeds flat, in a single series, often margined. (= ○). Garden perennials, mostly European. Leaves undivided. Fig. 55.

C. Cheìri. St. somewhat shrubby and decumbent at base; lvs. lanceolate, glabrous. pet. obovate. long-clawed. yellow; stig. capitate. ♃ S. Eur. 2f. June.

ORDER 11.—CRUCIFERÆ.

9. BARBÀREA, R. Br. WINTER-CRESS. (Dedicated to *Sta. Barbara.*) Sepals erect. Siliques columnar, 2 or 4-angled, valves carinate with a midvein. Seeds in a single row (━○). Leaves lyrate-pinnatifid. Fls. yellow.

1 **B. vulgàris** R. Br. Upper lvs. toothed or pinnatifid at base; siliques obtusely 4-angled, pointed with the style. ② Brooksides: common. 1—2f. Racemes dense, showy-panicled. Pod 9″. May, June.
2 **B. prècox** R. Br. *Scurvy-grass.* Upper lvs. pinnatifid, with the lobes all linear oblong; silique 2-edged. ♃ § ‡ South. Pod 2—3′. May, June.

10. ERÝSIMUM, L. FALSE WALL-FLOWER. ('Ερύω, to cure; from its salutary medicinal properties.) Calyx closed. Siliques columnar, 4-sided, valves with a strong mid-vein. Stigma capitate. Seeds in a single series. Cotyledons oblong, ‖ ○. Lvs. narrow, undivided. Fls. yellow.

1 **E. cheiranthoìdes** L. Pubescence minute, appressed, branched; lvs. lanceolate, denticulate, or entire; fls. small; siliques short (8—10″), on slender, spreading pedicels; stig. small, nearly sessile. ① Wet grounds. 1—2f. Rac. long. July.
2 **E. Arkansànum** N. *Yellow Phlox.* Simple, scabrous; lvs. linear-lanceolate, remotely dentate; rac. corymbed at top; pod long (3′), erect; stig. capitate. ②
Bluffs, O. to Ark. 2—3f. Flowers large, orange-yellow. June, July.
3 **E. orientàle** R. Br. Glabrous and glaucous; radical lvs. obovate, stem lvs. cordate-clasping, obtuse, entire; fls. white. ① Near Phila (A. H. Smith). § Eur.

11. SISÝMBRIUM, Allioni. (An ancient Greek name.) Calyx half-spreading, equal at base. Petals unguiculate, entire. Silique subterete, valves concave, marked lengthwise with 1—3 veins. Style very short. Seeds in a single series, ovoid, ‖ ○. Flowers small, yellow.

1 **S. officinàle** Scop. *Hedge Mustard.* Leaves runcinate; racemes slender, virgate; siliques subulate, erect, closely appressed to the rachis. ① A common weed, with branches at right angles. 1—3f. June—Sept. §
2 **S. Sòphia** L. *Flixweed.* Lvs. bipinnatifid, lobes linear-oblong, acute; sep. longer than pet.; pod linear, erect, longer than the spreading pedicel. ① N. Y. Can. §
3 **S. canéscens** Nutt. *Tansey Mustard.* Lvs. bipinnatifid, canescent, lobes oblong, subdentate, obtuse; pet. about equalling the calyx; pod oblong-linear, 3—6″, ascending, shorter (or never longer) than the spreading pedicel. ① U. S. 1—2f. Mar.—June.

12. WÀREA, N. (Named for *Mr. Ware,* the discoverer.) Sep. colored, ligulate. Pet. with very slender claws. Silique flattened, long and slender, raised on a slender stipe. Cotyledons oblong, ‖ ○. ① Glabrous, entire-leaved. Flowers white or purple, in short racemes. Siliques curved and declinate.

1 **W. cuneifòlia** N. Lvs. oblong, obtuse, cuneate at base, and subsessile. Ga. Fla. 1—2f. Pet. obovate, white. September.
2 **W. amplexifòlia** N. Lvs. oblong-ovate, partly clasping. Sand hills, Fla 1—2f. Pet. oval, purple. September.

13. HÉSPERIS, L. ROCKET. ('Εσπερα, evening, when the flower is most fragrant.) Calyx closed, shorter than the claws of the petals. Pet. bent obliquely, linear or obovate. Silique subterete. Seeds not margined. Stig. forked, with the apices converging (‖ ○). Flowers white or purple.

1 **H. matronàlis** L. Simple, erect; lvs. lance-ovate, denticulate; pet. obovate; pod torulous, elongated (8′), erect. ② Shores of L. Erie (Hankenson) and Huron. § †

ORDER 11.—CRUCIFERÆ.

14. SCHIZOPÉTALON, Sims. ($\Sigma\chi\iota\zeta\omega$, to cut, as the petals appear to be.) Sep. erect. Pet. pinnately lobed, involute in the bud. Silique linear, compressed. Stig. lobes erect, connate. Seeds oblong or globular, cotyl. twisted (|| ○). ① Lvs. sinuate-pinnatifid. Fls. white or purple.

S. WÁLKERI. Stem slender, erect, branching, 2f. Lvs. canescer.t. Fls. racemed. Chili. Raised from seed. Flowers large, curious, soon perishing.

15. BRÁSSICA (and Sinapis) L. CABBAGE, MUSTARD, &c. (The ancient names.) Silique long, terete, or 4-sided, pointed with a stout style or an ensiform 1-seeded beak. Valves 1-3-veined. Seeds in 1 row, globular, ((○. Root lvs. pinnatifid. Rac. elongated. Fls. yellow. Figs. 230, 192, 429.

§ SÍNAPIS. Sep. spreading. Pet. ovate. Pod with an acute beak ....Nos. 1, 2, 3
§ BRÁSSICA. Sep. erect. Pet. obovate. Pod squarish, with a blunt style...Nos. 4, 5, 6
1 **B. nigra** L. *Black Mustard.* Smooth; pod 1', smooth, somewhat 4-angled, appressed to the rachis, and beaked with a slender, 4-sided style. ① 3—6f. §
2 **B. arvénsis** (L.) *Field Mustard.* St. and lvs. hairy; pod 1½', smooth, many-angled, torulous, spreading, thrice longer than the slender ancipital style. ① § June, July.
3 **B. alba** (L.) *White Mustard.* Lvs. smoothish; siliques hispid, torulous. shorter than the ensiform beak; seeds large, pale yellow. ① Eur. 3—5f. Pod 4-seeded.
4 **B. campéstris** (L.) *Cole.* Lvs. somewhat fleshy and glaucous, the lower lyrate-dentate, subciliate, upper cordate-amplexicaul, acuminate. ① Fields. 2f. July. §
β. *Rutabàga.* *Swedish Turnip.* Root tumid, napiform, subglobous, yellowish. ‡
5 **B. Rapa** L. Radical lvs. lyrate, rough, not glaucous, cauline ones incised, upper entire, smooth.
β. DEPRÉSSA. *Common Turnip.* Root depressed, globous or napiform, contracted below into a slender radicle. ② Long cultivated for its root. ‡
6 **B. OLERÀCEA** L. *Cabbage.* Lvs. very smooth and glaucous, fleshy, repand-toothed or lobed. ② Europe, on rocky shores, forming no head.
β. BULLÀTA. *Savoy Cabbage.* Lvs. curled, subcapitate, finally expanding. ‡
γ. BOTRYTIS-CAULIFLÒRA. *Cauliflower.* Stem low; heads thick, compact, terminal; flowers abortive, on short, fleshy peduncles. ‡
δ. BOTRYTIS ASPARAGOÏDES. *Broccoli.* Stem taller; heads subramous; branches fleshy at the summit, consisting of clusters of abortive flower-buds. ‡
ε. CAPITÀTA. *Head Cabbage.* Stem short; leaves concave, packed in a dense head before flowering; raceme paniculate. ‡

16. ALÝSSUM, L. MADWORT. (Gr. α, privative, $\lambda\upsilon\sigma\sigma\alpha$, rage; supposed by the ancients to allay anger.) Calyx equal at base. Pet. entire; some of the stamens with teeth. Silicle orbicular or oval, with valves flat, or convex in the centre. Seeds 1—4 in each cell (= ⊃). Showy European herbs, half shrubby at base.

1 **A. marítimum** Lam. *Sweet A.* Lvs. lance-linear, acute, entire, some hoary; pods oval, smooth, 2-seeded; fls. white, small, sweet. ♃ 1f. Escaped from gardens. §
2 **A. calycínum** L. Calyx persistent; lvs. linear-spatulate, canescent; pods orbicular, lens-shaped, with a thin border, 4-seeded; fls. yellowish. ① 1f. Fields: rare. Mass. N. Y. (Wayne Co., Hankenson). §
3 **A. SAXÁTILE.** *Rock A.* Lvs. lanceolate, entire, downy; pods round-obovate, 2-seeded; flowers yellow, corymbed, abundant and brilliant. ♃ Candia. 9'. April.

17. LUNÀRIA, L. HONESTY. (Lat. *luna*, the moon; from the broad, round silicles.) Sep. somewhat bisaccate at base. Pet. nearly entire. Stam. without teeth. Silicle pedicellate, elliptical, or lanceolate, with **flat**

valves; funiculus adhering to the dissepiment (═ c). European. Leaves cordate. Flowers lilac.

1 L. REDIVIVA L. *Perennial Satin-flower.* Lvs. ovate, petiolate, mucronately serrate; silicles lanceolate, narrowed at each end. ♃ 2—3f. June.
2 L. BIÉNNIS DC. *Honesty.* Lvs. with obtuse teeth; silicles oval, obtuse at both ends. ⓐ Flowers large, purple. May, June.

18. DRABA, L. WHITLOW GRASS. ($\Delta\rho\acute{\alpha}\beta\eta$, acrid, biting; from the taste of the plant.) Calyx equal at base. Pet. equal. Fil. without teeth. Silicle oval or oblong, entire, the valves flat or slightly convex, veined. Seeds not margined, 2-rowed in each cell (═ c). Flowers white, rarely yellow. Plants small.

§ ERÓPHILA. Petals 2-parted..No. 1
§ DRABA *proper.* Petals entire or only emarginate. (a)
    a Style distinct, long or short. Pods twisted when ripe. Perenn..Nos. 2, 3, 4
    a Style none. Pods straight, plane. Plants annual or bienn. (b)
        b Pedicels as long as or longer than the pods.................Nos. 5, 6
        b Pedicels shorter than the pods............................Nos. 7, 8

1 D. (Eróphila) vérna L. *Whitlow Grass.* Scape naked; lvs. oblong, acute, subserrate, hairy; pet. bifid; stig. sessile; silicle oval, flat, shorter than the pedicel. ① A little Spring flower, in rocky places. Can. to Va. 1—3′.
2 D. ramosíssima Desv. Minutely pubescent, diffuse; lvs. linear-lanceolate, with remote and slender teeth; rac. panicled; silicle lanceolate, about the length of the pedicel, the style half as long. ♃ Va. Ky. 5—8′. May.
3 D. arábisans Mx. Slightly pubescent; root leaves in tufts, wedge-lanceolate, toothed; stems leafy, erect, its lvs. oblong; silicle glabrous, lance-oblong (6″), spreading; style very short. ♃ Lake shores, Vt. N. Y. Mich. 6—10′. White. May.
4 D. incàna L. Hoary pubescent; root leaves in tufts, wedge-lanceolate, slightly toothed; st. nearly naked, branches and ped. very erect; silicle oblong (5″), twisted, sty. very short. ♃ or ⓐ Mts. N.Vt. and N. 6—8′. Lvs. 6″. Fls. very small, white. June.
5 D. nemoràlis Ehrh. Pubescent, branched; lvs. oval, the cauline lanceolate, toothed; pet. emarginate; silicles half the length of the spreading pedicels. ⓐ Mich. Mo. 8—10′. Flowers small, white or yellowish. May.
6 D. brachycárpa N. Minutely pubescent; lvs. ovate, the cauline oblong; rac. ∞-flowered; pet. obovate, entire; silicle as long as the ped. 6-seeded. ① Mo. and South. 3—4′. Pod 2″. April.
7 D. cuneifòlia N. Hirsute, pubescent, branching and leafy below, naked above; lvs. cuneate-oblong, sessile, denticulate; rac. elongated in fruit; silicles twice longer (1″) than the pedicels. ① Ky. to La. 3—8′. March.
8 D. Caroliniàna Walt. Hispid, branching and leafy below, naked above; lvs. entire, obovate and oval; rac. short; silicles oblong-linear, longer than the pedicels (5″). ① R. I. to Ga. and W. 1—3′. Much like No. 7. April—June.
    β. *micrántha.* Silicles minutely hispid; pet. often wanting. (D. micrantha N.) W

19. ARMORÀCIA, Rupp. HORSE-RADISH. (*Armorica*, its native country, now the province Brittany, France.) Sep. spreading. Pet. entire, much exceeding the calyx. Silicles ellipsoid or globular, turgid, 1-celled from the incomplete partition. Style distinct. Seeds few (═ c). ♃ Lvs. oblong, undivided, or the lower pinnatifid. Flowers white.

1 A. rusticàna Rupp. Radical lvs. oblong, crenate; cauline long, lanceolate, incised; silicle roundish, ellipsoid, much longer than the style. § Eur.

2 A. Americàna Arn. Aquatic; immersed lvs. doubly pinnatifid with capillary segments, emersed, oblong, pinnatifid, serrate or entire; silicle ovoid, little longer than the style. Lakes and rivers, Can. to Ky. July, Aug.

20. VESICÀRIA, Lam. BLADDER-POD. (Lat. *vesica*, a bladder or blister; from the inflated silicles.) Pet. entire. Silicle globous or ovoid; inflated valves nerveless, hemispherical or convex. Seeds several in each cell, sometimes margined (═ ○). Flowers yellow. (See *Addenda*.)

V. Shórtii T. & G. Lvs. elliptical, sessile, entire; style twice as long as the globous silicle; seeds 2—4, not margined. ① Ky. rare.

21. CAMELÌNA, Crantz. FALSE FLAX. ($X\alpha\mu\alpha i$, dwarf, $\lambda i v o v$, flax.) Calyx equal at base. Pet. entire. Silicle obovate or subglobous, with ventricous valves and many-seeded cells. Styles filiform, persistent. Seeds oblong, striate, not margined (∥ ○). Flowers small, yellow.

C. satìva Crantz. Lvs. lanceolate, sagittate at base, subentire; silicle obovate-pyriform, margined, tipped with the pointed style. ① Fields. § Eur. ♃ June.

22. SUBULÀRIA, L. AWLWORT. (Named in reference to the linear subulate leaves.) Silicle oval, valves turgid, cells many-seeded. Stigma sessile; cotyledons linear, curved and incumbently folded on themselves. ① Aquatic acaulescent herbs.

S. aquática L.—Shores of ponds, Me. N. H. Lvs. all radical, entire, subulate, 1' Scape 2—3', with a few minute white flowers. July.

23. IBÈRIS, L. CANDYTUFT. (Most of the species are natives of *Iberia*, now Spain.) The 2 outside petals larger than the 2 inner. Silicles compressed, truncate, emarginate, the cells 1-seeded. Handsome herbs from the Old World, pretty in cultivation. Flowers white or purple.

1 **I.** UMBELLÀTA. *Purple C.* Herbaceous; lvs. lin.-lanceolate, acuminate. the lower serrate; silicles umbellate, acutely 2-lobed. ① Eur. 1f. Purple. June, July.
2 **I.** AMÀRA. *Bitter C.* Herbaceous; lvs. lanceolate, acute; fls. finally racemed; silicles obcordate, narrowly emarginate. ① Eng. 1f. White. June, July.
3 **I.** ODORÀTA. Herbaceous; lvs. linear, toothed, dilated at end; silicle round, with acute, spreading lobes. ① Alps. 1f. Sweet scented. Foliage pretty. July.
4 **I.** PINNÀTA. Lvs. pinnatifid, smooth. ① Eur. 1f. White, corymbed.
5 **I.** SAXÁTILIS. Shrubby; lvs. linear, entire. ♃ Eur. 1f. White, corymbed.

24. CAPSÉLLA, Vent. (Lat. *capsa*, a chest or box; alluding to the fruit.) Calyx equal at base; silicles triangular-cuneiform, obcordate, compressed laterally; valves carinate, not winged on the back; septum sublinear; style short; seeds ∞, oblong, small, ∥ ○. Fls. white. A common weed. Fig. 193.

C. Bursa-pastòris Mœnch. *Shepherd's Purse.* ① Grows everywhere. 6'—1'—2f. Root lvs. rosulate, cut-lobed; stem leaves lance-lin. clasping-sagittate; rac. long.

25. LEPÍDIUM, R. Br. PEPPER GRASS. ($\Lambda \epsilon \pi i \varsigma$, a scale; from the resemblance of the silicle.) Sepals ovate; petals ovate, entire; silicles oval-orbicular, emarginate; septum very narrow, contrary to the greater diameter; valves carinate, dehiscent; cells 1-seeded. Cotyledons ‖⊃, often —o. Flowers small, white, often incomplete.

* Stamens only 2. Petals often wanting. Leaves not clasping ....Nos. 1, 2
* Stamens 6. Silicles evidently winged........ ....Nos. 3, 4

1 **L. Virgínicum** L. *Tongue-grass.* Lvs. linear-lanceolate, the lower incisely serrate; pet. 4; silicles orbicular. emarginate; cotyledons = o. ① Dry places. 1f.
2 **L. ruderàle** L. Cauline lvs. incised, those of the branches entire; pet. none; pods broad-oval, notched, wingless. ① Dry fields. Rare. 10—15'. Always apetalous. §
3 **L. campéstre** R. Br. *Yellow-seed.* Cauline lvs. sagittate-clasping, denticulate; silicles ovate, notched, winged, rough. ① Dry fields. Rare. 6—10'. Jn. § Eur.
4 **L. satìvum** L. *Pepper-grass.* Lvs. oblong, variously incised and pinnatifid; silicles elliptic-ovate, notched and winged. ① Eur. 2f. A garden salad. July.

26. **SENEBIÈRA**, Poir. Carpet Cress. Swine Cress. (In honor of *Senebier*, a distinguished vegetable physiologist.) Silicle didymous, with the partition very narrow; valves ventricous, separating but indehiscent, and each 1-seeded, cotyledons incumbently folded on themselves. ① or ② Prostrate and diffuse, with minute white flowers.

1 **S. dídyma** Pers. Lvs. pinnate, with pinnatifid segments; silicles rugously reticulated, notched at the apex. Waste places coastward, Atlantic and Pacific.
2 **S. Coronòpus** DC. Lvs. pinnate, with the segm. entire, toothed, or pinnatifid; silicles tubercled, not notched at apex. R. Isl. (Robbins) to Car. Rare.

27. **ISÀTIS**, L. Woad. ('Ισάζω, to make equal; supposed to remove roughness from the skin.) Silicle elliptical, flat, 1-celled (dissepiment obliterated), 1-seeded, with boat-shaped valves, which are scarcely dehiscent (‖o). None North American.

1. tinctòria L. Silicles cuneate, acuminate at base, somewhat spatulate at the end, very obtuse, three times as long as broad. ① Eng. 4f. Yellow. May—July. Cultivated for the dye which is yielded by its leaves.

28. **CAKÌLE**, Tourn. Sea Rocket. (Named from the Arabic.) Silicle 2-jointed, the upper part ovate or ensiform; seed in the upper cell erect, in the lower pendulous, sometimes abortive. ① Maritime, fleshy herbs Flowers purple.

C. marítima Scop. Lvs. oblong, bluntly serrate, obtuse, often lobed; lower joint of silicle clavate, upper ovate-ensiform: racemes spike-like. Coasts, N. States. Prostrate. 6—12'. July, August.

29. **RÁPHANUS**, L. Radish. ('Pα, quickly, φαίνω, to appear; from its rapid growth.) Calyx erect. Pet. obovate, unguiculate. Siliques terete, torulous, not opening by valves, transversely 3-jointed, joints with 1 or several cells. Seeds large, subglobous, in a single series ((o.

1 **R. Raphanístrum** L. *Wild Radish.* Lvs. lyrate; silique moniliform, 3–8 seeded, becoming in maturity 1-celled, longer than the style. ① Fields: rare. 1—2f. Pet. yellow, blanching as they decay. June, July. § Eur.
2 **R. satìvus.** *Garden Radish.* Lower lvs. lyrate, petiolate; silique 2–3-seeded, acuminate, scarcely longer than the style. ① China. 2—4f. Root napiform or fusiform, red, black, or white. Flowers pink-white.

Order XII. CAPPARIDACEÆ. Capparids.

Herbs, shrubs, or even *trees,* destitute of true stipules. *Leaves* alternate, petiolate. *Flowers* cruciform, hypogynous. *Sepals* 4, *Petals* 4, unguiculate. *Stamens* 6—12, or some multiple of 4, never tetradynamous, on a disk or separated from the corolla by an internode of the torus. *Ovaries* often stipitate, of 2 united carpels. *Style* united. *Fruit* either pod-shaped and dehiscent, or fleshy and indehiscent. *Seeds* many, reniform. *Albumen* 0. *Embryo* curved. *Cotyledon* foliaceous.

§ Tribe CAPPAREÆ. Shrubs (or trees) with baccate or drupaceous fruit. S. Fla...Capparis
§ Tribe CLEOMEÆ. Herbs (or shrubs) with capsular 1-celled pods. (a)
    a Stamens 6, separated from the petals by an internode......Gynandropsis. 1
    a Stamens 6, not separated from the petals......Cleome. 2
    a Stamens 8—32, free. Torus not developed to an internode......Polanisia. 3

1. GYNANDRÓPSIS, DC. (*Gynandria,* a Linnæan class, ὄψις, appearance.) Sep. distinct, spreading. Stam. 6, separated from the 4 petals by a slender internode of the torus. Pod linear-oblong, raised on a long stipe which rises from the top of the torus. ① Lvs. digitate. Fls. racemed.
  G. pentaphýlla DC. Middle lvs. petiolate, 5-foliate, floral and lower ones 3-foliate, leaflets obovate, entire, or denticulate. Waste grounds, Va. to Ga. 2—3f. White. §

2. CLEÒME, L. Spider Flower. Sep. sometimes united at base. Pet. 4. Torus not developed between the petals and the stamens, which are 6—4. Pod stipitate more or less. Herbs or shrubs. Lvs. simple or digitate. Flowers racemed or solitary. (See Addenda.)
  1 **C. pungens** L. Stem simple, prickly; lfts. 5—9, elliptic-lanceolate, acute; flowers racemed; petals on filiform claws, half as long as the stamens. ② Gardens and fields. 3—4f. Flowers purple, curious. May—Aug. §
  2 **C. speciosíssima**. Stem branched below; lfts. 5—7, lanceolate, acuminate; petals as long as their claws, rose-purple. Mexico. 3—4f. June—Sept.

3. POLANÍSIA, Raf. (Πολύ, much, ἄνισος, unequal.) Sep. distinct, spreading. Pet. 4, unequal. Stam. 8—32, filaments filiform or dilated at the summit. Torus not developed, minute. Pods linear. ① Strong-scented herbs, with glandular, viscid hairs.
  1 **P. gravèolens** Raf. Viscid-pubescent; lvs. ternate, lfts. elliptic-oblong; fls. axillary, solitary; stam. 8—12; caps. oblong-lanceolate, attenuate at base. Gravelly shores, Vt. to Ark. 1f. Flowers in leafy racemes, yellowish-white. July.
  2 **P. tenuifòlia** T. & G. Viscid-glandular; lfts. 3, filiform-linear; pet. unequal, oval, on short claws; stam. 12—15; pod linear. Ga. Fla. 1—2f. White.

Order XIII. RESEDACEÆ. Mignonettes.

Herbs, with alternate, entire, or pinnate leaves. *Stipules* minute, gland-like. *Flowers* in racemes or spikes, small and often fragrant, 4–7-merous, unsymmetrical and open in bud. *Petals* unequal, entire or cleft. *Stamens* 8—20, inserted on the hypogynous, one-sided glandular disk. *Ovaries* sessile, 3-lobed, 1-celled, many-seeded. *Fruit* a capsule, 1-celled, opening between the stigmas before maturity. Illust. 40, 165.

RESÈDA, L. (Lat. *resedo*, to calm : the plants are said to relieve pain.) Sep. 4—7. Pet. of an equal number, often cleft. Torus large, fleshy, one-sided, bearing the 8—∞ stamens.

1 **R. lutèola** L. *Dyer's Weed.* Lvs. lanceolate, with a tooth on each side at base; sepals 4, united below; petals (greenish-yellow) 3-5-cleft. ① Roadsides, N. Y. 2f. Flowers numerous, in a tall raceme. § Eur.

2 **R.** ODORÀTA L. *Mignonette.* Lvs. cuneiform, entire or 3-lobed; sepals shorter than the 7-13-cleft petals. Egypt. 1f. Fragrant.

ORDER XIV. VIOLACEÆ. VIOLETS.

Herbs with simple (often cleft) alternate leaves with stipules. *Flowers* irregular, spurred, with the sepals, petals, and stamens in 5's. *Sepals* persistent, slightly united, elongated at base, the 2 lateral interior. *Petals* commonly unequal, the inferior usually spurred at base. *Stamens* 5, usually inserted on the hypogynous disk. *Filaments* dilated, prolonged beyond the anthers. *Ovary* of 3 united carpels, with 3 parietal placentæ. *Style* 1, declinate. *Stigma* cucullate. *Fruit* a 3-valved capsule. *Seeds* many, with a crustaceous testa and distinct chalaza. Illust. 50, 93, 137, 302, 515, 522.

Sepals not auricled at base. Filaments united into a tube..................................SOLEA. 1
Sepals more or less auriculate at base. Filaments scarcely cohering......................VIOLA. 2

1. SÒLEA, Gingins. GREEN VIOLET. (Dedicated to *W. Sole,* an English writer on plants.) Sep. nearly equal, not auriculate. Pet. unequal, the lowest 2-lobed and gibbous at base, the rest emarginate. Stam. united into a tube, sheathing the ovary and bearing a gland above the middle. Sds. 6—8, very large. ♃ An erect, leafy plant, with inconspicuous axillary fls.

S. cóncolor Gingins. *Green Violet.*—Woods, W. N-Y. (Hankenson) to Car. and Mo 1—2f. Lvs. large, lanceolate, acuminate. Fls. greenish. Pod 1'. May, June.

2. VÌOLA, L. VIOLET. PANSEY. (From the Latin.) Sep. 5, unequal, auricular at base. Pet. 5, irregular, the broadest spurred at base, the 2 lateral equal, opposite. Stam. approximate, anthers connate, 2 of them with appendages at the back. Caps. 1-celled, 3-valved, seeds attached to the middle of the valves. ♃ Low, herbaceous plants. Ped. angular, solitary, 1-flowered, recurved at the summit so as to bear the flowers in a resupinate position. Joints of the rhizome often bearing apetalous flowers. Figs. 50, 137, &c.

§ Acaulescent.—*a* Petals yellow...No. 1
  —*a* Petals white...Nos. 2, 3, 4
  —*a* Petals blue,—*b* beardless.....................................Nos. 5, 6, 7
    —*b* bearded.—*c* Lvs. divided..................Nos. 8, 9S, 9b
      —*c* Lvs. undivided ...9, 10, 11, and the Exot. 19

ORDER 14.—VIOLACEÆ.

§ Caulescent.--*d* Petals yellow. Stems leafy at the top only .... Nos. 12, 3, 14
—*d* Petals not quite yellow.—*e* Stipules entire .... No. 15
—*e* Stipules fringe-toothed .... Nos. 16, 17, 18
—*e* Stip. lyrate-pinnatifid, very large .. Nos. 20–22

1 V. rotundifòlia Mx. Fig. 50. Lvs. smooth, orbicular-ovate, cordate, with the sinus closed; petiole pubescent; sep. obtuse. Woods, N. E. to Tenn. Mar.—May.

2 V. lanceolàta L. Lvs. smooth, lanceolate, tapering at base into the long petiole, obtusish, subcrenate. Wet meadows. Lvs. 3–5'. Rt. stock creeping. Fls. white. May.

3 V. primulæfòlia L. Lvs. lance-ovate, abruptly contracted at base and decurrent on the petiole; pet. subequal, beardless. Damp soils, Mass. S. and W. White. Ap.May.

4 V. blanda Willd. Lvs. cordate, roundish, slightly pubescent; petiole pubescent; petals beardless. Meadows. Can. to Penn. Root creeping. Flowers fragrant. May.

5 V. palústris L. Lvs. reniform-cordate; stip. broadly ovate; sep. ovate, obtuse, spur very short; caps. oblong-triangular. White Mts. 3'. Pale blue. June.

6 V. Selkírkii Goldie. Lvs. orbicular-cordate, crenately serrate, the sinus deep and nearly closed; spur nearly as long as the petals, thick, very obtuse. Hills, N. Y. to Can. and Mich. 2'. Pale blue, with a large blunt spur. May.

7 V. pedàta L. Rt. premorse; lvs. pedately 5—9-parted, segments linear-lanceolate, entire; stig. large, obtusely truncate, scarcely beaked; spur short, obtuse. Hilly woods, 4—7'. Smooth and beautiful. Flowers large, violet-blue. April, May.

β. *bicolor*. Upper petals violet, the lower pale blue and yellow. Mass. to Ga.

8 V. delphinifòlia Nutt. Lvs. pedately 7—9-parted, with linear, 2—3 cleft segments all similar; stig. thick, distinctly beaked. Ill. Iowa, Mo. Deep blue. Mar. Apr.

9 V. cucullàta Ait. Lvs. reniform-cordate, cucullate at base, acute, crenate; stip. linear; inferior and lateral petals bearded. Common everywhere. 3—12'. Known by its broad hooded leaves and blue flowers. Varies much. April, May.

β. *palmàta*. Lvs. cordate, hastate-lobed, middle lobe largest. Fls. large. South, &c.

γ. *septemlòba*. Lvs. concave at base, deeply 5—7 lobed, mid. lobe lance. South.

10 V. villòsa Walt. Lvs. roundish-ovate, cordate, obtuse, flat, pubescent, sinus narrow or closed; pet. bearded; stig. beaked. Woods, Pa. to Ga.; com. 2—3'. Apr.

11 V. sagittàta Ait. Lvs. oblong-lanceolate, sagittate-cordate, subacute, often incised at base, serrate-crenate; pedicel longer than the leaves; pet. densely bearded. Dry hills. 3—5'. Lvs. varying to triangular-hastate. April—June.

β. *ovàta*. Lvs. ovate, incised and decurrent at base. N. J., southward.

12 V. hastàta Mx. Smooth; st. simple, erect, leafy above; lvs. deltoid-lanceolate or hastate, acute, dentate; stip. ovate, minute, ciliate-dentate; lower pet. dilated, obscurely 3-lobed; spur very short. Fla. to Tenn. 6—10'. April, May.

13 V. tripartìta Ell. Hairy. St. simple, erect, leafy above; lvs. deeply 3-parted, lobes lanceolate, dentate; stip. lanceolate. Upper Ga. 1f. Yellow.

14 V. pubéscens Ait. Villous-pubescent; st. erect, naked below; lvs. broad-cordate, toothed; stip. ovate, large, subdentate. Dry woods. 5—20'. May, June.

β. *eriocárpa*. Tall, pubescent; pods woolly. Westward.

γ. *scabriúscula*. Some scabrous; sts. decumbent, branched at base. Ct. to Ky.

15 V. Canadénsis L. Smooth; lvs. cordate, acuminate, serrate; ped. shorter than the leaves; stip. short, entire. Woods. 8—12'. Leafy all the way. Flowers large, subregular, white or light blue. Summer.

16 V. striàta Ait. Smooth, nearly erect; lvs. roundish-ovate, cordate, crenate-serrate; stip. large, ciliate-dentate, oblong-lanceolate; spur one-fourth as long as the corolla. Wet grounds. 6—12'. St. semi-terete. Flowers cream-white.

17 V. Muhlenbérgii Torr. St. weak, assurgent; lvs. reniform-cordate, upper ones rather acuminate; stip. lanceolate, somewhat fimbriate; spur half as long as the corolla, obtuse. Swamps. 6—8'. Pale purple. May.

18 V. rostràta L. Smooth; st. terete, diffuse, erect; lvs. cordate, roundish, serrate, upper ones acute; stip. lanceolate, deeply fringed; petals bearded; spur longer than the corolla. Moist woods, Can. to Ky. 6—8'. Pale blue. May.—Often beardless.

19 V. ODORÀTA L. *Sweet, or English Violet. Neapolitan.* Stolons creeping; lvs. cordate, crenate, nearly smooth; sep. obtuse. Eur. Flowers fragrant, blue, white, &c.
20 V. tricolor L. *Pansey, Heartsease.* St. angular, diffusely branched; lvs. oblong-ovate, lower ones ovate cordate, deeply crenate; stipules as large as the leaves; spur short, thick. Gardens. Flowers large, white-yellow-violet to black, in endless variety.
β. arvénsis. Slender, subsimple; petals scarcely longer than sepals. Fields. Perhaps this is the primary form. Abundant in Oregon.
21 V. GRANDIFLÒRA L. Stem 3-cornered, procumbent; leaves crenate, shorter than the peduncles, much larger than the stipules; flowers large, all violet.
22 V. CORNÙTA. Stems 3-cornered, ascending; lvs. cordate, crenate; stip. cut-toothed; fls. violet-purp., the spur subulate, longer than the sepals. From the Pyrenees. Hardy.

ORDER XV. CISTACEÆ. ROCK ROSES.

Herbs or low *shrubs* with simple, entire, opposite (at least the lower) leaves, with *flowers* perfect, regular, hypogynous, in one-sided racemes, very fugacious. *Sepals* 5, unequal, persistent. *Petals* 5 (sometimes 3 or wanting), convolute in bud. *Capsules* 1-celled, 3–5-valved, with as many parietal placentæ. *Seeds* albuminous. *Embryo* curved or spiral.

¶ Petals 3, linear-lanceolate, small, brown-purple. Stamens 3–12.................LECHEA. 1
¶ Petals 5,—a large, yellow, very fugacious, or none. Stamens ∞............HELIANTHEMUM. 2
—a small, bright yellow. Tufted shrublets. Stamens 9–30..............HUDSONIA. 3

1. **LECHÈA**, L. PINWEED. Sep. 5, the 2 outer minute. Pet. 3, lanceolate, small. Stig. 3, scarcely distinct. Caps. 3-celled, 3-valved, placentæ nearly as broad as the valves, roundish, each 1–2-seeded. ♃ Often shrubby at base, with numerous very small brownish purple flowers.

1 L. major Mx. Hairy; leaves elliptical, mucronulate; flowers minute, about as long as the pedicels. In dry woods. 1—2f, rigid, brittle, purple, much branched. Leaves 4″. Capsules the size of a small pin-head. July, August.
2 L. minor Lam. Smoothish; leaves linear, very acute; flowers small, on pedicels which are mostly twice longer. Dry, sandy grounds. Stems 8—16′, slender, red. Leaves 6—10″. Capsules the size of a large pin-head. Summer.
3 L. thymifòlia Ph. Shrubby; hoary with appressed hairs; leaves linear and linear-oblanceolate, rather acute, often verticillate; flowers small, on pedicels still shorter. Coasts, Mass. to N. J. 1f. Very bushy. Capsules size of a pin-head. Sum.
4 L. Novæ Cæsareæ Austin. Hairs minute, appressed; lvs. ellip., 6″, often opp.; pan. leafy, narrow; outer sep. lin., longer than the fl. or pedicels. N. J. (Prof. Porter).

2. **HELIÁNTHEMUM**, L. ROCK ROSE. Sep. 5, the 2 outer smaller. Pet. 5, or rarely 3, convolute contrary to the sepals, sometimes 0. Stam. ∞. Stig. 3, scarcely distinct. Capsules triangular, 3-valved, opening at top. Sds. angular. Fls. yellow, often of 2 kinds, the later ones being apetalous.

§ Flowers of 2 sorts, the later ones apetalous, and 3-10-androus.............Nos. 1, 2
§ Flowers all alike, pentepetalous and polyandrous........................Nos. 3, 4

1 H. Canadénse Mx. *Frost Plant.* Hoary pubescent; petaliferous flowers solitary, pedicellate, terminal; apetalous axillary, small, clustered, subsessile; sepals acute; leaves revolute on the margin, lanceolate, acute. In dry soils, Can. to Va. 8—12′.
2 H. corymbòsum Mx. Canescently tomentous; fls. in crowded, fastigiate cymes, the primary ones on elongated, filiform pedicels, and with petals twice longer than the calyx; sep. obtuse; leaves oblong-lanceolate, margins revolute. Sands, N. J. to Fla. 1f

3 H. Caroliniànum Mx. Villous, simple, erect; fls. all large, petaliferous and subterminal; sepals acuminate; lvs. oblong-oval, edges denticulate, not revolute. Dry woods, South. 8—12′. April, May.

4 H. arenícola Chapm. Hoary-tomentous; lvs. lance-oblong, obtuse, small (9″); fls. few. or solitary, pedicellate (7″), terminal. Fla. in sand. 3—6′. Apr. (H. Canadense, β. obtusum Wood. Ed. 5th.)

3. HUDSÒNIA, L. (In honor of *William Hudson*, author of Flora Anglica.) Sep. 3, united at base, subtended by 2 minute ones outside · pet. 5; sta. 9—30; style filiform, straight; cap. 1-celled, 3-valved, many-seeded ♃ with very numerous branches, minute leaves, and small, bright yellow flowers. May.

1 H. tomentòsa Nutt. Hoary tomentous; lvs. ovate, appressed-imbricate, acute; fls. subsessile; sep. obtuse. Coasts, Me. to N. J. and Wisc. In tufts, 7—10′.

2 H. ericoìdes L. Hoary-pubescent; lvs. subulate, a little spreading; pedicels exserted, as long as the calyx; sep. acutish. Shores, Vt. N. H. to Va. Delicate, 6′.

3 H. montàna Nutt. Minutely pubescent; lvs. filiform-subulate; pedicels longer than the flowers; sep. acuminate, the outer ones longer, subulate. Mts. Car. 5′.

ORDER XVI. HYPERICACEÆ. ST. JOHN'S WORTS.

Herbs or *shrubs* with opposite, entire, dotted, exstipulate *leaves*, with *flowers* perfect, regular, hypogynous, 4 or 5-merous, cymous and mostly yellow; *sepals* unequal, persistent; *petals* mostly oblique or convolute in the bud; *stamens* few or many, polyadelphous; *anthers* versatile; *ovary* compound, with styles united or separate, becoming in *fruit* a 1-celled capsule with parietal placentæ, or 3 to 5-celled when the dissepiments reach the centre. *Seeds* exalbuminous, minute. (Illust. 128, 129, 275.)

§ Sepals 4. Petals 4, oblique, contorted in æstivation, yellow..................ASCYRUM. 1
§ Sepals 5. Petals 5,—*a* oblique, contorted in æstivation, yellow..................HYPERICUM. 2
— *a* equilateral, imbricated in bud, purplish..................ELODEA. 3

1. ÁSCYRUM, L. ST. PETER'S WORT. Sep. 4, the two outer usually very large and foliaceous; pet. 4, oblique, convolute; fil. slightly united at base into several parcels; styles 2—4, mostly distinct; cap. 1-celled. ♃ Lvs. punctate with black dots. Fls. pale yellow, 1 or 3 terminating each branch.

The outer pair of sepals—*a* very large, ovate. Styles 1 or 2...................Nos. 1, 2
—*a* still larger, orbicular. Styles 3...................Nos. 3, 4
—*a* small, like the two inner. Styles 3, long, distinct.....No. 5

1 A. Crux-Andreæ L. Branches many, suberect, ancipital above; lvs. linear-oblong, obtuse; outer sep. twice longer than the pedicel; 2 bracteoles a little below the flower. Sandy woods, N. J. to Ga. and La. 1—2f. Lvs. 6—12″. Jn.—Sep.
β. *angustifòlia*. Lvs. smaller (3—6″), crowded; bractlets close to the fl. Car. Ga.

2 A. pùmilum Mx. Low, trailing at base; lvs. oval and obovate, obtuse, sessile; outer sepals shorter than the slender pedicel, inner sepal 0; bracteoles 0. Ga. Fla.

3 A. stans Mx. St. erect, ancipital; lvs. oblong, sessile and half-clasping, obtuse; caps. ovate, acute. Swamps, N. J. to Fla. and La. 1 to 3f. Lvs 10—15″. Jn.—Aug.

4 A. amplexicaùle Mx. St. erect, terete below; lvs. broadly ovate, cordate, clasping; caps. oblong; bracteoles 0. Ga. and Fla. 1 to 2f. Lvs. 8—12″. Apr.—Sep.

5 A. microsépalum Torr. and Gr. Lvs. oblong-linear, crowded; sep. much shorter than the obovate, unequal petals. Bushy, 1—2f. Lvs. 3—6″.

ORDER 16.—HYPERICACEÆ. 49

2. HYPÉRICUM, L. St. John's-wort. Sep. 5, connected at base, subequal. Pet. 5, oblique, contorted in bud. Stam. mostly ∞, generally cohering in 3—5 sets (polyadelphous), with no intervening glands. Styles 3—5, distinct or united. Caps. 1-5-celled. Herbs or shrubs. Flowers cymous, yellow. June—August. Figs. 128, 129, 275.

§ Stamens 25—100, more or less united into sets (polyadelphous)...(*a*)
§ Stamens 5—15, not at all united. Annuals. Flowers small. (*g*)
  a Carpels (and styles) 5 or more. Capsule 5-celled .... Nos. 1, 2
  a Carpels 3, capsule 3-celled (the placentæ meeting)...(*b*)
  a Carpels 3, capsule 1-celled (the placentæ not quite meeting)...(*c*)
    b Shrubby. Petals not dotted. Lvs. lanceolate or oblanceolate .... Nos. 3, 4, 5
    b Shrubby. Petals not dotted. Leaves linear .... Nos. 6, 7
    b Herbaceous. Petals sprinkled with black dots .... Nos. 8, 9, 10
  c Shrubs. Styles united into 1...(*d*)
  c Half-shrubby. Styles united into 1...(*e*)
  c Herbaceous. Styles distinct, at least at the top...(*f*)
    d Flowers solitary or in 3's, axillary. Stems 2-edged .... Nos. 11, 12
    d Flowers clustered in a compound terminal cyme.... .... Nos. 13, 14
    e Flowers in a leafless, stalked cyme. Leaves obtuse .... Nos. 15, 16
    e Flowers in a leafy (few-leaved) cyme. Leaves acute .... Nos. 17, 18
      f Stem and branches 4-cornered or square .... Nos. 19, 20
      f Stem and branches terete, not angular .... Nos. 21, 22
      g Flowers in corymbous cymes, orange-colored .... Nos. 23, 24
      g Flowers racemed on the slender branches .... Nos. 25, 26

1 **H. pyramidàtum** Ait. Herbaceous; lvs. sessile, oblong-ovate, acute; sty. 5; placentæ retroflexed. ♃ O. Pa. to Can. 3—5f. Flowers very large (2′).

2 **H. Kalmiànum** L. Shrubby; lvs. linear-lanceolate, very numerous, obtuse; caps. 5-celled, tipped with the 5 styles. Niagara, &c. 1f. Flowers 9″.

3 **H. Bucklèyi** Curtis. Low, diffuse, shrubby; lvs. obovate, very obtuse; fls. solitary, peduncled; caps. 3-celled, styles united. Mts. N. Car. to Ga. 8—12′.

4 **H. prolificum** L. Branches ancipital, smooth; lvs. oblong-lanceolate, obtuse; cymes compound, leafy; sepals unequal, leafy, ovate, cuspidate. M. W. 3—4f. †
  β. *densiflòrum.* Branches, lvs. and fls. crowded, and smaller. Lvs. 1′. Fls. 6″. South.

5 **H. galloìdes** Lam. Branches erect, terete; lvs. linear-lanceolate; cymules axillary and terminal, paniculate; sep. subequal, linear-lanceolate. S. Car. to Fla. 2—3f.

6 **H. rosmarinifòlium** Lam. Erect, sparingly branched; lvs. linear, shorter than the internodes, narrowed to a petiole. South. Handsome. 2f.

7 **H. fasciculàtum** Lam. Shrub much branched, bushy; lvs. linear, 1′, very narrow, longer than the internodes, sessile; cymules leafy. Pine-barrens, South. 1—2f.
  β. *abbreviàtum.* Lvs. very short (2—3″), tufted in the axils. Car. to Ga.

8 **H. perforàtum** L. Stem 2-edged, branched; lvs. with pellucid dots; sep. lanceolate, half as long as the petals. ♃ Dry pastures. 1—2f. Lvs. 6—10″. Flowers 1′.

9 **H. corymbòsum** Muhl. Stems terete, corymbously branched; lvs. oblong-ovate or oval, obtuse, marked with black (as well as pellucid) dots; sep. ovate, acute (very small), ⅓ as long as the petals. ♃ Can. to Pa. and Ark. 2f. Lvs. 1—2′. Flowers 9″.

10 **H. maculàtum** Walt. Stem terete, corymbously branched; lvs. oblong, thickly sprinkled with black dots; sep. lanceolate. ♃ S. Car. to Fla. 2—4f. Lvs. 1′. Fls. 10″.

11 **H. aùreum** Bartram. Branches spreading, ancipital; lvs. thick, lance-ovate, obtuse, sessile; flower (large) solitary, sessile. Ga. to Ark. 2—4f. Stamens 50′? †

12 **H. ambiguum** Ell. Branches ancipital; lvs. lance-linear, thin, acute; fls. solitary and in 3's in the axils of the upper leaves. Ga. 1—2f. Flowers 8″.

13 **H. myrtifòlium** L. St. terete; lvs. thick, ovate, or oblong, cordate-clasping; fls. in a leafy compound fastigiate cyme, the dichotomal sessile. Ga. Fla. 1—2f.

3

14 H. cistifòlium Lam. St. 2-winged, subsimple; lvs. linear-oblong, obtuse, sessile; flowers in a leafless, compound cyme. Ga. to Fla. and La. (No. 6, β. ?)

15 H. nudiflòrum Mx. St. and branches 4-angled and winged; lvs. ovate-lanceolate or oblong, obtuse, sessile; cyme leafless, peduncled; sep. linear; capsule almost 3-celled. ♃ Wet. Penn. to La. and Ga. 1—2f. Leaves 2', thin.

16 H. sphærocárpon Mx. St. obscurely 4-sided; lvs. linear-oblong, obtuse, with a minute callous tip; sep. ovate, mucronate; caps. globular. ♃ Rivers, W. 1f. Fls. 7".

17 H. adpréssum Bart. St. 2-winged above; lvs. linear-oblong, half erect; cymes few-leaved; petals obovate. ♃ R. I. to Ark.

18 H. dolabrifórme Vent. St. scarcely 2-edged above; lvs. linear-lanceolate, spreading; fls. in a leafy, fastigiate cyme; pet. very oblique (dolabriform). ♃ Ky.Tenr.

19 H. angulósum Mx. Herb smooth; st. acutely 4-cornered; lvs. oblong-lanceolate, acute; cymes leafless; style distinct, thrice longer than the ov. ♃ N. J. to Fla.

20 H. ellípticum Hook. Herb smooth; st. quadrangular, simple; lvs. elliptical, obtuse, somewhat clasping, pellucid-punctate; cyme pedunculate; sep. unequal; style united to near the summit, as long as the ovary. ♃ Can. to Pa. 1f. Flowers 6".

21 H. gravèolens Buckley. Stem terete, smooth; leaves oblong-ovate, clasping; sepals and petals narrow; styles 3. ♃ High Mts., N. Car. Strong-scented.

22 H. pilòsum Walt. Rough-downy; stem simple, terete, virgate; lvs. ovate-lanceolate, appressed, clasping, acute; styles distinct. ① Pine-barrens, South. 1—2f.

23 H. mùtilum L. Stem square, branched; lvs. ovate, 5-veined, clasping, obtuse; cymes leafy; pet. shorter (1") than sep.; sta. 6—12. ① Damp sandy soils. 3—9'. Com.

β. *gymnánthum*. Strict, simple or branched, cy. only bracted. Del., Penn. (Porter).

24 H. Canadénse L. Stem quadrangular, branched; lvs. linear, attenuated to the base, with pellucid and also with black dots, rather obtuse; petals shorter than the lanceolate, acute sepals; stamens 5—10. ① Wet sandy soils. Capsule red. 6—12'.

25 H. Saròthra Mx. Stem and branches filiform, erect, and parallel; lvs. very minute, subulate; flowers sessile; stam. 5—10. ① Sandy soils. 4—12'. Fls. minute.

26 H. Drummóndii T. & G. Branches alternate; lvs. linear, very narrow; flowers pedicellate; stamens 10—20. ① Dry. Ill. and South. 1f. Leaves 6".

3. ELODÈA, Adams. ('Ελώδης, marshy; from the habit.) Sep. 5, equal. Pet. 5, equilateral, imbricated in bud. Stam. 9 (rarely more), triadelphous, the sets alternating with 3 orange-colored glands. Styles 3, distinct. Capsule 3-celled. ♃ Herbs with pellucid-punctate leaves, the axils leafless. Flowers dull orange-purple. July—Sept.

1 E. Virgínica Nutt. Stem erect, somewhat compressed, subsimple; leaves oblong, amplexicaul: stamens united below the middle, with 3 in each set. Swamps. 1f.

2 E. petiolàta Ph. Leaves oblong, narrowed at base into a petiole; flowers mostly in 3's, axillary, nearly sessile; filaments united above the middle; caps. oblong, much longer than the sepals. Swamps, S. States, N. to N. J. Flowers smaller (4").

Order XVII. DROSERACEÆ. Sundews.

Herbs growing in bogs, often covered with glandular hairs, with *leaves* alternate or all radical, mostly circinate (rolled from top to base) in vernation. *Flowers* regular, hypogynous, 5-merous, the *Sepals, Petals,* and *Stamens* persistent (withering). *Ovaries* compound, 1-celled, with the *Styles* and *Stigmas* variously parted, cleft, or united. *Seeds* ∞ in the capsule, albuminous. *Embryo* minute.

* Stamens 5. Styles distinct. Seeds on the valves of the capsule.................DROSERA. 1
* Stamens 10—15. Styles united. Seeds all at the base of the cell.............DIONÆA. 2

ORDER 18.—ELATINACEÆ.

1. DRÓSERA, L. SUNDEW. (*Δρόσος*, dew; from the dew-like secretion.) Sep. 5, united at base, persistent. Pet. 5. Stam. 5. Sty. 3—5, each 2-parted, the halves entire or many-cleft. Caps. 3–5-valved, 1-celled, many-seeded. ② or ♃ Small marsh herbs. Lvs. covered with reddish, glandular hairs, secreting a viscid fluid. Flowers in a raceme on a slender scape which is at first coiled, uncoiling as the flowers open.

* Scapes 4—6 times as long as the spreading leaves..........................Nos. 1-3
* Scapes 1—2 times as long as the ascending leaves ....Nos. 4-6

1 **D. rotundifòlia** L. Lvs. orbicular, abruptly contracted into the hairy petiole; fls. white. ② A curious little plant, in bogs and muddy shores. Scapes 6—9′, 6-9-flowered. Leaves 1—2′, glistening as with dew-drops. June—Aug.

2 **D. capillàris** Poir. Lvs. obovate, cuneiform at base, the petioles naked; flowers purple; scape erect. ② Marshes, S. Car. to Fla. Scapes 3—12′, 6-12-flowered. May.

3 **D. brevifòlia** Ph. Lvs. cuneiform-spatulate, forming a small, dense tuft (1′ diam.); petioles very short, hairy; flowers few, rose-colored. ② N. Car. to Fla. 2—5′.

4 **D. longifòlia** L. Lvs. spatulate-oblong or obovate, ascending, alternate, tapering at base into a long, smooth petiole; scape declined at base; pet. wh. ♃ 4—7′. Lvs.2-3′.

5 **D. lineàris** Goldie. Lvs. linear, obtuse; petioles elongated, naked, erect; scapes few-flowered, about the length of the leaves (3′); calyx glabrous, much shorter than the oval capsule; seeds oval, smooth. ♃ Borders of lakes, North. White.

6 **D. filifórmis** Raf. Lvs. filiform, very long, erect; scape nearly simple, longer than the leaves, many-flowered; petals obovate, erosely denticulate, longer than the glandular calyx; style 2-parted to the base. ♃ Wet sand. 1f. Purple.

2. DIONÆA, L. VENUS' FLY-TRAP. (One of the names of Venus.) Stam. 10—15. Sty. united into 1, the stigmas many-cleft. Caps. breaking irregularly in opening, 1-celled. Seeds many, in the bottom of the cell. ♃ Glabrous herbs. Lvs. all radical, sensitive, closing convulsively when touched. Scape umbelled.

D. muscípula Ell.—A very curious plant. Sandy bogs in Car. Lvs. rosulate, lamina roundish, spinulose on the margins and upper surface, instantly closing upon insects and other objects which light upon it. Scape 6—12′, with an umbel of 8—10 white flowers. April, May. †

ORDER XVIII. **ELATINACEÆ.** WATER PEPPERS.

Herbs small, annual, with opposite leaves and membranous stipules *Flowers* minute, axillary. *Sepals* 2—5, distinct or slightly coherent at base, persistent. *Petals* hypogynous, as many as the sepals. *Stamens* twice as many as the petals, *anthers* introrse. *Ovaries* 2-6-celled. *Stigmas* 2—6, capitate; *placenta* in the axis. *Fruit* capsular. *Seeds* numerous, exalbuminous.

ELATINE, L. MUD PURSLANE. Fls. 2-, 3-, or 4-parted, symmetrical, all the parts distinct except the united ovaries. Stig. sessile. ☿ Very small plants growing in mud, with minute, axillary, sessile flowers.

1 **E. Americàna** Arn. Stems creeping, diffuse, in patches; branches ascending 1—2′; leaves wedge-obovate, 2″, obtuse; flowers 2-parted, rarely 3-parted; seeds 6- 8.

2 **E. Clintoniàna** (Peck). Stems erect, 4″, in very dense tufts, from matted roots; lvs spatulate, 1″; fls. 2-parted; seeds slightly curved. Sand Lake, N. Y. (C. H. Peck).

Order XIX. CARYOPHYLLACEÆ. Pinkworts.

Herbs with swollen joints, opposite, entire leaves, and regular ⚥ (rarely ⚲) flowers. *Sepals* persistent. *Petals* often unguiculate, or bifid, or 0. *Stamens* distinct, twice as many as the sepals, or fewer. *Torus* often some developed, separating the whorls. *Styles* 2—5, *ovary* 1. *Fruit* a 1–5-celled, 1 – ∞-seeded pod, opening by teeth or valves. *Embryo* curved around the albumen. Figs. 6, 41, 44, 45, 56, 131, 276, 330, 456.

§ Stipules present, dry (0 in No. 17). Calyx open. Petals sessile, minute, or 0. Tribe III...(*h*)
§ Stipules 0.—*a* Calyx a tube including the long claws of the petals. Pod ∞-seeded. Tribe I...(*c*)
—*a* Calyx open. Petals sessile (rarely 0 in No. 10). Pod 3 – ∞-seeded. Tribe II...(*e*)
—*a* Calyx open, *white*. Petals 0. Styles 3. Pod 3-celled. Tribe IV...Mollugo. 18
I. SILENEÆ.—*c* Calyx with scale-like bractlets at base. Styles 2................Dianthus. 1
    —*c* Calyx bractless.—*d* Styles 2......................Saponaria, 2, *or* Gypsophila, 2½
        —*d* Styles 3. Pod 6-toothed when open..........Silene. 3
        —*d* Styles 5. Pod 10-toothed or 5-valved..........Lychnis. 4
II. ALSINEÆ.—*e* Petals erose-denticulate at the end. Styles 3................Holosteum. 5
    —*e* Petals 2-parted (sometimes wanting in No. 7)...(*f*)
        f Styles 5. Capsule opening at the top by 10 teeth..............Cerastium. 6
        f Styles 3. Capsule opening to the base by half-valves..........Stellaria. 7
    —*e* Petals entire (often wanting in No. 10)..(*g*)
        g Styles 3, or if 5, opposite to the sepals. (No. 7 or)..........Arenaria. 8
        g Styles 4, opposite to the 4 sepals. Stamens 4..............Mœnchia. 9
        g Styles 4 or 5, and alternate with the sepals................Sagina. 10
III. ILLECEBREÆ.—*h* Styles or stigmas 3–5. Pod several-seeded. Pet. colored...(*k*)
        k Leaves opposite.—*l* Flowers axillary, solitary..........Spergularia. 11
            —*l* Flowers in terminal clusters........Stipulicida. 12
        k Leaves whorled.—*m* Styles 5, pod 5-valved..............Spergula. 13
            —*m* Styles 3, pod 3-valved............Polycarpon. 14
    —*h* Styles or stigmas 2 or 1. Utricle 1-seeded...(*n*)
        n Sepals distinct or nearly so, greenish..................Paronychia. 15
        n Sepals united into a tube below, white above..........Syphonychia. 16
        n Sepals united into an urn below, green above..........Scleranthus. 17

1. DIÁNTHUS, L. Pink. Calyx tube cylindrical, striated, with 2 or more pairs of imbricated scales or bracteoles at base. Pet. 5, with long claws, limb irregularly notched. Stam. 10, styles 2, recurved. Capsule cylindrical, 1-celled, 4-valved at top. Beautiful Oriental plants, everywhere cultivated. Figs. 6, 131, 276.

§ Bracts long-pointed, equalling the calyx tube (dry, obtuse, No. 2).........Nos. 1—4
§ Bracts much shorter than the calyx tube.................................Nos. 5—7

1 **D. Armèria** L. *Wild Pink.* Leaves linear-subulate, hairy; flowers aggregated, fascicled; bracteoles erect, lance-subulate. ① Sandy fields, E. 1—2f. Flowers small (6″ broad), pink-red sprinkled with white. August. § Europe.
2 **D. prólifer** L. Slender, strict, smooth; lvs. linear, erect, 1—2′; bracts dry, ovate, covering the calyx and pod; pet. small, pink; fl. mostly but 1. Penn. (Porter). § Eur.
3 **D.** BARBÀTUS. *Sweet-William,* or *Bunch P.* Leaves lanceolate; flowers in dense cymes; bracteoles erect, ovate-subulate. ♃ Europe. 1½f. Red-white. May—July.
4 **D.** CHINÉNSIS. Leaves lance-linear; flowers solitary; bracteoles spreading, linear. ② China. 1f. Evergreen, not glaucous. Flowers large, variegated.
5 **D.** CARYOPHÝLLUS. *Carnation P.* Glaucous; leaves linear; flowers solitary; bractlets very short, ovate; petals very broad, crenate. ♃ England. 2—3f. Fragrant.
6 **D.** PLUMÀRIUS. *Pheasant's Eye.* Glaucous; flowers solitary; bracts ovate, acute; petals many-cleft, hairy at throat. ♃ Europe. White-purple. June—August.
7 **D.** SUPÉRBUS. Leaves linear-subulate, green; cymes fastigiate; bracts ovate, mucronate; petals pinnatifid-fringed. ♃ Europe. White-roseate. July, August.

ORDER 19.—CARYOPHYLLACEÆ. 53

2. SAPONÀRIA, L. SOAPWORT. Calyx tubular, 5-toothed, without bractlets. Petals 5, unguiculate. Stamens 10. Styles 2. Capsules oblong, 1-celled. Flowers in cymous panicles. July, August. Fig. 45.

§ Calyx tube oblong, neither angled nor veined (SAPONARIA.).................No. 1
§ Calyx tube ovoid, 5-angled, at length 5-winged, very smooth. (VACCÀRIA.)...No. 2

1 S. officinàlis L. *Bouncing Bet.* Lvs. lanceolate; pet. crowned. ♃ 2f. White. §
2 S. Vaccària L. Lvs. lance-ovate; fls. cymous, pale red. ①1f. Waste grounds. §

2½. GYPSOPHILA, L. GYPSUM PINK. Sepals half united into a bell-form calyx. Pet. scarcely clawed. *Caps.* globular, 1-celled, 4-valved. —Neat, free-flowering exotics. Flowers panicled. June—Sept. Europe.

1 G. ÉLEGANS. Lvs. lance., thick; pan. loose, forked; pet. notched, wh. or pink. 1f. ①
2 G. MURÀLIS. Low, diffuse, with linear lvs. and a profusion of pinkish small fls. ① 6'.
3 G. PANICULÀTA. Tall; lvs. lance-lin.; fls. minute, numerous, white, in filiform pan. ♃
4 G. STÉVENI. Lvs. lance-lin., keeled; fls. white, in corymbs, fine for bouquets. ♃ 2f.

3. SILÈNE, L. CAMPION. CATCH-FLY. (*Silenus* was a drunken god of the Greeks, covered with slaver as these plants are with a viscid secretion.) Calyx tubular, swelling, without scales at the base, 5-toothed; pet. 5, unguiculate, often crowned with scales at the mouth, 2 or many-cleft, or entire; sta. 10; styles 3; capsule 3-celled, opening at top by 6 teeth, many-seeded. Figs. 41, 56, 330.

§ Acaulescent, low, tufted. Petals crowned. Perennial..................No. 1
§ Caulescent.—Petals fringe-cleft, white or rose-color, crownless. Perennial.Nos. 2—4
   —Petals bifid or entire.—Calyx inflated, veiny. Perennial.......Nos. 5, 6
                 —Calyx close on the pod. (*)
* Flowers spicate, alternate. Upper leaves linear, lower spat. Annual...Nos. 7, 8
* Fls. not spicate.—Petals pale, closed in sunshine. Upper lvs. linear...Nos. 9, 10
                   —Petals red, purple, &c.,—bifid......................Nos. 11, 12
                                              —entire....................Nos. 13—15

1 S. acaùlis L. *Moss Campion.* Low, moss-like; lvs. linear (6''); ped. solitary, short, 1-fld.; calyx bell-shaped; pet. obcordate, crowned. ♃ White Mts. 1—3'. Purp. Jl.
2 S. stellàta Alt. Erect, pubescent; lvs. in whorls of 4's, oval-lanceolate, acuminate; cal. loose and inflated; pet. fimbriate. ♃ Can. to Car. and W. 2—3f. White. July.
3 S. ovàta Ph. Erect, puberulent; lvs. opposite, lance-ovate, acuminate; cal. ovate, not inflated; pet. many-cleft, crownless. ♃ Car. Ga. 3f. White. July.
4 S. Baldwinii Nutt. Weak, hairy; lvs. obovate-spatulate; calyx not inflated; pet. cuneiform, divaricately fimbriate. ♃ Ga. Fla. 1f. Fls. 2', roseate. April.
5 S. nívea DC. Minutely puberulent, erect, subsimple; lvs. oblong-lanceolate, acuminate; fls. few, solitary, leafy; cal. inflated; pet. 2-cleft, with a small bifid crown; caps. shorter than its stipe. ♃ Penn. to Ill. Rare. 2f. Fls. few, white. July.
6 S. inflàta Smith. *Bladder Campion.* Glabrous and glaucous; lvs. ovate-lanceolate; fls. in cymous, leafless panicles, drooping; cal. ovoid-globular, much inflated; caps. on a short stype. ♃ Fields. 2f. White. July. §
7 S. quinquevúlnera L. Villous; spike somewhat one-sided; cal. very villous; pet. roundish, entire, crowned. ① S. Car. 1f. Pet. crimson, with a pale border. §
8 S. noctúrna L. Lvs. pubescent; fls. small, appressed to the stem in a dense 1-sided spike; cal. cylindrical, smoothish; pet. narrow, 2-parted. ① Ct. to Pa. Rare. 2f. Jl. §
9 S. Antirrhìna L. *Snap-dragon Catch-fly.* Sticky in spots; lvs. lanceolate, acute; fls. few, on slender branches; cal. ovoid; pet. emarginate. ① Waste pl. 1½f. Fls. r.
  β. linària. Very slender; lvs. all linear; cal. globular. Ga. and Fla.
10 S. noctiflòra L. Viscid-pubescent; lower lvs. spatulate; cal. cylindrical, teeth subulate, very long; petals 2-parted. ① Cult. grounds. Flowers large, white. §

11 S. Virgínica L. Slender, erect, branching; root-lvs. spatulate, cauline oblong lanceolate; flowers large, cymous, cal. large, clavate; pet. bifid, broad, crowned. ♃ Woods, Pa. to Ill. and S. 1—2f. Red. June.

12 S. rotundifòlia Nutt. Decumbent, branching; lvs. thin, roundish-oval; fls. solitary, very large; calyx cylindric-campanulate; petals bifid and toothed, deep scarlet, crowned. ♃ Rocks, W. States. Rare. June—August.

13 S. règia Sims. *Splendid Catch-fly.* Scabrous, somewhat viscid; st. rigid, erect; lvs. ovate-lanceolate; cyme paniculate; pet. oblanceolate, entire, erose at the end; sta. and stig. exserted. ♃ O. to Ill. and S. 3—4f. Bright scarlet. June, July.

14 S. Pennsylvánica Mx. *Wild Pink.* St. clustered, low, ascending; lvs. spatulate or cuneate, of the stem lanceolate; cyme few-flowered; pet. slightly emarginate, subcrenate. ♃ Dry soils, N. Eng., S. and W. 6—10′. Fls. pink-red. June.

15 S. Armèria L. *Garden Catch-fly.* Very smooth, glaucous; st. branching, glutinous below each node; lvs. ovate-lanceolate; flowers in flat cymes; pet. obcordate, crowned; cal. clavate, 10-striate. ① 12—18′. July, September. † §

4. LYCHNIS, L. (Λύχνος, a lamp; from fancied resemblance or use.) Cal. tube bractless, 10-veined, limb 5-lobed. Pet. 5, entire or cleft, often crowned. Stam. 10. Styles 5. Caps. more or less 5-celled at base, opening by 5 or 10 teeth. Handsome exotics, cultivated or §.

§ AGROSTÉMMA. Calyx limb of 5 leafy, deciduous lobes exceeding the petals.......No. 1
§ LYCHNIS *proper.* Calyx limb of 5 persistent lobes shorter than the petals...(a)
  a Fls. diœcious. Petals 2-lobed, white or purplish. Escaped from culture......No. 2
  a Fls. all perfect.—*b* Petals 2-lobed or entire..........................Nos. 3, 4
     —*b* Petals 4-parted or laciniate............................Nos. 5, 6

1 L. Githago Lam. *Corn Cockle.* St. forked; lvs. linear, hairy; fls. few, large, dull purple; seeds large, blackish. ① Fields. 2—3f. A handsome weed. July. §

2 L. diúrna L. Stem forked and panicled; fls. ♂ ♀; pet. half-2-cleft; pod ovoid or subglobous. ② Rare in cultivated grounds. 2f. June—August. § Eur.

3 L. CORONÀRIA DC. *Mullein Pink. Rose Campion.* Villous; stem dichotomous; ped. long, 1-flowered; petals broad, entire. ♃ Italy. 2f. Purple, &c.

4 L. CHALCEDÓNICA L. *Scarlet Lychnis* or *Sweet William.* Smoothish; fls. fasciculate; calyx cylindric-clavate, ribbed; petals 2-lobed. ♃ Russia. 2f. Scarlet.

5 L. FLOSCÙCULI L. *Ragged Robin.* Fls. fascicled; cal. campanulate, 10-ribbed; pet. in 4 deep, linear segments. ♃ Europe. 1—2f. Flowers pink.

6 L. CORONÀTA L. *Chinese Lychnis.* Fls. terminal and axillary, 1—3; calyx rounded, clavate, ribbed; petals laciniate. ♃ 1—2f. Flowers large, red, &c.

5. HOLÓSTEUM, L. ("Ολος, all, οστέον, bone; by antiphrasis, as the plant is *no bone*, but soft.) Sep. 5. Pet. 5, erose-denticulate at the end. Stam. 3—5, rarely 10. Styles 3. Caps. 1-celled, ∞-seeded, opening by 6 teeth. Fls. white, in an umbel.

H. umbellàtum L. Lvs. smooth and glaucous, oblong, sessile; ped. long, terminal, viscid, pedicels reflexed after flowering. ① Fields: rare. 6′. § Eur.

6. CERÁSTIUM, L. MOUSE-EAR CHICKWEED. (Κέρας, a horn; from the resemblance of the capsule.) Sep. 5, ovate, acute. Pet. 5, 2-cleft or lobed. Stam. 10, rarely fewer. Styles 5, opposite to the sepals. Capsule cylindrical or ovoid, elongated, opening at top by 10 teeth, ∞-seeded. Flowers cymous, white. Fig. 44.

§ Petals about as long as the sepals...Nos. 1, 2
§ Petals much longer than the sepals . . ..Nos. 3, 4, 5

ORDER 19.—CARYOPHYLLACEÆ.

1 **C. vulgàtum** L. Hairy, cæspitous; lvs. obovate or ovate, obtuse, attenuated at base; fls. in subcapitate clusters; sep. acute, longer than the pedicels; stam. often 5. ① Fields and waste grounds. 6—12′. June—Aug. §
2 **C. viscòsum** L. Hairy, viscid, spreading; lvs. oblong-lanceolate, rather acute; fls. in loose cymes; sep. obtuse, scarious on the margin and apex, shorter than the pedicels. ♃ Fields and waste grounds. 5—9′. Plant greener. June—Aug.
3 **C. arvénse** L. Pubescent; lvs. linear-lanceolate, acute; cyme on a long, terminal peduncle, 4-flowered; petals more than twice longer than the calyx; capsule scarcely exceeding the sepals. ♃ Rocky hills. 4—10′. May—Aug.
4 **C. oblongifòlium** Torr. Villous, viscid above; lvs. oblong-lanceolate; flowers numerous, in a spreading cyme; pet. twice as long as the sepals; capsule about twice as long as the calyx. ♃ Rocky places. Rare. 6—10′. Fls. large. April—June.
5 **C. nùtans** Raf. Viscid-pubescent, erect; lvs. lanceolate; fls. many, diffusely cymous, on long, filiform, nodding pedicels; pet. nearly twice as long as the calyx; capsule a little curved, nearly thrice as long. ① Low grounds. 8—12′. May.

7. **STELLÀRIA,** L. STAR CHICKWEED. (Lat. *stella*, a star; from the stellate or star-like flowers.) Sep. 5, connected at base. Pet. 5, 2-parted, rarely 0. Stam. 10, rarely fewer. Styles 3, sometimes 4. Caps. ovoid, 1-celled, valves as many as styles, 2-parted at top. Sds. many. Small herbs in moist, shady places. Fls. in forked cymes or axillary, small, wh. Fig. 456.

§ Stems hairy mostly in lines, leafy to the top. Leaves broad..............Nos. 1, 2, 3
§ Stems all glabrous,—*a* leafy to the top. Petals sometimes wanting......Nos. 4, 5, 6
—*a* leafless above, with scarious bracts...............Nos. 7, 8, 9

1 **S. mèdia** Smith. Lvs. ovate; st. procumbent, with an alternate, lateral, hairy line; pet. shorter than the sep.; stam. 3 to 5 or 10. ① A common weed. April—Nov.
2 **S. prostràta** Baldw. Lvs. ovate, the lower on long petioles; sts. procumbent, pubescent; fls. on long pedicels; pet. longer than sepals; stam. 7. ① Ga. Fla. Mar. Ap.
3 **S. pùbera** Michx. Stem ascending, pubescent in 1 lateral or 2 opposite lines; lvs. oblong, acute, sessile; pet. longer than the white-edged sep. ♃ Pa. S. and W. Apr. Ju.
4 **S. uniflòra** Walt. Smooth, erect from a prostrate base; lvs. linear-subulate, remote; ped. long, 1-flwd.; pet. obcordate, twice longer than cal. ② Swamps, S. 10—12′. May.
5 **S. boreàlis** Bw. Smooth, weak; lvs. veinless, lance-oblong; ped. at length axillary, 1-flwd.; pet. 2-parted (often 0), as long as calyx. ♃ Wet shades, N. Eng. to Wis. 6—15′.
6 **S. crassifòlia** Ehrh. Sts. weak; lvs. linear-oblong, thickish; pet. longer than the cal., or 0; sds. roughened. Wet rocky places, Ky. and N. (Sagina fontinalis Sh.& Pet.)
7 **S. uliginòsa** Murr. Decumbent; lvs. lance-oval and oblong, veiny; cymes lateral, sessile, leafless; sep. 3-veined, as long as the bifid pet. ♃ Springs, Md. to N. H., and W.
8 **S. lóngipes** Goldie. Smooth and shining; lvs. linear-lanceolate, broadest at base; ped. erect, filiform, cymous; sep. with membranous margins, shorter than the petals. ♃ Me. to Mich. and N. June.
9 **S. longifòlia** Muhl. Lvs. linear; cyme terminal, naked, at length lateral, the pedicels spreading; petals longer than the calyx. ♃ Common. July.

8. **ARENÀRIA,** L. SANDWORT. (Lat. *arena*, sand, in which most species grow.) Sep. 5, spreading. Pet. 5, entire, or notched, rarely 0. Stam. 10, rarely fewer. Styles 3, rarely more or fewer, opposite to as many sepals. Capsule 1-celled, ∞-seeded, opening by valves or half-valves. Slender herbs, mostly tufted, with white flowers. (The following sections have sometimes been regarded as genera.)

§ ARENÀRIA. Caps. splitting into 6 half-valves. Lvs. acute. Seeds naked.....Nos. 1, 2
§ MŒHRÍNGIA. Caps. as above. Lvs. and sep. obtuse. Sds. strophiolate..........No 3

§ Honkenya. Caps. splitting into 3 (—5) valves. Disk large, 10-lobed............No. 11
§ Alsine. Capsule splitting into 3 entire valves. Disk inconspicuous..(a)
    a Sepals 3 or 5-veined, acute, or acuminate.....................Nos. 4, 5, 6
    a Sepals veinless, obtuse.—*b* Leaves rigid, subulate, imbricated............No. 7
        —*b* Leaves soft, opposite, spreading........Nos. 8, 9, 10

1 **A. serpyllifòlia** L. St. dichotomous, spreading; lvs. ovate, acute, subciliate; pet. shorter than the acute sep.; pod ovate. ① Sandy pl. 2–5'. Lvs. 2–3''. Jn.-Aug. §
2 **A. diffùsa** Ell. St. long, diffuse; lvs. lance-ovate, acute at both ends; ped. 1-flwd.; pet. oval, much shorter than the calyx, or 0. ♃ Moist woods, S. 2–5f. Apr. June.
3 **A. lateriflòra** L. Upright, slightly pubescent; lvs. oval, obtuse; ped. lateral, 2 to 3-flwd.; seeds (strophiolate) appendaged at the hilum. ♃ Damp shades, N. 6–10'. Jn.
4 **A. pátula** Mx. Sts. divaricately branched, very slender; lvs. linear-filiform, obtuse; petals emarginate. ① Cliffs, Va. and Ky. 6–10'. June—July.
5 **A. Pítcheri** T. & G. Erect, fastigiately branched, almost glabrous; lvs. linear, obtuse, flat; pet. entire, twice as long as the 5-veined sepals. ① Tenn. and W. 3–6'.
6 **A. strícta** Mx. Glabrous, diffuse; st. branched from the base; lvs. subulate-linear, rigid, so fascicled in the axils as to appear whorled; cymes few-flowered, with spreading branches. ♃ Sterile grounds. 8–10'. May, June.
7 **A. squarròsa** Mx. Cæspitous; stem few-flowered; lower leaves squarrous-imbricate, crowded, upper ones few, all subulate, channelled, smooth; petals obovate, 3 times longer than the sepals. ♃ Barrens, L. I. to Ga. 6–10'. April-Aug.
8 **A. Greenlándica** Spr. Cæspitous; sts. numerous, filiform; lvs. linear, flat, spreading; ped. 1-flwd., elongated, divaricate. ♃ High Mts. N. 3'. Fls. 8'', numerous. Jl.Aug.
9 **A. brevifòlia** N. Erect (not tufted), few-leaved; stems many, filiform; lvs. minute, few, remote, ovate-subulate; sepals oblong. ① Rocks, Ga. 2–4'. May.
10 **A. glàbra** Mx. Cæspitous, glabrous; sts. filiform; lvs. linear setaceous, spreading; sep. oval, veinless, half as long as the petals. ♃ Mts. S. 4–6'. Fls. 6''. July.
11 **A. peploides** L. Sts. creeping, with upright branches, tufted; lvs. ovate, fleshy, half-clasping; fls. small, the veinless sepals exceeding the petals. ♃ Coast. 1f. May.

9. **MŒNCHIA**, Ehrh. (Dedicated to *Mœnch*, a German botanist.) Sep. 4, as long as the 4 entire petals and opposite to the 4 styles. Stam. 4. Caps. ovoid, not exceeding the calyx, opening by 8 teeth, ∞-seeded. ① Low, smooth, glaucous. Flowers white.

M. quaternélla Ehrh.—Dry places, Md. Stems simple, 2–3', with 1 or 2 flowers. Leaves lance-linear, acute. Apr. May. § Eur. (Sagina erecta L.)

10. **SAGÍNA**, L. Pearlwort. (Lat. *sagina*, food or nourishment; badly applied to these minute plants.) Sep. 4 or 5. Pet. 4 or 5, entire, often 0. Stam. as many or twice as many as the sepals. Styles 4 or 5, *alternate* with the sepals, but the valves of the pod are opposite. Diminutive herbs with linear leaves and small white flowers.

    * Petals 0, or 4, and much shorter than the 4 sepals. Stam. 4................Nos. 1, 2
    * Petals 5, equalling or much exceeding the 5 sepals. Stam. 10..............Nos. 3, 4

1 **S. procúmbens** L. Procumbent, glabrous; pet. about half as long as the roundish, obtuse sepals, sometimes 0; lvs. linear-filiform. ♃ Damp, N. 3–4'. June.
2 **S. apétala** L. Erect, puberulent; pet. very minute, or none; sep. oblong, acute; lvs. linear-subulate. ① Sandy, N. Y., N. J. and W. Stems filiform, 2–4'. May, Ju.
3 **S. subulàta** Wimmer. Smooth or puberulent, tufted; lvs. filiform-linear, mucronate, shorter than the erect ped.; pet. 5, as long as the ovate, obtuse sep., rarely 0. ② Sandy, S. 2–6'. Lvs. 6''. March, April. (S. Elliottii Fenzl.)
4 **S. nodòsa** Fenzl. Tufted, ascending, glabrous; lvs. subulate, the upper very short and fascicled; pet. much longer than the sepals. ♃ Sandy shores, N.

ORDER 19.—CARYOPHYLLACEÆ.   57

11. SPERGULÀRIA, Pers. SAND SPURRY. (Name derived from *Spergula.*) Sep. 5. Pet. 5, entire. Stam. 2—10. Styles 3. Caps. 3-valved, ∞-seeded.—Herbs low, spreading, with narrow opposite leaves and scarious stipules. Flowers red or rose-colored.

1 S. rubra Presl. Decumbent, divaricately branched, slender; stip. triangular-acuminate; lvs. linear; sep. lanceolate, with scarious margins; pet. as long, pink-red; seeds rough, marginless. ♃ Sandy, near the coast. 3—6'. May—October.

2 S. marìna. Plant thick and fleshy; caps. a third longer than the calyx, with the seeds nearly smooth and mostly margined. Otherwise like No. 1, and perhaps not distinct. ♃ Salt marshes. May—October. (Arenaria, L.)

12. STIPULÍCIDA, Michx. (Lat. *stipula, cædo;* the stipules being much cleft.) Sep. with scarious margins. Pet. 5, as long as the sepals, entire. Stig. 3, subsessile. Caps. subglobous, 3-valved, few-seeded. ① A slender, tufted, dichotomously branched herb, almost leafless, with the small flowers in terminal cymules.

S. setàcea Mx.—Dry sand, Ga. Fla. Stems almost setaceous, 6—10'. Joints distant, with a fringe of leaves and stipules ¼''. Root leaves roundish, 1''. Fls. reddish. May.

13. SPÉRGULA, L. SPURRY. (Lat. *spergo,* to scatter; from the dispersion of the seeds.) Sep. 5. Pet. 5, entire. Stamens 5 or 10. Styles 5. Caps. ovate, 5-valved, seeds ∞. Embryo coiled into a ring. ① Herbs with fls. in loose cymes. Leaves verticillate. Stipules scarious.

S. arvénsis L. Lvs. filiform; ped. reflexed in fruit; sds. reniform, angular, rough. Cultivated grounds. 1—2f. Lvs. 1—2', many in a whorl. May—August. §

14. POLYCÁRPON, L. ALL-SEED. ($\Pi o \lambda \acute{u} \varsigma$, much, $\varkappa a \rho \pi \acute{o} \varsigma$, fruit; the pods are many.) Sepals 5, carinate. Pet. 5, emarginate. Stam. 3—5. Style short, 3-cleft. Caps. 3-valved. ① Low, diffuse, with whorled lvs.

P. tetraphýllum L. Lvs. spatulate or oval, tapering to a petiole, some of them in whorls of 4; stam. 3. Around Charleston, S. Car. 3—6'. Lvs. 3—5''. Fls. minute. §

15. PARONÝCHIA, Tourn. NAILWORT. ($\Pi a \rho \acute{a}$, with, $\acute{o} \nu \upsilon \xi$, the nail; *i. e.,* the *whitlow;* supposed cure for.) Sep. 5, linear-oblong, connivent, mucronate or awned near the apex. Pet. or sterile filaments very narrow and scale-like, or none. Stam. 2, 3, or 5. Stig. 2, with the styles more or less united into 1. Utricle 1-seeded. Low herbs dichotomously branched, with scarious, silvery stips., and at least the lower lvs. opposite.

§ PARONYCHIA. Sepals evidently awned at apex. Lvs. linear and subulate....Nos. 1, 2
§ ANÝCHIA (Mx. partly). Sep. merely mucronate at apex. Lvs. lanceolate to oval.(*)
 * Stems procumbent, diffuse on the ground. Stamens 5....................Nos. 3, 4
 * Stems erect, with diffusely ascending branches. Stamens 2 or 3.........Nos. 5, 6

1 P. dichótoma Nutt. Glabrous, densely branched; lvs. accrose, mucronate; bracts like the leaves; cymes fastigiate, with no central flower; sepals 3-veined, cuspidate ♃ Rocks, Va. to Car. and Ark. 6—12'. Lvs. 1'. July—November.

2 P. argyrócoma Nutt. Pubescent, tufted, decumbent; lvs. linear, acute; cymes glomerate, terminal; fls. enveloped in dry, silvery bracts; sep. hairy, 1-veined, setaceously cuspidate. ♃ Mts. N. H. Va. to Ga. 4—10'. Lvs. 6—10''. July.

3 P. herniarioìdes Nutt. Scabrous, diffusely branched; lvs. oval or oblong, mucronate; the ramial alternate. Fls. sessile in the axils of the leaves; sep. 3-veined merely mucronate. ♃ Sand, S. Small, flat. Lvs. 1—3''. July- October.

4 P. Baldwinii Chapm. Diffusely branched, procumbent; leaves linear-lanceolate very acute, all opposite; flowers longer than the setaceous stipules, mostly terminal, stalked; stam. 5. ① Dry fields, Fla. Ga. 6—10'. Lvs. few. July—Oct.

5 P. Canadénsis. Stem erect, slender, pubescent, many times forked, with slender or capillary branches; lvs. lanceolate, the ramial alternate; style none; utricle equalling the sepals. ① Woody hills.

β. *púmila.* Dwarf (2—4'), tufted; fls. closely sessile; style as long as ovary, forked at apex. Dry hills, Md. (Mr. Shriver.)

16. **SIPHONÝCHIA**, Torr. and Gr. ($\Sigma i\varphi\omega\nu$, a tube; that is, *Anychia* with a tubular calyx.) Sep. linear, petaloid above, coherent into a tube below, unarmed. Pet. 5 setæ alternate with the stamens on the throat of the calyx. Style filiform, minutely bifid; utricle included. ① Diffuse and widely spreading. Fls. in glomerate, terminal cymes, white. Jn.—Oct.

§ Calyx tube bristly with hooked hairs. Stems prostrate, diffuse............Nos. 1, 2
§ Calyx smooth or merely pubescent. Stems erect............ ....Nos. 3, 4

1 S. Americàna T. & G. Sts. pubescent in lines; lvs. lanceolate; sep. rounded, incurved at apex; fls. solitary and clustered. ① S. Car. to Fla. 1—2f. Lvs. small.

2 S. diffùsa Chapm. Pubescent; lvs. lanceolate, obtuse; sep. linear, mucronate; fls. in dense cymes. ① Pine-barrens, Fla. 1f.

3 S. erécta Chapm. Sts. smooth, rigidly erect, subsimple; lvs. linear; sep. lanceolate, tube smooth, furrowed. ♃ Sands, Fla. 6—12'.

4 S. Rugèlii Chapm. Erect, dichotomous, pubescent; lvs. oblanceolate; sep. conspicuously mucronate, the tube hairy. ① E. Fla. 1f. (Paronychia, Shutt.)

17. **SCLERÁNTHUS**, L. Knawel. ($\Sigma\kappa\lambda\eta\rho\delta s$, hard, $\check{\alpha}\nu\vartheta os$; the calyx hardens in fruit.) Sep. 5, united below into a tube contracted at the orifice. Pet. 0. Sta. 10, rarely 5 or 2. Styles 2, distinct. Utricle very smooth, enclosed in the hardened calyx tube. ① A prostrate, diffuse little weed, exstipulate.

S. ánnuus L. Dry fields and roadsides, N. and M. 3—6'. Lvs. linear, acute, short, partially united at their bases. Fls. very small, green, in axillary fascicles. July.

18. **MOLLÙGO**, L. Carpet-weed. Calyx of 5 sepals, inferior, united at base, colored inside. Cor. 0. Sta. 5, sometimes 3 or 10. Fil. setaceous, shorter than and opposite to the sepals. Anth. simple. Caps. 3-celled, 3 valved, many-seeded. Seeds reniform. Lvs. at length apparently verticillate, being clustered in the axils.

M. verticillàta L. Lvs. cuneiform, acute; st. prostrate, branched; pedicels 1-flowered, subumbellate; sta. mostly but 3. ① Dry fields. 6—10'. White.

Order XX. PORTULACACEÆ. Purslanes.

Herbs succulent or fleshy, with entire leaves, no stipules, and regular flowers. *Sepals* 2, united at base. *Petals* 5, more or less imbricated. *Stamens* variable in number, but opposite the petals when as many. *Ovaries* free, 1-celled. *Styles* several, stigmatous along the inner surface. *Fruit* a pyxis, dehiscing by a lid, or a capsule, loculicidal, with as many valves as stigmas. *Seeds* with a coiled embryo. Figs. 122, 123.

ORDER 23.—MALVACEÆ. 59

* Stamens 8—20, perigynous. Capsule opening by a lid (a pyxis)......... ....PORTULACA. 1
* Stamens 10—30, hypogynous. Capsule opening by valves........................TALINUM. 2
* Stamens 5, each on the base of a petal. Capsule 3-valved. ....CLAYTONIA. 3
* Stamens 4—15. Capsule 3-valved. Leaves alternate........................CALANDRINIA. 4

1. PORTULACA, Tourn. PURSLANES. Sep. 2, the upper portion deciduous. Pet. 5 (4 to 6), equal. Stam. 8—20. Style 3-6-parted. Pyxis opening near the middle, ∞-seeded. Low, fleshy herbs.

1 **P. oleràcea** L. Stems reddish, prostrate; leaves cuneate. ① Cultivated grounds, especially gardens. 1f. Plant very smooth, succulent. Fls. small, yellow. June—Aug.
2 **P.** GRANDIFLORA. Upright; lvs. linear, acute; fls. large, rose-purple. ① S. Am. 8'.
3 **P.** GILLÉSII. Upright; lvs. short, terete, blunt; fls. large, deep purple. ① S. Am

2. TALINUM, Adans. Sep. 2, ovate, deciduous. Pet. 5, sessile, inserted with the 10—20 stamens into the torus. Style trifid. Caps. 3-valved, ∞-seeded.—Herbs fleshy, smooth.

T. teretifòlium L. Stem short, thick, with crowded linear lvs. at the ends of the short branches, with long (6') terminal, naked peduncles, bearing a cyme of purple, ephemeral flowers. ♃ Rocks, Penn. to Ga. June—Aug.

3. CLAYTÓNIA, L. SPRING BEAUTY. (In memory of *John Clayton*, one of the earliest botanists of Virginia.) Sep. 2, ovate or roundish. Pet. 5, emargined or obtuse. Stam. 5, inserted on the claws of the petals. Stig. 3-cleft. Caps. 3-valved, 2-5-seeded.—Small, fleshy, early flowering plants, arising from a small tuber. (Stem with 2 opposite leaves.)

1 **C. Caroliniàna** Mx. Lvs. ovate-lanceolate; sep. and pet. obtuse. ♃ Moist woods. Stem 3', bearing 2 (rarely 3 or 4) leaves; root leaves few; fls. white, with purple lines.
2 **C. Virgínica** L. Lvs. linear or lance-linear; sepals rather acute; petals obovate, mostly emarginate or retuse; ped. slender, nodding. ♃ In low, moist grounds, more common than the first, the 2 opposite leaves 3—5' long. Flowers roseate.

4. CALANDRÍNIA, H. B. K. (*Calandrini* was an Italian botanist.) Sep. 2. Pet. 3—5. Stam. 4—15, mostly hypogynous. Style short, stig. 3. Caps. 3-valved.—Herbs of Chili and California, smooth, with alternate leaves and purple flowers.

1 **C.** GRANDIFLORA. Leaves rhomboid; raceme terminal. ♃ Chili. 1f. Fls. near 2'.
2 **C.** SPECIÓSA. Leaves linear-spatulate; flowers axillary. ① Cal. 6'. Fls. 1' broad.

ORDER XXIII. MALVACEÆ. MALLOWS.

Herbs or shrubs with alternate, stipulate leaves and regular flowers, with 5 *sepals* united at base, valvate in the bud, often subtended by an involucel; 5 *petals* hypogynous, convolute in the bud, with the *stamens* ∞ monadelphous, hypogynous, and 1-celled reniform *anthers*. *Pistils* several, distinct, or united, and *stigmas* various. *Fruit* a several-celled capsule, or a collection of 1-seeded indehiscent carpels. *Seeds* with little or no *albumen*, and a curved *embryo*.

§ Calyx naked, i. e., having no involucel. (b)
§ Calyx involucellate.—Carpels (and styles) more than 5. (a)
—Carpels 3 to 5 only,—1-seeded. (c)
—3— ∞-seeded. (d)

ORDER 23.—MALVACEÆ.

```
a Involucel of 6 to 9 bractlets.  Carpels 1-seeded..................ALTHÆA.          1
a Involucel of 3 distinct bractlets.  Carpels 1-seeded.  Stigmas linear......MALVA.   2
a Involucel of 3 united bractlets.  Carpels 1-seeded..................LAVATERA.       3
a Involucel of 3 distinct bractlets.  Carpels 2-seeded..................MODIOLA.      4
a Involucels (of 2 or 3 distinct bractlets). Carpels 1-seeded. Stig. capitate....MALVASTRUM. 5
   b Flowers diœcious.  Stigmas 10, linear..................NAPÆA                     6
   b Flowers perfect.  Carpels 5 or more, 1-seeded..................SIDA.             7
   b Flowers perfect.  Carpels 5 or many, 3 to 9-seeded..................ABUTILON.    8
      c Stigmas 10.  Carpels 5, baccate, united..................MALVAVISCUS.         9
      c Stigmas 10.  Carpels 5, dry, distinct..................PAVONIA.              10
      c Stigmas 5.  Carpels 5, dry, united into a pod..................KOSTELETZKYA. 11
         d Involucre of many bractlets.  Calyx regular..................HIBISCUS.    12
         d Involucre of 3 incisely-toothed bractlets..................GOSSYPIUM.     13
```

1. ALTHÆA, L. MARSH MALLOW. ("Αλθω, to cure; the mucilaginous root is highly esteemed in medicine.) Calyx surrounded at base by a 6–9-cleft involucel. Styles ∞, with linear stigmas. Carpels ∞, 1-seeded, indehiscent, arranged circularly, and at maturity separating from the axis.

1 **A. officinàlis** L. Lvs. soft-downy on both sides, cordate-ovate, dentate, somewhat 3-lobed; ped. much shorter than the leaves, axillary, many-flowered. ♃ Salt marshes, North. 3f. Flowers large, pale purple. Sept. § Eur.
2 **A. ròsea** Cav. *Hollyhock*. St. erect, hairy; lvs. cordate, 5-7-angled, rugous; fls. axillary, sessile. ② Gardens, often sowing itself. 6f. Flowers of all colors. §

2. MALVA, L. MALLOW. (Μαλαχή, soft; on account of the soft mucilaginous properties.) Calyx 5-cleft, the involucel 3-leaved. Pet. obcordate or truncate. Styles ∞, with linear stigmas. Carpels ∞, 1-celled, 1-seeded, indehiscent, arranged circularly, and at maturity separating from the axis.

```
* Leaves triangular-hastate, crenate, scabrous.  Carpels acute..................No. 1
* Leaves orbicular, with 5—7 angular lobes.  Carpels obtuse.......... ....Nos. 2—4
* Leaves palmately 5-7-parted..................Nos. 5—7
```

1 **M. triangulàta** Lav. Rough-hairy; lvs. triang.-hastate, crenate; the lower cordate; panicle many-flowered; carp. 10—15, slightly beaked, at length 2-valved. ♃ Dry prairies, W. and S. 2—3f. Petals 1', purple. July, Aug. (Callirrhoë triang. Gr.)
2 **M. rotundifòlia** L. *Low Mallow*. St. prostrate; lvs. obtusely 5-lobed; cor. pale, twice as long as the calyx. ♃ Waste grounds. 1f. June, July. § Eur.
3 **M. sylvéstris** L. *High Mallow*. St. erect; lvs. 5-7-lobed, lobes rather acute; pet. purple, 3 times longer than sepals. ② Waysides. 3f. June, July. § Eur.
4 **M. crispa** L. St. erect; lvs. angular-lobed, dentate, crisped, smooth; fls. axillary, sessile, white. ① Gardens and waste grounds. 5f. June—Aug. § Syria.
5 **M. moschàta** L. *Musk Mallow*. Erect; radical lvs. reniform, incised, cauline 5-parted, the segments linear-cuneiform, incisely lobed; peduncles shorter than the leaves. ♃ Gardens and waysides. 2f. Flowers large, roseate. July. § Eur.
6 **M. A'lcea** L. Erect; rt. lvs. angular; st. lvs. 5-lobed, the lobes merely incised; stem and calyx velvety. ♃ Escaped from gardens: rare. 3f. Fls. purple. July. † § Eur.
7 **M. Papàver** Cav. *Poppy Mallow*. Lvs. 3-5-parted, segm. oblong or linear, entire or toothed; fls. on very long peduncles; bracteoles 1—3, subulate. ♃ Open woods, South. 12—18'. Flowers bright red. May, June. (Callirrhoë Papaver Gr.)

3. LAVATÈRA, L. (Named in honor of the two *Lavaters*, physicians of Zurich.) Calyx subtended by an involucel of 3 united bracteoles. Stigmas ∞, filiform. Carpels ∞, 1-celled, 1-seeded, indehiscent, arranged circularly as in Malva.

L. TRIMÉSTRIS. Annual; lvs. roundish-cordate, the upper angular; fls. large, red, solitary. Europe. 2f. The flowers vary to white. July, Aug.

Order 23.—MALVACEÆ.

4. MODÌOLA, Mœnch. (Lat. *modiolus*, a certain measure; from the fancied resemblance of the fruit to a basket.) Calyx 5-cleft, with an involucel of 3 bractlets at base. Stigmas 15—20, capitate. Carpels same number, 2-seeded, transversely 2-celled, 2-valved. ①② Prostrate, with cleft leaves and small flowers.

M. multífida Mœnch. Lvs. roundish, cordate, 3—5 cleft; segm. cut-toothed; ped. soon longer than the petioles. ♃ Car. Ga. and W. 1—2f. Fls 6″, red. July, Aug.

5. MALVÁSTRUM, Gray. (Name altered from *Malva*.) Involucel of 1—3 leaves, or 0. Styles 5—20. Stigmas capitate. Carp. 5—∞, often beaked or awned, each 1-seeded.

1 **M. angústum** Gr. Branched, erect, hairy; lvs. lanceolate, with bristle-form stip.; invol. bristleform; carps. 5, dehiscent. ① S. Car. Ga. 1f. Fls. yellow. (Sida, Ph.)
2 **M. tricuspidàtum** Gr. Shrubby; rough-hirsute; lvs. ov.-oblong; stip. lanceolate; invol. 3-leaved; carp. 10—12, 3-awned at apex. ♃ S. Fla. 1f. Yellow.

6. NAPÆA, Clayt. (*Νάπη*, a wooded valley between mountains, where Clayton discovered the plant.) Involucel none. Calyx 5-toothed; fls. diœcious. Styles 6—8, with filiform stigmas. Carpels as many, 1-seeded, indehiscent, beakless, circularly arranged. ♃ Tall, with large, palmately divided leaves and small white flowers in leafy panicles.

N. dioìca L.—Rocky thickets, Pa. Va. to Ill. Stem weak. 4—6f. Leaf segm. 5—11, lanceolate, acuminate, coarsely toothed. Flowers 4—5″. August.

7. SIDA, L. Involucel 0. Fls. perfect. Calyx 5-cleft. Styles 5 or more, with the stigmas capitate or truncate. Carp. 5—∞, 1-seeded, finally separable. Herbs or shrubs, mostly tomentous.

* Leaves palmately parted. Flowers rose-white. Carpels beaked..........Nos. 1, 2
* Leaves undivided. Flowers red or yellow.—*a* Carpels 5 or 7.............Nos. 3, 4
  —*a* Carpels 10—12.............Nos. 5—8

1 **S. Napæa** Cav. Nearly glabrous; lvs. palmately 5-lobed, lobes oblong, acuminate, coarsely-toothed; ped. many-flowered; carpels 10, acuminate-beaked. ♃ Woods, Penn. to Va. 3f. Fls. 8″. White. July.
2 **S. alcœoìdes** Mx. Strigoso-pubescent; lvs. palmately 5-7-parted, the segments laciniate; fls. corymbed, terminal; carp. 10, acute. ♃ In barren oaklands, Tenn. Ky. 1—2f. Fls. nearly as large as in the Musk Mallow. (Callirrhoë alcœoides Gr.)
3 **S. spinòsa** L. St. rigid; lvs. ovate-lanceolate, serrate, with a spinous tubercle at the base of the petiole; stip. setaceous, shorter than the petioles or axillary peduncles; carp. 5, birostrate. ② Sandy, M. and W. 8—16′. Yellow. July. §
4 **S. ciliàris** Cav. St. prostrate; lvs. elliptical, obtuse; stip. setaceous, and calyx ciliate; carp. 7, tipped with 2 spines; fls. red. ♃ S. Fla.
5 **S. stipulàta,** Cav. Smoothish; leaves rhombic-lanceolate, dentate; stip. subulate, longer than the petioles, persistent; carpels 10—12, pointed with 2 short spines. ♃ Sandy soils, S. 18′. Pet. 5″, yellow. July. (S. hispida C-B.)
6 **S. Elliòttii** Torr. & Gr. Lvs. linear-oblong, obtuse at base; ped. 1-flowered, a little longer than the petioles; caps. truncate. ♃ Sandy soils, S. 3f. Yellow.
7 **S. rhombifòlia** L. Leaves rhombic-oblong, serrate, cuneate and entire at base; ped. much longer than the petioles; caps. 2-beaked. ① S. Car. to Fla. 2f. Yellow.

8. ABÙTILON, Dill. Indian Mallow. Calyx 5 cleft, without an involucel, often angular. Styles 5 to 20, with capitate stigs. Carps. as many, arranged circularly, each 1-celled, 3 to 6-seeded, and opening by 2 valves

ORDER 23.—MALVACEÆ.

§ HERBACEOUS. Lvs. ovate, crenate, acuminate, velvety. Fls. erect........Nos. 1, 2
§ SHRUB. Leaves 3-5-acuminate-lobed. Fls. pendulous........................No. 3

1 A. Avicénnæ Gært. Tomentous; lvs. roundish, cordate; ped. shorter than the long petiole; carp. about 15, inflated, 2-beaked, 3-seeded. ① Waste places. 3f. Yel. Jl. §

2 A. Hulscànum Torr. Pilous-hispid; lvs. roundish; ped. 3–5-flowered; carpels about 12; fls. near 2′ broad, light purple. Fla. Lvs. small, whitish beneath.

3 A. STRIÀTUM. *Tassel-Tree.* Shrub with maple-like lvs. and tasselform fls., the column exserted. Greenhouse. 5—10f. Orange-red, scarlet-veined. Brazil.

4 A. VEXILLÀRIUM. Shrub with long, slender, drooping branches: leaves lance-ovate, cordate, crenate-serrate; flowers droop on filiform stalks, cylindric; calyx scarlet, corolla golden yellow, column exserted. Greenhouse. Flowers all Winter.

9. MALVAVÍSCUS DRUMMÓNDII. GLUE MALLOW. Shrub 4f, with showy, erect, axillary scarlet flowers. Involucel of many bractlets. Pet. erect. Styles 10, with capitate stigmas. Fruit fleshy. Leaves roundish, cordate, angularly 3-lobed, coarsely crenate-toothed. Column long-exserted. § About N. Orleans.

10. PAVÒNIA, L. (Latin *pavo*, peacock; suggested by the colors.) Involucel of 5 or more bracteoles. Calyx 5-cleft. Carpels 5, half as many as the branches of the style, 1-seeded. Stig. capitate. Fruit dry. ♄

P. Lecóntii T. & G. Shrubby; lvs. sagittate-oblong, obtuse, hoary-tomentous beneath; bractlets 5; carpels blunt, rugous. 5f. Ga. (Mr. Jones), rare. Fls. 18″ diam , rose-white, with a deep purple centre. (P. Jonesii C-B.)

11. KOSTELÈTZKYA, Presl. (In honor of *Kosteletzky*, a German botanist.) Calyx, involucel, styles, &c., as in Hibiscus. Fruit a 5-celled, depressed capsule, with a single seed in each cell.

K. Virgínica Presl. Lvs. acuminate, cordate, ovate, dentate, upper and lower ones undivided, middle 3-lobed; ped. axillary, and in terminal racemes; fls. nodding, pistils declinate. ♃ Marshes, L. I. to Ga. 3f. Fls. 2½′, rose-red. Aug.

12. HIBÍSCUS, L. Calyx 5-cleft, subtended by an involucel of many bractlets. Column long with the stamens lateral and the 5 stigmas capitate. Fruit a 5-celled capsule, loculicidal, the valves bearing the partitions in the middle. Seeds 3 or many in each cell. ♄ ♃ Flowers large and showy. Plants often cultivated.

§ HIBÍSCUS *proper.* Calyx equally 5-cleft or toothed, persistent...(a)
§ ABELMÓSCHUS. Calyx tube in flowering split down to the base on one side..Nos. 12, 13
  a Shrubs and trees. Leaves undivided, ovate, &c. Stip. persistent......Nos. 9—11
  a Herbs.—b Calyx, &c., tomentous. Lvs. undivided, angularly lobed ....Nos. 1, 2, 3
    —b Calyx, &c., hispid. Leaves palmately divided................Nos. 4, 5
    —b Calyx, &c., glabrous.—c Leaves strongly 3-5-lobed.............Nos. 6, 7
                    —c Leaves ovate, undivided..................No. 8

1 H. Moscheùtos L. Simple, erect, hoary-tomentous; lvs. ovate, obtusely dentate, some 3-lobed; ped. long, often cohering with the petiole; pod and seeds smooth; sepals abruptly pointed. Brackish marshes. 4—6f. Fls. 6′ diam., roseate. Aug.

β. *flavéscens.* Fls. larger (pet. 4′ long), of a light sulphur-yellow, with a purple centre. Marshes, Indiana to Fla. (H. incanus Wendl.)

3 H. grandiflòrus Mx. Lvs. cordate, acuminate, repand-dentate. downy both sides. hoary beneath; pods densely hirsute. S. and W. 5—7f. Pet. 4½′, flesh-color. Jl.-Oct

4 H. aculeàtus Walt. Prickly-hispid; lvs. 3-5-lobed, repand-toothed; bractlets of the involucel linear, forked at the end: sep. red-veined. S. 3—5f. Fls. 4½′, y-p. Jn. +

ORDER 24.—TAMARISCINEÆ. 63

5 **H. Triònum** L. *Flower-of-an-Hour.* Hispid; leaves 3-parted, middle segments long, all sinuate-lobed; bractlets entire; calyx inflated, membranous; flowers yellowish, dark-brown centre, ephemeral, numerous. Fields and gardens. § Italy.
6 **H. militàris** Cav. Glabrous; leaves hastately 3-lobed, lobes acuminate, serrate; corolla tubular-campanulate; capsules smooth, ovoid-acuminate; seeds hairy. ♃ Penn., S. and W. 4f. Petals flesh-color, purple at base, 3′. July, August.
7 **H. coccíneus** Walt. Very smooth; lvs. palmate, 5-parted, lobes lanceolate, acuminate; corolla expanding; caps. ovoid. ♃ South. 6f. Flowers 6′, scarlet. July, Aug.
8 **H. Caroliniànus** Muhl. Smooth; lvs. cordate, ovate, acuminate; ped. free from petiole; pet. downy inside, purple, 4′; pod globular. ♃ Wilmington Isl., Ga. (Elliott).
9 **H. Syrìacus** L. *Althæa. Tree Hibiscus.* Lvs. ovate, cuneiform at base, 3-lobed, dentate; ped. scarcely longer than petiole. Fls. wh.-purp. or roseate. 8—15f. § Syria.
10 **H. Floridànus** Shutt. Hispid; lvs. ovate-cordate, obtuse, small; fls. pendulous on long peduncles, scarlet or crimson; stamens exserted. S. Fla. 4—5f. Fls. 1′.
11 **H.** ROSA-SINÉNSIS. *Chinese H.* Shrub with very smooth ovate pointed lvs. coarsely dentate at end; fls. very large, dark red, varying to buff, yellow, striped, and double.
12 **H.** ESCULÉNTUS. *Okra.* Lvs. cordate, 5-lobed, obtuse, dentate; petiole longer than the fl.; involucel about 5-leaved, caducous. ① 5f. Cult. for its large, mucilaginous pods.
13 **H.** MÁNIHOT. Lvs. divided into 5—7 linear, pointed, few-toothed lobes; bractlets of the involucel 5—7, persistent. ♃ China. 4f. Fls. sulph.-yellow, purp. centre. Jl. +

13. **GOSSÝPIUM,** L. COTTON PLANT. Calyx obtusely 5-toothed, surrounded by an involucel of 3 cordate leaves, deeply and incisely toothed. Stamens very numerous, lateral. Stigmas 3, rarely 5, clavate. Seeds ∞, involved in cotton. Flowers yellow. Fig. 201.

1 **G.** HERBÀCEUM. Leaves 3–5-lobed, with a single gland below, lobes mucronate; seeds brownish, cotton white. ① 5f. Cultivated South. Yellow.
2 **G.** BARBADÉNSE. *Sea Island C.* Leaves with 3 glands on the mid-vein below; seeds black, cotton white, long and silky. ② Coasts, South. Planted in Autumn.

ORDER XXIV. STERCULIACEÆ. SILK COTTONS.

Large *trees* or *shrubs* with simple or compound leaves, with flowers similar to those of the Mallow, except that the *anthers* are 2-celled and turned outward. *Fruit* capsular, of 3, rarely 5 carpels.

* Involucel 0. Petals 0. Carpels 5. Stamens 10—20, all fertile, monadelphous..STERCULIA.
* Involucel 0. Petals 5, long-clawed. Carpels 5. Fertile stamens 5. S. Fla..AYENIA *pusilla.*
* Involucel 3-leaved. Petals 5. Carpel 1. Stamens 5, all fertile. S. Fla......WALTHÈRIA *Americàna.*

1. **STERCÙLIA,** L. Calyx 5-lobed, sub-coriaceous. Stam. monadelphous, united into a short, sessile cup. Anth. adnate, 10, 15, or 20. Carp. 5, distinct, follicular, 1-celled, 1 – ∞-seeded.—Trees with axillary panicles or racemes. (See *Addenda.*)

S. PLATANIFÒLIA L. Leaves cordate at base, palmately 3–5-lobed, smooth; calyx rotate, reflexed, greenish, in clusters. Cultivated South. 30f. Japan. A handsome tree.

ORDER XXIV. *bis.* TAMARISCINEÆ. TAMARISKS.

Shrubs or *herbs* with minute, scale-like leaves, dense slender racemes of small 4–5-parted flowers. *Stamens* definite, hypogynous. *Styles* 3. *Capsules* 3-valved, 1-celled, ∞-seeded. *Seeds* with a coma. *Albumen* 0. *Embryo* straight.

TÁMARIX GÁLLICA. Characters mainly as given in the Order. Pet. and sta. 5. A beautiful shrub, 10f, with virgate branches, bearing numerous exceedingly delicate racemes of flesh-colored fls. Lvs. lance-subulate, clasping. Eur. Nearly hardy

ORDER XXV. TILIACEÆ. LINDENBLOOMS.

Trees or *shrubs* (rarely *herbs*) with simple, stipulate, alternate, dentate leaves, with *flowers* axillary, hypogynous, usually perfect and polyadelphous; with the *sepals* 4 or 5, deciduous, valvate in bud, the *petals* 4 or 5, imbricated. *Stamens* ∞, with 2-celled, versatile *anthers*. *Ovary* of 2—10 united *carpels*, and a compound *style*. *Fruit* dry or succulent, many-celled, or 1-celled by abortion. *Embryo* in the axis of fleshy albumen.

1. CÓRCHORUS, L. Sep. and pet. 4 or 5. Stam. ∞, rarely as few as the petals. Style very short, deciduous, stig. 2 to 5. Caps. roundish or siliquose, 2-5-celled, many-seeded. ♄ Flowers yellow.

C. siliquòsus L. Lvs. ovate-lanceolate, acuminate, equally serrate, 4 times longer than the petioles; caps. siliquose, linear, 2-valved. La. to Fla. Flowers 4-merous.

2. TÍLIA, L. LINDEN or LIME TREE. Calyx of 5, united sepals, colored. Cor. of 5, oblong, obtuse petals, crenate at apex. Stam. ∞, somewhat polyadelphous, each set (in the N. American species) with a petaloid scale (staminodium) attached at base. Ov. superior, 5-celled, 2-ovuled. Caps. globous, by abortion 1-celled, 1-2-seeded. ♄ Lvs. cordate. Fls. cymous, cream-white, with the peduncle adnate to the vein of a large leaf-like bract.

§ Staminodia 5, petaloid, opposite the petals. Leaves mucronate-serrate....Nos. 1, 2
§ Staminodia none. Stamens scarcely cohering.................................No. 3

1 T. Americàna L. *Bass-wood.* Lvs. broad cordate, unequal at base, acuminate, coriaceous, smooth, and green on both sides; pet. truncate or obtuse at apex; sty. as long as the petals. Woods, N. and M. States. 70f. June. Timber valuable.
  β. *Walteri.* Lvs. pubescent (but green) beneath. A large tree. Va. to Fla.

2 T. heterophýlla Vent. *White Bass-wood.* Lvs. obliquely subcordate, scarcely acuminate, white and velvety beneath, shining, and dark green above; pet. obtuse, crenulate; sty. hairy at base, longer than the petals. River banks, W. 40f.
  β. *alba.* Lvs. whitish and minutely tomentous beneath, serratures fine and long-mucronate. Ky. and South along the mountains. 80f.

3 T. EUROPÆA L. *Lime Tree.* Lvs. suborbicular, obliquely cordate, abruptly acuminate, serrulate, twice as long as the petioles, glabrous except a woolly tuft in the axils of the veins beneath. Parks. 40f. † Eur.

ORDER XXVI. CAMELLIACEÆ. CAMELLIAS or TEAWORTS.

Trees or *shrubs* with alternate, simple, feather-veined, exstipulate *leaves*. *Flowers* regular, polyandrous, hypogynous, cyanic, with *sepals* and *petals* imbricated, the former often unequal in size. *Stamens* more or less coherent at base into one, three, or five sets. *Anthers* 2-celled. *Seeds* few, with little or no albumen. *Cotyledons* large.

§ Calyx of many imbricated sepals. Stamens monadelphous..................CAMELLIA. 1
§ Calyx simple.—Stamens united at the base into one set.......................STUARTIA. 2
                —Stamens in 5 sets, adhering to the base of the petals...............GORDONIA. 3

1. **CAMÉLLIA**, L. TEA ROSE. TEA. Sepals many, imbricated, the inner ones larger. Fil. ∞, shorter than the corolla, united at base, some of the interior free. Styles united. Stigmas 3—5, acute. ♄ ♃

1 **C. JAPÓNICA** L. *Japan Rose.* Leaves ovate, acuminate, acutely serrate, glabrous and shining; flowers terminal, solitary; petals obovate; stamens 50 (mostly transformed to petals); stigmas 5-cleft. Tree in Japan, here a beautiful greenhouse shrub.
2 **C.** (Thea) **BOHÈA.** Shrub 4f, lvs. elliptic-oblong, acute, some rugous, twice as long as broad; flowers axillary, white. Cultivated throughout China and Japan—rarely here.
3 **C.** (Thea) **VÍRIDIS.** Shrub 4f; lvs. lance-oblong, thrice longer than broad, flat, acute; fls. white, 1' broad. China. The leaf of these shrubs, variously *cured*, is the *Bohea, Black, Green*, or *Imperial Tea.*

2. **STUÁRTIA**, Catesby. Sepals 5 (or 6), ovate or lanceolate. Petals 5 (or 6), obovate, crenulate. Stamens monadelphous at base. Capsules 5-celled, 5- or 10-seeded, seeds ascending. ♄ Leaves large, deciduous; flowers showy, fragrant, axillary, nearly sessile.

§ STUÁRTIA *proper.* Styles united. Capsule globous. Seeds lenticular... ...No. 1
§ MALACHODÉNDRON. Styles distinct. Capsule ovoid. Seeds margined........No. 2
1 **S. Virgínica** Cav. Leaves oval, acuminate, thin, serrulate, downy beneath; sepals roundish; pet. white; fil. purple, anth. blue. Va. to Fla. and La. 6—12f. Apr., May.
2 **S. pentágyna** L'Her. Leaves ovate, acuminate; sep. lanceolate: one pet. smaller than the others, all cream-white; capsules 5-angled. Ky. to Ga. 10—15f. June, Jl

3. **GORDÒNIA**, Ellis. LOBLOLLY BAY. Sepals 5, roundish, strongly imbricated. Pet. 5. Sta. 5-adelphous, one set adhering to each petal at base. Styles united into one. Caps. woody, 5-celled. Seeds 2 or more in each cell, pendulous. ♄ With large, white, axillary, pedunculate flowers.

§ GORDONIA *proper.* Stam. inserted on a 5-lobed cup, as short as the style......No. 1
§ FRANKLÍNIA. Stam. inserted on the pet. at base, longer than the style.......No. 2
1 **G. Lasiánthus** L. Leaves coriaceous, perennial, glabrous, shining on both sides lance-oblong; peduncles half as long as the lvs.; fls. 3'. S. 70f. July, August.
2 **G. pubéscens** L'Her. Leaves thin, serrate, deciduous, oblong-cuneiform, shining above, canescent beneath; fls. on short peduncles; sep. and pet. silky. S. 30f. May.

ORDER XXVII. MELIACEÆ.

Trees or *shrubs* with exstipulate, often pinnate leaves. *Flowers* 4-5-merous. *Stamens* 6—10, coherent into a tube, with sessile anthers. *Disk* hypogynous, sometimes cup-like; *style* 1. *Ovary* compound, several-celled, cells 1—2—6-ovuled. *Fruit* fleshy or dry, often 1-celled by abortion. *Seeds* winged or wingless.

§ MELIEÆ. Cells of the ovary 2-ovuled. Seeds wingless, few (in a fleshy drupe)....MELIA. 1
§ SWIETENIEÆ. Cells of ovary many-ovuled. Seeds winged, many in the capsule..SWIETENIA. 2

1. **MELIA**, L. PRIDE OF INDIA. (Μέλι, honey; the name was first applied to the Manna Ash.) Sep. small, 5, united. Pet. spreading. Sta. tube 10-cleft at summit, with 10 anthers in the throat. Ovary 5-celled, 10-ovuled. Style deciduous. Drupe with a 5-celled, bony nut, cells 1-seeded. ♄ With bipinnate lvs. and panicles of delicate flowers.

M. AZÉDARACH L. Lvs. deciduous, glabrous, lfts. obliquely lance-ovate, acuminate, serrate. S. States. 30—40f. Fol. light; fls. lilac; drupes as large as cherries. † W. Ind

ORDER 29.—ZYGOPHYLLACEÆ.

2. SWIETÈNIA Mahógoni, L. Mahogany Tree. A large and beautiful tree growing in South Florida, Mexico, and the Isthmus. 80—100f. The reddish-brown ornamental wood is well known. Lvs. smooth, abruptly pinnate, with 6—10 lance-ovate lfts. Fls. small, yellowish, in panicles, 5-parted. Pod size of a goose-egg, ∞-seeded.

ORDER XXVIII. LINACEÆ. FLAXWORTS.

Herbs with entire, simple leaves, and no stipules; with *flowers* regular, symmetrical, and perfect, 5-(rarely 3 or 4)-parted. *Calyx* strongly imbricated in the bud, corolla contorted. *Stamens* definite, hypogynous, alternate with the petals. *Styles* distinct, with capitate stigmas, and each cell of the capsule more or less divided by a false dissepiment into two 1-seeded compartments. *Seeds* with little or no albumen, attached to axile placentæ. Figs. 10, 11, 130, 136, 469.

LINUM, L. FLAX. Sepals, petals, stamens, and styles 5, the latter rarely 3. Caps. 6–10-celled. Seeds 10, suspended, mucilaginous. Herbs with a bark of strong fibres, and simple, sessile leaves.

    Flowers yellow, small (2—7″ broad). Species ①, native. June—August...(a)
        a Sepals entire, 1-veined, as long as the depressed or globous capsule...Nos. 1—4
        a Sepals glandular-fringed, longer than the globular-ovoid capsule.......Nos. 5, 6
  * Flowers blue, large (1′ broad). In fields and gardens.....................Nos. 7, 8
  * Flowers large, showy, red or yellow. Garden exotics..................Nos. 9, 10

1 **L. Virginiànum** L. Sts. teretish, erect, corymbous above, branches short, spreading, terete; lvs. oblong to lanceolate, mostly scattered; fls. 4–5′ broad; caps. depressed, styles distinct. Woods and hills. 2f. *Prof. Porter* distinguishes No. 2 from this.
2 **L. striàtum** Walt. St. *striate*, often clustered; branches short, ascending, sharply about 4-angled; lvs. lance-oblong, the lower mostly opp. Fls. and fr. as in No. 1. Com.
3 **L. simplex** Wood. Stem single, terete, corymbed at top, branches subterete; leaves linear-subulate, erect, scattered; caps. globular; sty. distinct; fls. 3″, few. S-W. 18′.
4 **L. diffùsum** Wood. Stems very slender, ascending, with long, filiform, *diffuse*, angular branches; lvs. veiny, lance., spreading, 9–12″; fls. 2″ broad; pod depressed. W.
5 **L. sulcàtum** Riddell. St. and branches *sulcate*, strict, erect; lvs. lin., erect; sep. 3-veined, acuminate; sty. united below. Conn. to Ill., and S. 1—1½f. (L. rigidum C-B.)
6 **L. rigiduɪn** Ph. Stems low and branches rigidly erect, angular-sulcate; lvs. linear-subulate, erect; sepals lance-linear, twice longer than the pod. Iowa, Min., and W.
7 **L.** usitatíssimum L. *Common Flax.* ① Leaves lance-linear; panicle corymbous; flowers axillary; petals crenate. 2f. The strong bark yields *linen.* § Europe.
8 **L.** perénne L. ⚥ Leaves linear; flowers supra-axillary and terminal; petals retuse, light blue. California! and Europe. Flowers numerous and showy.
9 **L.** grandiflòrum. Leaves lance-elliptical; flowers red; styles 5. N. Africa. 10′.
10 **L.** trígynum. Leaves elliptical; flowers yellow; styles 3. E. India. 1f.

ORDER XXIX. ZYGOPHYLLACEÆ. BEAN CAPERS.

Herbs, shrubs, or *trees,* with leaves opposite, mostly pinnate (not dotted) and stipulate. *Flowers* 4- or 5-merous, corolla imbricate or convolute in bud. *Stamens* twice as many as the petals, hypogynous, distinct, each often with a scale. *Ovary* compound; style and stigma 1 · fruit and seeds as in Linaceæ.

ORDER 30.—GERANIACEÆ. 67

Herbs. Disk annular, 10-lobed. Fruit of 5—12 indehiscent carpels................TRIBULUS. 1
Trees. Disk inconspicuous. Fruit of 2—5 dehiscent, 1-seeded carpels.............GUIACUM. 2

1. TRÍBULUS, L. Sep. and pet. 5, imbricated. Stam. 10, the 5 alternate with the petals placed inside of hypogynous glands. Ov. sessile, cells 1-5-seeded, separating into nutlets.—Loosely branched, prostrate herbs, with abruptly pinnate leaves. Flowers solitary (yellow).

1 **T.** (Kallstrœmia) **máximus** L. Lfts. 3 or 4 pairs, oblong or oval, oblique, the terminal pair largest; nutlets 10, tubercled, 1-seeded. Ga. Fla. 1—2f.
2 **T. cistoídes** L. Lfts. 5—8 pairs, linear-lanceolate, subequal; ped. elongated, with one large flower; nutlets 5, spiny, 2-5-seeded. Fla. 2f.

2. GUAIÀCUM, Plm. LIGNUM-VITÆ. Sep. and pet. 4 or 5, deciduous, imbricated. Stam. 8—10. Ovary stipitate, 2-5-celled, cells many-ovuled, in fruit 1-seeded. ♄♄ Wood hard and resinous. Lvs. abruptly pinnate. Ped. in pairs, between the stipules, 1-flowered.

G. sanctum L. Branches jointed; lfts. 3 or 4 pairs, oblong, oblique, entire, mucronate; ped. short; pet. obtuse, blue. S. Fla. 20f. Bark white.

ORDER XXX. GERANIACEÆ. GERANIA.

Herbs or *shrubs* with perfect, hypogynous, symmetrical and regular, or irregular, 3-5-merous flowers. *Stamens* as many or twice as many as the sepals, often some of them abortive or rudimentary. *Carpels* as many as the sepals, 1-few-seeded, mostly separating from the persistent axis at maturity.—A large and rather incongruous order, as now constituted (by Bentham and Hooker), including the following tribes, heretofore regarded as orders. Figs. 27, 28, 172, 243, 265, 270, 315, 350, 497.

§ Flowers regular.—*a* Styles 5. Carpels several-seeded. TRIBE I.
　　　　　　　　　　—*a* Style 1.—*b* Sepals valvate. Fruit beakless. TRIBE II.
　　　　　　　　　　　　　　　　—*b* Sepals imbricate. Fruit beaked. TRIBE III.
§ Flowers irregular.—*c* Petals perigynous. Stamens 7 or 8. TRIBE IV.
　　　　　　　　　　—*c* Petals hypogynous. Stamens 5, short. TRIBE V.

I. OXALIDEÆ. Symmetrical. Stamens 10 +. Petals convolute. Pod 5-celled.....OXALIS. 1
II. LIMNANTHEÆ.—Symmetrical. Stamens (10 in LIMNANTHES, No. 3) 6 in........FLŒRKEA. 2
III. GERANIEÆ.—Stamens 10 +. 5 often sterile. Glands between the petals. Fruit a regma, &c)
　　e Stamens 10, all antheriferous. Tail of carpels beardless................GERANIUM. 4
　　e Stamens 5 antheriferous. Tail of the carpels bearded......................ERODIUM. 5
IV. PELARGONIEÆ.—Sepals spurred behind. Glands 0. Stamens declined..(*f*)
　　f Spur adnate to the pedicel. Fruit rostrate,—a regma......................PELARGONIUM. 6
　　f Spur free. Fruit not beaked. Carpels 1-seeded, separating................TROPÆOLUM. 7
V. BALSAMINEÆ.—Sepals spurred behind. Pod opening elastically........ ....IMPATIENS. 8

1. ÓXALIS, L. WOOD SORREL. ('Οξυς, acid; the herbage is sour.) Sep. 5, distinct or united at base. Pet. contorted, much longer than the calyx. Sty. 5, capitate. Caps. oblong or subglobous. Carp. 5, 1 to several-seeded. Mostly ♃, with palmately trifoliate leaves and inversely heart-shaped leaflets. Figs. 265, 270, 497. (See *Addenda*.)

1 **O. Acetosélla** L. Acaulescent; scape longer than the leaves, 1-flowered; leaflets broad-obcordate with rounded lobes; styles as long as the inner stamens; root dentate, scaly. ♃ Woods, Can. and N. States. 6'. Flowers white-purple. June.

2 O. violàcea L. Bulbous at base, acaulescent; scape umbelliferous; flowers nodding; tips of the calyx fleshy; styles shorter than the outer stamens. ♃ An elegant species in rocky woods. 5—8'. Flowers violet-purple. May.

3 O. stricta L. Caulescent; st. branching; ped. umbelliferous, longer than the petioles; style as long as the inner stamens; flowers yellow. ① Fields. 3—9'. Common.

4 O. flava. Scapes 6', 1-flowered; leaflets 6—10, linear; petals yellow, 1' long. S. Afr.

5 O. ròsea. Stem erect, 8'; lfts. 3, obcordate; pet. roseate, 1', toothed; fls. many. Chili.

6 O. versícolor. St. 3'; lfts. 3, linear, emarginate; pet. crimson-striped outside. S. Afr.

2. FLŒRKEA, Willd. FALSE MERMAID. Sep. 3, longer than the 3 petals. Glands 3. Stam. 6. Ovaries 3, tuberculate. Style 2-cleft. Fruit separating into 3 achenia. ① Small aquatics, with pinnately-divided leaves.

F. proserpinacoìdes Lindl.—By streams and lakes, Vt. to Penn., and W. 6—10'. Prostrate; lvs. alternate; lf. segm. 3—5; pet. white, shorter than the sepals; ach. 1—3.

3. LIMNÁNTHES, Br. Sepals 5, valvate. Pet. 5, convolute, with 5 glands. Stamens 10. Style 1. Ovary deeply 5-lobed, separating 5 achenia in fruit.—Herbs with pinnate leaves and cut-lobed leaflets. Summer.

L. Dougláshi. Stems low, diffuse, with numerous axillary flowers 1' broad; petals wedge-oblong, yellow, edged with white, notched at the end. California.

4. GERÁNIUM, L. CRANE'S BILL. Sep. and pet. 5, regular. Stam. 10, all perfect, the 5 alternate ones longer, and each with a gland at its base. Fruit at length separating from the axis into 5 achenia, and uplifted on the smooth curving styles.—Herbs. Ped. 1–3-flowered. Fig. 172.

* Petals entire, twice as long as the awned sepals, purplish... ....Nos. 1, 2
* Petals emarg. or 2-lobed, not longer than the sep., roseate. May—Aug...Nos. 3—6
European perennials, cultivated, hardy, ornamental..................No. 7

1 G. maculàtum L. Stem erect, angular, dichotomous, retrorsely-pubescent; leaves palmately 3-5-lobed, lobes cuneiform and entire at base, incisely serrate above, radical ones on long petioles. ♃ Woods. 2f. Flowers 1', purple. April—June.

2 G. Robertiànum L. *Herb Robert.* Stems weak, reddish, diffuse, hairy; leaves pinnately 2-parted to the base, the segments pinnatifid, and the pinnæ incisely toothed; capsule rugous, seeds smooth. ② Rocky places, Can. to Va. 1—2f. Jn.—Aug.

3 G. Carolinìànum L. Erect, at length diffuse, hairy; leaves 5-7-parted; segm. 3-lobed, lobes entire or incised; ped. short, clustered at the ends of branchlets; sepals awned; fruit hairy; seeds *obscurely* reticulated. ① Hills, dry or rocky. ½—2f.

4 G. disséctum L. Diffuse, pubescent; lvs. 5- or 7-parted, segm. linear, many-cleft; seeds *strongly* reticulated. ① Fields: rare. 6—12'. Fruit some hairy. § Europe.

5 G. pusillum L. Procumbent, puberulent; lvs. round-reniform, 7-parted, segments 3-cleft; sepals *awnless*; seeds smooth. ① Waste grounds, N. Y., Mass. 1f. § Eur.

6 G. columbìnum L. Slender, decumbent, with long, filiform flower-stalks; sep. awned, enlarged after flowering; fr. glab.; lvs. and sd. as in No. 4. Penn. (Porter). ①

7 G. sanguìneum. Erect, diffuse; leaf-lobes 3-cleft. linear; ped. 1-flowered; flowers red, large. β. LANCASTRIÉNSE is prostrate, with smaller (1') purple flowers, very elegant.

5. ERÓDIUM, L'Her. HERON'S BILL. Sep. and pet. 5, regular. Stam. 10, the 5 shorter ones sterile. Styles in fruit spirally twisted and bearded.

E. cicutárium Sm. Diffuse, hairy; leaves pinnately divided, segments sessile, pinnatifid, incised, acute; ped. several-flowered; petals equal, red. ① Lake shores, N. Y.: rare. In California it is one of the chief forage plants. May, June. § Europe.

6. PELARGÒNIUM, L'Her. STORK'S BILL. GERANIUM. Sepals 5,

ORDER 30.—GERANIACEÆ. 69

the upper one ending in a nectariferous tube extending down the pedicel.
Petals 5, irregular, longer than the sepals. Filaments 10, 3 or 5 of
them sterile. ♄ or herbs. A large and ornamental genus, chiefly S. African, everywhere cultivated. Lower leaves (in plants raised from the seed)
opposite, upper alternate. Figs. 243, 350.

§ Filaments 10, the alternate ones bearing anthers. Upper petals larger..........Nos. 1, 2
§ Filaments 10, of which 7 bear anthers, and 3 are sterile...(a)
    a The 2 upper petals smaller, all scarlet, 1-colored. Shrubby..............Nos. 3—5
    a Petals nearly equal in size, mostly variegated...(b)
        b Stemless. Root tuberous. Leaves laciniate. Flowers brown.......Nos. 6, 7
        b Stems shrubby.—c Lvs. cordate, palmate, lobed. Flowers small.....Nos. 8, 9
            —c Lvs. peltate or cordate, 5-lobed, smooth.............No. 10
    a Two upper petals longer and broader. Stems shrubby...(d)
        d Flowers white, the 2 upper petals striped with red................Nos. 11, 12
        d Flowers purple.—e Leaves undivided...................Nos. 13, 14
            —e Leaves divided below the middle..............Nos. 15—17

1 P. TRÍCOLOR. Lvs. lanceolate, cut-dentate; 3 lower pet. white, 2 upper purp.-blk. 18′. ♃
2 P. CORIANDRIFÒLIUM. Lvs. bipinnate; pet. white, upper purp.-veined, very large. 1f. ⊙
3 P. ZONÀLE. *Horse-shoe G.* Lvs. orbicular-cordate, slightly lobed, toothed, zoned; stem fleshy, shrubby; petals cuneiform; flowers umbelled. 2—3f. Numerous varieties.
    β. MARGINÀTUM. Silver-edged; the leaves bordered with white.
4 P. ÍNQUINANS. Lvs. round, reniform, scarcely lobed, crenate viscid; pet. obov. 2—3f.
5 P. FOTHERGÍLLII. Lvs. renifm., 5-lobed, crenate, zoned; stip. toothed, ciliate; pet. obov.
6 P. FLAVUM. *Carrot-leaved Geranium.* Lf. lobes many, lin., hairy: fls. brownish-yell.
7 P. TRISTE. *Mourning Ger.* Lf. lobes lin., acute; pet. dark-green, obl., obovate. 1f.
8 P. FRAGRANS. *Nutmeg G.* Branches thick velvety, lvs. very soft; stip. subulate. Fls. w.
9 P. ALCHEMILLOÌDES. Villous; lvs. 5-lobed; peduncle few-flowered; fls. pink-colored.
10 P. PELTÀTUM. *Ivy-leaved G.* Br. fleshy; lvs. more or less peltate; fls. purplish.
11 P. GLAUCUM. Glabrous, glaucous; lvs. lanceolate, entire; ped. 1-2-flowered. 3f.
12 P. GRANDIFLÒRUM. Glab., glaucous; lvs. 5-lobed, toothed at end; fls. very large. 3f.
13 P. BETULÌNUM. Smoothish; lvs. ovate, unequally serrate: ped. 2-4-flwd. Pale. 3f.
14 P. WATSÒNII. Lvs. orbicular, cordate, some lobed, dentate; fls. large, varieg. 3f.
15 P. GRAVÈOLENS. *Rose Ger.* Lvs. palmately 7-lobed; lobes toothed, revolute, very rough at the edge; umbels many-flowered, capitate. 3f. Very fragrant.
16 P. RÁDULA. Lvs. palmate, rough, lobes narrow, rolled at edge, pinnatifid with linear segments; umbels few-flowered. 3f. Fragrance mint-like.
17 P. QUERCIFÒLIUM. Hispid; lvs. sinuate-pinnatifid, often spotted, cordate at base. 3f.

7. TROPÆOLUM, L. INDIAN CRESS. NASTURTION. Fls. irregular.

Sep. 5, produced behind into a free spur. Pet. 5, the 2 upper exterior, different from the 3 lower. Stamens 8, free, unequal, perfect. Style 1. Ov. 3-celled, in fruit separating from the short axis into 3 hardened achenia. ♄ Leaves alternate. Stipule 0. Flowers showy. S. Am. (See *Addenda.*)

1 T. MAJUS L. *Nasturtion.* Lvs. peltate, roundish, repand on the margin; pet. obtuse, the 3 lower fringed and long-clawed at base. Flowers orange, scarlet, crimson, &c.
2 T. MINUS. *Smaller,* erect; petals pointed, yellow to white, or variegated. Peru.
3 T. LOBBIÀNUM. Leaves peltate, reniform, wavy, fixed near the base; petals crenate, rounded, the 2 lower fringe-toothed, all shades of red. Columbia.
4 T. PEREGRÌNUM. *Canary Bird.* Leaves deeply 5-7-lobed, lobes toothed, spur hooked; petals light yellow, 2 of them large and much lobed. A tall climber.

8. IMPÀTIENS, L. TOUCH-ME-NOT. Sepals colored, 4 (the upper one double), the lowest saccate and spurred. Petals apparently 2, each of them 2-lobed (double). Stamens 5, short, the anthers cohering at

apex; caps. often 1-celled by the obliteration of the dissepiments, 5-valved, bursting elastically.—Sts. smooth, succulent, tender, subpellucid, with tumid joints. Lvs. simple, alternate, serrate. Figs. 27, 28, 315.

1 I. pállida Nutt. Lvs. oblong-ovate; ped. 2–4-flowered, elongated; lower gibbous sepals dilated-conical, broader than long, with a very short, recurved spur; fls. pale yellow, sparingly dotted. ① Wet shades. 3–4f. Aug.

2 I. fulva Nutt. Lvs. rhombic ovate; ped. 2–4-flowered, short; lower gibbous sepal acutely conical, longer than broad, with an elongated, closely reflexed spur; fls. deep orange, spotted. ① Damp grounds. 2–3f. July.

3 I. BALSÁMINA L. *Balsamine.* Lvs. lanceolate, serrate, upper ones alternate; ped. clustered; spur shorter than the flower. ① E. India. Fls. large, white and red.

ORDER XXXI. RUTACEÆ. RUEWORTS.

Herbs or generally *shrubs* or *trees*, with the exstipulate leaves dotted with transparent glands containing aromatic or acrid oil. *Flowers* regular, 3–5-merous, hypogynous, perfect or polygamous. *Stamens* as many or twice as many as the sepals. *Pistils* 2—5, separate or united, styles united. *Fruit* capsular or separating into its component, 1-2-seeded *carpels*.

§ RUTEÆ. Flowers perfect. (Herbs. Stamens 10.)..(a)
    a Petals equal, concave. Capsule 5-lobed..............RUTA. 1
    a Petals unequal, clawed. Capsules separable............DICTAMNUS. 2
§ ZANTHOXYLEÆ. Flowers ♀ ☿ ♂. (Trees, shrubs)..(b)
    b Pistils 3—5, separate below. Stamens 3—6........ ZANTHOXYLUM. 3
    b Pistils 2, united. Samara 2-seeded........ ......PTELEA. 4

1. RUTA, L. RUE. Calyx of 4 or 5 sepals, united at base. Petals 4 or 5, concave, obovate, distinct, torus surrounded by 10 nectariferous pores. Stamens 10. Capsule lobed. ♃ ♭, mostly European.

R. GRAVÈOLENS L. *Common Rue.* Suffruticous, nearly glabrous; leaves 2—3 pinnately divided, segm. oblong, obtuse, terminal ones obovate-cuneate, all entire or irregularly cleft; fls. terminal, corymbous; pet. entire. 3f. Greenish.

2. DICTÁMNUS, L. FRAXINELLA. Calyx of 5, deciduous sepals; petals 5, unguiculate, unequal; filaments 10, declinate, with glandular dots; capsules 5, slightly united. ♃ Native of Germany.

D. ALBUS Willd. St. simple; lvs. pinnate, the rachis more or less winged; fls. in a large, terminal, erect panicle.—In gardens. 1—2f. Fls. showy.
    β. RUBRA. Fls. purple; rachis of the leaves winged.

3. ZANTHÓXYLUM, L. PRICKLY ASH. (Ξανθός, yellow, ξύλον, wood.) Sepals 4 or 5, rarely obsolete. Petals 4 or 5. Sta. as many as the petals in ♂, rudimentary in ♀. Pistils 3 to 5, distinct below, with coherent styles, in fruit crustaceous, 2-valved, 1 or 2-seeded. ♭ ♅ With sharp prickles, pinnate leaves, and small, greenish flowers.

1 Z. Americànum Mill. Prickly; lfts. 9—11, ovate, sessile, equal at base; umbels axillary; sep. obsolete, pet. 5. Woods. 10—12f. Flowers before leaves April.

2 Z. Caroliniànum Lam. Prickly; lfts. 7—13, falcate-lanceolate, very inequilateral, petiolulate; panicles terminal; sep. minute; bark warted around the prickles. S. States. Tree, 20—40f. Bark intensely pungent to the taste. May.

ORDER 34.—SIMARUBACEÆ. 71

β. *fruticosum.* Shrub; lvs. ovate-oblong, scarcely pointed; ovaries 2. S.

3 Z. Floridànum N. *Satin-wood.* Unarmed; lfts. 5—7, ♀ ovate-lanceolate, ♂ & liptical, obtuse; fls. minute; carp. 1—2, 1-seeded, obovoid. S. Fla.

4. PTÈLEA, L. SHRUB TREFOIL. (Πτελέα, the elm-tree; from the resemblance of the fruits.) ♀ ♂ ☿. Sepals 3 to 6, mostly 4, much shorter than the spreading petals. ☿ Stamens longer than the petals and alternate with them, very short and imperfect in ♀. Ovary of 2 united carpels. Stig 2. Fruit 2-celled, 2-seeded samaræ, with a broad, orbicular margin. ♄ Lvs 3–5-foliate. Fls. cymous.

1 P. trifoliàta L. Lvs. 3-foliate, lfts. sessile, ovate, short-acuminate, lateral ones in equilateral, terminal ones cuneate at base; cymes corymbous; stam. mostly 4; styl short. Rocky places, N. Y. S. and W. 6—8f. Fls. white, odorous. June.

β. *mollis.* Young branches, petioles and leaves beneath, soft-downy and hoary. S

2 P. Baldwinii T. & G. Lvs. glabrous, very small; lfts. sessile, oval, obtuse; stam 4; stig. sessile. E. Fla. 1f. Branches numerous and scraggy. Lvs. 1'.

ORDER XXXII. AURANTIACEÆ. ORANGEWORTS.

Trees or *shrubs*, glabrous, abounding in little transparent receptacles of volatile oil, with *leaves* alternate, 1-3-foliate or pinnate. *Flowers* regular, 3- 5-merous. *Stamens* with flat filaments, distinct or cohering in one or several sets. *Ovary* compounded of several united carpels. *Style* 1. *Fruit* (hesperidium) many-celled, pulpy, covered with a thick rind. *Albumen* 0 *Cotyledon* thick. Figs. 37, 363.

CITRUS, L. (Κίτρον, the citron; the fruit of one of the species.) Sepals and petals in 5's. Anthers 20, or some other and higher multiple of 5, versatile, the connectile articulated to the filament. Filaments dilated at base, polyadelphous. Berry 9–18-celled. ♄ ♄ A noble E. Indian genus Lvs. 1-foliate, entire, evergreen. Petiole often winged.

1 C. vulgàris Risso. *Bitter Orange.* Petiole winged; lvs. elliptical, acute, crenu late; stam. 20; fruit globular, with a thin rind and bitter pulp. S. Fla. 15—20f. § Asia.

2 C. AURÁNTIUM. *Sweet Orange.* Petiole scarcely winged; lft. oblong, acute, crenu late; sta. 20; fr. globous, with a thin rind and sweet pulp. 30f.

3 C. LIMÉTTA. *Lime.* Petioles not at all winged; lft. ovate-orbicular, serrate; stam. 30; fr globous, with a sweet pulp, and a protuberance at top. 15f.

4 C. LIMÓNUM. *Lemon.* Petioles somewhat winged; sta. 35; fr. oblong-spheroid, with a thin rind and very acid pulp. 20f. Fr. yellow.

5 C. DECÙMANA. *Shaddock.* Petioles broadly winged; lft. obtuse, emarginate; fr very large, with a thick rind. 15f. Fruit green-yellow. 5' diam.

ORDER XXXIV. SIMARUBACEÆ. QUASSIAWORTS.

Trees or *shrubs* with bitter bark, alternate, exstipulate, pinnate leaves, and small, diclinous, regular, hypogynous 3–5-merous flowers. *Stamens* as many or twice as many as the *petals*, inserted on the hypogynous disk. *Styles* 2—5. *Ovaries* 2-5-lobed or carpelled. *Fruit* 1—5 one-seeded drupes or samaras

ORDER 36.—ANACARDIACEÆ.

§ Leaves abruptly pinnate. Flowers diœcious. Styles united. Fruit baccate............SIMIRUBA. 1
§ Leaves odd-pinnate. Flowers polygamous. Styles distinct. Fruit a samara.........AILANTHUS. 2

1. SIMARÙBA, Aubl. QUASSIA. (Its name in Guiana.) ♄
S. glauca DC. Leaflets 4—8, alternate, entire, obtuse, coriaceous. S. Fla. Tree, 40f.

2. AILÁNTHUS, Desf. CHINESE "TREE-OF-HEAVEN." (*Ailanto*, its name in China.) ⚥ ♀ ♂ Sep. 5. Pet. 5. ♀ Stam. 2—3. Ov. 3—5. Sty. lateral. Fr. 1-celled, 1-seeded samaræ, with oblong margins. ♂ Stam. 10. ♀ Ovaries, styles, and samaræ as in ⚥. ♄ Oriental, with odd-pinnate leaves. Flowers in panicles.

A. GLANDULÒSUS Desf. Lfts. glabrous, 21—41, ovate or oblong-lanceolate, acuminate, with 1 or 2 obtuse, glandular teeth each side at base, terminal one long-petiolate. Parks, &c. 40—60f. Flowers greenish, ill-scented. June.

ORDER XXXV. BURSERACEÆ. BURSERIDS.

Trees and *shrubs* abounding in balsam or resin, with exstipulate, compound, dotted leaves, and small, regular, racemed or panicled flowers. *Calyx* 3–5-cleft. *Petals* 3—5. *Stamens* twice as many. *Ovaries* free, 1–5-celled. *Stigmas* 2–5-lobed, ovules 2 in each cell. *Fruit* drupaceous, indehiscent, rarely capsular. *Seeds* pendulous, exalbuminous.

* Flowers perfect, 4-parted. Stamens 8, hypogynous. Leaves opposite.........................AMYRIS. 1
* Flowers polygamous, 4 and 6-parted. Stamens 8—10; disk crenate. Leaves alternate....BURSERA. 2

1. AMÝRIS, L. BALM-OF-GILEAD. (Μύρρα, myrrh; from its perfumed gum.) ♄ Flowers in panicles, white.

A. Floridàna N. *Torch-wood*. Shrub; lvs. opposite, trifoliate, on short petioles, lfts. ovate, obtuse, entire, petiolulate; drupes small, globular. E. Fla.

2. BÚRSERA, L. (To *Joachin Burser*, an Italian botanist.) ♄

B. gummífera Jacq. Lfts. 3—9, petiolulate, ovate, acum., entire; fls. racemed. Fla.

ORDER XXXVI. ANACARDIACEÆ. SUMACS.

Trees or *shrubs* with a resinous, gummy, caustic, or even milky juice. *Leaves* alternate, simple, or ternate, or unequally pinnate, without pellucid dots. *Flowers* with bracts, commonly diœcious, small. *Sepals* 3—5, united at base, persistent. *Petals* of the same number (sometimes 0), imbricated. *Stamens* as many as petals, alternate with them, perigynous. *Ovary* 1-celled, free. *Ovule* 1. *Stigmas* 3. *Fruit* a berry or drupe, usually the latter, and 1-seeded. *Albumen* 0.

RHUS, L. SUMAC. (The ancient name, from Celtic, *rhudd*, red?) Calyx of 5 sepals united at base. Pet. and stam. 5. Sty. 3. Stig. capitate. Fruit a small, 1-seeded, subglobous, dry drupe.—Small trees or shrubs. Leaves alternate, mostly compound. Flowers often, by abortion, imperfect, greenish.

Order 37.—SAPINDACEÆ.

§ Leaves simple. Flowers perfect (or all abortive in cultivation)..........Nos. 10, 11
§ Leaves compound. Flowers diœcious. A tree. South Florida..............No. 9
§ Leaves compound. Flowers polygamous...(a)
    a Flowers in clustered spikes *preceding* the trifoliate leaves.................No. 8
    a Flowers in axillary panicles, *with* the 3-13-foliate lvs. Poisonous.....Nos. 5—7
    a Flowers in terminal thyrses, *with* the 9-31-foliate leaves...(b)
        b Common petiole winged between the leaflets........ ....No. 4
        b Common petiole not winged......................................Nos. 1—3

1 R. glabra L. Lvs. and branches glabrous; lfts. 11—31, lanceolate, acuminate, acutely serrate, whitish beneath; fr. red, with crimson hairs. Thickets and pastures. 6—15f. The fruit hairs are extremely acid, and dye red. June, July.

2 R. typhina L. Branches and petioles densely villous; lfts. 11—31, oblong-lanceolate, acuminate, acutely serrate, pubescent beneath; fruit red, with crimson hairs. Rocky soils. 10--20f. Branches thick, straggling. Drupes acid. Wood yellow. June.
    β. *laciniàta*. Lfts. irregularly gashed; panicles leafy. Hanover, N. H. (*Ricard.*)

3 R. pùmila Mx. Procumbent, villous-pubescent; lfts. 9--13, oval or oblong, coarsely toothed; drupes red, silky pubescent. N. Car. to Ga. Branches 1f high.

4 R. copallìna L. *Mountain Sumac.* Branches and petioles pubescent; lfts. 9—21, oval-lanceolate, mostly entire, unequal at base, common rachis winged; fls. in dense panicles; drupes red, hairy. Rocky hills. 2—8f. Thyrse sessile. July.

5 R. venenàta DC. *Poison Sumac. Dog-wood.* Very glabrous; lfts. 7—13, oval, abruptly acuminate, very entire; panicles loose, axillary, pedunculate; drupes greenish-yellow, smooth. Swamps. 10—15f. Flowers green. Very poisonous. June.

6 R. Toxicodendron L. *Poison Oak. Poison Ivy.* Erect, or decumbent; lvs. pubescent; lfts. 3, broadly oval, acuminate, angular, or sinuate-dentate; drupes smooth, roundish. Thickets, Can. to Ga. Perhaps runs into the next. June.

7 R. radicans L. *Climbing Ivy.* Stems climbing by means of innumerable radicating tendrils; leaflets ovate, smooth, entire. Ascending trees. 20—30f. Drupes dull white. Stems 1—2′ in thickness. June.

8 R. aromática Ait. *Sweet Sumac.* Lfts. sessile, incisely crenate, pubescent beneath, lateral ones ovate, terminal one rhomboid; fls. in close aments, preceding the leaves; drupe globous, villous. Copses. 2—6f. Flowers yellowish. May.

9 R. Metòpium L. Lfts. 3—7, smooth, entire, ovate, acumin.; drupes smooth. 30f.

10 R. cotinoìdes N. Smooth; lvs. oval, obtuse, entire, acute at base, thin, long-stalked; fls. minute, in loose, erect panicles; drupes smooth. Mts. Car. to Ark.

11 R. Cótinus. *Venetian Sumac. Smoke-tree.* Lvs. obovate, entire, thick; flowers mostly abortive, pedicels diffusely branched and hairy. Italy.

Order XXXVII. SAPINDACEÆ. Mapleworts.

Trees, shrubs, or rarely *herbs,* with simple or compound, alternate or opposite leaves. *Flowers* mostly unsymmetrical, often irregular, 4 or 5-merous, with the *sepals* and *petals* both imbricated in the bud, with the *stamens* 5 to 10, inserted on a hypogynous or perigynous disk. *Ovary* 2 or 3-celled, lobed, and with 1 or 2 (rarely more) ovules in each cell. *Embryo* mostly curved or convoluted, with little or no albumen. Figs. 100, 224, 230, 236, 237, 308, 312, 444, 515.

I. ACERINEÆ.—Leaves opposite. Flowers regular, diclinous. Fruit a double samara...(a)
    a Disk annular. Petals 4 or 5 or 0. Leaves simple, lobed......................Acer. 1
    a Disk obsolete. Petals none. Leaves pinnately compound..................Negundo. 2
II. STAPHYLEÆ.—Leaves opposite. Flowers regular, perfect. Stamens 5.......Staphylea. 3
III. HIPPOCASTANEÆ.—Leaves opposite. Flowers irregular. Stamens 7.......Æsculus. 4
IV. SAPINDEÆ.—Leaves alternate. Flowers polygamo-diœcious...(b)

ORDER 37.—SAPINDACEÆ.

b Petals 5, regular. Ovules solitary. Fruit baccate. Trees.................SAPINDUS.
b Petals 5 or 4, regular. Ovules 2 or 3 in each cell. Trees. South Florida....HYPELATE.
b Petals 4, irregular. Trees. KŒLREUTERIA, No. 7..........Vines...........CARDIOSPERMUM.
b Petals 0. Ovules 2 in each cell. Capsules winged. Shrub. South Fla....DODONÆA.

1. **ACER.** MAPLE. (The ancient name, meaning sharp, vigorous.) Fls. polygamous. Cal. 5 (4–9)-cleft. Cor. 5 (4–9)-petalled or 0. Stam. 8 (4–12). Sty. 2. Samaræ 2-winged, united at base, by abortion 1-seeded. Leaves simple, palmately 5 (rarely 3–9)-lobed. (See *Addenda*.)

§ Flowers in dense, umbellate clusters, appearing *before* the leaves..........Nos. 1, 2
§ Flowers in pendulous corymbs, yellowish, appearing *with* the leaves.......Nos. 3, 4
§ Flowers in terminal racemes, greenish, appearing *after* the leaves...(*a*)
　　a Shrubs or small trees, native. Leaves 3-lobed................... ....Nos. 5, 6
　　a Large trees, exotic, cultivated. Leaves 5–7-lobed....................Nos. 7, 8

1 A. **rubrum** L. *Red Maple. Swamp Maple.* Lvs. cordate, acutely and incisely toothed, the sinuses acute, glaucous beneath; ped. elongated in fruit; pet. linear-oblong; ovaries and fruit smooth. Swamps. 30—80f. Flowers red. April.
　β. *tridens.* Lvs. 3-lobed, rounded at base; flowers yellowish. N. J. to La. 20f.
2 A. **dasycárpum** Ehrh. *White Maple.* Lvs. truncated at base, unequally and incisely toothed, with obtuse sinuses, white and smooth beneath; fls. greenish, with downy ovaries; petals 0; fruit divergent. Woods. 50f. Mar. April. (Fig. 308.)
3 A. **saccharinum** L. *Sugar Maple. Rock Maple.* Lvs. subcordate at base, acuminate, remotely toothed, with rounded and shallow sinuses, glaucous beneath; fls. pedunculate, pendulous. Rocky hills, N. 40—70f. A noble tree.
4 A. **nigrum** Mx. *Black Maple. Sugar Tree.* Lvs. cordate, with the sinus closed, lobes divaricate, sinuate-dentate, paler beneath, with the veins beneath and the petioles pubescent; flowers on long, slender pedicels. Vt. to Ind. 30—70f. April.
5 A. **Pennsylvánicum** L. *Striped Maple. Whistle-wood.* Lvs. with 3 acuminate lobes, rounded at base, sharply denticulate, smooth; rac. simple, pendulous. Can. to Ga. and Ky. 10—15f. Bark striped, green and black. May.
6 A. **spicátum** Lam. *Mountain Maple-bush.* Lvs. 3–5-lobed, acute, dentate, pubescent beneath; racemes erect, compound. Woody hills. 5—8f. Flowers greenish.
7 A. PSEUDO-PLÁTANUS L. *Sycamore.* Lvs. cordate, glabrous, glaucous beneath, lobes acute, unequally dentate; raceme pendulous; fruit smooth. Europe. 40f.
8 A. MACROPHÝLLUM Ph., with large, very deeply 5-lobed leaves, nodding racemes, and hispid fruit. Oregon. 30—50f.

2. **NEGÚNDO**, Mœnch. BOX ELDER. ASH MAPLE. Flowers ♀ ♂. Corolla 0; ♀ flowers racemed, ♂ fascicled. Disk 0. Stam. 3—5. Fruit as in the last genus. Leaves compound, pinnately 3–5-foliate.

N. **aceroìdes** Mœnch. Lfts. ovate, acuminate, remotely and unequally dentate; ♀ rac. long and pendulous; fruit oblong, with large wings dilated upward. A handsome tree, 20—40f. N. Y. to Car. and Cal.! April.

3. **STAPHYLEA**, L. BLADDER-NUT. (A Greek word, meaning a cluster of grapes; from the form of the fructification.) Fls. ☿. Calyx of 5, colored, persistent sepals. Pet. and sta. 5. Styles 3. Caps. 2—3, membranous and inflated, slightly cohering. Seeds not arilled. ♄ With opposite, 3–7-foliate lvs. and caducous stipules. Fig. 444.

S. **trifòlia** L. Lfts. 3, ovate, acuminate, serrate; fls. in drooping cymous panicles, white; pet. ciliate at base. Can. to Car. and Tenr. 6—10f. Caps. large. May.

4. **ÆSCULUS**, L. HORSE CHESTNUT. BUCKEYE. Calyx 5-toothed;

cor. irregular, 4 or 5-petalled; sta. 7 (6 to 8), distinct, unequal. Style filiform, ov. 3-celled, with 2 ovules in each cell. Fruit coriaceous, 2–3-valved, containing but one or very few large, smooth seeds. Cotyledons thick, bulky, inseparable. ♄ With opposite, digitate, 5–7-foliate leaves. Fls. paniculate, terminal. Fig. 100.

§ PAVIA. Fruit smooth. Petals 4, erect, the two upper clawed. *Buckeye*..Nos. 1—3
§ ÆSCULUS *proper*. Fruit prickly. Petals 4 or 5, spreading.............Nos. 4, 5

1 Æ. Pàvia L. Lfts. 5–7, shining, oblong-lanceolate; cuneate at base, short-acuminate, finely serrate; fls. red, very irregular in a lax, thyrsoid raceme; pet. as long as stamens; cal. half as long as the two shorter petals. S. 3–10f. Mar. April.

2 Æ. parviflòra Walt. Lfts. 5–7, obovate, acuminate, serrate, velvety canescent beneath; petals 4 white, somewhat similar and spreading, thrice shorter than the capillary stamens. S. 2–9f. Fls. very numerous.

3 Æ. flàva Ait. *Sweet Buckeye*. Lfts. 5–7, oblong or elliptic-ovate, acuminate, serrulate, pubescent beneath; fls. in thyrsoid, pubescent panicles; pet. very unequal, longer than the stamens. W. and S. 6–70f. Yellowish. April, May.

4 Æ. glàbra Willd. *Ohio Buckeye*. Lfts. 5, oval or oblong, acuminate, serrate or serrulate; fls. in lax thyrsoid panicles; pet. 4, half as long as the stamens. River banks, W. Tree 20–40f, ill-scented, with small, yellowish flowers. June.

5 Æ. HIPPOCÁSTANUM L. *Horse Chestnut*. Lvs. of 7 obovate lfts.; pet. 5, spreading; fruit prickly. Tartary. A noble tree, in parks, &c. June.

5. SAPÍNDUS, L. SOAP-BERRY. (That is, by syncope, *Sapo Indicus*, Indian soap.) Sep. 4 or 5. Pet. as many, or one less by abortion, appendaged inside with a gland, scale, or beard. Sta. 8—10. Stig. 3. Fruit 3, connate, globular, fleshy carpels, often by abortion 2 or 1. Seed large, solitary. ♄ Lvs. alternate, pinnate, exstipulate.

S. marginàtus Willd. Common petioles wingless; lfts. 9–13, ovate-lanceolate, long-pointed, very inequilateral, short-stalked, entire, glabrous, shining above; flowers in white, dense panicles. Ga. to Ark. 20–40f. Fruit globular.

6. CARDIOSPÉRMUM, L. HEART-SEED. ($K\alpha\rho\delta i\alpha$, heart, $\delta\pi\epsilon\rho\mu\alpha$, seed.) Sep. 4, two of them smaller. Pet. unequal, each with a scale at base. Sta. 8. Style 3-fid. Caps. membranous, inflated. ♃ Leaves biternate. Pedicels changed to tendrils.

C. Halicàcabum L. Lfts. ovate-lanceolate, incisely lobed and dentate; fr. pyriform-globous, large, bladder-like. Banks of streams, S. and W. 4–6f. July. §

7. KŒLREUTÈRIA, Lam. (To *J. G. Köhlreuter*, a Russian botanist and author, 1755.) Sep. 5. Pet. 4, irregular. Sta. 8. Sty. exserted. Caps inflated, 3-celled, cells 2-seeded. ♄ Lvs. alternate, pinnate, lfts. about 13, cut-serrate. Flowers yellow, in large panicles.

K. PANICULÀTA.—China. 20–30f. Odd leaflet cut-lobed. A curious tree.

ORDER XXXVIII. CELASTRACE.Æ. STAFF TREES.

Shrubs with simple leaves alternate or opposite, with *flowers* small, regular, 4 or 5-merous, perigynous, *sepals* and *petals* both imbricated in æstivation, *stamens* alternate with the petals, and inserted on a disk which fills the bottom of the calyx. *Carpels* 2—5, *styles* united. *Fruit* free from the calyx, with 2—5 cells. *Seeds* arilled, few, albuminous.

ORDER 40.—RHAMNACEÆ.

* Leaves alternate.—a Capsule dehiscent. Cells 2-ovuled. Vine..................CELASTRUS. 1
     —a Capsule dehiscent. Cells 1-ovuled. Erect. S. Fla............MAYTENUS.
     —a Drupe dry, 2-celled, 2-seeded. Erect. S. Fla..............SCHAEFFERIA.
* Leaves opposite.—b Capsule 3–5-celled. Cells 2-ovuled............................EUONYMUS. 2
     —b Drupe 1-celled, 1-seeded (ovary 2–4-celled.) S. Fla............MYGINDA.

1. CELÁSTRUS, L. STAFF-TREE. Fls. often imperfect. Sep. and pet. 5. Disk 5-lobed, bearing the 5 stamens on its edge. Caps. subglobous, or 3-angled, 3-celled. Seeds with an arillus, 1 or 2 in each cell. ♃ With alternate, deciduous lvs. and minute, deciduous stipules.

    C. scándens L. St. twining; lvs. oblong, acuminate, serrate; rac. terminal; flowers diœcious. Woods. 20—40f. Arilled seeds scarlet, persistent in winter. June.

2. EUÓNYMUS, Tourn. BURNING BUSH. (Εὖ, good, ὄνομα, name.) Fl. perfect; calyx flat, of 5 (sometimes 4 or 6) united sepals. Corolla flat, inserted on the outer margin of the broad disk. Stamens 5, with short filaments. Caps. colored, 5-angled, 5-celled, 5-valved. Seeds wholly invested with a scarlet aril. ♃♃ Lvs. opposite, serrate. Flowers purple.

  1 E. atropurpùreus Jacq. Lvs. elliptic-ovate, petiolate, acuminate, finely serrate, puberulent beneath; ped. compressed, many-flowered; fls. usually 4-merous; capsule smooth, lobed. Woods. 4—10f. Fruit crimson. June. Varieties in cultivation have orange-red or even whitish fruit.

  2 E. Americànus L. Branches 4-angled; lvs. oval and elliptic-lanceolate, acuminate, acute, or obtuse, smooth, subsessile; ped. round, about 3-flowered; fls. mostly pentamerous; caps. warty. Woods. 2—5f. Fruit dark red. June.
    β. *obovàtus.* Trailing; lvs. obovate, obtusish, petiolate. Ohio, &c.
    γ. *angustifòlius.* Lvs. linear-lanceolate, inequilateral, acute at each end. South.

  3 E. EUROPÆUS, has smooth, shining, lance-oblong, serrate leaves, the flattened ped 3-flowered; fls. 4-parted. Europe. Not hardy North. (See Addenda.)

ORDER XL. RHAMNACEÆ. BUCKTHORNS.

Shrubs or small *trees*, often spiny, with simple, alternate, stipulate *leaves*, with *flowers* regular, sometimes apetalous or otherwise imperfect; with the *stamens* perigynous, as many (4 or 5) as the valvate sepals, alternate with them, and opposite to the petals when they are present. *Disk* perigynous. *Capsule* or *drupe* with one albuminous seed in each cell.

* Leaves opposite or subopposite, with opposite branches...a
   a Flowers small, in axillary clusters or umbels. S. Fla.....................SCUTIA.
   a Flowers minute, spicate, in terminal panicles.........................SAGERETIA. 1
* Leaves alternate.—b Shrubs climbing by twining. Petals sessile...............BERCHEMIA. 2
    —b Shrubs climbing by tendrils. Pet. short........................GOUANIA. 3
* Leaves alternate.—c Clusters of (white) flowers terminal. Pet. unguiculate........CEANOTHUS 4
    —c Clusters axillary. Pet. 4, 5, or 0, on the margin of disk..........RHAMNUS. 5
    —c Clusters axillary. Pet. 5, under the 5-lobed disk. S. Fla........COLUBRINA.

1. SAGERÈTIA, Brongn. (Named for *M. Sageret*, a French florist and veg. physiologist.) Calyx 5-cleft. Petals 5, cucullate. Sta. 5. Ovary immersed in the entire disk, with a 3-lobed stigma. Drupe 3-celled. ♃ With slender branches. Fls. in rigid, interrupted spikes.

  S. Michàuxii Brongn. Branches at length spiny; leaves ovate or oblong-ovate, subsessile, shining, subentire. Sandy coasts, Car. to Fla. Trailing, 6—15f. October.

2. **BERCHÈMIA**, Necker. SUPPLE JACK. Calyx 5-parted. Pet. 5, convolute, enclosing the 5 stamens. Ovary half immersed in the disk, but free from it, 2-celled. Style bifid. Drupe oblong, with a bony, 2-celled nut. ♂ ♀ Unarmed. Lvs. pinnate-veined. Panicles terminal, small.

B. volùbilis DC. Climbing, glabrous; lvs. ovate, straight-veined, repandly serrate; drupe dark purple. Damp soils, S. Stem supple, 10—20f. May, June.

4. **CEANÒTHUS, L.** JERSEY TEA. RED-ROOT. Calyx tubular-campanulate, 5-cleft. Petals 5, saccate, arched, with long claws. Sta. mostly exserted. Style 3-cleft. Capsule obtusely triangular, 3-celled, 3-seeded, surrounded at base by the persistent tube of the calyx. ♂ ♀ Thornless. Fls. small, aggregated at the end of the branches.

1 **C. Americànus** L. Leaves oblong-ovate, or ovate, serrate, 3-veined; flowering branches leafy or leafless, elongated. Dry woods. 2—4f. June.
2 **C. ovàlis** Bw. Lvs. oval-lanceolate or narrowly oblong, with glandular serratures, 3-veined, veins pubescent beneath; thyrse corymbous, abbreviated. Vt. to Mich. 2—3f. Less common than No. 1. Lvs. smooth, shining. May.
3 **C. microphýllus** Mx. Diffusely branched, branches very slender; leaves minute, obovate, rigid, glabrous, strigous beneath. Pine-barrens, S. 1—2f. April.
. β. *serpyllifòlius*. Very slender; branches filiform; lvs. oval (2—3″ long). S.

5. **RHÁMNUS, L.** BUCKTHORN. (The Greek name.) Calyx urceolate, 4 or 5-cleft. Pet. 4 or 5, notched, lobed, or entire, or sometimes wanting. Ov. free, not immersed in the thin torus, 2—4-celled. Styles 2—4, more or less united. Drupe containing 2—4 cartilaginous nuts. ♀ Lvs. alternate, rarely opposite. Fls. in axillary clusters.

§ Flowers tetramerous. Leaves with arcuate veinlets..................Nos. 1, 2
§ Flowers pentamerous. Leaves with the veinlets nearly straight..........Nos. 3, 4

1 **R. cathárticus** L. Thorny; lvs. ovate, denticulate-serrate; fls. fascicled; polygamo-diœcious, mostly tetrandrous; sty. 4, at apex distinct and recurved: fr. globular, 4-seeded. Hedges, rarely wild. 10—15f. Drupes black, cathartic. May +. § Eur.
2 **R. lanceolàtus** Ph. Thornless; lvs. lanceolate or oblong, acute at each end, the earlier ones obtuse; fls. 1—3 together; pet. 4, minute; sty. 2 at apex, distinct; drupe 2-seeded. Pa. to Iowa (Colman). Rare. 4—8f. May.
3 **R. alnifòlius** L'Her. Unarmed; lvs. oval, acute, serrate; ped. aggregate, 1-flowered; fls. mostly pentandrous and apetalous; sep. acute; styles 3, united, very short; fruit 3-seeded. Pa. to Can. 2—4f. June.
4 **R. Caroliniànus** Walt. Unarmed; leaves oblong-oval, serrulate, acute, paler beneath; fls. perfect, in short, axillary umbels, petals minute; stigmas 3; fr. 3-seeded. River banks, Va. to Fla. 7—15f. June.

ORDER XLI. VITACEÆ. VINES.

Shrubs with a watery juice, tumid nodes, and usually climbing by tendrils. *Flowers* small, regular, racemous, often polygamous or diœcious. *Calyx* minute, truncated, the limb obsolete or 5-toothed. *Petals* hypogynous, valvate in æstivation, as many as and opposite to the stamens. *Stamens* inserted on the disk which surrounds the 2-celled, 1-styled ovary. *Fruit* a berry, usually 4-seeded. *Seeds* bony. *Albumen* hard. Figs. 187, 250.

VITIS, L. GRAPE-VINES. (Celtic *gwyd*, a tree or shrub.) Petals 4 or

5, deciduous, cohering at the top, or distinct and spreading. Ovaries 2-celled, cells 2-ovuled. Fruit a globular berry, 1–4-seeded. ♃ Lvs. simple or compound. Ped. opposite the lvs. often changed to tendrils. Fls. small, clustered.

§ VITIS *proper*. Petals cohering at the top, and falling without expanding...*a*
§ CISSUS. Petals free, expanding before falling. Tendrils coiling, or 0...*b*
§ AMPELÓPSIS. Petals free, expanding. Tendrils with an adhesive foot.......No. 9
  a Leaves beneath clothed with a whitish or rusty wool................Nos. 1, 2, 3
  a Leaves glabrous except the veins, and green both sides.............Nos. 4, 5, 10
    b Leaves simple, angular or entire..No. 6
    b Leaves pinnately compound. ....Nos. 7, 8

1 **V. labrúsca** L. *Fox Grape. Isabella, Catawba.* Leaves broad-cordate, angular-lobed, hoary tomentous beneath; berries large. Woods. 30–80f. Fr. p. gr. or amb.
2 **V. æstivàlis** L. Lvs. broadly cordate, 3–5-lobed or palmate-sinuate, coarsely dentate, with scattered ferruginous hairs beneath; fertile racemes long, panicled, berries small. Shady banks. Fruit deep blue, small, ripe in September.
3 **V. Caribǽa** DC. Hoary; lvs. round-cordate, 3-lobed or entire, smooth above. Fla.
4 **V. cordifòlia** Mx. *Frost Grape.* Lvs. cordate, acuminate, somewhat equally toothed, smooth, or pubescent beneath the veins and petioles; rac. loose, many-flwd.; berries small. River banks. 10–20f. Fruit blackish, ripe in November.
5 **V. vulpìna** L. *Muscadine. Scuppernong.* Lvs. (small) cordate, slightly 3-angled or lobed, shining on both sides, coarsely toothed, the teeth not acuminate; rac. composed of many capitate umbels. Va. to Fla. Fruit large, purple, few.
6 **V. indivìsa** Willd. Lvs. simple, cordate or truncate at the base, often angular-lobed; flowers 5-merous; berry 1 or 2-seeded. Swamps, S. Fruit small (2 ?).
7 **V. bipinnata** T. & G. Lvs. bipinnate, lfts. incisely serrate, glabrous; flowers 5 merous. S. States along rivers. Fruit small, black. No tendrils.
8 **V. incìsa** N. Lvs. 3-foliate, thick; lfts. 2–3-lobed; berry 1-seeded. Fla. to La.
9 **V. quinquefòlia** Lam. *Virginia Creeper.* Lvs. digitate, lfts. 5, oblong, acuminate, dentate; berries dark blue, smaller than peas, acid. Woods, thickets. 20–40f.
10 **V. vinífera** L. *European Wine-grape.* Lvs. cordate, sinuately 5-lobed, glabrous; flowers all perfect. Europe. Many varieties.

ORDER XLII. POLYGALACEÆ. MILKWORTS.

Herbs or *shrubs*, with the leaves mostly simple and without stipules. *Flowers* irregular, unsymmetrical, hypogynous, perfect. *Sepals* 5, unequal, distinct, some or all of them colored. *Petals* 3, often 5, and 2 of them scale-like. *Stamens* 4 to 8, distinct, or cohering in a tube which is split on the upper side. *Ovary* superior, compound, with suspended ovules, united styles and stigmas. *Fruit* a 2-seeded pod. *Seeds* pendulous, with or without a caruncle and albumen.

Sepals 5, unequal, 2 larger, wing-shaped, petaloid. Petals 3. Stamens 8..............POLYGALA. 1
Sepals 5, nearly equal. 3 of the 5 petals long-clawed. Stamens 4..................KRAMERIA. 2

1. POLÝGALA, Tourn. MILKWORT. (Πολύς, much, γάλα, milk; said to favor the lacteal secretions of animals.) Fls. very irregular. Sep. 5, 2 of them wing-shaped and petaloid. Pet. 3, cohering by their claws to the filaments, lower one carinate and often crested on the back. Stam. 6 or 8, filaments united into a split tube. Anth. 1-celled. Caps. obcordate,

ORDER 42.—POLYGALACEÆ.

2-celled, 2-seeded, loculicidal. Sd. appendaged with a various caruncle at the hilum. Mostly herbs, bitter, and with simple leaves. Flowers often of two forms, the subterranean apetalous.

* Leaves alternate.—*a* Fls. purple, solitary, 2—4. Perennial........................ No. 1
    —*a* Fls. purple, racemed, many. Biennial.................. ....Nos. 2, 3
    —*a* Fls. white. Spike slender. Seeds hairy. Perennial.....Nos. 4, 5
    —*a* Fls. purple. Spike capitate.—Caruncle double...........Nos. 6—8
            —Car. appears simple. ①...Nos. 9—11
    —*a* Fls. xanthic.—*b* Spikes solitary, large. Biennial........Nos. 12, 13
      —*b* Spikes ∞, corymbed, small. Bien.....Nos. 14, 15
* Lvs. vertic. on the stem.—*c* Spikes acute, slender. Fls. greenish-white...Nos. 16, 17, 18
      c Spikes obtuse, thick..(Sbrubs, †. No. 22—25)..Nos. 19, 20, 21

1 P. paucifòlia L. St. simple, erect, naked below; lvs. ovate, acute, smooth; terminal fls. large, crested, radical ones apetalous. ♃ Woods. 3—4′. Flowers few, large (10″), very showy. May, June.

2 P. grandiflòra Walt. Ascending, pubescent; lvs. ovate-lanceolate to lance-linear, acute; fls. distant, pendulous after blooming, wings large, roundish, covering the fruit, keel as long as the wings (3′), crestless. ②? Dry soils, S. 9—12′. May—Aug.

3 P. polýgama Walt. Sts. simple, numerous, glabrous; lvs. linear-oblong, mucronate, obtuse; fls. racemed, short-pedicelled, those of the stem winged, those of the root wingless; keel cristate. ② Fields. 6—12′. Rac. showy. Fls. 2″. June, July.

4 P. Sénega L. *Seneca Snake-root.* St. erect, smooth, simple, leafy; lvs. lanceolate, tapering at each end; fls. slightly crested, in a terminal spike-form, slender raceme. ♃ Woods, W. States, rare in E. 8—14′. Spike 1—2′. Leaves 1—2′. July.
 β. *latifòlia.* Leaves ovate, acuminate at each end. Leaves 2—3′. Ind.

5 P. alba N. St. angular, branched above; lvs. linear; spike lance-linear, pointed, on a long stalk. ♃ Ala. to La. 6—12′. Spikes 1—3′.

6 P. setàcea Mx. Sts. filiform, simple, apparently leafless (lvs. minute, deltoid-acum.); spike (small) oblong, acute; wings short-pointed, shorter than the petals; caruncle enclosing the short stipe of the hairy seed. ♃ South. 1f. Leaves 1″. June.

7 P. incarnàta L. Glaucous; st. erect, slender, mostly simple; lvs. few, scattered, linear-subulate; spike oblong; wings lanceolate, cuspidate; claws of the petals united into a long, cleft tube; seed very hairy. ① N. J. to Fla. 1—2f. June.

8 P. Chapmánii T. & G. Very slender, simple, or nearly so; lvs. linear-subulate; spike loose, roundish-oblong, rather acute; wings obovate, slightly clawed; caruncle lateral on the thin-haired seed. ① South. 1f.

9 P. Nuttállii T. & G. St. erect, somewhat fastigiate; lvs. linear; spikes acute, roundish-oblong, dense; wings elliptical, attenuate at base; crest minute; caruncle notched, lateral on the thick seed-stipe. ① Mass., R. I., to La. 6—10′. August.

10 P. fastigiàta Nutt. Slender and much branched above; lvs. linear; spikes roundish, loose-flowered; wings ovate-oblong, distinctly clawed; caruncle broad, nearly embracing the small seed-stipe (immature). ① N. J. to Fla. 8—12′. July+.

11 P. sanguínea L. St. branching at top; lvs. linear and lance-linear; spikes oblong, obtuse, dense; wings oval or ovate, obtuse, subsessile; caruncle mostly simple, nearly as long as the hairy seed. ① Wet grounds. 10′. Leaves 1′. July+.

12 P. lùtea L. St. mostly simple; root leaves spatulate, obtuse, attenuate at base; cauline ones lanceolate, acute; rac. ovate-globous, obtuse, dense; fls. pedicellate; wings ovate, mucronate, keel with a minute crest. ② Sands, N. J. to Fla. 1f. June +.

13 P. nana DC. Low, ascending; lvs. obovate and spatulate, mostly radical; heads ovate, becoming oblong, dense; wings lance-ovate, cuspidate-acuminate, twice longer than the slightly-crested keel. ② Pine woods, S. 4′. April, May.

14 P. ramòsa Ell. Erect, corymbously branched above; spikes loose, oblong, numerous, forming det se, level-topped cymes; radical lvs. few, spatulate, cauline oblong-linear; seed oval carunclel. ② Swamps, Del. to Fla. 1f. June.

15 P. cymòsa Walt. Tall, corymbously branched at top; lvs. mostly radical, linear, pointed, crowded; stem lvs. very few, linear-subulate; racemes spike-like, forming a dense, fastigiate cyme; seed globular, naked. ② Swamps, S. 2—5f. June+.

16 P. verticillàta L. St. branched above, erect; lvs. linear, verticillate both on the stem and opposite branches; fls. crested; calycine wings roundish; seed oblong, smooth, caruncle hardly half as long. ① Dry hills. 6—8′. July+.

β. *ambigua*. Branches and upper lvs. alternate; spikes long; fls. scattered.

17 P. Boykínii T. & G. Sts. erect from an ascending base, simple; lvs. obovate and lanceolate; spike slender, pointed, dense; caruncle two-thirds the length of the very hairy seed. ♃ South. 12—18′. June—Aug.

18 P. leptóstachys Shuttl. Sts. filiform, strict; lvs. setaceous, in 4's or 5's, remote; spikes linear; seed smooth. ① Dry sands, Fla. 1f. Greenish.

19 P. Hoókeri T. & G. Sts. weak, 4-angled; lvs. in 4's, linear; spikes lance-ovate, pointed. Pine woods, Fla. to Tex. 1f. Flowers pale red.

20 P. cruciàta L. St. erect, winged at the angles, fastigiate; lvs. in 4's, linear-oblong, punctate; spikes ovate, dense, obtuse, subsessile; caruncle as long as the ovoid smooth seed. ① Wet grounds. 3—12′. July, Aug.

β. *cuspidàta*. Lvs. linear; heads squarrous with the wing-cusps. South.

21 P. brevifòlia Nutt. Slender, branched above; lvs. linear, short, remote, in 4's, or on the branches scattered; spike oblong, dense, obtuse, on long peduncles; wings ovate-lanceolate, acute; seed just as in No. 20. ① N. Y. to Fla. 1f. August.

22 P. SPECIÒSA. Shrub 6f; lvs. cuneate-oblong, alternate; fls. purple, in terminal rac.

23 P. MYRTIFÒLIA. Shrub 3—4f; lvs. oblong-obovate, altern.; fls. purple, in lateral rac.

24 P. OPPOSITIFÒLIA. Shrub 3f; lvs. opp., sessile, cordate, smooth; fls. roseate, large.

25 P. LATIFÒLIA. Shrub 3f; lvs. opposite, ovate, glaucous, downy beneath; fls. purple.

2. **KRAMÈRIA**, L. Ovary 1-celled, with 2 collateral ovules. Seed with no caruncle and no albumen. ♄ Racemes terminal.

K. lanceolàta Torr. Prostrate; lvs. lance-lin., acute, longer than ped.; fr. spiny. Fla

ORDER XLIII. LEGUMINOSÆ. LEGUMINOUS PLANTS.

Herbs, shrubs, or *trees. Leaves* alternate, usually compound, margins entire. *Stipules* 2, at the tumid base of the petiole. *Stipels* commonly 2. *Sepals* 5, more or less united, often unequal, the odd one always anterior. *Petals* 5, either papilionaceous or regular, perigynous, the odd one (when present) posterior. *Stamens* diadelphous, monadelphous, or distinct. *Anthers* versatile. *Ovaries* superior, single, and simple. *Style* and *stigma* simple. *Fruit* a legume, either continuous (1-celled), or (a loment) jointed into 1-seeded cells. *Seeds* solitary or several, destitute of albumen. Figs. 59, 60, 102, 157, 190–1, 203–4, 214, 233, 308, 354–6, 361–2, 397, 401–2, 480.

A vast and important order, containing 400 genera and 6,500 species, of which 350 are native in the United States.

I. MIMOSEÆ. Corolla regular, valvate in bud. Stamens exserted, hypogynous. Lvs. bipinnate...(§)
II. CÆSALPINEÆ. Corolla irregular, upper petal *interior* in bud. Stamens 5—10, perigynous...(§§)
III. PAPILIONACEÆ. Corolla papilionaceous, upper petal (the banner) larger and *exterior*...(*)

* Stamens 10, all distinct to the base. Plants erect. (Tribe PODALYRIEÆ)...(1)
* Stamens 10, monadelphous or diadelphous...(**)
　** Leaves cirrhous, ending with a tendril. Stamens 9 and 1. Vines. (Tribe VICIEÆ)...(2)
　** No tendrils. Pod a loment (§ 165), or rarely 1-seeded. Lvs. pinnate. (Tr. HEDYSAREÆ)...(3)
　** No tendrils. Pod a legume (§ 165), rarely 1-seeded...(***)
　　*** Erect (or if prostrate, with palmately 3-foliate leaves). (Tribe LOTEÆ)...(4)
　　*** Twining or trailing vines, with pinnately compound leaves. (Tribe PHASEOLEÆ)..(5)

ORDER 43.—LEGUMINOSÆ. 81

§ Pods flat, composed of 1 or more 1-seeded joints. Petals united. Stamens 4—10..MIMOSA. 1
§ Pods continuous,—m prickly, 4-sided and 4-valved. Petals united. Sta. 8—10..SCHRANKIA. 2
— m smooth,—n Petals distinct. Pod linear. Stamens 5 or 10..DESMANTHUS. 3
— n Petals distinct. Pod oblong. Stamens 10....NEPTUNIA. 4
— n Petals united. Trees, shrubs. Sta. ∞, monadel..ALBIZZIA. 5
— n Petals distinct, ylw. Shrubs. Stamens ∞ ..ACACIA. (5 a) 58
§§ Flowers perfect, red or yellow, showy. Trees or shrubs. Lvs. bipinnate..POINCIANA.(9 a) 59
§§ Flowers perfect, red or rose-colored. Trees with simple broad leaves......CERCIS. 9
§§ Flowers perfect, yellow (in our species). Herbs with pinnate leaves.......CASSIA. 8
§§ Flowers imperfect, greenish.—Trees thornless, with bipinnate leaves.......GYMNOCLADUS. 6
—Trees thorny. Lvs. pinnate and bipinnate...GLEDITSCHIA. 7
L PODALYREÆ.—c Trees. Leaves pinnate. Pod flat and thin.................CLADASTRIS. 10
— c Trees or shrubs. Lvs. ternate...CALLISTACHYS, 60, or pinnate in..SOPHORA.(10 a) 61
— c Shrubs in the greenhouse, with simple, spiny-toothed leaves....CHORIZEMA. (10 b) 62
— o Herbs.—p Pod inflated, stipitate. Leaves 1-3-foliate.............BAPTISIA. 11
— p Pod flattened, sessile. Leaves 3-foliate...............THERMOPSIS. 12
L VICIEÆ.—d Erect. Tendrils obsolete. Fls. white, with a black spot on each wing..FABA. 13
— d Climbing.—q Leaflets serrate. Pods 2-seeded...................CICER. 14
— q Lfts. entire.—r Sty. grooved on the back. Sds. 3—9 glob..PISUM. 15
— r Sty. flattened on the bk. Sds. 3-9, flattish.LATHYRUS. 16
— r Sty. flattish. Seeds 1 or 2, lens-shaped....LENS. (17 a) 64
— r Style filiform. Seeds 2—7, roundish......VICIA. 17
L HEDYSAREÆ.—s Fls. yellow.—s Leaves palmately 4-foliate. Stam. monadelphous..ZORNIA. 18
— s Leaves pinnate, 7-49-foliate. Stam. diadelphous..ÆSCHYNOMENE.19
— s Lvs. pinnately 3-7-foliate. Stam. monadelphous..CHAPMANIA. 20
— s Leaves pinnately 3-foliate. Pod slender at base...STYLOSANTHES.21
— s Leaves pinnately 4-foliate. Pod gibbous at base..ARACHIS. 22
s Fls. cyanic.—u Lvs. pinnate, 5-21-foliate.—t umbels pedunculate...CORONILLA. 23
— t rac. pedunculate.......HEDYSARUM. 24
— u Lvs. pin. 3-foliate.—t stipellate. Pod 3-7-jointed...DESMODIUM. 25
— t exstipellate. Pod 1-jointed..LESPEDEZA. 26
LOTEÆ—(including GENISTEÆ, Gen. 27—30, TRIFOLIEÆ, 31—34, and GALEGEÆ, 35—48).
f Leaves wanting ; if present, simple. Flowers yellow...................SPARTIUM. 27
f Leaves present, simple. Flowers yellow.—o Keel oblong, straight.............GENISTA. 28
— v Keel falcate, pointed.............CROTALARIA. 29
f Leaves palmately 5-15-foliate (rarely simple). (Genus 35, or)..................LUPINUS. 30
f Leaves palmately 3-foliate.—w Small tree with yellow hanging racemes......LABURNUM. 31
— w Shrubs. Fls. ylw., axil. Some of the lvs. simple..CYTISUS. (31 a) 65
— w Herbs with straight, small pods. Fls. capitate...TRIFOLIUM. 32
f Lvs. pinnately 3-foliate.—x Pods curved or spiral. Fls. in spikes, heads, &c......MEDICAGO. 33
— x Pods long and long-pointed. Flowers axillary..TRIGONELLA, (33 a) 66
— x Pods 1-2-seeded. Rac. (red, Gen. 50) white or yellow..MELILOTUS. 34
— x Pod 1-seeded,—y Fls. yellow. Lvs. resinous-dotted..(Genus 48
— y Fls. cyanic.—z Lvs. dark-dotted. .PSORALEA. 35
— z Lvs. not dotted....(In Genus 26
f Lvs. pinnate, with no odd leaflet.—* 15 to 25 pairs. Tall. Fls. yellow. S......SESBANIA. 36
—* 1 to 6 pairs. Flowers purple. Cult......OROBUS. (13 a) 63
f Lvs. odd-pinnate,—h dotted with dark glands.—k Shrub. Fls. spicate.........AMORPHA. 37
— k Herbs 10-androus.. ....DALEA. 38
— k Herb 5-androus...............PETALOSTEMON.39
— h dotless.—i Herbs. Style glabrous. Pod partly 2-celled...ASTRAGALUS. 40
— i Herbs. Style hairy. Pod 1-celled.............TEPHROSIA. 41
— i Herbs. Style glabrous. Pod 1-celled.........INDIGOFERA. 42
— i Trees or shrubs. Flowers white or roseate......ROBINIA. 43
— i Shrubs with yellow flowers....................COLUTEA. 44
— i Shrubs with scarlet flowers................CLIANTHUS.(44 a) 67
L PHASEOLEÆ.—g Lvs. pinnate, 5-15-foliate.—m Vine shrubby. Keel falcate.......WISTARIA. 45
— m Herbs. Keel (straight, Gen. 41) spiral..APIOS. 46
— g Leaves pinnately 3-(rarely 1)-foliate...(n)
n Flowers yellow. Legumes 5-seeded..... ....VIGNA. 47
n Flowers yellow Legumes 1-2-seeded.................RHYNCHOSIA. 48
n Flowers cyanic. .(*)

* Keel with stamens and style spirally twisted. Bushy or twining .... PHASEOLUS. 49
* Keel straight or merely incurved...(o)
o Shrubby at base. Flowers and seeds scarlet. Wings and keel very short. S..ERYTHRINA. 50
o Herbs.—x Calyx ebracteolate. Style beardless. Petals suberect, pale.....AMPHICARPÆA 51
—x Calyx 2-bracteolate,—y 4-cleft. Style beardless. Fls. pale......GALACTIA. 52
—y 4-toothed. Style bearded at top.........DOLICHOS. 53
—y 5-cleft, long. Style bearded inside........CLITORIA. 54
—y 5-cleft, short. Style bearded at top......CENTROSEMA. 55
—y 5-lobed. Style beardless. Cultivated...KENNEDYA. 56
—y 4-toothed. Style beardless. Cult......HARDENBERGIA. 57

1. MIMÒSA, L. SENSITIVE PLANT. (*Μίμος*, a buffoon: the leaves seem sporting with the hand that touches them.) Fls. ♀ ☿ ♂. ☿ Calyx valvate, 5-toothed. Cor. 0, or 5-toothed. Stam. 4—15. Legume separated into 1-seeded joints. ♂ Like the perfect, but without ovaries or fruit. ♃ ♄ Tropical. Leaves bipinnate.

1 **M. strigillòsa** T. & G. Nearly unarmed, prostrate, diffuse, strigous; stip. ovate; petioles and peduncles very long; pinnæ 4 to 6 pairs; lfts. 10 to 15 pairs, oblong-linear; heads oblong. ♃ Fla. to La. Flowers rose-color.

2 **M.** PÙDICA L. St. prickly, more or less hispid; lvs. digitate-pinnate, pinnæ 4, of many (30 or more) pairs of linear leaflets. Brazil. 1f. Leaflets 3′′.

2. SCHRÁNKIA, Willd. SENSITIVE BRIER. (In honor of *Francis de Paula Schrank*, a German botanist.) Fls. ☿ ♂. Cal. minute, 5-toothed. Pet. united into a funnel-shaped, 5-cleft corolla. Stam. 8—10. Pod long and narrow, echinate, dry, 1-celled, 4-valved, many-seeded. ♃ Prickly. St. procumbent. Lvs. sensitive, bipinnate. Fls. in spherical hds., purplish.

S. uncinàta Willd. St. angled, grooved; pinnæ 6 to 8 pairs; lfts. numerous, minute, elliptic-oblong or linear; heads axillary, 1 to 2 together, on peduncles shorter than the leaves. S. States. 2—4f. Leaflets 2′′. May—July. (& S. angustata T. & G.)

3. DESMÁNTHUS, Willd. (*Δέσμη*, a bundle, *ἄνθος*, flower.) Cal. valvate, 5-toothed. Pet. 5, distinct. Stam. 5 or 10, distinct. Pod dry, flat, 2-valved, 4–6-seeded, smooth. ♃ ♄ With bipinnate lvs. and white fls. in axillary, pedunculate heads. Petioles with 1 or more glands.

D. brachýlobus Benth. Erect, smoothish; pinnæ 6 to 13 pairs; lfts. minute, 20 to 30 pairs; stam. 5; pods short 1′, 2–4-seeded. ♃ Ill. to La. 2f. June—Aug.

4. NEPTÙNEA, Lour. Anthers 10, crowned with a stipitate gland. Pod oblong, oblique, deflexed on the stipe, 2-valved. Otherwise as in Desmánthus.

N. lùtea Benth. Sts. ascending, strigous; pinnæ 4—5 pairs; lfts. linear-oblong, ciliate, crowded; ped. longer than the leaves; pod 5–8-seeded. ♃ Prairies, Fla. to La. The leaves similar to those of Mimosa. Flowers yellow. Pods stiped. (Acacia lutea C-B.)

5. ALBÍZZIA, Durazz. Calyx 4- or 5-toothed. Petals united into a funnel-form corolla. Stamens ∞, monadelphous at base, very long. Pod linear and flat, jointless, dry, 2-valved, many-seeded. ♄ ♄ Tropical, with the leaves twice pinnate. Flowers in dense heads or spikes, roseate or white, polygamous.

A. JULIBRÁSSIN. *Silk Tree.* Tree about 20f, glabrous, thornless; pinnæ 8—12 pairs, each with 20—30 pairs of *halved* leaflets (being one-sided), acute; heads pedunculate, forming a terminal panicle; corollas white, with the innumerable long silky stamens purplish; pods some contracted between the seeds. Very ornamental, hardy South, sparingly naturalized in the Gulf States.

6. GYMNÓCLADUS, Lam. COFFEE TREE. (Γυμνός, naked, κλάδος, a shoot; for its coarse, naked shoots in winter.) Fls. ♀ ♂. Cal. tubular, 5-cleft, equal. Pet. 5, inserted into the summit of the tube. ♂ Stam. 10, distinct. ♀ Style 1. Leg. 1-celled, oblong, very large, pulpy within. ♄ Unarmed, with unequally bipinnate lvs. Lfts. ovate, acuminate. Fig. 480.

G. Canadénsis Lam.—Woods, N. Y. to Ill. and Tenn. 50f. Rac. greenish; seeds round, polished, brown, very hard, ½′ diam. May—July.

7. GLEDÍTSCHIA, L. HONEY LOCUST. (To *John G. Gleditsch*, a botanical writer, Leipzig.) Fls. ♀ ☿ ♂. Sep. equal, 3—5, united at base. Pet. 3—5. Stam. 3—5, distinct, opposite the sepals. Style short. Legume continuous, compressed, often intercepted between the seeds by a sweet pulp. ♄ With branched spines. Lvs. abruptly pinnate and bipinnate, often in the same specimen. Fls. small, green, racemous. Figs. 362, 401.

1 **G. triacánthus** L. Branches armed with stout, triple, or multiplex spines; lfts. alternate, oblong-lanceolate, obtuse; leg. linear-oblong, compressed, many-seeded. Pa. to Mo. and La. 40—70f. Wood very heavy. Pods 8—18′. May—July.

2 **G. monospérma** Walt. *Water Locust.* Spines few, mostly simple; lfts. ovate-oblong; pod broadly oval, without pulp, 1-seeded. Swamps, S. 30f.

8. CÁSSIA, L. SENNA. (Hebrew *Katzioth.*) Sep. 5, scarcely united at base, nearly equal. Pet. 5, unequal, but not papilionaceous. Stam. distinct, 10, or by abortion fewer, anth. opening by terminal pores, the three upper often sterile. Pod many-seeded, 1-celled or many-celled transversely. ♄ ♄ or herbs. Lvs. abruptly pinnate. Fls. mostly yellow. Fig. 357.

§ Stam. 5 or 10, all perfect. Sep. acute. Lfts. small. Stip. persistent.......Nos. 1, 2
§ Stam. 10, the 3 upper abortive. Sep. obtuse. Lfts. large. Stip. deciduous..(*a*)
a Gland on the petiole at or near the base..............................Nos. 3, 4
a Gland on the rachis between the two lowest leaflets.................Nos. 5, 6

1 **C. Chamæcrísta** L. *Sensitive Pea.* Lfts. 8—12 pairs, oblong-linear, obtuse, mucronate; fls. large, pedicellate, 2 or 4 in each fascicle; anth. 10, unequal, all fertile. ① Dry soils. 12—18′. Flowers large, 2 petals spotted. August.

2 **C. níctitans** L. *Wild Sensitive Plant.* Lfts. 6—15 pairs, oblong-linear, obtuse, mucronate, sessile; fls. small, 2 or 3 in each subsessile fascicle; stam. 5, subequal. ① Sandy soils. 1f. Flowers small (3″), pale yellow. July.

3 **C. Marilándica** L. *American Senna.* Lfts. 6—9 pairs, oblong-lanceolate, mucronate, an obovoid gland near the base of the common petiole: fls. racemed; pod curved, 12-20-seeded. ♃ Stony places. 4—5f. Flowers showy. August.

4 **C. occidentàlis** L. Lfts. 3—6 pairs, ovate or lance-ovate, sharply acuminate; fls. in short racemes; pod nearly straight, 25-40-seeded. ① Va. to Ga. 5—6f. July. §

5 **C. obtusifòlia** L. Lfts. about 6, obovate, obtuse; pod long (6′) and narrow, recurved, 20-40-seeded; seeds longitudinal. ① Dry soil, S. 1—4f. July, Aug.

6 **C. melanocárpa** Vogel. Shrubby; lfts. 2—3 pairs, narrowly lanceolate, acute, coriaceous; rac. as long as the leaves. Ga. §

9. CÉRCIS, L. JUDAS-TREE. RED-BUD. Calyx 5-toothed. Petals scarcely papilionaceous, distinct, wings longer than the banner and smaller than the keel petals. Stamens 10, distinct. Pod compressed. Seeds obovate. ♄ Leaves simple, appearing after the roseate flowers. Fig. 308.

1 **C. SILIQUÁSTRUM.** Lvs. round-reniform; flowers more open than in No. 2. Eur. ♄

2 C. Canadénsis L. Lvs. broadly ovate-cordate, acuminate, villous on the veins beneath. Mid. and W. States. 20—30f. Flowers covering the branchlets. April.

10. CLADÁSTRIS, Raf. YELLOW-WOOD. Cal. 5-toothed, teeth short, obtuse. Pet. of nearly equal length, those of the keel distinct and straight like the wings. Vex. large, roundish, reflexed. Stam. 10, distinct. Fil. glabrous, incurved. Leg. flat and thin, short-stiped, 5 or 6-seeded. ♄ With yellow wood, pinnate leaves, and pendulous clusters of white flowers.

   C. tinctòria Raf.—Hills, Ky. and Tenn. 20—40f. Lfts. 7—11, oval, pointed, 3′; rac. 6—10′, resembling Robinia. April, May.

11. BAPTÍSIA, Vent. WILD INDIGO. ($Bάπτω$, to dye; a use to which some species are applied.) Cal. 4–5-cleft half way, persistent. Pet. of about equal length, those of the keel nearly distinct and straight. Vex. orbicular, emarginate. Stam. 10, distinct, deciduous. Pod inflated, stipitate, many (or by abortion few)-seeded. ♃ Lvs. palmately 3-fol. or simple.

  § Leaves simple. Flowers yellow...Nos. 1, 2
  § Leaves 3-foliate.—Flowers blue, in few elongated racemes............ ....No. 3
            —Flowers white, in few elongated racemes..(a)
            —Flowers yellow, solitary or in short racemes..(b)
    a Stipules leaf-like, longer than the petioles. Hairy. Cream-white....Nos. 4, 5
    a Stipules much shorter, or not longer than the petioles. Glabrous....Nos. 6, 7
      b Pedicels not longer than the calyx. Drying dark.............Nos. 8—10
      b Pedicels much longer than the calyx. Drying bright........Nos. 11—13

1 B. perfoliàta R. Br. Glabrous and glaucous; lvs. large, oval-orbicular, perfoliate; fls. solitary, axillary. Pine woods, S. Car. Ga. 1–2f. Pod inflated. May—July.

2 B. simplicifòlia Croom. Lvs. broadly ovate, obtuse, sessile; rac. terminal, elongated, many-flowered. Quincy, Fla. 2—3f. Pod ovate. 6″. June.

3 B. austràlis R. Br. Petioles short; lfts. obovate or oblong, obtuse; stip. lanceolate; rac. long, erect; pod oblong-oval. Ohio River and S. 2—3f. Flowers large and showy, indigo blue. June—Aug.

4 B. leucophæa Nutt. Lfts. oblanceolate, varying to obovate; stip. triangular-ovate; rac. nodding, the many flowers turned to the upper side on their long pedicels; pod ovoid, inflated. Prairies, W. and S. 2—3f. Flowers large. April.

5 B. villòsa Ell. Lfts. lance-oblong, or oblanceolate; stip. lance-linear, persistent; rac. long, declining; bracts minute, deciduous; ped. not secund; leg. oblong. N. Car. to Ga.; rare. 2—3f. Plant of rough aspect, as well as No. 4. June, July.

6 B. leucántha T. & G. Lvs. petiolate; lfts. cuneiform-obovate, obtuse; stip. lance-linear, about as long as petioles; rac. elongated, erect; bracts caducous; pod inflated, stipitate. Prairies, &c. W. and S. 2—3f. Flowers large. May—July.

7 B. alba R. Br. Fastigiate-branched above; petioles slender; lfts. elliptic-oblanceolate, acute at base; stip. and bracts minute, caducous; rac. erect or nodding, on a long peduncle. In rich soils, Va. to Fla. 2—3f. March, April.

8 B. lanceolàta Ell. Much branched, bushy; lvs. subsessile; lfts. narrowly elliptic to oblanceolate, obtuse, petiolulate; fls. axillary, subsolitary, short-pedicelled; pod ovate-globous. Pine woods, S. 1½f. Flowers large, dull yellow. April, May.

  β. *stricta.* Erect, strict; lfts. obovate, very obtuse: rac. few-flwd., termin. La. Fla.

9 B. tinctòria R. Br. Glabrous, branching; lvs. subsessile; lfts. small, roundish-obovate, acute at base, very obtuse at apex; stip. setaceous, caducous; rac. loose, terminal; pod subglobous. Dry woods. 2f, bushy. Pod size of a pea. July—Sept.

10 B. microphýlla N. Smooth, bushy; lvs. small, 2-3-foliate below. simple, sessile above; stip. and bracts large, persistent; fls. small, axillary, and in terminal racemes. S. Car. to Fla. 2—3f. (B. stipulacea Ravenel.)

Order 43.—LEGUMINOSÆ.

11 B. Lecóntii T. & G. Pubescent; lvs. short-petioled; lfts. obovate-oblong; pedicels with 2 bractlets; bracts persistent; pod short-stiped; branches, stipules, and racemes as in No. 9. Ga. Fla. 2f. May.

12 B. Scrènae Curtis. Smooth, diffuse; lfts. oblong-obovate, cuneate; fls. in terminal racemes, the central longest. S. Car. 1—2f. Pod oblong.

13 B. megacárpa Chapm. Glabrous, slender; lvs. petioled; lfts. oval; rac. short and short-stalked; stip. and bracts minute, caducous; fls. nodding; pod large, globular, and much inflated. Ga. Fla.: rare. 2—3f. Pods 1½'.

12. THERMÓPSIS, R. Br.

(Named for its resemblance to the Egyptian Lupine—*L. Thermis*.) Vex. roundish, sides reflexed. Sta. persistent. Pod subsessile, linear-oblong, many-seeded. ♃ Rhizome creeping, stems with sheathing bracts at base. Leaves 3-foliate. Flowers large, yellow.

1 T. mollis M. A. Curtis. Pubescent, diffusely branched; lfts. obovate-oblong; stip. leafy, as long as the petioles; ped. shorter than calyx. Woods, N. Car. 2f. April.

2 T. fraxinifòlia Curt. Smoothish, slender, branching; petioles longer than the stipules; lfts. wedge-oblong; ped. as long as the flower. Mts. Tenn. Car. 2f. May.

3 T. Caroliniàna Curt. St. stout, simple; petioles as long as the ovate clasping stipules; lfts. obl.-obov.; fls. on short ped. with decid. bracts. Mts. N. Car. 4f. June.

13. FABA, Mœnch. COFFEE BEAN.

Fls. as in Vicia. Seeds oblong, with a long scar (hilum) on the narrower end, and leathery, tumid legumes. ① Lvs. equally pinnate, with the tendril obsolete (in the following species). Peduncle shorter than the flowers.

F. vulgàris Mœnch. St. rigidly erect, with very short axillary racemes; lfts. 2—4, oval entire; stipules semisagittate Gardens. From Egypt. 2—3f. Glaucous.

14. CICER ARIETÌNUM,

the CHICK PEA, rarely cultivated, may be known by its serrated leaflets, a character quite strange in this Order.

15. PISUM, L. PEA.

(Celtic *pis*, Lat. *pisum*, Eng. *pea*, Fr. *pois*.) Style dilated above, grooved on the back, villous and stigmatic on the inner side. Otherwise as in Lathyrus. ① Figs. 59, 60, 190.

P. satìvum L. Lfts. ovate, entire, usually 4; stip. ovate, semicordate at base, crenate, ped. several-flowered. Nativity unknown. Many varieties.

16. LÁTHYRUS, L.

Calyx campanulate, the two upper sepals shortest. Stam. diadelphous (9 and 1). Style flat, dilated above, ascending, bent at a right angle with the ovary, pubescent or villous along the inner side next the free stamen. Pod oblong, several-seeded. ♄ ♃ Leaves abruptly pinnate, of 1 to several pairs of leaflets. Petioles produced into tendrils. Peduncles axillary. Fig. 497.

* Native.—*a* Leaflets a single pair. Southern... No. 1
  —*a* Leaflets commonly 3 pairs. Perennial............... ......Nos. 2, 3, 4
  —*a* Leaflets commonly 5 pairs. Perennial.........................Nos. 5, 6
* Exotic.—*b* Leaflets a single pair..Nos. 7—9
  —*b* Leaflets 3 to 6 pairs. (Species of Orobus).................Nos. 10—12

1 L. pusillus Ell. St. winged; lfts. 2, linear-lanceolate, acute at each end; stip. conspicuous, lance-falcate, half-sagittate; ped. long. S. Car. to La. Purple. May.

2 L. ochroleùcus Hook. St. slender; lfts. broadly ovate; stip. semicordate, large; ped. 7-10-flowered, shorter than the leaves; fls. cream-white. Shades. N. 3f. June.

3 L. palústr s L. St. winged; stip. semisagittate, mucronate; lfts. 2 or 3 pairs, lance-linear or oblong, mucronate; ped. 3–5-flowered, equalling the leaves. Wet thickets, N. Eng. to Oreg. 1—2f. Blue-purple. June—July.

4 L. myrtifòlius Muhl. St. slender, 4-angled; lfts. elliptic-oblong, obtuse; stip. ovate, entire; ped. longer than lvs., 5-flwd. N. E. to Va. and Ind. 2—4f. Pale purp. Jl.

5 L. venòsus Muhl. St. 4-angled; stip. semisagittate, lanceolate, very small; ped. 8–16-flowered, shorter than the leaves; lfts. 4—7 pairs, somewhat alternate, obtusish, mucronate. Shady banks. 2—3f. Flowers large, purple. June, July.

6 L. marítimus Bw. *Beach Pea.* St. 4-angled, compressed; petioles flat above; stip. cordate-hastate, nearly as large as the 8—12 ovate leaflets; ped. many-flowered. Sandy shores, N. Y. to Oreg. 1—2f. Leaves pale green. Flowers blue. May, June.

7 L. LATIFÒLIUS. *Everlasting Pea.* Ped. many-flowered; lfts. 2, lanceolate, internodes membranous-winged. ♃ Eur. 6f. Flowers large, pink. July, Aug.

8 L. ODORÀTUS. *Sweet Pea.* Ped. 2-flowered; lfts. 2, ovate-oblong; leg. hirsute. ① Sicily. Flowers very large, fragrant, red-white. June.

9 L. SATÌVUS. *Chick Pea.* Ped. 1-flowered; lfts. 2—4; leg. ovate, compressed, with 2 winged margins at the back. ① S. Eur. An unhealthy food.

10 L. VERNUS. Lfts. 6, ovate, acuminate; fls. red-purple-blue. Europe. 1f. April.

11 L. NIGER. Lfts. 12, ovate-oblong; fls. dark purple. Europe. 3f. July.

12 L. ATROPURPÙREUS. Lfts. linear, 3 pairs, acute; fls. dark purple. Algiers. 1f. May.

17. **VÍCIA**, L. VETCH. (Celtic *gwig*, whence Gr. βικίον, Lat. *vicia*, Fr. *vesce*, and Eng. *vetch*.) Style filiform, bent at right angles with the ovary, villous beneath the stigma on the outside (next the keel). Otherwise nearly as in Lathyrus.

* Peduncles 1-2-flowered, shorter (in flower) than the leaves....................Nos. 1—3
* Peduncles 3-20-flowered.—*a* Leaflets 3—6, very narrow.......................No. 4
　　　　　　　　　　—*a* Leaflets 8—20.—*b* Stipules long-toothed..............No. 5
　　　　　　　　　　　　　　—*b* Stipules entire................Nos. 6—8

1 V. satìva L. *Vetch. Tares.* Fls. solitary or in pairs, subsessile; lfts. 10—12, oblong-obovate, often linear, retuse, mucronate; pod linear, erect, 4-8-seeded. ① Fields. 2—3f. Fls. 6″, pale purple. June. § Eur.

2 V. tetraspérma Loisel. Ped. 1-2-flowered, in fl. shorter (in fr. longer) than the lvs.; pod 4-seeded; lfts. 4—6, small, linear, obtuse. Fields, Can. to Penn. St. very slender, 1—2f. Fls. bluish-white. Pod 5″. July.

3 V. micrántha N. Lfts. 4—6, linear, acute, obtuse or retuse; fls. mostly solitary, minute, pale; pod 1′, sabre-shaped, erect, 6-10-seeded; seeds black. S. 2—3f.

4 V. acutifòlia Ell. Leaflets 3—6, linear, acute; stip. lance-linear; tendrils mostly simple; rac. 3-9-flowered, longer than the leaves. Ga. Fla. 2—4f. Whitish.

5 V. Americàna Muhl. Ped. 4-8-flowered, shorter than the lvs.; stip. semisagittate, deeply dentate; lfts. 10—14, elliptic-lanceolate, obtuse; pod oblong-linear, compressed, reticulated. N. Y. westward. 1—3f. Blue-purple. May.

6 V. Caroliniàna Walt. Pedicel 6-12-flowered, rather shorter than the leaves; fls. loose; calyx teeth very short; stip. lance-linear; lfts. 8-12, linear-oblong or linear, smoothish; pod oblong. Woods and banks. 4—6f. Pale purple. May.

7 V. Cracca L. *Tufted Vetch.* Fls. imbricated, 12—20 or more in the raceme; lfts. 12—24, oblong, puberulent; stip. semisagittate, linear-subulate, entire. Thickets. 2—3f. Flowers blue-purple, 4″. July.

8 V. hirsùta Koch. Hairy; lfts. 8—20, linear, truncate, mucronate; ped. 3-6-flwd. shorter than leaves; leg. hirsute, 2-seeded. Fields. 1—3f. June. §

18. **ZÓRNIA**, Gmel. (For *John Zorne*, M. D., of Bavaria.) Calyx bilabiate, upper lip obtuse, emarginate, lower 3-cleft. Vex. orbicular, with the sides revolute. Sta. monadelphous, the alternate anthers different. Pod

ORDER 43.—LEGUMINOSÆ. 87

compressed, of 2–5 roundish joints. ♃ Lvs. palmately 2–4-foliate with sagittate stip., which are enlarged above and supply the place of bracts.

Z. tetraphýlla Mx. Lfts. 4; stip. or bracts oval, acute: pod aculeate, about 3-jointed. ♃ N. Car. to Fla. and Tex. 1–2f. Deep yellow. Pods adhesive. June—Aug.

19. **ÆSCHYNÓMENE**, L. (Αἰσχύνομαι, to be modest; alluding to its sensitive property.) Calyx bilabiate, bibracteolate; upper lip bifid, lower trifid. Vex. roundish. Stamens diadelphous, 5 in each set. Pod exserted, composed of several truncated, separable, 1-seeded joints.—Lvs. odd-pinnate. Stip. semisagittate. Rac. axillary (yellow). August.

1 **Æ. híspida** Willd. Erect, scabrous; lfts. very smooth, 27—37, oblong-linear, obtuse; rac. 3–5-flowered; pod 6–9-jointed. ① Marshes, Pa. and S. 2—3f.

2 **Æ. viscídula** Mx. Slender, procumbent, viscidly pubescent; lfts. 7—11, obovate; ped. filiform, 1 or 2-flowered; pod 2 or 3-jointed. ① Sandy fields, S.

20. **CHAPMÁNIA**, T. & G. (To *A. W. Chapman*, M. D., author of "Flora of the Southern States.") Fls. nearly as in Stylosanthes. Cor. inserted on the throat of the calyx. Keel 2-cleft at apex. Anth. alike, oblong. Leg. hispid, 1–2-jointed.—A viscid-hirsute branching herb. Leaves pinnately 3–7-foliate. Fls. small, yellow, in terminal racemes.

C. Floridàna T. & G.—E. Fla. 2—3f. Lfts. oblong.

21. **STYLOSÁNTHES**, Swartz. (Στῦλος, a style, ἄνθος.) Fls. of two kinds. ♂ Calyx bibracteolate at base, the tube slender and stalk-like, with the corolla inserted on its throat. Vex. orbicular. Sta. 10, monadelphous. Ov. sterile, with a filiform style. ♀ Cal. and corolla 0. Ov. between 2 bracteoles. Leg. 1–2-jointed, uncinate with the short, persistent style.—Lvs. pinnately trifoliate.

S. elàtior Swartz. *Pencil Flower.* St. pubescent on one side; lfts. lanceolate, smooth, acute; spikes 3-4-flowered; loment 1-seeded (lower joint abortive). ♃ Dry, gravelly woods, Long Isl. to Fla. 1f. Fls. yellow. July, August.

22. **ÁRACHIS**, Willd. PEANUT. (Lat. *aracos*, used by Pliny to designate some subterranean plant.) Calyx bilabiate. Cor. resupinate. St. monadelphous. Pod gibbous at base, coriaceous, veiny, turgid, and indehiscent, the joints not separating.—S. American herbs, with equally pinnate leaves and yellow flowers.

A. hypogæa Willd. Leaflets 2 pairs, oval or roundish, cuneate at base; stip. entire, lance-subulate, as long as the leaflets; fruit subterranean. Cult. South.

23. **CORONÍLLA**, L. (Lat. *corona*, a crown; from the inflorescence.) Calyx bilabiate. Petals unguiculate. Loment somewhat terete, jointed. Seeds mostly cylindrical. ♄ ♃ Lvs. unequally pinnate. Fls. in simple, pedunculate umbels, rose-colored.

1 **C. ÉMERUS.** *Scorpion Senna.* St. woody, angular; ped. about 3-flowered; claws of the petals thrice longer than the calyx. France. 3f. May.

2 **C. VÀRIA.** Herbaceous; lfts. 11—19, oblong; ped. 10-15-flwd. Eur. ♃—1f. Jl.—Sept.

24. **HEDÝSARUM**, L. (ἡδύς, sweet, ἄρωμα, smell.) Calyx cleft into 5 linear-subulate, subequal segments. Keel obliquely truncate, longer

than the wings. Sta. diadelphous (9 and 1), and, with the style, abruply bent near the summit. Pod (loment) of several 1-seeded joints connected by their middle. ♃ Leaves unequally pinnate.

H. boreàle N. Erect; lfts. 13—21, oblong; stip. united, sheathing; flowers deflexed, spiked on the long peduncle, violet-purple; pod of 1—4 lens-shaped, veiny joints. Rocks, Willoughby Lake, Vt. and N. 1—2f. Flowers large. June, July.

25 DESMÒDIUM, DC. Bush Trefoil.

Calyx more or less bilabiate. Vex. roundish, keel obtuse. Sta. diadelphous (9 and 1) sometimes monadelphous. Pod (loment) compressed, jointed, constricted most on the lower (dorsal) suture, the joints 1-seeded, separable, mostly aculeate and adhesive. ♃ ♭ Leaves pinnately trifoliate. Flowers in racemes or often large, loose panicles, purplish, in Summer. Figs. 191, 355.

§ Legumes distinctly stiped, the stipes about as long as the joints...(a)
    a Stems prostrate, creeping. Leaflets round or oval..............Nos. 1—2½
    a Stems erect. Leaflets broadly ovate, or (in No. 6) narrowly...(b)
        b Calyx teeth shorter than the tube......................Nos. 3—5
        b Calyx teeth longer than the tube,—upper one notched...Nos. 6—8
                          —upper one entire.........No. 9
§ Legumes subsessile, the stipes, if any, not exceeding the calyx...(c)
    c Bracts large, covering the flower buds, caducous...(d)
    c Bracts inconspicuous, smaller than the flower buds...(e)
        d Stipules large (6—9" long), ovate-lanceolate...........Nos. 10, 11
        d Stipules quite small, subulate ....Nos. 12, 13
            e Leaflets large (2—3' by 1—2'), oblong-ovate...Nos. 14, 15
            e Leaflets small, orbicular or oval.............Nos. 16—18
            e Leaflets long, linear...No. 19. Lfts. oblong.†...No. 20

1 **D. rotundifòlium** DC. Plant prostrate, downy; leaflets suborbicular; bracts and stipules broadly ovate, acuminate; racemes few-flowered; loment constricted on both margins nearly alike. Rocky woods. 2—3f. Purplish. August.

2 **D. ochroleùcum** Curt. Plant decumbent, smoothish; lfts. ovate, rarely single; stip. ovate, pointed; raceme long, fls. white; loment twisted. Woods, Md. & S. (Porter).

3 **D. nudiflòrum** DC. Lfts. roundish ovate, bluntly acuminate, slightly glaucous beneath; scape radical, panicled, smooth; joints of the loment obtusely triangular. Woods, com. St. 1f, scape 2—3f, with many small purple flowers.

4 **D. acuminàtum** DC. Plant erect, simple, pubescent, leafy only at top; leaflets ovate, long-acuminate, the odd one round-rhomboidal; pan. terminal, on a very long peduncle. Woods, com. 8—12', the panicle 2—3f. Fls. small, flesh-color. Pod 3-jointed.

5 **D. pauciflòrum** DC. St. assurgent, leafy all the way, retrorsely hairy; lfts. thin, obliquely ovate, acutish, terminal one rhomboidal; rac. terminal, the flowers few, in pairs; petals all distinct, spreading. Woods, N. Y. to Ill. and La. 1f. Whitish.

6 **D. paniculàtum** DC. Erect, slender, nearly glabrous; lfts. oblong-lanceolate, obtuse; stip. subulate, deciduous; fls. on slender pedicels in panicled racemes; loment of about 3 triangular joints. Woods, common, 2—3f. Purple.

7 **D. viridiflòrum** Beck. Densely pubescent; lfts. ovate, scabrous above, whitened beneath; stip. lance-ovate, acuminate; pan. naked, very long; pod of 3 or 4 triangular joints. Alluvion, N. Y. and S. 3—4f, rigid. Violet, fading to green.

8 **D. lævigàtum** DC. Glabrous, or nearly so; lfts. ovate; panicle subsimple, pedicels slender, in pairs. Woods, N. J., and S. 2—3f. Purple.

    β. *monophýllum*. Dwarf, simple; lower lvs. 1-foliate. Uxbridge. Ms. 1f. (Ricard.)

ORDER 43.—LEGUMINOSÆ. 89

9 **D. glabéllum** DC. St. smoothish; lfts. ovate, small, rough-pubescent on both sides; pod of 3 or 4 triangular, minutely hispid joints. Shades, Car.

10 **D. cuspidàtum** T. & G. Smooth; lfts. oblong-oval, or ovate, sharply acuminate; bracts deciduous, ovate, acuminate; joints of the loment suboval. Woods. 3—5f. Stipules and bracts 9″. Flowers 8″, purple.

11 **D. canéscens** DC. St. striate, scabrous; lfts. ovate, rather obtuse, scabrous on the upper surface, soft-villous beneath; pan. densely canescent, naked; joints of the loment 4, obliquely oval, hispid. Woods. 3f.

12 **D. Canadénse** DC. St. pubescent; lfts. oblong-lanceolate, obtuse, nearly smooth; stip. filiform; bracts ovate, long-acuminate; joints of the loment obtusely triangular, hispid. Woods, Can. to Pa. and W. 3f. Flowers 8″, purple.

13 **D. sessilifòlium** T. & G. St. tomentous-pubescent; lvs. sessile; lfts. linear or linear-oblong, obtuse at each end, scabrous above, softly tomentous beneath; stip. subulate; pod of 2—3 semiorbicular joints. Woods, W. 2—3f. Fls. small, crowded.

14 **D. Dillènii** Darl. Branching, hairy; lfts. oblong, villous beneath; stip. subulate; rac. panicled; joints of the loment 3, rhomboidal, reticulate, a little hairy, connected by a narrow neck. Moist soils, N. and W. 2—3f. Purple.

15 **D. rígidum** DC. Scabrous, pubescent; lfts. ovate-oblong, obtuse; petioles short, hairy; stip. ovate-acuminate, ciliate, caducous; leg. with 2—4 obliquely obovate joints. Hills and woods, Mass. to La. 2—3f. (D. Floridanum Chapm.)

16 **D. ciliàre** DC. Erect, slender, scabrous-pubescent; lvs. crowded, on short, hairy petioles; lfts. small, ovate, ciliate on the margin; joints of the short-stiped loment 2 or 3. Woods. 2f. Purple.

17 **D. Marilándicum** Boott. Erect, slender, nearly smooth; lfts. ovate, obtuse, subcordate at base, the lateral ones as long as the petioles; loment stipe as long as the calyx, joints 1 or 2. Woods. 2—3f. Violet.

18 **D. lineàtum** DC. Slender, reclining; st. finely striate with colored lines; lfts. small, roundish oval, smoothish, green both sides; pod quite sessile in the calyx, joints about 2. Dry woods. 2 or 3f.

19 **D. strictum** DC. Slender, nearly glabrous; lvs. petiolate; lfts. linear, elongated; pan. few-flowered; pod hispid, incurved, of 1—3 lunately triangular joints, with a filiform isthmus. Pine woods, N. J. and S. 3f.

20 **D. gyrans.** *Moving-plant.* Lateral lfts. very small; pods pendulous. From Bengal Wonderful for the leaves, which in warm weather are always in motion.

26. **LESPEDÈZA**, Mx. BUSH CLOVER. Calyx 5-parted, bibracteolate, segments nearly equal. Keel of the corolla very obtuse, on slender claws. Pod (loment) lenticular, compressed, small, unarmed, indehiscent, 1-seeded. ♃ Leaves pinnately trifoliate, reticulately veined. Summer.

§ Fls. all complete. Calyx villous, long. Cor. whitish with a purple spot....Nos. 1, 2
§ Fls. partly apetalous. Calyx short. Corolla violet.—a Stems upright......Nos. 3, 4
—a Stems prostrate........No. 5

1 **L. capitàta** Mx. *Bush Clover.* Lfts. elliptical to linear, silky beneath; stip. subulate; fascicles of flowers ovate, subcapitate, shorter than the leaves, axillary; loments hairy, shorter than the villous calyx. Dry soils, Can. to Car. 2—4f.

2 **L. hirta** Ell. Stem villous; lfts. roundish oval, pubescent beneath; rac. capitate, axillary, oblong, longer than the leaves; corolla and pod about as long as the calyx. Dry woods. 2—4f. Flowers reddish-white.

3 **L. Steùvi** Nutt. Branched and bushy, tomentous or pubescent; lfts. oval-obovate or roundish, longer than the petiole; rac. axillary, capitate or loose; pod villous-pubescent. Dry soils, Mass. to Ga. 2f. Variable.

4 **L. violàcea** Pers. Erect or diffuse, sparingly pubescent; lfts. oval, varying to oblong and linear, obtuse, mucronate, as long as the petioles; rac. axillary, few-flowered, the apetalous ones generally below. Dry woods. Leaflets 1′.

ORDER 43.—LEGUMINOSÆ.

β. *sessiliflòra.* Flowers many, in clusters shorter than the leaves.
γ. *reticulàta.* Leaflets linear, rigid; flowers in short fascicles. Erect.
δ. *divérgens.* Leaflets ovate; upper peduncle longer than the leaves.

5 L. procúmbens Mx. St. prostrate, diffuse, tomentous-pubescent; lfts. oval or obovate-elliptical, smooth above, on very short petioles; ped. filiform, few-flowered; pod roundish. Dry soils. Leaflets 5—9″. Ted. 2—5′.
β. *repens.* Nearly smooth and very slender; leaflets oval or elliptical.
γ. *Feayàna.* Decumbent; leaflets obovate upper ped. apetalous. South.

27 SPÁRTIUM, L. COMMON BROOM. (Σπάρτον, a rope; formerly made of the Broom.) Calyx spathe-like, split behind, teeth very short. Keel incurved, acuminate, longer than the wings. Otherwise like Genista.

S. SCOPÀRIUM.—Shrub native of Spain, 6f, with rush-like erect branches often leafless. Leaves simple (if any), oblong. Flowers showy, yellow or white.

28. GENÍSTA, L. DYER'S BROOM. WOAD-WAXEN. (Celtic *gen*, Fr. *genet;* a small shrub.) Calyx with the upper lip 2-parted and the lower 3-toothed. Vex. oblong. Keel oblong, scarcely including the stamens and style. Stigma involute. Stamens monadelphous. ♄ With simple leaves and yellow flowers.

G. tinctòria L. Branches round, striate, unarmed, erect; lvs. lanceolate, smooth pod smooth. Dry hills, Mass. N.Y. 1f. August. § Europe.

29. CROTALÀRIA, L. RATTLE-POD. (Κρόταλον, a rattle; from the rattling of the loose seeds in the horny pod.) Calyx 5-cleft, somewhat bilabiate. Vex. cordate, large. Keel acuminate. Stam. 10, monadelphous. Filamentous sheath cleft on the upper side. Pod pedicellate, turgid.— Herbs or shrubs. Lvs. simple or palmately compound. Flowers yellow.

1 C. sagittàlis L. Annual, erect, branching, hairy; lvs. lance-oval to lance-linear; stip. acuminate, decurrent; rac. 3-flowered, opposite to the leaves; cor. shorter than the calyx. Sandy fields. 6—12′. Cor. small. July.
2 C. ovàlis Ph. Perennial, hairy, diffuse; lvs. oval and elliptic; stip. small or minute, partly decurrent; pedicels long, 3-6-flowered; corolla longer than the calyx. Sandy woods, S. 4—12′. Flowers showy. April, May.
3 C. Púrshii DC. Perennial; slender, assurgent, nearly smooth; lvs. oblong-linear or linear, subsessile; stip. narrowly decurrent through the whole internode; pedicels 5-7-flowered; corolla as long as the calyx. Damp shades, S. 1—1½f.

30. LUPÌNUS, Tourn. LUPINE. (Lat. *lupus*, wolfish as a weed?) Cal. deeply bilabiate; upper lip 2-cleft, lower entire or 3-toothed. Wings united at the summit. Keel falcate, acuminate. Stam. monadelphous, the sheath entire. Anth. alternately oblong and globous. Pod compressed. ①♃ ♄ Leaves palmately 5-15-foliate, rarely unifoliate. Raceme terminal.

1 L. villòsus Willd. Unifoliate, densely silky-tomentous; sts. decumbent-assurgent; lvs. large, elliptic-oblong, long-petioled; rac. terminal, long, dense-flwd. Pine woods, S. 1—2f. Flowers roseate, with a purple spot. Pods very woolly. April—June.
2 L. diffùsus N. Diffusely branched from the base; lvs. oval-oblong, obtuse, soft-silky, on short petioles; pods very silky. Sands, S. Blue-purple. April.
3 L. perénnis L. Minutely pubescent, 5-7-foliate; lfts. oblanceolate, mucronate: fls. alternate; calyx without appendages, upper lip emarginate, lower entire. Sandy hills. 1L Flowers blue, varying to white. May, June. Cultivated.

ORDER 43.—LEGUMINOSÆ. 91

4 **L.** POLYPHÝLLUS. Lfts. 11—15, lanceolate; calyx lips subentire. Oreg. 3f. Purp.-wh.
5 **L.** NOOTKATÉNSIS. Villous; lfts. 5—9, oblong; cal. lips subentire. N.W. Coast. 2f. Pur.
6 **L.** HARTWÉGII. Hairy; lfts. 7—9, obl. obtuse; stip. and bracts setaceous. Mex. Blue
7 **L.** VÀRIUS. Small and delicate; calyx appendaged, lips 2-fid and 3-fid. Blue.

31. LABÚRNUM, Benth. Calyx campanulate, bilabiate, upper lip 2-, lower 3-toothed. Vex. ovate, erect, as long as the straight wings. Fil. diadelphous (9 and 1). Leg. continuous, tapering to the base, several-seeded. ♄♃, Leaves palmately trifoliate. Flowers mostly yellow.

1 **L.** VULGÀRE L. *Golden Chain.* Arborescent; lfts. oblong-ovate, acute at base, acuminate; raceme elongated (1f), pendulous; legume hirsute. Europe. 15f.
2 **L.** ALPÌNUM L. Arborescent; lfts. oblong-ovate, rounded at base; raceme long, simple, pendulous; legume glabrous. Alps. 30f.

32. TRIFÒLIUM, Tourn. CLOVER. ($T\rho\iota\varphi\acute{\upsilon}\lambda\lambda o\nu$ (three-leaved), Lat. *trifolium;* Fr. *trèfle;* Eng. *trefoil.*) Calyx 5-toothed. Pet. united at the base, withering. Vex. reflexed. Alæ oblong, shorter than the vexillum. Carina shorter than the alæ. Stam. 10, diadelphous (9 and 1). Legume mostly indehiscent, covered by and scarcely longer than the calyx, 1-4-seeded. Seeds roundish.—Herbs. Leaves palmately trifoliate. Leaflets with straight veinlets. Flowers in heads or spikes. Figs. 233, 354.

§ Flowers yellow, in small, dense, roundish heads. Legume 1-seeded..........Nos. 1, 2
§ Flowers cyanic,—*c* pedicellate, finally deflexed...(*a*)
　　　　　　　　—*c* subsessile, never deflexed...(*b*)
　a Heads small, on stalks some ten times longer. Legume 4-seeded......Nos. 3, 4
　a Heads large, on stalks two or three times longer......................Nos. 5, 6
　　b Calyx teeth plumose, longer than the whitish corolla..................No. 7
　　b Calyx teeth shorter than the purple or roseate corolla............Nos. 8—10

1 **T. procúmbens** L. *Yellow Clover.* St. procumbent or ascending; lfts. denticulate, terminal one stalked; stip. ovate-lanceolate, acuminate, much shorter than the petioles; heads small, subglobous; style short. ① Dry soils, N. H. to Va. 1—2f. Jn.
2 **T. agràrium** L. St. ascending or erect; lfts. denticulate, all subsessile; stipules linear-lanceolate, cohering with and longer than the petiole; heads ovoid-elliptic; style equalling the pod. ① Dry fields, N. H. to Va. 1f. July.
3 **T. Caroliniànum** Mx. Slender, diffuse; lfts. cuneate-obovate, the m'dle one obcordate; stip. ovate-acuminate, foliaceous; cal. teeth thrice longer than its tube. ① Fields, S.
4 **T. repens** L. *White Clover. Shamrock.* St. creeping, diffuse; lfts. obcordate, denticulate; stip. narrow, scarious; cal. teeth shorter than the tube. ♃ Pastures, &c.
5 **T. reflexum** L. *Buffalo Clover.* Pubescent; ascending or procumbent; lfts. obovate, serrulate; stip. leafy, semicordate; cal. teeth nearly as long as the corolla; leg. 4-seeded. ② Prairies, W. and S. 8—16'. April—June.
6 **T. stoloníferum** Muhl. Glabrous, creeping; lfts. broadly obcordate, denticulate; stip. leafy, ovato-lanceolate; cal. teeth not half the length of the corolla; legume 2-seeded. W. States. 6—12'. May, June.
7 **T. arvénse** L. Hds. cylindrical, very hairy; cal. teeth setaceous, longer than the cor.; leaflets narrow-obovate. (i) Dry, sandy fields. 5—10'. June—Aug. § Eur.
8 **T. praténse** L. *Red Clover.* Ascending, thinly hirsute; lfts. spotted, oval, entire; stip. ovate, cuspidate-acuminate; heads sessile; lower tooth of the cal. longer than the four others which are equal. ♃ Fields and meadows. 2f.
9 **T. médium** L. *Zig-zag Clover.* St. suberect, branching, flexuous, nearly glabrous; lfts. not spotted, oblong, subentire; stip. lanceolate, acuminate; heads ovoid-globous, peduncilate; cal. teeth setaceous, hairy. ♃ Hills, N. § Eur

10 T. incarnàtum. St. erect, flexuous; lfts. round-ovate, obtuse o₁ obcordate, villous; spike dense, oblong, pedunculate. Italy. 2f. Red.

33. MEDICÀGO, L. MEDICK. Calyx 5-cleft. Cor. deciduous. Vex. free and remote from the keel. Leg. variously curved, or spirally coiled or twisted.—Lvs. pinnately 3-foliate, denticulate. European.

* Pods smooth..........Nos. 1, 2, 3.   ** Pods spiny..........Nos. 4, 5, 6.

1 M. lupulìna L. *None-such.* Procumbent, pubescent; lfts. wedge-obovate; fls. yellow; pod reniform, 1-seeded. ① Waste grounds. 6—20'. May—July. §
2 M. satìva L. *Lucerne.* Erect, glabrous; lfts. oblong-lanceolate; stip. lance-linear; fls. violet-purple, large; pod spiral. ⚰ Fields : rare. 2—3f. June, July. §
3 M. scutellàta L. *Snails.* Lfts. elliptical and obovate; ped. 1-3-flowered, shorter than the leaf; pod coiled like a snail-shell. Gardens. July. §
4 M. denticulàta Willd. Lfts. obovate; stip. bristly-gashed; ped. with 1—3 yellow flowers; pod loosely spiral, border doubly echinate. ① 1—2f. June. §
5 M. maculàta Willd. Lfts. obcordate, with a purple spot; ped. 2-3-flowered; pod compactly spiral, outer edge grooved and doubly spiny. ① §
6 M. intertéxta L. *Hedgehog.* Lfts. rhomboidal; stip. gashed; pod spirally coiled in 5 or 6 turns, bordered with bristly prickles. Rare. §

34. MELILÒTUS, Tourn. MELILOT. Legume ovoid, wrinkled, longer than the calyx, 1-2-seeded. Fls. as in Trifolium. ①② Leaves pinnately trifoliate, leaflets toothed. Flowers in racemes. June, July.

1 M. officinàlis Willd. Fls. yellow; lfts. obovate-oblong, obtuse; stem erect, with spreading branches. Alluvion. 3f. Raceme slender, one-sided. §
2 M. alba Lam. *Sweet-scented Clover.* Fls. white; lfts. ovate-oblong, truncate, mucronate; vex. longer than the other petals. Fields. 4—6f.

35. PSORÀLEA. Cal. 5-cleft, campanulate. Segm. acuminate, lower one longest. Stam. diadelphous, rarely somewhat monadelphous. Pod as long as the calyx, 1-seeded, indehiscent. ⚰ ♄ Often glandular-dotted. Stip. cohering with the base of the petiole. Flowers cyanic.

* Leaves, at least the upper ones, 1-foliate, lowest 3-foliate...................Nos. 1, 2
* Leaves all pinnately 3-foliate..Nos. 3, 4, 5
* Leaves pinnately 19-21-foliate.. ....No. 6
* Leaves palmately 3-7-foliate.—*a* Silky or smooth. Fls. loosely spicate...Nos. 7, 8, 9
  —*a* Villous. Flowers densely capitate......Nos. 10, 11

1 P. canéscens Mx. Bushy, downy-canescent; lower lvs. palmately 3-foliate; lfts. roundish obovate, dotted, upper simple. Woods, S. 2f.
2 P. virgàta N. Virgate, smoothish; *lowest* lvs. pinnately 3-foliate; lfts. linear or oblong, often all simple; spikes rather dense. Ga. 2f.
3 P. stipulàta T. & G. Smoothish; lfts. elliptic-ovate, obtuse; stipules large, ovate; ped. as long as the leaves; spikes capitate. Falls of Ohio, Ky.
4 P. melilotòides Mx. Smoothish; lfts. lance-oblong, obtuse; stip. lanceolate; ped. much longer than the leaves. Dry soils, S. and W. 2f.
5 P. Onóbrychis N. Pubescent; lfts. ovate, acuminate; stipules filiform; ped. long, with slender spikes. Thickets, W. 3—5f. June, July.
6 P. multijùga Ell. Lfts. numerous, oblong-lanceolate, obtuse; spikes oblong; calyx villous, with long teeth. Upper country. Car. Ga.
7 P. Lupinéllus Mx. Slender, glabrous; lfts. 5—7, linear-filiform; rac. **elongated**; fls. violet; pod S-shaped. Woods, S. 2f. May. June.

ORDER 43.—LEGUMINOSÆ 93

8 P. floribúnda N. Canescent; lfts. 3, rarely 5, dotted, oblong to linear; rac. slender; ped. as long as the flowers (3''); pod smooth. Ill. and W. 3f. June.
9 P. argophýlla Ph. Erect, silky-white; lfts. elliptic, obtuse, 5, rarely 3; ped. much longer than the leaves; fls. whorled. Wis. to Dakota (Matthews.)
10 P. subacaúlis T. & G. Nearly stemless, hirsute; lvs. 7-foliate on very long petioles; lfts. obovate-oblong; ped. long, rigid; cal. teeth obtuse. Tenn. April.
11 P. esculénta Ph. Erect, rigid, diffuse, white-haired; lfts. 5, oblanceolate; petioles long, ped. longer (3'); head oblong; sep. and bracts long, pointed. Minn. to Dakota (Matthews, Colman.) 1f. Tubers farinaceous.

36. SESBÀNIA, Pers. Calyx bell-shaped. Vex. spreading or reflexed. Keel incurved, with long claws. Leg. linear or oblong, ∞- or few-seeded. Seeds transverse.—Lvs. abruptly pinnate, with many leaflets. Raceme axillary, loose (yellowish). Fig. 356.

1 S. macrocárpa Muhl. Tall, glabrous; lfts. oblong-linear, 20—30; pod linear, long, jointed, many-seeded. ① Damp. S. 3—9f. Pods 1f. Aug.—Oct.
2 S. platycárpa Pers. Tall, glabrous; lfts. as above; pod oblong-elliptic, valves double, the inner membranous, 2-seeded. ① S. 10f. Aug. (Glottidium Flor. DC.)

37. AMÓRPHA, L. LEAD PLANT. Calyx 5-cleft. Vex. concave, unguiculate, erect. Wings and keel none. Stam. exserted. Leg. oblong, somewhat curved at the point, scabrous with glandular points, 1 or 2-seeded. ♭♃ American. Lvs. unequally pinnate, punctate. Fls. bluish-white, small, in virgate racemes.

 * Leaves stalked (lowest leaflets remote from base). Legume 2-seeded..........No. 1
 * Leaves sessile or nearly so. Lfts. 16—20 pairs. Legume 1-seeded.........Nos. 2, 3

1 A. fruticósa L. Scarcely pubescent; lfts. 9—19, oval, obtuse (1'); cal. teeth short, obtuse, the lowest pointed. W. and S. to Rocky Mts. 6—16f. May, June.
2 A. herbàcea Walt. Pubescent or not; lfts. 41—51, oblong, obtuse (7''); cal. teeth subequal, villous, upper obtuse, lower acute. South. 2—4f. June, July.
3 A. canéscens N. Villous-canescent; lfts. small (4''), crowded, ovate-oblong; vex. bright blue; calyx teeth equal, acute. Wis. to Ga. and W. 2—4f. July, Aug.

38. DÀLEA, L. Calyx subequally cleft or toothed. Pet. unguiculate, claws of the wings and keel adnate to the staminate tube half way up. Vex. free, the limb cordate. Sta. 10, united into a cleft tube. Ov. 2-ovuled. Pod enclosed in the calyx, indehiscent, 1-seeded.—Glandular-punctate. Lvs. odd-pinnate. Stipels 0. Stip. minute, setaceous. Spikes mostly dense.

D. alopecuroides Willd. Glabrous and much branched; lfts. 8—11 pairs, linear-oval, obtuse or retuse, punctate beneath; spike pedunculate, oblong-cylindric, silky-villous. ① Ill. to Ala. and W. 2f. Flowers white and violet. August.

39. PETALOSTÈMON, Mx. Calyx 5-toothed, nearly equal. Pet. 5, on filiform claws, 4 of them nearly equal, alternate with the stamens and united with the staminate tube. Stam. 5, monadelphous, tube cleft. Leg. 1-seeded, indehiscent, included in the calyx. ♃ Leaves unequally pinnate, exstipellate. Flowers in dense, pedunculate, oblong spikes or heads.

 § KUHNISTERA Lam. Heads corymbed, each with an involucre of scales; calyx teeth long, plumous, pappus-like, setaceous............................No. 5
 § PETALOSTEMON *proper*. Spikes solitary, not involucrate. Calyx teeth short...(*a*)
  a Bracts awn-pointed, longer than the calyx. West................Nos. 1, 2
  a Bracts not awned, short, acute or obtuse. South........... . Nos. 3, 4

1 P. cándidum Mx. Glabrous, erect; lfts. 7—9, all sessile, linear-lanceolate, mucronate, glandular beneath; spikes on long peduncles; bracts longer than the white petals. Dry prairies, S. and W. Slender. 3f. Leaflets 1'. July.

2 P. violàceum Mx. Minutely pubescent, erect; lfts. 5, linear, glandular beneath; spikes pedunculate; bracts shorter than the violet petals. Prairies, West. 2f. Leaflets 1'. Heads 1' long, brilliant. July, Aug.

3 P. cárneum Mx. Glabrous, erect; lfts. 5—7, lance-linear; spikes oblong, pedunculate; bracts obovate; pet. oblong. Ga. and Fla. Slender. 1—2f. Rose-wh. Aug.

4 P. grácile Nutt. Glabrous, decumbent at base; lfts. 7, lance-linear; spikes somewhat sessile; bracts acute; petals ovate. Pine woods, Fla. and W. 1—2f. White.

5 P. corymbòsum Mx. St. corymbously branched; spikes capitate, sessile; bracts broad, colored, the outer leaf-bearing; lfts. linear, 5—7. South. 2f. White. Sept.

40. ASTRÁGALUS, L. MILK VETCH.
Calyx 5-toothed. Pet. elongated, erect, clawed. Vex. narrow, equalling or exceeding the obtuse keel. Stam. diadelphous (9 and 1). Legume mostly turgid, 2-valved, 1-celled, or 2-celled partly or completely by the intrusion of the sutures. Seeds 1— ∞, funiculus slender. ♃ chiefly. Leaves unequally pinnate. Flowers in spikes or racemes. (Including Phaca, L.)

§ Legume abruptly stipitate, oblong, straight-pubescent..................... Nos. 1, 2
§ Legume sessile in the calyx.—*a* Fls. white or yellowish...(*b*)
  —*a* Fls. blue or tipped with blue...(*c*)
  b Legume straight, ovoid-oblong, smooth, dry, turgid............. ....Nos. 3, 4
  b Legume curved, oblong, woolly or veiny, dry, flattened.................Nos. 5, 6
  c Legume curved, crescent-shaped, 1-celled, smooth....................No. 7
  c Legume globular, fleshy; when dry splitting into two.... ....Nos. 8, 9

1 A. Robbínsii Oakes. Erect; lfts. 5—11, elliptical; cor. white, twice longer than the calyx; pod puberulent, 1-celled. Rocky shores, Vt. Rare. 8—14'. Cor. white, 5". May, June.

2 A. alpìnus L. Diffuse; lfts. 13—21, ovate; cor. blue above, thrice longer than the calyx; pod pubescent with black hairs, 2-celled. Mts. Vt. Me. Can. June, July.

3 A. Canadénsis L. Canescent, tall; lfts. 21—31, elliptical; bracts as long as the calyx; fls. greenish; pod 2-celled. Banks. 2—3f. Pod 6". July, August.

4 A. Coóperi Gray. Smoothish; lfts. 13—27, elliptical; rac. exceeding the leaves; fls. white; pod inflated, 1-celled, roundish-ovate, with a deep groove at the ventral suture. Banks, N. Y. and W. 1—2f. June, July. (Phaca neglecta T. & G.)

5 A. glaber Mx. Erect, smoothish; lfts. 15—23, lance-oblong or linear; spikes loose; pod smooth, flattened, 2-celled. Pine woods, S. 1—2f. Flowers greenish. July.

6 A. villòsus Mx. Low, villous; lfts. 9—15, oblong-oval; rac. ovoid, dense; pod 3-angled, 1-celled, clothed with long hairs. Dry, S. 3—6'. Fls. dull yellow. Mar. Apr.

7 A. obcordàtus Ell. Low, assurgent, smoothish; lfts. 7—12 pairs, 4", oblong to obovate, cordate at apex; ped. as long as the leaves, 8-15-flowered; pod deflexed, incurved, pointed. Ill. to Ga. 6—10'. April—June. (A. distortus T. & G.)

8 A. caryocárpus Ker. Low, diffuse, whitish, downy or nearly smooth; leaves stalked; lfts. 15—21, obovate; ped. longer; fls. 8—10", capitate; pod as large as a grape, smoothish, eatable. Ill. W. and S. May. (A. Mexicanus DC.)

9 A. Platténsis N. Villous, diffuse; lfts. 8—12 pairs, oblong; stip. lanceolate; rac. capitate; pod ovoid, villous. Gravel, Ill. Tenn. and W. May.

41. TEPHRÒSIA, L. GOAT'S RUE. CAT-GUT.
Calyx with 5, nearly equal, subulate teeth. Bracteoles 0. Vex. large, orbicular. Keel obtuse, cohering with the wings. Sta. diadelphous (in the following species) or monadelphous. Legume linear, much compressed, many-seeded. ♃ Lvs. unequally pinnate. Leaflets mucronate. Flowers white-purple.

Order 43.—LEGUMINOSÆ.

§ Flowers large (9–10″ long) in a leafy terminal cluster. Lfts. 15–27 .... No. 1
§ Fls. small, spicate, on long peduncles.—*a* Lfts. 9–17. Pods downy .... Nos. 2, 3, 4
— *a* Lfts. 5–9. Pods smoothish .... Nos 5, 6

1 **T. Virgínica** Pers. Erect, villous; lfts. oblong; fls. subsessile, axillary and terminal, variegated with white, rose, and purple; pod villous. Dry. 1—2f. July.
2 **T. spicáta** T & G. Rusty-villous, diffuse; lfts. oval-oblong, obtuse or retuse; ped very long; calyx teeth longer than tube. S. 1—3f. July.
3 **T. hispídula** Ph. Minutely hispid or pubescent, slender, decumbent; lfts. elliptic-oblong, acute; cal. teeth not longer than tube. S. 1—2f. May—July.
4 **T. ambígua** M. A. Curt. Smoothish, decumbent; lfts. 7—15, oblong-oblanceolate, truncate, brownish beneath; ped. angular, 2-3-flowered, as long as the leaves; calyx teeth shorter than tube. S. 1f. June, July.
5 **T. grácilis** Wood. Slender, diffuse, subglabrous; lvs. stalked; lfts. oblong-obovate, emarginate; ped. twice longer than the leaves; fls. on slender pedicels; cal. teeth very short; pod smooth. Fla. to La. 6—12′.
6 **T. chrysophýlla** Ph. Prostrate, rust-pubescent; lvs. sessile; lfts. round-obovate, acutish, wavy, yellowish; pedunc. much longer than the leaves; calyx teeth subulate Dry woods, Ga. Fla. to Tex. 10—20′. May—July.

42. **INDIGÓFERA**, L. INDIGO-PLANT. Calyx with 5 acute segments. Vex. roundish, emarginate. Keel spurred each side, at length reflexed. Legume 2-valved, 1 to ∞-seeded. ♄ ♃ Stip. small, distinct from the petiole. Leaves odd-pinnate. Legume pendulous.

§ Racemes longer than the leaves. Leaflets obovate-oblong, obtuse.......... Nos. 1, 2
§ Racemes shorter than the leaves. Leaflets oval. Naturalized South .... Nos. 3, 4

1 **I. Carolinàna** Walt. Erect, branched; lfts. 11—15, petiolulate; fls. yellowish-brown; pod oblong, veiny, rugous, 2-seeded. Sandy woods, S. 3—7f. July—Sep.
2 **I. leptosépala** N. Decumbent, strigous; lfts. 7—9, subsessile; calyx teeth subulate; fls. pale-scarlet; pod linear, 6-9-seeded. Ga. Fla. to Ark. 2—3f.
3 **I.** TINCTÒRIA L. Erect; lfts. 9—11; pod terete, torulous, curved. Waste pl. § E. Ind.
4 **I.** ANIL L. Erect; lfts. 7—11; pod flattened, even, with thick edges. Waste. § W. Ind.

43. **ROBÍNIA**, L. LOCUST. Calyx 5-cleft, the 2 upper segments more or less coherent. Vex. large. Alæ obtuse. Sta. diadelphous (9 and 1). Style bearded inside. Legume compressed, elongated, many-seeded. ♄ ♄ With stipular spines. Lvs. odd-pinnate. Fls. showy, in axillary racemes. Fragrant. Fig. 402.

1 **R. Pseudacàcia** L. *Common Locust.* Branches armed with spines; lfts. ovate and oblong-ovate; rac. pendulous, white, smooth, as well as the pods. Penn. S. and W. Introduced everywhere. 30—80f. Wood very durable. April, May.
2 **R. viscòsa** Vent. *Clammy Locust.* Spines very short; branchlets, petioles, and pods glandular-viscid; lfts. ovate; rac. crowded, erect, roseate. Mts. S. 40f. Ap. Ju. †
3 **R. hispida** L. *Rose Acacia.* Spines almost wanting, shrub mostly hispid; rac. oose, mostly pendulous; fls. large, rose-red. Mts. S. 3—8f. May, June. †

44. **COLÙTEA**, L. BLADDER SENNA. Calyx 5-toothed. Vex. with 2 callosities, expanded, larger than the obtuse carina. Stig. lateral, under the hooked summit of the style, which is longitudinally bearded on the back side. Legume inflated, scarious. ♄ Leaves odd-pinnate.

C. ARBORÉSCENS L. Lfts. elliptical, retuse; vex. shortly gibbous behind. Mt Vesuvius 8—12f. Leaflets about 9. Flowers large, yellow. June— Aug.

45. WISTÀRIA, Nutt. Cal. bilabiate, upper lip emarginate, the lower one 3 subequal teeth. Vex. with 2 callosities ascending the claw and separating above. Wings and keel falcate, the former adhering at top. Legume torulous. Seeds many, reniform. ♄ Leaves odd-pinnate. Raceme large, with large, colored bracts. Flowers lilac-purple.

1 **W. frutéscens** DC. Pubescent when young, at length glabrous; lfts. 9—13, ovate or elliptic-lanceolate, acute; raceme densely ∞-flowered; calyx teeth obtuse; ovary glabrous. Swamps, S. 15—30f. Woody. April, May.
2 **W.** consequàna Benth. Pubescent; lfts. 9—13, ovate or oblong-lanceolate, acuminate; raceme loose, pendulous, 1f long; calyx teeth acuminate. China. April.

46. APIOS, L. Ground Nut. Calyx obscurely bilabiate, the upper lip of 2 very short, rounded teeth, the 2 lateral teeth nearly obsolete, the lower one acute and elongated. Keel falcate, pushing back the broad, plicate vex. at top. ♄ Glabrous. Root bearing edible tubers. Leaves pinnately 5–7-foliate.

A. tuberòsa Ph. St. twining; lfts. ovate-lanceolate; rac. shorter than the lvs. Thickets and shady woods. 2—8f. Rac. 1—3′ long. Fls. brownish-purple. Handsome. Jl.Aug.

47. VIGNA, Savi. (To *Dominic Vigna*, commentator on Theophrastus.) Calyx of 4 lobes, the upper twice broader, the lower longer. Vex. broad, with 2 callosities near the base of the limb. Keel not twisted. Stigma lateral. Legume terete. ♄ Leaves pinnately trifoliate.

V. hirsùta Feay. Plant hirsute, the stem retrorsely so; cal. with 1 bractlet at base, segm. all acute, the lower acuminate; lfts. ovate-lanceolate, pointed. Marshes, S. Car. Fla. to La. 6—10f. Flowers pale yellow, 6″. Pod 2′, 4–6-seeded. July—Sept.

48. RHYNCHÒSIA, DC. Calyx somewhat bilabiate, or 4-parted, with the upper segment 2-cleft. Vex. without callosities. Keel falcate. Style glabrous. Legume oblique, short, compressed, 1–2-seeded. Seeds carunculate. ♃ ♄ Leaves resinous-dotted beneath, pinnately 3-foliate, sometimes reduced to a single leaflet. Flowers yellow.

§ Phaseoloìdeæ. Twining. Raceme long, ∞-flowered. Calyx teeth short....No. 1
§ Arcypufllum. Low, or twining. Flowers in fascicles or short racemes. Calyx teeth leafy, as long as the corolla............ ....Nos. 2, ♂ 4
§ Orthodànum. Erect. Ped. 1-flowered, axillary. Calyx teeth subulate......No. 5

1 **R. mínima** DC. Scrambling; lfts. thin, rhomboidal; rac. with about 12 remote, reflexed fls.; pod torulous, 6″ long. Banks, S. Car. to Fla. and La. 3—5f, delicate.
2 **R. simplicifòlia** (Ell.) Low, erect, pubescent; lvs. reduced to a single leaflet, orbicular or reniform, obtuse. Sandy woods, S. 1—3′. Leaves 1½′. April, May.
3 **R. volùbilis** Wood. Twining, pubescent; lvs. 3-fol.; lfts. oval or orbicular; rac. 3-10-flwd.; calyx teeth ovate, cuspidate. Dry woods, S. 3—4f. Lfts. 1′. June, July.
4 **R. erécta** Wood. Tall, velvety pubescent; lvs. 3-foliate; lfts. oval, acute; sepals scarcely united, lance-ovate to linear. Dry. Md. to Fla. 2—5f. June—Aug.
5 **R. galactoìdes** Chapm. St. erect, rigid, branched; lfts. small, elliptic or oval, margins revolute; ped. half as long as the flowers. Ala. Fla. 2—3f.

49. PHASÈOLUS, L. Kidney-bean. Cal. upper lip 2-toothed, lower 3-toothed. Keel with the stamens and style spirally twisted. Leg com-

ORDER 43.—LEGUMINOSÆ. 97

pressed and falcate, or cylindric, many-seeded. Seeds compressed, reniform. Leaves pinnately trifoliate. Leaflets stipellate. Figs. 157, 203–4, 214.

§ Flowers arranged in racemes. Legume falcate. July—Sept....................No. 1
§ Flowers few. capitate on long stalks. Legume straight, linear..........Nos. 2–4
                                  Exotic.—*a* Stems climbing............Nos. 5, 6, 7
                                    —*a* Stem erect, bushy................No. 8

1 **P. perénnis** Walt. *Wild Bean Vine.* Twining, pubescent; rac. paniculate, mostly in pairs, axillary; lfts. ovate, acuminate, 3-veined; leg. pendulous, falcate, broad-mucronate. Dry woods: common. 4—7f. Pod 2′.

2 **P. diversifòlius** Pers. St. prostrate, scabrous; lfts. angular, 2–3-lobed or entire; ped. longer than leaf; pod pubescent, broadly-linear, cylindric. ② Sandy shores. 3–5f.

3 **P. hélvolus** L. St. slender; lfts. between oblong-ovate and lance-ovate, not lobed; ped. slender, several times longer than the leaves; pod straight, cylindric, 8–10-seeded. ♃ Sandy fields. 3—5f.

4 **P. paucifiòrus** Benth. Stem slender, retrorsely hirsute; lfts. linear-oblong, not lobed, as long as the petiole, hirsute; pod hirsute, 5–8-seeded. Prairies, Ill. (Mead) and W. 2—4f.

5 **P. vulgàris**. Lfts. ovate, acuminate; rac. solitary; pod pendulous, long-pointed; seed reniform, variously colored. ① E. Ind. Flowers white. 3—8f.

6 **P. lunàtus**. *Lima B.* Lfts. ovate-deltoid; pod broad, flat, falcate, with large, flat, white seeds; flowers whitish. ① E. Ind. 6—12f.

7 **P.** multiflòrus. *Scarlet Pole B.* Lfts. ovate, acute; rac. as long as the lvs.; fls. scarlet; pod pendulous, seeds reniform. ① S. Am. 6—10f.

8 **P.** nanus. *Bush B.* Lfts. broad-ovate, acute; pod torulous; flowers and seeds white. ① India. 1f. There are many varieties.

50. **ERYTHRÌNA**, L. Calyx truncate or lobed. Vex. long, lanceolate, with no callosities. Wings and keel much smaller. Stam. straight, nearly as long as the vexillum. Style glabrous. Legume torulous. ♄ ♃ Often prickly. Leaves pinnately trifoliate. Flowers racemed.

1 **E. herbàcea** L. Glabrous; lfts. rhombic-hastate, with 3 rounded, shallow lobes, petioles with here and there a small hooked prickle; rac. terminal; flowers slender, deep scarlet, 2′. Rich soils, S. Rhizome thick. 3—4f. April.

2 **E.** crista-galli. Shrub or tree; lfts. ovate or elliptical, with hooked prickles beneath; banner recurved; fls. scarlet, in large racemes. Planted South.

51. **AMPHICARPÆA**, Ell. Pea-vine. Calyx with 4 or 5 nearly equal segments. Pet. oblong. Vex. with the sides appressed. Stig. capitate. Ovary on a sheathed stipe. Leg. flat, 2–4-seeded. ① Slender, twining. Leaves pinnately trifoliate. The upper flowers complete, but usually barren, the lower apetalous and fruitful.

1 **A. monoìca** Nutt. St. retrorsely pubescent; lfts. ovate, thin; cauline racemes pendulous; cal. segm. very short; bracts minute. Woods. 4—8f. Very slender. Flowers pale purple. Upper pods 4-seeded, lower 1-seeded. July—Sept.

2 **A. Pitcheri** T. & G. Stem rusty-villous; lfts. rhomboid-ovate; rac. erect, often branched; bracts broad, conspicuous. N. Orleans and W. Seeds blackish.

52. **GALÁCTIA**, L. Cal. bibracteolate, 4-cleft, the segments of nearly equal length, upper one broadest, entire. Pet. oblong. Vex. broadest and incumbent. Keel petals slightly cohering at top. Legume many-seeded. ♄ ♃ Lvs. pinnately compound. Rac. axillary. Fls. purplish. Aug. Sept.

ORDER 43.—LEGUMINOSÆ.

§ Leaves pinnate, 7-9-foliate. Sts. prostrate, twining. Lvs. coriaceous.........No. 1
§ Leaves pinnately 3-foliate. Sts. prostrate, twining. Pods 12—18" long...Nos. 2—4
§ Leaves pinnately 3-foliate. Sts. erect or ascending. Petioles longer than lfts..Nos. 5, 6

1 **G. Ellióttii** N. Lfts. elliptic-oblong, obtuse; ped. longer than the lvs., few-flwd. at the top; upper sep. (double) broad-ovate. ♃ Ga. Fla. 3—7f. Rose-white. May, Jn.
2 **G. glabélla** Mx. St. nearly glabrous; lfts. elliptic-oblong, emarginate at each end, shining above, a little hairy beneath; rac. pedunculate, about the length of the leaves; flowers 6", pedicellate. ♃ Arid soils, N. J. to Fla. 2—4f. Rose-purple.
3 **G. mollis** Mx. St. softly pubescent; lfts. oval, obtuse, nearly smooth above, softly villous and whitish beneath; rac. longer than the leaves, pedunculate, fasciculate; fls. 4", on very short pedicels; pod villous. ♃ Dry soils, Md. to Ga. 2—4f.
   β. *microphýlla*. Lfts. small (4—6"). oval; fls. solitary, and nearly sessile in the upper axils; pods 5 or 6-seeded. Ga. Fla. (Miss S. Keen.)
4 **G. pilòsa** N. St. pubescent or smoothish; lfts. thin, oblong-ovate or oval, obtuse or retuse at both ends; rac. very slender, twice or thrice longer than the leaves, with scattered, distant flowers. ♃ Dry soils, S. 3—7f. Leaflets 1—2'. Flowers 4".
5 **G. brachýpoda** T. & G. Slender, branching; lfts. oblong, odd one petiolulate; rac. stalked, shorter than the leaves. ♃ Sandy woods, W. Fla. 2—3f, ascending.
6 **G. sessiliflòra** T. & G. St. simple, flexuous; lfts. oblong-linear, odd one subsessile; rac. very short, sessile. Sandy woods, S. 1—2f. Lfts. 1'—20". Pod erect.

53. **DÓLICHOS**, L. Calyx 4-lobed, the upper lobe 2-toothed or entire. Vex. with 2 or 4 callosities at the base of the limb. The free stamen spurred at base. Legume flattened with a few oval, flattened seeds. ♄ Leaves pinnately 3-foliate.

1 **D. multiflòrus** T. & G. Lfts. ample, orbicular, acute, thin, pubescent; racemes equalling the petioles, densely ∞-flwd. at the top of the stout peduncle; calyx upper lip entire; pod 4-5-seeded. ♃ Banks, Ga. to La.
2 **D. Hàlei** Wood. Lfts. ample, round-ovate, acuminate; petioles 3 times longer than the few-(3-8)-flwd., stalked raceme; pod broad, 2-3-seeded, the point incurved. ♃ N. Orleans and W. (Dr. J. Hale.) Pod 2'.
3 **D.** SESQUIPEDALIS. Pods smooth, subterete, very long (1f). W. Ind. † South.
4 **D.** CAT-IANG. Pods linear, erect, twin at top of the long ped. E. Ind. † South.

54. **CLITÒRIA**, L. Calyx bibracteolate, 5-toothed, segm. acuminate. Vex. large spreading, roundish, emarginate, not spurred. Keel smaller than the wings, acute, on long claws. Legume linear-oblong, torulous, several-seeded. ♄ Leaves pinnately 3-5-foliate. Flowers very large, solitary, or several together.

C. **Mariàna** L. Glabrous; lfts. 3, oblong-ovate or lanceolate, obtuse, lateral ones petiolulate; ped. short, 1-3-flwd.; bracteoles and bracts very short; pod 3-4-seeded ♃ Dry soils, N. J. to Fla. 1—3f. Flowers pale purple. July, Aug.

55 **CENTROSÈMA**, DC. Sep. lance-linear, slightly united, the lower longest and with 2 broad bractlets. Vex. very large, with a short spur on the back near the base. Keel and stamens much shorter, incurved Legume long, linear, margined and long-pointed. ♄ Leaves pinnately 3 foliate. Flowers very large, purple.

C. **Virginiàna** Benth. St. very slender; lfts. oblong-ovate to oblong-linear, firm, very veiny, the veins incurved; ped. 1-4-flowered, bractlets larger (not longer) than the calyx; pod veined along the margin. ♃ Dry soils, S. 2—5f. July, August.

ORDER 43.— LEGUMINOSÆ.

56. KENNEDYA, Vent. Two upper lobes of calyx half-united. Banner broad, spreading, keel as long as the wings, incurved. Legume linear. ♂ Australian twiners with brilliant flowers in clusters. Leaves 3-nate.

1 K. COMPTONIÀNA. Smoothish; lfts. 3, ovate, retuse, veiny; peduncle bearing an erect raceme of many bright blue flowers, very ornamental in the conservatory. 12f.
2 K. RUBICÚNDA. Hairy; lfts. ovate; ped. 3-flwd., fls. dark-red or crimson, to scarlet. 5f.

57. HARDENBERGIA, Benth. Two upper teeth of calyx united. Banner broad, spreading, keel much shorter than wings. Legume linear. ♄ Australian. Flowers in racemes, very delicate. Leaflet mostly but 1.

H. MONOPHÝLLA. Plant very smooth; lft. lance-ovate; rac. erect; fls. blue-purple. 10f.

58. ACÀCIA, Necker. Calyx valvate, 4- or 5-toothed. Pet. 4 or 5, small, distinct or nearly so. Sta. numerous, distinct. Legume not jointed, dry, 2-valved, ∞-seeded. Beautiful trees or shrubs, native of warm climates. Lvs. twice pinnate, or reduced to phyllodia (§ 321). Fls. yellow or yellowish, in spikes or heads, very numerous and showy.

§ Leaves bipinnate. Flowers collected in heads or spikes..................Nos. 1—3
§ Leaves abortive—reduced to flattened petioles (phyllodia) with their edges
    vertical. Flowers yellow,—x in globular, solitary heads..................Nos. 4, 5
                              —x in globular, racemed heads..................Nos. 6, 7
                              —x in cylindrical spikes..................Nos. 8—11

1 A. FARNESIÀNA L. *Sponge Tree.* Tree armed with straight stipular spines; lvs. with 4—8 pairs of pinnæ, leaflets 15—20 pairs, oblong, crowded; ped. 2 or 3 together. Naturalized along the Gulf, Fla. to N. Orleans. Pods 2—3' long. (Vachellia, C-B.)
2 A. ÁLBICANS. Shrub from Mexico, 5f, with stipular spines, silvery-pubescent; leaves with 8 or 9 pairs of pinnæ, leaflets 19—30 pairs, linear-oblong, glabrous; flowers white, the heads in axillary racemes, 2—5 together.
3 A. DEALBÀTA. Shrub thornless, 5f, from N. Holland, all velvety-pubescent; pinnæ 15 pairs, leaflets 30—35 pairs, linear, crowded; heads in axillary racemes.
4 A. JUNIPERÌNA. Shrub from N. Holland, spinescent; phyllodia linear-subulate, pungent; branches terete, hairy or downy; heads solitary; petals 5.
5 A. ARMÀTA. Shrub 5—8f, downy or hairy, with spinescent stipules; phyllodia half-oblong-ovate, entire, 1-veined; heads solitary; pods velvety. N. Holland.
6 A. VESTÌTA. Shrub 6f, clothed with a soft down; leaves (phyllodia) halved, elliptic-oblanceolate; heads loosely racemed along the ped., one being terminal. N. Holland.
7 A. CULTRIFÓRMIS. Shrub 5f, smooth and glaucous; leaves curved, triangular-lanceolate, coriaceous; heads in racemes, panicled at the end of the branches.
8 A. VERTICILLÀTA. Shrub bushy, leafy, with the phyllodia and leaf-like stipules crowded and irregularly whorled; spikes oblong, solitary, axillary. New Holland.
9 A. LONGIFÒLIA. Shrub 5f, unarmed, with the phyllodia long, linear-lanceolate, 3-veined at base, veiny above; spikes axillary, in pairs; flowers 4-parted. N. S. Wales.
10 A. LINEÀRIS. Shrub 5f, unarmed, with phyllodia very long (7') and narrowly linear, 1-veined; spikes axillary, many, often branched; calyx 4-parted.
11 A. FLORIBÚNDA. Shrub or small tree, 6—10f; phyllodia linear-lanceolate, attenuate both ways, 3-5-veined; spikes simple, axillary, solitary; calyx 4-toothed. N. Holland.

59. POINCIÀNA, L. Sepals 5, united just at base. Petals broad, unguiculate, spreading. Stam. 10, very long, decurved with the slender style. Legume flat. ♄ Tropical. Leaflets very many, no odd one. Fls. large.

1 P. PULCHÉRRIMA. Shrub prickly (used in the W. Ind. for hedges, hence called *Flower fence*); leaflets oval-oblong; fls. 2' broad, orange, with crimson filaments 2' long. 10f.

Order 43.—LEGUMINOSÆ.

2 P. Gillèsii. From S. Am. Thornless; lfts. very small; fls. 2′, ylw., the pet. subequal, subsessile, glandular-ciliate at apex. [one spotted. From Madagascar. 10f.
3 P. regia, has crimson flowers 3′ broad, the petals long-clawed, crenate-edged, upper

60. CALLÍSTACHYS, Vent. Calyx 2-lipped. Banner erect, keel and wings deflexed. Stam. 10, separate, as in Baptisia. Style incurved. Pod woody before ripening, many-seeded. ♭ From New Holland. Leaves 3-foliate but sessile. Flowers yellow, in a terminal cluster.

1 C. lanceolàta. Hairy, half-shrubby; leaflets lanceolate, apparently *whorled* in 3's.
2 C. ovàta. Pubescent; leaves ovate, acute; spike short and broad, many-flowered.

61. SOPHÓRA, L. Keel obtuse, not shorter than the wings or roundish banner. Pod stipitate, many-seeded, moniliform, indehiscent. Seeds globular. ♭♭ Leaves odd-pinnate. Panicles terminal.

1 S. tomentòsa L. Shrub 4—6f, hoary-tomentous; lfts. about 15. oblong, thick; fls. in long racemes, yellow, handsome; calyx obscurely 5-toothed; pod 6′. Coast, Fla.
2 S. Japónica. Tree 30—40f, from Japan, hardy from Philadelphia south. Leaflets about 13, smooth; panicles large, erect, open, white, in July and August.

62. CHORÓZEMA ilicifolia. Shrub from N. Holland, 3f, bushy, with thick spinescent, holly-like, simple leaves, and a profusion of deep orange or scarlet racemes. Calyx 2-lipped. Keel shorter than the wings. Pod inflated, many-seeded.

63. ÓROBUS, Tourn. Bitter Vetch. Calyx obtuse at base, deeper cleft on upper side. Cor. long, keel incurved, shorter than wings or banner. Sty. terete, downy above. ♃ Lfts. 2—12, rachis ending in a short point.

1 O. vérnus. Lfts. 6, ovate, pointed; stip. ½-sagit., entire; fls. blue and purp. Apr. 1f.
2 O. niger. Branched, 3f; lfts. 12, ovate to oblong; flowers dark purple. June—Aug.
3 O. atropurpùreus. Leaflets 6, linear; flowers dark purple, in long 1-sided racemes.

64. LENS esculénta. Lentil. Herb cultivated for food at the East since the times of Esau, seldom seen here. Stem weak, 1f. Leaves of many pairs of oblong leaflets, ending in a branched tendril. Raceme of 2 or 3 pale flowers succeeded by a short broad pod. Seed exactly lens-shaped, giving the name. ①

65. CÝTISUS, L. Cal. 2-lipped, with 5 teeth, keel obtuse, straightish. Style incurved or at length involute. Seeds with a *scale* at the hilum (strophiolate). ♭ Leaves of 3 leaflets, the upper becoming simple.

C. scopàrius. *Scotch Broom.* Shrub with smooth angular, virgate branches; lfts. oblong, pedicels solitary, axillary; flowers yellow, showy; pods hairy at edge. Europe.

66. TRIGONÉLLA Fœnum-Græcum. Fenugreek. Herb from Europe, in gardens. Cult. for its strong-scented herbage. 2f. Lfts. 3, cuneate at base. Fls. axillary, sessile, small, white. Pods linear, long, slightly falcate at point, 2 or 3 together.

67. CLIÁNTHUS, Soland. Cal. bell-form, 5-cleft. Banner lance-ovate, acuminate, reflexed, keel boat-shaped, decurved, as long as the banner, longer than the narrow wings. Pod oblong, inflated. ♭ From New Zealand. Leaves odd pinnate. Flowers large and splendid.

1 C. puníceus. Shrub smoothish, 4f; leaflets about 17, oblong, retuse, alternate, flowers 3′ wide, crimson-red, in dense hanging racemes of superb appearance.
2 C. Dampièrii. Shrub hairy, 4f; leaflets about 17, oval, acute; flowers very large, scarlet, with a black prominence at the base of the banner. Flowers freely.

Order XLIV. ROSACEÆ. Roseworts.

Herbs, shrubs, or *trees,* with alternate, stipulate leaves and regular flowers. Sepals 5, rarely fewer, united, often re-enforced by as many bractlets. Petals 5, rarely 0, distinct, inserted on the disk which lines the calyx tube. Stamens ∞, rarely few, distinct, inserted with the petals (perigynous) Ovaries 1, 2, 5, or ∞, distinct, or often coherent with each other, or immersed in the tube of the calyx. *Fruit* a drupe, or achenia, or a dry or juicy etærio (§ 158), or pome. *Seeds* 1 or few in each carpel, anatropous, exalbuminous. *Embryo* straight. Figs. 5, 35, 38, 117, 139, 158, 183–5, 188, 197, 244, 251, 285, 297, 300–1, 307, 358, 365–6, 400, 428.

A Ovary superior, and the fruit not enclosed in the tube of the calyx...(a)
A Ovary inferior, and the fruit enclosed in the calyx tube...(m)
  a Carpel 1, forming a drupe in fruit. Calyx deciduous. Trees or shrubs...(b)
  a Carpels 2— ∞. Calyx persistent, bractless. Shrubs or herbs...(c)
  a Carpels 4— ∞. Calyx persistent, with 5 bractlets added. Herbs mostly...(f)
    b Tribe I. CHRYSOBALANEÆ.—Style lateral. Ovules 2, ascending......CHRYSOBALANUS. 1
    b Tribe II. AMYGDALEÆ.—Style terminal. Ovules 2, pendulous.........PRUNUS. 2
    c Tribe III. SPIRÆEÆ.—Carpels 2–8, several-seeded follicles in fruit...(d)
      d Petals obovate, equal, imbricate in the bud............................SPIRÆA. 3
      d Petals lance-linear, convolute in the bud..............................GILLENIA. 4
    e Tribe IV. RUBEÆ.—Carpels 2— ∞, 1-seeded drupes or achenia....(e)
      e Shrubs unarmed. Carpels 5—8. Petals 5 (or ∞), yellow...............KERRIA. 5
      e Shrubs unarmed. Carpels 2—4. Petals 0. Calyx leafy.................NEVIUSIA. 6
      e Shrubs prickly. Carpels ∞, drupaceous and juicy....................RUBUS. 7
      e Herbs not prickly. Carpels 5—10 (2—6 in No. 12), dry...............DALIBARDA. 8
  f Tribe V. FRAGARIDEÆ.—Carpels 4— ∞, 1-seeded achenia in fruit...(g)
    g Style persistent on the dry achenia.—h Petals 8 or 9................DRYAS. 9
      —h Petals 5...........................GEUM. 10
    g Style deciduous.—k Torus pulpy, globular, red.....................FRAGARIA. 11
      —k Torus spongy or dry.—l Bractlets minute or 0.....WALDSTEINIA. 12
      —l Bractlets 5...............POTENTILLA. 13
  m Tribe VI. SANGUISORBEÆ.—Carpels 1–3, acheninato. Petals 0 or 5...(n)
    n Stamens 1—4. Style lateral. Flowers apetalous, scattered..........ALCHEMILLA. 14
    n Stamens 4— ∞. Style terminal. Flowers apetalous, spicate.........POTERIUM. 15
    n Stamens 10—15. Styles 2. Petals 5, yellow........................AGRIMONIA. 16
  m Tribe VII. ROSEÆ.—Carpels ∞, 1-seeded, free in the calyx tube......ROSA. 17
  m Tribe VIII. POMEÆ.—Carpels 2–5, consolidated with the calyx tube...(o)
    o Petals oblong-spatulate. Carpels half-2-celled....................AMELANCHIER. 18
    o Petals roundish.—p Carpels 1-seeded.............................CRATÆGUS. 19
      —p Carpels 2-seeded.............................PYRUS. 20
      —p Carpels ∞-seeded ....CYDONIA. 21

1. **CHRYSOBALANUS**, L. Cocoa Plum. Calyx 5-cleft. Pet. 5. Sta. about 20, in a single series. Ov. solitary, sessile, the style arising from the base. Ovules 2, collateral. Drupe 1-seeded, with thin pulp. ♄ With entire, veiny leaves, minute stipules, and terminal panicles.

  C. oblongifolius Mx. Lvs. oblong, varying to oblanceolate, subsessile, pedicels and calyx tomentous-hoary; filaments and ovary glabrous; drupe as large as a plum. Pine-barrens, Ga. Ala. Fla. 8—12f. Leaves shining. Flowers small, white.

2. **PRUNUS**, Tourn. Plum, &c. Calyx 5-cleft, the tube bell-shaped or cup-shaped, deciduous. Pet. 5, spreading. Sta. 15—30. Ov. solitary, with 2 pendulous ovules. Drupe fleshy, with a bony nucleus. ♄ ♃ Fruit mostly edible. Fls. white or purplish. Figs. 51, 119 21, 124 5, 158, 285, 297

§ Prunus. Drupe smooth, more or less glaucous with a bloom. Stone smooth, more or less flattened. Leaves mostly convolute (rolled) in vernation. Plums...(a)
   a Umbels 2-5-flowered. Leaves conspicuously acuminate...............No. 1
   a Umbels 2-5-flowered. Leaves acute or obtuse ....Nos. 2, 3, 4
   a Umbels 1-2-flowered. Leaves acute, obovate, or oval............Nos. 5, 6
§ Cérasus. Drupe smooth, without bloom. Stone smooth, globular. Leaves conduplicate (folded §254) in vernation. Cherries...(b)
   b Flowers in lateral leafless umbels. Drupes small. Native.........Nos. 7, 8
   b Flowers in lateral leafless umbels. Drupes large. Exotic.........Nos. 9, 10
   b Flowers in racemes—c terminating the leafy branches...........Nos. 11, 12
   —c in the axils of the evergreen leaves. ....No. 13
§ Armeniaca. Drupe soft-velvety. Stone smooth. compressed. Lvs. convolute in bud, expanding after the flowers. Apricots.............Nos. 14, 15
§ Amýgdalus. Drupe tomentous or smooth. Stone rugous-furrowed, compressed. Leaves conduplicate in vernation...(d)
   d Fruit with a soft juicy pulp. Small trees. Peach, &c..............No. 16
   d Fruit with a hard dry pulp. Trees or low shrubs. Almond...Nos, 17, 18, 19

1 **P. Americàna** Marsh. *Red Plum. Yellow Plum.* Somewhat thorny; lvs. oblong-oval and obovate, abruptly and strongly acuminate, doubly serrate; drupes roundish oval, reddish orange, with a tough skin. Low woods. 10—15f. May. †

2 **P. marítima** Wang. *Beach Plum.* Lvs. oval or obovate, slightly acuminate, sharply serrate; petioles with 2 glands; umbels few-flowered; ped. short, pubescent; fruit nearly round. Sea beach, Me. to Va. 3—4f. Fruit size of a grape. May.

3 **P. umbellàta** Ell. Lvs. lanceolate or lance-oval, acute or barely acuminate, obscurely serrulate; petioles glandless; umbels 3-5-flowered, precocious; fruit oval, small, glaucous, red. Dry soils, South. 10—15f. Fruit pleasant. May.

4 **P. Chícasn** Mx. *Chickasaw Plum.* Branches spinous; lvs. oblong-lanceolate or oblanceolate, ₊landular serrulate, not at all acuminate; pedicels short, smooth; drupe globous. Thickets, South. 6—12f. Fruit red or yellowish. April.

5 **P. spinòsa** L. β. *insititia. Bullace Plum.* Branches thorny; lvs. pubescent beneath; obovate-elliptical, varying to ovate, sharply and doubly dentate; umbels 1-2-flowered; fruit globular, black, glaucous. Roadsides. 15—20f. §

6 **P.** doméstica L. *Common Garden Plum. Damson Plum.* Branches unarmed; lvs. oval or ovate-lanceolate, acute; pedicels nearly solitary; drupe globous, oval, ovoid, and obovoid. Long cultivated. 15f. Italy.

7 **P. pùmila** L. *Sand Cherry.* Lvs. oblanceolate or obovate, acute, subserrate, smooth, paler beneath; umbels few-flowered, sessile; drupe ovoid. Shrub trailing in sandy soils. 1—2f. Fruit small, dark red, pleasant. May.

8 **P. Pennsylvánica** L. *Wild Red Cherry.* Lvs. oblong-ovate, acuminate, finely serrate, thin, smooth; umbels corymbous, with elongated pedicels; drupe small; ovoid-subglobous. Woods. N. 25f. Bark red-brown. May.

9 **P.** Avium L. *Ox-heart. English Cherry.* Branches erect or ascending; lvs. oblong-obovate, acuminate, hairy beneath; umbels sessile, with rather long pedicels; drupe ovoid-globous, subcordate at base. Gardens, parks. 30—50f. †

10 **P.** Cérasus L. *Sour Cherry. Large Red. Morello, &c.* Branches spreading; lvs. ovate-lanceolate, acute at apex, narrowed at base, nearly smooth; fls. with short pedicels; drupes globous. Tree 15—20f. †

11 **P. serótina** Ehr. *Black* or *Wild Cherry.* Lvs. firm, oval-oblong or elliptic, acuminate, smooth, shining above, unequally glandular-serrate; petioles with 2—4 glands; raceme long; drupes black. Woods. 50—80f. Bark black. May.

12 **P. Virginiàna** L. *Choke Cherry.* Lvs. smooth, oval or obovate, short-pointed, thin, not shining, with sharp, subulate serratures, veins bearded at base; petioles with 2 glands; raceme short. Thickets. 5—20f. Fruit blackish, astringent. May.

13 **P. Caroliniàna** Ait. *Cherry Laurel.* Lvs. oblong-oblanceolate, acuminate, on short petioles, entire, coriaceous; fls. small, in numerous, dense racemes shorter than the leaves; drupes persistent, poisonous. Banks, S. 30—50f. April. †

Order 44.—ROSACEÆ.

14 P. ARMENÌACA Willd. *Apricot.* Lvs. broadly ovate, acuminate, subcordate at base, denticulate; stip. palmate; fls. sessile, subsolitary; drupe large, subglobous. From Armenia. 10—15f. Fruit purple-yellow, 1—2'.

15 P. DASYCÀRPA Ehrh. *Black Apricot.* Lvs. ovate, acuminate, doubly serrate; petioles with 1 or 2 glands; fls. pedicellate; drupe subglobous. From Siberia. 10—15f. Fruit dark purple, in July. Flowers white, April.

16 P. VULGÀRIS Mill. *Peach.* Lvs. lanceolate, serrate, with all the serratures acute; fls. solitary, subsessile, preceding the leaves; drupe tomentous. Persia. 8—15f. Fls. rose-color, with the odor of prussic acid. Fruit yellow-purple.

β. LÆVIS. *Nectarine.* Drupes glabrous, yellow, purple, red, large.

17 P. COMMÙNIS. *Almond.* Lvs. lanceolate, serrate, with the lower serratures glandular; flowers sessile in pairs. Barbary. 15f. Varies with flowers double.

18 P. NANA. *Dwarf single-flowering Almond.* Lvs. ovate, attenuate at base, simply and finely serrate; flowers subsessile, Russia. 3f. May, June.

19 P. LANCEOLÀTA. *Dwarf double-flowering Almond.* Lvs. lanceolate, doubly serrate; fls. pedicellate, covering the stems. China. 2—3f. Roseate. (Amygd. pumila, Ait.)

3. SPIRÆA, L. Calyx 5-cleft, persistent. Pet. 5, roundish. Stam. 10—50, exserted. Carp. distinct, 3—12, follicular, 1-celled, 1-2-valved, 1-10-seeded. Styles terminal. ♄ ♃ Branches and leaves alternate. Flowers white or rose-colored. Fig. 244.

§ Shrubs, with stipulate, simple, lobed leaves. Carpels inflated... ....No. 1
§ Shrubs, with stipulate, pinnate leaves. Carpels 5, united. Exotic............No 2
§ Shrubs, without stipules. Leaves simple. Ovaries distinct...(a)
   a Flowers in umbels or corymbs.—b Corymb compound, terminal. Mts....No. 3
     —b Clusters many. Gardens. Exotic...Nos. 4—7
   a Flowers in a terminal panicle,—c roseate-purple........................Nos. 8, 9
     —c white, rarely blush-colored.......Nos. 10—12
§ Herbs, without stipules. Leaves tripinnate. Ovaries 5, drooping.... ....No. 13
§ Herbs, stipulate. Leaves pinnately divided.—d Flowers rose-purple.........No. 14
     —d Flowers white........Nos. 15—17

1 S. opulifòlia L. *Ninebark.* Lvs. roundish, 3-lobed, doubly serrate; fls. white, in pedunculate corymbs; carp. 3—5. By streams. Rare. 4f. June.

2 S. SORBIFÒLIA. Lvs. odd-pinnate; lfts. lanceolate, acuminate, doubly serrate, terminal one lobed; fls. white, in terminal panicles. Siberia. 6f. May.

3 S. corymbòsa Raf. Lvs. ovate, cut-serrate above, whitish beneath; fls. innumerable, white or roseate, in a dense, level-topped corymb; styles and carpels generally 3. Penn. Ky. and S. 1—2f. May, June.

4 S. HYPERICIFÒLIA. *St. Peter's Wreath.* Lvs. obovate-oblong, subentire; fls. in many lateral clusters, on short branches, white, mostly double. Europe. 3f. May.

5 S. PRUNIFÒLIA. Branches virgate; lvs. ovate, petiolate, serrate, 5-veined, silky beneath; fls. in 3's—5's (very double), white. Japan. Beautiful.

6 S. REEVESIÀNA. Lvs. lanceolate, serrate, 3-lobed or pinnatifid, glaucous beneath; rac. capitate, pedunculate, often forming long wreaths. June.

7 S. TRILOBÀTA. Lvs. roundish, lobed, crenate, veiny; fls. corymbed. Alps.

8 S. tomentòsa L. *Hardhack.* Rusty tomentous; lvs. lance-ovate, smoothish above, serrate; rac. short, dense, aggregated into a dense thyrse-like, terminal panicle; carp. 5. Pastures, thickets. Common. 2—3f. July, Aug. †

9 S. DOUGLÀSII. Much like No. 8, but larger, smoother, and with redder fls. Oregon.

10 S. salicifòlia L. Nearly smooth; lvs. lanceolate to oblanceolate, serrate; rac. panicled, dense or lax, white, often with a blush; carp. 5. Meadows, thickets. Common. Stem purplish. 3—4f. Stam. conspicuous as in other species. July. †

11 S. ARIÆFÒLIA. Lvs. elliptic-oblong, crenately lobed and toothed; fls. innumerable in large, terminal panicles, white. Oreg. 6—12f. Stems virgate. June, July.

12 S. lævigàta. Lvs. obovate-oblong, very smooth and entire, sessile. Siberia.
13 S. Arúncus L. *Goat's Beard.* Lvs. tripinnate; lfts. oblong-lanceolate, acuminate, straight-veined, doubly serrate, odd ones lance-ovate; pan. large, of numerous slender racemes; carpels 3—5, glabrous, 1″. Mts. N. Y. to Ga. 3—5f. July.
14 S. lobàta L. *Queen-of-the-Prairie.* Lvs. pinnatifid, the term. lobe largest, pedately 7-9-parted, lobes all doubly serrate; stip. reniform; panicle large, roseate, exceedingly delicate; carpels 6—8. Low prairies, W. & S. 4—8f. June, July. †
15 S. Ulmària. *Double Meadow-sweet.* Lvs. interruptedly pinnate, white-downy beneath; lfts. lance-ovate, the terminal one large, palmately 3–5-lobed. Eur. July.
16 S. Filipéndula. *Pride-of-the-Meadow.* Lfts. 9—21, pinnatifid-serrate, minute ones between; stip. clasping, large; corymbs lax; sep. reflexed. Europe. Root tuberous.
17 S. Japónica. Lvs. biternate; lfts. oblong, acuminate, cordate, their stalks bearded at base; panicle terminal; flowers with 10 stamens and 2 styles, pure white. 3—4f

4. GILLÈNIA, Mœnch. Indian Physic. Calyx tubular-campanulate, contracted at the orifice, 5-cleft. Pet. 5, linear-lanceolate, long. Sta. 10—15, very short. Carpels 5, connate at base. Styles terminal. Follicles 2-valved, 2-4-seeded. ♃ With trifoliate, doubly-serrate leaves.

1 G. trifollàta Mœnch. Lfts. ovate-oblong, acuminate; stip. linear-setaceous, entire; fls. on long pedicels, in pedunculate, corymbous panicles. In woods, W. N. Y. to Ga. 2—3f. Flowers axillary and terminal, rose-white, 1½′ broad. June, July.
2 G. stipulàcea Nutt. *Bowman's Root.* Lvs. lanceolate, deeply incised; radical leaves pinnatifid; stipules leafy, ovate, doubly incised, clasping; flowers large, in loose panicles. W. N. Y. to Ala. Flowers rose-color. June.

5. KÉRRIA, DC. Calyx of 5, acuminate, nearly distinct sepals. Cor. of 5 petals. Ov. 5—8, smooth, globous, ovules solitary. Sty. filiform. Ach. globous. ♄ Stems virgate. Lvs. simple, ovate, acuminate, doubly serrate, with stipules. Flowers terminal on the branches, solitary or few together, orange yellow.

K. Japónica. *Japan Globe-flower.*—Gardens. 5—8f. Flowers double.

6. NEVIÙSIA, Gray. Calyx 5-parted, the lobes leafy, cut-serrate, persistent. Cor. 0. Sta. ∞, filiform. Ov. 2—4, 1-ovuled. Ach. drupaceous. ♄ Lvs. simple, ovate, petiolate. Stipules subulate, free. Flowers terminal, numerous, showy.

N. Alabaménsis Gr.—Tuscaloosa, Ala. (Rev. R. D. Nevius.) 2—3f.

7. RÙBUS, L. Bramble. Calyx spreading, 5-parted. Pet. 5, deciduous. Stam. ∞, inserted into the border of the disk. Ovaries many, with 2 ovules, one of them abortive. Achenia pulpy, drupaceous. ♄ With ② stems, armed with prickles. Inflorescence imperfectly centrifugal. Fruit esculent, July—Sept. Flowers in May, June. Fig. 185.

§ Fruit inseparable from the juicy, deciduous receptacle. Blackberries. .(a)
   a Stems (mostly) erect, stout, armed with stout, recurved prickles.......Nos. 1, 2
   a Stems procumbent, trailing, mostly with slender, minute prickles.....Nos. 3—5
§§ Fruit separating from the dry, persistent receptacle. Raspberries...(b)
   b Leaves simple, lobed. Not prickly.......... ....Nos. 6—8
   b Leaves compound.—Stems not prickly, herbaceous...................... No. 9
           —Stems prickly, shrubby.—Corollas single.......Nos. 10—12
                        —Corollas double............No. 13

1 R. villòsus Ait. *High Blackberry.* Pubescent, viscid, and prickly; st. recurved at top, angular ; lfts. 3—5, ovate, acuminate, serrate ; petioles prickly ; calyx acuminate ; raceme leafless, ∞-flowered ; fruit ovoid, small-grained, sweet. Thickets. 3—6f. Fruit black, in August.

β. *frondòsus*. *Lawton B.* Smoothish ; rac. leafy at base, short ; fr. subglobous, large-grained, very acid. Fields and gardens.

γ. *humifùsus*. Trailing ; leaves smaller ; peduncles few-flowered.

2 R. cuneifòlius Ph. *Sand B.* Pubescent ; lvs. 3-foliate ; lfts. wedge-obovate, entire at base, dentate above ; racemes few-flowered, loose. Sandy woods, L. I. to Fla. 2—3f. Pet. white, thrice longer than calyx. May, June.

3 R. hìspidus L. Hispid with retrorse bristles; lvs. 3-foliate, smooth, green both sides; lfts. obovate, thickish, persistent ; fls. and fr. small, corymbed, on filiform pedicels. Damp woods. 3—7f long. Fruit sour. May, June.

β. *setòsus*. Lvs. oblanceolate ; fruit red. (R. setosus Bw.)

4 R. Canadénsis L. *Northern Dewberry.* Slightly prickly ; lvs. 3 (rarely 5)-foliate ; lfts. elliptic or rhomb-oval, acuminate, thin ; ped. long, hardly in clusters ; fruit large, black, very sweet in August. Stony fields, North.

5 R. triviàlis Mx. *Southern Dewberry.* Prickly and bristly; lvs. 3–5-foliate, thick, ovate-oblong or oval; ped. 1-3-flowered ; sep. obtuse, reflexed. South.

6 R. odoràtus L. *Mulberry.* St. erect or reclining, unarmed, glandular-pilous ; lvs. palmately 3-5-lobed, middle lobe longest, unequally serrate ; fls. large, in terminal corymbs ; pet. orbicular, purple. Woods: common. 3—5f. Fr. red, sweet, in Aug.

7 R. Nutkànus Mocino. Somewhat pilous; lvs. broad, 5-lobed, lobes nearly equal, coarsely serrate ; ped. few-flowered ; sep. long-acuminate, shorter than the very large, round-oval, white petals. Mich., Wis. to Oreg. 5—7f.

8 R. Chamæmòrus L. *Cloudberry.* Herbaceous, diœcious ; st. decumbent at base, erect, unarmed, 1-flwd. ; lvs. mostly but 2, cordate reniform, rugous, with 5-rounded lobes, serrate ; sep. obtuse ; pet. obovate, white. White Mts. 1f. June.

9 R. triflòrus Rich. Branches herbaceous, green ; lvs. 3- or 5-foliate; lfts. nearly smooth, thin, rhombic-ovate, acute, odd one petiolulate ; stip. ovate, entire ; pet. erect, oblong-obovate. Hilly woods, N. Fruit few-grained, dark red.

10 R. strigòsus Mx. *Wild Red Raspberry.* St. strongly hispid ; lvs. pinnately 3- or 5-foliate ; lfts. oblong-ovate or oval, obtuse at base, canescent tomentous beneath, odd one stalked ; cor. cup-shaped, white. Old fields, N. Common. Fruit red.

11 R. occidentàlis L. *Black Raspberry. Thimble-berry.* St. glaucous with bloom, long, recurved, prickly ; lvs. pinnately 3-foliate ; lfts. ovate, acuminate, hoary-tomentous beneath, lateral ones sessile ; pet. shorter than sep. ; fr. blk. Rky. fields and gard.

12 R. Idæus. *Garden R.* Hispid or prickly ; lvs. pinnately 3-5-foliate ; lfts. rhomb-ovate, acuminate, hoary-tomentous beneath ; sep. hoary-tomentous, pointed, longer than the white petals ; fruit red, white, or yellow. § ? ‡

13 R. rosæfòlius. *Bridal Rose.* Prickles straight ; lvs. pinnately 3-7-foliate ; lfts. lance-ovate, doubly serrate, velvety ; flowers large, white. Mauritius.

8. DALIBÁRDA, L. FALSE VIOLET.

Calyx inferior, deeply 5-6-parted, spreading, 3 of the segm. larger. Pet. 5. Sta. ∞. Sty. 5—8, long, deciduous. Ach. nearly dry. |., Lvs. undivided. Scapes 1-2-flowered.

D. rèpens L. Low, pubescent, bearing creeping shoots ; lvs. simple, roundish-cordate, crenate ; stipule linear-setaceous ; calyx spreading in flower, erect in fruit. ⅔ Damp woods, Penn. to Can. 2—12′. Scapes with 1 small white flower. June.

9. DRYAS integrifolia, Vahl.

—On the White Hills of N. H. *Prof. Peck* (Pursh). On Pike's Peak, Colorado. (A. H. Thompson.)

10. GEUM, L. AVENS.

Calyx 5-cleft, with 5 alternate segments or bractlets smaller and exterior. Pet. 5. Sta. ∞. Ach. ∞, aggregated on a dry receptacle, and caudate with the persistent, mostly jointed, geniculate and bearded style. ♃ Leaves pinnately divided.

§ SIEVÉRSIA. Style straight, jointless, all of it persistent. Flowers large...Nos. 1, 2
§ GEUM *proper.* Style bent and jointed in the middle, upper part deciduous...(a)
  a Head of fruits raised on a stipe. Flowers yellow or purple..........Nos. 3, 4
  a Head of fruits sessile (no stipe).—*b* Flowers yellow.................Nos. 5, 6
  —*b* Flowers white..............Nos. 7, 8

1 G. triflòrum Ph. Villous, erect, about 3-flowered; lvs. mostly radical, interruptedly pinnate, of numerous cuneate, incisely dentate, subequal lfts.; bractlets linear, longer than the sepals; styles plumous, very long in fruit (2—3′). N-W. States, rare in the North. 8—12′. Flowers purplish-white. May, June.

2. G. radiàtum Mx. Hirsute or smoothish; stem erect, nearly leafless; root lvs. lyrate, the terminal leaflet large, reniform, lobed and toothed, lateral ones minute; bractlets minute; pet. obcordate, yellow, large; styles hairy at base. White Mts. N. H., Roan Mt. N. Car. 9—15′. (G. Peckii Ph.)

3 G. vernum T. & G. Smoothish; lvs. pinnately divided, incisely lobed and toothed, the lowest often simple; fls. small, yellow; sep. reflexed; torus conspicuously stipitate. W. and S-W. 12—20′. Stipules large. April—June.

4 G. rivàle L. Pubescent; st. subsimple; radical lvs. lyrate; stip. ovate, acute; fls. nodding, purple; pet. as long as the erect cal. segments, purplish-yellow; upper joint of the persistent style plumous. Wet meadows, N. and M. 1—2f. June.

5 G. strictum Ait. Hirsute; lvs. interruptedly pinnate; lfts. ovate, lobed and toothed; pet. roundish, longer than the reflexed sepals; torus densely pubescent. Fields, N. States and Can. 2—3f. Terminal leaflet largest. July, August.

6 G. macrophýllum Willd. Hispid; lvs. interruptedly lyrate-pinnate, the terminal lft. much the largest, roundish cordate, 3—5′, all unequally dentate; petals longer than the calyx; recept. nearly smooth. White Mts. and Can. 1—2f. June, July.

7 G. album Gmel. Smoothish or pubescent; root lvs. ternate or often simple, upper lvs. simple; lfts. ovate, lobed and dentate; pet. as long as calyx; torus white-bristly. Thickets. Common. 2—3f. July. (G. Virginianum T. & G. &c.) (See Addenda.)

8 G. Virginiànum L. Hirsute; lvs. pinnate below, then ternate, the upper simple; lfts. incisely lobed, wedge-lanceolate, very acute, cut-toothed; pet. shorter than calyx; torus nearly naked. Wet thickets. 2—3f. Stout. July.

11. FRAGÀRIA, L. STRAWBERRY. Cal. concave, deeply 5-cleft, with an equal number of alternate, exterior segments or bractlets. Pet. 5, obcordate. Sta. ∞. Sty. ∞, lateral. Ach. smooth, affixed to a large, pulpy, deciduous receptacle. ♃ ∟ Stems stoloniferous. Leaves trifoliate. Fruit red. Flowers white, in Spring. Figs. 5, 117, 184, 251, 428.

§ Bractlets entire; petals white. Stemless, stoloniferous....................Nos. 1, 2
§ DUCHÉSNIA. Bractlets 3-lobed; petals yellow. Stems trailing................No. 3

1 F. Virginiàna Ehrh. Pubescent; lvs. thick; cal. of the fruit erect-spreading; acu. imbedded in pits in the globous receptacle; ped. commonly shorter than the lvs. Fields and gardens. 6—12′. Some of its varieties are polygamo-diœcious.
  β. *Illinoënsis.* Larger, very villous in the stems. Prairies. Westward.

2 F. vesca L. *Alpine, Wood,* or *English Strawberry.* Villous-pubescent; cal. of the fruit spreading or reflexed; ach. superficial on the conical or hemispherical receptacle, which is without pits; lvs. thin. Fields and woods.
  β. *pallida.* Fruit white. A var. well established in Wayne Co. N.Y. (Hankenson.)

3 F. Índica Ait. Pubescent, rooting at the joints; lfts. ovate, obtuse, incisely crenate-serrate; stip. lanceolate, free; pedicels axillary, solitary 1-flowered; bractlets leafy in fruit. ♃ Damp places, Penn. and S. § India.

ORDER 44.—ROSACEÆ.   107

12. WALDSTÉINIA, Willd.   DRY STRAWBERRY.   Cal. 5-cleft, with 5 alternate, sometimes minute and deciduous bractlets.   Pet. 5 or more, sessile, deciduous.   Sta. ∞.   Sty. 2—6.   Ach. few, dry, on a dry receptacle.   ♃ Acaulescent, with lobed or divided leaves, and yellow flowers.

1 W. fragarioìdes Traut. Lvs. trifoliate ; lfts. broad-cuneiform, incisely dentate-crenate, ciliate ; scapes bracteate, many-flowered. Hilly woods. 8'. June.
2 W. lobàta T. & G. Lvs. simple, roundish, cordate, 3-5-lobed, incisely crenate; scapes filiform, bracted, 3-7-flowered. Hills, South. 6'. May, June.

13. POTENTÍLLA, L.   CINQUEFOIL.   Calyx concave, deeply 5-cleft, with 5 bractlets added.   Pet. 5, roundish.   Sta. ∞, slender.   Ovaries collected into a head on a small, dry, hairy torus.   Sty. terminal and lateral, deciduous.   Achenia ∞.   ① ♃ ♄ Leaves compound.   Flowers solitary or cymous, mostly yellow.   Figs. 365–6.

§ SIBBÁLDIA. Stamens 5. Achenia 5—10, styles lateral. Low herbs. Mts.....No. 1
§ CÓMARUM. Sta. ∞. Flowers brown-purple. Torus in fruit ovoid, spongy....No. 2
§ POTENTÍLLA *proper*. Sta. ∞. Flowers yellow to white. Torus not enlarged..(a)
   a Leaves palmately 3-foliate..Nos. 3, 4, 5
   a Leaves palmately 5-foliate. Flowers yellow...................Nos. 6, 7
   a Leaves pinnate.—*b* Shrubs, with the flowers axillary above...............No. 8
      —*b* Herbs, with the flowers axillary, solitary..........Nos. 9, 10
      —*b* Herbs, with the flowers in terminal cymes.......Nos. 11, 12
         Exotic species, with fls. roseate and purple..Nos. 13, 14

1 P. procúmbens Clairv. Lfts. 3, obovate, 3-toothed at apex, hairy beneath ; fls. corymbed. White Mts.? (Pursh), and N. (Sibbaldia L.)
2 P. palústris Scop. Lvs. pinnate ; lfts. 3—7, lance-oblong, obtuse, sharply serrate, hoary beneath ; sep. much longer than the purple petals ; torus persistent, large, tasteless. ♃ Swamps, N. 1—2f. June. (Comarum L.)
3 P. Norvégica L. Hirsute; st. erect, dichotomous above ; lfts. 3, elliptical or obovate, dentate-serrate, petiolulate ; cymes leafy ; cal. exceeding the emarginate pale-yellow petals ; sty. terminal. ② Old fields, thickets, Can. to Car. 1—4f. July—Sept.
4 P. tridentàta Ait. Smooth ; st. ascending, woody and creeping at base ; lfts. 3, obovate-cuneate, evergreen, entire, with 3 large teeth at the apex; cymes nearly naked ; petals white, obovate. ♃ High Mts. N. Eng. 6—12'. June.
5 P. mínima Haller ? St. pubescent, ascending, mostly 1-flowered ; lfts. 3, obovate, obtuse, incisely serrate with 5—9 teeth above ; petals yellow, longer than the sepals. ♃ White Mountains. 1—3', tufted. June, July.
6 P. Canadénsis L. Villous-pubescent, procumbent, producing runners ; lfts. 5, obovate, cut-toothed above ; pedicels axillary, solitary, 1-flowered.
   α. *pùmila.* Small and delicate, flowering in Apr. May, everywhere.
   β. *símplex.* Subsimple, ascending. 8—14', smoothish ; fls. June—Aug. Common.
7 P. argéntea L. St. ascending, tomentous ; lfts. 5, oblong-cuneiform, with a few, large, incised teeth, smooth above, silvery canescent beneath, sessile ; flowers in a cymous corymb, small (3''). ♃ Rocky hills, N. 6—10'. June–Sept.
8 P. fruticòsa L. St. fruticous, very branching, hirsute, erect ; lfts. 5—7, linear-oblong, all sessile, margin entire and revolute ; petals large, much longer than the calyx. A low, bushy shrub, N. States. 1—2f. Flowers 1'. June–Aug.
9 P. anserìna L. *Silver-weed. Goose-grass.* St. slender, prostrate, rooting ; lvs. interruptedly pinnate ; lfts. many pairs, oblong, deeply serrate, canescent beneath ; peduncle solitary, 1-flowered, very long. ♃ Wet, N. Eng. N. and W. 1—2f. Ju.–Sept.
10 P. paradóxa N. Decumbent at base, pubescent ; lvs. pinnate ; lfts. 7—9, ovate-obl. incised, upper ones confluent ; ped. solitary, recurved in fruit ; ach. 2-lobed. ① Shores of Sodus Bay (Hankenson), W. to Oreg. 1f. June—July.

11 P. Pennsylvánica L. Erect, whitish-downy; lfts. 5—9, oblong, obtuse, pinnatifid, upper ones larger; cyme fastigiate, at length loose. ♃ N. Eng.: rare.

12 P. argùta Ph. Erect, grayish, pubescent and villous; radical lvs. on long petioles, 7-9-foliate, cauline few, 3-7-foliate; lfts. broadly ovate, cut-serrate, crowded; fls. in dense terminal cymes. ♃ By streams, N. and W. 2—3f, stout. May, June.

13 P. NEPALÉNSIS. Root lvs. quinate; stem ternate; lfts. wedge-oblong, serrate; stip. large, adnate, entire. ♃ Nepal. 1½f. Flowers large, rose, scarlet, orange, &c.

14 P. ATROSANGUÍNEA. Lvs. ternate; lfts. obovate, cut-serrate, white-downy beneath; sep. elliptic; pet. obcordate. ♃ Nepal. 1½f. Flowers crimson, often double.

14. ALCHEMÍLLA, L. LADIES' MANTLE.
Calyx 4-toothed, with 4 external bractlets. Petals 0. Sta. 1—4. Carp. (1—4) mostly solitary, with the style lateral. Stig. capitate. Seed suspended. Low herbs, with palmately lobed or incised leaves and small green flowers. Fig. 38.

1 A. arvénsis Scop. *Parsley Piert.* Lvs. crenate at base, incisely 3-lobed or parted, the segm. 2-3-cleft, pubescent; fls. axillary. ① E. Va. A small weed. § Europe.

2 A. alpìnus L. Lvs. radical, silky beneath, 5-7-parted, cut-serrate at apex; fls. corymbed. High Mts. of N. Eng. (Pursh, 1816.) † Europe.

15. POTÈRIUM, L. BURNET.
Calyx tube contracted at the top. Lobes 4, imbricated, petaloid, deciduous. Pet. 0. Sta. 4—∞, exserted. Styles slender, 1—3. Stig. penicillate. Ach. included in the hardened, 4-angled calyx tube. ♃ Lvs. unequally pinnate, with long stalks and adnate stipules. Lfts. petiolulate, serrate. Fls. in a spike or head, on a long peduncle or scape, often ☿. (Includes Sanguisorba L.)

1 P. Canadénse (L.) Glabrous; lfts. many, ovate or oval, obtuse, cordate, with serrate stipels and stipules; spikes cylindric (3'); stam. 4, long exserted. Wet meadows along the mountains. Can. to Ga. 2—4f. Flowers green-white. Aug.

2 P. Sanguisórba L. Glabrous; leaflets many, ovate or roundish, deeply serrate, heads subglobous; sta. ∞, in the lower fls. L. Huron (*Hooker*) and W. Purp. † Aug.

16. AGRIMÒNIA, L. AGRIMONY.
Calyx tube turbinate, contracted at the throat, muricate, limb 5-cleft, connivent in fruit. Pet. 5. Sta. 12—15. Ov. 2. Styles terminal. Ach. included in the indurated tube of the calyx. ♃ Lvs. pinnately divided. Fls. yellow, in long, slender racemes.

1 A. Eupatòria L. Lfts. 5 to 7, lance-oval or obovate, with small ones interposed, coarsely dentate; stip. large, dentate; pet. twice longer than the reflexed calyx. Dry soils, common. 1—3f. Rac. spicate, 6'—1f. Fls. 3—4'' broad. July, Aug.

2 A. parviflòra Ait. Lfts. 9—17, crowded, pubescent beneath, lanceolate, cut-serrate, with smaller ones interposed; pet. small. Woods, &c., Pa. S. and W. Plant fragrant, 3—4f, with spreading brownish hairs. July, Aug.

β. *incisa.* Lfts. incisely pinnatifid. South. (A. incisa T. & G.)

17. ROSA, Tourn. ROSE.
Calyx tube urceolate, contracted at the orifice, lined with the fleshy disk. Petals 5 (greatly multiplied by cultivation). St. ∞, inserted into the rim of the disk. Ach. ∞, bony, hispid, borne free within the calyx tube. ♄ Prickly. Lvs. odd-pinnate. Stip. mostly adnate to the petiole. Figs. 35, 139, 197, 301.

Obs. Our innumerable varieties of garden Roses have mostly originated with the few species mentioned below. To define these varieties in order to their recognition would generally be impossible, for their forms are as evanescent as their names are arbitrary. All that we propose is to aid the learner in tracing back each form to the *species* whence it sprung. This will be easily done in all cases except with the hybrids.

Order 44.—ROSACEÆ. 109

* Wild Roses, with simple, 5-petalled flowers, open in June and July...(§)
  § Leaflets 3, rarely 5, smooth. Branches long, climbing or trailing..........Nos. 1, 2
  § Leaflets 5—9,—*a* rusty glandular and fragrant beneath....................Nos. 3, 4
    —*a* not glandular. Erect.—*b* Prickles stout, falcate.............No. 5
                                  —*b* Prickles weak, straight......Nos. 6, 7, 8
* Garden Roses, with either simple or double flowers...(§§)
  §§ Styles cohering in an exserted column. Climbers...(*a*)
    a Leaflets 3—5, mostly 3. Prickles stout, deflexed.........................No. 1
    a Leaflets 5—9.—*b* Stipules and sepals mostly entire....................Nos. 9, 10
                     —*b* Stipules, or sepals, dissected. Prickles slender....Nos. 11, 12
  §§ Styles separate.—*c* Stipules nearly free, and caducous................Nos. 2, 13, 14
                     —*c* Stipules adnate to the petiole.—*d* Prickles falcate...(*e*)
                                                         —*d* Prickles straight...(*f*)
      e Leaflets not at all glandular. Shrubs erect, often slender.....Nos. 15, 16, 17
      e Leaflets glandular and fragrant beneath, downy or not.......Nos. 3, 18, 19
      f Lvs. and often the calyx, glandular. Fls. roseate or yellow..Nos. 20, 21
      f Lvs. not at all glandular. Prickles numerous, weak, or 0..Nos. 22, 23, 24

1 R. setígera Mx. *Prairie Rose.* Spines strong, straightish; lfts. ovate; stip. adherent; fls. in corymbs, deep roseate, becoming pale, scentless; styles united in an exserted column. Prairies, &c., N. Y. W. and S. 12—20f. June, July. †
  Var. *Prairie Queen, Baltimore Belle, Rosa Superba,* &c.
2 R. lævigàta Mx. *Cherokee R.* Prickles very strong, recurved; lfts. elliptical, evergreen, polished; stip. free, setaceous; fls. solitary, large, white; calyx bristly; styles separate. Tenn. to Fla. 15—30f. § ? In hedges and gardens.
3 R. rubiginòsa L. *Sweet Brier. Eglantine.* Prickles strong, recurved, many weak ones intermixed; lfts. broad-oval; fls. solitary; fruit obovoid and, with the pedicels, glandular hispid. Fields, roadsides. 4—8f. Fls. light red, single or double.
  Var. *Clementine, Maiden, Royal, Scarlet, Tree-double, White,* &c.
4 R. micrántha Smith. Prickles strong, recurved, few and equal; lfts. ovate; fls. solitary, small (15″), mostly white. Pastures, &c. N. Eng. 6—8f. June.
5 R. Carolìna L. *Swamp R.* Tall, erect, glabrous; lfts. elliptical, glaucous beneath, not shining; fls. corymbed; fr. depressed-globous, dark red, with hispid peduncles. Damp woods. 4—8f. Fls. varying from red to white. June, July.
6 R. lùcida Ehrh. *Wild R.* Prickles scattered, setaceous; lfts. elliptical, simply serrate, shining above; fls. in pairs (1—3); fr. depressed-globous and, with the pedicels, glandular-hispid. Dry woods. 1—3f. Branches greenish. Fls. red.
7 R. nítida Willd. *Wild R.* Stems reddish with very numerous reddish prickles; lfts. narrow-lanceolate, smooth and shining; fls. solitary; calyx hispid. Swamps, N. Eng. 1—2f. Fls. red. Fr. scarlet. Perhaps a variety of No. 6.
8 R. blánda Ait. *Thornless Wild R.* Prickles few, slender, deciduous; lfts. oblong, obtuse, not shining; stip. broad; ped. short, and with the calyx smooth and glaucous; fr. globous. Dry hills, N. and M. 2—3f. Petals reddish.
9 R. sempérvirens. Prickles subequal; lfts. thick, evergreen; fls. clustered, mostly white; fr. round-ovoid, yellow, glandular-hispid. S. Eur. 6—12f.
10 R. arvénsis. *Ayreshire R.* Prickles unequal, falcate; lfts. ovate, acute, deciduous; glaucous beneath; fls. solitary or clustered, white to purple. Eur. 20f.
  Var. *Dundee Rambler, Virginia Lass, Weeping-tree R.,* &c.
11 R. moschàta. *Musk R.* Lfts. lanceolate, acuminate; stip. very narrow; sep. long-appendaged, pinnatifid; fls. panicled, peculiarly fragrant, white. Asia. 10—12f.
12 R. multiflòra. *Japan R.* Lfts. lance-ovate, rugous, soft; stip. pectinate-fringed; fls. corymbed; sep. short and ped. tomentous. South. 15—20f. Pet. wh. to purp. § †
  Var. *Boursault, Seven Sisters, Russel's,* &c.
13 R. bracteàta. *Macartney R.* Erect; prickles recurved; lfts. 5—9, obovate, shining; stip. bristle-fringed; fls. solitary, with large bracts under the tomentous calyx. China. 2—3f. Fls. white, creamy, &c. § S.

14 R. Bánksiæ. *Thornless R.* Prickles none; lfts. lanceolate, 3—5, subentire; fls. small, in umbels; fruit globular, nearly black. China.

15 R. Indica. *Chinese Monthly R. Bengal R.* Lfts. 3—5, ovate, pointed, shining; stip. very narrow; sep. subentire; stam. inflexed; fruit top-shaped. China. 1—8f. Fls. white to crimson. April to November.

   β. Lawrenciàna. *Miss Lawrence's R.* Aculeate; fls. small (1?). pink-purple. Other var. *Noisette, Youland of Aragon, Giant of Battles, Cloth of Gold* (sulphur-yellow), and the favorite *Tea Roses*.

16 R. canìna. *Dog R.* Prickles strong, compressed; lfts. 5—9, with acute, incurved serratures; stip. rather broad, serrulate; sep. deflexed after flowering, deciduous; fr. ovoid, red. Eur. 4—8f. Fls. often simple, red. Often runs wild.

   β. Bourboniàna. Lfts. ovate, subcordate, glossy; fls. double and semidouble, purple. Numerous subvarieties, everblooming.

17 R. cinnamòmea. *Cinnamon R.* Lfts. 5—7, oval-oblong, grayish-downy beneath; stip. broad, involute, pointed; ped. and cal. glabrous; sep. as long as the petals, closed and persistent on the fruit. Eur. 6—12f. Purple.

18 R. damascèna. *Damask R.* Prickles broad, unequal; lfts. large, broad-elliptic, whitish-downy; sep. reflexed. Levant. 3—4f. Fls. pale roseate, very fragrant. The common Monthly is a variety.

19 R. alba. *White R.* Erect, tall; prickles slender, or 0; lfts. round-ovate; petioles and veins downy, glandular; sep. pinnatifid; fr. ovoid. Eur. Stout, 4—8f. Flowers large, clustered, sweet-scented, pure white, semidouble.

20 R. centifòlia. *Provens R. Cabbage R.* Very prickly; leaflets 5—7, ovate, edges gland.-ciliate; cal. and ped. gland.-hispid, viscid and frag. S. Eur. 2—4f. Fls. pink, &c. Var. very numerous, among which is the incomparable *Moss Rose*.

21 R. eglantèria. *Yellow R. Austrian Eglantine.* Branches red, all prickly; lfts. 5—7, small, broad-oval, or obovate; sep. smooth, entire; pet. large, yellow. Aust. 3f. Var. The *Copper Austrian*, single; *Persian Yellow*, double, and others.

22 R. alpìna. *Boursault R.* Climbing; lfts. 5—11, ovate or obovate, sharply serrate; ped. deflexed after flowering, and sep. connivent on the ovoid hip. Alps. 10—20f. Older stems thornless. Fls. clustered, pink, blush, crimson, &c.

23 R. Gállica. *Common French R.* Erect; leaflets 5—7, oval to lanceolate, thick; fls. erect, with large spreading red petals; sep. ovate, some viscid. Eur. 2—5f. Var. 300 or more; as the *Velvet, Carmine, Carnation.* Some are variegated, as *York-and-Lancaster, Tricolor, Picotée, Nosegay,* &c.

24 R. pimpinellifòlia. *Scotch R. Burnet R.* Very prickly, erect; lfts. 5—9, round-ovate, obtuse, smooth; sep. entire, finally convergent on the fruit; fls. small, roseate; but there are varieties with purple and even yellow flowers.

18. AMELÁNCHIER, Medic. Shad-flower. Wild Service.
Cal. 5-cleft. Pet. 5, oblong-obovate or oblanceolate. Sta. short. Sty. 5, somewhat united at base. Pome 3–5-celled, cells partially divided, 2-seeded. ♄ ♄ Leaves, simple, serrate. Flowers racemous, white.

A. Canadénsis T. & G. Lvs. oval or oblong-ovate, sharply serrate. smooth; raceme loose; calyx segments lance-triangular; fruit globous, purplish. Woods: common. 5—35f. Flowers showy, in early Spring. Fruit pleasant, ripe in June.

   β. *oblongifòlia.* Shrub; lvs. oblong-oval, mucronate; pet. oblong-obovate.
   γ. *rotundifòlia.* Lvs. broad-oval; pet. linear-oblong. Shrub 10—20f.
   δ. *alnifòlia.* Lvs. round-oval, serrate near apex; pet. linear-oblong. 15—30f.
   ε. *oligocárpa.* Shrub; lvs. elliptic-oblong, cuspidate; rac. 2–4-flowered. North.

19. CRATÆGUS, L. Thorn. Hawthorn.
Calyx urceolate, limb 5-cleft. Pet. 5. Sta. ∞. Ov. 1—5, with as many styles. Pome fleshy, containing 1—5 bony, 1-seeded carpels, and crowned at the summit by the

ORDER 44.—ROSACEÆ. 111

persistent calyx and disk. ♄♄ Armed with thorns. Lvs. simple, often lobed. Bracts subulate, deciduous. Fls. corymbous, white or purplish.

§ Corymbs 6–30-flwd., appearing with the leaves. Fruit red or yellowish...(a)
    a Villous or pubescent. Leaves plicate or sulcate along the veins.....Nos. 1, 2
    a Pubescent. Leaves plain, not at all plicate, cleft or not............Nos. 3, 4
    a Glabrous throughout.—*b* Leaves abrupt at base, lobed, petioled.....Nos. 5—7
               —*b* Leaves attenuate at base, seldom lobed....Nos. 8, 9
§ Corymbs 1–6-flowered,—*c* appearing before the downy leaves.................No. 10
        —*c* appearing with the leaves,—*d* pubescent..........No. 11
                     —*d* glabrous........Nos. 12, 13

1 **C. tomentòsa** L. *Black Thorn.* Lvs. broad-ovate or oval, abrupt at base, doubly serrate or cut-lobed, villous beneath when young, and plicate; fls. large, in compound pubescent corymbs; fruit oval, large (8″), 2–5-seeded, red. Can. to Ky. and Car. Mts. 15—25f. Flowers white, April, May. Fruit July, Aug. Varies greatly.
  β. *plicàta.* Lvs small, glabrous, strongly plicate. Vt., N. H., N. Y.
  γ. *pyrifòlia.* Lvs. elliptic, acute at base, thinly pubescent. Styles 3. W.
  δ. *flabellàta.* Lvs. fan-shaped; corymbs glandular-pubescent. W.
  ε. *móllis.* Lvs. large, soft-villous, subcordate, many-lobed; corymbs canescently-villous; fruit downy when young. Ohio to Iowa.

2 **C. punctàta** Jacq. Lvs. cuneiform-obovate, doubly and often incisely serrate, entire at base, and narrowed to a short, winged petiole, veins straight and prominent, corymbs villous-downy; styles 3; fruit globous, punctate. Woods. 12—25f. April—June. (See *Addenda.*)

3 **C. arboréscens** Ell. Thornless; lvs. lanceolate, acute at each end, deeply serrate; calyx hairy; segments subulate, obtuse, entire; corymbs very numerous; styles 5; fruit ovoid, red, 3″. Ga. Fla. and W. 20—30f. March, April.

4 **C. apiifòlia** Mx. Thorny. Lvs. deltoid, truncate at base, cut-lobed and toothed; petioles slender; styles 2 or 3. Woods, S. 8—12f. March, April.

5 **C. Oxyacántha** L. *Hawthorn.* Lvs. wedge-obovate, 3–5-lobed at apex; corymbs glabrous, white to purple; styles 1—3; fruit small, red. Hedges, &c. 8—18f. §

6 **C. coccínea** L. *White Thorn.* Lvs. broadly ovate, acutely serrate, 7–9-lobed (lobes shallow), thin, abrupt at base; petioles long, slender, and (with the calyx) subglandular; styles 3—5. Thickets: common. 10—20f. May.

7 **C. cordàta** Ait. *Washington Thorn.* Lvs. cordate-ovate, somewhat deltoid, incisely and often deeply 3–5-lobed, serrate, with long petioles; sep. short; sty. 5; fr. small, globous-depressed. Banks, Va. to Fla. 15—20f. ‡

8 **C. Crus-galli** L. *Cock-spur Thorn.* Lvs. obovate-cuneiform, tapering to a short petiole, serrate, coriaceous, shining above; spines very long; corymbs glabrous; sep. lanceolate, subserrate; styles 1 (2 or 3). Thickets. 10—20f. Fruit pyriform. June.

9 **C. spathulàta** Mx. Lvs. small, coriaceous, shining, oblong-spatulate, attenuated to the subsessile base, crenate above, sometimes lobed; corymbs numerous, lateral, 20–25-flowered; sepals very short; fruit very small, scarlet. South. 10—15f. June.

10 **C. æstivàlis** T. & G. *Apple Haw.* Young lvs. rust-downy, older smooth above, elliptic, repand, short-stalked; corymbs glabrous, 2–5-flowered; fruit large (8—9″), globular, red. Wet shores, S. 20—30f. Fruit pleasant, in May. (See Addenda.)

11 **C. parviflòra** Ait. Thorns straight and slender; lvs. cuneate-obovate, subsessile; fls. subsolitary, villous-tomentous; sep. incised, leafy, as long as the petals; sty. 5; fr. large, roundish, yellowish. Sandy woods, N. J. and S. 4—7f. April, May.

12 **C. flàva** Ait. *Summer Haw.* Thorns straight or arcuate; lvs. rhombic-obovate, attenuate into a glandular petiole; corymbs 1 (often 2 or 3)-flowered; styles 4 or 5; fruit large, pear-shaped. Va. to Fla. 15—25f. April, May.

13 **C. víridis** L. Thorns few and short; lvs. roundish or oval, acute at each end, sharply and doubly toothed above; petioles glandless; corymbs 3–6-flowered; styles 2 or 3; fruit large, globular. Iowa to Fla. 12—18f. April, May.

20. PYRUS, L. PEAR, APPLE, &c. Calyx urceolate, limb 5-cleft. Pet. 5, roundish. Styles 5 (2 or 3), often united at base. Pome closed, 2–5-carpelled, fleshy or baccate. Carp. cartilaginous, 2-seeded. ♄ ♄ Lvs. simple or pinnate. Flowers white or rose-colored, in cymous corymbs.

§ PYRUS. Leaves simple, glandless. Styles distinct. Pome pyriform..............No. 1
§ MALUS. Leaves simple, glandless. Styles united below. Pome globous .Nos. 2—4
§ ARÒNIA. Leaves simple, glandular on the midvein. Styles united, &c..........No. 5
§ SORBUS. Leaves pinnate. Styles 2—5, distinct. Pome small (scarlet).......Nos 6, 7

1 **P. commùnis.** *Pear-tree.* Lvs. ovate-lanceolate, obscurely crenate, glabrous and polished above, acute or acuminate; corymbs racemous; cal. and pedicels pubescent; styles 5, distinct and villous at base. Europe. 20—35f.

2 **P. Malus.** *Common Apple-tree.* Lvs. ovate or oblong-ovate, serrate, not lobed, downy, the veins all incurved; corymbs subumbellate; pet. with short claws; styles 5, united and villous at base. Europe. 20—30f. Nearly §.

3 **P. coronària** L. *Wild Crab-tree.* Lvs. ovate. rounded at base, cut-serrate, often sublobate, straight-veined, soon smoothish; sep.subulate; fls. large, roseate, corymbed, fragrant; pome large (18″), sour. Glades. 10—20f. May.

4 **P. angustifòlia** Ait. Lvs. lanceolate, often acute at base, crenate-serrate or subentire, short-stalked; sep. ovate; styles distinct. Pa. and S. 20—30f. March.

5 **P. arbutifòlia** L. *f. Choke Berry.* Downy; lvs. oblong or obovate, crenate-serrulate, narrowed at base into a short petiole; fruit pyriform or subglobous, dark red. Damp woods. 5—8f. Fruit size of currants. May, June.

β. *melanocàrpa.* Nearly smooth; fruit blackish purple. Swamps. 2—4f.

6 **P. Americàna** DC. *Mountain Ash.* Lfts. oblong-lanceolate, acuminate, mucronately serrate, smooth, subsessile; cymes compound, with numerous flowers; pome small, globous; styles 3—5. Mountain woods, Can. to Ga. 15—20f. May. †

7 **P. Aucupària.** *English Mountain Ash.* Lfts. as in P. Americana, except that they are always smooth on both sides, and, with the serratures, less acute at apex, flowers corymbous; fruit globous. Europe. 20—40f. †

21. CYDÒNIA, Tourn. QUINCE. Flowers and leaves as in Pyrus. Carpels cartilaginous, many-seeded. Seeds covered with mucilaginous pulp. ♄ ♄ Flowers mostly solitary.

1 **C. vulgàris.** Lvs. oblong-ovate, obtuse at base, acute at apex, very entire, smooth above, tomentous beneath; fls. solitary, large, roseate; pome tomentous, obovoid. Europe. 8—12f. Stems crooked. April, May.

2 **C. Japónica.** *Japan Quince.* Lvs. glabrous, shining, coriaceous, ovate-lanceolate, acute at each end, serrulate; stip. reniform; spines short, straight; fls. axillary, subsessile, crimson. Japan. 5—6f. Very bushy. April, May.

ORDER XLV. SAXIFRAGACEÆ. SAXIFRAGES.

Herbs or *shrubs. Leaves* alternate or opposite, sometimes - stipulate. *Sepals* 4 or 5, cohering more or less, and partly or wholly adherent. *Petals* as many as the sepals, inserted between the lobes of the calyx. *Stamens* as many. as the petals, and alternate with them, or 2 to 10 times as many. *Ovary* mostly inferior, usually of 2 (2—4) carpels cohering at base and distinct or united above. *Fruit* generally capsular, 1-2-celled. *Seeds* small, many, albuminous. Figs. 25, 52, 53, 132, 250, 273.

A large order, now including Ribes and Parnassia, each often regarded as constituting separate orders.

Order 45.—SAXIFRAGACEÆ.

I. SAXIFRAGEÆ. Herbs. Stipules none or adnate. Petals imbricate, rarely convolute in the bud. Calyx free or partly adherent...(a)
  a Petals wanting. Ovary adherent, 1-celled. Stamens 10.................CHRYSOSPLENIUM. 1
  a Petals pinnatifid. Ovary half adherent, 1-celled. Stamens 5 or 10..........MITELLA. 2
  a Petals entire.—d Stam. 10.—e Ovary 1-celled, nearly free.................TIARELLA. 3
         —e Ovary 2-celled. Fls. perfect. Lvs. simple.....SAXIFRAGA. 4
         —e Ovary 2-celled. Fls. polyg. Lvs. compound...ASTILBE. 5
    —d Stam. 5.—f Ovary 2-celled, adherent. Seed rough........BOYKINIA. 6
         —f Ovary 2-celled, free. Seed wing-margined....SULLIVANTIA. 7
         —f Ovary 1-celled.—g Styles and carpels 2.......HEUCHERA. 8
               —g Styles and carpels 5.......LEPUROPETALON. 9
               —g Stigmas and carpels 4......PARNASSIA. 10
II. ESCALLONIEÆ. Shrubs with alternate leaves, no stipules, and a valvate corolla bud...(b)
  b Calyx free from the 2-celled ovary. Stamens 5. Capsule ∞-seeded.........ITEA. 11
  b Calyx adherent to the ovary. Stam. 5. Berry ∞-seeded. (From S. Am.).....ESCALLONIA. 12
III. HYDRANGEÆ. Shrubs with opposite, simple leaves, and no stipules...(c)
  c Corolla valvate in the bud.—h Cymes radiate. Shrub erect................HYDRANGEA. 13
         —h Cymes naked. Shrub climbing.............DECUMARIA. 14
  c Corolla convolute in the bud.—k Stamens 20—40 Petals 4.................PHILADELPHUS. 15
         —k Stamens 10. Petals 5. (Asiatic)...........DEUTZIA. 16
IV. RIBESIEÆ. Shrubs with alternate, palmately-lobed leaves, and baccate fr...RIBES. 17

1. CHRYSOSPLÉNIUM, Tourn. WATER CARPET. Calyx adnate to the ovary, 4–5-lobed, colored inside. Cor. 0. Sta. 8—10, short. Sty. 2. Caps. obcordate, 1-celled, 2-valved, many-seeded. ♃ Prostrate, small.

C. **Americànum** Schw. Lvs. opposite, roundish, slightly crenate, tapering to the petiole; cal. 4-cleft. Cool springs, Northward. 3—6′. Calyx yellowish. Apr. May.

2. MITÉLLA, Tourn. MITRE-WORT. Calyx 5-cleft, adherent to the base of the ovary. Pet. 5, pectinately pinnatifid, inserted on the throat of the calyx. Sta. 5 or 10, included. Sty. 2, short. Caps. 2-beaked, 1-celled, with two equal valves. ♃ Flowers small, in a slender raceme or spike.

1 **M. diphýlla** L. Lvs. cordate, acute, sublobate, serrate-dentate, radical ones on long petioles, the cauline 2, opposite, subsessile; fls. white, in a long, loose spike. Woods, N. Eng. to Car. 1f. May, June. Curious.
2 **M. nuda** L. Lvs. orbicular-reniform, doubly crenate, with scattered hairs above; scape filiform, few-flwd., naked or with a single leaf; pet. pinnatifid with filiform segments. Damp woods, N. Eng. N. Y.; rare. 6′. Very delicate. June.

3. TIARÉLLA, L. BISHOP'S CAP. Calyx 5-parted, the lobes obtuse Pet. 5, entire, the claws inserted on the calyx. Sta. 10, exserted, inserted into the calyx. Sty. 2. Caps. 1-celled, 2-valved, one valve much larger. ♃ Flowers white.

T. cordifòlia L. Lvs. cordate, acutely lobed, mucronate-dentate, pilous; scape racemous; stolons creeping. Rocky woods, Can. to Ga. Common North. 1f.

4. SAXÍFRAGA, L. SAXIFRAGE. Sep. 5, more or less united, often adnate to the base of the ovary. Pet. 5, entire, inserted on the tube of the calyx. Sta. 10. Anth. 2-celled, with longitudinal dehiscence. Caps. of 2 connate carpels, opening between the 2 diverging, acuminate beaks (styles). Seeds ∞. ♃

§ Leaves opposite (small) on the prostrate stem. Flowers purplish......... No. 1
§ Leaves alternate on the ascending stem. Flowers yellow or white..... Nos. 2, 3, 4
§ Leaves tosulate at the base of the mostly leafless scape...(a)

Order 45.—SAXIFRAGACEÆ.

a Calyx entirely free from the ovary (inferior) .... Nos. 5, 6, 7
a Calyx adherent to the base of the ovary (half superior)......... Nos. 8, 9, 10
                Exotic species, cultivated .... Nos. 11, 12

1 **S. oppositifòlia** L. Lvs. opposite, obovate, carinate, obtuse, punctate, persistent; fls. solitary; cal. free; pet. large, obovate, 5-veined, longer than the stamens. Rocky cliffs, Willoughby Lake, Vt. June.

2 **S. aizoìdes** L. Cæspitous, leafy; lvs. linear-oblong, thick, flat; sep. ovate, slightly adherent; pet. oblong, yellow, longer than the sepals; capsules as long as the styles. With No. 1, and N. W. June.

3 **S. rivulàris** L. St. weak, ascending, 3-5-flowered; radical lvs. petiolate, reniform, crenately lobed, cauline lanceolate. subentire; cal. lobes broad-ovate, nearly as long as the white, ovate petals. White Mts. and N.

4 **S. tricuspidàta** Retz. St. thick, erect; lower lvs. crowded, oblong, 3-cuspidate; fls. few, large, somewhat corymbed; sep. thick, ovate, shorter than the oblong-obovate, yellow, dotted petals. Lake shores, Can. and N.

5 **S. leucanthemifòlia** Mx. Viscid-pubescent; lvs. radical, spatulate, cut-dentate, tapering to a petiole; scape diffusely paniculate; calyx free, reflexed; pet. unequal, white, 3 of them spotted. Mts. S. 18′.

6 **S. eròsa** Ph. Viscid-pubescent; lvs. radical, thin, oblong-lanceolate, acute, with erose teeth; panicle oblong, loose, with leafy bracts; cal. free, with reflexed, obtuse sepals as long as the equal, obtuse white petals. Mts. Pa. to Car. 15′.

7 **S. Careyàna** Gr. Lvs. round-ovate to deltoid, coarsely dentate, abrupt at base; panicle diffuse; pet. equal, ovate or oblong, white, dotted, twice longer than the recurved sepals. Mts. S. (and S. Caroliniana Gray).

8 **S. aızòon** Jacq. Lvs. spatulate, obtuse, bordered with white cartilaginous teeth, and a marginal row of impressed dots; flowers corymbous paniculate; pet. obovate, white. Rocky shores, N. Ver. to Mich. and N. 5—10′. July.

9 **S. Virginiénsis** Mx. *Early Saxifrage.* Lvs. spatulate obovate, crenately toothed, shorter than the broad petiole; scape nearly leafless, paniculately branched; petals white, oblong, much exceeding the calyx. Rocks, common. 4—12′. April, May.

10 **S. Pennsylvánica** L. Lvs. oblong-lanceolate, rather acute, tapering at base, denticulate; scape forming a diffuse panicle; fls. pedicellate; pet. greenish, linear-lanceolate, but little longer than the cal. Wet meadows, N. Eng. to O. 1—2f. May, Jn.

11 **S. sarmentòsa.** With creeping runners; leaves roundish; pet. white, 2 longer than the other 3; scapes naked; plant hairy. China. Pretty for baskets.

12 **S. crassifòlia.** No runners; lvs. thick, oval; sc. naked; fls. pk. Siberia. Jn. Jl.

5. **ASTILBE**, Don. ☆ ☿ ♀ Calyx obconic, with 4 or 5 erect segments. Pet. 4 or 5, spatulate. St. 8 or 10, exserted. Ov. 2-celled. Carpels in fr. separating and dehiscing lengthwise inside. Seeds 1—4 in each cell. ♃ Coarse, weed-like plants. Leaves bi- or tri-ternate. Fls. small, yellowish-white, in spicate rac. forming a compound panicle (like Spiræa Aruncus).

A. decándra Don. St. tall, angular; lfts. subcordate, incisely lobed, mucronate-serrate; sterile flowers mostly apatelous; sta. 10. Mts. South. 4—6f. June—August.

6. **BOYKÍNIA**, Nutt. Calyx turbinate, adherent, 5-cleft. Pet. 5, deciduous. Sta. 5. Ov. 2-celled, 2-beaked. Capsule invested with the calyx, dehiscent between the beaks. ♃ Lvs. alternate, petiolate, palmate. Fls. cymous, white.

B. aconitifòlia Nutt. St. viscid-glandular; lvs. smoothish, deeply 5-7-lobed (like those of Aconitum); cyme fastigiate, the fls. secund. Mts. S. 1—2f. July.

7. **SULLIVÁNTIA**, T. & G. Calyx adherent to the base of the ovary

Segm. ovate, acute. Pet. oval-spatulate, twice as long as the calyx. Sta. 5, shorter than the calyx. Capsule 2-beaked, 2-celled. Seeds wing-margined. ♃ Lvs. mostly radical, palmate-veined. Fls. in a loose pan., small, wh.

S. Ohiònis T. & G.—Ohio, Wisc. Stem weak, ascending, 6—12′. Lvs. roundish, cordate, lobed and toothed. May, June.

8. HEÙCHERA, L. ALUM ROOT.
Calyx of 5 obtuse segm. Cor. of 5 small, entire petals, inserted with the 5 stamens on the throat of the calyx. Cap. 1-celled, 2-beaked, dehiscent between the beaks. Seeds many, with a rough, close testa. ♃ Lvs. radical, long-petioled, petioles with adnate stipules at base.

§ Fls. small (1—2″ long), regular; stamens and style much exserted.........Nos. 1—3
§ Fls. larger (3—5″ long), oblique; stamens and style short.................Nos. 4, 5

1 H. Americàna Willd. Viscid-pubescent; leaves roundish, cordate, somewhat 7-lobed; pan. elongated, loose, divaricate; cal. obtuse, short, about equalling the spatulate petals; stam. much exserted. Shades, W. and S., rare N. 2—4f. May, June.

2 H. villòsa Mx. Villous, with rusty, spreading hairs; radical lvs. round-cordate, thin, glabrous above, 7-9-lobed; pan. loose, filiform; pet. white, about as long and narrow as the filaments. Mts. Md. to N. Car. and Ky. 1—3f. June, July.

3 H. cauléscens Ph. Smooth or nearly so; lvs. 5-7-lobed, dentate; pan. loose, slender; scape bearing one or two leaves below; pet. linear-spatulate, twice longer than the calyx. Mts. Car. Tenn. Ky. 1—2f. (H. Curtisii Gr.)

4 H. pubéscens Ph. Lvs. glabrous, round-cordate, 7-9-lobed; panicle dichotomous, geniculate; style exserted, stam. included; pet. white. Mts. Middle States.

5 H. hìspida Ph. Lvs. hispid-rough, 5-7-lobed, lobes very obtuse; fls. scattered; pet. spatulate, purple; sta. a little exserted. Mts. S. and prairies W. June.

9. LEPUROPÉTALON, Ell.
Calyx 5-parted, lobes obtuse, tube turbinate, adherent to the base of the 3-carpelled ovary. Petals 5, minute, spatulate, persistent. Sta. 5, short. Capsule globous, 1-celled, 3-valved, many-seeded. Placentæ opposite the stigmas. ① A minute, succulent herb, growing in tufts. Lvs. entire, dotted. Fls. terminal.

L. spatulàtum Ell.—Hard soils S. Stems scarcely 1′; leaves spatulate, veinless; fls. large in proportion, white. March, April.

10. PARNÁSSIA, Tourn. GRASS OF PARNASSUS.
Sep. 5, united at base, persistent. Pet. 5, persistent, with a bundle of sterile fil. at the base of each, and 5 perfect stamens alternating. Caps. 1-celled, 4-valved. Placentæ opposite the stigmas, in the middle of each valve. Seeds winged ♃ Glabrous. Lvs. radical. Scape 1-flowered, often with one sessile leaf. Pet. white, with green veins.

1 P. Caroliniàna L. Sterile filaments 3 in each group, each with a little round head; pet. sessile; lvs. broad-oval, rounded at base, one sessile on the scape. Wet meadows. 10—15′. Flower handsome, 1′ broad. June—August.

2 P. asarifòlia Vent. Sterile fil. 3 in each set; pet. abruptly clawed; lvs. reniform. Mts. Va. and Car. 10′. Lvs. large (1—2′).

3 P. palustris L. Sterile fil. pellucid, setaceous, 9—15 in each set; cauline leaf, if any, sessile; radical lvs. all cordate. Bogs, Mich. N. and W. 6′. Fls. 1′. August.

11. ÌTEA, L.
Calyx small, with 5 subulate segm. Pet. 5, lance-linear, inflexed, inserted with the 5 stam. on the calyx. Styles united. Caps. 2-

celled, 2-furrowed, 8–12-seeded. ♄ With alternate, simple leaves, and a simple, spicate, terminal raceme of white flowers.

1. Virgínica L.—Swamps, Pa. to Fla. 6f. Lvs. oval, acuminate, short-stalked. May, Jn.

12. ESCALLÒNIA RUBRA and E. GLANDULOSA are handsome shrubs, with evergreen leaves and scarlet flowers, prized in the greenhouse. S. Am.

13. HYDRÁNGEA, L. HYDRANGEA. Marginal fls. sterile, neutral— an enlarged, rotate 5-lobed, colored calyx only. ☿ Calyx tube hemispherical, adherent. Limb 4–5-toothed, persistent. Pet. ovate, sessile. Stamens twice as many as the petals. Caps. 2-beaked, opening between the beaks. Seeds ∞. ♄ With opposite leaves. Fls. cymous, generally radiant.

§ Cymes paniculate. Lvs. sinuate-lobed. Fls. rose-white......................No. 1
§ Cymes corymbous, level-topped. Leaves undivided....................Nos. 2, 3, 4

1 H. quercifòlia Bartram. Lvs. deeply sinuate-lobed, dentate, tomentous beneath, and on the petioles and veins above; cymes paniculate, radiant, the sterile fls. very large and numerous. Shady banks, S. 4—8f. A superb plant. †

2 H. arboréscens L. Lvs. ovate, obtuse or cordate at base, acuminate, serrate-dentate, paler beneath, nearly smooth; fls. white-red. Banks, S. and W. 5—6f.

3 H. radiàta Walt. Lvs. ovate, abrupt or cordate at base, acuminate, serrate, silvery-tomentous beneath; fls. white. Uplands, S. 6—8f.

4 H. HORTÉNSIS L. *Changeable Hydrangea.* Lvs. elliptical, narrowed at each end, dentate-serrate, strongly veined, smooth. China? 1—3f. In cultivation the fls. are generally *all* neutral, of varying hues, white, blue, pink, &c.

14. DECUMÀRIA, L. Calyx 7–10-toothed, tube adherent to the 5–10-celled ovary. Pet. as many as calyx teeth, valvate in the bud. Sta. 3 times as many as the petals, in one row. Stig. radiate. Caps. many-ribbed, crowned with the style, ∞-seeded. ♄ With rootlets, opposite leaves and cymes of white, fragrant flowers.

D. bárbara L.—A beautiful climber, in damp woods, S. 15—30f.

15. PHILADÉLPHUS, L. FALSE SYRINGA. Calyx 4–5-parted, half superior, persistent. Cor. 4–5-petalled. Sty. 4-cleft. Sta. 20—40, shorter than the petals. Caps. 4-celled, 4-valved, with loculicidal dehiscence. Sds. many, arilled. ♄ Handsome. Leaves opposite, exstipulate.

1 P. inodòrus L. Lvs. ovate, acute or pointed, 3 (rarely 5)-veined, smooth, entire or with remote slender teeth; calyx lobes ovate, acute, as long as the tube; styles united; fls. scentless, 1 or several together, pure white, 1'. Uplands, S. 5—8f. May–Jl.
 β. *grandiflòrus.* Pubescent; flowers larger (1½'); sepals acuminate. Cultivated.
 γ. *hirsùtus.* Hairy; leaves and flowers smaller, the latter 7''. Mt. woods.

2 P. CORONÀRIUS. *Mock Orange.* Glabrous; lvs. ovate, remotely serrate above, 5–7-veined; flowers in dense clusters, cream-white, very fragrant; styles separate. S. Europe. 5—8f. June, July.

16. DEÙTZIA, Thunb. Pet. 5, valvate or imbricate in bud. Sta. 10, the alternate longer, fil. dilated, 3-toothed, middle tooth antheriferous. Ov inferior. Caps. 3–5-celled. ♄ Leaves opposite. Fls. numerous, white.

1 D. SCABRA. Lvs. ovate, acute, serrate, rough-hairy; racemes terminal, dense; styles 3; flowers bell-shaped. Japan. 5—8f. Very fragrant. June.

2 D. GRÁCILIS. Foliage similar to the other but smooth er. Shrub only 2–3f, branches covered with flowers in June.

17. RIBES, L. CURRANTS. Calyx tube ovoid, adherent to the one-celled ovary, limb tubular or bell-shaped, 4–5-cleft. Pet. 4—5, small, inserted with the 4—5 stamens on the top of the calyx tube. Sty. 2. Berry filled with pulp, with 2 parietal placentæ. Seeds ∞, albuminous. ♂ ♀ Leaves alternate, palmately lobed. 3—6f. Styles often united.

§ RIBÉSIA. *Currants.* Stems and berries not prickly. Flowers in racemes...(a)
  a Flowers greenish or red. Lvs. plicate in the bud.—*b* Fruit smooth....Nos. 1, 2, 3
    —*b* Fruit hairy......Nos. 4, 5, 6
  a Flowers bright yellow. Leaves convolute in the bud................No. 7
§ GROSSULÀRIA. *Gooseberries.* Stems spinescent. Leaves plicate...(c)
  c Peduncles 5-8-flowered. Style 2-cleft. Berries small, hispid..............No. 8
  c Peduncles 1-3-flowered.—*d* Calyx tube and fruit prickly.................Nos. 9, 10
    —*d* Fruit smooth.—*e* Leaves cordate at base........No. 11
    —*e* Leaves not cordate....Nos. 12, 13, 14

1 **R. rubrum** L. *Common Red C.* Lvs. obtusely 3-5-lobed, pubescent beneath, subcordate; rac. smoothish, pendulous; calyx limb rotate; bracts short; fr. globous, glabrous, red, rarely amber. Woods, Vt. Wisc. †

2 **R. flóridum** L'Her. *Wild Black C.* Lvs. acutely 3-5-lobed, resinous-dotted, subcordate; rac. pubescent, pendulous; cal. cylindrical; bracts long; fruit obovoid, smooth, black. Copses, Can. to Ky. 3—4f. May, June.

3 **R.** NIGRUM. *Black C.* Lvs. 3-5-lobed, resinous-dotted beneath, not cordate; rac. lax, hairy; calyx bell-shaped; fruit roundish, black. Eur. 4—5f.

4 **R.** SANGUÍNEUM. Lvs. 3-5-lobed, white-downy beneath, cordate; rac. long, lax, all rose-red; calyx segments spreading; styles united; fruit blue. Oregon.

5 **R. prostrátum** L'Her. *Mountain C.* Stems reclined; lvs. 5-7-lobed, rugous, cordate; rac. erect, lax; cal. rotate; berries globous, glandular-hispid, red, ill-scented. Rocks, N. Eng. to Car. Raceme becoming erect. May.

6 **R. resinósum** Ph. Clothed with resinous-glandular hairs; lvs. 3-5-lobed, roundish; raceme erect; calyx spreading. Mts. Car. (Lost.)

7 **R.** AÙREUM Ph. Glabrous; lvs. 3-lobed, subentire, shorter than their stalks; raceme lax; calyx limb tubular, longer than the pedicels; fruit oval, yellow, soon brown. Mo. to Oreg. 6—10f. Flowers fragrant.

8 **R. lacústre** Poir. Spiny and prickly; lvs. deeply 3-5-lobed and incised, cordate; raceme hairy; style 2-cleft; fruit hispid. Swamps, Northward.

9 **R. Cynósbati** L. *Prickly G.* Spines in pairs, prickles few or none; lvs. cordate, lobed, pubescent, cut-dentate; styles united to the top; fruit brown-purple, with long spines, catable. Thickets, Northward. May.

10 **R.** SPECIÒSUM. Glabrous; lvs. roundish, lobed, crenate, polished; spines long, in 3's; flowers nearly solitary, pendulous, scarlet. California. Very handsome.

11 **R. hirtéllum** Mx. Spines few and short, prickles 0; lvs. roundish, lobed, toothed; calyx limb bell-shaped, lobes twice longer than the petals; stamens exserted; style 2-cleft. Rocky woods, N. Eng. to Wisc. Fruit purple.

12 **R. rotundifòlium** Mx. Spines few and short; prickles few or 0; lvs. roundish, lobed, cut-crenate-dentate, smooth or downy; calyx lobes linear, reflexed; stamens and styles much exserted. Rocky woods. May.

13 **R.** UVA-CRISPA. *English G.* Spiny; lvs. roundish, short-stalked, hairy beneath; peduncle hairy, 1-flowered; fruit oval or globous, large (8—12''), red, green, amber, white, &c. Europe.

ORDER XLVI. CRASSULACEÆ. HOUSE-LEEKS.

Plants herbaceous or shrubby, succulent. *Leaves* entire or pinnatifid. *Stipules* 0. *Flowers* sessile, usually in cymes and perfectly symmetrical.

Order 46.—CRASSULACEÆ.

Sepals 3—20, more or less united at base, persistent. *Petals* as many as the sepals. *Stamens* as many as the petals, and alternating with them, or twice as many. *Ovaries* as many as the petals. *Filaments* distinct. *Anthers* 2-celled, bursting lengthwise. *Fruit* distinct follicles or a capsule, many-seeded. Figs. 8, 9, 468.

§ Carpels distinct, forming a circle of follicles...(*)
* Petals distinct.—*a* Flowers all 3- or 4-parted. Stamens 3 or 4...................TILLÆA. 1
—*a* Flowers 5-, or 4- and 5-parted. Stamens 8 or 10.............SEDUM. 2
—*a* Flowers all 5-parted. Stamens 5.......................CRASSULA. 3
—*a* Flowers 6-12-parted, with cleft hypogynous scales..........SEMPERVIVUM. 4
Petals united at base.—*b* Flowers 4-parted. Stamens 8........................BRYOPHYLLUM. 5
—*b* Flowers 5-parted. Stamens 5......................ROCHEA. 6
—*b* Flowers 5-parted. Stamens 10.....................ECHIEVERIA. 7
§ Carpels united into a many-seeded capsule...(x)
x Flowers 4-parted, with 8 stamens..............................DIAMORPHA. 8
x Flowers 5-parted, with 10 stamens. Petals often wanting.............PENTHORUM. 9

1. TILLÆA, Mx. PIGMY-WEED. Calyx of 3 or 4 sepals united at base. Petals 3 or 4, equal. Sta. 3 or 4. Caps. 3 or 4, distinct, follicular, opening by the inner surface, 2- or many-seeded. ≋ Very small. Lvs. opposite.

T. simplex Nutt. St. ascending or erect, rooting at base; lvs. connate at base, linear-oblong, fleshy; flowers axillary, solitary, subsessile, their parts in 4's; pet. greenish; carpels 8-10-seeded. ① Muddy banks, Ct. to Md. 1–2'. July–Sep

2. SEDUM, L. STONE-CROP. Sep. 4 or 5, united at base. Pet. 4 or 5, distinct, spreading. Sta. 8—10. Carp. 4—5, distinct, many-seeded, with an entire scale at the base of each. ♃ Lvs. fleshy. Inflorescence cymous.

§ Fls. in scorpoid racemes or spikes, or axillary, the latter often 4-parted....Nos. 1—4
§ Fls. in corymbous cymes, all 5-parted.—*a* Leaves mostly alternate........Nos. 5—7
—*a* Leaves opposite, and whorled......No. 8

1 S. ternàtum Mx. Leaves scattered, flat, obovate, the lower mostly in whorls of 3, the upper spatulate; spikes 3, rarely 2—4, radiating, secund; central flower 5-parted, the rest 4-parted, white. Damp woods. 3—8'. May, June.

2 S. Nèvii Gr. Stem weak, branched, 3—5'; leaves alternate, imbricated, small, obovate-spatulate; petals lance-linear, white. Mts., Va. (Porter), and S. June, July.

3 S. pulchéllum Mx. Leaves linear, alternate, crowded; spikes radiating, dense flowered, secund, central flower 5-, the others 4-parted, rose-purple. Rocks, Va. to Tex. 4—12'. May—July. Very pretty in gardens.

4 S. acre L. *English Moss.* Procumbent, diffuse; leaves very small, fleshy, crowded, alternate, appressed; cyme leafy, somewhat trifid; fls. yellow. Gardens. Jl. § Eur.

5 S. Rhodiola DC. Stems clustered, erect, 5—10'; leaves mostly scattered, obovate, with several angular teeth or entire, crowded; flowers 4-parted, in a small cyme at top, yellowish, *diœcious*. Rocks, Penn. (Prof. Porter), Me., and Can.

6 S. telephioìdes Mx. Ascending, tall; lvs. round-oval to lance-oval, narrowed to the base, subdentate, alternate; pet. acuminate, pink. Rocks, Md., and S. Stems 1f, leaves 1—2'. Flowers numerous, in a terminal branching cyme. June.

7 S. Telèphium L. *Live-forever.* Clustered, erect, very leafy; lvs. ob·, ng-ovate, obtuse, dent-serrate; corymb dense, leafy, blue-purple. Waste grounds, &c. Stems 1-2f, round, simple, with a compact pale-purple cyme at top. August. § Europe.

8 S. Sieböldii. Lvs. opposite, or in 3's, roundish, glaucous, sessile; cymes dense, leafy fls. 5-parted, small, bluish-purple, blooming in October. Japan. In dense tufts. A pretty plant, and one of the last to flower in the garden. Like most of the Sedums its severed stalks will grow even if suspended in air.

3. **CRÁSSULA**, Haw. Parts of the flower all in 5's, distinct and free Scales at base of ovaries 5. ♄ ♃ Fleshy plants, from S. Africa, remarkable for the perfect symmetry of their flowers.

1 C. ARBORÉSCENS. Stem shrubby, terete, erect; lvs. opposite, fleshy, roundish, cuspidate, flattish, glaucous, dotted above; cyme 3-parted; flowers handsome, roseate.

2 C. LÁCTEA. Stem erect, twisted below, branched; lvs. ovate, narrowed to the connate bases, dotted along the margin; cyme panicled, with many white star-like flowers. Leaves bright green. From S. Africa, as are many other species.

4. **SEMPERVÌVUM**, L. LIVE-FOREVER. HOUSE-LEEK. Sep. 6—20, nearly distinct. Petals and pistils as many, and stamens twice as many. Scales lacerated. ♄ ♃ Leaves thick and fleshy, crowded.

S. TECTÒRUM. Lvs. oval-obovate, ciliate-fringed, densely packed at the ends of the offsets, scattered on the stems; flowers purplish, usually 12-parted. Europe. Will grow on walls and on the roofs of houses (*tectorum*), or in borders.

5. **BRYOPHÝLLUM** CALYCÌNUM. ♄ Evergreen, fleshy, 2f. Leaves opposite, 3-5-foliate, with thick, oval, crenate leaflets. Flowers in a loose, terminal panicle, with an inflated calyx and a tubular, exserted, purplish corolla, which has a 4-lobed limb. The plant is propagated from the leaves, which produce buds on their margins becoming new plants,—like ovules from a carpellary leaf.

6. **RÒCHEA**, DC. Corolla funnel-form, 5-cleft. Sepals, stamens, ovaries, and hypogynous scales each 5. ♄ Fleshy. S. African.

1 R. FALCÀTA. Shrub 2f; leaves opposite, the pairs some united at base, glaucous, oblong, deflexed-falcate; flowers in corymbous cymes, red, open, fragrant.

2 R. COCCÍNEA. Leaves connate-sheathing, ovate-oblong; cymes scarlet. Beautiful.

7. **ECHEVÈRIA**, DC. Corolla tubular to bell-form, 5-lobed or parted. Calyx 5-cleft. Stamens 10. Ovaries 5, with 5 scales. ♄ ♃ Fleshy.

1 E. GRANDIFÒLIA. Plant 2f, erect, glaucous with a bloom; lvs. spatulate to obovate, acute, the lowest large, rosulate; flowers urn-shaped, panicled, orange-red. From Mexico.

2 E. COCCÍNEA. Plant 2f, erect; leaves obovate-cuneate, acute, scattered; flowers carmine outside, yellow within, in a tall leafy spike. Mexico.

8. **DIAMÓRPHA**, N. Fls. 4-parted, with 8 stamens. Carp. 4, united below, at length spreading, opening by an irregular valve on the back, 4-8-seeded. ② Small, fleshy, tufted, with cymes of white or pink flowers.

D. **pusilla** N.—Sunny rocks, S. 1—3'. Leaves oval, sessile, 1''. March, April.

9. **PENTHÒRUM**, L. VIRGINIA STONE-CROP. Calyx of 5 sepals united at base. Pet. 5 or 0. Sta. 10. Caps. of 5 united carpels, 5-angled, 5-celled, 5-beaked, dehiscent by an obliquely-terminal valve. Seeds ∞, minute. ♃ Not succulent. Lvs. alternate. Fls. yellowish, cymous.

P. **sedoìdes** L. Stem branched and angular above; leaves nearly sessile, lanceolate, acute, serrate; fls. in secund, radiating racemes. Wet places. 10-16'. July—Sept.

ORDER XLVII. HAMAMELACEÆ. WITCH HAZELWORTS.

Shrubs or *trees* with alternate simple leaves and deciduous stipules. *Flowers* in heads or spikes, often ♂ ☿ ♀ or ⚥. *Calyx* adherent. *Petals* linear, or 0. *Stamens* twice as many as the petals, the opposite sterile and scale-like, or ∞. *Ovaries* of 2 carpels, 2-celled, 2-styled, *ovules* 2 or ∞. *Fruit* a woody capsule, 2-beaked, 2-celled, 1–2-seeded.

§ Petals 4. Calyx 4-lobed. Stamens 4. Flowers mostly ☿ .... HAMAMELIS. 1
§ Petals 0. Calyx truncate. Stamens 20—28. Flowers ☿ .... FOTHERGILLA. 2
§ Petals 0. Calyx 0. Stamens ∞. Flowers ⚥, in globular heads .... LIQUIDAMBAR. 3

1. HAMAMÈLIS, L. WITCH HAZEL. Calyx with an involucel of 2—3 bracts at base. Pet. very long, linear. Sterile stamens scale-like, opposite the petals, alternating with the 4 fertile ones. Caps. nut-like, 2-celled, 2-beaked. ♄ ♄ Flowers yellow.

H. Virginiàna L. Lvs. oval or obovate, acuminate, crenate-dentate, obliquely cordate; fls. sessile, 3—4 together, blooming in late autumn and winter. Woods. Stems crooked, 10—15f. Pet. twisted, 9″ long.

2. FOTHERGÍLLA, L. *filius*. Calyx campanulate, truncate and obscurely 5–7-toothed, bearing the stamens in one marginal row. Styles distinct. Caps. 2-lobed. ♄ Lvs. oval or obovate, expanding after the dense spikes of flowers.

F. alnifòlia L. *f*.—Swamps, Va. to Fla. 2—4f. Calyx white, fringed with the long white or pink filaments. Styles long, recurved. March, April.

3. LIQUIDÁMBAR, L. SWEET GUM TREE. Involucre 4-parted deciduous. ♂ Ament conical. ♀ Ament globular. Calyx a scale, if any. Fruit a globular sorosis (§ 171), woody, consisting of the scales, and capsules which open between their beaks. Ovules ∞, 1 or 2 maturing. ♄ Leaves and gum fragrant. Twigs winged with corky bark.

L. styracíflua L. Lvs. palmate, with 5 acuminate, serrate lobes; veins villous at their bases. A large and handsome tree, Conn. to Ill. and S. 60f. May.

ORDER XLVIII. HALORAGEÆ. THE HIPPURIDS.

Herbs mostly aquatic, with incomplete or minute ☿—☿ flowers. *Calyx tube* adherent. *Petals* 0—4. *Stamens* 1—8. *Pollen* 4-grained. *Ovary* 1–4-celled. *Styles* 1—4, distinct, one pendulous ovule in each cell. *Fruit* indehiscent, 1–4-celled, 1–4-seeded. *Seed* pendulous, anatropous, albuminous (Formerly joined to Onagraceæ.)

* Flowers 3-parted, apetalous, perfect .... PROSERPINACA. 1
* Flowers 4-parted, monœcious; petals 4 or 0 .... MYRIOPHYLLUM. 2
* Flowers 1-parted, apetalous, perfect .... HIPPURIS. 3

1. PROSERPINÀCA, L. MERMAID WEED. Calyx tube adherent to the ovary, 3-sided, limb 3-parted. Pet. none. Sta. 3. Stig. 3. Fruit 3-angled, 3-celled, bony, crowned with the calyx. ♃ Roots creeping. Lvs. alternate. Fls. greenish.

Order 51.—MYRTACEÆ.

1 P. palústris L. Lvs. linear-lanceolate, sharply serrate above the water, those be low (if any) pinnatifid. ♃ Swamps: common. 6—20′. Lvs. 1—2′. June, July.
2 P. pectinácea Lam. Lvs. all pectinate, with linear-subulate segm.; fr. obtusely 3-angled. ♃ Sandy swamps, Ms. (rare) to Fla. 5—10′; long creepers at base. Jl. Aug

2. MYRIOPHÝLLUM, Vaill. WATER MILFOIL. Flowers ☿, or frequently ⚥. Calyx 4-toothed in the ⚥ and ♀ flowers, 4-parted in the ♂. Pet. 4, often inconspicuous or none. Sta. 4—8. Stig. 4. pubescent, sessile. Fr. of 4 nut-like carpels, cohering by their inner angles. ≈ ♃ Submersed lvs. parted into capillary segments. Upper fls. usually ♂, middles ones ⚥, lower ♀, greenish, emerging in summer.

§ Stamens 8. Carpels smooth and even. Leaves whorled in 3's, rarely in 4's..Nos. 1, 2
§ Stamens 4.—Carpels ridged on the back. Leaves whorled in 4's and 5's....Nos. 3, 4
—Carpels smooth and even. Leaves alternate or wanting........Nos. 5, 6

1 M. spicátum L. Floral lvs. ovate, entire, shorter than the flowers, the rest all pinnately capillary; fls. in term. spikes. Deep waters, fls. emerging. 10f.
2 M. verticillátum L. Floral lvs. pectinate-pinnatifid, much longer than the flowers, the lower pinnately-setaceous. Spikes leafy, terminal. Slow waters.
3 M. heterophýllum Mx. Floral lvs. ovate-lanceolate, serrate, longer than the fls. crowded, the rest pinnately or pectinately capillary. Ponds: rare.
4 M. scabrátum Mx. Floral lvs. linear, pectinately toothed; fr. roughened, sharply angled; verticils axillary. Shallow waters. 6—12′. Capillary segments few.
5 M. tenéllum Bw. Erect and almost leafless; floral leaves or bracts alternate, minute, entire, obtuse; fls. ☿; petals linear. Water edges, N. Eng. N. Y. and N. Scapes 4—12′, from long creeping rhizomes. Fls. purplish-white, sessile.
6 M. ambíguum Nutt. Lvs. many, submersed ones pinnate, with capillary segments, middle ones pectinate, upper linear; fls. mostly ⚥. Floating in ponds and ditches. Ms. to Ga.
β. *limòsum*. Small, procumbent, rooting, in muddy places; lvs. all linear.
γ. *capilláceum*. Very slender; lvs. all immersed and capillary, in ponds.

3. HIPPÚRIS, L. MARE'S TAIL. Calyx with a minute, entire limb crowning the ovary. Cor. 0. Sta. 1, inserted on the margin of the calyx. Anth. 2-lobed, compressed. Style 1, longer than the stamen, stigmatic the whole length. Seed 1. ≈ ♃ St. simple. Lvs. verticillate, entire. Fls. axillary, greenish.

H. vulgáris L. Lvs. in verticils of 8 to 12, linear, acute, smooth, entire; fls. solitary, minute. Borders of ponds, marshes. N. and W.: rare. 1—2f. Dakotah (Matthews)

Order LI. MYRTACEÆ. MYRTLEBLOOMS.

Trees and *shrubs*, without stipules. *Leaves* opposite, entire, punctate, usually with a vein running close to the margin. *Calyx* adherent below to the compound ovary, the limb 4- or 5-cleft, valvate. *Petals* as many as the segments of the calyx. *Stamens* numerous. *Anthers* introrse. *Style* and *stigma* simple. *Fruit* with many seeds. *Albumen* none.

Our Myrtleblooms are either tender exotics, or indigenous far South The following table must suffice for their recognition.

* Calyx truncate. Petals connate into a caducous calyptra or lid...(a?
  a Fruit a capsule. Stam. free. Australian trees, alternate-leaved................EUCALYPTUS.
  a Fruit a berry. Stam. free. Leaves opposite. Small trees in S. Florida ......CALYPTRANTHES
* Cal. 4-lobed. Pet. 4, spreading. Fr. bac. Lvs opp. Trees, shrubs. S. Fla. *Allspice, &c* .EUGENIA. 1
* Cal. 5-lobed. Pet. 5, spreading. Stam. long-exserted. Shrubs. Cultivated...(b?

ORDER 52.—MELASTOMACEÆ.

§ Stamens united into 5 sets. Fruit capsular. Lvs. alternate or opposite. Austrl..**Melaleuca.** 2
§ Stamens distinct.—*c* Flowers in dense lateral cymes. (Lvs. alternate.) Austrl..**Callistemon.** 3
—*c* Flowers solitary, axillary. Sepals equal. Lvs. opposite...**Myrtus.** 4
—*c* Flowers solitary, axillary. Sep. unequal. Opp. *Guava*...**Psidium.** 5

1. EUGÈNIA JAMBOS. *Rose Apple.* Tree (20—30f in India), with lanceolate leaves. Flowers white, in terminal showy cymes. Fruit round-ovoid, crowned with the calyx, 1½' diam., yellow, with a thick rind, which has a sweetish, *rose-like* flavor.

2. MELALEUCA HYPERICIFÒLIA. Shrubby, 5f, with opposite, elliptic-oblong, shining, 3-veined leaves on the drooping branches. Flowers of a splendid red, in slender spikes, with innumerable stamens (1' long) radiating in all directions.— M. LEUCADÉNDRON, the famous *Cajeput Tree* of the East, has long lance-linear leaves, white fls. spiked on the pendent branchlets. The trunk is black and the branches white.

3. CALLISTÈMON LANCEOLÀTUM. *Bottle-brush.* Beautiful shrub, with long, thick, lanceolate leaves, and the flowers in dense, cylindric spikes, crimson stamens innumerable, radiant at right angles, suggesting the English name. Often cultivated.

4. MYRTUS COMMUNIS. *Myrtle.* Evergreen shrub or tree of S. Europe, emblematic of victory in honorable contests. The leaves are long, ovate, shining, the flowers pure white or rose-tinged, with innumerable stamens, and the berries black.

ORDER LII. MELASTOMACEÆ. MELASTOMES.

Trees, shrubs, or *herbs,* with square branches and usually no stipules. *Leaves* opposite, undivided, dotless, and 3–5-veined. *Calyx tube* urceolate, adherent, at least to the angles of the ovary. *Petals* 4—6, convolute in bud. *Stamens* definite. *Anthers* opening by terminal pores. *Fruit* capsular or baccate.—Genera more than a hundred, all tropical except the following.

1. RHÉXIA, L. DEER-GRASS. Calyx 4-cleft, swelling at the base. Petals 4. Stamens 8, 1-celled. Styles declined. Capsules 4-celled, nearly free from the investing calyx tube. Seeds numerous. ♃ Leaves opposite, exstipulate, 3-veined. Flowers showy. June—September.

§ Anthers curved, saccate and appendaged at base. Flowers purplish...(a)
  a Stem square, winged. Leaves ovate to lanceolate, bristly-serrate.....Nos. 1, 2
  a Stem terete or teretish. Leaves lanceolate to linear............. .....Nos. 3, 4
§ Anthers straight, oblong.—*b* Stems simple, with purple flowers.....Nos. 5, 6
  —*b* Stems brachiate, with yellow flowers.......... No. 7

1 R. Virgínica L. *Meadow Beauty.* Stem narrowly 4-winged; leaves sessile, and with the stem clothed with scattered hairs; calyx hispid. Wet grounds, E. Mass., S. and W. 12—16'. Cymes corymbed. Flowers purple. July, August.

2 R. stricta Ph. Stem tall, strongly 4-winged, glabrous; leaves acuminate, glabrous; calyx glabrous, tube very short. Bogs, S. 3—4f. Purple. June, July.

3 R. Mariàna L. Hairy; leaves lanceolate and lance-linear, acute, bristly-serrate, tapering to a short petiole. Sandy bogs, N. J. to Fla. 1—2f. Purple.
  β. *lineàris.* Diffusely branched; lvs. almost linear. South. (R. lanceolata Walt.)

4 R. glabélla Ph. Glabrous, glaucous; lvs. lanceolate, subserrulate, acute, sessile; cal. glandular-hispid. Damp woods, S. 2—3f. Fls. few, large, purple. June—Aug.

5 R. ciliòsa Mx. Stem 1—2f, squarish; leaves broad-ovate, sparsely hispid above, margin ciliate with long bristles; flowers few, subsessile, terminal; calyx glabrous, lobes acute. Damp woods, Md. to Fla. Petals roundish. June—August.

6 R. serrulàta N. Stem 6—8', square; leaves small, roundish-oval, glabrous both sides, serrulate-ciliate; calyx glandular-hispid, lobes obtuse. Swamps, S.

7 R. lútea Walt. Leaves obloug-linear; flowers panicled; calyx much constricted above the ovary, limb bell-form, with cuspidate teeth. Damp woods, S. 18′.

2. CENTRADÈNIA ROSEA, from Mexico, is often seen in conservatories. A small shrub, with opposite, lanceolate leaves (one of each pair much smaller or obsolete). Fls. 4-parted, roseate, in numerous hanging clusters. Sta. 8, anthers appendaged.—**C.** GRANDIFÒLIA has the large lanceolate leaves crimson beneath, and cymes erect.

ORDER LIII. LYTHRACEÆ. LOOSESTRIFES.

Plants with entire, exstipulate, mostly opposite leaves, with a tubular *calyx* bearing the (4—7) petals and stamens in its throat, and a compound ovary and style. *Stamens* 4—14, rarely ∞. *Fruit* capsular and free, or baccate, 2—6-, or by abortion, 1-celled, ∞-seeded. *Albumen* 0.

§ Shrubs, with alternate leaves, ∞ stamens, and a bell-shaped calyx.............LAGERSTRŒMIA. 1
§ Shrubs, with opposite leaves, ∞ stamens, and a tubular, adherent calyx.........PUNICA. 2
§ Herbs—*a* Flowers irregular   Calyx inflated, gibbous at base....................CUPHEA. 3
 —*a* Flowers regular.—*b* Calyx cylindrical, striate, with 5 minute horns......LYTHRUM. 4
  —*b* Calyx campanulate,—*c* 5 teeth with 5 long horns....NASÆA. 5
   —*c* 4 teeth with 4 short horns...AMMANNIA. 6
   —*c* 4 teeth. Horns 0. Petals 0..DIDIPLIS. 7

1. LAGERSTRŒMIA INDICA. CRAPE MYRTLE. Petals 6, crisped, on claws inserted into the calyx tube. Sta. ∞. Lvs. round-ovate, thick, smooth. Branches winged. Flowers blue-purple, in panicles. Common S. † and §. From E. India.

2. PÙNICA GRANATUM. POMEGRANATE. Lvs. lanceolate. Pet. 5, oval, obtuse, erect, scarlet, large. Fr. large, crim., crowned with the calyx, eatable, of singular structure, being 3-celled below and 5-celled above, 10—20f. Hardy in Fla. and La. (Eur.)

3. CÙPHEA, Jacq. Calyx tubular, 12-veined, gibbous at base, with 6 erect teeth, and often as many intermediate processes. Pet. 6 or 7, unequal. Stam. about 12, unequal. Sty. filiform. Caps. thin, 1-2-celled, few-seeded.

1 **C. viscosíssima** Jacq. ① Viscid-pubescent; branches alternate; lvs. opp., lanceovate; flowers violet-purple, short-stalked, 1 in each axil; capsules bursting laterally before ripe. Wet grounds, Mass., W. and S. Not common. 9—18′. August.
2 **C. PLATYCÉNTRA.** Low, bushy perennial; leaves lanceolate; fls. with a scarlet calyx tube and short, purple petals, produced in profusion all Sum. From Mex. Not hardy.
3 **C. STRIGULÒSA.** Shrubby, hispid and viscid; lvs. oblong-ovate; cal. scarlet, gibbous at base; petals 6, subequal, large, violet-purple, varying to yellow; sta. 11, hairy.
4 **C. SILENOÌDES.** Lvs. lanceolate; cal. green and red; pet. 5, purple, 2 large and 3 small.

4. LÝTHRUM, L. LOOSESTRIFE. Calyx cylindrical, striate, limb 4-6-toothed, with as many intermediate, minute processes. Pet. 4—6, equal. Stam. as many or twice as many as the petals, inserted in the calyx. Style filiform. Capsule 2-celled, many-seeded. ♃ Mostly with entire leaves and purple or pale flowers. June—Aug.

§ Stamens as many as the petals. Flowers axillary, solitary.................Nos. 1—3
§ Stamens twice as many as the petals. Flowers spicate or racemed........Nos. 4, 5

1 **L. hyssopifòlium** L. *Grass-poly.* Glabrous, slender; branches square; lvs. alternate or opposite, linear or oblong-lanceolate, obtuse; fls. solitary, axillary, subsessile: pet. and stam. 5 or 6. Low grounds, coastward, Ms., N. Y. Rare. 6—10′.
2 **L. alàtum** Ph. Glabrous, erect, branched; stem winged below; lvs. lance-ovate acute, sessile, broadest at base, alternate and opposite; flowers axillary, solitary with 6 wavy petals and 6 short stamens. Damp. S. and W. 1—2f.
3 **L. lineàre** L. St. slender, somewhat 4-angled, branched above; lvs. linear, mostly opposite, obtuse; fls. nearly sessile; pet. and sta 6. Swamps, N J to Fla 2 ff

4 L. Salicària L. More or less pubescent; lvs. lanceolate, cordate at base; fls. nearly sessile, in a long, somewhat verticillate, interrupted spike; pet. 6 or 7; stam. twice as many. Wet meadows, N. Eng., N. Y. Rare. 2—5f. Fls. showy, purple. †
β. ROSEUM. Flowers rose-red, in many spikes, all summer. A fine garden variety.

5. NESÆA, Juss. Calyx short, broadly campanulate, with 5 erect teeth, and 5 elongated, spreading, hornlike processes. Sta. 10, alternate ones very long. Sty. filiform. Caps. globous, included, ∞-seeded. ♃ Lvs. opposite or verticillate. Flowers axillary, purple.

N. verticillàta Kunth. Swamps, common. Stems woody at base, stoloniferous, 2—4f, angular; lvs. lanceolate, acuminate, opposite or in whorls of 3's; fls. in a long, leafy, showy, slender panicle of umbels. (Decodon verticillatum Ell.)

6. AMMÁNNIA, L. Calyx campanulate, 4–5-toothed or lobed, generally with as many hornlike processes, alternating with the lobes. Pet. 4 or 5. Sta. as many, rarely twice as many as the calyx lobes. Capsule globular, 2–4-celled, ∞-seeded. ① Stems square and leaves opposite, entire. Flowers axillary.

A. hùmilis Mx. St. branched from the base, ascending; lvs. lanceolate, obtuse, tapering at base into a short petiole: fls. solitary, closely sessile, all the parts in 4's; sty. very short. Ditches. A low herb, with inconspicuous flowers. Aug., Sept.
2 A. latifòlia L. St. erect, branching; lvs. linear-lanceolate, acute, dilated and auricled at the sessile base; cal. 4-angled, 4-horned; fls. crowded. Wet, W. 1—2f. Purp.

7. DÍDIPLIS, Raf. Calyx 4-lobed, without accessory teeth. Pet. 0. Sta. 2—4. Ov. 2-celled. Stig. 2-lobed, subsessile. Caps. globous, bursting irregularly, ∞-seeded. ♒ Leaves opposite, crowded, linear. Flowers axillary, sessile, minute. (Hypobrichia, Curt.)

D. diándra.—Ponds and sluggish streams, Ill. and S. 10—20′ long. Jn.-Aug.

ORDER LIV. ONAGRACEÆ. ONAGRADS.

Herbs, rarely *shrubs*, with the flowers 4-(sometimes 2 or 3)-parted, with the *calyx tube* adhering to the 2–4-celled ovary, and teeth valvate in the bud; the *petals* convolute in the bud, sometimes obsolete as well as the calyx teeth. *Stamens* as many or twice as many as the petals or calyx teeth. *Ovary* 2–4-celled, *styles* united, and *stigmas* capitate or 4-lobed. *Fruit* capsular or baccate, 2–4-celled. *Seeds* with little or no albumen Figs. 13, 54, 188, 317, 385.

* Stamens 8, or twice as many as the petals or sepals...(a)
  a Calyx tube not prolonged above the ovary.—b Seeds comous...................EPILOBIUM. 1
                              —b Seeds glabrous..................JUSSLÆA. 2
  a Calyx tube prolonged,—c the free summit slender.—d Seeds comous, ∞......ZAUSCHNERIA. 3
                                  —d Seeds glabrous, ∞.....ŒNOTHERA. 4
                                  —d Seeds glabrous, 1—4....GAURA. 5
            —c the free summit enlarged,—e short. Pet. clawed....CLARKIA. 6
                                —e long. Pet. sessile.....FUCHSIA. 7
* Stamens 4 or 2, as many as the sepals.—d Flowers 4-parted.....................LUDWIGIA. 8
                            —d Flowers 2-parted........................CIRCÆA. 9

1. EPILÒBIUM, L. WILLOW-HERB. ROSE BAY. Cal. tube not prolonged beyond the ovary, limb deeply 4-cleft, deciduous. Sta. 8. Stig often with 4 spreading lobes. Ov. and caps. linear, 4-cornered, 4-celled, 4-valved. Seeds ∞, comous with long silky hairs. ♃ Flowers purple to white. July—Sept.

* Lvs. alternate. Fls. showy, expanding. Stig. with 4 long lobes. Sty. declined .No. 1
* Lvs. opposite. Fls. small. Stigma undivided.—*a* Petals entire...........Nos. 2, 3
                                              —*a* Petals 2-lobed............Nos. 4, 5

1 **E. angustifòlium** L. St. simple, erect; lvs. lanceolate, subentire with a marginal vein; rac. long, terminal, spicate; pet. unguiculate, purple; stig. with 4 linear, revolute lobes. In newly-cleared lands, fence-rows, &c., E. and W. 4—6f.
   β. *canescens*. Flowers pure white throughout; ovaries silvery canescent.
2 **E. alpìnum** L. St. creeping at base, usually with 2 pubescent lines, few-flwd.; lvs. glabrous, oblong-ovate, obtuse; caps. glabrous. High Mts. N. 6—12′. Fls. pale-roseate.
   β. *nutans*. Taller (1f), nodding at the summit; lvs. oblong, denticulate. White Mts.
3 **E. palústre** L. β. *albiflòrum*. Minutely downy, branching; lvs. sessile, linear or narrowly lance-lin.; caps. pubescent. Swamps, Pa., N. & W. 6′—2f. Fls. nearly wh.
4 **E. molle** Torr. Velvety-pubescent, strict, branched above; lvs. sessile, crowded, lanceolate- to linear-oblong, subentire; pet. deeply-emarginate, rose-color. Swamps. E. and W. 1—2f. Varies to nearly smooth, and less leafy. (N. Y., Hankenson.)
5 **E. coloràtum** Muhl. Nearly smooth, much branched; lvs. lance-oblong, dentserrulate, some petiolate, often with reddish veins; pet. 2-cleft, rose-color. Wet. 1-3f

2. **JUSSIÆA**, L. Calyx tube long, but not produced beyond the ovary; the lobes 4—6, leafy, persistent. Pet. 4—6, spreading. Sta. 8—12. Pod 4-6-celled, long, opening between the ribs. Seeds very numerous.— Herbs with alternate leaves and yellow flowers.

1 **J. decúrrens** DC. Glabrous; fls. 4-parted, 9″; st. erect, branched, winged by the decurrent, lanceolate lvs.; pod clavate, 4-angled. ♃ Wet. Pa., and S. 6-20′. Jl.-Sep.
2 **J. rèpens** L. Smooth, or hairy above, creeping, with erect branches; fls. 5-parted, 2′; lvs. oblanceolate to oblong, narrowed to the slender pet.; ov. much shorter than the pod. ♃ Ponds, ditches, Pa. to Ill., and S. 2—3f. May—Aug. (J. grandiflora Mx.)
3 **J. leptocárpa** N. Hairy; fls. mostly 6-parted, small (9″); lvs. lanceolate, subsessile; pod slender, much longer than the ped. ① Marshes, Fla. to La. 1—2f. June

3. **ZAUSCHNÈRIA** CALIFÓRNICA. ♃ Bushy, hairy-viscid, with lanceolate leaves and scarlet (varying to white) flowers resembling Fuchsias. Sta. exserted.

4. **ŒNOTHÈRA**, L. EVENING PRIMROSE. Calyx tube prolonged beyond the ovary, deciduous. Segm. 4, reflexed. Pet. 4, equal, obcordate or obovate. Sta. 8. Caps. 4-celled, 4-valved. Stig. 4-lobed. Seeds many, without a coma.—Herbs with alternate leaves. Summer.

* Native. Fls. nocturnal, yellow. Pods sessile, oblong, terete.............Nos. 1—3
* Native. Fls. diurnal, yellow. Pods clubshaped, 4-angled and 4-ribbed..(*a*)
  a Calyx tube not longer than the ovary. Fls. 5″ or 6″ diameter..........Nos. 4, 5
  a Calyx tube about twice longer than the ovary. Fls. 15″—18″.........Nos. 6—8
  a Calyx tube 3 or 4 times longer than the ovary. Fls. 2′—1′............Nos. 9, 10
* Exotic.—*b* Fls. yellow, large. Tube much longer than the ovary........Nos. 11, 12
  —*b* Fls. white, very large. Pods 4-winged and 4-ribbed.... .....Nos. 13, 14
  —*b* Fls. purple or roseate. Tube short, funnel-form. GODETIA..Nos. 15—18

1 **Œ. biénnis** L. St. erect, hirsute; lvs. ovate-lanceolate, repand-denticulate; fls. in a terminal, leafy spike; cal. tube 2 to 3 times longer than the ovary; stam. shorter than the obcordate or obtuse petals; pod oblong, obtusely 4-angled. Com. 2—5f.

β. **murìcàta.** Stem rough-hirsute; petals but little longer than the stamens.
γ. **grandiflòra.** St. branching; pet. much longer than stam., deeply obcordate. †
δ. **parviflòra.** Calyx tube elongated; petals small, as long as the stamens.
ε. **cruciàta.** Petals linear-oblong, shorter than the stamens.
ζ. **canéscens.** Petals enlarged; whole plant canescently hairy.

2 Œ. rhombipétala N. St. erect, tall, smooth; lvs. lance-linear: pet. rhombic-elliptical, pointed; cal. tube 3—4 times longer than ovary. ② Prairies, W. 2—3f. †

3 Œ. sinuàta L. Pubescent, decumbent at base; lvs. oval-oblong, sinuate-dentate, or incised; fls. axillary, solitary; tube twice longer than ovary. ① N. J. and S. 3—8'.
β. **mínima.** Low, simple, 1-flowered; lvs. subentire. Pine-barrens, N. J. and S.

4 Œ. pùmila L. Low, pubescent, half-erect; lvs. lanceolate; fls. 6", in a leafy spike; calyx tube shorter than the oblong-clavate ovary. ② Meadows, Can. to Car. 6—10'.

5 Œ. chrysántha Mx. Ascending, slender; fls. small (5") crowded, spicate; lvs. lanceolate; cal. tube as long as the ovary; pet. emarginate. ② N. Y. to Wis. 12—18'.

6 Œ. fruticòsa L. St. rigid, hairy or downy; lvs. lance-oblong; rac. corymbed; fls. 18" diam.; pod oblong-clavate, 4-winged, 4-ribbed, pedicellate. ♃ Hard soils. 1—3f.

7 Œ. ripària N. St. slender, branched, purple, and polished; lvs. lin.-lanceolate, petiolate, denticulate; rac. corymbed; fls. large (18"). Banks, N. J., and S. 1—2f. May+.

8 Œ. lineàris Mx. Hoary-puberulent, subsimple; lvs. linear, subentire, obtuse; fls. large, corymbed; pod obovoid. ♃ Montauk Pt. to Tenn., and S. 1—1½f. May, June.

9 Œ. glaùca Mx. Smooth, glaucous; lvs. ovate, sessile, pointed; fls. large, clustered at the ends of the branches; pod oval. ♃ Va. to Ky., and S. 2—3f. May—July.

10 Œ. MISSOURIÉNSIS Sims. Simple, decumbent; lvs. thick, lanceolate, petiolate; fls. very large (4'), tube very long; pod very large, 4-winged. Dry hills, Mo. July—Oct.

11 Œ. NOCTÚRNA. St. erect, downy; lvs. lanceolate, repand-dentate. ③ S. Af. 2f.

12 Œ. LONGIFLÒRA. Simple, hairy; lvs. lanceolate, denticulate; pet. 2-lobed. ③ S. Am.

13 Œ. SPECIÒSA. Lvs. pinnatifid below; fls. diurnal, white, fading red. ♃ Ark. 18'.

14 Œ. TETRÁPTERA. Lvs. pinnatifid below; fls. nocturn., large, pure wh. ① Mex. 1-2f.
 Œ. RUBICÚNDA. Erect; lvs. lance-linear; pet. rose-purp., orange at base. ① Cal. 2f.

6 Œ. LÍNDLEYI. Diffusely branched; lvs. lance-lin.; pet. lilac, red at base. ① Cal. 1f.

17 Œ. VINÒSA. Erect; lvs. linear-oblong; pet. white-roseate; fls. 2' broad. ① Cal. 2f.

18 Œ. LÉPIDA. Erect, simple; lvs. lance-obl.; pet. pale-purp., crimson-spotted at edge.

5. GAURA, L. Calyx tube much prolonged above the ovary, cylindric, limb 4-cleft. Pet. 4, unguiculate, somewhat unequal. Sta. 8, declinate, alternate ones a little shorter. Ovary oblong, 4-celled, but usually by abortion, 1-celled, 1—4-seeded.—Herbaceous or shrubby. Lvs. alternate. Flowers white and red, in slender spikes. July, August.

1 G. biénnis L. St. branched, pubescent; lvs. lance-oblong, spikes dense; cal. tube as long as the segments, the pet. rather shorter. ② Dry bluffs, rare, handsome. 3—5f.

2 G. filipes Spach. Paniculate and naked above; lvs. linear-oblong, tufted at the base of the slender racemes; calyx segments longer than the tube or petals; pods obovoid-clavate, on slender pedicels. Dry soils, S. and W. 3—5f.

3 G. angustifòlia Mx. Pubescent; lvs. linear, very acute; calyx seg. much longer than tube or pet.; pod sessile, ovoid, sharply 4-angled. S. Car. to Fla. Fls. small, wh

4 G. LINDHEIMERI. Erect, much branched; lvs. lin.; cal. red; pet. blush, long in bloom.

6. CLARKIA, Ph. Calyx tube slightly prolonged beyond the ovary, limb 4-parted, deciduous. Pet. 4, unguiculate, 3-lobed or entire, claws with 2 minute teeth. Sta. 8. Sty. 1, filiform. Stig. 4-lobed. Capsule largest at base, 4-celled, 4-valved, many-seeded. —① Herbs (from Oreg. and Cal.) with showy, axillary flowers.

1 C. PULCHÉLLA. Lvs. lin.-lanceolate; pet. 3 parted; 4 sterile sta. Fls. wh., rose, or lilac

ORDER 54.—ONAGRACEÆ.   127

2 C. ÉLEGANS. Lvs. lance-ovate; pet. rhombic-ovate; sta. all fertile. Purple to white.
3 C. RHOMBOÌDEA. Lvs. ovate-obl.; pet. rhomb.-ovate, 2-toothed, lilac, with purple spots.

7. FÚCHSIA, L. LADIES' EARDROP. Calyx tubular-funnel-form, colored, deciduous, limb 4-lobed. Pet. 4, in the throat of the calyx. Sta. 8, exserted. Disk glandular, 8-furrowed. Baccate capsule oblong, obtuse, 4-sided. ♄ S. American, beautiful. Fls. drooping, axillary. Figs. 54, 138.

1 F. COCCÍNEA. Smooth; lvs. opp. or 3-whorled, ovate, denticulate; pet. convolute, violet-purple, half as long as the scarlet sepals, quarter as long as the purple stamens.
2 F. GRÁCILIS. Half-shrubby; lvs. ovate, glandular-dentate; pet. nearly as long as sep.
3 F. FULGENS. Lvs. cordate-ovate; cal. tube long, trumpet-shaped, bright red.—Many hybrid varieties of the above three species are in cultivation.

8. LUDWÍGIA, L. BASTARD LOOSESTRIFE. Calyx tube not prolonged beyond the ovary, limb 4-lobed, mostly persistent. Pet. 4, equal, obcordate, often minute or none. Sta. 4, opposite the sepals. Sty. short. Caps. short, 4-celled, 4-valved, many-seeded, and crowned with the persistent calyx lobes. ♃ and mostly ☿. Leaves entire. Flowers in summer.

§ Leaves opposite. Stems creeping.—*a* Petals none. Flowers very small...Nos. 1, 2
         —*a* Petals yellow, showy..................Nos. 3, 4
§ Leaves alternate, sessile. Stems mostly erect...(*b*)
 b Petals large, yellow. Pods pedicellate, short.........................Nos. 5—7
 b Petals small, yellowish. Pods sessile, elongated, smooth........ ......Nos. 8, 9
 b Petals 0 or minute.—*c* Pods elongated, hairy or smooth..............Nos. 10, 11
     —*c* Pods short, rounded, shorter than the sepals..Nos. 12, 13
     —*c* Pods short, square,—*d* axillary................Nos. 14—16
         —*d* capitate....................No. 17

1 L. palústris Ell. *Water Purslane.* Creeping or floating, smooth, some fleshy; lvs. ovate-spatulate, on winged petioles; fls. sessile, solitary, apetalous; pod oblong (2″), with 4 green angles. Stem 10—18′, round, reddish.
2 L. spatulàta T. & G. Ascending, branched, downy, not fleshy; lvs. obovate-spat., on winged petioles; fls. very small, sess.; pod ovoid, 4-sided, downy. Fla. 6-12′.
3 L. natans Ell. Creeping or floating, smooth; lvs. oblong, on margined petioles; fls. sessile; pet. as long as the calyx; ov. with 2 bractlets at base. Swamps, S. Pod 4″.
4 L. arcuàta Walt. Creeping, smoothish; lvs. linear-oblanceolate, tapering to the slender base; fls. solitary, on ped. twice longer than the lvs.; petals bright yellow, longer than the narrow sepals; pod clavate, finally arcuate. Va. to Fla. 3—10′.
5 L. alternifòlia L. *Seed Box.* Erect, glabrous; lvs. lanceolate, acute; ped. axillary, 2-bracted; sep. large, purplish, crowning the 4-winged pod. Swamps. 1—3f.
6 L. hirtélla Raf. Erect, hairy; lvs. ovate-oblong, obtuse; ped. axillary, 2-bracted; sep. shorter than the yellow petals; pod 4-winged, subglobous. Wet. N. J. to Fla. 1 3f
7 L. virgàta Ph. Erect, with virgate branches, smoothish; lvs. oblong to linear, obtuse; fls. large; pet. longer than the leafy calyx, which is finally persistent and reflexed on the roundish-cubical 4-winged pod. Dry soils, S. 2—3f. Flowers 1′.
8 L. lineàris Walt. Slender, with erect branches; lvs. lance-linear, acute; fls. axillary, sessile; pet. obovate-obl.; pod clavate, 4-sided, longer than sep. N. J. and S. 2f
9 L. hiulfòlia Poir. Simple, erect from a creeping base; lvs. spreading, lin., attenuate at base; sep. ovate, pointed, equalling the pet. and oblong pods. Mid. S. 1f. Lvs. 1′.
10 L. cylíndrica Ell. Smooth; lvs. lanceolate; fls. minute, 1—3 together, apetalous; pod slender, cylindrical, blunt, longer than the calyx segm. S. Car. to Fla and La. 3f
11 L. pilòsa Walt Villous-pubescent; lvs. lanceolate; fls. axillary and spiked above pod villous, oblong, 4-sided, as long as the ovate, pointed sepals. Swamps, S. 2f.

12 L. sphærocàrpa Ell. Lvs. lanceolate, attenuate to base; ped. subsol., bractless, short; sep. as long as the small subglobous pod. Wet swamps, Mass. to Ga.; rare. 3f.
13 L. microcárpa Mx. Ascending from a creeping base; lvs. spatulate-obovate · sep. roundish, acuminate, larger than the very small obovoid pod. Wet, S. 1f.
14 L. alàta Ell. St. slender, strongly 4-angled; lvs. wedge-lanceolate; fls. in the upper axils few, white, apet.; pod cubic-obconic, winged; sds. ovoid. Marshes, S. 2–3f.
15 L. lanceolàta Ell.? (Chapm.) St. stout, terete; lvs. lanceolate; fls. in all the axils green, apetalous; pod cubical, with sharp angles. Swamps, Ga. Fla. 1–2f, bushy.
16 L. polycárpa Short & Peter. Lvs. lance-linear, on the runners oblanceolate; fls solitary, with 2 subulate bractlets at base; pod cubical-obconic. Swamps, W. 1–3f.
17 L. capitàta Mx. Erect; lvs. lance-linear to lance-obl., obtuse at the sessile base; flowers sessile, crowded in a terminal bracted head or spike. Wet barrens, S. 2–3f.

.9. CIRCÆA, L. ENCHANTER'S NIGHTSHADE. Calyx slightly produced above the ovary, deciduous, limb 2-parted. Pet. 2, obcordate. Sta. 2. Caps. obovoid, uncinate-hispid or pubescent, 2-celled, 2-seeded. Sty. united. ♃ Leaves opposite. Flowers small, racemed. Figs. 13, 317, 385.

1 C. Lutetiàna L. St. erect, pubescent above; lvs. ovate, subcordate, acuminate, slightly repand-dentate, opaque, longer than the petioles; bracts none; fr. reflexed, hispid-uncinate. Damp shades. 1—2f. Rac. slender. Fls. rose-colored. June, Jl.
2 C. alpìna L. Smooth; st. ascending at base, weak; lvs. broad-cordate, diaphanous, dentate, as long as the petioles; bracts setaceous; caps. pubescent. Wet, rocky woods, N. Eng. to Oreg. 6—10′. Fls. white. Plant small and delicate. July, Aug.

ORDER LV. LOASACEÆ. LOASADS.

Herbs often hispid with stinging hairs, with *leaves* opposite or alternate and no stipules. *Calyx* adherent to the ovary, 4 or 5-parted, lobes persistent, equal. *Petals* 5, or 10 in 2 circles. *Stamens* ∞. *Ovary* 1-celled, with several parietal placentæ.

1. MENTZÈLIA, L. Calyx tubular, limb 5-parted. Pet. 5—10, flat, spreading. Sta. ∞, 20 to 200. Ov. inferior. Sty. 3, filiform, connate, and often spirally twisted. Stig. simple, minute. Caps. 1-celled, many-seeded. —Branching herbs. Leaves alternate.

1 M. oligospérma Nutt. Very rough, with barbed hairs; stem dichotomous; lvs. ovate-lanceolate, lobed or incisely toothed; pet. entire, cuspidate, longer than the 20+ sta.; caps. 3-5-seeded. ♃ Dry rocks, Ill. Mo. and S. 1f. Fls. deep yellow, 9″. May–Jl.
2 M. Floridàna N. Slightly roughened; lvs. deltoid-ovate, unequally toothed, petiolate; pet. wedge-oval, obtuse: sta. 30; caps. 6-seeded. Fla. 1f. Fls. small, yellow.
3 M. LINDLEYI. *Golden Bartonia.* Hispid; lvs. lance-ovate, pinnatifid, lobes often dentate; pet. broad obovate; seeds ∞; stamens 200. ① California. Fls. golden, 2—3′.

2. LOÁSA, Adans. Cal. 5-parted. Pet. 5, concave. Scales 5, petaloid, 2-3-lobed, connivent, with 2 sterile filaments inserted at base. Sta. ∞, in many fascicles. Style 3-fid. Caps. 1-celled, half 3-valved.

L. LATERÍTIA. *Brick-red L.* Climbing, stinging; leaves palmately lobed, cordate; fls. large, on long stalks, brick-red to orange. Chili. 20f. June–October.

ORDER LVI. TURNERACEÆ.

Herbs with alternate, exstipulate leaves, solitary, 5-parted flowers, a free calyx bearing the 5 petals and 5 stamens in its throat. *Ovary* 1-celled, with

3 parietal placentæ. *Styles* 3, distinct. *Fruit* a 3-valved capsule. *Seeds* albuminous, strophiolate.

TURNÈRA, L. Calyx campanulate. Styles 3. Stigmas 2–5-∞-parted or fringed. Caps. of 3 valves separating to the base. Herbs pubescent or tomentous. Flowers on jointed pedicels, yellow. (Piriqueta, Aub.)

1 **T. cistoïdes** L. Hairy, erect; lvs. lanceolate, obtuse, denticulate; the upper bractlike, shorter than the peduncles; pet. obovate, cor. 1'. Dry. S. 1f. June, July.
2 **T. tomentòsa**. Tomentous; lvs. oblong (1'), longer than the peduncles. Fla. 1f.
3 **T. glabra** (Chapm.) Smooth, branched; ped. 2–3 times longer than lin. lvs. Fla

Order LVII. PASSIFLORACEÆ. Passionworts.

Plants often woody, climbing by tendrils, with alternate leaves and leafy stipules. *Flowers* perfect, 5-parted. *Calyx* tubular, the throat crowned with several rows of sterile filaments, and the corolla above them. *Stamens* 5, monadelphous, sheathing the stipe of the ovary. *Fr.* fleshy, ∞-seeded. Figs. 111, 112, 348.

PASSIFLÒRA, L. Passion-flower (*i. e.*, emblematic of our Saviour's passion). Cal. colored, deeply 5-parted, the throat with a complex filamentous crown. Ov. raised on a stipe. Stig. 3, with 5 large anthers. Fr. a pulpy berry. ♄♃ Fls. large, wonderful and beautiful. May—July.

1 **P. lùtea** L. Lvs. glabrous, cordate, 3-lobed, obtuse; petioles glandless; ped. mostly in pairs; pet. gr.-yel., narrower and much longer than sep. ♃ Woods, O., and S. 10f.
2 **P. incarnàta** L. Lvs. deeply 3-lobed, serrate; petioles with 2 glands above; involucre 3-leaved; crown triple, roseate. ♃ Dry fields, Va. to Fla. 20–30f. Pet. wh.
3 **P. cœrùlea**. Shrubby; lvs. palmately 5-parted, entire; invol. 3-bracted; petioles glandular; pet. longer than the crown, blue, purple, and white. Brazil. Not hardy

Order LVIII. CUCURBITACEÆ. Cucurbits.

Herbs succulent, creeping or climbing by tendrils, with alternate leaves. *Flowers* monœcious or polygamous, never blue. *Calyx* 5-toothed, adherent. *Petals* 5, often united, inserted on the calyx. *Stamens* 5, generally cohering in 3 sets. *Anthers* united, contorted. *Ovary* 1-celled, with 3 parietal placentæ often filling the cells. *Fruit* a pepo or membranous. *Seeds* flat, with no albumen, often arilled. Figs. 186, 476, 482.

§ Corolla white,—*a* 6-cleft. Stigmas 2. Fruit echinate .... Echinocystis. 1
    —*a* 5-petalled. Pepo smooth, many-seeded .... Lagenaria. 2
    —*a* 5-parted. Berry smooth, few-seeded .... Bryonia. 3
    —*a* 5-lobed. Fruit prickly, 1-seeded .... Sicyos. 4
§ Corolla yellow,—*b* 5-lobed. Berry small, smooth, ∞-seeded .... Melothria. 5
    —*b* 5-lobed. Pepo large. Seeds thick at edge .... Cucurbita. 6
    —*b* 5-cleft. Pepo large, Seeds colored, thick-edged .... Citrullus. 7
    —*c* Seeds white, acute-edged .... Cucumis. 8

1. **ECHINOCÝSTIS**, T. & G. Flowers ♂. Calyx of 6 filiform-subulate segments, shorter than the corolla. Petals 6, united at base into a rotate-campanulate corolla. ♂ Sta. 3, diadelphous. ♀ Abortive fil. 3, dis-

tinct, minute. Style very short. Stig. 2, large. Fruit roundish, inflated, echinate, 4-seeded. ① Climbing, with branched tendrils.

E. lobàta T. & G. Alluvion, Can. to Penn. and W. Smoothish. Lvs. thin, palmately 5-lobed. Fls. small, white, the barren in large racemes, fertile few below. Jl.—Sep.

2. **LAGENÀRIA**, Ser. GOURD. Fls. ☿. Calyx campan., 5-toothed. Pet. 5, obovate. ♂ Sta. 5, triadelphous. ♀ Stig. 3, thick, 2-lobed, subsessile. Pepo ligneous, 1-celled. Seeds arilled, obcordate, compressed, margin tumid.—Mostly climbing by tendrils.

L. VULGARE. Stem soft-pubescent; tendrils branched; lvs. roundish, cordate, 2 glands beneath at base; fls. solitary, peduncled, white; pepo bottle-shaped. ① Gardens.

3. **BRYÒNIA**, L. BRYONY. Fls. ☿ or ♂ ♀. Cal. 5-toothed, teeth short. Cor. 5-cleft or -parted. ♂ Stamens 5, triadelphous, with flexuous anthers. ♀ Sty. trifid. Berry small, globular. ♄ Fls. greenish-wh. June.

B. Boykínii T. & G. Scabrous pubescent; lvs. deeply 3-5-lobed, cordate; flowers small, axillary, mixed, on short pedicels; berries 3-seeded, bright red. Ga. to La. 10f.

4. **SÍCYOS**, L. SINGLE-SEED CUCUMBER. Fls. ☿. Cal. 5-toothed. Pet. 5, united at base. Anthers cohering, contorted. Styles 3, united at base. Fruit ovate, membranous, hispid or echinate, with one large, compressed seed. ♄ With compound tendrils. Flowers axillary, mixed.

S. angulàtus L. Hairy, branched; lvs. roundish, 5-angled or lobed, lobes pointed, fls. wh. with gr. veins, the ♂ in long rac., the ♀ smaller, capitate. Thickets. Jl.—Sep.

5. **MELÒTHRIA**, L. Fls. ♀ ☿ ♂ or ☿. Calyx bell-form, limb in 5 subulate segments. Pet. 5, united into a bell-form corolla. Sta. 5, triadelphous. Style 1, stig. 3. Berry ovoid, small, ∞-seeded. ♄ Tendrils simple.

M. péndula L. Lvs. roundish, small, 5-lobed or angled, pointed; fls. axillary, ♂ in small rac., ♀ solitary, on long peduncles. N. Y. to Ga. Delicate. Fls. yellowish. Jl.

6. **CUCÚRBITA**, L. SQUASH. Fls. ☿. Cal. 5-toothed, limb deciduous after flowering in ♀. Cor. bell-shaped, cohering with the calyx. Stam. 5, anth. connate, straight. Stig. 3. Pepo fleshy. Seeds thick at margin, smooth. ♄ Flowers yellow.

1 **C. PEPO.** *Pumpkin.* Rough-hispid; lvs. very large, cordate, 5-lobed or angled; fls. large, ♂ long-stalked; fr. very large, rounded, smooth, torulous, finally yellow. ①

2 **C. MELOPÈPO.** *Flat Squash.* Hairy; lvs. cordate, 5-lobed; fr. depressed-orbicular, margin torulous, smooth or warty, whitish. ① Hybridizes with No. 1.

3 **C. VERRUCÒSA.** *Crookneck S.* Hairy; lvs. cordate, deeply 5-lobed; fr. oblong or clavate, often elongated and curved at base. ① The varieties are numerous.

β. MEDULLÒSA. *Vegetable Marrow.* Lvs. triangular in outline, deeply 3-lobed; fr. oblong or club-form, dark-green and wh., 10—20′ long. Highly prized in England.

4 **C. máxima.** *Mammoth S. Winter S.* Rough-hairy; lvs. round-reniform, obtusely 5-lobed; fruit 10′—3f! diam., with a lobed, yellowish-white surface and dense pulp.

7. **CITRÚLLUS**, Neck. WATERMELON. CITRON. Cal. deeply 5-cleft, segm. linear-lanceolate. Pet. 5, united at base. Sta. triadelphous. Style trifid. Stig. reniform-cordate. Fr. rounded or oblong, the succulent placentæ filling the cell. Seeds colored, truncate at base. ♄

C. **vulgāris** Schrad. Hirsute; lvs. somewhat 5-lobed, the lobes sinuate-pinnatifid, glaucous beneath; fls. with a bract; fr. dark-spotted. ① India. Africa.

8. CUCUMIS, L. Fls. ♂ or ☿. Cal. tubular-campanulate, with subulate segments. Cor. deeply 5-parted. Sta. triadelphous. Style short. Stig. 3, thick, 2-lobed. Pepo elongated. Seeds lance-oblong, white, acute, not margined at the edge. ♃ Fls. axillary, solitary, yellow.

* Leaves angular, not lobed, subcordate. Tendrils simple................Nos. 1, 2
* Leaves deeply-lobed or cleft. Tendrils simple or forked................Nos. 3—5

1 C. **satīvus**. *Cucumber.* Rough; lf. angles acute; fr. oblong, prickly when young. ①
2 C. **Mělo**. *Musk Melon.* Hairy; lf. angles obtuse; fr. globular, torulous. ① Asia.
3 C. **Angŭria**. *Prickly C.* Lvs. sinnate-lobed; tendrils simple; fr. ovoid, echinate.
4 C. **Colocýnthis**. *Colocynth.* Lvs. cut-lobed; tend. short; fr. round, yel., very bitter.
5 C. **anguīnus**. *Serpent C.* Lvs. 3-5-lobed; tendrils forked; fr. long, coiled, snake-like.

ORDER LIX. BEGONIACEÆ. BEGONIADS.

Herbs or *shrubby plants*, with alternate, inequilateral leaves, and dichinous, unsymmetrical flowers. *Perianth* of 2—∞ lvs., all petaloid or the inner only. *Stamens* ∞, anth. connate. *Ovary* inferior, 3-angled or winged, 3-celled, the placentæ in the angles. *Styles* united at base. *Albumen.*0, or thin.

BEGÒNIA, L. ♂ Sepals 2. Pet. 2, rarely more, or 0. ♀ Sepals 2, larger than the 4 petals. Cap. with 3 angles unequally winged, opening below the apex. Sds. ∞, minute. ♃ ♄ Lvs. alternate, stipulate, with the sides unequal, margins toothed or lobed. Fls. often showy. Species 320, mostly tropical, often found in the greenhouse. Much mixed.

§ Leaves feather-veined, and glabrous as well as the whole plant............Nos. 1—4
§ Leaves palmi-veined, with 5—9 veins from near the base...(a)
  a Plant glabrous throughout. Leaves toothed or crenulate............Nos. 5—7
  a Leaves hairy, at least on the deeply 5–9-lobed margins...............Nos. 8, 9
  a Leaves hairy, at least on the undulate or toothed margins...(b)
    b Staminate flowers with 2 sepals only, the petals usually 0........Nos. 10, 11
    b Staminate flowers with 2 sepals and 2 petals. E. India.........Nos. 12—15

1 B. **maculāta**. Very smooth; lvs. ovate-oblong, wavy, cordate, white-spotted above, purple beneath; fls. white or flesh-colored, in forked cymes. Brazil. (B. argentea.)
2 B. **fucusioīdes**. Smooth; lvs. oblong to obovate, obtuse at base, serrulate; fls. bright red, drooping like Fuchsias, in many terminal cymes, very handsome. N. Granada.
3 B. **sempérvirens**. Leaves bristly on the crenate edges, ovate, subcordate; fls. white to rose-colored, 1'—18", in an open panicle, with scarious, persistent bracts. Brazil.
4 B. **incarnāta**. Leaves bristly-serrate, ovate to oblong; fls. roseate, large, in compound, pendulous cymes, with caducous bracts or 0. Mexico. (B. insignis.)
5 B. **nítida**. Leaves ovate, half-cordate, subcrenate, shining, green as well as the stipules; flowers purplish-white, with caducous bracts, on axillary peduncles. W. Ind.
6 B. **sanguínea**. Leaves oblique-ovate, deeply cordate, crenulate, red beneath, large; flowers white, small, many, in cymes longer than the leaves. Stalks red.
7 B. **coccínea**. Leaves oblique oblong, half cordate, dentate; stipules obovate, caducous; flowers scarlet, pendulous, 8" broad, in cymes equalling the leaves (5).
8 B. **hieracifòlia**. Leaves roundish, palmately 7-cleft, lobes toothed; fringed scales on the petiole above; scape long, with many roseate flowers, 1' diameter. Mexico.
9 B. **parviflòra**. Shrub rusty-downy; leaves ample, roundish, subcordate, 7-9-lobed, lobes serrulate; cymes 1f long, with numerous small pale flowers. Peru.

10 B. manicàta. Leaves oblique-ovate, cordate, angular, toothed, with purple-fringed scales on the petioles; flowers flesh-colored, in open cymes, on long peduncles. Mex.
11 B. phyllomanicàta. Stem covered with leaf-like bulblets; leaves broad-ovate, cordate, doubly dentate; peduncles longer than the leaves; flowers roseate. Brazil.
12 B. Evansiàna. Leaves ovate, subcordate, bristly denticulate, purple beneath; flowers rose-colored, 1', in cymes on long stalks. Our oldest species, from China.
13 B. Rex. Leaves ample, ovate, cordate, variegated with zones of dark-green, silvery-gray, and purple, sinuate-crenate; scape 1—2f, with large roseate flowers. E. Ind.
14 B. Griffìthii. Like No. 13, but densely downy all over, even the large whitish fls.
15 B. xanthìna. Lvs. like No. 13, but varied with metallic spots; scape with *yellow* fls.

Order LX. CACTACEÆ. Indian Figs.

Plants with a green fleshy caudex or stock, angular or jointed, mostly leafless, armed with numerous prickles and terrible spines. *Flowers* solitary, mostly very showy. *Sepals* ∞ on the surface. *Petals* and *stamens* ∞ on the top of the ovary or calyx tube. *Fruit* fleshy, 1-celled, with parietal placentæ. *Style* filiform, with stellate *stigmas*. Figs. 472, 487.

* Calyx tube not produced above the ovary. Stock jointed, branching.................Opuntia. 1
* Calyx tube produced above the ovary.—a Joints flat, leaf-like, spineless...(x)
　　　　　　　　　　—a Stocks 3 - ∞-angled or grooved, spiny...(y)
　x Flowers rose-red, oblique, from the top of the short truncated joints...............Epiphyllum. 2
　x Flowers pink to red, regular, from the notches of the long joints.....................Phyllocactus. 3
　y Stock long-cylindrical, many-ridged. Flowers lateral, long-tubed............. Cereus. 4
　y Stock depressed-globular to oblong. Flowers subterminal, short-tubed.........Echinocactus.
　y Stock globular to conical. Flowers terminal, small, woolly-tubed............ Melocactus. 5
　y Stock globular to cylindrical, covered with tubercles. Flowers lateral.... .....Mammillària.

1. OPÚNTIA, Mill. Indian Fig. Sep. and pet. ∞ adnate to the ovary, not produced into a tube above it, longer than the stamens, the inner obovate. Stig. 4—10. Berry smoothish or prickly. ♄ Branches composed of fleshy, mostly flattened joints. Lvs. small, deciduous, alternate, with tufts of prickles in their axils. Flowers large, yellow.

　§ Joints obovate or broadly oval. Stigmas 8—10. Seeds many. Fr. eatable...Nos. 1—4
　§ Joints oblong or nearly cylindrical. Stigmas 4—6. Seeds 1—6.............Nos. 5, 6

1 O. Ficus-Indica Haw. Stock branches stout, erect-spreading, pale-glaucous; lvs. subulate, with pungent bristles, no spines; fr. bristly, obovoid, purple. Florida! to San Diego! 3—20f. Joints 1f. Fruit pleasantly acid. § Trop. Am.

2 O. vulgàris Mill. Stock prostrate, pale-glaucous; lvs. minute, scale-like, with ∞ bristles and few spines; fr. nearly smooth, ovoid, eatable, crimson when ripe. Dry rocks, &c., Ct. to Fla. 1—2f, the joints 4—6'. Flowers 2½—4' broad. Pet. 7—10. Jn.

3 O. Rafinesquii Eng. Stock prostrate, bright green; lvs. spreading, subulate, longer (3—4''); spines 1-5 in each axil; petals 10-12, often purplish at base. Ky. to Ill., and W.

4 O. Missouriénsis DC. Stock prostrate; leaves minute, the axils bristly and with whorl of many spines; fruit prickly, dry. Wis., along the rivers, and W. June.

5 O. polyántha Haw. Erect; joints oblong, the upper bearing many flowers at top; spines strong, yellow, unequal; stigmas 6; fruit small, 6-seeded. Waysides, Fla. Jn.

6 O. Pes-Corvi Leconte. Stk. prostrate; joints compressed-cylindric, small (2'); spines in pairs, unequal; pet. few, spatulate; stig. 4; fr. small, prickly, 1-4-seeded. Ga., Fla.

7 O. Braziliensis. Stock cylindrical, 6—10f; branches short, bearing ovate joints, which are thin and somewhat leaf-like; spines 1—3 together, sharp and strong. Brazil.

2. EPIPHÝLLUM truncàtum. Stock consisting of short, flat, notched joints, truncate at top; flowers at top of the joints, 2—3' long, conspicuously oblique. Style longer than the stamens or 6—8—10 reflexed petals. From Brazil. 1f.

3. PHYLLOCÁCTUS PHYLLANTHOÌDES. Stock consisting of narrow, ensiform, crenate joints, fleshy but leaf-like. Flowers 4' long, open by day, with many rose-colored petals and sepals longer than the tube, gradually spreading. Mexico.

2 P. ACKERMÁNNI. Fls. scarlet; pet. channelled, pointed, very many, 3—4'. Mexico.
3 P. PHYLLÁNTHUS. *Spleenwort.* Joints ensiform, serrate; fls. 9—12', the white funnel-form cor. much shorter than the slender tube, opening by night, fragrant. S. Am.

4. CÉREUS, DC. Sep. and pet. imbricated, adnate to and prolonged into a long tube above the ovary. Sta. and style filiform, adnate to the tube. Stig. 10. Berry scaly with the remains of the sepals. ♄ ♃ Stock fleshy, green, prismatic, often jointed, with fascicles of spines on the ridges.

1 C. GRANDIFLÒRUS. *Night-blooming C.* Stock long, about 5-angled; flowers very large, nocturnal; pet. spreading 6—8', pearl-white; sep. yellow. Mex. A magnificent flower.
2 C. TRIANGULÀRIS. Stock 3-angled, prickles bristly; fl. very large, white; sep. green.
3 C. FLAGELLIFÓRMIS. Stock slender, long, prostrate, 10-angled, hispid; fls. pink-color, smaller, open by day many days in succession; tube longer than the petals.
4 C. SERPENTÌNUS. Stock 12-angled, 4f; spines white, bristly; fls. pale, open by night.
5 C. SPECIOSÍSSIMUS. Stock 3- or 4-angled, erect, 4f; angles winged, undulate; fls. large (4' long), with many red or crimson petals and white stamens, diurnal. Common.
6 C. SENÌLIS. *Old-Man C.* Stk. erect, oblong, with tufts of long, white, hair-like bristles.

5. MELOCÁCTUS COMMÙNIS. Stock very succulent, roundish ovate, 1f, 12-18-ribbed, surmounted by a sort of spadix, consisting mostly of dense wool, in which at the top the small red flowers are imbedded. W. Indies.

ORDER LXI. FICOIDEÆ. MESEMBRYANTHS.

Plants fleshy, of forms variously singular, with entire, mostly opposite leaves, and solitary, regular flowers, remarkable for their profusion and duration. *Calyx lobes* 4 or 5. *Petals* ∞—5, or rarely 0. *Stamens* ∞, distinct, perigynous. *Ovary* more or less adherent. *Stigmas* 2—∞. *Capsules* 1-∞-celled, ∞-seeded. *Embryo* curved.

§ Petals and stamens ∞, in several rows. Capsule fleshy, valvate............MESEMBRYANTHEMUM. 1
§ Petals none, stamens ∞—5. Capsule 3-5-celled, circumsessile..............SESUVIUM..2 (& p. 446)

1. MESEMBRYÁNTHEMUM, L. ICE PLANT. Calyx lobes 5. Pet. linear, inserted with the filiform stamens on the calyx tube. ♃ ♄ Air bubbles beneath the epidermis appear like dew or frost.

1 M. CRYSTALLÌNUM. Procumbent, fleshy; lvs. large, ovate, acute, wavy at the margin, 3-veined beneath. ♃ Greece. Stem 1f. Flowers white, all summer. Not hardy.
2 M. GRANDIFLÒRUM. Procumb.; lvs. cord. ovate; cal. 4-cleft, 2-horned; pet. pink. Afr.

2. SESÚVIUM, L. SEA PURSLANE. Sep. 5, united at base, colored inside. Sta. 5—50, inserted on the calyx tube. Ov. free, 3-5-celled. Sty. 3—5. Pyxis opening transversely by a lid. ♃ Prostrate sea-side herbs.

S. **Portulacustrum** Tourn. Lvs. linear-spatulate; fls. on short peduncles; sta. ∞. Sandy coasts, N. C. to Fla. 1f +. Plant very smooth and fleshy. Fls. axil., roseate. Jl. +.

ORDER LXIII. UMBELLIFERÆ. UMBELWORTS.

Herbs with hollow, striate stems, sheathing petioles, and flowers in um-

ORDER 63.—UMBELLIFERÆ.

bels. *Calyx* adherent to the ovary. *Petals* 5, usually inflected at the point. *Stamens* 5. *Ovaries* 2-carpelled, surmounted by the fleshy disk which bears the petals and stamens. *Styles* 2, distinct, or united at their thickened bases. *Fruit* a cremocarp (§151), consisting of 2 coherent achenia called *mericarps*, which separate along the middle space, which is called the *commissure*.

Carpophore, the slender, simple, or forked axis attached to and supporting the mericarps at top, enclosed between them at the commissure.
Ribs, 5 ridges traversing each mericarp lengthwise, and often 4 intermediate or secondary ones, some, all, or none of them winged.
Vittæ, little tubular receptacles of colored volatile oil imbedded in the substance of the pericarp, just beneath the intervals of the ribs, and also sometimes in the face of the commissure.
Embryo in the base of abundant, horny albumen.
Figs. 42, 177, 235, 238, 303, 334–5, 360, 442–3.

A large and well-defined Order. As the flowers in all are nearly alike, the genera are best distinguished by characters taken from the fruit—the number and form of the ribs, the presence or absence of vittæ, the form of the albumen at the commissure, &c. These parts, therefore, minute as they are, will require the special attention of the student.

§ Flowers in simple umbels, sometimes spicate. Leaves simple...(*a*)
§ Flowers in capitate umbels, *i. e.*, sessile, forming dense heads...(*b*)
§ Flowers in regularly-compound umbels, not sessile in heads...(2)
  2 Fruit flattened on the back, singly-winged on the margin only...(*c*)
  2 Fruit flattened on the back, doubly-winged on the margin only...(*d*)
  2 Fruit flattened on the sides, or terete and not flattened either way...(3)
    3 Fruit slender, teretish, 2–3 times longer than wide. Flowers white...(*e*)
    3 Fruit nearly as broad as long.—*m* Flowers yellow...(*f*)
                  —*m* Flowers white...(*i*)
      4 Ribs of the fruit either muricate, or crenulate-winged...(*g*)
      4 Ribs smooth, entire, winged or sharply prominent...(*h*)
      4 Ribs obtuse or obsolete.—*n* Calyx teeth obsolete or 0...(*k*)
                  —*n* Calyx teeth prominent...(*l*)

a Fruit flat, orbicular. Leaves round or roundish..............................HYDROCOTYLE. 1
a Fruit globular. Leaves linear, fleshy phyllodia..........................CRANTZIA. 2
b Flowers partly sterile. Fruit densely muricate, few........................SANICULA. 3
b Flowers all fertile. Fruit scaly, many in the head.......................ERYNGIUM. 4
c Flowers of two sorts, the marignal with enlarged corollas, radiant..............HERACLEUM. 5
c Flowers all alike.—*o* Fruit with a thick, corky margin. Vittæ ∞ ....POLYTÆNIA. 6
        —*o* Fruit with a thin margin. Vittæ single.................PEUCEDANUM. 7
d Seed adherent to the pericarp. Intervals with single vittæ...............ANGELICA. 8
d Seed loose in the pericarp. Intervals with numerous vittæ..........ARCHANGELICA. 9
e Beak slender, longer than the fruit, all without vittæ. South.................SCANDIX. 10
e Beak short or none.—*p* Fruit clavate, upwardly hispid...................OSMORRHIZA. 11
        —*p* Fruit smooth, linear-oblong. Styles very short......CHÆROPHYLLUM.12
        —*p* Fruit smooth, elliptical. Styles very slender..........CRYPTOTÆNIA. 13
f Involucels of 5 ovate, entire bracts. Leaves simple, entire..................BUPLEURUM. 14
f Involucels of 3 subulate bracts.—*r* Fruit laterally compressed.............CARUM. 15
        —*r* Fruit subterete transversely...............THASPIUM. 16
f Involucra none.—*s* Fruit laterally compressed. Vittæ ∞.....................PIMPINELLA. 17
        —*s* Fruit transversely subterete. Vittæ single.................FŒNICULUM. 18
  g Calyx teeth prominent. Ribs of the fruit muricate......................DAUCUS. 19
  g Calyx teeth obsolete. Ribs of the fruit crenulate-undulate.................CONIUM. 20
  h Marginal wings twice broader than the dorsal.......................SELINUM. 21
  h Marginal and dorsal ribs alike sharp,—*u* with ∞ vittæ ....LIGUSTICUM 22
        —*u* with single vittæ....................ÆTHUSA. 23
  k Fruit a double globe. Petals not inflected. Low, early-flowering........ ERIGENIA 24
  k Fruit ovate-oblong. Petals emarginate-inflected. Involucra 0........ .. CARUM. 15

‡ Fruit round-ovate.—v Petals concave, not emarginate. Vittæ single.......AP OM. 25
—v Petals inflected, emarginate. Vittæ ∞............PIMPINELLA. 17
‡ Ribs of the carpels obsolete. Fruit ovate, covered with large vittæ............EULOPHUS. 26
‡ Ribs of each carpel 9. Fruit globular. Outer flowers radiant..................CORIANDRUM. 30
‡ Ribs of each carpel 5.—x Fruit round, didymous.............. ....CICUTA. 27
—x Fruit oval. Leaves pinnate...........................SIUM. 28
—x Fruit ovate. Leaves capillaceous.....................DISCOPLEURA. 29

1. HYDROCÓTYLE, L. PENNYWORT. Calyx limb obsolete. Pet. spreading, the point not inflected. Fr. laterally flattened, the commissure narrow. Carpels 5-ribbed, without vittæ. ≈ Low, smooth, creeping. Umb. simple. Invol. few-leaved. Fls. small, white. June—Aug. Figs. 334-5.

* Leaves reniform or cordate, the base lobes not united............... .....Nos. 1—3
* Leaves peltate, orbicular, the base lobes united... ....Nos. 4, 5

1 **H. Americàna** L. St. filiform; lvs. round-reniform, slightly lobed, crenate; umb. sessile, 3-5-flwd.; fr. orbicular. ♃ Damp shades. 2—6′. Plant very smooth and shining.

2 **H. ranunculoides** L. *f.* Lvs. round-reniform, deeply 3-5-cleft, lobes crenate; ped. 1—2′, branched; umbels 5-9-flwd., capitate. ♃ Waters, Pa., and S. Lvs. veiny, 4-8′.

3 **H. repánda** Pers. Lvs. broad-ovate, cordate, rounded, margin repand-dentate; ped. 2—3′, simple; umb. capitate, 3 or 4-flwd.; invol. 2-bracted. ♃ Muddy shores, S.

4 **H. umbellàta** L. Lvs. crenate, with a notch at base, long-stalked (4—6′); scapes 4—6′, bearing a simple (rarely proliferous) umb. of 20-30 fls. ♃ Ponds, bogs. Ms. to La.

5 **H. interrúpta** Muhl. Lvs. crenate; umb. proliferous, 5-flwd. ♃ Wet. Ms. to Ga.

2. CRÁNTZIA, Nutt. Calyx margin obsolete. Pet. obtuse. Fr. subglobous. Carpels unequal, 5-ribbed, with a vitta in each interval. ≈ Small, creeping, with linear or filiform, entire lvs. Umbels simple, involucrate.

C. lineàta Nutt. Lvs. cuneate-linear, sessile, obtuse at apex, and with transverse veins, shorter than the peduncles. ♃ Muddy banks, coastward. Umb. 4-8-flowered.

3. SANÍCULA, Tourn. SANICLE. Fls. ♀ ☿ ♂. Cal. segm. acute, leafy. Pet. obovate, erect, with a long, inflected point. Fr. subglobous, armed with hooked prickles. Carpels without ribs. Vittæ numerous. ♃ Umbel nearly simple. Rays few, with many-flowered, capitate umbellets. Involucre of few, often cleft leaflets, involucel of several entire.

1 **S. Marilándica** L. Lvs. 5-7-parted, digitate, mostly radical; segm. thick, oblong, incisely serrate; sterile fls. many, pedicellate, fertile ones sessile; cal. segm. entire; styles slender, conspicuous, recurved. Woods: common. 2—3f. May—July.

2 **S. Canadénsis** L. Lower lvs. 5-parted, upper 3-parted; segm. cuneate-obovate, mucronate-serrate; sterile fls. few, much shorter than the fertile; sty. shorter than the prickles. Woods, thickets: com. 1-3f. Lvs. thin, 1-3′. Umb. few-flwd. Jn.-Aug.

4. ERÝNGIUM, Tourn. Fls. sessile, collected in dense heads. Cal lobes somewhat leafy. Pet. inflexed. Sty. filiform. Fr. scaly or tuberculate, obovate, terete, without vittæ or ribs. ♃ ② Fls. blue or white, bracteate; lower bracts involucrate, the others smaller and chaffy. Summer.

* Scales and chaff of the heads entire, often spinescent.....................Nos. 1-3
* Scales and chaff of the heads tricuspidate.—a Flowers white.... ... .Nos. 4, 5
—a Flowers blue..... ....Nos. 6, 7

1 **E. yuccæfòlium** Mx. Erect; lvs. broadly linear, parallel-veined, ciliate with remote, soft spines; invol. bracts entire, spinescent, shorter than the ovoid-glob. heads. ♃ Prairies and pine-barrens, W. and S. 2—5f. Fls. white, inconspicuous. Jl., Aug.

2 E. Baldwinii Spr. Sts. prostrate, filiform; rt. lvs. wedge-oblong, st. lvs. 3-parted, segm. lance-lin., cut-toothed; invol. scales and chaff alike; hds. oblong. Fla. 10'. Blue.

3 E. prostràtum Baldw. Sts. prostrate. filiform, rooting: lvs. of two forms at the same node, small, some ovate, some 3-parted with lance-linear segm.; invol. scales linear, longer than the small oblong heads; fls. blue. ♃ Swamps, Ga. Fla. 6–12'. Jn. +
  β. *foliosum*. Bracts of the invol. leafy, twice longer than the heads. Fla. La.

4 E. aromáticum Baldw. Sts. assurgent; vs. short (1'), pinnate, with cuspidate segm., the 3 terminal largest; hds. globons (6–8''); invol. scales 5. Dry. Fla. 9–18'.

5 E. Mettaùeri. Erect, tall; lvs. linear-terete, consisting chiefly of the fistulous, jointed midvein, barely winged and toothed; bracts 8–10. leafy. Wet. Fla. 4–6f.

6 E. Virginiànum Lam. Erect; lvs. lance-oblong to linear, flat, the lower long stalked, upper uncinate-serrate; bracts longer than the roundish head. ♃ Swamps. 2–4f. Hds. in umbel-like cymes, numerous, 5–6''. Varies with lvs. all linear. Jl. Aug.

7 E. virgàtum Lam. Erect; lvs. oval or oblong, thin, petiolate, dentate, the upper sessile; bracts 6–8, longer than the depressed, cymous heads. ♃ Wet, S. 2–4f.

5. HERACLEUM, L. Cow Parsnip. Calyx 5-toothed. Pet. often radiant in the exterior flowers, and apparently deeply 2-cleft. Fruit compressed, flat, with a broad, flat margin, and 3 obtuse, dorsal ribs to each carpel; intervals with single vittæ. Seeds flat. ♃ Stout, with large umbels. Involucre deciduous. Involucels many-leaved.

H. lanàtum L. Villons; lvs. ternate, petiolate, tomentous beneath; lfts. petioled, round-cordate, lobed; fr. orbicular. Can. to N. Car. and W. 4f. Lvs. very large. June.

6. POLYTÆNIA, DC. Calyx 5-toothed. Fruit oval, glabrous, compressed on the back, with a thickened, corky margin. Commissure with 4 to 6 vittæ. Seeds plano-convex. ♃ A smooth herb, with bipinnately-divided leaves. Involucre 0. Involucel of setaceous bracts.

P. Nuttállii DC.—Prairies, W. 2–3f. Smoothish. Lower leaves long-stalked. Umbels 2'. Fruit 3''. May.

7. PEUCEDÁNUM, L. Fruit ovate, oval, or roundish, compressed on the back, the margin acute or broadly winged, carpels plane or convex, intervals with single vittæ. Seeds plano-convex. ♃ ② Smooth, rarely pubescent. Lvs. pinnately or ternately divided or decompound. Umbels compound, with or without involucra. Fls. yellow or white. Fig. 238.

§ Eupeucedànum. Cal. 5-toothed. Lvs. pinnatisect. Fr. narrowly winged. Yellow...1, 2
§ Archemora. Cal. 5-toothed. Lfts.1–11, narrow. Fr. narrowly winged. Fls. white. 3–5
§ Pastinàca. Calyx teeth 0. Lfts. oval. Fruit broadly winged. Flowers yellow....No. 6

1 P. fœniculàceum N. and other species with radical, pinnatisect leaves grow in Kansas, and W. (Rev. J. H. Carruth.)

2 P. graveolens. *Dill.* Lvs. cauline, tripinnate; seg. capillary; umb. on long stalks; fr. oval, flat, brown, aromatic, pungent, medicinal. ② Spain. 2f. (Anethum, C-B.)

3 P. rígidum *Cowbane.* St. rigid, striate: lvs. pinnate; lfts. 3–11, lance-ovate, sub entire; umb. 2 or 3, spreading, with slender rays; fr. with large purp. vittæ. ♃ Swamps, N. Y., W. and S. 2–5f. August.
  β. *ambígua*, has the leaflets linear and entire.

4 P. ternàtum. Stem slender, smooth; lvs. on long petioles. ternate; segm. very long, linear, entire, 3-veined; invol. 0–3-leaved; involucel 4–6-leaved. Swamps, in pine-barrens, S. 2–3f. Sept.—Nov. (Neurophyllum longifolium. C-B.)

5 P. teretifòlium. Tall, slender, smooth; lvs. reduced to fistular, jointed phyllodia, terete tapering. 6–16' long; fr. 3''; invol. 5–6-leaved. ♃ Wet, S. (Tiedmannia, C-B.)

6 **P. sativum.** Root fusiform; stem furrowed; lvs. pinnate, downy beneath; lfts. oblong, incisely toothed, the terminal 3-lobed; umbels large; involucra near.y 0. ② Fields, gardens. 3—4f. July—Sept. ‡ Wild and Common Parsnip.

8. **ANGÉLICA,** L. Calyx teeth obsolete. Fruit dorsally compressed, doubly winged. Carpels 5-ribbed, the 3 dorsal ribs filiform, the 2 marginal winged, intervals with single vittæ. Carpophore 2-parted. Seed semiterete. ♃ Leaves bi- or tri-ternate, sessile. Umbels terminal. Invol. 0 or few-leaved. Involucels many-leaved.

A. Curtísii Buckley. Lvs. biternate or with 3 quinate divisions; lfts. thin, ovate or lance-ovate, acuminate, incisely toothed; fr. broadly winged. Mts. Pa., & S. Aug.

9. **ARCHANGÉLICA,** Hoffm. ANGELICA. Calyx teeth short. Fr. dorsally compressed, with 3 carinate, thick ribs upon each carpel, and 2 marginal ones dilated into membranous wings. Seed loose in the ripe carpel, covered with vittæ. ♃ Petioles usually large, inflated and 3-parted. Umbels perfect. Involucels many-leaved. Fls. greenish white. Fig. 177.

* Involucels less than half the length of the pedicels. Fruit 3″ long, winged....No. 1
* Involucels about as long as the pedicels.—a Fruit scarcely winged............No. 2
　　　　　　　　　　　　　—a Fruit broadly winged........Nos. 3, 4

1 **A. atropurpùrea** Hoffm. St. dark purple, furrowed; petioles 3-parted, the divisions quinate; lfts. incisely toothed, terminal lft. rhomboidal, sessile, the others decurrent; involucels setaceous. Meadows, E. and W. 4—6f. Stout, aromatic. June.

2 **A. peregrìna** N. St. striate; lf. divisions ternate, segm. incisely serrate; involucel of many bracts, as long as the pedicels; fruit ribs corky, thick. Sea-coast, Mass. to Labrador. 2—3f. July. (A. Gmelini DC.)

3 **A. hirsùta** T. & G. Stem striate, the summit with the umbels tomentous-hirsute; lvs. bipinnately divided, the divisions quinate; segm. oblong, acutish, the upper pair connate, but not decurrent at base. Dry woods, N. Y. to Car. 2—5f. July.

4 **A. dentàta** Chapm. Slender, smooth; lvs. 1-2-ternate; segm. lance-ovate, incised; umbels few-rayed; involucel 5-6-leaved, as long as the pedicels. Ga. Fla. 2—3f. Jl. +

10. **SCANDIX,** L. VENUS'S COMB. Cal. limb obsolete. Fr. laterally compressed or nearly terete, attenuated into a beak which is longer than the seed. Carpels with 5 obtuse, equal ribs. Vittæ 0, or scarcely any. ① or ② Lvs. finely dissected. Invol. 0. Involucel 5–7-leaved. Flowers white.

S. apiculàta Willd. Petioles and peduncles slender; lvs. finely dissected into subulate segments; umbels 3-rayed; fruit with beak and forked style 9″. Ga. 1f. § Eur.

11. **OSMORHÌZA,** Raf. SWEET CICELY. Calyx margin obsolete. Sty. conical at base. Fr. linear, very long, clavate, attenuate at base. Carpels with 5 equal, acute, bristly ribs. Vittæ 0. Commissure with a deep, bristly channel. ♃ Leaves biternately divided, with the umbels opposite. Involucels 4–7-leaved. Flowers white. May, June. Figs. 42, 442–3.

1 **O. longistylis** DC. Sty. filiform, nearly as long as the ovary; fr. clavate; rt. spicy and sweet-flavored; st. and lvs. smoothish. Rich woods, Can. to Va. 1–3f. Fruit 1″.

2 **O. brevistylis** DC. Sty. conical, scarcely as long as the breadth of the ovary; fr. somewhat tapering at the summit; root nauseous; plant hairy. Woods. 1—3f.

12. **CHÆROPHÝLLUM,** L. CHERVIL. Calyx limb obsolete. Fruit laterally compressed, linear or oblong, contracted above but scarcely

beaked. Carpels with 5 obtuse, equal ribs, intervals with single vittæ. Commissure deeply sulcate. ① ② Leaves 2–3-pinnately divided. Segm. incisely cleft or toothed. Invol. 0, or few-leaved. Involucel many-leaved. Flowers mostly white. Umb. mostly sessile.

1 C. procúmbens Lam. Slender, spreading, smoothish; lf. segm. trifid and pinnatifid, lobes oblong, obtuse; umb. few-rayed, sessile or pedunculate; fr. acute, ribs narrower than the intervals. Damp woods, Ill. to Penn., and S. 1—2f. April, May.

2 C. Tainturièri Hook. Ascending or erect, some hairy; lf. segm. crowded, again pinnatifid or bipinnatifid, ultimate segm. acute; fr. short-beaked, ribs broader than the intervals. Ga. to Fla. and La. 10—20′. Much branched. Fruit 4″. March, Apr.

3 C. sativum. *Garden C.* Lf. segm. ovate, cut or cleft; fr. smooth, shining. Eur. 18′.

13. CRYPTOTÆNIA, DC. Honewort. Margin of the calyx obsolete. Fruit elliptical, with slender styles. Carpels with 5 obtuse ribs. Carpophore free, 2-parted. Vittæ very narrow, twice as many as the ribs. ♃ Leaves 3-parted, lobed and doubly-serrate. Umbels compound, with very unequal rays. Invol. 0. Involucels few-leaved. Flowers white.

C. Canadénsis DC.—Common in moist woods. Plant smooth, 2–3f, with large lfts. (3′ by 2′). Umb. panicled, slender, involucels minute. Fr. 2″ long, styles 1″. Jn.-Sept.

14. BUPLEÙRUM, Tourn. Thorough-wax. Calyx teeth 0. Fruit laterally compressed. Carpels 5-ribbed, lateral ones marginal. Seed teretely convex, flattish on the face.—Herbaceous or shrubby. Lvs. (or phyllodia) entire. Involucra various. Flowers yellow.

B. rotundifòlium L. Lvs. (phyllodia) roundish-ovate, entire, perfoliate; invol. 0, involucels of 5, ovate, mucronate bracts. ① Fields, N. Y. to Va. Rare. § Europe.

15. CARUM, L. Caraway. Alexanders. Cal. teeth minute or 0. Disk broad-conic. Fr. ovate or oblong, laterally compressed. Carpels 5-angled, with 5—10 prominent, filiform, equal ribs, the two lateral bordering the commissure. Intervals with a single, rarely 2, vittæ. Seeds sub-terete.—Leaves ternate to decompound. Involucra various.

§ Zizia. Lvs. simple, or 1-2-ternate, ovate. Cal. teeth minute. Pet. yellow...Nos. 1. 2
§ Carum. Lvs. pinnately or ternately dissected. Cal. teeth 0. Pet. white...Nos. 3, 4

1 C. aùreum. *Golden Alexanders.* Lvs. 1-2-ternate; lfts. thin, lance-oblong, sharply serrate; umb. rays 1′; invol. 0; involucels 3-lvd.; fr. oval, the ribs acute or winged. ♃ Meadows and banks. 1—2f. Smooth throughout. Fls. deep yel. Jn. (Thaspium, N.)

2 C. cordàtum. Root lvs. simple, cordate, crenate, on long stalks; st. lvs. becoming 3-parted, ternate, or quinate, serrate: fr. roundish-oval, with acute or winged ribs; fls. yellow, varying to brownish. Rocky shades. 2—3f. May, June. (Thaspium, N.)

3 C. Petroselìnum B. & H. *Parsley.* Leaf segm. numerous, wedge-ovate to lance-oblong, acute, incised; invol. lvs. few or 0; involucels subulate. ♃ Greece. 2—3f. Jn.

4 C. Carvi. *Caraway.* Lf. segm. numerous, linear to filiform; invol. 1-lvd. or 0; involucels 0. ♃ Europe. 2—3f. Lvs. large. Fls. white. Fr. oblong, aromatic. June.

16. THÁSPIUM, Nutt. Golden Alexanders. Calyx margin 5-toothed. Fruit ovoid, transversely subterete. Carpels semiterete, with 5 prominent or winged ribs, the lateral margined. Intervals with single vittæ. ♃ Umbels without an invol. Involucels 3-lvd., lateral. Fls. yellow

ORDER 63.—UMBELLIFERÆ.

1 T. barbinòde N. St. pubescent at the nodes; lvs. triternate and biternate; lfts. wedge-ovate, cut-serrate: fr. large (3″), elliptical, 6-winged. River banks. St. 2—3f, angular and grooved. Rays 2′, each 20-flowered. Flowers deep yellow. June.

2 T. Wálteri Shntt. Stem rough-puberulent above; lvs. triternate to ternate; lfts. pinnatifid with linear-oblong segments; fruit oblong, narrowly 8-10-winged. Barrens, Ky. to E. Tenn. and W. Car. (Zizia pinnatifida Buckley.)

17. PIMPINÉLLA, L. ANISE. ZIZIA.
Calyx teeth obsolete. Fruit ovate, oval, or roundish, laterally compressed and contracted at the commissure, ribs very slender, with many vittæ. Styles slender. Seeds teretely 5-angled. ♃ Leaves decompound. Involucra 0, or scarcely any.

1 P. integrifòlia (B. & H.) Smooth, glaucous; lvs. bi- or tri-ternate, with elliptic-oblong, entire, acute lfts. (1′); umb. (yellow) with 13 very slender (2—3′) rays; fr. oval, with 3 vittæ in each interval. Rocky woods. 1—2f. May—July. (Zizia, DC.)

2 P. ANÌSUM. *Anise.* Smooth, shining; root lvs trifid, cauline multifid, with narrowly-linear segments; umbels large, many-rayed. Egypt. Richly aromatic.

18. FŒNÍCULUM, Adans. FENNEL.
Fruit elliptic-oblong, subterete. Carpels each with 5 carinate ribs, intervals with single vittæ. Involucra 0. Leaves biternately dissected. Flowers yellow.

F. VULGÀRE. Leaf segm. linear-subulate, elongated, or filiform; umb. of 15—30 unequal rays. ② Europe. 3—5f. The turgid seeds are warmly aromatic. (Anethum, C-B.)

19. DAUCUS, Tourn. CARROT.
Calyx limb 5-toothed. Pet. the 2 outer often largest and deeply 2-cleft. Fr. oblong. Carpels with 5 primary, bristly ribs, and 4 secondary, the latter more prominent, winged, and divided each into a single row of prickles, and having single vittæ beneath. ② Invol. pinnatifid. Involucels of entire or 3-cleft bracts. Fls. white, the central one abortive.

1 D. Caròta L. Stem hispid; lvs. tripinnatifid, the segm. linear, cuspidate-pointed; umbels dense, concave; invol. pinnate. Fields, waysides: common. 3f. § Eur.—In cultivation the root becomes conical, fleshy, red to yellow, and nutritious. Jl.—Sept.

2 D. pusíllus Mx. Slender, retrorsely hispid; lvs. bipinnatifid, divisions deeply lobed with linear-oblong, merely acute segments; invol. bipinnatifid. Dry soils, S. Car. to Fla., and W. 1—3f. June.

20. CONIUM, L. POISON HEMLOCK.
Calyx margin obsolete. Fruit ovate, laterally compressed. Carpels with 5 acute, equal, undulate-crenulate ribs, lateral ones marginal. Vittæ 0. Seeds with a deep, narrow groove on the face. ② Poisonous. Leaves decompound. Involucra and involucels 3-5-leaved, the latter unilateral. Flowers white.

C. maculàtum L. St. spotted; lvs. tripinnate; lfts. lanceolate, pinnatifid; involucel short; fruit smooth. Waste grounds, waysides. 4f. Much branched. An ill-scented narcotic. July. § Europe.

21. SELINUM, L.
Calyx teeth obsolete. Fr. ovoid to oblong, terete. Carpels slightly compressed on the back, semiterete, with 5 winged ribs, the lateral wings broadest, intervals with 1 (rarely 2) vittæ. ♃ Glabrous, tall, branched. Lvs. pinnately decompound. Um.b. rays ∞. Invol. bracts 0- -few. Involucels ∞-bracted. Fig. 303.

S. Canadénse B. & H. Petioles large, sheath-like, inflated; lf. segm. linear-oblong, very acute, or acuminate; umb. 12-rayed, long-stalked; bracts lin.-filiform; fls. white, conspicuous. Wet woods, Me. to Va. and Wis., rare, 3–5f. Ang., Sept. (Conioselinum.)

22. LIGÚSTICUM, L. Lovage. Calyx teeth minute. Fruit as in Selinum, except that the intervals are filled with numerous vittæ. ♃ Glabrous. Lvs. ternately divided. Involucra few-∞-bracted. Fls. white.

1 L. Scóticum L. *Sea L.* Lvs. 2-1-ternate; lfts. rhombic-ovate, cut-dentate, some oblique; invol. bracts ∞-linear; fr. oblong. Sea-coast. northward. 2f. Fruit 5″. July.

2 L. actæfòlium Mx. *Angelico.* Lvs. triternate, with ovate, dent-serrate leaflets; umbels panicled or triply compound; involucra about 3-bracted; fruit short. Woods, Ms. to Tenn. 3—6f. May—July.

23. ÆTHÚSA, L. Fool's Parsley. Calyx margin obsolete. Fruit globous-ovate. Carpels with 5 acutely-carinated ribs, lateral ones marginal, broader. Intervals acutely angled, with single vittæ, commissure with 2. ① Poisonous herbs. Leaves ternately or pinnately decompound. Involucra 0. Involucels one-sided, 3-leaved, deflexed. Flowers white.

Æ. Cynàpium L.—Waste grounds, N. Eng. to Penn.: rare. 2f. Stem green. Leaf segm. numerous, wedge-shaped, uniform. Plant ill-scented, dark green. Jl. § Eur.

24. ERIGENÌA, Nutt. Daughter-of-Spring. Calyx limb obsolete. Pet. not inflexed, entire. Fr. contracted at the commissure. Carpels 8-ribbed, ovate-reniform. ♃ Rt. tuberous. Radical leaf triternately decompound. Involucrate lvs. solitary, biternately compound. Involucels of 3—6 entire, linear-spatulate bracts. Figs. 235, 369.

E. bulbòsa Nutt. A small, early-flowering herb, 4—6′. Shady banks, Penn., W. N.Y. and W. Tuber roundish, deep in the ground. Pet. white, anth. brown-purple (hence called *Pepper-and-Salt*). March, April.

25. ÀPIUM, L. Celery, &c. Calyx teeth obsolete. Pet. not emarginate. Fr. ovate or globular, laterally compressed, often didymous. Carpels 5-angled, ribs equal, obtuse. Vittæ single in each interval. Carpophore undivided. Seed terete. ① ♃ Smooth. Leaves pinnately decompound. Involucra various. (Flowers white.)

§ Helosciàdium. Lvs. simply pinnate. Involucels ∞-bracted. Fr. roundish..Nos. 1—3
§ Euàpium. Lvs. pinnately decompound. Involucels 0. Involucre 1-leaved...Nos. 4—6

1 A. lineàre. Stem angular, tall; lfts. 9—11 (3 above), linear-oblong or linear, tapering to a very acute point, serrate; umb. pedunculate; invol. ∞-bracted; fr. globular with very prominent ribs. ♃ Wet. 2—4f. July, Aug. (Sium, C-B.)

2 A. Cársoni (Durand). Erect, branched; lfts. 3—7, lin. to ovate, serrate to gashed fr. broadly ovate, the ribs filiform, with broad intervals. Wet. Conn. to Penn. Jn., Jl.

3 A. nodiflòrum. Stems procumbent; lvs. pinnate; lfts. lance-oblong, equally serrate; umb. opposite the lvs., subsessile; invol. 0-2-lvd. ① Wet. S. Car. 1—2f. Apr. §

4 A. leptophýllum. Erect or diffuse; lf. segm. linear to filliform; umb. opp. the lvs., sessile; fr. very small (½″), globular, with thick ribs. ① Ga. to La. Jn. (Helosc.)

5 A. divaricàtum. Small and slender; lf. segm. filiform or capillary, obtuse; umb. very small, pedunculate, 3-5-rayed; fr. rough with minute scales. ① Dry sands, S. 2—8′. March, April. (Leptocaulis, N.)

6 A. graveolens. *Celery.* Lvs. on long petioles, segm. broad-cuneate, incised, upper lvs. 3-parted and cut-lobed; invol. 0; fr. roundish. ② Eur. Well known as a salad.

Order 63.—UMBELLIFERÆ.

26. EULOPHUS, N. Calyx limb 5-toothed, deciduous. Fr. contracted laterally, somewhat double. Carpels surrounded with large vittæ, ribs obsolete. Seed channelled on the inner face. ♃ Smooth, branched. Lvs. ternately decompound. Invol. nearly 0. Involucel setaceous. Fls. white.

E. Americàna N. Lvs. mostly radical; segm. lance-lin., 1' long, acute, upper lvs. in 3 long, entire seg.; umb. long-stalked, 3-10-rayed. Prairies, O. to Ill. and Tenn. 3—4f.

27. CICÙTA, L. WATER HEMLOCK. Calyx margin of 5 broad segments. Fr. subglobous, didymous. Carpels with 5 flattish, equal ribs, 2 of them marginal, intervals filled with single vittæ. Seeds terete. ⚓ ♃ Poisonous. Leaves compound. Stems hollow. Umbels perfect. Invol. few-leaved or 0. Involucels many-leaved. Flowers white.

1 C. maculàta L. St. streaked with purple; lower lvs. triternate and quinate, upper biternate; segments lanceolate, mucronately serrate, the veins running to the notches. Wet meadows. 3-6f. Smooth, glaucous. Leaflets 1—3'. Fruit 1¼'', 10-ribbed. Umbels 3'. July, August.

2 C. bulbífera L. Lvs. biternate; lfts. linear, with remote, divergent teeth; lvs. of the branches 3-cleft or simple, subopposite, bearing bulblets in their axils. Swamps, Can. to Penn. and W. 3—4f. Leaflets 2—4' by 1—4''. Umbels few. August.

28. SIUM, L. WATER PARSNIP. Calyx teeth acute. Pet. obcordate, with an inflexed point. Fr. nearly oval, laterally compressed. Carpels with 5 obtusish ribs, and several vittæ in each interval. Carpophore undivided. ♃ Leaves pinnate, dentate. Umbels perfect, with many-leaved involucra. Flowers white.—Stout herbs.

S. latifòlium L. St. angular, sulcate; lfts. oblong-lanceolate, acutely and coarsely serrate, barely acute; cal. teeth conspicuous. Swamps, Ind. (Green Co. !) and Can. 3—4f. Lfts. 4—6' by 1—2', 2-10-toothed. Umb. with 20—30 long (3—4') rays. Jl., Aug.

29. DISCOPLEÙRA, DC. BISHOP-WEED. Cal. teeth subulate, persistent. Fr. ovate, often didymous. Carp. 5-ribbed, the 3 dorsal ribs filiform, subacute, prominent, the 2 lateral united with a thick, accessory margin; intervals with single vittæ. Sds. subterete. ① Lvs. capillaceous dissected. Umbels compound. Bracts of the invol. cleft. Fls. white.

1 D. capillàcea DC. Erect or procumbent; umbels 3-10-rayed; lfts. of the invol. 3—5, mostly 3-cleft; fr. ovate. Swamps near the coast, Mass. to Ga. 1—2f. June+.

2 D. costàta Hale (1850). Branched, erect; umbels 7-15-rayed; bracts of the invol. 10—12, 2-5-parted; lf.-segm. filiform, numerous, apparently verticillate; fr. with ribs and vittæ strongly contrasted. Swamps, Ogeechee R. and W. 1—2f. stout. Oct., Nov.

3 D. Nuttàllii DC. Erect, tall; umbel 15-20-rayed; invol. few bracted, bracts entire; fr. broadly cordate-ovate. Wet prairies, Ky. and S. Slender, 2—4f.

30. CORIÁNDRUM, L. CORIANDER. Cal. with 5 conspicuous teeth. Outer petals radiant, inflex-bifid. Fr. globous. Carp. cohering, with the 5 depressed, primary ribs, and 4 secondary more prominent ones, seeds concave on the face. ① Smooth. Invol. 0 or 1-leaved. Involucels 3 leaved, unilateral.

C. sativum L. Lvs. bipinnate, lower ones with broad-cuneate lfts., upper with linear lfts.; carp. hemispherical. Europe. 2f. Cultivated for its spicy fruit.

Order LXIV. ARALIACEÆ. Araliads.

Trees, shrubs or *herbs* closely allied to the Umbelworts in the leaves, inflorescence and flowers, but the *styles and cells of the* ovary are usually more than 2 (3 to 5), cells 1-ovuled. *Fruit* baccate or dry, 3–5-celled, with 1 albuminous seed in each cell, and the petals not inflected. Fig. 242.

§ Styles and carpels 5. Umbels ∞. Flowers perfect. Leaves alternate, pinnate... ......Aralia. 1
§ Styles and carpels 2—3. Umbel 1. Flowers dioecious. Leaves verticillate, palmate......Ginseng. 2
§ Styles 5, united into 1. Umb. ∞. Flowers polygamous. Lvs. simple. Climbing.......Hedera. 3

1. ARALIA, L. Wild Sarsaparilla. Cal. tube adherent, limb 5-toothed. Pet. 5, ovate, spreading. Stam. 5, epigynous. Styles 5, recurved above, persistent. Fr. a berry, 5-celled, 5-seeded, and 5-angled when dry, ♃ ♄ Lvs. pinnately compound, alternate. Umbels several or many, white or greenish, in summer.

* Plants low (1—2f), with few (3—7) umbels corymbously arranged..........Nos. 1, 2
* Plants tall (3—12f), with numerous umbels in racemes...................Nos. 3, 4

1 A. nudicaulis L. Nearly stemless, with 1 ternate-pinnate leaf longer than the scape, which bears 3 umbels at top; plant smooth. ♃ Rich, Rocky wds. E. & W. 1f.
2 A. hispida L. *Wild Elder.* Stem shrubby and hispid-prickly at base, herbaceous above; lvs. 1-2-pinnate; lfts. ovate, cut-serrate, often lobed; umbels about 5, long-stalked, forming a terminal corymb. ♃ Dry fields. N. Eng. to Va. 1—2f. Fr. blue-blk.
3 A. racemòsa L. *Pettymorrel.* Herbaceous, smooth, branched; lvs. large, bi-ternate-pinnate, lfts. ovate, serrate; umb. small, ∞, in a panicle of racemes. ♃ Rocky woods. 3—5f. Root aromatic, an ingredient in *small-beer.*
4 A. spinòsa L. *Angelica-tree. Hercules' Club.* Shrub prickly; lvs. bi- and tri-pinnate, lfts. thick, ovate, cusp-pointed, glaucous beneath. Damp woods, O. to Fla. 8—12—20f. Trunk usually simple, bearing all the lvs. and panicles at the top.

2. GINSENG. (Panax, L. *in part.*) Dioecious-polygamous. Cal. tube adherent, limb obsolete. Pet. 5, ovate, obtuse. Stam. 5, epigynous. Sty. 2 or 3, distinct, erect. Fruit baccate, 2- or 3-seeded. ♂ Styles obsolete. ♃ Root tuberous. Stem simple, bearing 3 leaves in a whorl and one umbel. Flowers white. Fig. 242.

1 G. trifòlium. *Ground-nut.* Root a round tuber; stem low (3—6′); lvs. palmately 3-5-foliate, lfts. lance-oblong, serrate, subsessile; peduncle longer than the petioles; sty. 3; berries 3-lobed, greenish-yellow. Low woods: com. May. Root farinaceous.
2 G. quinquefòlium. *True Ginseng.* Root fusiform, fleshy; st. taller (1f+); lvs. palmately 5-foliate, lfts. ample, obovate, petiolulate, acuminate, serrate; peduncle shorter than the petioles; sty. 2; berries 2-seeded, bright red. Rocky woods. Jn.-Aug.

3. HÉDERA. L. European Ivy. Calyx 5-toothed. Pet. 5, valvate. Sta. 5. Sty. united into 1. Fr. ovoid, baccate, 5-seeded. ♄ Lvs. coriaceous, simple. Flowers green.

¶. Helix. Stems woody, slender, climbing high by radicating fibres; lvs. dark green, with whitish veins, roundish ovate, 5-angled; umbels corymbed; fr. black. Europe.

Order LXV. CORNACEÆ. Cornels.

Trees and *shrubs*, seldom *herbs*, without stipules. *Leaves* opposite or rarely alternate, simple, with pinnate veinlets. *Calyx* adherent to the

Order 65.—CORNACEÆ.

ovary, the limb minute, toothed or lobed. *Petals* distinct, alternate with the calyx teeth, valvate in the bud, often 0. *Stamens* same number as petals, inserted on the margin of the epigynous disk (in the ☿ flowers.) *Ovary* 1- or 2-celled. *Fruit* a baccate drupe, crowned with the calyx. Fig. 43C.

1. CORNUS, L. Dogwood. Flowers perfect. Calyx limb of 4 minute segments. Pet. 4, oblong, sessile. Sta. 4. Style somewhat club-shaped. Drupe baccate, with a 2- or 3-celled nut. ♄ ♄ ♃ Lvs. entire. Flowers in cymes, often involucrate. Floral envelopes valvate in æstivation. Bark bitter, tonic. Fig. 430.

§ Cymes subtended by a 4-leaved, white involucre. Fruit red............Nos. 1, 2
§ Cymes naked.—*a* Lvs. alternate, clustered at the ends of the branches........No. 3
—*a* Lvs. opposite.—*b* Twigs and cymes pubescent ....Nos. 4, 5
—*b* Twigs, &c., glabrous.—*c* Drupes blue.Nos. 6, 7
—*c* Drupes wh...Nos. 8, 9

1 C. Canadénsis L. *Low Cornel.* Herbaceous, low; upper lvs. whorled, veiny, on short petioles; st. simple; invol. lvs. ovate. ♃ Damp woods, N. 4—8′. May, June.

2 C. flórida L. *Flowering Dogwood.* Arboreous; lvs. opposite, ovate, acuminate, entire; fls. small, in a close, cymous umbel or head, surrounded by a very large, 4-lvd. obcordate involucre. Tree in woods, 20—30f. Invol. showy. May. Bark tonic.

3 C. alternifòlia L. Lvs. alternate, oval, acute, hoary beneath; branches alternate, warty; drupes purple, globous. Shrub or tree, 8—20f, with a flattened top. June.

4 C. seríеca L. Branches spreading, purplish, branchlets woolly; lvs. ovate or elliptical, acuminate, silky-pubescent beneath; cymes depressed, woolly; cal. teeth lanceolate; drupes light blue. Shrub 5—9f. Flowers yellowish white, crowded. June.

5 C. asperifòlia Mx. Branches erect, brownish, branchlets rough-downy; lvs. lance-oval, scabrous above, downy beneath; cymes hispid; sep. minute. W. and S. May+.

6 C. stricta Lam. Branches erect, brown, smooth; lvs. ovate to lanceolate, smooth and green both sides, long-acuminate; cymes loose, smooth; sepals subulate, half as long as the ovary; anth. and fr. pale blue. Swamps, Va. to Fla. 8—12f. April.

7 C. circinàta L. Branches warty; lvs. round-oval, white-tomentous beneath; cymes spreading, depressed; drupes light blue. Shrub 5—10f, E. and W. Lvs. large. June.

8 C. paniculàta L'Her. Branches erect, grayish, smooth; leaves ovate-lanceolate, acuminate, hoary beneath; cymes and drupes small, paniculate, white. 6f. May, Jn.

9 C. stolonífera Mx. *Red Osier.* St. often stoloniferous; branches smooth; shoots virgate, reddish-purple; lvs. broad-ovate, acute, pubescent, hoary beneath; cymes naked, flat; berries bluish-white. Small tree, E. and W. 8—10f. May, June.

2. NYSSA, L. Tupelo. Gum-tree. Fls. diœcious or polygamous. ♂ Calyx tube very short, limb truncate. Pet. 5, oblong. Sta. mostly 10, inserted in the bottom of the calyx. Ov. 0. ♀ Calyx tube oblong, adherent to the 1-celled ovary, limb as in ♂. Pet. 2—5, oblong, often 0. Sty. large, stigmatic on one side. Drupe oval, 1-seeded. ♄ with small green fls. clustered on axillary peduncles, the sterile more numerous. Apr. June.

1 N. multiflòra Wang. Lvs. oblong-obovate, acutish or obtuse at each end, entire; the petiole, midvein, and margin villous; fertile peduncles 3-(2–5)-flowered; sty. revolute; nut short, obovate, striate, obtuse. Tree 30—70f. Drupe blue-black. †

2 N. uniflòra Walt. *Swamp Tupelo.* Lvs. green, oblong-ovate or ovate, long-petiolate; fertile fls. solitary, 3-bracted, on slender peduncles; sty. nearly straight; sterile fls. 5—10; drupe oblong, as large as a plum. Tree 50—80f, in swamps, S.

3 N. capitàta Walt. *Ogeechee Lime.* Leaves oval or oblong, short-petiolate, entire, whitened beneath, obtuse at apex, acute at base; fertile fls. solitary, on short peduncles, downy, 3–4-bracted, with 5 petals and 10 stamens; sterile fls. 20—30 in each dense globular head; fruit large, oblong. River banks, S. 20—30f.

COHORT 2, GAMOPETALÆ,

Or MONOPETALOUS EXOGENS.—Plants having a double perianth, consisting of both calyx and corolla, the latter composed of petals partially or wholly united. (Cohort 3, page 278.)

ORDER LXVI. CAPRIFOLIACEÆ. HONEYSUCKLES.

Shrubs, rarely *herbs*, often twining with opposite leaves; no stipules. *Flowers* clustered and often fragrant, 5-parted and often irregular. *Corolla* monopetalous, tubular or rotate. *Stamens* inserted on the corolla tube, rarely one less than the lobes. *Ovary* adherent to the calyx. *Style* 1, *stigmas* 3 to 5. *Fruit* a berry, drupe, or capsule. *Embryo* small, in fleshy albumen. Figs. 67, 383, 390, 466, 471, 477.

I. LONICEREÆ. Corolla tubular, with a filiform style...(a)
  a Herbs.—*b* Corolla 5-lobed, the stamens but 4 ....LINNÆA. 1
    —*b* Corolla 5-lobed, the stamens 5..TRIOSTEUM. 2
  a Shrubs.—*c* Corolla bell-shaped, regular. Berry 4-celled. 2-seeded..........SYMPHORICARPUS. 3
    —*c* Corolla tubular, lobes unequal. Berry 2–3-celled.. ....LONICERA. 4
    —*c* Corolla funnel-form. Capsule 2-celled, ∞-seeded. (Addenda.)..DIERVILLA. 5
II. SAMBUCEÆ. Corolla rotate, deeply 5-lobed. Stigmas sessile...(b)
  b Shrubs with pinnate leaves. Berry 3-seeded......... ....SAMBUCUS. 6
  b Shrubs with simple leaves. Drupe 1-seeded........................... . .....VIBURNUM. 7

1. LINNÆA, Gron. TWIN-FLOWER. Calyx tube ovate, limb 5-parted, deciduous. Bractlets at base 2. Cor. campanulate, limb subequal, 5-lobed. Sta. 4, two longer than the others. Berry dry, 3-celled, indehiscent, 1-seeded (two cells abortive). ♄ Lvs. roundish, petiolate. Ped. filiform, erect, 2-flowered. Inhabits the N. temperate zone of both hemispheres.

L. boreàlis Gron.—Moist rocky shades, N. J. to Oreg. and N. Filiform stems 3—6f. Ped. 3, bearing at top a pair of nodding, bell-shaped, roseate, fragrant flowers. June.

2. TRIÓSTEUM, L. FEVERWORT. Calyx tube ovoid, limb 5-parted, segm. linear, nearly as long as the corolla. Cor. tubular, gibbous at base, limb 5-lobed, subequal. Sta. 5. Included. Stig. capitate, lobed. Fr. drupaceous, crowned with the calyx, 3-celled, containing 3 ribbed, bony seeds. ♃ Coarse, hairy, with large, connate leaves and axillary flowers.

1 **T. perfoliàtum** L. Hirsute; lvs. oval, acuminate; fls. verticillate or clustered, sessile, brownish-purple. Rocky woods. 2—4f. Fruit orange-colored. 6″. June.
2 **T. angustifòlium** L. Hispid: lvs. lanceolate, acuminate. scarcely connate: fls. mostly solitary, short-stalked, yellowish or straw-colored. L. I., W. & S, 2—3f. May.

3. SYMPHORICÁRPUS, Dill. SNOWBERRY. Calyx tube globous, limb 4–5-toothed. Cor. funnel- or bell-shaped, the limb in 4—5 equal lobes Sta. 4 or 5. Stig. capitate. Berry globous, 4-celled, 2-seeded (two opposite cells abortive). ♄ Leaves oval, entire. Flowers small, roseate.

ORDER 66.—CAPRIFOLIACEÆ.

1 **S. racemòsus** Mx. Fls. in terminal, loose, interrupted, often leafy rac.; cor. campanulate, densely bearded within; sty. and sta. included; berries snow-white. W. Vt. to Wis. and Pa., on rocky banks. 2—3f. A smooth, handsome shrub. July—Aug. †
2 **S. occidentàlis** R. Br. *Wolfberry.* Lvs. ovate, obtusish; spikes dense, axillary and terminal, nodding; cor. densely bearded inside; sta. and bearded style exserted: berries white. Woods, Mich. Wis. and N. 2—4f. July.
3 **S. vulgàris** Mx. Lvs. roundish-oval; spikes axillary, subsessile, capitate, and crowded; cor. lobes nearly glabrous; sta. and bearded style included; berries dark red. River banks, Penn. to Iowa, and S. 2—3f. Flowers greenish-red. July.

4. LONICÈRA, L. HONEYSUCKLE. WOODBINE.
Calyx 5-toothed, tube subglobous. Cor. funnel- or bell-form, limb 5-cleft, often labiate. Sta. 5, exserted. Ov. 2–3-celled. Berry few-seeded. Stig. capitate. ♄ ♭ Lvs. entire, often connate. Fls. fragrant and beautiful. May–Jl. Figs. 67, 390.

§ XYLÓSTEON. Shrubs erect. Leaves never connate. Flowers in pairs...(*a*)
  a Corolla gibbous at base, lobes somewhat irregular..................Nos. 1—3
  a Corolla not gibbous, lobes spreading, equal, roseate.....................No. 4
§ CAPRIFÒLIUM. Shrubs climbing. Flowers sessile, mostly whorled...(*b*)
  b Leaves all distinct. Corolla ringent. Cultivated exotics..............Nos. 5, 6
  b Leaves (the upper pair) connate-perfoliate...(*c*)
    c Corolla subequal, both tube and limb scarlet.........................No. 7
    c Corolla limb ringent,—*d* tube equal (not gibbous) at base.........Nos. 8—10
                           —*d* tube gibbous at the base................Nos. 11, 12

1 **L. ciliàta** Muhl. *Fly Honeysuckle.* Lvs. ovate, subcordate, ciliate; cor. limb with short and subequal lobes, tube saccate at base; sty. exserted; berries distinct, red. Woods, Mc. to O. and N. 3—4f. Flowers straw-yellow, on short ped. May.
2 **L. oblongifòlia** Hook. Lvs. oblong or oval, velvety beneath; cor. limb deeply bilabiate; ped. long, filiform, erect; berries connate or united into one, globous, purple. Swamps, N. Y., W. and N. 2—3f. Purple-yellow. †
3 **L. cœrùlea** L. Lvs. oval-oblong, ciliate, obtuse, villous both sides, at length smoothish; ped. short, reflexed in fruit; bracts longer than the ovaries; cor. lobes short, subequal; berries connate, deep blue. Rocky woods, Ms. N. Y. and N. 2—3f.
4 **L. TARTÁRICA.** *Tartarian Honeysuckle.* Much branched; lvs. ovate, cordate, polished; cor. segm. oblong, obtuse, purple-white. Russia. 4—10f.
5 **L. JAPÓNICA.** *Chinese Honeysuckle.* Sts. soft-pubescent; lvs. ovate and oblong; ped. axillary, 2-bracted and 2-flowered; flowers orange, &c. China. 15f.
6 **L. PERICLÝMENUM** Tourn. *Woodbine.* Lvs. deciduous, elliptical, acute, on short petioles; fls. in dense, terminal heads, red, yellow. Europe. 15f.
  β. QUERCIFÒLIUM. Leaves sinuate-lobed.
7 **L. sempérvirens** Ait. *Trumpet Honeysuckle.* Lvs. oblong, evergreen; flowers in nearly naked spikes of distant whorls; cor. trumpet-shaped, nearly regular, ventricous above. Moist groves, N. Y., W. and S. 15f. May–Sept. †
8 **L. flava** Sims. *Yellow Honeysuckle.* Lvs. ovate, glaucous both sides; spikes terminal, of about 2 close whorls; cor. smooth, slender, bright yellow; stam. exserted. N. Y., W. and S. Shrub scarcely twining. Corolla 15″. †
9 **L. gràta** Ait. *Evergreen Honeysuckle.* Lvs. evergreen, obovate, smooth, glaucous beneath; fls. in sessile, terminal, and axillary whorls; cor. ringent, long, slender, reddish without, yellowish within. Damp woods, M. and W. States. 12f.
10 **L. CAPRIFÒLIUM.** *Italian Honeysuckle.* Lvs. deciduous; fls. in a single, terminal verticil; lips of corolla revolute, red, yellow, white. Europe.
11 **L. parviflòra** Lam. Lvs. smooth, shining above, glaucous beneath, oblong, all sessile or connate, the upper pair perfoliate; fls. in heads of 4 or more approximate whorls; cor. glabrous, short, yellow-red; fil. bearded. Rocky woods. 8—10f.
  β. *Douglàsii.* Lvs. large, pubes. beneath, lower petiolate; fls. pubes. O., and W

12 L. hirsùta Eaton. Lvs. hairy above, soft-villous beneath, veiny, broad-oval, abruptly acuminate; fls. in verticillate spikes, greenish-yellow; fil. bearded. Woods. N. Eng. to Mich. and N. 15—20f.

5. DIERVÍLLA, Tourn. BUSH HONEYSUCKLE.

Calyx tube oblong, limb of 5 linear segm. Cor. twice as long, funnel-shaped, limb 5-cleft and nearly regular. Sta. 5. Capsular fr. 2-celled, 2-valved, crowned with the cal., many-seeded. ♄ Lvs. acuminate, serrate, deciduous. Ped. axillary. Ju.

1 D. trífida Mœnch. Lvs. ovate, on distinct petioles; ped. 1-3-flwd.; pod attenuate at top beneath the calyx limb. Thickets, Can. to Car. 2f, bushy. Fls. greenish-yellow.
2 D. sessilifòlia Buckley. Lvs. lance-oblong, sessile or subamplexicaul; peduncles 3-5-flwd., crowded in the axils above; caps. short-beaked. High Mts. N. Car. 2—4f.

6. SAMBÚCUS, L. ELDER.

Calyx small, 5-parted. Cor. 5-cleft, segm. obtuse. Sta. 5. Stig. obtuse, small, sessile. Berry globous, pulpy, 3-seeded. ♄ ♃ Lvs. odd-pinnate or bipinnate. Fls. in cymes, white. Figs. 466, 477.

1 S. Canadénsis L. Woody, with large pith; lfts. 7—11, oblong-oval, acuminate, smooth; cymes fastigiate; berries dark-purple. Hedgerows, thickets: common. 9—12f. Cymes broad, white. May—July.
2 S. pubens Mx. Woody; lfts. lance-oval, acuminate, 5—7, downy beneath; cymes paniculate; berries scarlet. Copses. Can. to Car. 5—10f. June.—Berries rarely white. Catskill Mountains.

7. VIBÚRNUM, L.

Calyx small, 5-toothed, persistent. Cor. rotate, limb 5-lobed, seg. obtuse. Stam. 5. Stig. 1—3, sessile. Fr. a drupe, 1-celled, 1-seeded,—a stony nut covered with soft pulp. ♄ ♃ Lvs. simple, petioles often minutely stipulate. Fls. white, in compound flat cymes, which are often radiant. Fig. 388.

§ Cymes radiant,—the outer flowers sterile and showy. Leaves stipuled .... Nos. 1, 2
§ Cymes not radiant,—the flowers all alike perfect..(a)
    a Leaves 3-lobed, palmately 3-5-veined, with setaceous stipules .... Nos. 3, 4
    a Leaves not lobed,—*b* coarsely toothed, straight-veined. Cyme stalked ... Nos. 5—7
        —*b* finely and sharply serrate. Cymes sessile. June .... Nos. 8, 9
        —*b* entire, or nearly so.—*c* Species native .... Nos. 10, 11
        —*c* Species exotic .... Nos. 12, 13

1 V. lantanoìdes L. *Hobble-bush.* Leaves round-cordate, abruptly acuminate, unequally serrate; petioles and veins rusty-downy; cyme sessile; fruit ovate. Rocky woods, N. 5f. Shoots often reclined and rooting. Handsome. May.
2 V. Ópulus L. *High Cranberry.* Smooth; lvs. 3-lobed, 3-veined, broader than long, rounded at base, lobes acuminate, crenate dentate; petioles glandular; cymes pedunculate. Borders of woods, N. 8—12f. Fruit bright red, very acid. June.
    β. ROSEUM. *Snow-ball.* Fls. all neutral, in globous cymes. †
3 V. acerifòlium L. *Dockmackie.* Leaves subcordate, 3-veined, lobes acuminate, acutely dentate, downy beneath; stam. exserted; fr. purple. Woods. 4—6f. June.
4 V. pauciflòrum Pylaie. Lvs. roundish, 5-veined at base, with 3 short lobes, serrate; cymes few-flowered; stamens included; fr. red. Mts. N.: rare. 2—3f.
5 V. dentàtum L. *Arrow-wood.* Smooth; lvs. round-ovate, acutely-toothed, often with downy tufts in the axils of the stout veins beneath; petioles slender; fr. blue; nut concavo-convex. Damp woods, Can. to Ga. 8—12f. Branches virgate. June.
6 V. pubéscens Ph. Lvs. ovate, acuminate, broadly dentate, hairy most beneath; petioles short, downy; fr. black, nut plano-convex, grooved. Rocks, Can. to Car. 2—3f.
7 V. molle Mx. *Poison Haw.* Downy throughout, with forked or stellate hairs; lvs. broad oval, acute, crenate dentate; fr. blue, nut grooved. Woods. Ky. to Fla. 10f. May

Order 67.—RUBIACEÆ. 147

8 V. Lentàgo L. *Sweet Viburnum.* Lvs. ovate and oval, long-acuminate, acutely and finely uncinate-serrate; petiole with undulate margins; fr. glaucous-black, oval, eatable. Rocky woods, Can. to Ga. and Ky. 10—20f. A small, handsome tree. June.

9 V. prunifòlium L. *Black Haw. Sloe.* Lvs. shining, oval or ovate, obtuse, sharply uncinate-serrulate; petioles slightly margined; cymes sessile; fr. blackish, oval, sweet. Woods, N. Y. to Ga. and Ill. 10—20f. A small tree. Lvs. 2—3'.
  β. **ferrugíneum.** *Possum Haw.* Lvs. lance-oval, rusty beneath; fr. tasteless. S.

10 V. nudum L. Smooth; lvs. oval-oblong, or lance-oval, subrevolute at edge, entire or subcrenulate, not shining, veiny and dotted beneath; petioles not winged; cymes on short stalks. Thickets. 10—20f. Lvs. 3—4'. Drupes blue, eatable. Apr.–Jn.
  β. **angustifòlium.** Lvs. lance-oblong, acute at both ends, subentire. S.
  γ. **cassinoìdes.** Lvs. ovate or oval, denticulate, obtuse, acute, &c. N.
  δ. **ovàle.** Lvs. small (15″), oval, obtuse, very entire. South.

11 V. obovàtum Walt. Lvs. small (6—12″), obovate, obtuse, entire or nearly so, subsessile, dotted; cymes small, many, sessile. River banks, S. 12f. Fruit black. Ap.

12 V. Tinus. *Laurestine.* Lvs. lance-ovate, entire, thick, shining. Eur. 5f.

13 V. odoratíssimum. Lvs. elliptic-oblong, repand-dentate, thick. China.

Order LXVII. RUBIACEÆ. Madderworts.

Plants with opposite or verticillate, entire leaves. *Stipules* between the petioles sometimes leaflike or 0. *Calyx tube* adherent to the ovary; limb 4- to 5-cleft. *Corolla* regular, inserted upon the calyx tube, and of the same number of divisions. *Stamens* inserted upon the tube of the corolla, equal in number and alternate with its segments. *Ovaries* 2-(rarely more)-celled. *Style* single or partly divided. *Fruit* various.

§ STELLATÆ. Herbs with the leaves in whorls of 4—8 and no stipules..(a)
  a Flowers 4-parted. Fruit twin, separating into 2 nutlets. ....Galium. 1
  a Flowers 5-parted. Fruit twin, separable, baccate, smooth................. ....Rubia. 2
CINCHONEÆ. Leaves opposite or in whorls of 3, with stipules.—*b* Herbs..(c)
                                         —*b* Shrubs or trees..(d)
  c Flowers in pairs, with a double ovary. Berry double......................Mitchella. 3
  c Flowers separate. Carpels 2,—*e* each 1-seeded, separating in fruit..(*f*)
                   —*e* each ∞-seeded, forming a capsule..(*g*)
  f Fls. in clusters.—*h* Both carpels open after separating....................Borreria. S. Fla.
            —*h* One carpel open, the other indehiscent ....Spermacoce. 4
  f Flowers subsolitary. Both carpels indehiscent,—*k* dry........ ....Diodia. 5
                —*k* baccate........Ernodea *littoralis.* S. Fla.
  g Corolla funnel-form. Seeds 16+, cup-shaped..............................Houstonia. 6
  g Corolla wheel-shaped. Seeds 80+, angular...............................Oldenlandia. 7
  d Flowers capitate, in round, dense heads. Leaves often ternate..(*l*)
  d Flowers not capitate.—*m* Carpels 2—10, each 1-seeded. In S. Florida..(*n*)
          —*m* Carpels 2—5, each ∞-seeded. Florida..(*o*)
    n Carpels 2—4, fewer than the lobes of the corolla. Fruit fleshy..(*p*)
    n Carpels 4—10, symmetrical with the corolla lobes..(*q*)
      l Flowers 4-parted, white. Fruit compacted but distinct, dry............Cephalanthus. 8
      l Flowers 5-parted, red. Drupes united into a compact berry.......Morinda *Roio.* S. Fla.
        p Leaves opposite. Racemes axillary. Carpels flattened.........Chiococca *racemosa.*
        p Leaves opposite. Corymbs terminal. Carpels angular............Psycotria.
        p Leaves in 3's, linear, rigid. Racemes axillary. Shrub........,Strumpfia *maritima.*
      q Spikes axillary, forked. Anthers on the throat of corolla.............Guettarda.
      q Panicles axillary. Filaments inserted on the base of corolla........Erithalis *fruticosa.*
        o Fruit baccate, 5-celled. Corolla tubular. Stigma entire........Hamelia *patens.*
        o Fruit baccate, 2-celled. Cor. funnel-form, white. †...*Cape Jessamine.* Gardenia, p. 145.
        o Fruit capsular.—*s* Flowers in radiant cymes. A slender tree.........Pinckneya. 9
            —*s* Flowers in cymes, not radiant, red. Shrub. †...Bouvardia. 10
            —*s* Flowers solitary, axillary. Shrub 6—10f. . Exostemma. S. Fla.

Order 67.—RUBIACEÆ.

1. GÀLIUM, L. CLEAVERS. BEDSTRAW. Calyx limb minutely 4-toothed. Cor. rotate, 4-cleft. Sta. 4, short. Sty. 2. Carpels 2, united, separating into 2 1-seeded, indehiscent nutlets.—Herbs with slender, 4-angled stems. Verticels of 4, 6, or 8 leaves, rarely of 5.

 a Flowers yellow. Leaves in whorls of about 8. Fruit smooth.................No. 1
 a Flowers dull-purple. Leaves (large) in whorls of 4. Fruit hispid or not..Nos. 2– 4
 a Flowers white.—*b* Leaves in 4's only. Fruit dry. Panicle terminal..........No. 5
      —*b* Leaves in 4's only. Fruit smooth, purple berries.......Nos. 6, 7
      —*b* Leaves in 4's—6's.—*c* Fruit hispid with hooked hairs......No. 8
          —*c* Fruit smooth or nearly so, dry..Nos. 9—11
      —*b* Leaves in 8's, long and narrow. Fruit hispid.............No. 12

1 G. verum L. *Yellow Bedstraw.* Erect; lvs. in 8's, grooved, entire, rough, linear; fls. densely paniculate. ♃ Dry soils, Mass. 1—2f. Branches short. June. § Eur.

2 G. pilòsum Alt. Hirsute; lvs. in 4's, oval, punctate with pellucid dots; ped. several times 2- or 3-forked; fls. pedicellate, densely hispid. ♃ Dry thickets. 1—2f. June.

3 G. circæzans Mx. *Wild Liquorice.* Smoothish; lvs. oval or ovate-lanceolate, obtuse, 3-veined, ciliate on the margins and veins; ped. divaricate, few-flowered; fr. subsessile, nodding, hispid. ♃ Woods: common. 8—12'. July.

 β. *lanceolàtum.* Very smooth; leaves lanceolate, 2' long; fruit sessile.
 γ. *montànum.* Dwarf; leaves obovate. White Mountains. (Oakes.)

4 G. latifòlium Mx. St. erect. smooth; lvs. lanceolate, 3-veined, very acute; ped. axillary (leafy) and terminal, about twice 3-forked; purple flowers and smooth fruit on filiform pedicels. ♃ Woody hills, Pa. S. and W. 2f. July.

5 G. boreàle L. Erect, smooth; lvs. linear-lanceolate, rather acute, 3-veined, smooth; fls. in a terminal pyramidal panicle. ♃ Shaded rocks, N. 1f. July.

6 G. hispídulum Mx. Diffuse, minutely hispid; lvs. oval, thickish, mostly acute; ped. axillary, 1-3-flwd.; fr. large, bluish-purple. ♃ Sandy. S. 2f. May—Oct.

7 G. uniflòrum Mx. Glabrous, cæspitous, slender; lvs. linear, acute; ped. axillary, solitary, mostly 1-flwd. bracted; fr. purple. ♃ Damp woods, S. 1f. May.

8 G. triflòrum Mx. Stems weak, rough on the angles; lvs. in 5's and 6's, lance-elliptic, cusp-pointed, 1-veined; ped. mostly 3-flowered. ♃ Moist woods. 1—3f. July.

9 G. aspréllum Mx. *Rough Cleavers.* St. diffuse, very branching, rough backward, lvs. in 6's, 5's, or 4's, lanceolate, acuminate, or cuspidate, margin and midvein retrorsely aculeate; ped. short, in 2's or 3's. ♃ Thickets, N. 2—5f. July.

10 G. trífidum L. *Dyer's Cleavers. Goose-grass.* St. decumbent, very branching, ronghish with retrorse prickles; lvs. in 6's and 4's, linear-oblong or oblanceolate, obtuse, rough-edged; flowers mostly 3-parted. ♃ Swamps. 6'. July.—Variable.

 β. *tinctòrium.* Ped. 3-6-flowered; parts of the flower in 4's. The root dyes red.
 γ. *latifòlium.* Lvs. in 4's, oblanceolate; ped. 3-flowered; fls. 4-parted.

11 G. concínnum T. & G. St. decumbent, diffuse, scabrous; lvs. in 6's, linear, glabrous, 1-veined, scabrous upward on the margins; ped. filiform, twice or thrice 3-forked, panicled. ♃ Dry woods, Pa. Va. Ill. 1f. June.

12 G. Aparìne L. St. weak, procumbent, retrorsely prickly; lvs. in 8's, 7's, or 6's, linear-oblanceolate, mucronate; ped. axillary, 1-2-flwd. ① Wet thickets, N. 3-5f. Jn.

2. RÙBIA, Tourn. MADDER. Like Galium, but its flowers are mostly 5-merous, and its fruit always smooth and berry-like.

R. tinctòrum L. Stem weak, rough backward; lvs. in 6's, lanceolate, aculeate; fls. brownish-yellow, paniculate above, with 3-forked peduncles. Europe. 3—5f.

3. MITCHÉLLA, L. PARTRIDGE BERRY. Flowers 2 on each double ovary. Cal. 4-parted. Cor. funnel-shaped, hairy within. Stam. 4, short,

ORDER 67.—RUBIACEÆ.

inserted on the corolla. Stig. 4. Berry composed of the 2 united ovaries, each 4-seeded. ⌊, Smooth. Leaves opposite.

M. repens L.—Woods: com. Sts. creeping, 6-18′. Lvs. roundish-ovate, petiolate, evergreen. Cor. reddish-white, fragrant. Berry red, seeds (nutlets) bony. Very pretty. Jn.

4. SPERMACOCE, L. Cal. 2-4-parted. Cor. tubular, limb 4-lobed. Stam. 4. Stig. 2-cleft. Fr. dry, 2-celled, crowned with the calyx, separating into 1 open and 1 indehiscent carpel. Sds. 2.—Low herbs. Stip. bristly. Flowers small, in dense, axillary, sessile whorls, or clusters, white.

1 **S. glabra** Mx. Glabrous; lvs. lanceolate; cal. 4-toothed; cor. funnel-form, short, throat hairy; anth. included in the tube; stig. subsessile. ♃ River banks, W. 1—2f.
2 **S. Chapmánii** T. & G. Nearly glabrous; lvs. oblong-lanceolate; cor. funnel-form, thrice longer than the calyx; stam. and sty. exserted. Fla. Ga. 10′.
3 **S. involucràta** Ph. Hispidly hairy; lvs. ovate-lanceolate; heads terminal, involucrate; stam. exserted. Carolina (Fraser). 1f. Leaves oblique.

5. DIÒDIA, L. Carpels 2, rarely 3, separating, each 1-seeded and indehiscent. Fls. otherwise as in Spermacoce.—Herbs. Stip. fringed with bristles. Fls. few or solitary, axillary, sessile, small, white; the tube often slender. Summer.

1 **D. Virgínica** L. Procumbent; lvs. lanceolate, sessile; corolla tube slender, with a broad, spreading limb; sta. exserted. ♃ Damp places. 1—2f. Varies with the lvs. ovate-lanceolate; also with the leaves more or less hairy.
2 **D. teres** Walt. Erect or ascending, nearly terete; lvs. lance-linear, rigid, sessile; bristles long; cor. reddish-white, with a wide tube and short limb; sta. scarcely exserted. ① Sandy fields, N. J. to Ill., and S. 5—18′.

6. HOUSTÒNIA, L. BLUETS. Cal. 4-toothed or cleft, persistent. Cor. tubular, the 4 lobes spreading. Fil. 4, inserted on the corolla. Style 1. Anth. and stig. dimorphous, that is, in some plants the former exserted and the latter included—in others the style exserted and anthers included. Caps. 2-lobed, the upper half free, cells few- (8-20)-seeded.—Herbs. Stip. connate with the petiole, entire. Fls. solitary or in cymes, white, bluish, &c.

§ Corolla salver-form, glabrous. Peduncles 1-flowered—*a* terminal............Nos. 1, 2
—*a* axillary.............Nos. 3, 4
§ Corolla funnel-form. Peduncles ∞-flowered, cymous.—*b* Lvs. lance-ovate....No. 5
—*b* Lvs. lance-linear..Nos. 6,7

1 **H. cœrùlea** L. *Dwarf Pink. Innocence.* Cæspitous; radical lvs. ovate-spatulate, petiolate; sts. erect, numerous, dichotomous; ped. filiform, 1-2-flowered. ⓖ Moist soils. 3—5′. Flowers 5″, pale blue, with a yellow centre. May, June. Pretty.
β. *minor.* Branches divaricate; flowers smaller (3—4″ wide). South.
2 **H. serpyllifòlia** Mx. Cæspitous; sts. filiform, procumbent; lvs. roundish-ovate, petiolate, ciliolate; ped. terminal, very long. ♃ Mts. of Car., Tenn. 6—12′. May—Jl.
3 **H. mínima** Beck. Glabrous; lvs. linear-spatulate; ped. at first nearly radical, at length axillary, often not longer than the leaves; seeds concave, smooth. ⓖ Prairies, Ill. to La. 1—3′. Flowers rose-color, 3—4″. March—May.
4 **H. rotundifòlia** Mx. Procumbent, creeping, leafy; lvs. roundish-oval, abrupt at base, petiolate; ped. mostly longer than the leaves; caps. emarginate, few-seeded. ♃ Sandy, damp places, S. In patches. 2—5′. Flowers white. Mar.—Dec.
5 **H. purpùrea** L. Erect; lvs. 3-5-veined, closely sessile; cymes 3-7-flowered, often clustered; calyx segm. lance-linear, longer than the pod. ♃ Penn., S. and W. 1f. White-purple. May—July. Very pretty.

6 H. longifòlia Gaert. Radical leaves oval-elliptic, cauline linear or lance-linear, 1 veined; fls. in small, paniculate cymes; sepals shorter than the pod.

β. *tenuifòlia.* Much branched; leaves very narrow; ped. filiform.

γ. *ciliolàta.* Leaves oblong-linear, obtuse, often ciliate; branches erect. N. and W.,—all the forms, on river banks and prairies. 1f. June, July.

7 H. angustifòlia Mx. Slender, tall, strictly erect; lvs. narrowly linear, 1-veined; fls. very numerous, short-pedicelled, in compact, terminal cymules; cal. lobes subulate; caps. obovoid or top-shaped. ♃ Prairies, Ill. to La. 1—2f. June—July.

7. OLDENLÁNDIA, L.

Calyx 4- or 5-lobed, persistent. Cor. funnelform, with a short tube, little longer than the calyx, 4–5-lobed. Sta. 4—5. Sty. short or 0. Stig. 2. Caps. wholly adherent. Seeds very numerous and minute (40—60 in each cell).—Herbs erect or prostrate. Stipules with 2—4 subulate points each side. Flowers small, axillary, white.

1 O. glomeràta Mx. *Creeping Greenhead.* Stems assurgent; lvs. ovate-lanceolate, pubescent, narrowed at the base; fls. glomerate in the axils and terminal; cor. shorter than the leafy calyx teeth. Swamps, N. Y. to La. 1—12′. June—Sept.

2 O. Bóscii. Erect, much branched; lvs. lance-linear, acute; fls. subsolitary, axillary, sessile. ♃ Banks of rivers, S. 6—10′. Corolla purplish. July, Aug.

3 O. Hàlei. Weak, diffuse, succulent; lvs. oval-oblong, acute; fls. subsolitary, white, pentamerous. ♃ River banks, Fla. to La. 8—10′.

8. CEPHALÁNTHUS, L. BUTTON BUSH.

Calyx limb 4-toothed. Cor. tubular, slender, 4-cleft. Sta. 4. Sty. much exserted.—Shrubs with opposite lvs. and short stip. Fls. in globous heads, without an involucre.

C. occidentàlis L. Lvs. opposite and in 3's, oval, acuminate, entire, smooth; heads pedunculate. Margins of streams. 6f. Heads nearly 1′ diam. July.

9. PÍNCKNEYA, Mx.

Calyx 5-parted, one of the segm. in the outer flowers changed to a large, rose-colored bract. Cor. tubular, lobes 5, spreading. Sta. 5, exserted. Stig. 2-lobed. Caps. 2-valved, ∞-seeded. ♄ Lvs. large, ovate. Cymes corymbous, terminal, splendidly *radiant.* Cor. purplish.

P. pubéscens Mx.—Swamps, S.: common. 15—25f. Pods size of a hazel-nut. May, June.—In cultivation it is a shrub, flowering when 8—12f high.

10. BOUVÁRDIA, H. K.

Calyx toothleted between its 4 lobes. Cor. tubular. Anth. 4, included. Caps. 2-partible, ∞-seeded. Sds. margined. ♄ Glabrous. Leaves lanceolate, coriaceous. (See p. 445.)

1 B. TRIPHÝLLA. Lvs. in whorls of 3's; cymes corymbed; fls. scarlet. Mexico. 2f.

2 B. VERSICOLOR. Lvs. opp.; cymes racemed; cor. clavate, curved, red and purp. S. Am.

ORDER LXVIII. VALERIANACEÆ. VALERIANS.

Herbs with opposite leaves and no stipules. *Calyx* adherent, the limb either membranous or resembling a pappus. *Corolla* tubular or funnelform, 4–5-lobed, sometimes spurred at base. *Stamens* distinct, inserted into the corolla tube alternate with, and generally fewer than its lobes. *Ovary* inferior, with one perfect cell and two abortive ones. *Seeds* solitary, pendulous, in a dry, indehiscent pericarp.

1. VALERIÀNA, L. VALERIAN.

Calyx limb at first very small, involute, at length evolving a plumous pappus. Cor. funnel-form, regular, 5-cleft. Sta. 3. Fruit 1-celled, 1-seeded. ♃ Leaves opposite, mostly pinnately divided. Flowers in close cymes. June, July.

§ Stems climbing and twining. Leaves ternately divided, long-stalked.........No. 1
§ Stem erect.—*a* Leaves and leaflets broad, somewhat ovate. Root fibrous...Nos. 2, 3
—*a* Leaves and leaflets narrow, nearly linear. Root fusiform.......No. 4
† Garden exotics, native of Europe.............Nos. 5—8

1 V. **scandens** L. Glabrous; lfts. ovate, thin, entire, pointed; cymes diffusely panicled, axillary and terminal; corolla very short. E. Fla. 4—6f, slender.
2 V. **pauciflora** Mx. Rt. lvs. ovate, cordate, crenate-serrate; cauline of 3—7 ovate, toothed lfts.; cor. tube long (7—8″) and slender, rose-white. O. to Va. and W. 1—2f.
3 V. **sylvatica** Richd. Rt. lvs. ovate or oblong, never cordate, entire; cauline of 5—11 lance-ovate, entire lfts.; cor. short (3—4″), roseate. Swamps, Vt. and W.
4 V. **édulis** N. Smooth, thickish; root lvs. linear-spatulate, entire; cauline of 3—7 lance-linear, acute segm., the margins ciliate; cor. white, short (2—3″), in a dense panicle. Low grounds, O. Wis. and W. The thick root is edible. 1—3f.
5 V. **dioica**. Root lvs. undivided; cauline pinnatifid; fls. panicled, ♂ ♀, blush. 1f.
6 V. **Pitu**. Root lvs. undivided; cauline pinnate; fls. corymbed, ♂, white. 3f.
7 V. **officinalis**. Lvs. all pinnate and toothed; fls. corymbed, blush-colored. 3f.
8 V. **Pyrenaica**. Lvs. cordate, toothed, upper pinnate; fls. corymbed, pink-red. 1—2f.

2. **VALERIANÉLLA**, Mœnch. DC. Calyx limb obsolete. Cor. tube short, not spurred, limb 5-lobed, regular. Sta. 3. Stig. 3-cleft or entire. Fr. 3-celled, 1-seeded, 2 cells empty. ① Stems forked above. Lvs. opposite, oblong or linear, entire or toothed, sessile. Fls. in dense, terminal cymelets. The specific characters are afforded mainly by the fruit. (Fedia, Gaert. T. & G.)

* Flowers pale blue. Fruit orbicular, fertile cell larger than the empty..........No. 1
* Flowers white.—*a* Fruit ovoid, fertile cell larger than the 2 empty........ Nos. 2, 3
—*a* Fruit subglobous, empty cells larger than the fertile....Nos. 4, 5

1 V. **olitòria** Mœnch. *Lamb Lettuce.* Fr. finally broader than long; fertile cell with a corky back, seed laterally compressed. Fields, N. Y. to Va.: rare. 8—12′. June.
2 V. **Fagopyrum**. Fruit smooth, ovoid-triangular, the empty cells at the obtuse angle, and no groove between; fls. large (1½″). W. N-Y. to Wis. 1f. June.
3 V. **radiata** Dufr. Fruit pubescent, ovoid, somewhat 4-angled, 1-toothed at apex; empty cells with a groove between; fls. small (½″). N. Y. (*Howe*) to Mich., and S.
4 V. **umbilicàta**. Fr. inflated, apex 1-toothed, the anterior face deeply umbilicate and perforated into the empty cells, which are much larger. Ohio (Sullivant).
5 V. **patellària**. Fruit orbicular, flattened, the empty cells widely divergent, at length forming a winged margin to the fertile cell. N. Y. to O. (*Howe, Sullivant.*)

ORDER LXIX. DIPSACEÆ. TEASELWORTS.

Herbs with whorled or opposite leaves and no stipules. *Flowers* in dense heads, surrounded by an involucre as in Compositæ. *Calyx* adherent, pappus-like, surrounded by a special scarious involucel. *Corolla* tubular. *Stamens* 4, alternate with the lobes of corolla, and distinct. *Ovary* inferior, 1-celled, 1-ovuled. *Style* 1, simple. *Fruit* dry, indehiscent, with a single suspended seed. Fig. 441.

1. **DÍPSACUS**, L. TEASEL. Fls. in heads. Involucre many-leaved.

Involucel 4-sided, closely investing the calyx and fruit. Cor. 4-cleft, lobes erect. Fruit 1-seeded, crowned with the calyx. ② Stout, prickly. Leaves connate at base. Hds. oblong, the middle zone of florets first expanding.

1 D. sylvéstris Mill. *Wild T.* Lvs. sinuate or jagged; bracts slender, erect, pungent, longer than the heads; chaff pungent, with a straight point. Waysides and hedges, Mass. to Cal.! 5f. Flowers bluish. July. § Europe.

2 D. fullònum. *Fullers' T.* Leaves serrate or entire; bracts of the involucre spreading; chaff rigid, erect, with sharp, hooked points. Europe. 4f. July.

2. SCABIOSA, L. Scabish. Fls. in heads. Involucre many-leaved. Involucel nearly cylindrical, with 8 little excavations. Calyx limb consisting of 5 setæ, sometimes partially abortive. ♃ Mostly European.

S. atropurpùrea. *Mourning Bride.* Leaves pinnatifid and incised; heads radiant, receptacle cylindric. India? 3f. Purple. Beautiful.

S. candidíssima. Flowers pure white.—There are many other varieties.

Order LXX. COMPOSITÆ. Asterworts.

Plantæ herbaceous or shrubby, with compound flowers (of the old botanists), *i. e.*, the flowers in dense heads (capitula) surrounded by an involucre of many bracts (scales), with 5 united anthers, and the fruit an achenium (cypsela). *Leaves* alternate or opposite, exstipulate, simple, yet often much divided. *Flowers* (florets) ∞, crowded, sessile, on the receptacle with or without pales (chaff). *Calyx* adherent, the limb wanting or divided into bristles, hairs, &c. (pappus). *Corolla* tubular, of 5 lobes with a marginal vein, often ligulate or bilabiate. *Stamens* 5, alternate with the lobes of the corolla, anthers cohering into a tube. *Ovary* 1-celled, with 1 erect ovule. *Style* single, with 2 stigmas at summit. *Fruit* a cypsela (§ 151), dry, indehiscent, 1-seeded, often crowned with a pappus. (See § 104, 348, 362.)

Figs. 68, 72–7, 103, 146, 160, 178, 261, 319, 341–6, 387–8, 433–4, 446–8, 492.

An immense and perfectly natural assemblage, of about 1000 genera and 9000 species. In the United States very few are shrubby.

The flowers are perfect or variously diclinous. If the head has all its flowers of one kind, whether ☿, or ♂, or ♀, it is *homogamous;* if of different kinds, it is *heterogamous.*—The following are De Candolle's Suborders and Tribes, with a convenient artificial analysis appended.

I. TUBULIFLORÆ.—*Corolla of the perfect flowers tubular, 5-lobed.* (A.)
  Tribe 1, VERNONIACEÆ. Branches of the style long, slender, terete, and hispid all over. Heads discoid. Flowers all alike, perfect................................Nos. 1–3
  Tribe 2, EUPATORIACEÆ. Branches of the style clavate, obtuse, flattened, minutely pubescent. Heads discoid. Flowers all alike, perfect...............................Nos. 4–15
  Tribe 3, ASTEROIDEÆ. Branches of the style flat, linear, downy above and opposite the distinct, stigmatic lines, appendaged at top. Heads discoid or radiate ....Nos. 16–34
  Tribe 4, SENECIONIDÆ. Branches of the style linear, fringed at the top, truncate or extended into a conical, hispid appendage..Nos. 35–89
  Tribe 5, CYNAREÆ. Style thickened or node-like at top. Branches not appendaged, the stigmatic lines not prominent, reaching the apex......................Nos. 90–98

II. LIGULIFLORÆ.—*Corollas all ligulate (radiant), flowers all perfect.* (B.)
  Tribe 6, CICHORACEÆ. Branches of the style long, obtuse, pubescent all over; stigmatic lines commencing below their middle. Juice milky........................Nos. 99–115

ORDER 70.—COMPOSITÆ.   153

III. LABIATIFLORÆ.—*Corolla of the perfect flowers bilabiate.* (C.)
TRIBE 7, MULISIACEÆ. Style nearly as in Cynareæ, the branches obtuse, very convex outside, minutely downy at the top...... .... No. 118

A. SUBORDER TUBULIFLORÆ.

§ Heads discoid, that is, without rays...(1)
§ Heads radiate, *i. e.*, the outer flowers ligulate...(8)
  1 Receptacle naked, *i. e.*, with no pales or bristles among the flowers...(2)
  1 Receptacle chaffy, bearing pales among the flowers...(6)
  1 Receptacle bearing bristles, or deeply alveolate (honeycombed)...(7)
    2 Pappus a circle of 5—20 chaffy scales...(*a*)
    2 Pappus none, or a short, toothed margin...(*b*)
    2 Pappus composed of many capillary bristles...(3)
      3 Leaves opposite. (Heads homogamous)...(*d*)
      3 Leaves alternate...(4)
        4 Heads homogamous,—flowers all perfect...(*c*)
        4 Heads heterogamous,—flowers not all perfect...(5)
          5 Scales herbaceous, often deciduous...(*e*)
          5 Scales scarious, persistent, often colored...(*f*)
    6 Leaves alternate...(*g*)
    6 Leaves opposite...(*h*)
      7 Pappus none, or consisting of scales...(*i*)
      7 Pappus composed of many bristles...(*j*)
  8 Receptacle naked (not chaffy), or (in No. 67) deeply honeycomb-celled...(9)
  8 Receptacle chaffy, with pales among the flowers...(13)
    9 Pappus of 5—12 scales, which are 1-awned or (in No. 62) cleft-bristly...(*k*)
    9 Pappus none, or of a few short awns...(*l*)
    9 Pappus of many capillary bristles...(10)
      10 Rays cyanic, in a single row...(*m*)
      10 Rays cyanic, in several rows...(*n*)
      10 Rays yellow, in about one row...(11)
        11 Pappus double, or of very unequal bristles...(*o*)
        11 Pappus simple, the bristles all similar...(12)
          12 Involucre scales imbricated, the outer shorter...(*p*)
          12 Involucre scales equal, not imbricated...(*r*)
    13 Disk and ray flowers both fertile, the latter pistillate...(14)
    13 Disk flowers sterile, ray flowers fertile...(*u*)
    13 Disk flowers fertile, ray flowers sterile...(15)
      14 Rays yellow...(*s*)
      14 Rays cyanic...(*t*)
      15 Achenia obcompressed, often beaked...(*v*)
      15 Achenia compressed laterally, or not at all...(*x*)

a Corolla lobes one-sided. Head large, many-flowered .... STOKESIA. 2
a Corolla lobes one-sided. Heads 4–5-flowered, aggregated..................... ELEPHANTOPUS 3
a Corolla lobes equal.—Leaves opposite. Pappus awned..................... AGERATUM. 4
      —Leaves whorled. Pappus obtuse..................... SCLEROLEPIS. 5
      —Leaves alternate.—Pappus scales 8—10................ PALAFOXIA. 65
        —Pappus scales 12—20................ HYMENOPAPPUS. 66
b Leaves opposite. Flowers diœcious, obscure... .... AMBROSIA. 47
h Leaves alternate.—Flowers yellow. Disk conical.................. MATRICARIA. 73
      —Flowers yellow. Disk convex. .... TANACETUM. 74
      —Flowers whitish.—Erect, leafless above.................... ADENOCAULON. 15
        —Fls. 8 ....ARTEMISIA. 76......Fls. 0 ....HUMEA. (82 *a*) 116
        —Low and depressed................. SOLIVA. 77
c Scales of the involucre in one row.—Flowers cyanic......... .... CACALIA. 86
      —Flowers yellow.—Receptacle flat........... SENECIO. 87
        —Receptacle convex...... RUGELIA. 88
e Scales imbricated.—Flowers yellow..................(No. 89, or)........ BIGELOVIA. 77
      —Flowers whitish.............. EUPATORIUM, 10, and...... KUHNIA. 8
      —Flowers purple.—Pappus simple. Involucre not radiate...LIATRIS. 7
        —Pappus simple. Involucre dry, radiate...RHODANTHE. 82
        —Pappus double .... VERNONIA. 1

Order 70.—COMPOSITÆ.

d Achenia 10-striate. Flowers purple..BRICKELLIA. 9
d Achenia 5-angled.—Receptacle conical. Flowers blue.....................CONOCLINIUM. 12
— Receptacle flat.—Scales 4 or 5....................................MIKANIA. 11
— Scales 8—20...EUPATORIUM. 10
e Shrubs. Flower diœcious, the ♀ and ♂ in different heads.............BACCHARIS. 34
e Herbs.—Stem winged. Heads spicate.....................................PTEROCAULON. 35
— Stem wingless.—Heads, corymbous, purplish........................PLUCHEA. 33
— Heads paniculate.—Pappus reddish.............................CONYZA. 31
— Pappus white..................................ERECHTITES. 85
f Receptacle chaffy except in the centre................................FILAGO. 80
f Receptacle naked.—Heads diœcious....................................ANTENNARIA. 79
— Heads heterogamous.—Involucre erect...........................GNAPHALIUM. 78
— Involucre radiate.........................HELICHRYSUM. 83
g Scales dry, fadeless. Pappus 4 teeth. Stem winged....................AMMOBIUM. 81
g Scales dry, fadeless. Pappus of scale-like awns......................XERANTHEMUM. 84
g Scales herbaceous.—Flowers heterocephalous. Fruit a burr ....XANTHIUM. 43
— Flowers all perfect.—Pappus of 5 or 6 scales.................MARSHALLIA. 69
— Pappus of many bristles.............CARPHEPHORUS. 6
h Flowers yellow. Pappus 2 inversely hispid awns......................BIDENS. 59
h Flowers yellow. Pappus 2 erectly hispid awns.......................COREOPSIS. 58
h Flowers whitish,—heterocephalous, Anthers yellowish.................AMBROSIA. 47
— monœcious. Anthers yellow..................................IVA. 46
— all perfect. Anthers black.................................MELANTHERA. 49
i Outer scales of the invol. leafy. Pappus none........................CARTHAMUS. 91
i Outer scales pectinate or ciliate-fringed, or entire.................CENTAUREA. 93
j Pappus plumous. Achenia obovate......................................CYNARA. 90
j Pappus plumous. Achenia oblong.......................................CIRSIUM. 97
j Pappus scabrous,—triple, each row by 10's..........................CNICUS. 95
— simple.—Scales spinescent, ...ONOPORDON. 96
— Scales hooked...LAPPA. 98
k Leaves opposite. Pappus scales deeply cleft into bristles............DYSODIA. 62
k Leaves alternate.—Receptacle with deep horny cells..................BALDWINIA. 68
— Receptacle with shallow fringed cells.........................GAILLARDIA. 63
— Receptacle areolate.—Rays all yellow..........................HELENIUM. 67
— Rays spotted at base †.....................GAZANIA. 64
l Leaves opposite. Involucre double, outer 8 united....................DAHLIA. 23
l Leaves opposite. Involucre single. Scales united....................TAGETES. 91
l Leaves alternate.—Pappus of a few short awns or bristles............BOLTONIA. 24
— Pappus a membranous margin.....................................MATRICARIA. 73
— Pappus 0.—Rays fertile, disk sterile..........................CALENDULA. 91
— Flowers all fertile.—Involucre scales equal.......BELLIS. 22
— Invol. broad, flat..........LEUCANTHEMUM. 72
— Invol. hemispherical......CHRYSANTHEMUM. 75
m Rays 4 or 5 Involucre oblong, imbricated. Cypsela very silky........SERICOCARPUS. 17
m Rays 5—75 Involucre loosely or closely imbricated. Pap. simple, copious...ASTER. 18
m Rays 8—12 Involucre imbricated. Pappus double, the outer very short......DIPLOPAPPUS. 19
m Rays 40—200. Involucre scarcely imbricated, scales nearly equal.........ERIGERON. 20
n Flowers diœcious, purplish. Leaves all radical......................NARDOSMIA. 14
n Flowers all fertile.—Native. Scales subequal, flat. Fruit smoothish.....ERIGERON. 20
— Exotic. Scales subequal, keeled. Fruit hairy........AGATHEA. 16
— Exotic. Scales imbricated. Pappus double..........CALLISTEPHUS. 21
o Pappus double in the disk flowers, none in the rays.................HETEROTHECA. 29
o Pappus double in both disk and ray flowers..........................CHRYSOPSIS. 30
p Heads large, about 20-rayed. Pappus in one row......................INULA. 32
p Heads very small, 1-15-rayed.—Pappus 1 row, shorter than achenia....BRACHYCHÆTA. 75
— Pappus 1 row, tawny, longer than achenia......ISOPAPPUS. 28
— Pappus irregularly 2-rowed, white.............SOLIDAGO. 26
r Head solitary, on a scape with alternate bracts.......................TUSSILAGO. 13
r Heads corymbed, &c.—Leaves alternate................................SENECIO. 87
— Leaves opposite..ARNICA. 88
s Shrubby. Pappus 4-toothed, obscure...................................BORRICHIA. 36
s Herbaceous.—Scales (the 4 outer) united into a cup..................TETRAGONOTHECA. 52

Order 70.—COMPOSITÆ.

—Scales distinct.—Cypselæ 4-angled. Pappus 0..................HELIOPSIS	51
—Cypselæ flattened. Pappus 0.... :...........SPILANTHES.	60
—Cypselæ flat, with a 2-awned pappus..........VERBESINA.	61
i Leaves alternate. Pappus none. Achenia tereto........................ANTHEMIS.	70
i Leaves alternate. Pappus none. Achenia obcompressed..............ACHILLEA.	71
i Leaves opposite.—Pappus none..............................ECLIPTA.	37
—Pappus of fringed scales.....................GALINSOGA.	38
—Pappus of the disk a single awn, of the ray 0..............ZINNIA.	50
a Leaves opposite. Rays yellow. Pappus none.......................POLYMNIA.	39
u Leaves opposite. Rays yellow. Pappus a 2- or 3-toothed crown. Gen. 41, & CHRYSOGONUM.	40
u Leaves alternate.—Rays whitish, very short, 5 only..........................PARTHENIUM.	45
—Rays yellow, disk dark-purple. Leaves entire...:..........MADIA.	43
—Rays yellow, disk brown. Leaves cut....................SPHENOGYNE.	44
—Rays and disk yellow.—Fruit winged..................SILPHIUM.	41
—Fruit wingless.....................BERLANDIERA.	42
v Cypsela with erectly hispid awns, or awnless, never rostrate...............COREOPSIS.	58
v Cypsela with retrorsely hispid awns, often attenuated above..............BIDENS.	59
x Rays white, spreading. Pappus none............................ANTHEMI'S.	70
x Rays purple, pendent. Pales sharp, elongated...............................ECHINACEA.	53
x Rays yellow.—Pappus none. Cypsela quadrangular.......... ....RUDBECKIA.	54
—Pappus none. Cypsela compressed.........................LEPACHIS.	55
—Pappus of 2 awns.—Fruit wingless.................... ....HELIANTHUS.	56
—Fruit broad-winged......................ACTINOMERIS.	57

B. Suborder LIGULIFLORÆ.

§ Pappus none, or consisting of little scales...(*a*)
§§ Pappus double (of scales and bristles), or simple and plumous...(*b*)
§§§ Pappus composed of capillary bristles, not plumous...(*)
　* Achenia terete or angular, not flattened...(*c*)
　* Achenia evidently flattened...(*d*)

a Flowers yellow. Pappus none. Heads paniculate......................LAMPSANA.	99
a Flowers yellow. Pappus none. Heads solitary or umbellate... ....APOGON.	100
a Flowers blue.—Pappus of many little scales. Receptacle naked......... ......CICHORIUM.	101
—Pappus of 5 scales. Receptacle chaffy......................CATANANCHE.	107
b Flowers purple. Feathery pappus on a long filiform beak...............TRAGOPOGON.	105
b Flowers yellow. Feathery pappus on a short beak or sessile.................LEONTODON.	104
b Flowers yellow.—Pappus of many bristles with the scales....................CYNTHIA.	103
—Pappus of 5 bristles and 5 scales.....................KRIGIA.	102
c Flowers whitish or purplish, mostly nodding. Stem leafy.....................NABALUS.	108
c Flowers rose-purple, erect. (Stem almost leafless)...........................LYGODESMIA.	109
c Flowers yellow.—Achenia long-beaked. Pappus white...............TARAXACUM.	112
—Achenia long-beaked. Pappus reddish....................PYRROPAPPUS.	111
—Achenia not beaked.—Pappus dull-white or tawny..........HIERACIUM.	106
—Pappus bright white...................TROXIMON.	110
d Achenia contracted into a slender beak. Flowers mostly yellow...............LACTUCA.	113
d Achenia scarcely beaked.—Flowers mostly blue........................MULGEDIUM.	114
—Flowers yellow. Pappus silky...................SONCHUS.	115

C. Suborder LABIATIFLORÆ.

§§§ Head radiate, solitary, nodding in bud. Pappus capillary......................CHAPTALIA. 117

1 VERNONIA, Schreb. IRON WEED. Fls. all tubular, perfect. Invol. of ovate, imbricated scales, the inner longest. Recept. naked. Pap. double, the exterior chaffy, the interior capillary. ♃ ♄ Leaves alternate. Fls. purple (in our species). Cymes corymbed. Figs. 446–8.

§ Scales of the involucre all obtuse and closely appressed. Stem tall, grooved. ...No. 1
§ Scales of the invol. (usually all)—*a* with slender, flexuous points... ....Nos. 2, 3
　　　　　　　　　　　　　　　—*a* with acute or mucronate points. South...Nos. 4, 5

1 V. fasciculàta Mx. Lvs. narrowly lanceolate, serrulate; cyme fastigiate; invol. ovoid-bell-shaped, half as long as the showy, dark-purple fls. Com. W. 3—10f. Jl. Aug.
2 V. Noveboracénse Willd. Lvs. many, lanceolate, serrulate, rough; cyme fastigiate; invol. scales filiform at the ends, or the upper cuspidate. Com. 3—6f. Aug.
3 V. scabérrima N. Lvs. all sessile, lanceolate and lance-linear, margins revolute, subentire; hds. 20–30-flowered; scales lanceolate, ciliate, protracted into long, flexuous points. Pine-barrens, S. 2—3f. June—August.
4 V. angustifòlia Mx. Lvs. linear and lance-linear, margins revolute; hds. 10–15-flowered; lower scales some filiform-pointed. Barrens, S. 2f. September.
5 V. ovalifòlia T. & G. Lvs. many, the lower oval or oblong; invol. bell-form, 20-flowered; scales acute or mucronate, short. Dry woods, Fla. 2—3f. June, July.
6 V. oligophýlla Mx. Lvs. mostly radical, oblong-obovate, the 2 or 3 cauline bract-like, lanceolate; scales spreading, acuminate. S. 2f. June, July.

2. **STOKÈSIA**, L'Her. Fls. all tubular, the marginal larger, ray-like, irregular; scales of the invol. imbricated, in several rows, the outer spinulous and leaf-like. Recept. naked. Cypsela 4-angled. Pap. of 4 or 5 awn-like, rigid, deciduous scales. ♃ Erect, with a downy stem, alternate lvs., and large terminal heads of showy blue flowers.

S. cyama L'Her.—Wet woods, S. Car. and W.; very rare. 2f. Lvs. glabrous, entire. Bracts spinulous at base, gradually becoming scales. †

3. **ELEPHÁNTOPUS**, L. ELEPHANT'S-FOOT. Heads 3–5-flowered, glomerate into a compound head with leafy bracts. Fls. all ☿ and equal. Invol. scales about 8, in 2 series. Cor. deeply cleft on one side. Fr. ribbed. Pap. chaffy-setaceous. ♃ Erect, with large, alternate, subsessile lvs. Cor. purple or white. July—September.

1 E. Caroliniànus Willd. St. much branched, leafy, hairy; lvs. somewhat hairy, ovate or oval-oblong, obtuse, crenate-serrate. Dry soils, Pa. S. and W. 2f.
2 E. tomentòsus L. St. hirsute, nearly leafless, simple or dichotomous above; root lvs. hirsute-tomentous, oblong-obovate. Woods, S. 1—2f. Flowers whitish.

4. **AGERÀTUM**, L. Heads ∞-flowered, ☿, discoid. Scales linear, imbricated, pointed. Recept. naked. Corollas all tubular. Cyp. 5-angled, narrowed at base. Pap. 5 or 10, chaffy, awned scales. ①② Mostly tropical, with opposite, petioled lvs. and corymbed heads. Fig. 75.

A. conyzoìdes L. Branching; lvs. ovate, tooth-crenate, acute or cordate at base, somewhat rugous; pap. scales 5, as long as the corolla, but much shorter than tho conspicuous styles. Wet places, near Savannah. 1—1½f. Blue or white. Apr.—Jn.
β. MEXICANA. Lvs. all, or nearly all, cordate. Fls. light blue, perpetual. †

5. **SCLERÓLEPIS**, Cass. Head ∞-flowered, ☿, discoid. Scales equal, linear, in 2 series. Recept. naked. Cor. 5-toothed. Styles much exserted. Cyp. 5-angled, crowned with a cup-shaped pappus of 5 obtuse, horny scales. ♃ Glabrous, simple, with 1—3 terminal hds. Lvs. verticillate. Flowers purple.

S. verticillàta Cass. ♃ In shallow water, N. J. to Fla. Erect, 1—2f, from a decumbent base. Lvs. lin., entire, 1', in whorls of 5's and 6's. Hds. mostly solitary. Jl.—Sep.

6. **CARPHÉPHORUS**, Cass. Heads (about 20-flowered), involucre, flowers, and fruit as in Liatris. Recept. chaffy. Pales narrow, 3-veined

rigid, shorter than the flowers. ♃ Sts. simple, leafy, corymbous at top, with middle-sized heads of purple flowers in Autumn. (Liatris, Mx. Ell.)

* Scales of the involucre acute, downy-tomentous. Leaves acute ....Nos. 1, 2
* Scales of the involucre rounded-obtuse, nearly glabrous. Leaves obtuse ..Nos. 3, 4

1 **C. pseudo-liàtris** Cass. Lvs. linear-subulate, rigid, closely appressed to and covering the stem; hds. few, rac. or cor.; plant downy, erect. W. Fla. to La. 2f.
2 **C. tomentòsus** T. & G. Lvs. lanceolate, petiolate, the cauline lance-ovate, sessile, small, erect; plant tomentous, corymb loose. Swamps, S. 2f.
3 **C. bellidifòlius** T. & G. Low, nearly smooth, tufted; lvs. spatulate below, linear above; hds. few, in a loose corymb; scales herbaceous. Sand hills, N. Car. 1f.
4 **C. corymbòsus** T. & G. St. single, stout, erect, hairy; lvs. oblanceolate, the upper oblong, sessile; corymb dense; scales scarious-edged. Swamps, S. 3f.

7. **LIÀTRIS**, L. Fls. all ☿, tubular. Invol. oblong, imbricate. Recept. naked. Pap. of ∞ capillary bristles. Cyp. tapering to the slender base, 10-striate. Styles much exserted. ♃ With simple, erect stems, alternate, entire lvs., and handsome rose-purple flowers in spicate, racemed, or paniculate heads. August—November.

§ Heads in a corymb or thyrse-like panicle. Root fibrous, no tuber............Nos. 1-3
§ Heads in a spike or a simple raceme. Root a roundish tuber..(a)
    a Scales of the involucre colored and petaloid at their lengthened ends...... No. 4
    a Scales not petaloid, green or slightly tinged at the end..(b)
        b Pappus evidently plumous. Corollas (13 to 60) hairy within..........Nos. 5, 6
        b Pappus evidently plumous. Cor. (3 to 5) smooth within. South....Nos. 7, 8
        b Pappus only barbellate (smooth to the naked eye)..(c)
            c Heads 20-40-flowered, roundish, with rounded scales.... ....No. 9
            c Heads 7-15-flowered.—d Scales all similar, obtuse. .. ..Nos. 10, 11
                —d Scales all, or the inner only, acute....Nos. 12, 13
            c Heads 3-7-flowered,—e in a regular spike, raceme (or panicle)...Nos. 14-16
                —e in one-sided spikes or racemes........ ...No. 17

1 **L. odoratíssima** Willd. *Vanilla Plant. Deer's Tongue.* Smooth; lvs. obovate-spatulate, obtuse, thick, the cauline oblong; heads 7-8-flowered, in a loose, compound corymb. Pine-barrens, Va. to Fla. 1—3f. Used to perfume tobacco.
2 **L. paniculàta** Willd. Viscid-tomentous; lvs. lance-spatulate, the cauline small, pointed; hds. 5-flwd., in an oblong, dense panicle, white-purple. Damp. S. 2—3f.
3 **L. fruticòsa** N. Shrubby, smooth; lvs. obovate, fleshy, veinless, the lowest opposite; hds. corymbed, 5-flowered; scales lanceolate, acute, dotted. E. Fla. Lvs. 1′.
4 **L. élegans** Willd. Hairy above; lvs. oblanceolate, cauline linear; rac. dense, 1f; hds. 4-5-flowered, scales longer and more showy than the flowers. Woods, S. 4f.
5 **L. squarròsa** Willd. *Blazing Star.* St. 2—3f; lvs. linear, the lower narrowed at base; rac. leafy; hds. few, 15-40-flowered, 9—12″ long, scales squarrous-spreading, the outer leafy, inner sharp-pointed. Dry soils, Penn. to Fla and W.
6 **L. cylindràcea** Mx. St. low (6—18′), slender; lvs. linear, rigid; hds. few, cylindrical, 15-20-flowered; scales short, rounded, appressed. Dry. N. Y. and W.
7 **L. Boykínii** T. & G. Lvs. linear, dotted; hds. 3 or 4-flowered in a close, virgate spike; scales pointed and spreading at the tips. Near Columbus, Ga. 1—2f.
8 **L. tenuifòlia** L. Lvs. narrowly linear or filiform; hds. 5-flwd., crowded in a long raceme; scales oblong, obtuse-mucronulate. Woods, S. 2—4f. Fine.
9 **L. scariòsa** L. *Gay Feather.* Scabrous-pubescent; lvs. lanceolate, the lower on long petioles, upper linear; hds. remotely racemed; invol. hemispherical, with obovate, very obtuse scales. Dry soils. 4—5f. Beautiful.
10 **L. spicàta** Willd. Lvs. lance-linear, the lower narrowed at base; hds. sessile, in a long spike; scales oblong, obtuse, narrow-margined. N. J., W. and S. 2—5f.

11 **L. graminifòlia** Willd. Leaves linear, 1-veined; hds. mostly pediceliate. rac. rarely paniculate below; invol. acute at base, scales obovate-spatulate, obtuse, appressed; cyp. hairy. Sandy soils, N. J. and S. Variable.
12 **L. pilòsa** Willd. Downy and hairy, stout; lvs. linear and lance-linear; hds. loosely racemed, scales linear-oblong, obtuse, the inner linear. N. Car. Rare.
13 **L. heterophýlla** R. Br. Glabrous; lvs. lanceolate, the upper greatly diminished; hds. spiked, scales lance-acuminate, spreading. N. Car. to Ga. Rare.
14 **L. grácilis** Ph. Pubescent; lvs. linear, 1-veined, the lower lanceolate; heads on slender stalks, in a long virgate rac.; scales oblong, obtuse. Dry. Ga. Fl. 2—3f.
15 **L. pychnostáchia** Mx. Hirsute; lvs. rigid, lanceolate, the upper narrow-linear; spike dense, thick, of numerous cylindric heads; scales appressed, with acute, scarious, colored and spreading tips. Prairies. Ill. to Tex. 3—5f. Spike 10—20′.
16 **L. Chapmánii** T. & G. Tomentous; lvs. linear, obtusish, the upper very short; hds. cylindric. 3-flowered, densely spiked; scales acum.; fr. hairy. Fla. 1—2f.
17 **L. pauciflòra** Ph. St. pubescent, recurved; lvs. linear, short, the lowest lance-linear; rac. recurved, with the hds. all turned to the upper side; hds. 4–5-flowered; scales lance-oblong, acute. Dry sand-hills, S. 1—3f. (L. secunda Ell.)

8. **KÚHNIA**, L. Heads 10–25-flowered, ☿. Scales lanceolate, loosely imbricated. Recept. naked. Cor. slender, 5-toothed. Pap. in a single series, plumous. Fr. cylindrical, striate, pubescent. ♃ With alternate, resinous-dotted lvs., and corymbed heads of pale yellow florets. September.

K. eupatorioìdes L. St. somewhat viscid-pubescent; lvs. lance-ovate to lance-lin., resinous-dotted, petiolate, toothed or entire. Dry soils, N. J., W. and S.

9. **BRICKÉLLIA**, Ell. Heads many-flowered, ☿. Scales imbricated, lanceolate or linear, striate. Receptacle naked, flat. Cor. tube slightly expanded above, 5-toothed. Branches of the style clavate. Fr. 10-striate, contracted above. Pap. setaceous, in one series. ♃ With opposite, 3-veined leaves and large heads of purple florets in corymbs.

B. cordifòlia Ell. Pubescent; lvs. triangular, truncate or cordate, crenate, petiolate; hds. 30–40-flowered, scales obtuse; pap. purple. Ga. Fla. 2—4f. August.

10. **EUPATÒRIUM**, Tourn. BONESET. Fls. all tubular, ☿. Invol. imbricate, oblong. Style much exserted, deeply cleft. Anth. included. Recept. naked, flat. Pap. capillary, simple, scabrous. Cyp. 5-angled. ♃ Generally with opposite, simple lvs. and corymbous hds. Fls. of the cyanic series—that is, white, blue, red, &c., never yellow. July—September.

§ Leaves mostly alternate, pinnately dissected. Heads paniculate, very ∞......Nos. 1, 2
§ Leaves mostly opposite or verticillate,—*c* pinnately dissected. Hds. corymbed...No. 3
—*c* undivided. Heads corymbed..(*)
  * Scales imbricated in several rows, the outer gradually shorter...(*a*)
    a Flowers bluish. Leaves opposite. Scales strongly striate... ....No. 4
    a Flowers purplish. Lvs. whorled. Scales streaked and flesh-colored..Nos. 5—7
    a Flowers white, 5 only in each head. Lvs. subsessile. (exc. No. 18)..(*b*)
      b Leaves acute at base. Scales with acute white points... ....Nos. 8—10
      b Leaves acute at base. Scales obtuse, short, downy.............Nos. 11—14
      b Leaves obtuse, roundish or truncate at the base.................Nos. 15—18
    a Flowers white, 7—15 in each head. Leaves various........ ....Nos. 19—22
  * Scales all of equal length, in about 1 row. Leaves petiolate.... ....Nos. 23—25

1 **E. foeniculàceum** Willd. Very branching; lvs. all alternate, compoundly pinnate, in linear-filiform segments, the upper setaceous, simple; heads 3–5-flowered. Fields, Pa. (rare) to Fla. 3—10f. Flowers yellowish-white. 1—2″ long.

Order 70.—COMPOSITÆ.

2 E. coronopifòlium Willd. Much branched, pubescent; leaves mostly alternate (the lower opp.), twice pinnatifid, with lance-linear lobes and segm., the upper linear, simple; hds. 5-flowered, scales 10. Dry soils, S. 3—5f. Flowers white, 2″.
3 E. pinnatifidum Ell. Pubescent; lvs. laciniate-pinnatifid, segm. linear, toothed or entire, the lower whorled in 4's, middle opp., upper altern.; corymb fastigiate; hds. small, ∞, 5-9-flowered; scales oblong, mucronate. Pine woods, S. 3—4f.
4 E. ivæfòlium L. Lvs. opposite, lanceolate, tapering to each end, 3-veined; heads pedicellate, 15-20-flowered; scales 20, imbricated, erect, obtuse, with 3—5 distinct striæ. Woods, Miss. and Fla. 3—5f. Blue.
5 E. purpùreum L. Stem solid, purple at the joints; lvs. feather-veined, in whorls of 3's—5's, thin, ovate to lanceolate, coarsely serrate. Dry. 3—6f.
6 E. fistulòsum Barratt. *Trumpet-weed*. Stem hollow, striate, glabrous, glaucous-purple; lvs. lance-oblong, in 5's, 6's, finely serrate; corymb globous, with whorled rays. Thickets. 6—10f. Lvs. 8′. Corymbs 1f. (E. purpureum. β. T. & G.)
7 E. maculàtum L. Stem solid, marked with purple glands and lines; leaves 3-veined, ovate, in 3's—5's. Low grounds: common. 3—5f. (E. purpureum. β. Darl.)
8 E. scábridum Ell.? (Chapm.) St. stout, tomentous; lvs. lance-ovate, acute, scr., 3-veined from base; scales lance-obl., cuspidate, edged, shorter than fls. Car. Fla. 2f.
9 E. album L. Rough-downy; lvs. lance-oblong, acutish; hds. oblong, 5-flowered; scales white-scarious at the point, longer than the fls. Sands, N. J. and S. 2f.
10 E. leucólepis T. & G. Nearly smooth; lvs. lance-linear, obtuse; heads 5-flwd.; scales white-scarious at the tip, as long as the fls. Sands, L. I. and S. 2—3f.
11 E. hyssopifòlium L. Lvs. linear-lanceolate, 1-3-veined, punctate, lower ones subserrate, upper ones entire; scales oval. Dry. Mass., W. and S. 2f. Hds. 3″.
12 E. parviflòrum Ell. Lvs. lanceolate, sessile, acutely serrate above, 3-veined; heads 2″, crowded; outer scales very short, inner linear. Damp. Va. to Fla. 2—3f.
13 E. altíssimum L. Tall, downy; lvs. lanceolate, few-toothed above, conspicuously 3-veined; scales 8—12, elliptical, 2½″; fls. 5″. Dry. Pa. to Car., and W. 3—7f.
14 E. cuneifòlium Willd. Downy; lvs. small, glaucous, obovate-oblong, 3-veined, apex *obtuse* and subserrate; scales oval, 2″; fls. 4″. Rich shades, S. Car. to Fla. 2f.
15 E. teucrifòlium Willd. Rough-downy; leaves sessile, ovate, veiny, the lower doubly serr.; scales elliptical, faintly striate, rather acute. Damp. Mass. to La. 2—3f.
16 E. sessilifòlium L. Smooth; leaves half-clasping, lance-ovate, serrate; inner scales oblong-obovate, obtuse. Rocky woods, Mass. to Ind., and S. 2—4f. Lvs. 3—5′
17 E. rotundifòlium Willd. *Hoarhound*. Downy; lvs. roundish ovate, subcordate, 3-veined, sessile, coarsely toothed; inner scales acuminate, as long as the fls. Dry fields, N. J. and S. A compact, bushy plant. 3f.
18 E. mikanioides Chapm. St. creeping at base, ascending; lvs. deltoid, truncate at base, petioles subconnate; scales lanceolate, acute. Isl. St. Vincent, Fla. 1—2f.
19 E. pubéscens Muhl. Hairy; lvs. distinct, sessile, ovate, acute, blunt-toothed; hds. about 8-flwd.; scales lanceolate, acute, short. Dry. N. H. to N. J., and Ky. 3—4f.
20 E. resinòsum Torr. Viscid-resinous; leaves distinct, closely sessile, lin.-lanceolate, long-pointed; hds. 10-15-flwd.; scales obtuse, white-downy. Barrens, N. J. 2-3f.
21 E. perfoliàtum L. *Thoroughwort. Boneset.* Hairy; lvs. lanceolate, each pair united at base around the stem; heads about 12-flowered, in a large, dense corymb; scales lance-oblong, acute. Low grounds: common. 3—4f. A powerful tonic.
22 E. serótinum Mx. Soft-puberulent; lvs. petiolate, lance-ovate, sharp-serrate, 3-veined; hds. 12-15-flwd.; scales 9—11, similar, very downy, obtuse. Md., S. and W. 5f.
23 E. ageratoides L. Smooth; lvs. long-petiolate, ovate, acuminate, sharp-serr., 3-veined; hds. 10-20-flwd., in a compound corymb; scales oblong, obtuse. Woods. 3f.
24 E. aromáticum L. Rough-downy; lvs. petiolate, lance-ovate, acute, 3-veined, blunt-serr.; hds. 10-15-flwd., in small corymbs; scales lance-linear. Low woods. 2f.
25 E. incarnàtum Walt. Diffusely branched; leaves long-petioled, deltoid-ovate, pointed, coarsely crenate-dentate; hds. on slender ped., 15-20-flwd.; scales lin.-subulate, 3-striate; lobes of the corolla pale purple. Damp soils, N. Car. to Fla. 3f.

11. **MIKÀNIA**, Willd. CLIMBING BONESET. Fls. all tubular, ☿. Involucre 4-leaved, 4-flowered. Receptacle and flowers as in Eupatorium. ♃ Climbing and twining. Leaves opposite.

M. scandens Willd. Smooth; lvs. cordate, repand-toothed, acuminate, the lobes divaricate; hds. in pedunculate, axillary corymbs. Thickets, Ms. to Ga. Not common. Clusters on the short, lateral branches, of white or pink-colored flowers. Aug. Sept.

12. **CONOCLÍNIUM**, DC. Heads many-flowered. Receptacle conical. Character otherwise as in Eupatorium. ♃ ♄ Leaves opposite, petiolate, serrate. Flowers sky-blue, in crowded corymbs.

C. cœlestìnum DC. Much branched; lvs. deltoid-ovate, truncate or subcordate, crenate-serrate, petiolate; scales linear. ♃ Copses, Pa., S. and W. 1—2f. Aug. Sept.

13. **TUSSILÀGO**, Tourn. COLT'S-FOOT. Head radiate, many-flowered. Flowers of the ray ♀, those of the disk ♂. Invol. simple. Recep. naked. Pappus capillary. ♃ Lvs. radical. Fls. yellow, with very narrow rays.

T. Fárfara L.—Cold, clayey banks, N. and M. Scape 5′, appearing with its single head of yellow flowers in March and April, before the large angular leaves.

14. **NARDÓSMIA**, Cass. Heads radiate, ∞-flowered, somewhat ♀ ♂. Fls. of the ray ♀, of the disk ☿, but abortive in the sterile plant. Invol. simple. Recep. flat, naked. Pappus capillary. ♃ Leaves radical. Fls. cyanic. The ray flowers of the sterile heads are in a single row; of the fertile in several rows, but very narrow.

N. palmàta Hook. Scape with a thryse or corymb; lvs. roundish-cordate, 5-7-lobed, woolly beneath, coarsely dentate. Swamps, N. Eng. and W. Rare. May.

15. **ADENOCAÙLON**, Hook. Fls. few, all tubular, of the margin ♀, of the disk ♂. Scales equal, in one series. Recep. naked. Cyp. clavate, exserted, bearing stalked glands above. Pap. 0. ♃ Nearly acaulescent, with alternate leaves, and small, paniculate heads, also gland-bearing.

A. bìcolor Hook. Lvs. deltoid, cordate, angular-toothed, decurrent on the petioles, white-downy beneath. Shores of Lake Superior, and W. (Common in Oregon.) 2f.

16. **AGATHÆ̀A**, Cass. Heads as in Erigeron, but the scales are 1-veined, keeled or channelled, and the cypselæ rough-haired. ① ♄ S. Afr. Leaves opposite. Disk flowers yellow, rays blue. (Cineraria, L.)

A. AMELLOÌDES. Lvs. ovate or oval, petiolate, entire, scabrous. Not hardy. A beautiful shrub, often cultivated in the greenhouse. 1—2f. Heads solitary.

17. **SERICOCÁRPUS**, Nees. WHITE-TIPPED ASTER. Ray fls. 4—6, ♀: disk fls. 6—10, ☿. Invol. oblong, imbricated. Scales appressed, white with green, spreading tips. Recep. alveolate. Cyp. obconic, very silky. Pap. simple. ♃ With alternate lvs. and close corymbs. Rays white.

1 **S. solidagíneus** Nees. Smooth: lvs. linear-oblanceolate, obtuse, entire, sessile; heads subsessile; scales obtuse; pap. white. Woods: com. 2f. Rays long. Jl. Aug.

2 **S. conyzoìdes** Nees. Some pubescent; lvs. lance-oval, acute, serrate, the lower narrowed into a petiole; rays short; pappus rusty. Woods, Ms. to Fla. 1-2f. Jl. Aug.

3 **S. tortifòlius** Nees. Grayish pubescent; lvs. short, oblong-obovate, sess., twisted to a vertical position, both sides alike; pappus white. Woods, Va. to Fla. 2f. Sept.

18. **ASTER**, L. Invol. oblong, imbricate. Scales loose, often with green tips, the outer spreading. Disk fls. tubular, ☿, ray fls. ♀, in one row, ligulate, 3-toothed at apex, finally revolute. Recep. flat, alveolate. Pap. simple, capillary. Cypsela compressed. ♃ Very abundant in the U. S., flowering in late summer and autumn. Lvs. alternate, diminishing gradually upward. Disk-flowers yellow, changing to purple; ray-flowers blue, purple, or white, never yellow. Figs. 146, 388. (See also p. 446.)

A Scales of the invo.ucre tipped with green or wholly green...(§ 1, 2, 3)
B Scales destitute of green tips, white or scarious. Lvs. never cordate...(§ 4-p)
 § 1. Biòtia. Heads corymbous, large. Rays 6—15, white. Lvs. cordate....Nos. 1, 2
 § 2. Calliástrum. Heads corymbous or few, large. Rays 12—30, violet-blue. Pap. bristles unequally thickened. Lvs. rigid, not cordate....(a)
  a Lvs. ovate to lanceolate, serrate more or less. Fr. smoothish.....Nos. 3—5
  a Leaves lance-linear to linear,—b entire, merely acute.............Nos. 6, 7
   —b bristly-fringed, pungent........ ..Nos. 8, 9
 § 3. Astèria. Hds. panicled or racemed, rarely few. Pap. equal, soft...(c)
  c Leaves petiolate, the lower cordate,—d evidently serrate............Nos. 10, 11
   —d entire or obscurely serrate...Nos. 12—15
  c Leaves all sessile, entire, silky-canescent both sides. Pap. tawny....Nos. 16, 17
  e Lvs. not silky,—d clasping with a cordate or auriculate base...(f)
   —d clasping with a broad base not cord. or auric...(h)
   —d sessile with a narrow base, not clasping...(m)
  f Lvs. very small (1″—3″), entire. Scales with spreading tips..........Nos. 18, 19
  f Leaves ordinary (1′—6′).—e Scales with abrupt, appressed tips........Nos. 20, 21
   —e Scales loosely spreading. Lvs. entire....Nos. 22—25
   —e Scales very loose. Lvs. long, serrate....Nos. 26, 27
  h Scales of the involucre closely imbricated (obtuse, No. 20), acute....Nos. 28—31
  h Scales loose, or spreading, or recurved.—k Pappus bright-colored...Nos. 32—34
   —k Pappus tawny-brown.....Nos. 35, 36
  m Scales squarrous-spreading at the tips.—o Hds. large (6″-1′), purple..Nos. 37, 38
   —o Hds. small (2—4″),whitish..Nos 45-47
  m Scales loosely divergent, straight. Heads medium size, rays pale........ No. 42
  m Scales erect, straight, in 1 row. Heads 2—3, or solitary, rays white.....No. 43
  m Scales closely imbricated.—n Hds. medium (3-6″), purp. or pale...Nos. 43, 44, 31
   —n Heads small (2—3″), white or pale.....Nos. 39—41
 § 4. Scariòsi.—p Lvs. lanceolate, broadly or narrowly. Scales obtusish....Nos. 49—51
   —p Lvs. subulate or lin. Scales very acute.—s Hds. large, few..Nos.52, 53
   —s Hds. small, many...54—56

1 **A. corymbòsus** Ait. Nearly smooth; lvs. thin, ovate-acuminate, serrate, the petioles wingless; rays 6—9. Dry woods, N., M. 1—2f. Heads oblong, 4″. Lvs. large.
2 **A. macrophýllus** Willd. Rough-pubescent; leaves thickish, ovate, serrate with close teeth, petioles some winged; rays 8-15. Woods, N. 1-2f. Lvs. very large. Hds. 6″.
. **A. miràbilis** T. & G. Lvs. ovate, serrate, the lowest petiolate, the ramial roundish; invol. hemispherical, scales obtuse; rays about 20. S. Car. Very rare.
4 **A. radula** Ait. Lvs. lanceolate, acuminate, sessile, sharp-serrate, rough and rugous, invol. squarrous with the spreading scale-tips; rays 20. N. 1—3f.
5 **A. spectábilis** Ait. Lvs. lance-oblong. sessile, entire, the lower subserrate; invol hemispherical, scales linear-spatulate, ciliate. Sands, Mass. to Fla. 1—2f.
6 **A. surculòsus** Mx. Root a creeping, knotted rhizome; lvs. lance-linear and linear heads 1—5; scales linear-oblong, ciliate, inner obtuse. Wet. N. J. to Car. 1f.
 β. **grácilis**. Heads 8—12, smaller; rays 12; scales but slightly spreading.
7 **A. paludòsus** L. Slender, glabrous; lvs. long, linear; hds. 1—6; scales lance-linear rays 30, longer than the (6″) invol. Swamps, S. 2—3f. Heads very large

8 A. spinulòsus Chapm. Bristly-hairy, rigid; lvs. narrowly linear, pungent, bristle-fringed; heads few, spicate; scales spine-pointed; rays 13, blue. Fla. 1f.

9 A. eryngiòlius T. & G. Hairy, rigid; lvs. lance-linear, pungent, fringed with spiny teeth; heads very large, 1—4, loosely racemed; scales green, rigid, lanceolate, long-pointed; rays many, white. Fla. 1—2f. (Prinopsis Chapmanii, C-B.)

10 A. cordifòlius L. Stem paniculate; leaves sharply serrate, acuminate; petioles winged; scales appressed, with short green tips. Woods and glades, N. and W.: com. 1—3f. Heads numerous, rather small, blue varying to white, in a large panicle.

11 A. sagittifòlius Willd. Branches racemed; lvs. lance-obl., some arrow-shaped; petioles winged; scales loose, lin.-subulate. Low woods, N. and W. 2—4f. Wh.-blue.

12 A. undulàtus L. Racemous-paniculate, rough, grayish; lvs. ovate-oblong, undulate-crenate, the base, or the winged petioles, cordate-clasping, the upper acute, entire, sessile: scales appressed. Dry woods. 2f. Blue. (A. diversifolius Mx.)
  β. *aspérulus.* Lowest petioles slender, not clasping; lvs. scarcely cordate. Com.

13 A. azùreus Lindl. Slender, rigid, rough; lvs. below on slender petioles, cordate-lanceolate, the others successively lanceolate, linear, and subulate, acute at each end; rac. paniculate, heads obconic; scales acute, appressed. Woods, prairies, W. 2f.

14 A. Shórtii Hook. Smoothish, subsimple; lvs. lance-ovate, deeply cordate, petiolate, long-pointed, entire, the upper sessile; rac. paniculate; scales green-tipped, shorter than the disk, Rocky banks, O. to Wis. and Ark. 3f.

15 A. anómalus Eng. Lvs. as in No. 13; scales with linear, spreading, leafy tips; hds. large; rays spreading, 15—18″, bright blue. Rocks, Ill. Mo. (Mr. J. Wolf.) 2—4f.

16 A. seríceus Vent. Bushy; lvs. silvery-silky both sides, lance-oblong, sessile; hds. large. terminal on the short, leafy branches; scales spreading at tip; fr. glabrous; rays 15—25. violet blue. Banks, Mich. (H. Mapes) to Iowa, and S. 1—2f.

17 A. cóncolor L. Subsimple; lvs. grayish-silky, lance-oblong, the upper cusp-pointed; heads in a terminal, virgate raceme; scales lanceolate, appressed; fruit silky; rays purple. Pine-barrens, N. J. to Fla. 2—3f. Aspect of Liatris.

18 A. squarròsus Walt. Slender, with simple, 1-flowered branches; leaves very small, triangular, heart-clasping, reflexed-squarrous; scales with spreading green tips; fr. pubescent. Dry soils, S. 2—3f. Rays 20, blue.

19 A. adnàtus N. Slender, rough; lvs. oblong to lanceolate, erect, adhering to the stem by the midvein, the summit only free. Sands, Fla. to La. 2—3f.

20 A. turbinnéllus Lindl. Smooth, subcorymbed; lvs. lance., tapering both ways; hds. club-top-shaped (6″); sc. tips short, blunt. Ill. Mo. to La. B!ue. Pap. brown.

21 A. lævis L. Very smooth; branchlets 1-flwd.; lvs. oblong, entire, shining, lowest lanceolate, subserrate, upper auriculate; scales with a broad, acute, appressed tip; heads large, rich blue, showy. Low woods. 2—3f.
  β. *lævigàtus.* Not glaucous; leaves linear-lanceolate; scales linear.
  γ. *cyáneus.* Plant glaucous; leaves thickened, very entire. Beautiful Asters.

22 A. patens L. Pubescent; rac. paniculate; lvs. ovate-oblong, cordate-clasping, ciliate at edge; heads large, terminal on the leafy branchlets; scales lax, green-tipped; rays 20, violet-blue. Dry soils, Mass. to Ga. 2—3f.
  β. *phlogifolius.* Leaves lance-ovate, auriculate-clasping, very acute.

23 A. amethystìnus N. Hoary-puberulent; rac. paniculate; lvs. lin.-oblong, acute, some auricled at the clasping base; heads broad-bell-shaped (3″); scales erect, with only the green tips spreading. Damp, Mass. to Ill. (J. Wolf.) 2—3f.

24 A. Novæ-Angliæ L. Corymbous-paniculate, pubescent; lvs. lanceolate and lance-linear, auriculate-clasping; scales equal, lax, glandular-viscid, green their whole length; rays 70+, deep purple. Damp. 4—6f.—Varies with the rays rose-purple, or rarely, white. Fine in cultivation.

25 A. Caroliniànus Walt. Rough-downy; branches divaricate; lvs. lance-ovate, entire, clasping with small auriculate lobes; heads very large, scattered; scales with spreading green tips; rays rose-purple. Damp, S. 6—13f.

26 A. puníceus L. Hispid, panicled; lvs. lance-oblong, auriculate-clasping, ap pressed-serrate; scales 2-rowed, long, revolute; heads large, showy, with 30—60 narrow, pale-purple rays. Swamps, Can. to Car., and W. 4—6f. Stem often red.

β. *vimineus.* Tall, slender, smoothish; heads few, very large; leaves narrow.

γ. *glaber.* Low (2f), subsimple, smoothish; leaves narrow, erect, entire; scales loose, not recurved; rays large, about 20, white? Ill. (J. Wolf.)

δ. *firmus.* Low (2—3f), scabrous, stout; leaves thick, subentire; heads many.

ε. *candidus*—the common form, with white rays. N. Y. (Hankenson.)

27 **A. prenanthoides** Muhl. Hairy or downy, corymbous-paniculate; lvs. lance-oval, pointed, serrate, the long petiole winged and auriculate-clasping; scales spatulate, the green tips spreading. Wet banks, N. Y. to Va., and W. 2—3f.

28 **A. concinnus** Willd. Pubescent, subsimple; lvs. lanceolate and lance-linear, remotely serrate, narrowed to the clasping base, the upper entire; scales appressed-imbricate; heads medium, rays blue. Woods, &c. 2—3f.

29 **A. gracilléntus** T. & G. Very smooth, slender, simply panicled: leaves long-linear, the lower toothed, upper clasping, erect; scales short; rays blue. S. Rare.

30 **A. mutábilis** Ait.? Stem smooth, paniculate-branched from base, dense-flwd.; leaves linear-lanceolate, serrulate, clasping, thickish, upper lance-oblong, entire; heads medium; scales lanceolate, loose, much shorter than the disk; rays pale? Wet. Ill. (J. Wolf.) 2—3f.—Varies with leaves serrate, heads loose, &c.

31 **A. cárneus** Nees. Smoothish; branches leafy, ascending, racemed with 1-headed branchlets; lvs. uniform, linear-lanceolate, pointed, only the upper clasping; scales acute, shorter than the disk. Moist, E. and W. Heads larger than in No. 30, purple to rose, showy. Stem often red, 2—3f high.

32 **A. virgátus** Ell. Smooth, virgate branches racemed; leaves linear-lanceolate, ciliate-serrulate, half-clasping, graded above into numerous subulate bracts and spreading, pointed scales; fruit glabrous. Ga. to La. 3—4f.

33 **A. Novi-Bélgii** L. St. smoothish, branches pubescent; lvs. subclasping, lance-obl. to linear, pointed, the lower subserrate; heads large, racemed or subcorymbed; scales subequal, loose, equalling the disk. N. Y. to Ill. 2—4f. Blue. (A. æstivus Ait.)

β. *lætiflorus.* Branches slender, corymbed at end; lvs. very narrow. W. Showy.

34 **A. longifòlius** Lam. Stem glabrous, paniculate-spreading; lvs. lance-linear to linear, long, pointed, subclasping, nearly or quite entire, upper subulate; heads large; scales linear-subulate, the outer spreading. E. and W. 2—6f. Blue.

β. *præáltus.* Tall, strict, with thyrsoid panicles, medium heads; lvs. serrulate.

35 **A. Elliottii** T. & G. Stout, smooth, corymbous-branched; lvs. ample, lanceolate, subclasping, subserrate; ped. naked; scales attenuate. Swamps, S. 2—4f. Purple.

36 **A. oblongifòlius** N. Hairy, bushy; branches spreading; leaves obl.-lanceolate, acute, entire, clasping, graded above into subulate bracts and subequal spreading scales. Va. (Harper's Ferry) to Iowa and Mo. Rays purple. 1—2f.

37 **A. grandiflòrus** L. Rough, bristly-hairy; branches some corymbed, 1-flowered; lvs. small, linear-oblong, obtuse; hds. very large, blue-purple; scales obtuse. S. 2f.

38 **A. Curtisii** T. & G. Smooth, racemous; lvs. thin, sessile, lanceolate, acuminate, subentire; scales with green spreading tips; heads large, showy. Mts. N. Car.

39 **A. dumòsus** L. Rac. paniculate; lvs. linear to oblong, sessile, lowest subserrate; invol. obtuse at base, closely imbricated; scales obtuse; heads small, rays 20+, purplish-white. Dry woods, &c.: common. 1—2f. Lvs. very numerous, 3'—3''.

β. *cordifòlius,* is a starved, attenuate form, very slender every way.

40 **A. Tradescánti** L. Smoothish, slender, much branched; lvs. lance-linear, long, remotely serrulate, teeth sharp, upper leaves entire, all sessile; heads many, subsecund; scales close; rays small, pale. Fields, copses. 2—4f. Leaves 5'—5''.

β. *fragilis.* Leaves nearly linear, minutely serrulate; heads scattered.

41 **A. miser** L. Hairy or downy, very leafy; branches spreading, racemous; lvs. al. lanceolate, tapering both ways, sessile, sharply serrate in the middle, the ramial smaller, entire; scales acute, close; rays whitish, short. Old fields, N- 30'.—Varies greatly. Lvs. 5'—1', broad or narrow. Hds. dense or scattered. Rays 15+, 2—3''.

ORDER 70.—COMPOSITÆ.

42 A. simplex Willd. Loosely corymbous-paniculate, smoothish; lvs. .auceolate. acuminate, the lower serrate; heads scattered; scales loosely imbricated, linear-subulate. Low grounds: common. 3—6f. Heads twice larger than No. 41, blue to white.
  β. *divérgens.* Diffusely branched, loosely racemous; branches hairy in lines.
43 A. tenuifòlius L. Paniculate-branching, with 1-flowered branchlets; lvs. linear-lanceolate to lance-linear, slender-pointed, sessile, remotely serrulate, upper entire; scales linear-subulate, equalling the disk. Moist fields. 2—6f.
  β. *bellidiflòrus.* Leaves scabrous, slightly clasping; scales loosely imbricated.
  γ. *distichus.* Leaves and strict ascending branches in 2 rows! Ill. (Mr. J. Wolf.)
44 A. subásper Lindl.? Pubescent above; racemous-branched, branches short, dense-flwd.; lvs. lance-acuminate, appressed-serrate, rough, attenuate to a petiole, upper reduced, entire, sessile; invol. closely imbricated; rays purp. Dry. Ill. 2f. (Wolf.)
45 A. ericoìdes L. Smoothish; branches virgate, branchlets secund, 1-headed; lvs. lance-lin. to subulate; hds. small; sc. as long as disk, with subulate-mucronate spreading tips. Rocky fields. 1-3f. Lvs. 4'-4'', attenuate-mucronate. Rays white or purplish
46 A. racemòsus Ell. Rough-downy; branches slender, erect; hds. very small (2'') spicate-racemous, crowded above; lvs. linear, sessile, rigid, 3'—3''. Coast, S. Car. 2f.
47 A. multiflòrus L. Grayish-downy, diffusely branched; lvs. linear, entire, sess., obtuse-mucronate; hds. small; sc. with obtusish spreading tips. Dry fields. 1f. Very bushy, with crowded racemes. Rays about 12, pale, 2—3'' long.
48 A. graminifòlius Ph. Slender, with filiform erect branches, 6- -12' lvs. linear, crowded below; ped. slender, leafless, 1-flwd.; sc. subulate-linear; rays about 20, white or rose. Rocks, Vt. N. H.: rare. (Willoughby Lake, Vt., Bradford, Vt., Whi'e Mts.)
49 A. acuminàtus Mx. St. simple, flexuous, angular, branching into a corymbous panicle above; lvs. broad-lanceolate, narrowed and entire at the base, serrate and acuminate; scales lax, linear. Wooded hills, N. 1f. Rays 12+, long, white.
50 A. nemoràlis Ait. Branches corymbed or 0; ped. 1-flwd., nearly naked, filiform; lvs. narrowly lanceolate. acute at each end, veinless, subentire; sc. very acute, loose, shorter than the disk; rays long, about 20. Wet woods. 1f. White-purple.
51 A. ptarmicoìdes T. & G. St. corymbous-fastigiate above; lvs. lin.-lanceolate, acute, rough-margined, entire, lower ones dentate, attenuated into a short petiole, rays short, snow-white. Rocky shores, Vt. to Mo. Rare. Heads rather large.
52 A. flexuòsus N. Smooth, slender, flexuous; branches leafy, 1-flwd.; lvs. fleshy, long-lance-linear to subulate; hds. large; rays short, many, purple. Marshes. 1f.
53 A. Chapmúnii T. & G. Smooth, slender, strict; branches filiform, 1-flwd.; lvs. linear-subulate; rays longer than invol., 20—30, purp.; cyps. glabrous. Swamps, Fla.
54 A. linifòlius L. *Sea Aster.* ① Smooth, much branched, paniculate; lvs. lance-linear to subulate; scales in 3 rows; rays minute, scarcely exserted. Marshes. 1f.
55 A. subulàtus Mx. ① Smooth, slender, much branched, corymbed; lvs. linear-subulate; rays many, narrow, in 1 row, longer than the disk, blue. Wet. S. 1—3f.
  β. *ixilis.* Taller (2—4f), less branched; heads few, rays pale purple. Ga.

19. DIPLOPÁPPUS, Cass. DOUBLE-BRISTLED ASTER.

Ray-flowers about 12, ♀. Disk-flowers ∞, ☿. Invol. imbricate. Scales narrow, destitute of green tips. Recep. flat, subalveolate. Pap. double, the exterior very short (about ½'' long), interior copious, capillary. Fruit compressed. ♃ Lvs. entire, alternate. Heads corymbous or few, rays cyanic, disk yellow.

§ Rays violet. Achenia silky. Bristles of the inner pappus alike. Sept. Oct.....No. 1
§ Rays whitish. Some of the longer bristles clavellate.—Ach. smoothish. Aug..Nos. 2, 3
    —Ach. villous. Sept. Oct...No. 4

1 D. linariifòlius Hook. St. clustered, leafy; branches 1-flwd., fastigiate; lvs. lin., entire, 1-veined, obtuse, rigid, rough. Dry places. 1f. Heads rather large, showy.
2 D. umbellàtus Hook. Smooth, simple, strict, with ∞ heads in a level corymb;

lvs. long (4–6'), lanceolate, acuminate; sc. obtuse; fr. pubes. in lines. Low grounds 2–4f. Stems purplish. Rays about 12, 3–4" long. Handsome

β. *amygdalinus*. St. roughish above; lvs. ovate-lanceolate; sc. rather loose. 2–3f
3 **D. cornifòlius** Less. Rough above, some hairy in lines; hds. few, corym.-panicu late; lvs. elliptical, thin, long-pointed both ways, entire; scales shorter than the disk obtuse; cypsela glabrous. Woods, Can. to Car. 1–2f. Rays about 10, white.
4 **D. obovàtus** (Ell.) Cinereous-pubescent: heads corymbed; lvs. obovate-oblong acute; sc. lin.-subulate, rusty yellow; fr. villous; rays white. Damp shades, S. 2–3f

20. **ERÍGERON,** L. FLEABANE. WHITE-WEED. Heads subhemispherical. Ray-flowers ♀ (40—200), narrow, linear. Fls. of the disk ☿, ∞ Recep. flat or convex, naked. Invol. scales nearly in one row and equal Pap. generally simple. Herbs with alternate lvs., rays cyanic, disk yellow

§ Rays minute, shorter than the cylindrical involucre, white. Pappus simple....1, 2, 10
§ Rays long, showy, 30—40. Pappus simple. Lvs. all radical. Hds. corymbous..No. 3
§ Rays long, showy, 50—200.—a Pappus simple. Leaves clasping. Corymbous...Nos. 4—6
—a Pappus double. Leaves sessile. Corymbous....Nos. 7—9

1 **E. Canadénse** L. Erect; invol. oblong; rays 40—50, crowded, minute; pap. simple; stem hairy, paniculate; leaves lanceolate. ① A common weed. 6'—6f. Jl.—Oct.
2 **E. divaricàtum** Mx. Decumbent and diffusely branched, hirsute; lvs. linear anc subulate; hds. very small, loosely corymbous. ② Dry soil, W. and S-W. 6'—2f. Purp
3 **E. nudicaùle** Mx. Glabrous; lvs. obovate or spatulate, radical, rosulate, entire hds. few; rays narrow, white. ♃ Pine-barrens. S. Scape bracted, slender. 18'. Jn. Jl.
4 **E. bellidifòlium** Muhl. *Robins' Plantain*. Hirsute; radical lvs. obovate, obtuse, subserrate; stem lvs. remote, mostly entire, clasping; hds. 3—7; rays 50—60, purple, linear-spatulate. ♃ Dry soils: common. 1—2f. May, June. Handsome.
5 **E. Philadélphicum** L. Pubescent or hirsute; lvs. thin, lower spatulate, crenate-dentate, upper clasping, sometimes cordate-auriculate; heads few, on long, slender ped.; rays 150-200, filiform, reddish. ♃ Damp: com. 2f. St. lvs. various. Jn.-Aug.
6 **E. quercifòlium** Lam. Pubescent; root lvs. oblong-obovate, lyrate-pinnatifid, or deeply sinuate-toothed, the cauline sharply serrate, clasping; heads ∞, small, with innumerable filiform flesh-colored rays. ♃ Low grounds. S. May.
7 **E. ánnuum** Pers. *Common Fleabane. White-weed.* Hirsute, branching; leaves coarsely serrate, ovate to lanceolate, the lower on winged stalks; rays very numerous, narrow, white. ①② Fields: common. 2—4f. June—Aug.
8 **E. strigòsum** L. Rough, with short, appressed hairs, or nearly smooth; lvs. lanceolate, tapering to each end, entire, or with a few large teeth in the middle, lower ones 3-veined and petiolate; pan. corymbous, white. ② Grass lands: com. 2f. Jn.—Oct.
9 **E. glabéllum** Nutt. Lvs. smooth, entire, spatulate, long-tapering at base, upper lanceolate and lance-linear, sessile, acuminate; heads 4—6, pubescent; rays very numerous, pale blue. Wis. to Dak. 12'—18'. July, Aug.
10 **E. ncre** L. Erect, 1f; lvs. entire, oblong to lanceolate; heads few or many, hemispherical, with bluish-purple rays as long as the pappus. Lake Superior (Porter).

21. **CALLÍSTEPHUS,** Cass. CHINA ASTER. Ray-flowers ♀, ∞, disk-flowers ☿. Involucre hemispherical. Recep. subconvex. Pappus double, each in – series, outer series short, chaffy-setaceous, with the setæ united into a crown; inner series of long, filiform, scabrous, deciduous bristles.

C. CHINÉNSIS. Stem hispid; branches divergent, 1-flwd.; leaves ovate, coarsely dentate, petiolate, cauline ones sessile, cuneate at base. China? Cultivation has produced innumerable varieties, double and semi-double, of every color. Aug., Sept. ①

22. **BELLIS,** L. GARDEN DAISY. Rays ∞, ♀. Disk ☿. Involucel

hemispherical, of equal scales. Recep. subalveolate, conical. Pap. none.
① ♃ Heads solitary.

1 B. integrifòlia Mx. Annual, diffusely branched; lvs. entire, spatulate-obovate to lance-obl.; sc. with scarious margins; rays violet-purp. Ky. to Tex. 6–12'. Mar.-May.
2 B. perénnis. Perennial, acaulescent; root creeping; scape naked, single-flwd.; lvs. obovate, crenate. Europe. 3—4'. Fls. white, double, quilled, &c. June–Aug.

23. DÀHLIA, L. Rays ♀. Disk ☿. Invol. double, the outer series of many distinct scales, the inner of 8 scales united at base. Recep. chaffy. Pappus none. ♃ Splendid Mexican herbs. Leaves opposite, pinnate.

D. variàdilis. Lfts. ovate, acuminate, coarsely serrate, 3—7 in number; stems stout, widely branched; heads solitary, very large; root tuberous. Colors exceedingly variable and splendid. Heads about 3' diameter; but a variety (the *bouquet Dahlia*) has the heads from 1½ to 2' broad.

24. BOLTÒNIA, L'Her. Ray-flowers ♀, in a single series, those of the disk tubular, ☿. Scales in 2 series, appressed, with membranous margins. Recep. convex, punctate. Cyp. flat, 2- or 3-winged. Pap. of minute setæ, 2 (to 4) of them usually lengthened into awns. ♃ Glabrous, loosely branching. Leaves sessile. Rays white. Aug.—Oct.

1 B. asteroìdes L'Her. Lvs. lanceolate, all entire; heads corymbed; fruit broadly-oval with a few minute setæ,—no awns. Swamps, Pa. to Ga. 1—3f. Rays 13—20.
2 B. glastifòlia L'Her. Lvs. linear-lanceolate, the lowest serrate; heads in a loose paniculate corymb; fruit obovate, with 2 long awns. Prairies, W. & S. 3—7f. Rays 30.
3 B. decúrrens. Lvs. lance-oblong, the broad base decurrent on the green, winged stem; heads corymbed, globular in fruit; fruit obovate, with 2 awns and several minute bristles; rays purple. Bottoms. Ill. (J. Wolf.) (B. glastifolia. β. ? T. & G.)
4 B. diffùsa Ell. Lvs. lance-linear to subulate, entire; hds. small, in a diffuse panicle; fruit obovate, with 2 short (half its own length) awns. Prairies, W. & S. 3—6f.

25. BRACHYCHÆTA, T. & G. False Goldenrod. Pap. a single row of scale-like bristles, shorter than the obconic cypsela. Otherwise as in Solidago. The golden yellow heads arranged in little clusters, forming 1 or more unilateral racemes.

B. cordàta T. & G.—Woods, E. Ky. (at Cumberland Gap) to Ga. along the mountains. 2–4f. Lvs. ovate, cordate, the lower petiolate, serrate. Hds. small (3" long). Aug.-Oct.

26. SOLIDÀGO, L. Goldenrod. Fls. of the ray about 5, ♀, remote; of the disk ☿. Invol. oblong, imbricate, with appressed scales. Recep. punctate, narrow. Pap. simple, capillary, scabrous. ♃ Very abundant in the U. S. Stem erect, branching near the top. Lvs. alternate. Hds. small, with 1—15 (very rarely 0) small rays. Fls. yellow (one species whitish), expanding in the autumnal months. Fig. 319. (Addenda.)

§ Shrubs 1—3f. Leaves punctate, veinless, entire. Rays 1—3. Chrysoma......No. 1
§ Herbs. Scales of involucre with spreading herbaceous tips. Chrysàstrum.. Nos. 2—4
§ Herbs. Scales imbricated, erect, scarious, seldom herbaceous...(a)
 a Inflorescence chiefly axillary, in clusters or short racemes...(b)
 a Inflorescence terminal, virgate or paniculate...(d)
 a Inflorescence terminal, in a fastigiate corymb...(s)
  b Rays white or cream-white. Clusters approximate above... ... ...No. 5

ORDER 70.—COMPOSITÆ.

   b Rays golden yellow.—*c* Cypsela glabrous. Scales acute..............Nos. 6, 7
                        —*c* Cypsela pubescent. Scales obtuse..........Nos. 8—10
  d Clusters or racemes erect, not secund. Leaves feather-veined...(*e*)
  d Clusters or racemes recurved and secund (one-sided)...(*g*)
    e Heads large, with loose scales. Alpine plants....................Nos. 11—13
    e Heads not large.—*f* Plants glabrous. Rays 4—7..................Nos. 14—16
                   —*f* Plants soft-downy. Rays 9—12................Nos. 17, 18
    g Leaves evidently feather-veined, mostly serrate...(*m*)
    g Leaves evidently 3-veined. Herbs inland, not maritime...(*h*)
    g Leaves 3- or 1-veined, fleshy. Very smooth, salt-marsh herbs......Nos. 19, 20
    g Leaves not veiny, thick, subentire. Herbs some downy, inland...Nos. 21—23
      h Leaves entire or very nearly so.................................Nos. 24—26
      h Leaves serrate.—*k* Stem smooth and glabrous................Nos. 27—29
                 —*k* Stem roughish-pubescent....................Nos. 30, 31
    m Heads discoid, rays none. Southern..........................Nos. 32, 33
    m Heads radiate.—*n* St. hairy or downy. Lvs. rough or smooth.. 24, 34—37
               —*n* St. glab. Lvs. glab. or not.—*o* Rays 1–5....Nos. 38—40
                                         —*o* Rays 6—12..(*p*)
      p Racemes distant, loosely if at all panicled...................Nos. 41, 42
      p Racemes close, forming a compact panicle................Nos. 43—45
    s Hds. large, rays fewer than the disk fls.—*x* St. and lanc. lvs. smooth..Nos. 46—49
                            —*x* Plant hairy. Lvs. oblong..Nos. 50, 51
    s Hds. small, rays more numerous than the disk flowers. EUTHAMIA..Nos. 52, 33

1 S. paucifloscúlosa Mx. Bushy, glabrous, glaucous and some viscid; lvs. lance linear, entire, sessile; rac. erect, panicled; fls. 5—7, rays 1—3, large. Coast, S.

2 S. discoidea (Ell.) Downy-canescent; hds. about 12-flwd., with no rays; rac. erect, in a long, narrow panicle; lvs. ovate to lanceolate, serrate. Ga. Fla., and W. 3f.

3 S. squarròsa Muhl. Pubescent; hds. very large, ∞-flwd., rays 9—12; panicle long, spike-like; lvs. smooth, broad-oval to elliptic, serrate. Hills, Can. to Ga. 2—5f.

4 S. petiolàris Ait. Pubescent, striate; hds. 20-25-flwd., rays 6—10; rac. long, compound; lvs. rough, small, oval to elliptic, the upper subpetiolate; scales subulate, the outer herbaceous, loose, spreading. Uplands, S. and W. 1—3f. (S. squarrulosa, C-B.)

5 S. bícolor L. Hairy, simple; leaves elliptical, the lower serrate; heads glomerate, virgate-panicled above; scales obtuse; rays about 8, *whitish*. Hills. 2f.
  β. *hirsúta*. Rays yellow, as well as the disk flowers. Penn. (S. hirsuta N.)

6 S. Búckleyi T. & G. Villous-pubescent; leaves oblong, serrate, acute at each end; clusters shorter than the leaves; fls. 15—20, rays 4—6; scales glabrous, rather acute; fruit compressed, glabrous. Interior of Alabama. 2—3f. Leaves 3'. October.

7 S. montícola (T. & G.) Stem terete, slender, puberulent above; lvs. oblong-lanceolate, pointed, subserrate; rac. approx.; fls. 12—15; fr. glabrous. Mts. N. Car. (Curtis).

8 S. latifòlia Muhl. Stem flexuous, angular, downy above; lvs. broad-ovate or oval, acuminate both ways, deeply serrate; racemes axillary and terminal, dense or loose; cypsela silky-pubescent; flowers 9—12, rays 3—4. Woody vales. 2f.
  β. *pubens*. Pubescent, becoming woolly above. Mts. N. Car. (M. A. Curtis).

9 S. ambigua Ait. Smooth or smoothish; st. tall, angled; lvs. long-lanceolate, acuminate, finely serrate, the upper reduced and shorter than the racemes; heads large; scales obtuse, oblong; fruit hairy. Mts. N. Car. 3f. Leaves 4—5'.
  β. *Curtísii* (T. & G.) Rac. shorter than the lvs.; sc. lin.-oblong; fr. silky. N. Car

10 S. cǽsia L. Stem slender, recurved at top, terete, smooth, glaucous; lvs. lin.-lanceolate, pointed, the lower serrate; fls. 6—10, rays 3—5, oval; racemes axillary, usually short; fruit puberulent. Hilly woods. 2—4f. Very elegant, wreath-like.

11 S. thyrsoídea Meyer. St. stout, simple, angular; lvs. ovate, acute, sharply and unequally toothed, the lower on long petioles; hds. large, in a narrow, downy raceme or panicle, rays 8-10; cyp. glabrous. Mt. woods, Me. to N Y. 1—4f. Coarse and showy.

168 ORDER 70.—COMPOSITÆ.

12 S. virgaùrea L. *β. alpina* (Bw.) St. dwarf, furrowed, simple; lvs. oval, subserrate or entire, narrowed to a petiole, upper lanceolate; hds. few (1—9), large, rays 10—12; sc. acute, very thin. Tops of high mts. Me. to N. Y., shores of L. Sup. 3—6'.
γ. glomerata. Taller; lvs. ovate-oblong, serrate; hds. very large. Mts. N. Car.

13 S. hùmilis Ph. Glabrous, simple; lvs. oblanceolate, crenate-serrate, acute, the lower obtuse, petiolate; rac. paniculate; hds. middle-size, about 12-flwd.; sc. obtuse. Mt. streams, N. H. and N. 6—12'—2f.—Varies with the branches pubescent above.

14 S. virgàta Mx. Tall, virgate, with a simple raceme at top; lvs. thickish, entire, oblanceolate, the lower subserrate, petiolate: hds. about 15-flwd., rays 6—7; fr. pubescent. Damp pine-barrens, N. J. to Fla. 3—5f. Rac. 6'—1f. long, of small clusters.

15 S. stricta Ait. Strict, simple; lvs. lanceolate, lower serrate, very long-petiolate, upper entire, panicle slender; heads 10-12-flowered; scales obtuse; rays 5 or 6. Wet woods, N. 2f.

16 S. speciòsa N. Stout, simple; lvs. lanceolate, entire, thick, lower very broad, subserrate, petiolate; panicle thyrsoid; ped. pubescent; rays, 6—8, large. Thickets; not common. 3—6f. Very handsome.—Varies with the panicle slender or virgate.

17 S. verna Curtis. Hoary-pubescent; stem few-lvd., loosely paniculate; lvs. ovate to lance-ovate, the lower finely serrate; rays, 10—12. Barrens, S. Fls. in May, June.

18 S. pubérula N. Puberulent as if dusty, strict, simple; lvs. oblanceolate to lanceolate, the lower subserrate; pan. dense, compound; sc. linear-subulate; fls. 20—25, rays about 10, elongated. In woods. Stem purplish, 2—3f. Heads rather large.

19 S. sempérvirens L. Lvs. thick, lanceolate, entire, obscurely 3-veined; hds. paniculate, 25-30-flwd., rays 8—10; ped. scabrous-pubescent. Marshes. 3—6f. Handsome.

20 S. angustifòlia Ell. Lvs. thick, entire, erect, 1-veined, the lower lanceolate; pan. dense, virgate; hds. 15-20-flowered, rays 7; ped. glabrous. Swamps, S. 2—4f.

21 S. pilòsa Walt. Hirsute, tall, stout; lvs. lance-oblong to lance-ovate, remotely serrulate, rough; rays minute, 2—10, disk-fls. 5—6. Damp barrens, N. J. and S. 4-7f.

22 S. odòra Ait. St. terete, smoothish, slender; lvs. lin.-lanceolate, abrupt at base, acute, pellucid-punctate; rays 2—4, disk-fls. 3—4. Dry hills and woods. 2—3f. The plant is yellowish-green, fragrant, and yields by distillation a fragrant oil.
β. retrórsa. Lvs. linear to subulate, acute, often twisted; rays 1—3. Ga.

23 S. tortifòlia Ell. St. rough-pubescent; lvs. many, linear, small, subentire, not punctate, often twisted at base; sc. obtuse; rays 3-5, disk-fls. 3-5. Dry fields, S. 2-3f.

24 S. nemoràlis Ait. Dusty-subtomentous; lvs. obscurely 3-veined, roughish, acute, attenuate at base; hds. small; fls. 10—15, rays 5—6, conspicuous. Dry fields, roadsides. 1-2f.—Varies with stem much branched, or with stem and panicle simple and slender.

25 S. rupéstris Raf. Smooth, slender; lvs. linear-lanceolate, plainly 3-veined; hds. small, in a simple panicle; fls. 15, rays very short. Rocky banks, Ind. Ky. 2—3f.

26 S. Leavenwórthii T. & G. St. minutely downy, very leafy; lvs. smooth, lin.-lanceolate, entire above; panicle open; heads rather large; ray and disk flowers each 10-12. Damp soils, South. 2—3 feet high.

27 S. Missouriénsis N. Low, simple; lvs. lance-lin., tapering both ways, shining, the lowest oblanceolate, with slender serratures; rac. small, dense; pedicels glabrous; hds. small, 12-15-flwd.; sc. with greenish tips; rays about 8. Dry prairies, Ill. Mo. 1-2f.

28 S. seròtina Willd. St. terete, striate, tall; lvs. slightly serrate, lin.-lanceolate, veins beneath pubescent; ped. pubescent; hds. small, 15-20-flwd. Low grounds. 3-6f.

29 S. gigántea Ait. St. striate, tall; lvs. lanceolate, with sharp, spreading serratures; strongly 3-veined; pan. downy-hirsute; hds. 15-20-flwd. 4-7f. Generally much branched.

30 S. Canadénsis L. St. downy; lvs. lanceolate, acuminate, rough; hds. very numerous and small; fls. 12—17, rays short and obscure, about 7. Copses, hedges: com. 2-5f.
β. pròcera. St. and lvs. beneath villous; hds. and rays larger. Low grounds. 4—7f.

31 S. Shórtii T. & G. St. minutely rough-downy; lvs. lance-oblong, acute, smooth pan. contracted, elongated; sc. with greenish tips; fls. 10-15, rays 5-7. O. Ky. 2f.

32 S. gracíllima T. & G. Smooth, slender; lvs. lance-spatulate, obtuse, to linear, entire; panicle narrow, hds. 9-12-flowered, scales obtuse; rays 0. Barrens, Fla. 2f.

ORDER 70.—COMPOSITÆ.

33 **S. brachyphylla** Chapm. Pubescent; leaves spatulate to round-oval, serrulate; rac. spreading; scales obtuse, rigid; disk-fls. 3—5, rays 0. Dry soils, Ga. Fla. 3f.
34 **S. altíssima** L. Hairy, tall; lvs. lanceolate, very veiny, rough and wrinkled, the lower serrate; scales acute; rays 6—8. Fields: common. 3—5f. Variable.
35 **S. Drummóndii** T. & G. Minutely velvety; lvs. ovate or broad-oval, acute both ways, sharply serrate, veiny; scales oblong-obtuse; rays 4—5. Ill. opp. St. Louis. 1—2f.
36 **S. Rádula** N. Rough-downy, simple; lvs. oblong-spatulate, tapering to base, serrate above, very rough and rigid; hds. small, rays 5, disk-fls. 3—6. Ill. to La. 1—2f.
37 **S. amplexicaùlis** T. & G. Rough-pubescent, subsimple; lvs. broad-cordate to ovate, serrate; petioles wing-clasping; rays 1—3. Dry woods, W. Fla. to La. 2—3f.
38 **S. ulmifòlia** Willd. Stem glabrous, with hairy branches; lvs. thin, elliptic-ovate, acuminate, serrate, tapering to base, smooth above, villous beneath; raceme recurved-spreading; hds. small, scales acute, rays 3—4, disk-fls. 3—4. Thickets, N. and W. 3f.
39 **S. Boottii** Hkr. Stem glabrous, with hairy branches; lvs. ovate to lance-ovate, pointed at both ends, serrate; pan. long, loose; hds. middle-size, scales oblong, obtuse; rays 2—5, disk-flowers 8—12. Sandy soils, S. 2—3f.—Varies with stem downy.
40 **S. linoìdes** Sol. Smooth throughout, slender, simple; lvs. lanceolate, finely serrate; scales oblong-linear, obtuse; hds. small, rays 1—4, disk 4—5. Bogs, near Boston to N. J. 12—20′. Racemes of the panicle short, secund, at length spreading.
41 **S. Muhlenbérgii** T. & G. St. furrowed; lvs. smooth both sides, strongly serrate, ovate to lanceolate, pointed both ways; rac. axillary, remote, spreading; hds. 15-20-flowered, scales linear, obtuse. Damp woods, N. H. to Pa. 2—3f.
42 **S. pátula** Muhl. St. angular-striate; lvs. elliptic, acute, serrate, very rough above, the lower oblong-spatulate; panicle loose; scales obtuse, flowers 12—15. N. and W. 3f.
43 **S. elliptica** Ait. Glabrous, leafy; lvs. elliptical, acute both ways, subserrate; pan. pyramidal; rays very short, 5—8, disk-fls. 6—7; scales obtuse. Marshes, R. I. to Ga.
   β. *Elliottii.* Panicle more widely spreading. South. (S. Elliottii T. & G.)
44 **S. argùta** Ait. Strict; lvs. smooth, unequally serrate with divergent teeth, oblong-ovate to elliptical; pan. corymbous; rays about 10, disk-fls. 9—10; cyp. smooth Woods, meadows: common. 3f. Plant smooth and shining.
   β. *júncea.* Leaves lanceolate, upper entire; rays twice longer than involucre.
45 **S. neglécta** T. & G. St. striate; leaves lanceolate to linear, the lower divergent-serrate, long-stalked; panicle oblong or pyramidal; rays 6—10, disk-flowers 7—12; cypsela smooth. Swamps, Me. to Penn., and W. 3—4f. Root leaves 6—12′.
46 **S. Ohiénsis** Riddell. Entirely smooth; lvs. entire, lanceolate, flat, obtuse, to oblong-lanceolate, abruptly-acute, the lower on long stalks; hds. numerous, large, 15-20-flowered, rays about 6. Meadows and prairies, West N-Y. to Ind. and Wis. 2—3f.
47 **S. Riddéllii** Frank. Stout, nearly smooth; root lvs. very long, lance-linear, long-pointed, on long petioles, the cauline clasping, carinate, acute; heads 20-24-flowered, densely clustered in the level corymb. Wet prairies, O. to Mo., and N. 15—30′.
48 **S. corymbòsa** Ell. Glabrous, with the corymbous branches hirsute; lvs. sessile, lance-obl thick, rigid, smooth; hds. large, rays 10, disk-fls. 20; fr. smooth. Ga. 4—6f.
49 **S. Houghtónii** T. & G. Low, smooth; lvs. lin.-lanceolate, acutish, flat, entire, tapering to base or petiole; hds. few, large, 20-30-flwd., rays 9 or 10. N. Y. Mich. 1—2f.
50 **S. rìgida** L. Stout, rough-hairy; lvs. rigid, ovate to oblong, serrate, upper minute; hds. very large (4—5″), scales obtuse, rays 7-10, disk-fls. 25+. Dry Ct., S. and W. 3—5f.
51 **S. spithamæa** Curt. Low, villous; lvs. lance-oval to oblong, thin, sharply serrate; hds. middle-size; scales lanceolate, acute; rays 6—8, disk-fls. 15—20. High mts. N. Car.
52 **S. lanceolàta** Ait. St. angular, hairy, much branched; lvs. lin.-lanceolate, entire, 3-veined; rays minute, about 17, disk-ds. 10. Meadows, copses: com. 2—4f. Fragrant.
53 **S. tenuifòlia** Ph. St. angular, smooth, much branched; lvs. narrowly linear, 1-veined, the axils leafy; corymb open, loose; rays about 10. Dry fields, coastward.

27. **BIGELÓVIA**, DC. Fls. 3—4, all tubular, ♀. Rays 0. Invol. cylindrical, as long as the flowers. Scales rigid, linear, closely imbricated.

Recep. pointed by a scale-like cusp. Fr. obconic, hirsute. Pap. bristles in one row. ♃ Glabrous, slender. Leaves alternate, entire. Heads fastigiately corymbous, with yellow flowers and colored scales.

B. virgàta DC.—Swamps, N. J. to Fla. and La. 1—2f. With virgate branches from base. Lvs. narrowly lin., 1-veined, the cauline lin.-spatulate. Sc. glutinous. Aug.-Oct.

28. **ISOPÁPPUS,** T. & G. Ray-fls. 5—12, ♀; disk-fls. 10—20, ☿. Scales of the invol. lance-subulate, closely imbricated. Recep. alveolate. Fr. terete, silky-villous. Pap. a single row of equal capillary bristles. ② Rough-hairy, branching, with alternate leaves and loose panicles. Aug.—Oct.

I. divaricàtus T. & G. Scabrous, hispid; lvs. lin.-lanceolate, taper-pointed each way; ped. slender, naked; rays 6–8, disk-fls. 10–13; pappus tawny. Dry. Ga. Fla. to Tex.

29. **HETEROTHÈCA,** Cass. Hds. ∞-flowered. Rays in one series, ♀; disk-fls. ☿. Scales imbricated, appressed. Recep. alveolate, fringed. Fr. minutely canescent, of the ray without pappus (naked), of the disk with a double pap., the outer very short, scale-like, the inner of capillary bristles. ♃ Hairy, corymbously branched, with alternate leaves and yellow flowers.

H. scàbra DC. St. flexuous, striate; lvs. scabrous, oblong-ovate, dentate; pet. wing-clasping; hds. large, rays 15–20; pap. tawny red, the outer white. S. 2–3f. Sept. Oct.

30. **CHRYSÓPSIS,** Nutt. Hds. ∞-flowered. Ray-fls. ♀; disk-fls. ☿. Invol. imbricate. Recep. subalveolate, flat. Pap. of the ray and disk similar, double, the exterior short, interior copious, capillary, brownish. Cyp. hairy, compressed. ♃ ② Hairy, with alternate and entire leaves and yellow flowers. Heads corymbous.

§ Leaves linear and lance-linear, grass-like, veined. Cypsela linear..........Nos. 1—4
§ Leaves oblong. Cypsela clavellate.—*a* Corymbs simple, umbel-like... ...Nos. 5—7
   —*a* Corymbs compound or paniculate..Nos. 8–10

1 **C. graminifòlia** N. Canescent with long, silky hairs; stem leafy to the top; lvs. linear, the upper reduced; hds. many, large, loosely corymbed. Del. to Fla. 2f. Sept.

2 **C. oligántha** Chapm. Canescent with silky hairs; st. almost leafless above; hds. quite large, few, on slender peduncles; lvs. lance-lin. Damp sands, Fla. 2f. Apr. May.

3 **C. pinifòlia** Ell. Glabrous; lvs. narrowly linear to setaceous, rigid, erect; hds. solitary, few; cyp. villous; pap. reddish-brown, the outer whitish. Hills, Ga. 1–2f. Sept.

4 **C. falcàta** Ell. Villous; lvs. somewhat falcate, spreading, narrow; hds. small, in axillary corymbs; rays 3-toothed. Dry sands, Ms. to N. J. St. 8', stout, leafy. Sep. Oct.

5 **C. Mariàna** N. Silky-arachnoid, simple; lvs. oblong-lanceolate, smooth when old, the lower spatulate, rather obtuse, upper reduced, acute; hds. about 7, large, 15–20-rayed; ped. and acute scales glandular. ♃ Barrens, N. J. to Fla. 2f. Sept.

6 **C. gossýpina** N. Cottony-tomentous, simple; lvs. uniform, ovate-oblong, obtuse, the lower tapering to base; hds. few, large; ped. short, glandular. ③ Md. to Fla. in barrens. 1—2f. Lower leaves rarely sinuate-toothed. (C. dentata Ell.) Sept.

7 **C. villòsa** N. Villous-pubescent, leafy to top; lvs. acute, lower oblong-spatulate, upper oblong-linear, bristly-ciliate; hds. large, umbel expanded. Ill. to Ala. 2f.

8 **C. trichophýlla** N. Silky-villous, branching, leafy; lvs. oblong to lance-linear, the lower obtuse; corymb large; ped. and scales smoothish. ② Barrens, S. 2–3f. Sept.

9 **C. scabrélla** T. & G. Dusty-scabrous, stout, branched; lvs. oblong-lanceolate, the lower narrowed to base, upper acute; corymb large; ped. glandular. Fla. 2f. Oct.

10 **C. decúmbens** Chapm. Silky-villous, decumbent; lvs. lance-oblong, obtuse. with leafy axils, lower spat.-oblong; hds. very large, paniculate, glandular. Fla. 3–4f. Nov.

31. CONYZA, L. GNATBANE. Fls. all tubular, those of the margin ♀, of the centre ♂ or ☿. Scales in several rows. Recep. flat or convex. Cyp compressed. Pap. 1 row of (red) capil. bristles.—Herbs chiefly trop. Fls. yel

C. ambígua DC. Cinereous-pubescent; lower lvs. sinuate-lobed, acute, middle repand-dentate, upper linear, entire; hds. panicled. Ga. S. Car. Ap.–Jl. § (C. sinuata Ell.)

32. ÍNULA, L. ELECAMPANE. Hds. many-flowered. Invol. imbricate. Ray-fls. numerous, ♀; disk-fls. ☿. Recep. naked. Pap. simple, scabrous. Anthers with 2 bristles at base. ♃ Coarse European herbs, with alternate leaves and very large yellow heads.

I. Helènium L. Lvs. amplexicaul, ovate, rugous, downy beneath; hds. solitary, terminal; sc. ovate. Pastures and roadsides, N. Eng. to Ill. 4–6f. Root lvs. 1–3f. Jl. Aug. §

33. PLÙCHEA, DC. MARSH FLEABANE. Hds. ∞-flowered; fls. of the margin ♀, of the centre ☿, but sterile. Invol. imbricated. Recep. flat, naked. Sty. undivided. Pap. capillary, simple.—Strong-scented herbs, with alternate leaves and corymbs of purple fls., and copious, reddish pappus.

1 **P. bìfrons** DC. Pubescent, leafy; lvs. oval-oblong, acute, finely serrate, cordate-amplexicaul, veiny; heads in compound, corymbous clusters. ♃ Damp, S. 2f.

2 **P. camphoràta** DC. Lvs. ovate-lanceolate, somewhat pubescent, acute, sessile or short-petioled, serrate; fls. in crowded corymbs; sc. viscid-downy, pointed. ① Salt marshes, Mass. to Fla. 1—3f. Stout, some fleshy, with upright branches. Aug. Sept.

3 **P. purpuráscens** DC. Glandular-tomentous; lvs. ovate-lanceolate, serrate, on slender petioles; hds. on slender ped.; sc. downy, acute. ① Swamps. 1–2f. Fla. Sept.

4 **P. fœtida** DC. Nearly glabrous, very leafy; lvs. broadly lanceolate, acute or acuminate at each end, petiolate, obtusely subserrate; heads numerous, in paniculate corymbs; scales smoothish, acute. ♃ Open hills, W. & S. 1—2f. Aug.—Oct.

34. BÁCCHARIS, L. GROUNDSEL TREE. Hds. discoid, ♂ ♀. Invol. imbricate, cylindric, or ovate, with subcoriaceous, ovate scales. ♂ Sta. exserted. Recep. naked. Pap. capillary. ♄ With alternate leaves and white flowers in Autumn.

1 **B. halimifòlia** L. Whitish-scurfy; lvs. obovate, incisely- or repand-dentate above, the highest lanceolate; panicle compound, leafy; fascicles pedunculate, terminal, in a dense panicle. Sea-coast, Conn. to Fla. 6—12f. A handsome shrub.

2 **B. glomerulifòra** Pers. Minutely scurfy; lvs. all obovate, very obtuse, repand-few-toothed; heads in sessile, axillary glomerules. Coast, Va. to La. 3—6f.

3 **B. angustifòlia** Mx. Diffusely branched; lvs. linear, sessile, entire; hds. small, 15-20-flowered, cylindrical, axillary, loosely paniculate. Marshes, S. 6—10f.

35. PTEROCAÙLON, Ell. BLACK-ROOT. Hds. many-flowered, the fertile flowers ♀, in several rows, the sterile flowers central, mostly ☿. Sc. imbricated, caducous with the fruit, ♀ corollas 3-toothed, ☿ 5-cleft. Cyp. angular, hispid. Pap. of equal capillary bristles longer than the involucre. ♃ Rhizome tuberous. Leaves alternate, decurrent, and the stem winged. Heads sessile, crowded in a thick woolly spike.

P. pychnostáchyum Ell. Simple; lvs. lanceolate, smooth above, cream-white-tomentous beneath, as well as one side of the wings of the stem. Sandy soils, S. 2-3f. Spike 2—3'. May—Aug. A curious plant.

36. BORRÍCHIA, Adans. SEA OX-EYE. Ray-fls. ligulate, ♀, fertile

Scales imbricated, the outer leafy. Recep. flat, chaffy, the chaff rigid, persistent. Fr. 4-angular, crowned with a 4-toothed pappus. ♂ ♀ Maritime, with opposite leaves and solitary yellow heads.

1 B. frutéscens DC. Canescent, downy; lvs. oblanceolate, repand, obtuse-cuspidate, subconnate at base; chaff of the recep. rigidly cuspidate. Marshes, Va. to Fla. 1—3f.

2 B. arboréscens DC. Smoothish; lvs. spatulate, entire; chaff obtuse. S. Fla. 8f.

37. **ECLÍPTA,** L. Ray-fls. ♀, numerous, narrow; disk ☿, mostly 4-toothed. Scales 10—12, in two rows, leafy, lance-ovate. Recep. flat. Chaff bristly. Cypsela somewhat angular or 2-edged. Pap. 0. ① Strigous. Lvs. opposite. Heads axillary and terminal, solitary. Flowers white. Fig. 72.

E. alba (L.) Erect or diffuse, with short, appressed hairs; lvs. lance-oblong, tapering to each end, subserrate; ped. longer than the hds.; scales lanceolate. Damp soils, Ill. to Md., and S. 1–3f. Rays minute. (E. erecta L. E. procumbens Mx. Cotula alba L., &c.)

38. **GALINSÓGA,** R. & P. Rays 4 or 5, small, obtuse, ♀. Invol. scales 4 or 5, ovate, thin. Recep. conical, chaffy. Cyp. angular. Pappus of small, fringed scales, or 0. ① Leaves opposite, 3-veined. Heads small, with white rays and yellow disk-flowers.

G. parviflòra Cav. Lvs. ovate, acute, subserrate; pap. scales 8—16. A weed in cultivated grounds, coastward, Mass. to Penn. 1—3f. Summer. § S. America.

39. **POLÝMNIA,** L. Leaf-cup. Involucre double, outer of 4 or 5 large, leafy scales, inner of about 10 leaflets, concave. Ray-flowers pistillate, few; disk sterile. Receptacle chaffy. Pappus none. ♃ Coarse and clammy. Leaves opposite. Flowers yellow.

1 P. Canadénsis L. Viscid-villous; lvs. petiolate, acuminate, lower pinnatifid, upper 3-lobed or entire, rays shorter than the invol. Can. to Car. and Ill. 3—5f. June.

2 P. uvedàlia L. Hairy and rough, stout; lvs. 3-lobed, acute, decurrent into the petiole, lobes sinuate-angled; rays 7—12, much longer than the involucre. In highland woods, N. Y. to Ill., and S. 3—6f. Lvs. very large (as also in No. 1). Hds. showy.

40. **CHRYSÓGONUM,** L. Rays about 5, ♀, fertile; disk ☿ but sterile. Scales in two rows of about 5 each, the outer leafy, the inner chaffy. Recep. flat, chaffy. Cyp. of the ray obcompressed, obovate, each embraced by a chaff scale, of the disk abortive. Pappus a small, 2-3-toothed crown ♃ A little prostrate herb, with opposite leaves and solitary, pedunculate, bright yellow vernal flowers.

C. Virginiànum L.—In rich shady soils, Md. to Ill., and South. Acaulescent, finally caulescent. One of the earliest flowers of Spring.

41. **SÍLPHIUM,** L. Rosin-weed. Ray-fls. numerous, in 2 or 3 rows, fertile, outer row ligulate; disk-fls. sterile. Invol. campanulate. Scales in several series, leafy and spreading at summit. Recep. small, flat, chaffy. Cyp. broad, flat, obcompressed, crowned with a 2-toothed pappus. ♃ Stout, coarse, resinous herbs. Heads large. Flowers yellow. Summer (p. 447).

* Stem nearly leafless, scape-like. Lvs. very large, alternate, mostly radical....Nos. 1—3
* Stem leafy.—*a* Leaves verticillate, in whorls of 3's, rarely 4's............ ...Nos. 4, 5 β.
    —*a* Leaves opposite, rarely the highest scattered.................Nos. 5—7
    —*a* Leaves alternate (the lowest opposite or verticillate or alternate)..No. 8
    —*a* Leaves connate-perfoliate.......................................No. 9

1 **S. laciniàtum** L. *Polar Plant.* Very rough, with white, hispid hairs; leaves (18′) pinnately parted, petiolate, segments sinuate-lobed or entire; heads spicate, distant; scales ovate, appendaged and squarrous at apex. Prairies, W. 5—10f. July—Sept.

2 **S. terebinthinàceum** L. *Prairie Burdock.* St. glabrous; lvs. ovate to oblong, cordate, tooth-serrate, obtuse (1—2f); hds. panicled; scales round-oval; rays about 20; fr. winged. Prairies, W. and S. 4—8f. Exudes much resin. Hds. 1′ broad, rays 1′ long.
  β. *pinnatifidum.* Lvs. more or less deeply lobed or pinnatifid. Prairies.

3 **S. compósitum** Mx. Glabrous throughout; slender, glaucous; lvs. cordate, variously sinuate-pinnatifid with lobed segments; hds. corymbed; fr. roundish-obcordate; rays about 10. Barrens, S. 3—6f. July, Aug. Varies with leaves only toothed. Hills.

4 **S. trifoliàtum** L. St. glabrous, terete or 6-angled; lvs. lanceolate, acute, short-petioled, in 3's or 4's, upper opp.; cyme loose; fr. oval, 2-toothed. Dry, O. to Fla. 4—6f.

5 **S. integrifòlium** Mx. Scabrous; st. 4-angled; lvs. opp., sessile, ovate-lanceolate, entire, cordate; corymb close; fr. broad-winged, 2-toothed. Prairies, W. and S. 2—3f.
  β. *ternàtum.* Stem 6-angled; lvs. verticillate in 3's. With the common form.

6 **S. scabérrimum** Ell. Rough-hispid; lvs. rigid, oval, some pointed, serrate, petiolate, scales ciliate-serrulate; fr. roundish, broad-winged, deeply notched at apex. W. Ga. to La. 3—4f. Corymbed. Rays 20, spreading 2′. Fruit 6″. Aug. Sept.

7 **S. lævigàtum** Ell. Glabrous; lvs. lance-oblong, acute, serrate, petiolate; scales ciliate; fruit, large, oval, narrowly winged, emarginate. W. Ga. Ala. 2—3f. Heads small, loosely corymbed. Rays spreading, 1½′. Fruit 4″. Aug. Sept.

8 **S. Asteríscus** L. Hispid or hairy; lvs. lanceolate, crenate-serrate, petiolate; scales leafy; fruit broad-obovate, 2-toothed. Dry soils, Va. to Fla. 2—4f. June—Aug.
  β. *pùmilum.* Downy, low; leaves elliptical; heads small; fruit truncate.

9 **S. perfoliàtum** L. *Cup-plant.* Stem square; leaves large, thin, ovate, forming a cup with their connate bases; heads on long peduncles; fruit broad-obovate, winged, notched. By streams, W. and S. 4—7f. Heads large. July, Aug.

42. **BERLANDIÈRA**, DC. Ray-fls. ♀, fertile, in one series; disk ☿ but sterile. Scales in three series, leafy, subequal. Recep. chaffy. Pales obtuse. Cyp. all marginal, in one row, obcompressed, wingless, obovate, adherent to the inner scales. Pap. minute. ♃ Velvety-canescent, with alternate, cordate, petiolate leaves and yellow rays.

1 **B. tomentòsa** T. & G. Caulescent, simple, white-tomentous; lvs. oblong, obtuse, crenate; heads in small, dense corymbs. Barrens, S. 1—2f. April—Aug.

2 **B. subacaùlis** N. Acaulescent, at length some caulescent, roughish canescent; lvs. sinuate-pinnatifid; scapes tall, bearing a single head. Ga. Fla. May, June.

43. **MÀDIA**, Molina. Invol. scales as many as the rays, complicate and embracing the compressed cypselæ. Recep. chaffy at its border. Rays 5—15, ♀; disk-fls. ☿, but often sterile. Pap. 0. ① Hairy and glandular.

M. élegans. Lvs. lance-linear, sessile; heads corymbed; rays linear-cuneate, 3-toothed at apex, yellow, with a purple base. From California, very showy. (*Madaria*, DC.)

44. **SPHENÓGYNE**, Br. Invol. imbricate. Sc. with broad scarious tips. Recep. chaffy, pales embracing the flowers. Rays neutre; disk-fls. ☿. Cyp. hairy. Pap. of obtuse, contorted, chaff-scales.—S. Afr. Lvs. alternate.

S. speciòsa. Leaves pinnatifid, with oblong cut segments; rays linear-oblong, spreading 2′, yellow, disk dark purple. ① 1f. Blooms profusely from July to Oct.

45. **PARTHÈNIUM**, L. Rays 5, very short, fertile; disk-fls. ∞, tubular, sterile. Invol. hemispherical. Sc. in two series, outer ovate, inner

orbicular. Recep. conical, chaffy. Cyp. 5, compressed, cohering with 2 contiguous pales. American herbs with alternate leaves. (Flowers white.)

1 P. integrifòlium L. Pubescent, rigidly erect; lvs. lance-ovate, coarsely dentate-crenate, coriaceous; hds. many, corymbed. ♃ Dry. Md., W. and S. 3—5f. Jl.—Sept.

2 P. Hysteróphorus L. Puberulent, decumbent; lvs. bipinnatifid, the upper linear; heads numerous, very small, in a diffuse panicle. River banks, Fla. to La.

46. IVA, L. MARSH ELDER. HIGHWATER SHRUB. Hds. discoid, monœcious. Invol. of 3—9 scales, distinct or partly united. Marginal fls. 1—5, fertile, the others sterile. Recep. chaffy. Cyp. obconic, obtuse. Pap. none. Herbs or shrubs. Lower lvs. opposite. Hds. small, greenish white

1 I. frutéscens L. Shrubby; lvs. fleshy, lanceolate, coarsely serrate, upper lance linear, entire; hds. axillary; scales 5, distinct, rounded; cypselæ 5. Borders of salt marshes, Mass. to Fla. 3—8f, bushy. Racemes paniculate, hds. drooping. July—Sept

2 I. cillàta Willd. Annual, hairy; lvs. lance-ovate, acuminate, coarsely toothed; hds. spicate; sc. 3, distinct, roundish, ciliate; cyp. 3. Wet. Ill. to La. 3—7f. Aug.—Oct.

3 I. imbricària Walt. ♃ Terete, glabrous; lvs. fleshy, linear-lanceolate, 3-veined, sessile; heads drooping, in leafy racemes; scales 6—9, obtuse, imbricated in 2 rows, with torn edges. Sea-coast, S. 1—2f.

47. AMBRÒSIA, Tourn. HORSE-WEED. Monœcious. Sterile involucre of several scales united into a depressed, hemispherical cup, many-flowered. Anth. approximate, but distinct. Fertile involucre 1-leaved, entire or 5-toothed, 1-flowered. Cor. 0. Sty. 2. Sta. 0.—Herbaceous plants with mostly opposite leaves and unsightly flowers. July—Sept. Figs. 73, 342.

§ Sterile heads sessile, densely spicate, chaffy. Leaves alternate............No. 1
§ Sterile heads pedicellate, racemed, not chaffy.—*a* Leaves opposite............No. 2
—*a* Leaves alternate.........Nos. 3, 4

1 A. bidentàta Mx. Hairy and leafy, with simple branches; lvs. sessile or clasping, oblong, with a single tooth on each side near the base; fertile hds. axillary; fr. 4-angled, acutely pointed, the ribs produced into 4 short spines. ① Prairies, Ill. to La. 1–3f.

2 A. trífida L. Rough-hairy; lvs. 3-lobed, serrate, lobes oval-lanceolate, acuminate; fr. with 6 ribs ending below the conical top. ① Along streams, &c. 5—10f. Aug.

β. integrifòlia. Leaves ovate, acuminate, often some of them 3-lobed.

3 A. artemisiæfòlia L. *Hog-weed.* Lvs. twice-pinnatifid, smoothish, petioles ciliate; sterile hds. in panicled racemes, fertile axillary, sessile. ① Gardens, fields. 2–3f.

4 A. psilostàchya DC. Whitish, woolly, branching and leafy; lvs. rigid, the lower opp., bipinnatifid, upper pinnatifid; rac. spike-like; fr. hairy. ① Prairies, Wis. to Tex.

48. XÁNTHIUM, Tourn. CLOT-WEED. Monœcious. ♂ Hds. spicate above. Scales distinct, in one row. Anth. approximate, but distinct. Recep. chaffy. ♀ Invol. clustered below, 2-lvd., clothed with hooked prickles, 1- or 2-beaked, enclosing 2 fls. Sta. 0. ① Coarse weeds with alternate leaves.

1 X. Strumàrium L. Rough, unarmed, branching; lvs. cordate, lobed, 3-veined, unequally serrate; fruit elliptical, armed with stiff, hooked thorns, and ending with 2 spreading, straight horns. Fields, waysides, N., M. 2—3f. Aug. Unsightly.

2 X. spinòsum L. Whitish-downy, armed with triple, slender, subaxillary spines; lvs. lance-ovate, 3-lobed, dentate, or entire; ♀ invol. oblong Waysides, &c. 2f. Sept.

49. MELÁNTHERA, Cass. Fls. all tubular, ☿. Scales in 2 subequal series. Recep. chaffy, the pales partly investing the fls. Cyp. short, truncate.

angular. Pap. a few minute caducous awns or bristles. ♃ Scabrous, with square stems, opposite, petioled, 3-veined leaves and long peduncled heads. Corolla white. Anthers black, tipped with a white appendage.

1 **M. hastàta** Mx. Lvs. hastately 3-lobed, acuminate, dentate; sc. lance-ovate, acuminate, pales rigid, cusp-pointed. Dry soils, S. Car. to Fla., and W. 3—6f. Jl.—Sept.
2 **M. deltoìdea** Mx. Lvs. ovate-deltoid; scales ovate; pales or chaff obtuse. S. F.s

50. **ZÍNNIA**, L. Ray-fls. ligulate, ♀; disk tubular, ☿. Sc. oval, margined, imbricate. Recep. chaffy, conical. Pap. of the disk of 1 or 2 erect, flat awns. ① American herbs, with opposite, entire leaves and solitary terminal heads. Rays bright-colored, showy.

1 **Z. multiflòra** L. Lvs. lance-oblong, sess.; peduncles scarcely longer than the lvs.; rays oval, shorter than the invol.; fr. 1-awned; pales entire. Fields, S. 6'-2f. May,Ju. §
2 **Z. élegans** L. Lvs. ovate, cordate, sessile and clasping; peduncles much longer than the leaves; pales serrated; fruit 2-awned. Mexico. 2—4f. Fls. single or double, of all colors, often brilliant, blooming in gardens throughout the Summer.

51. **HELIÓPSIS**, Pers. Ox-eye. Invol. imbricate, with ovate, subequal scales. Rays linear, large, ♀; disk ☿. Recep. chaffy, conical, the pales lanceolate. Fruit 4-sided. Pappus 0. ♃ Leaves opposite. Heads large. Flowers yellow, like Helianthus.

H. lævis Pers. St. smooth; lvs. ovate-oblong to lanceolate, coarsely serrate, petiolate, 3-veined, smooth beneath. Hedges and thickets: common. 3—5f. June, July.
β. *gràcilis.* Slender, 2f; lvs. lance-ovate, scabrous, acute at base.
γ. *scabra.* Stem and leaves scabrous, yellowish; leaves truncate at base. W. 6f.

52. **TETRAGONOTHÈCA**, Dill. Hds. radiate. Invol. double, the outer of 4 leafy bracts united at base, the inner of 8 small scales similar to the chaff of the conical receptacle. Ach. smooth, truncate, destitute of pappus. ♃ Clothed with viscid hairs, opposite leaves, with 1 or few yellow-flowered, large heads, on long peduncles.

T. helianthoìdes L.—Sandy soils, Va., and S. 3f. A stout, coarse, unsightly herb. Leaves ovate, sessile, repand-toothed. Rays spreading nearly 3'. April—June.

53. **ECHINÀCEA**, Mœnch. Purple Cone-flower. Scales of the invol. in 2 or 3 rows. Ray-fls. neutral; disk-fls. ☿. Recep. conic, bristling with stiff, spiny pales. Cyp. 4-angled. Pap. a few teeth. ♃ Branches each with 1 large head. Leaves alternate. Rays rose-purple, drooping.

1 **E. purpùrea** Mœnch. Very rough; lower lvs. broad-ovate, 5-veined, cauline lance-ovate, acuminate, nearly entire; rays 12—15, very long (2—3'), bifid. Thickets, W. and S. 4f. July-Sept.—Varies in roughness, and with white rays. (See *Addenda.*)
2 **E. angustifòlia** DC. St. hispid, slender; lvs. all entire, hispid-pubescent, 3-veined, lanceolate to lance-linear; rays 12—15, narrow, 1—2' long. Prairies and marshes, Ill. Mo., and S. 2—3f. Rays sometimes white. May—July. (See Addenda.)
3 **E. atrórubens** N. Smooth or rough; stem simple, furrowed; lvs. lance-linear to linear, rigid, the lower 3-veined; rays 8—11, shorter than the disk (1'); scales in 3 rows; pappus of 4 teeth. Damp barrens, Ga. Fla., and W. 2f. June—Aug.

54. **RUDBÉCKIA**, L. Invol. scales nearly equal, leafy, in a double row, 6 in each. Ray-fls. neutral; disk ☿. Recep. conic or columnar, with

unarmed pales or chaff. Cyp. 4-angled. Pap. a lacerate or toothed margin, or 0. ♃ Leaves alternate. Heads large. Rays yellow.

§ Rays large, drooping.—*a* Leaves divided. Disk ovoid or rounded..........Nos. 1, 2
—*a* Leaves undivided. Disk columnar............ ..Nos. 3, 4
§ Rays spreading. Disk dark purple, conical or rounded...(*b*)
    b Leaves deeply lobed or parted, the upper undivided....................Nos. 5, 6
    b Leaves undivided.—*c* Pales of the disk whitish downy........ ....Nos. 7, 8
        —*c* Pales dark purple as well as the flowers........Nos. 9—12

1 **R. laciniàta** L. Glabrous: lower leaves pinnate, segments 3-lobed, upper leaves ovate; disk ovoid, yellowish, pales truncate. Swamps. 3—5f. Rays near 2′. Aug.
2 **R. heterophýlla** T. & G. Downy; lvs. coarsely toothed, 3-5-lobed or parted, the lowest often round-cordate, highest ovate; disk globous; pales acute. Fla. 4f. Ang.
3 **R. máxima** N. Glabrous; leaves thin, ample, oval to oblong, subentire, the upper clasping; head solitary, on a long ped.; rays 2′. Wet barrens, Fla. to La. 7f. Aug.
4 **R. nítida** N. Glabrous and shining; leaves thick, lanceolate, acute, 3-5-veined; heads few or solitary; disk brown; rays 9—12, near 2′. Swamps, S. 4f. July.
5 **R. subtomentòsa** Ph. Tomentous-downy, corymbous; leaves serrate, the lower 3-parted or lobed, upper ovate; disk globular; pales bearded, obtuse; rays 10—15, orange-yellow, 1′. Prairies, W. and S-W. 3—5f. July, Aug.
6 **R. tríloba** L. Hairy, paniculately branched; lvs. coarsely serrate, 3-lobed to ovate-lanceolate, the lowest cut-pinnate or undivided; hds. rather small, disk conical, dark purple; pales smooth, awned. Fields. M., W. 3—4f. Aug. Sept.
7 **R. mollis** Ell. Soft-woolly all over; lvs. oblong, sessile or clasping; sc. reflexed; disk dark purp., with canescent pales; rays 15-20, 1′. W. Ga. 2-3f. Lvs. small. Aug.-Oct.
8 **R. Heliópsidis** T. & G. Slightly downy; lvs. ovate or oval, 5-veined, petiolate; sc. obtuse, squarrous, rays 10—12; pales canescent. W. Ga. and Ala. 1-2f. Aug. Sept.
9 **R. hirta** L. Very rough-hairy; ped. leafless; lvs. ovate-spatulate, 3-veined, petiolate, mostly entire, upper ones sessile, lance-ovate; scales in 3 rows; rays oval, 12—15; disk rounded, dark brown; pales bearded. Fields. 2f. Showy. July—Sept.
10 **R. fùlgida** Ait. Rough-hirsute; branches leafless above; lvs. ovate to lance-oblong, remotely dentate, lower petiolate; scales oblong, spreading as long as the 12—14 orange rays; pales glabrous, lin.-oblong, obtuse. Mts. Pa. to O., and S. 1-3f. July-Oct.
11 **R. speciòsa** Wend. Hairy and downy; branches slender, leafless above; lvs. strongly dentate, acuminate, ovate to lanceolate, 5-3-veined, lower long-petiolate; sc. much shorter than the 18 rays; pales smooth, acute. Ill. to Va. 2—4f. Aug.—Oct.
12 **R.** AMPLEXIFÒLIA. ① Branching, glabrous; lvs. cordate-clasping; rays spotted at base, brilliant. La. (Dracopsis.)

55. **LEPACHYS**, Raf. Invol. in one series of linear scales. Ray-fls. few, neutral; disk ☿. Recep. columnar, chaffy. Chaff obtuse, and bearded at apex. Pap. 0. Fertile achenia compressed, 1-2-winged. ♃ Lvs. alternate, pinnately divided. Hds. with long, drooping, yellow rays. June-Sept.

1 **L. pinnàta** T. & G. Rough; lvs. all pinnate, divisions 5—7, 2-parted or entire; rays light yellow, twice longer than the ovoid yellowish disk. W. N-Y., W. and S. 2-4f.
2 **L. columnaris.** Rough, branching; root lvs. undivided, oblanceolate; stem lvs. pinnatifid; disk nearly 2′ long, longer than the 5—8 broad rays, which, in Variety pulcherrima, are crimson, tipped with yellow. Montana. 2f.

56. **HELIÁNTHUS**, L. SUN-FLOWER. Ray-fls. neutral; disk ☿. Sc. of the invol. imbricated in several series. Recep. flat or convex, the chaff persistent, embracing the fruit. Pap. of 2 or 4 chaffy awns, mostly deciduous. Fruit compressed or 4-angled. ① ♃ Rough. Lvs. opposite, the up-

ORDER 70.—COMPOSITÆ. 177

per often alternate, mostly tripli-veined. Rays yellow; disk yellow or purple: in late Summer and Autumn. Figs. 74, 261, 433–4.

§ HELIANTHÉLLA (T. & G.) Pap. persistent. .vs. scattered, 1-veined.....Nos 24, 25
§ HELIÁNTHUS *proper*. Pappus deciduous. Lower leaves opposite...(*)
 * Disk (its corollas and pales) dark purple, mostly convex...(a)
    a Herbs annual. Leaves chiefly alternate..........................Nos. 1, 2
    a Herbs perennial. Leaves opposite.—e Scales acuminate..........Nos. 3—5
                                —e Sc. obtuse or barely acute..Nos. 6, 7
 * Disk (its corollas and pales) yellow...(b)
    b Leaves chiefly alternate and feather-veined.....................Nos. 8- ·11
    b Leaves chiefly opposite and 3-veined or tripli-veined...(c)
        c Scales erect, closely imbricated.—f Plants green, rough......Nos. 12, 13
                                —f Plants whitish, downy...Nos. 14, 15
        c Scales loosely spreading. Heads large, 9-15-rayed...(d)
            d Scales lance-linear, longer than disk. Leaves thin......Nos. 16, 17
            d Scales lance-ovate, as long as the disk. Leaves thick...Nos. 18—21
        c Scales loosely spreading. Heads small, 5-8-rayed...........Nos. 22, 23

1 H. ánnuus L. *Great Sunflower.* Erect, stout; lvs. all cordate, only the lowest opposite; hds. very large (6—12'), nodding; fr. glabrous. Gardens and fields. 2—10f. § S. America.—A variety with the flowers all ligulate is sometimes found in gardens.
2 H. débilis N. Decumbent, slender; leaves mostly alternate, ovate, serrulate, petiolate; hds. small; scales with slender points; fr. pubescent. Shores, E. Fla. to La. 1–2f.
3 H. Rádula T. & G. Hirsute, simple, bearing a single head; lvs. roundish-obovate or ovate, obtuse; scales and pales lanceolate, acuminate, erect; rays 7—10, rarely 0. ⁄ Barrens, Ga. Fla. Ala. 1—3f. Often growing in clusters. Hds. near 1'. Aug. Sept.
4 H. heterophýllus N. Slightly hispid, slender, bearing a single head; lvs. entire, the lower oval, upper linear-lanceolate; scales acuminate, erect, ciliate; pales acute; rays 12—18. ⁄ S. 1—2f. Heads 6'' diam., rays spreading 2¼'. Aug. Sept.
5 H. angustifòlius L. Erect, slender, scabrous or hispid; lvs. lance-linear, tapering to a long point, 1-veined, rigid; heads few; scales lance-linear, the long point spreading; pales linear, 3-toothed. Dry soils. N. J., Ky. and S. 2—3f. Aug.—Oct.
6 H. rígidus Desf. Rigid, subsimple; lvs. lanceolate, pointed, rough both sides; hds. few; scales ovate, acute, short; rays 12—20. Prairies, Wis. Mo. to La. 2—3f.
7 H. atrórubens L. Ped. few, long, leafless; st. hirsute below; lvs. ovate or oval, obtusish, on winged petioles; sc. oblong, obtuse, 3-veined. Dry soils. S. 2—4f.
8 H. gigánteus L. Rough or hairy; lvs. lanceolate, serrate, pointed, on ciliate, winged petioles; scales lance-linear, ciliate; rays 12—20; pappus of 2 short, fringed scales Can. to Car. and Ky. 4—10f.—Varies with the leaves mostly opposite.
9 H. tomentòsus Mx. Stout, pubescent, branched; lvs. ovate to long-lanceolate, acuminate, subentire, the lower petiolate; scales long-pointed, villous, spreading; pales hairy and 3-toothed at top. Dry hills, Ill. to Ga. 4—8f. Rays 15''.
10 H. grosse-serràtus Martens. St. smooth and glaucous; lvs. lanceolate or lance-ovate, long-acuminate, sharply serrate, downy beneath, on winged stalks; scales loose, subulate, as long as the disk; rays 15—20. W. and S. 4—6f.
11 H. tuberòsus L. *Jerusalem Artichoke.* Root bearing oblong tubers; lvs. cordate-ovate to ovate, acuminate; petioles ciliate. Fields, hedges. 4f. § Brazil.
12 H. lætiflòrus Pers. St. branched above; lvs. thick, lance-oval, pointed, serrate, on short stalks; scales ovate-lanceolate; rays 12—20, 2'. Woods, W. and S-W. 3—4f.
13 H. occidentàlis Riddell. Slender, simple, nearly naked above; lvs. oval, subserrate, on long hairy petioles; hds. 1—5, small; scales lance-oval. Sandy. W. 3f.
14 H. mollis Lam. Canescent-tomentous, subsimple; lvs. ovate, sessile, cordate-clasping, acumⁱnate; sc. lanceolate; pales entire, acute; rays 15—25. O. to Mo. 2—4f.
15 H. cinèreus, β. Sullivántii (T. & G.) Cinereous-pubescent; stem virgate, branched above; lvs. ovate-oblong, narrowed to the sessile base, the lower to a winged petiole; pales pointed, with 2 lateral teeth; rays about 20. Ohio. 2—3f.

16 H. decapétalus L. Lvs. all opposite, thin, ovate, acuminate, toothed, on winged stalks, scabrous above, smoothish beneath.—Varies with the invol. scales enlarged and leaflike, or only lance-linear. Can. to Penn. 3—4f.

17 H. trachellifòlius Willd. Branch lvs. alternate, thin, appressed-serrate, acuminate, all ovate to lance-linear; pales 3-toothed; rays 12—15. Thickets, W. 3—8f.

18 H. doronicoìdes Lam. Branching; lvs. ovate to lance-ovate, acuminate, serrate; scales lance-linear; rays 12—15, 1½', very showy. W. and S. 4—7f.

   β. **plena-flora.** Flowers all ligulate. Gardens. Very handsome.

19 H. strumòsus L. Smooth below; lvs. all similar, ovate-lanceolate, acuminate, serrulate; heads few, about 10-rayed; scales ciliate, squarrons. Swamps. 3—5f.

20 H. hirsùtus Raf. St. simple or forked, hirsute; lvs. petiolate, ovate-lanceolate, subserrate, hirsute beneath; scales lance-ovate, hairy; rays 11—15. Dry, W. and S. 6f.

   β. **pubéscens.** Leaves tomentous beneath, subsessile. (H. pubescens Hook.)

21 H. divaricàtus L. St. smooth, simple, or forked; lvs. rough, lance-ovate, long-pointed from an abrupt sessile base; heads few, corymbous. Woods, &c. 4—5f.

   β.? **scabérrimus.** Stem subsimple; leaves thick, exceedingly rough and rigid, opposite or ternately verticillate, rounded at base. W.

22 H. microcéphalus T. & G. St. smooth or hispid, branched; lvs. lanceolate, acuminate, narrowed to a short petiole, rough above, whitish-downy beneath; scales lanceolate; rays 5—8, spreading 1'. Dry, W. and S. 3—5f. (H. Schweinitzii T. & G.)

23 H. longifòlius Ph. Smooth throughout, branching; lvs. lance-oblong to lance-linear, acute, the lowest petiolate, serrulate; heads few; scales ovate-lanceolate; rays 6—10, spreading 1½—2'. Damp. S. 3—5f. (H. lævigatus T. & G.)

24 H. grandiflòrus. Rough-downy; simple, leafy; lvs. 1—2', lance-linear, sessile; scales lanceolate, loose; rays 15—20, near 2'; pappus 2 fringed scales. E. Fla. 3f.

25 H. tenuifòlius. Rough-hairy, simple; lvs. narrow-linear; scales lance-subulate, loose; rays 10—13 (15''); pappus 2—4 awns. W. Fla. 1—2f. Leaves 2—3'. July.

57. ACTINÓMERIS, Nutt. Heads many-flwd.; ray-fls. 4—14, rarely 0. Invol. scales foliaceous, subequal, in 1—3 series. Recep. conical or convex, chaffy. Ach. compressed, flat, obovate, mostly winged and 2-awned. ♃ Plants tall, with 3-veined, serrate leaves. Heads corymbous. Rays when present yellow. Autumn.

  § ACTÍMERIS. Pappus of 2 awns. Stems tall, corymbous...(a)
  § ACHÆTA. Pappus wanting. Cypsela winged. Stems low, simple. Jn. Jl....No. 1
    a Rays wanting. Disk corollas white. Stem narrowly winged..............No. 2
    a Rays 4—14, flowers all yellow. Scales in 2 or 3 series.................Nos. 3—5

1 A. pauciflòra N. Lvs. opp. or alternate, lanceolate to elliptical, rigid, obtuse; hds. 1—3, discoid, yellow; fr. narrowly winged, the disk cupshaped. Barrens, Fla. 1—2f.

2 A. alba T. & G. Lvs. narrow-lanceolate, acute both ways, serrulate; scales lance-linear, few, in one series; fruit broadly winged. S. Car. to Fla. and La. 7f.

3 A. helianthoìdes N. Stem winged; lvs. alternate, ovate-lanceolate, decurrent, acuminate, serrate, rough, hairy; rays 1' long, 6—14. unequal; scales erect; fruit narrowly winged. Copses, prairies, Ohio to Ga., and W. 2—4f. June, July.

4 A. squarròsa N. Stem winged, tall (6—10f); lvs. alternate, some opposite, lance-oblong, long (6—14'), pointed both ways. decurrent; heads small; scales spreading or deflexed; rays 4—8, regular, short. Alluvion, N. Y., W. and S. Homely.

5 A. nudicaulis N. Stem wingless, branched and leafless above; lvs. oblong, unequally serrate, closely sessile; rays 7—12, broadly winged. Ga. Fla. Ala. 2—3f.

58. COREÓPSIS, L. TICK-SEED. Rays about 8, rarely 0. Involucre double, each 6—12-leaved. Recep. chaffy. Cyp. obcompressed, emarginate, each commonly with a 2-toothed, upwardly-hispid pappus, sometimes

ORDER 70.—COMPOSITÆ. 179

...ione. Leaves mostly opposite. Rays usually yellow; disk-flowers yellow or dark purple.

§ Corollas of the disk dark purple...(a)
   a Ray-flowers yellow with a purple base. Achenia incurved.............Nos. 1—3
   a Ray-flowers wholly yellow. Achenia not incurved, 2-awned. Summer....Nos. 4—6
§ Corollas of the disk yellow. Rays rose-colored. Leaves simple.............Nos. 7, 8
§ Corollas of the disk and ray all yellow (disk brownish in No. 9)...(b)
   b Leaves sessile, divided often so as to appear verticillate................Nos. 9—12
   b Leaves petiolate, never serrate,—*c* pinnate with lance-linear segments..Nos. 13, 14
                                        —*c* simple, or rarely auricled below....Nos. 15, 16
   b Leaves petiolate, serrated,—*d* simple. Achenia awns obsolete........Nos. 17, 18
                                   —*d* compound.—*e* Rays about 8............Nos. 19—21
                                          —*e* Rays wanting............Nos. 22, 23

1 C. Drummóndii. ② Pubescent; lvs. pinnately (1-5)-divided; segm. oval or oblong, entire; sc. lance-acuminate; rays unequally 5-toothed. Tex. 1-2f. Rays ample, showy.
β. *atrosanguinea*. A garden variety, with the rays wholly dark purple. July-Oct.

2 C. tinctòria. ② Glabrous; lvs. alternate, some pinnate; lobes lin.-oblong and linear; scales very short, acute; rays 3-lobed at apex. Nebraska. 1-3f. Beautiful. Summer.

3 C. Atkinsoniàna. ♃ Lf. lobes linear-spatulate to linear; sc. oblong, obtuse; rays 3-lobed; fr. distinctly winged. Columbia River. Oreg. Hds. handsome, like C. tinctoria.

4 C. gladiàta Walt. St. terete; lvs. alternate, thick, some ternately divided, lance-oblong to lance-linear; outer scales lance-ovate; fr. fringed, awns 2, slender; rays 3-toothed at the dilated apex. Moist barrens, S. 2—3f. Heads several, corymbed.

5 C. angustifòlia Ait. St. square; lvs. opposite (mostly), undivided, spatulate to linear, obtuse; outer sc. ovate, obtuse; fr. wing-fringed, awns 2, short; rays 3-lobed. S.

6 C. Æmleri Ell. St. angular above; lvs. opp., lance-ovate to lanceolate; outer scales oblong, obtuse; fruit margined, ciliate, the 2 awns very short. Ga. (Elliott) and Fla.

7 C. nudàta Nutt. Very slender; lvs. few, terete, rush-like, alternate, the lower very long; hds. few; rays wedge-obovate, crenate-lobed at apex. ♃ Swamps, Ga. Fla. 2f.

8 C. ròsea N. Branching; lvs. opp., 1-veined, linear; ped. short; outer sc. very short; rays oblong, obscurely tridentate. ♃ Wet grounds, Ms. to Ga. 8-16′. Delicate. Jl. Aug.

9 C. senifòlia Mx. Minutely downy or glabrous; lvs. opposite, ternate, sessile, appearing in whorls of 6; lfts. ovate-lanceolate, varying to linear-lanceolate or even to linear; scales downy, obtuse; rays entire. ♃ Dry, Va. Ky. to Ga. 1—2f. July, Aug.

10 C. delphinifòlia Lam. Lvs. opp., sessile, divided into lfts. which are each again 2-5-parted; seg. linear, entire, acute; disk-fls. brown at the tips. ♃ Va. to Fla. 2f. Aug.

11 C. verticillàta L. Branched; lvs. 3-divided, closely sessile, the divisions 1-2-pinnately-parted; seg. filiform-lin.; rays 1-3-toothed. ♃ Moist, Md. to Ga, 1-3f. Ju.-Aug.

12 C. palmàta N. St. angled, striate, leafy to top; lvs. sessile, deeply 3-cleft, rigid; lobes linear, acutish, entire or again cleft; fr. linear-elliptic. ♃ Prairies, W. 1-2. July.

13 C. tripteris L. St. simple tall, corymbous; lvs. opp., stalked, thick, 3-5-divided; seg. lin.-lanceolate, entire, acute; hds. small; rays obtuse. ♃ Dry, W. and S. 4-8f. Jl.

14 C. grandiflòra N. St. low; hds. solitary, large, on long naked stalks; lvs. lance-olate, mostly divided into lance-lin. seg.; rays 4-5-cleft. ♃ Mo. to Tex. Much like No.15.

15 C. lanceolàta L. Ascending; lower lvs. oblanceolate, upper lanceolate, all entire; heads solitary, on long naked peduncles; rays 4-5-toothed. ♃ Damp soils, West and So..th. Head showy. Rays about 8, spreading 2′ or more. June—Aug. †

16 C. auriculàta L. Lower lvs. round-ovate, petiolate, some of them with 2 small lateral segm. (auriculate) at base, the upper oblong, subsessile; hds. few, on long ped., outer scales oblong-linear. Dry soils, Ill. to Va., and S. 1—3f. May—Aug.

17 C. latifòlia Mx. Very glabrous, tall; lvs. thin, opp., ovate to oblong, acuminate, unequally toothed; hds. small, rays 5 or 6, entire, large; sc. lin., spreading. Mts. S. Aug

18 C. argùta Ph. Stem strict; lvs. simple, ovate to lanceolate, petiolate, acuminate, sharply serrate; scales oblong; rays 9—12, 3-toothed; awns obsolete. Hills, S. 2-5f

19 C. aurea A.t. Lower lvs. pinnately divided, upper ternately, or simple; lfts. ovate to lance-linear, serrate; rays 6—9, obtuse; fruit toothed. Ditches, S. 2-4f. Aug.-Oct.

20 C. aristòsa Mx. Sparingly pubescent; lvs. pinnately 5-9-parted, segm. lance-lin., incised; hds. small, rays large; outer scales 10—12, linear; awns slender, spreading, as long as the fruit. ② Low woods, W. 2—3f. Rays expanding 18″. Aug.—Oct.— Varies with the outer involucre leafy; and with the awns short, &c.

21 C. trichospérma Mx. Stem glabrous, square, dichotomous; lvs. pinnately 5-7-parted, segm. lanceolate, cut; rays entire, large; cyp. narrowly cuneate, with 2 short stout awns. ① Wet grounds, Mass. to Ill. (J.Wolf), and Car. 1-2f. Fls. showy. Jl. Aug.

22 C. discoìdea T. & G. Glabrous, much branched; leaves ternate, long-petiolate; lfts. lance-ovate, dentate; hds. small (2—3″); fr. linear-cuneate, the 2 stout awns (upwardly hispid) half as long and equalling the corolla. ① Wet, W. and S. 1-3f. Jl.-Sept.

23 C. bidentoìdes N. Glabrous, paniculate; lvs. simple, lanceolate, serrate; heads 7-1 y ; fr. lin.-oblong, the slender (up-hispid) awns longer than cor. ① Pa. Del.: rare.

59. BIDENS. L. BURR-MARIGOLD. Invol. double. Scales somewhat similar, or the outer foliaceous. Rays 4—8 (sometimes none), neutral; disk-flowers perfect. Recep. chaffy, flat. Pap. of 2—4 awns, rough backwards. Cypsela obcompressed, obscurely quadrangular. Leaves opposite, incised. Flowers yellow. July—October. (See Addenda.)

§ Cypsela linear-subulate, tapering to the top, 3-4-angled, 2-6-awned........Nos. 1—3
§ Cypsela oblanceolate, broader at the top, flat, 2-4-awned..................Nos. 4—7

1 B. leucántha Willd. Lvs. in 3—5 serrate lobes: hds. with 5 white rays. S. Fla. 1f.

2 B. bipinnàta L. *Spanish Needles.* Lvs. bipinnate, lfts. lanceolate, pinnatifid; rays very short, obovate, 3, 4, or 0; sc. all equal in length. ① Waste grounds, Ct. to Ill. 2-4f.

3 B. Beckii Torr. St. subsimple; submersed lvs. capillaceous-multifid, emersed lvs. lanceolate, connate, acutely serrate or cut; rays longer than the involucre. ♃ Slow waters, Vt. (rare), W. and N. Stem 2—3f. Heads solitary, terminal.

4 B. frondòsa L. *Beggar-ticks.* Rays 0; outer sc. leafy, 6 times longer than the fls.; lower leaves pinnate, ternate, upper lanceolate, serrate; awns 2. ① Fields: com. 2f.

5 B. connàta Willd. Rays 0; outer sc. leafy, longer than the head; lvs. lanceolate, serrate, subconnate at base, lower some trifid; awns 3. ① Swamps, E. and W. 1—3f.

6 B. cérnua L. Rays 0—4—8, small; hds. *cernuous;* outer scales as long as the disk; leaves all lanceolate, subconnate, dentate. ① Swamps, ditches, E. and W. 1—2f.

7 B. chrysanthemoìdes L. Lvs. oblong, attenuate at each end, connate at base, regularly serrate; rays thrice longer than the involucre. ① Ditches: common. 6′-2f.

60. SPILÁNTHUS, L. Invol. shorter than the disk, double, appressed. Recep. conical, chaffy, the pales embracing the flowers. Cyp. of the disk compressed, with 1—3 bristly awns or awnless, of the ray (when present) 3-angled. Herbs with acrid taste, opposite leaves, and solitary, yellow heads. Chiefly tropical. Aug.—Oct. (Acmella, Rich.)

1 S. repens Mx. Diffuse, rooting at the lower joints; lvs. lanceolate, subserrate, acute at each end, petiolate; rays about 12; fr. awnless, not ciliate. ♃ Wet, S. Car. to Fla.

2 S. Nuttállii T. & G. Ascending, diffuse; lvs. ovate to oblong, coarsely serrate abruptly petiolate; fruit ciliate on the margins; rays 10-12. Bogs, E. Fla. 1-2f.

61. VERBESÌNA, L. CROWN-BEARD. Rays ♀, few or none; disk ☿. Sc. in 2 or more series, imbricated, erect. Chaff concave or embracing the flowers. Achenia compressed, 2-awned. ♃ ♭ Leaves often decurrent serrate or lobed. Heads solitary or corymbous.

1 V. Siegisbeckii Mx. Stem 4-winged; lvs. opposite, ovate, serrate, acuminate, ♂

Order 70.—COMPOSITÆ.

veined, tapering to the winged petiole; hds. corymbous, yellow; rays 1-5; fr. wingless. ♃ Dry, W. and S. 5f. Aug. Sept.

2 V. Virgínica L. Stem narrowly winged; lvs. alternate, lance-ovate, subserrate, feather-veined, tapering to the sessile base; rays 3—4, white; fruit narrowly winged. ♃ Dry woods, Pa. to La. 4f. August.

3 V. sinuàta Ell. St. wingless, striate-angled; lvs. alternate, ovate, acuminate, contracted to a long slender base and petiole, irregularly repand-toothed or lobed; rays 3—5, white; fr. broadly winged. ♃ Sandy fields, S. 2-4f, with ample lvs. Sept.—Nov.

62. DYSÒDIA, Cav. FALSE DOG-FENNEL. Rays ♀, disk ☿. Invol.

a single series of partially united scales, usually calyculate. Cyp. elongated, 4-angled, compressed. Pap. scales chaffy, in 1 series, fimbriately and palmately cleft into bristles. ① With large, pellucid glands. Lvs. mostly opp., pinnately parted or toothed. Hds. paniculate or corymbous. Fls. yellow.

D. chrysanthemoìdes Lagasca. Smooth, much branched; lvs. pinnately-parted, lobes linear, toothed; hds. with few very short rays. Prairies and waysides, W., migrating E. 1f. An ill-scented plant. Aug. Sept.

63. GAILLÁRDIA, Foug. Rays neutral. Scales in 2 or 3 series, acute,

leafy, spreading, outer largest. Recep. convex, fimbrillate (naked in No. 1). Rays cuneiform, 3-cleft. Cyp. villous with long hairs from its base. Pappus of 6—10 long awns, which are membranous at base.—Leaves alternate, entire, often dotted. Heads on long, naked peduncles. May—Aug.

1 G. lanceolàta Mx. Lvs. lanceolate to linear; sc. as long as the dark purple disk; rays 8—10, small, yellow; *recep. naked.* ② Barrens, S. Car. to Fla. and Tex. 1-2f.
2 G. PICTA. Lvs. lanceolate; sc. hairy, longer than disk; rays 10-12, violet-purple with yellow teeth; recep. fimbrillate with slender awns. ① ♃ Dak. to Tex. 2f. Handsome.

64. GAZÀNIA, Gært. Rays neutral, disk-fls. ☿. Sc. in several rows,

united at base. Cyp. wingless, densely hairy. Pap. chaffy. Recep. alveolate. —From S. Africa. Hds. solitary, showy, on naked stalks. Rays tricolored.

G. SPECIÒSA. Trailing, half-shrubby; leaves oblong, entire or pinnatifid, smooth above, white-tomentous beneath; rays (1?) orange-yellow, each with an eye of white and chocolate at its base. Singularly beautiful.

65. PALAFÓXIA, Lagasca. Rays ♀ or 0. Sc. 8—15, scarious at tip,

shorter than the disk. Recep. flat, naked. Cyp. 4-angled, slender at base. Pap. of 6—12 membranous, denticulate, pointed scales. ♃ ♭ With scattered, narrow, entire lvs. and cyanic fls. in a corymb. (Polypteris, N.) Jl.-Sept.

P. Integrifòlia T. & G. Rough; lvs. lance-linear, 1-veined; rays none; pap. of 8—10 pointed scales with fringed edges. Barrens, Ga. and Fla. 3—5f. Heads purplish.

66. HYMENOPÁPPUS, L'Her. Fls. all ☿, tubular. Sc. 6—12, in 2

series, oval, obtuse, colored. Recep. small, naked. Anth. exserted. Cyp. broad at the summit, attenuate to the base. Pap. of many, short, obtuse, membranous scales in 1 series. ② ♃ Hoary-villous. Stem grooved and angled. Leaves alternate, pinnately divided.

H. scabiosæus L'Her. Leaf segm. linear-oblong; corymb simple; sc. obovate, white, greenish at base, longer than the disk; fr. pubescent. W. and S. 1—2f. Apr.—June.

67. HELÈNIUM, L. Rays ♀ or neutral, 3-5-cleft at the expanded

summit. Disk-fls. ☿. Invol. small, scales linear to filiform, reflexed. Recep. naked, convex to oblong. Cyp. angled, clavate or turbinate. Pap. of 5—12 silvery, thin scales.—Herbs with alternate, often decurrent leaves, punctate, resinous. Heads corymbous or solitary, showy, yellow.

§ HELLENIÁSTRUM. Rays pistillate. Pappus awned. Heads corymbed...(a)
  a Disk globular, its corollas 5-toothed. Pappus awned..................Nos. 1—3
  a Disk oblong, its corollas 4-toothed. Pappus scales obtuse............No. 4
§ LEPTÓPODA. Rays mostly neutral and fruitless...(b)
  b Heads corymbed, on short peduncles. Pappus awned. Disk globous......No. 5
  b Head solitary, on a long ped. Disk convex.—c Cypsela glabrous......Nos. 6, 7
                                                —c Cypsela hairy........Nos 8—10

1 **H. autumnàle** L. St. strongly winged; lvs. lanceolate, serrate, decurrent, heads loosely corymbed. ♃ Damp. 2-3f. Hds. large, with drooping rays. Sept. Very bitter.
2 **H. parviflòrum** N. St. scarcely winged; lvs. lanceolate, subentire, slightly decurrent; sc. filiform, shorter than the globular disk; hds. small, few. Ga. (Nuttall.) Scarce.
3 **H. tenuifòlium** N. St. and numerous fastigiate branches wingless; lvs. crowded, linear or filiform, fascicled; sc. subulate. ③ Fields, Ga. to La. 1-2f. Rays spread 10″.
4 **H. quadridentàtum** Lab. Much branched, strongly winged; lvs. oblong, some lobed or toothed; disk oblong, longer than the rays. Swamps, S-W. 1-3f. June-Aug.
5 **H. Brachýpoda**. St. strongly winged, branches few, corymbous, 1-headed; hds. small (4″), rays 8-12, short (3-4″); disk brown-purp., globular. Damp, Ill. to Ga. 1-2f.
6 **H. Leptópoda**. Smooth; st. simple, clustered, naked above; lvs. lanceolate to oblong-linear, some decurrent; rays 20—30, spreading 1½′; disk convex. Moist soils, S. Car. to Fla. 2f. March, April.
7 **H. incìsum**. Smooth; lvs. lanceolate, sessile, not decurrent, sinuate-pinnatifid or incised; rays about 40, in 2 or 3 rows; fruit glabrous. Low barrens, Ga., and W. 2f.
8 **H. pubérulum**. Downy; sts. much clustered; lvs. lance-linear, sessile, not decurrent; rays 20-30, broad, spreading 1½-2′; fr. hairy. Wet pine-barrens, S. 2f. Ap.,May.
9 **H. brevifòlium**. Pubescent above, single, often some branched; lvs. lance-obl. to linear, obtuse, the radical spatulate, cauline subdecurrent. Wet. S. 2f. May, June.
10 **H. fimbriàtum**. Smooth; often branched; leaves lance-linear, subentire, acute, decurrent; pap. scales deeply cleft into a fringe of bristles. Barrens, Fla. 1-2f. Apr. +

68. **BALDWÍNIA**, N. Invol. scales closely imbricated in 2—4 rows. Recep. convex, deeply honeycombed, with horny walls. Rays 8—20, neutral, in 1 row, 3-toothed. Disk ☿. Cypsela silky-villous, immersed in the cells. Pappus of 9—12 oblong scales. ♃ Simple or corymbed. Leaves alternate, linear, punctate. Heads yellow. July—Sept.

  B. uniflòra N. St. simple, puberulent, with 1 large head; rays about 20; lvs. below linear-spatulate; pap. scales 9. Swamps, Va. and S. 1—2f. Rays spreading 2′.
  ↙ **B. multiflòra** N. Glabrous, corymbously branched; rays about 10; lvs. crowded, narrow-linear; fruit truncate and ray-marked at summit, crowned with 12 obovate scales. Sand hills, Ga. Fla. 1—3f. Rays 1½′. (Actinospermum, T. & G.)

69. **MARSHÁLLIA**, Schreb. FALSE SCABISH. Invol. scales lance-linear, subequal, erect, in 1 or 2 rows. Recep. convex, with linear, rigid pales. Fls. all tubular, ☿. Cor. lobes slender, spreading. Cyp. 5-angled. Pappus of 5 or 6 membranous, awned scales. ♃ Simple or branched, with alternate, entire, 3-veined leaves, and solitary, long-stalked heads of purplish flowers, resembling a Scabish. Ornamental.

1 **M. latifòlia** Ph. St. simple, leafy; lvs. ovate-lanceolate, acuminate, sessile; scales

rigid, acute; pales narrowly linear; pappus triangular-acuminate. Dry soils, Va. to Ala. 1f. Stem purple, smooth. Corollas 6—7″, slender. May, June.
2 M. lanceolàta Ph. Stem simple, leafy below; leaves oblanceolate to lanceolate, mostly obtuse and petiolate; scales obtuse. Uplands, S. 1—2f. April—June.
3 M. angustifòlia Ph. Mostly branching, leafy; lvs. narrow-lanceolate to linear, all acute; scales acute. Swamps, S. 1f. Very handsome. July, Aug.

70. ÁNTHEMIS, L. Снамоміle, &c. Invol. hemispherical, with subequal, small imbricated scales. Rays numerous, generally ⚥. Recep. chaffy (at least at summit), convex or conical. Disk-flowers ⚥. Cypsela ribbed, smooth, linear or clavate. Pappus a slight border, or 0.—Herbs with 1-3-pinnatifid leaves, usually strong-scented. (Rays white.)

§ Chamæmelum. Rays pistillate. Cyp. teretish. Lvs. mostly alternate.... Nos. 1, 2
§ Marùta. Rays neutral. Cypselæ clubshaped or obovoid. Lvs. alternate....No. 3
1 A. arvénsis L. *Corn C*. St. erect, bushy, whitish-downy; lvs. bipinnatifid, segm. lance-lin.; branches naked above, 1-headed; pales cuspidate, longer than the flowers. ① Fields: not common. Resembles Mayweed, but inodorous. 8—15′. § Eur. July.
2 A. nóbilis L. *Garden C*. St. prostrate, branched from base, woolly; lvs. hairy, decompound-pinnatifid, seg. lin.-subulate; pales some shorter than the fls. ⚇ Gardens, rarely in fields. Aromatic. § Eur.—Var. with fls. double (florets all radiate). Jl.-Sept
3 A. Cótula L. *Mayweed*. Nearly smooth, erect, bushy; lvs. bipinnatifid, seg. linear subulate; pales bristly, shorter than the flowers. ① Waysides: com. 1f. Hds. terminal, corymbed, disk yellow, rays white, showy. Ill-scented. Jn.-Sept. (Maruta, DC.)

71. ACHILLÆA, L. Millfoil. Yarrow. Invol. ovoid, of unequal, imbricated scales. Rays 5—10, short, ♀. Recep. flat, chaffy. Cyp. without a pappus. ⚇ Leaves much divided, alternate. Heads small, corymbous.

1 A. Millefòlium L. Lvs. bipinnatifid, with linear, dentate, mucronate segments; stem furrowed, corymbed at top; sc. oblong, rays 4—5, short. Fields, waste grounds, everywhere. 1—2f. June—Sept.—A variety with rose-purple flowers, is very pretty.
2 A. ptármica L. *Sneezewort*. Leaves linear, acuminate, sharply serrate, smooth; hds. loosely corymbed; rays 8—12, longer than invol. (double in cult.) Rare. 15′. §

72. LEUCÁNTHEMUM, Tourn. Whiteweed. Invol. broad, depressed, imbricated. Rays ♀, numerous. Recep. flat, naked. Cyp. striate, without pappus. ⚇ Lvs. alternate. Hds. solitary, disk yellow, rays white.

1 L. vulgàre Lam. St. simple or branched; cauline lvs. clasping, few, lance-oblong, obtuse, cut-pinnatifid at base; scales brown at the edge. Too common in our fields and pastures. 2f. Rays spreading 1½′. July—Sept. § Europe. [N. Y. (Gerard.)
β. *tubuliförme* (Tenney). Ray-fls. tubular, very slender, 5-3-lobed. Po'keepsie.
2 L. Parthènium Godron. *Feverfew*. Branched; lvs. petiolate, 2-3-pinnate, segm. ovate, cut; hds. corymbed. Gardens, rarely in fields. 2f. Often double. (Matricaria, C-B.)

73. MATRICÀRIA, Tourn. Mother-Carey. Invol. scales imbricated, with scarious margins. Recep. conical or convex, naked. Rays ♀ or 0. Pap. a membranous border on the cyp., or 0.—Herbs with alternate leaves.

1 M. discoìdea DC. Hds. discoid, few, terminal; lvs. sessile, 2-3-pinnately-parted, lobes small, linear-oblong, acute; sc. oval, obtuse, white-edged, much shorter than the conical disk. Ill. and W. Common in Cal. 3-8′. Disk 3″ broad and high. Jl.—Sept.
2 M. Tanacètum. *English Mint*. Downy; leaves oval, serrate, lower petiolate; heads small, corymbed, discoid. Europe. 1—2f. Aromatic. Jl. Aug. (M. Balsamita C-B.)

74. TANACÈTUM, L. Tansy. Invol. hemispherical, imbricate, the

scales all minute. Recep. convex, naked. Pap. a slight membranous border. Cyp. with a large, epigynous disk.—Lvs. alternate, much dissected. Flowers yellow, discoid.

1 T. vulgàre L. Lvs. pinnatifid, segm. oblong-lanceolate, pinnatifid and cut-serrate; hds. fastigiate-corymbous, ray-fls. terete, tubular, 3-toothed. ♃ Waysides. 2–3f. Aug.

2 T. Huronénse Nutt. Lvs. bipinnatifid, lobes oblong, often again pinnatifid; heads large, corymbed; ray-fls. flattened, unequally 3–5-cleft. ♃ Sandy shores, W. 2–3f.

75. CHRYSÁNTHEMUM, L.

Invol. bell-shaped, sc. imbricated, scarious at the edges. Recep. flat or convex, naked in the disk. Rays ♀, disk-fls. ☿, 5-toothed. Cyp. angular or compressed. Pap. 0 or tooth-like.—Plants ornamental, from E. Asia, with alternate, lobed lvs. and large rays. Fig. 387.

§ PYRÈTHRUM. Cypselæ wingless, angular, all alike. Plants perennial.....Nos. 1–3
§ CHRYSÁNTHEMUM. Cyp. of the ray 3-angled, of the disk compressed. ①.....Nos. 4, 5

1 C. SINÉNSE. Shrubby; lvs. sinuate-pinnatifid, thick, glaucous; rays much longer than the obtuse scales. Beautiful flowers of all colors, late in Autumn. 2–3f.

2 C. INDICUM. Shrubby; leaves incisely-pinnatifid, thin, flaccid; rays little longer than the obtuse scales, spreading about 1'. Heads much smaller than in No. 1.

3 C. RÒSEUM. Perennial, glabrous; lvs. 2-3-pinnatisect; hds. solitary, terminal; scales brown-edged; rays rose-colored or white, often double. ♃ Heads 1' broad.

4 C. CORONÀRIUM. Annual; lvs. clasping, bipinnatifid, lobes dilated at summit; flowers large, terminal, yellow; pappus none. S. Europe. 3f. Varieties double, &c. Aug.

5 C. CARINÀTUM. *Tricolored C.* Annual; lvs. thick, bipinnatifid; scales carinate; rays white, yellow at base, disk purple. Barbary. 1–2f. Flowers all Summer.

76. ARTEMÍSIA, L. WORMWOOD, &c.

Invol. ovoid, imbricate, with dry, connivent scales. Recep. without pales. Disk-fls. numerous, ☿, tubular; ray-fls. few, often without stamens and with a subulate corolla or none. Cypsela with a small disk. Pappus 0.—Bitter herbs. Leaves alternate. Heads yellow or purplish, discoid. Aug., Sept.

§ ABSÍNTHIUM. Recep. villous or hairy. Fls. all fertile, heterogamous......Nos. 1,
§ ABRÓTANUM. Recep. naked. Fls. all fertile.—*a* Lvs. or segm. lanceolate..Nos. 3, 4
 —*a* Lvs. or segm. linear......Nos. 5—7
§ DRACÚNCULUS. Recep. naked. Disk-fls. sterile.—*b* Lvs. trifid or entire...Nos. 8, 9
 —*b* Lvs. pinnatisect.....Nos. 10—12

1 A. Absínthium L. *Common W.* Leaves multifid, clothed with short silky down both sides; seg. lanceolate; hds. hemispherical, drooping. Waysides, N. 1-2f. § Eur.

2 A. frígida Willd. Lvs. silky canescent, the cauline pinnatifid; seg. linear, 3-5-cleft; hds. small, glob., drooping; inner sc. woolly. Rocky hills, Minn. Dak., and W. 6-12'.

3 A. Ludoviciàna N. Canescent-tomentous; leaves lanceolate, the lower serrate or pinnatifid, upper entire; heads ovoid, in a slender, leafy panicle. ♃ Shores, Mich. and W. 2—5f. Heads small and crowded.

4 A. vulgàris L. *Mugwort.* Lvs. canescent-tomentous beneath, pinnatifid with lanceolate segments, upper entire; heads erect, ovoid, subsessile, in a branched panicle. Waysides, N. and W. 3f. § Europe.

5 A. ABRÓTANUM. *Southernwood.* Hoary; leaves bipinnatisect; heads hemispherical, nodding, downy. From S. Europe. 3f. [ding. Eur. 3f.

6 A. PÓNTICA. *Roman W.* Lvs. hoary beneath, 2-3-pinnatisect; heads globular, nod-

7 A. biénnis Willd. Erect, glabrous, simple; lvs. 1-2-pinnatifid, lobes sharply serrate or cut, those above subentire; hds. globular, erect, spicate, in a virgate, leafy panicle. ② Common westward, migrating E. to Po'keepsie (Gerard), and to Pa. 1–3f. Aug ⁓.

ORDER 70.—COMPOSITÆ. 185

8 A. DRACÚNCULUS. *Tarragon.* Glabrous; lvs. lin.-lanceolate, lower trifid; heads globous. From Siberia. 3f. A garden salad. Give a rich fragrance to vinegar.
9 A. dracunculoìdes Ph. Canescent when young, branched; lvs. lin.-filiform, the radical often trifid; hds. small, globular; inner scales roundish, outer oblong. ♃ N-W.
10 A. boreàlis Pal. Tufted, silky-villous, simple; lower lvs. petiolate, lance-linear, entire at base, ternately, pinnately, or bipinnately parted at apex with lin. lobes; hds. hemispherical; scales colored. ♃ Shores of Lake Superior, N. and W. 6–12′.
11 A. Canadénsis Mx. *Sea W.* Glabrous (mostly); lvs. 1-2-pinnatisect with linear seg.; hds. roundish, sessile, in a pan. of glomerules. ♃ Lake shores, N. 2–4f. Hds. 1″.
12 A. caudàta Mx. Glabrous, simple, densely paniculate; lvs. 3-2-1-pinnatisect with alternate, filiform segm.; heads globous, pedicellate, erect. ⓢ Coast, N. H. to Ga. 4f.

77. SOLÌVA, R. & P. Invol. of 5—15 scales in 1 row. Recep. flat, naked. Fertile fls. in several rows, apetalous; ♀ fls. few, interior, with a 3–5-toothed corolla. Cyp. obcompressed, tipped with the persistent style and no pappus.—Little matted herbs with pinnatifid lvs. and sessile heads.

S. nasturtiifòlia DC. Lf. lobes 5—9, oblong, obtuse; sc. 10—15; fr. obconic, rugous, crowned with a dense tuft of wool instead of a pappus. Sandy shores, S. 1—3′.

78. GNAPHÀLIUM, L. CUDWEED. EVERLASTING. Heads discoid, heterogamous. Invol. imbricate with scarious, colored scales. Marginal fls. ♀, subulate, mostly in several rows; central fls. ☿. Recep. flat, naked. Pappus a single row of scabrous, hair-like bristles.—Herbs generally clothed with whitish wool. Leaves alternate, entire.

* Heads in terminal corymbous clusters. August. . . . . .Nos. 1–3
* Heads in axillary, somewhat spicate clusters. . . . . Nos. 4, 5

1 G. decúrrens Ives. Lvs. decurrent, linear-lanceolate, very acute, naked above, white and woolly beneath; fls. in dense, roundish, terminal clusters. ♃ Hilly pastures, N. Eng. to Penn. and Mich. 2f. Lvs. green above. Fls. yellow, scales white.
2 G. polycéphalum Mx. Woolly; lvs. sessile, linear-lanceolate, acute, scabrous above; hds. capitate, corymbous; sc. ovate-lanceolate, acute. ① Dry. 1-2f. Fragrant.
3 G. uliginòsum L. *Cudweed.* St. diffusely branched, woolly; lvs. sessile, linear-lanceolate; hds. small (1″ wide), in terminal, crowded, leafy clusters; scales obtuse, yellowish or brownish; fruit smooth. ① Moist hollows, N. M. W. 4—6′.
4 G. purpùreum L. Erect; lvs. linear- or obovate-spatulate, canescent beneath, green above; hds. sessile, crowded; sc. acuminate, purplish. ① Dry fields. 8—12′. June.
5 G. supìnum Villars. Cœspitous, woolly; lvs. linear; hds. few, oblong, in a spicate raceme or solitary; scales acute, brown. White Mountains. 2—4′. Rare.
6 G. FŒTIDUM, from S. Africa, has yellow heads, entire, clasping leaves. 2f. Hardy.

79. ANTENNÀRIA, Br. EVERLASTING. Heads ♀ ♂. Invol. of imbricate, colored scales. ♀ Cor. filiform. Recep. subconvex, alveolate. Pap. a single row of bristles. ♃ Tomentous. Lvs. alternate, entire. Hds. corymbous, with white or brownish, never yellow scales.

1 A. margaritàcea Br. Woolly-white, erect, corymbed above; lvs. lin.-lanceolate, 3-veined; scales elliptic, obtuse, pearl-white, fadeless. Fields. 1—2f. July.
2 A. plantaginifòlia Br. *Mouse-ear E.* Simple, with running stolons; leaves oval to spatulate, the cauline small, bract-like; hds. in a close terminal cluster, purplish, all ♀ in some plants, all ♂ in others, in early Spring. Borders of woods. 5—8 ′.

80. FILÀGO, Tourn. COTTON ROSE. Heads heterogamous. Recep. columnar, naked at top, chaffy below, with pales resembling the scales, each with a ♀ fl. in its axil. Cyp. terete, the central with a hairy pappus. —Herbs canescent downy. July, Aug. §.

F. Germánica L. Lvs. erect, crowded, linear-lanceolate; hds. in capitate clusters, which are successively proliferous; scales cuspid., straw-color. ① Fields, E. 6—10'.

81. **AMMOBIUM**, Br. Invol. imbricated, sc. with broad, scarious, spreading tips. Recep. broad-conic, chaffy. Fls. all tubular, ☿. ·*Cyp. 4-angled, 4-toothed. ♃ Australia. Stems winged with the decurrent leaves.

A. ALÀTUM. ① In gardens. 1—2f. Villous-canescent. Root lvs. oblong-petioled. Involucre white, flowers yellow. Summer.

82. **RHODÁNTHE**, Lindl. Involucre top-shaped, imbricate, sc. dry, ovate, acute, the inner radiate-spreading. Recep. naked. Fls. all tubular, 5-toothed, ☿. Cyp. woolly. Pap. of plumous bristles. ① Australia.

R. MANGLÉSII. Lvs. oblong, clasping, entire; hds. large. fadeless, rose-colored, variegated. A splendid "Everlasting," with many beautiful varieties. Hds. 1—2' diam.

83. **HELICHRYSUM**, Cass. IMMORTAL FLOWER. Invol. imbricate, with scarious, colored scales. Recep. flat, naked of pales. Pap. a row of bristles, often cohering.—Herbs or shrubs, chiefly S. African. Lvs. alternate. A vast genus of 200 species.

1 **H. BRACTEÀTUM.** Branching, puberulent; lvs. lanceolate to linear, repand, acuminate; hds. terminal, bracted at base; outer scales brownish, the inner radiant, ylw. to wh.
2 **H. MACRÁNTHUM.** Subsimple, scabrous; lvs. spatulate to lance-oblong, obtuse, en tire; hd. 1 or few, large, white outside, roseate within; inner scales radiant. ①—♃.
  β. COMPÓSITUM. Hds. composite (or double), purple. carmine, yellow, white.
  γ. ATROSANGUÍNEUM. Hds. composite, with deep crimson scales and pappus. 18'.

84. **XERÁNTHEMUM**, Tourn. Hds. discoid, heterogamous. Invol. hemispherical, imbricated, dry, with radiant, colored scales. Recep. with 3-toothed, dry pales. Pap. chaffy-bristly. ① S. Eur. Lvs. entire. Hds. white or rose-colored.

X. RADIÀTUM. *Eternal Flower.* Erect, branched. Lvs. linear-oblong; hds. 1–2' diam.

85. **ERÉCHTITES**, Raf. FIRE-WEED. Fls. all tubular, those of the margin ♀, of the disk ☿. Invol. cylindrical, simple, slightly calyculate. Recep. naked. Pap. of numerous, fine, capillary bristles. ① Lvs. simple, alternate. Fls. corymbous, whitish. A rank weed.

E. hieracifòlius Raf. St. virgate, paniculate; lvs. oblong, acute, clasping. unequally and deeply cut-toothed; invol. smooth; fr. hairy. Burnt grounds, &c. 3f. Aug.+.

86. **CACÀLIA**, L. TASSEL-FLOWER. Fls. all tubular, ☿. Involucre cylindric, oblong, in one series, often calyculate with small scales at the base. Recep. not chaffy. Pap. capillary, scabrous. ①♃. Smooth. Lvs. alternate. Heads of flowers corymbed, mostly cyanic.

§ Scales of the invol. cohering, about 12. Flowers 60—80. scarlet............ ....No. 8
§ Scales of the invol. distinct,—a about 12. Flowers 20—30. white...............No. 1
  —a 5 only. Fls. 5.—b Lvs. cordate or lobed....Nos. 2—4
  —b Lvs. never cordate.......Nos. 5—7

ORDER 70.—COMPOSITÆ.

1 **C. suavèolens** L. Glabrous; st. striate-angular; lvs. on winged petioles, hastate-sagittate, dentate, green on both sides; fls. white. ♃ Ct., W. and S.: rare. 4—5f. Aug.
2 **C. reniförmis** Muhl. St. sulcate-angled; lvs. palmately-veined, nearly smooth, green, petiolate, lower reniform, upper flabelliform. ♃ Woods, Ill. to Car. 3—6f. Jl.
3 **C. atriplicifòlia** L. St. terete; lvs. petiolate, smooth, glaucous beneath, palmate-veined, angularly-lobed and dentate, the lower subcordate. N. Y., S. and W. 3—5f. Jl.
4 **C. diversifòlia** T. & G. Not glaucous; st. striate-angled; lower lvs. cordate-ovate, obtuse, repand-dentate, upper 3-5-lobed, subhastate. ♃ Swamps, Fla. 2—3f. May +.
5 **C. tuberòsa** N. St. angular-sulcate; lvs. oval or ovate, strongly 5-7-veined, not glaucous, petiolate, lower petioles very long. ♃ Swamps, W. and S. 2—5f. May—Jl.
6 **C. ovàta** Ell. St. terete; lvs. glaucous beneath, 3-5-veined, ovate and oval, entire or undulate-margined, contracted at base into petioles. ♃ Swamps, S. 3—4f. July +.
7 **C. lanceolàta** N. St. terete; lvs. 3-veined, glaucous beneath, lanceolate to lance-linear, the lower tapering to petioles, upper sessile; corymb simple. ♃ Ga. Fla. 5f.
8 **C. coccínea.** *Tassel-flower.* Root leaves ovate-spatulate, cauline clasping-auriculate; invol. much shorter than the scarlet fls., finally reflexed. E. Ind. 1—2f. June—Sept.

87. **SENECIO**, L. GROUNDSEL. Invol. of many equal scales, or invested with a few shorter ones at base. Fls. all tubular, ☿, or usually radiate and rays ♀. Recep. not chaffy. Pap. simple, capillary and copious. —A vast genus of herbs and shrubs. Lvs. alternate. Fls. mostly yellow, exceeding the invol. Fig. 160.

§ Rays none. Root annual. (A perennial climber, No. 11.)....................No. 1
§ Rays yellow.—*a* Radical leaves undivided. Achenia glabrous............Nos. 2, 8
—*a* Radical leaves undivided. Achenia pubescent............Nos. 4, 5
—*a* Radical leaves divided, as well as the cauline..............Nos. 6, 7
§ Rays purple, &c. Species of Cineraria, L. &c. in the greenhouse........Nos. 8—10

1 **S. vulgàris** L. St. paniculate, erect, angular; lvs. sinuate-pinnatifid, dentate, am plexicaul. ① A weed in gardens, &c. 1f. 18′. Flowers all Summer.
2 **S. aùreus** L. Radical lvs. ovate, cordate, crenate-serrate, petiolate, cauline ones lyrate-pinnatifid, dentate, terminal segments lanceolate; ped. subumbellate, thick; rays 8—12; fr. glabrous. ♃ Woods, meadows. 1—2f. Rays spread 1′. May—Aug.
β. *Balsdmitæ.* Pubescent; lvs. few, small, the radical lance-oblong. Rocks.
γ. *gracilis.* Root lvs. roundish, on long petioles, cauline linear-oblong, dentate.
δ. *obovàtus.* Root leaves obovate to oblong-spatulate; peduncles long.
ε. *lanceolàtus.* Lvs. lanceolate, the cauline pinnatifid at base. Vt. Rare.
ζ. *discoideus* (Porter). Rays none; lvs. obov.-spatulate, cauline pinnatifid. Penn.
3 **S. obovàtus** Ell. Tomentous, becoming glabrous; root lvs. obovate or roundish, crenate, with an attenuated, sessile base; cauline few, small, cut-pinnate; corymb small, rays 10—12, spreading 1′. ♃ Va. to Fla. 1f. Stem nearly leafless. May.
4 **S. tomentòsus** Mx. Cottony-tomentous; st. lvs. obovate to oblanceolate, obtuse, long-petioled, crenate, upper sessile or none; rays 12—15, spreading 16″. ♃ Va. & S.
5 **S. anónymus** Wood. Cottony-tomentous; root lvs. oblong, obtuse, creately toothed or lobed, cauline pinnatifid, the lobes dentate; hds. small, rays 6—9, spreading 6″. ♃ Thickets, Ala. (Montgomery). 2f. Corymbs compound. May, June.
6 **S. Canadénsis** L. Lvs. glabrous, bipinnatifid; seg. lobed, obtuse, the few upper pinnatifid; corymbs compound; rays 9—12. ♃ Canada (Kalm), Mts. N. Car.
7 **S. lobàtus** Pers. *Butterweed.* Glabrous; leaves all pinnatifid, the lower lyrately, lobes crenate; invol. subcalyculate; rays 10—12. ① Wet. S.; com. 2—3f. Mar. Apr.
8 **S. ÉLEGANS.** *Purple Jacobæa.* Lvs. pinnatifid, hairy, viscid; scales scarious at tip, calyculate with an outer row of short green ones. ① S. Afr. Purp., varying to white.
9 **S. LANÀTA.** Lvs. roundish, angular, cordate, woolly beneath; rays vivid purple in side, wh. outside; disk white or blue. ♃ Canaries. 3f. Shrubby.—Many var'eties.

10 S. cruéntus. Lvs. angular, cordate, cut-toothed, purple beneath, the petioles wing-ed, ear-shaped at the base; heads in a broad corymb, crimson, purple, blue, white. ♃ Canaries. A common handsome greenhouse plant.

11 S. scándens. *German Ivy.* Climbing and twining; leaves smooth, roundish-cordate, 5–7 angled or lobed; corymbs axillary, of small rayless yellow heads. ♃ S. Africa. Blooms freely in California, rarely in our greenhouses.

88. ÁRNICA, L. Involucre of equal, lanceolate scales, 1- or 2-rowed. Ray fls. ♀, disk ☿. Receptacle flat, with scattered hairs. Pap. single, rigid, and serrulate. ♃ Stem simple. Leaves opposite. Flowers yellow.

1 A. mollis Hook. Pubescent; stem leafy; lvs. becoming nearly glabrous, dentate, lance-oblong, radical ones petiolate; hds. few; fr. hairy. Mts. &c. N. H., N. Y. July.

2 A. nudicaùlis Ell. Hairy; st. nearly naked; lvs. all sessile, ovate, subentire, tho cauline bract-like; heads large, rays 12, spreading 2'; fruit glabrous. Wet sands, Va. to Fla. 1f. April, May.

89. RUGÈLIA, Shutt. Invol. as in Arnica. Fls. all tubular, ☿. Recep. convex, naked. Cyp terete, striate. Pap. of rough bristles. ♃ Lvs alternate. Heads large.

R. nudicaùlis Shutt. St. simple, erect; branches 1-flwd.; root lvs. ample, ovate, narrowed to long winged petioles; stem lvs. small, subsessile. Mts., Tenn. 1f.

90. CÝNARA, L. Heads discoid, homogamous. Invol. dilated, imbricate, scales fleshy, emarginate, pointed. Receptacle fibrillate. Pap. plumous. Cypselæ not beaked. ♃ Spiny. Leaves not decurrent.

1 C. Scólymus. *Garden Artichoke.* Leaves subspinous, pinnate, and undivided; invol. scales ovate. Gardens. The heads are used as asparagus. Coarse plants.

2 C. cardúnculus. *Cardoon.* Leaves spiny, all pinnatifid; invol. scales ovate. S. Eur. The petioles, blanched by culture, are used as celery.

91. TAGÈTES, L. Marigold. Heads heterogamous. Invol. simple, tubular, of 5—10 united scales. Ray-fls. 5, persistent. Receptacle naked. Pap. of 5 erect awns. ① Tropical America. Leaves pinnately divided.

1 T. pátula. *French Marigold.* Stem erect, with widely-spreading, 1-headed branches, lf. segm. linear-lanceolate; ped. long; invol. terete. Yel. and dark purp. Handsome.

2 T. erécta. *African Marigold.* Stem stout, erect; lf. segm. lanceolate; ped. 1-flwd., thickened at top; involucre angular. Yellow and orange.

3 T. flórida. Erect, corymbously branched; lvs. lanceolate, opposite, aristate-serrate; rays mostly 3, large, yellow. Mexico. 18'.

92. CALÉNDULA, L. Pot Marigold. Heads radiate. Invol. of many equal leaves, in about 2 series. Rays ♀, disk ♂. Receptacle naked. Cyp. of the disk membranaceous. Pap. 0. ① Oriental. Lvs. alternate.

C. officinàlis. Viscid-pubescent; stem branched; lvs. oblong, acute, mucronate, sessile; hds. terminal, solitary; large, brilliant, orange, lemon, double, &c. June—Sep.

93. CENTAÙREA, L. Knap-weed. Bachelor's-button. Hds. discoid. Invol. imbricate. Fls. all tubular, the marginal often enlarged, ray-like, neutral. Pappus filiform, scale-form, or 0. ①♃ Lvs. alternate.

* Scales of the involucre with a fringed or pectinate appendage............Nos. 1, 2
* Scales not appendaged,—*a* merely ciliate or spinescent..................Nos. 3, 4
    —*a* nor ciliate nor spinescent (Amberboa).....   Nos. 5, 6

ORDER 70.—COMPOSITÆ. 189

1 **C. Americàna** N. Erect, sparingly branched; leaves sessile, glabrous, repand-toothed, ovate-oblong to lanceolate; hds. few, very large, with the marginal fls. much enlarged, pale-purple. ① Ark. La. and § in Ill. 2—4f. Appendages straw-yellow.
2 **C. nigra** L. Erect, branched, pubescent; lvs. angular-lyrate to lanceolate, dentate; sc. ovate; marg. fls. not enlarged, all pnrp. ♃ Fields. Append. dark brown. § Eur.
3 **C. Cýanus** L. *Bachelor's-button.* Erect, branched, downy; lvs. linear; sc. ciliate-serrate; outer fls. much enlarged. ① Fields, gardens. Purple, blue, white.
4 **C. Calcítrapa** L. *Star Thistle.* Hairy, diffusely branched; lvs. pinnately lobed, lobes lin.; scales tipped with spreading spines. ② Pa. to N. Car. Purple. § Eur.
5 **C. moschàta**. Lvs. lyrate, dentate; invol. subglobous, smooth; sc. ovate; ray-fls. scarcely enlarged; pap. 0. ① Persia. Purple, varying to white. July—October.
6 **C. suavèolens**. *Yellow Sweet Sultan.* Lvs. oblong, toothed, the upper pinnatifid at base; ray fls. much enlarged, yellow; pap. chaff-like. ① Levant. 1—2f. July—Sept.

94. CÁRTHAMUS, L. SAFFRON. Hds. discoid. Invol. imbricated, outer bracts foliaceous. Fls. all tubular and ☿, filaments smooth. Pap. 0. Receptacle with setaceous pales. Cypselæ 4-angled.—Oriental herbs.

C. tinctòrius. St. smooth; leaves ovate-lanceolate, sessile, spinous-denticulate, half-clasping. ① Egypt. Heads large, with long, slender, orange-colored flowers. July.

95. CNÍCUS, Vaill. BLESSED THISTLE. Heads discoid. Invol. ventricous, imbricate with doubly spinous scales. Ray-fls. sterile. Receptacle very hairy. Pappus in 3 series, the outer 10-toothed, the 2 inner each 10-bristled.—Oriental herbs.

C. benedíctus L. Lvs. somewhat decurrent; dentate and spiny; invol. doubly spinous, woolly, bracteate. Fields, &c.: rare. 2f. Heads large, yellow. §

96. ONOPÓRDON, Vaill. COTTON THISTLE. Heads discoid, homogamous. Involucre ventricous, imbricate with spreading, spinous scales. Recep. deeply alveolate. Pappus copious, capillary, scabrous. Cypselæ 4-angled.—Large, branching herbs, with decurrent leaves.

O. acánthium L. Plant cottony-white; involucre scales spreading, subulate; leaves ovate-oblong, sinuate, spinous. ② Waste grounds: rare. 3f. Fls. purp. Jl., Aug.

97. CÍRSIUM, Tourn. THISTLE. Hds. discoid, homogamous. Invol. subglobous, of many rows of spinous-pointed, imbricated scales. Recep. bristly. Style scarcely divided. Pap. copious, plumous. Cyp. compressed, smooth.—Herbs with alternate lvs., generally armed with spinous prickles. Flowers in Summer. Figs. 178, 345.

* Leaves decurrent on the stem more or less, floccous-woolly beneath..........Nos. 1, 2
* Leaves not decurrent,—*a* white-tomentous both sides. Plants low, stout...Nos. 3, 4
 —*a* white tomentous beneath only. Plants slender..Nos. 5—7
 —*a* green oth sides.—*b* Hds. leafy-bracted at base...Nos. 8, 9
 —*b* Hds. naked, few, large (1′). Nos. 10, 11
 —*b* Hds. naked, many, small........No. 12

1 **C. lanceolàtum** Scop. *Common Thistle.* Lvs. decurrent, pinnatifid, hispid, the segments divaricate and spinous; hds. several, ovoid, villous; scales lanceolate, tipped with a spine, spreading. ② N. and M.: common. 3—tf. Heads purple.
2 **C. Lecóntii** T. & G. Slender, subsimple, with few hds.; lvs. lin.-lanceolate, more or less decurrent, hoary beneath, teeth few, spinous; scales not spinous. cuspidate heads large (1′ diameter), purple. Swamps, Ga. Fla. to La. 2f.

ORDER 70.—COMPOSITÆ.

3 **C. Pítcheri** T. & G White-tomentous; lvs. pinnatifid, segm. linear, spinous, margins revolute; scales spine-pointed; flowers ochroleucous. ♃ Lake shores, W. June, July.
4 **C. undulàtum** Spr. White-tomentous; lvs. lance-oblong, sinuate-pinnatifid, wavy, prickly; scales scarcely prickly; flowers purple. ③ Mich., and N. 1—2f.
5 **C. díscolor** Spr. Slender, much branched; lvs. pinnatifid, segm. 2-lobed, divaricate, spinous; scales ovate, tipped with a spreading spine. ② N. 3—5f. July+.
6 **C. altíssimum** Spr. Tall, branching, villous, leafy to the top; lvs. lance-oblong, often sinuate-dentate, or pinnatifid, spinescent; scales lance-ovate, the outer with a spreading spine. Fields, M. and W. 3—8f. Purple. August.
7 **C. Virginiànum** Mx. Slender, subsimple, naked above; lvs. lanceolate, margins revolute, spinescent, lobed or dentate, white-downy beneath; heads small (6″); scales bristle-tipped. Woods, W. and S. 3—4f. Purple. April—Sept.
8 **C. horrídulum** Mx. Cottony when young; leaves cut-pinnatifid, spinous; heads large, invested by a whorl of very spiny bracts; scales sharp-pointed. ③ Uplands, N. Eng. to Fla. Flowers purple or cream-color. 1—3f. April—August.
9 **C. pùmilum** Spr. Hairy; lvs. few above, green, clasping, lance-oblong, pinnatifid, segm. lobed, spinous; heads few, very large, subtended by 1—5 bracts; invol. round-ovate, spinous. ③ Pastures, waysides, N. Eng. to Pa., and W.: com. 1—2f, stout. Flowers purple, fragrant. July, August.
10 **C. mùticum** Mx. Lvs. pinnatifid; heads on naked peduncles, bractless; invol. unarmed, with webbed and glutinous scales. ③ Damp. 3—7f. Hds. 1′. Aug., Sept.
β. **glabrum.** Nearly glabrous; lvs. lance-lin., lobed; scales with minute spines. S.
11 **C. repándum** Mx. Lvs. crowded to top, at length green both sides, clasping, lin.-oblong, wavy, spinous-ciliate; hds. 1 or 2; inner scales subulate. Barrens, S. 1—2f.
12 **C. arvénse** Scop. *Canada Thistle.* Lvs. sinuate-pinnatifid, wavy, lance-oblong; hds. panicled, small (5″), numerous; scales with minute prickles. ♃ Waysides, fields, N. and W. A pernicious weed, hard to extirpate. 3f. Very prickly, except its heads.
13 **C. pulchérrimum** with yellow flowers, 3f high, is rarely planted in borders.
14 **C. Lánthium.** A greenhouse shrub, covered with pale blue flowers. From Mexico.

98. LÁPPA, Tourn. BURDOCK.
Heads discoid, homogamous. Invol. globous, the scales imbricated and hooked at the extremity. Recep. bristly Pap. bristly, scabrous, caducous. ② European herbs. Lvs. alternate, large cordate, petiolate. Hds. panicled, pink-purple, very adhesive by the hooks

L. officinàlis Allioni.—A coarse weed, in waste and cultivated grounds, E. and W. 3f (L. major Gært.)—Varies with small hds. and lvs somewhat pinnatifid. (L. minor DC.)

99. LAMPSÁNA, Tourn. NIPPLEWORT.
Hds. radiant, 8—12-flwd. Invol. cylindrical, angular. Scales 8, erect, in one row, with 2 or 3 minute bractlets at base. Recep. naked. Cyp. glabrous. Pap. 0.—Slender, oriental herbs, with small, yellow heads, in paniculate corymbs.

L. commùnis L. Stem leafy; lvs. ovate, petiolate, dentate; ped. cylindrical; invol. angular in fruit. ① Waysides, Quebec, Boston, and W. Rare.

100. APOGON, Ell.
Heads radiant. Invol. scales ovate, acuminate, about 8, in two rows. Recep. naked. Ach. glabrous, oval, longitudinally 12-striate. Pappus 0. ① Herbs glabrous and glaucous, branched from the base. Leaves alternate, lanceolate. Heads small, yellow.

A. hùmilis Ell.—Woods, S. Car. to Fla. and La. 3—12′. Slender, smooth; lvs. varying to linear, entire or lyrately lobed. Heads 3″ broad. March—June.

101. CICHÓRIUM, Tourn. CHICORY.
Invol. double, the outer of 5

ORDER 70.—COMPOSITÆ. 191

leafy scales, the inner of about 8 linear ones. Receptacle chaffy. Pappus scaly. Cypselæ not rostrate, obscurely 5-sided.—Oriental herbs with bright blue flowers, about 20 in a head.

1 C. Íntybus L. Root lvs. runcinate, cauline bract-like; heads axillary, subsessile, mostly in pairs. ♃ Dooryards, waysides, E. 2—3f. Rays large, showy, 5-toothed. The root, or its extract, is often mixed with coffee. July—Sept. § Europe.

2 C. Endívia. *Endive.* Root leaves sinuate-dentate or pinnatifid, cauline auricled at base; heads axillary, 3—5 together. ① India. Cultivated as a salad.

102. KRÍGIA, Schreb. Dwarf Dandelion. Involucre many-leaved, nearly simple, equal. Recep. naked. Cypselæ turbinate, striate, 5-angled. Pappus double, consisting of 5 broad, membranous scales, alternating with as many slender, scabrous bristles. ① Acaulescent, small. Leaves lyrately lobed. Scapes simple. Heads solitary, with 20—30 yellow flowers.

1 K. Virgínica Willd. Early lvs. round-spatulate, subentire, the later toothed and pinnatifid; scapes 1—5 or more, 1'—10' high. Rocks and sands. Hds. 5—6''. May +.
2 K. Caroliniàna N. Early lvs. lin.-oblanceolate, few-toothed, later lvs. lyrate-pin natifid, or angular-lobed; scapes 1—5 or more, 2'—12'. Sands. S. Feb.—May.

103. CÝNTHIA, Don. Invol. nearly simple, of equal, narrow scales. Recep. flat, alveolate. Pap. double, the outer minute, scaly, inner copious, capillary. Cyp. short. ♃ Lvs. alternate or all radical. Fls. 15—20, yellow.

1 C. Virgínica Don. St. few-leaved, subumbellate; lvs. lance-obl., repand-dentate, rarely lobed, petiolate. N. Y. to Ill., and S. Very smooth. 1—2f. Hds. 9''. June.
2 C. Dandèllon DC. Acaulescent; scapes leafless, simple, 1-flwd.; lvs. spatulate-obl. to lance-lin., entire or toothed, rarely pinnatifid. Md. to Ga. and Tex. 6—18'. Mar.—Ju.

104. LEÓNTODON, L. Autumn Dandelion. Invol. imbricate, the outer sc. very short. Recep. naked. Pap. plumous, persistent on the somewhat rostrate cypsela.—Acaulescent herbs with yellow fls., many in a head.

L. autumnàlis L. Scape branching; ped. scaly-bracted; lvs. lanceolate, dentate-pinnatifid, smoothish. Waysides, meadows, &c. E. N. Eng. 6'—20'. Hds. several, near 1' in diameter. July—Oct. § Europe.

105. TRAGOPÒGON, L. Vegetable Oyster. Invol. simple, of many leaves. Recep. naked. Pap. plumous. Cyp. longitudinally striate; contracted into a long, filiform beak. ② European, with long, grass-like lvs.

T. porrifòlius L. Invol. much longer than the corolla; lvs. lance-linear; ped. thickened upward; pappus tawny. Waysides, &c. N. Y. (Hankenson). 3f. June. § ⚹

106. HIERÀCIUM, Tourn. Hawkweed. Invol. more or less imbricated, ovoid, many-flwd. Sc. very unequal. Cyp. not rostrate. Pap. a single row of copious, tawny, fragile bristles. ♃ Lvs. alternate, entire or toothed.

* Heads 40-50-flwd. Invol. more or less imbricated. Cyp. blunt at top......Nos. 1, 2
* Heads 12-30-flwd. Involucre simple.—*a* Achenia contracted at the top....Nos. 3, 4
—*a* Achenia not contracted upward...Nos. 5, 6

1 H. Canadénse Mx. St. erect, subvillous, leafy, corymbed; lvs. sessile, ovate obl. to lanceolate, acute, with few acute teeth; invol. strongly imbricated; fruit brown. Rocky woods, N. Eng. to Wis., and N. 1—2f. Stout. Hds. near 1' broad Aug., Sept

2 H. scàbrum Mx. Leafy. rough-hirsute, glandular above; lvs. obovate to elliptic subentire; invol. scarcely imbricated; fr. red. Hilly woods. 1—3f. Hds. 9″. Aug

3 H. longípilum Torr. Clothed with *long.* erect. shaggy *hairs*; lvs. lance-oblong, entire; hds. glandular. 20-30-flwd. in a small naked panicle. W. 1—2f. July. Aug.

4 H. Gronòvii L. Hairy, paniculate. glandular at top; lvs. obovate to lance-oblong, slightly toothed, the cauline sessile. often few; fr. 20—30, *narrowed above.*—Varies with stems leafy or subnaked; pan. close or diffuse. Dry hills: com. 1—3f. Aug.+.

5 H. venòsum L. Scape or stem leafless, or with one leaf. paniculate. smooth: lvs. obovate. entire. nearly glabrous, with purple veins; scales smooth; fls. 20; *fr. linear.* Woods. E. and W. 1—2f. Hds. on slender ped., broader (9″) than in No. 4. Jl., Aug.

6 H. paniculàtum L. Slender, leafy, diffusely paniculate: lvs. lanceolate, glabrous; ped. very slender: fls. 10—20; fr. short-cylindric. black. Woods: com. 2—3f. Aug.

107. CATANÁNCHE, L. Invol. imbricated, scarious. Recep. paleaceous. Pap. paleaceous, 5-leaved. Pales awned. ① Oriental herbs, with alternate, lanceolate leaves.

C. CŒRÙLEA L. Lvs. villous, somewhat bipinnatifid at base; invol. lower scales ovate, mucronate. S. Europe. 2—3f. Heads on long peduncles. Blue. July+.

108. NÁBALUS, Cass. DROP FLOWER. Invol. cylindric, of many linear scales in one row, calyculate with a few short, appressed scales at base. Recep. naked. Pap. copious, capillary, brownish, 2-rowed, persistent. Cyp. not beaked, smooth, striate. ♃ Erect, with a tuberous, bitter root. Heads 5–18-flowered, not yellow, although often straw-colored.

§ Heads pendulous, glabrous. Leaves variously lobed or shaped...(*a*)
   a Dwarf species (6—10′ high) native of high mountains......................Nos. 1, 2
   a Tall (2—5f high).—*c* Heads 5–6-flowered.................................No. 3
       —*c* Heads 8–12-flowered.—*d* Pappus tawny................No. 4
          —*d* Pappus straw-colored....Nos. 5, 6
§ Heads nodding or erect. hairy. Leaves mostly undivided...(*b*)
   b Heads about 12-flowered. Pappus straw-color......................Nos. 7, 8
   b Heads about 25-flowered. Pappus tawny or dusky.......................No. 9

1 N. Boottii DC. St. simple. dwarf; lvs. hastate-cordate to lanceolate. mostly entire; heads racemed; flowers 10—18, inner scales 10—15. High mountains. N. July+.

2 N. nanus DC. Smooth. simple; lvs. deltoid-hastate and variously lobed, upper lanceolate, all petiolate; hds. clustered-paniculate; sc. 8. fls. 10—12. White Mts. Aug.

3 N. altíssimus Hook. Smooth. strict. paniculate. tall. leafy; lvs. petiolate. palmately 3–5-cleft. or lobed. varying to hastate. cordate. or even ovate. dentate; hds. 6″ long. yellowish. forming a slender. leafy panicle; sc. 5. Woods. N. 3—5f. August.

4 N. albus Hook. *Lion's-foot. White Lettuce.* Smooth. glaucous. corymb.-paniculate; lvs. hastate-ovate to ovate. petiolate, the lobes or leaves obtuse; heads 6—7″, with 9 scales, 9—12 fls., and brown pappus. Moist woods. 2—4f. Purplish in spots. Aug.
   β. *Serpentària. Snake-root.* Lvs. deeply 3-lobed, the middle lobe 3-parted.

5 N. Fràseri DC. *Earth-gall.* Smoothish. corymb.-paniculate; lvs. hastate or deltoid. rarely 5–7-lobed, on winged stalks. upper lanceolate.—Varies with the leaves all lanceolate and merely toothed. Hard soils. Conn. to Fla. 2—4f. August.

6 N. virgàtus DC. Glaucous. simple, strict; lvs. sinuate-pinnatifid. narrow, the upper toothed or entire; panicle or raceme virgate. Sands. N. J. to Fla. 2—4f. Sept., Oct.

7 N. racemòsus Hook. Smooth (exc. the invol.). simple. slender; lvs. lance-oval to lance-ovate. denticulate; hds. suberect. spicate-paniculate. Swamps. N. J. to Iowa, and N. 2—4f. Flowers pale red.—Varies with the lower leaves cut-pinnatifid. Sept.

8 N. asper T. & G. Rough-downy. simple, strict; leaves oval-oblong to lance-oblong, dentate; hds. erect. fascicled in a spicate panicle; fls. yellowish. W. 2—4f. Sept.

ORDER 70.—COMPOSITÆ. 193

θ **N. crepidíneus** DC. Smoothish, tall, stout, corymb.-paniculate; lvs. broadly tri-ang.-ovate to lanceolate, toothed, petiolate; hds. nodding, of 12 sc. and 25—35 ochroleucous fls. Fields, thickets, W. States. 5—8f. Larger than any of the foregoing. Sept.

109. **LYGODÉSMIA,** Don. Invol. fls., &c., as in Nabalus. Pappus whitish. Corollas rose-colored. ♃ With linear-subulate leaves and erect heads on long, naked peduncles.

1 **L. aphýlla** DC. St. scape-like, erect, slender, forked above; lvs. nearly all radical, short, linear-filiform; heads 5-flowered. Pine woods, Ga. Fla. 2f. May.
2 **L. júncea** N. St. much branched; lvs. lance-linear; fls. 5. Min. (Matthews), and W.

110. **TROXIMON,** Nutt. Hds. many-flowered. Invol. campanulate, scales loosely imbricate, in 2—3 rows. Cyp. oblong-linear, compressed, glabrous, not rostrate. Pap. setaceous, copious, white. ♃ Lvs. all radical. Scape bearing a single, large, showy head, with yellow flowers.

T. cuspidátum Ph. Rt. fusiform; lvs. linear-lanceolate, woolly at the edge; scales lanceolate, cuspidate-pointed. Prairies, Ill. Wis., and West. April—June.

111. **PYRRHOPÁPPUS,** DC. FALSE DANDELION. Invol. double, the outer row numerous, loose and spreading. Receptacle naked. Cyp. 5-grooved, at length long-beaked, bearing a copious, soft, capillary, reddish pap. ① ♃ Hds. solitary on long ped., large, with numerous deep yel. fls.

P. Caroliniànus DC. St. simple or branched, scape-like; lvs. mostly radical, lanceolate, acute, sinuate-toothed, lobed, or pinnatifid. Fields, Md. to Fla. May—July.

112. **TARÁXACUM,** Desf. DANDELION. Invol. double, the outer of small scales, much shorter than the inner appressed row. Recep. naked. Cyp. produced into a long beak crowned with the copious, white, capillary pappus.—Acaulescent herbs, with runcinate leaves. Figs. 68, 346, 492.

T. Dens-leònis Less. Outer scales of the involucre reflexed; lvs. runcinate, smooth, dentate; scape short in fl., long in fr.—a globe of pappus. ♃ Fields: common. § Eur.

113. **LACTÙCA,** Tourn. LETTUCE. Invol. few-flowered, scales imbricated in 2 or more unequal rows. Cyp. obcompressed (flattened same way as the scales), glabrous, abruptly narrowed to a long, filiform beak. Pappus copious, soft, capillary, white.—Herbs with leafy stems and paniculate heads of variable colors. Fig. 77.

1 **L. Canadénsis** L. β. *elongàta.* *Trumpet Milkweed.* St. tall, hollow; lvs. pale beneath, clasping, rnnc.-pinnatifid, upper lance., entire; heads racemous-paniculate, with few scales and 12 + fls. ② Rich soils, thickets. 3—6f. Yel. to purplish. Jl., Aug.
β. *sanguínea.* Stem; lf. veins, and fls. purple; lvs. some hairy, glaucous beneath.
γ. *graminifòlia.* Lvs. long, linear, the lower few-lobed, upper entire. South.
δ. *integrifòlia.* Lvs. lanceolate, all entire, lower some sagittate at base.
2 **L. satíva.** *Garden Lettuce.* Stem corymbous; lvs. roundish, the upper cordate; fls. white. ① Said to be § in some places, when its lvs. become dentate-lobed and prickly.

114. **MULGÈDIUM,** Cass. WILD LETTUCE. Involucre somewhat double, outer scales short and imbricated. Recep. naked, faveolate. Pap. capillary, crowning the short-beaked, *compressed* cypsela.—Leaves mostly spinulous. Hds. paniculate, small, ∞-flwd. Jl.—Sept. Figs. 76, 448-50.

* Pappus bright white. Corollas blue......................... ....Nos. 1, 2
* Pappus tawny. Corollas cream-colored, turning purplish................ .....No. 3

1 **M. Floridànum** DC. Smooth; lvs. runcinately pinnate-parted, segm. few, sinuate-dentate or angular; pan. loose, hds. 9″. ② Thickets, N. Y., W. and S. 3—6f.
  β. *acuminàtum.* Lvs. lance-ovate, acuminate. toothed, or the lower subruncinate.
2 **M. pulchéllum** N. Smooth and glaucous, strict; lvs. lance-oblong to lin., entire, or the lowest runcinate; pan. corymbed; fls. bright blue. L. Huron to Oreg. 2—7f.
3 **M. leucophæum** DC. Tall, leafy; lvs. lyrate-runcinate, coarsely-toothed; ped. scaly-bracted; pan. long, compound; fr. scarcely beaked. ② Moist thickets. 5 –10f.

115. SÓNCHUS, L. Sow Thistle. Invol. many-flowered, imbricate, of numerous unequal scales, at length tumid at base. Recep. naked. Pap. of white-silky hairs, in many series. Cypselæ compressed, not rostrate.— Leaves mostly spinulous. Heads with many yellow flowers. Europe.

§ Flowers bright yellow, in showy heads. Achenia angular. Perennial.........No. 1
§ Flowers pale yellow, in large heads. Achenia flat. Annual. Aug., Sept..Nos. 2, 3

1 **S. arvénsis** L. Smooth, erect, hispid above; leaves runcinate-pinnatifid, spinulous-dentate, clasping with short auricles at base; hds. subumbellate. Fields, waysides, N. Eng., N. Y. 2f. §.
2 **S. asper** Vill. Leaves cordate, amplexicaul, oblong-lanceolate, undulate, spinulous dentate; ped. subumbellate; fruit oval-obovate, 3-ribbed on each side. 1—2f. §
3 **S. oleràceus** L. Lvs. sagittate-amplexicaul, runcin.-pinnatifid, subspinulous, den tate; ped. downy; involucre at length smooth; fruit many-striate. Rubbish. 2—3f. §

116. HÙMEA Elegans. Tall, 4f, branching above into an ample capillary panicle; lvs. lance-ovate, clasping; heads numerous, small, drooping, with dry, loose scales, and 3 or 4 carmine-red florets, with no pales or pappus. N. Hol. July—Oct.

117. CHAPTÀLIA, Vent. Invol. campanulate. Scales in few series, linear, acute. Recep. naked. Ray-fls. ♀, ligulate, disk-fls. ☿, but sterile, oilabiate, lips equal, outer 3-, inner 2-parted. Cypselæ glabrous. Pappus capillary. ♃ Acaulescent. Lvs. all radical. Head cyanic. Mar., Apr.

C. tomentòsa Vent. White-tomentous; lvs. oblong-ovate to lance-oval; hd. nodding in bud, erect in fl., on the scape. Moist barrens, S. 6—12′. Rays 20, rose-colored.

Order LXXI. LOBELIACEÆ. Lobeliads.

Herbs or *shrubs* with a milky juice, alternate, exstipulate leaves and scattered flowers. *Calyx* 5-lobed or entire. *Corolla* monopetalous, irregular, split down to the base on one side. *Stamens* 5, free from the corolla, united into a tube at least by their anthers. *Ovary* adherent to the calyx tube. *Style* 1. *Stigma* surrounded by a fringe. *Fruit* a capsule 2-3-(rarely 1-) celled. *Seeds* numerous, albuminous.

1. LOBÈLIA, L. Cor. tubular, irregular, cleft nearly to the base on the upper side, upper lip of 2 separate lobes, lower 3-lobed. Anth. united above into a curved tube. Stig. 2-lobed. Caps. opening at the summit. Seeds minute. ① ♃ Flowers axillary and solitary, or in terminal, bracted racemes. July—Sept.

¶ Corollas scarlet or bright crimson, large...* Exotic, Nos. 15, 16.. ...* Native, No. 1
¶ Corollas blue, or blue and white. ..† Exotic, Nos. 17, 18 ...† Native..(*a*)

a Calyx lobes auricled at base, denticulate, shorter than corolla tube....Nos. 2–4
a Calyx lobes auricled at base, entire, linear, long as corolla tube........Nos. 5, 6
a Calyx lobes not auricled, entire,—*b* very slender and long............Nos. 7–9
    —*b* much shorter than corolla...(*c*)
  c Leaves cauline, entire, few. Racemes loose, few-flowered......Nos. 10–12
  c Leaves radical, entire. Racemes strict, few-flowered...........Nos. 13, 14

1 **L. cardinàlis** L. *Cardinal Flower.* Tall, simple, glabrous; lvs. oblong-lanceolate, slightly toothed, acute at each end, sessile; fls. in a terminal, bracted, second raceme; stam. longer than the corolla. ♃ Swamps. 2—4f. Splendid.
  β. *integérrima.* Leaves all very entire; stem naked above. Northern N. Y.
  γ. *cándida.* Flowers white, the segments narrower. Mass.
2 **L. syphilítica** L. *Great Lobelia.* Stem erect, angular; leaves oblong-lanceolate, acute or acuminate, unequally serrate, some hairy; raceme leafy; calyx hispidly ciliate, with the sinuses reflexed. ♃ By streams. 1—3f. Flowers 1'.
  β. *alba.* Flowers pure white. N. Y. (E. L. Hankenson; G. M. Wilbur).
3 **L. glandulòsa** Walt. Subsimple, leafless above; lvs. lance-lin., acutish, and with the lanceolate, auricled sepals some glandular-toothed; fls. few, remote, large (9''); cal. hispid or smoothish, short. ♃ Damp barrens, Va., and S. 1½—2f. Sept.—Oct.
4 **L. brevifòlia** N. Erect, simple, hispid; lvs. 1', crowded, oblong-lin., denticulate; sep. ovate, fringe-toothed, half as long as cor. Damp, Fla. to La. 18'. (L. Ludov. C-B.)
5 **L. leptóstachys** A. DC. Glabrous, erect, simple, virgate; lvs. lance-oblong; fls. small (4''), spike not secund; auricles awl-shaped, long. ♃ Prairies, W. and S. 1—2f.
6 **L. pubérula** Mx. Downy or smoothish, erect, simple; lvs. elliptic-ovate, denticulate; fls. large (7—8''), in a long, secund spike; auricles ear-shaped. N. J., W. and S. 2f.
7 **L. amǽna** Mx. Erect, simple, smooth; lvs. lanceolate, pointed both ways; fls. large (8—9''), secund, numerous, in a long rac.; bracts very small. ♃ Swamps, Va., and S. 2f.
8 **L. spicàta** Lam. Erect, simple, puberulent; lvs. oblong, mostly obtuse; fls. small (3—4''), crowded in a slender rac.; pedicels and bracts as long as the fl. Dry soils. 1-2f.
9 **L. inflàta** L. *Indian Tobacco.* Erect, branching, hairy; lvs. ovate-lanceolate, serrate; fls. short (4''), with leafy bracts; caps. inflated, large. ① Fields. 1f. Narcotic.
10 **L. Boykínii** T. & G. Slender, smooth; branches erect; lvs. awl-shaped, erect; fls. small (4''), on filiform ped. in long, loose racemes. Wet sands, Ga. Fla. 2f. Lvs. 6''.
11 **L. Nuttállii** DC. Erect, very slender, smooth; lvs. few, linear, remote; fls. few, small (3''); ped. as long as cor.; cal. tube almost none. ⚥ Swamps, L. I., and S. 1-1½f.
12 **L. Kálmii** L. Simple or branched; rt. lvs. spatulate, st. lvs. lance-lin. to lin., all obtuse; rac. loose, leafy; ped. about equalling the showy blue-wh. fls., minutely bracted, or naked (in same specimen); cor. 5''. lobes obovate. Rocky swamps, E. &W. 6–18'.
13 **L. paludòsa** N. Lvs. lin.-spat., thickish, obtuse, petiolate; scape simple, nearly naked; rac. loose, ped. about as long as the cal. ⚥ Bogs, Del., and S. 2–3f. Lvs. 5–10'.
14 **L. Dortmánna,** L. Lvs. submerged, tufted, linear, entire, hollow with 2 longitudinal cells, short, obtuse; scape simple, nearly naked; fls. in a terminal raceme, remote, pedicellate, nodding. ♃ In ponds, N. States. 2—3f. Only the fls. emerging.
15 **L. fúlgens.** Downy, erect, simple; lvs. narrow-lanc., revolute at edge. ♃ Mex. 3f.
16 **L. spléndens.** Smooth, erect; lvs. narrow-lanc., flat; fls. large, in long rac. Mex. 3f.
17 **L. Erýnus.** Slender, diffuse; lvs. toothed, ellip. to lin.; fls. scattered, bluish. S. Afr.
18 **L. cœlestìna,** a garden variety, with larger blue flowers, yellow in the centre.

2. **DOWNÍNGIA**, Torr. Sep. 5, linear. Cor. 2-lipped, *tube not split*, upper lip 2-parted, erect, lower lip 3-lobed. Stam. tube incurved. *Caps. silique-form*, 1-celled, ∞-seeded, opening by 3 linear valves. ① Low, with axillary, solitary flowers. (Clintonia, Doug.)

1 **D. élegans.** Stem few-branched, angular; lvs. ovate, acute; ovary curved, 3-angled, longer than the lvs.; corolla blue with a white palate. Oregon! 6—12'. July, Aug.
2 **D. pulchélla.** Stem much branched; lvs. obtuse; fls. 8'', middle lobe longest. Cal. !

Order LXXII. CAMPANULACEÆ. Bellworts.

Herbs with a milky juice, alternate leaves, and without stipules. *Flowers* mostly blue, showy. *Calyx* superior, generally 5-cleft, persistent. *Corolla* regular, campanulate, generally 5-cleft, withering, valvate in æstivation *Stamens* 5, free from the corolla. *Anthers* distinct, 2-celled. *Pollen* spherical. *Ovary* adherent to the calyx, 2 or more celled. *Capsule* crowned with the remains of the calyx, loculicidal. *Seeds* many. Figs. 62, 63.

§ Calyx tube short. Pod roundish, opening at the sides. Cor. bell or wheel form... ...CAMPANULA. 1
Calyx tube elongated. Pod prismatic, opening at the sides. Corolla wheel-form......SPECULARIA. 2
§ Calyx tube short. Pod ovoid, opening at the top. Corolla bowl-form................PLATYCODON. 3

1. CAMPÁNULA, Tourn. Calyx mostly 5-cleft. Cor. campanulate, or subrotate, 5-lobed, closed at base by the broad, valve-like bases of the 5 stamens. Stig. 3–5-cleft. Caps. 3–5-celled, opening by lateral pores. Mostly ♃. Flowers in racemes or spikes, or few and axillary.

§ Native or naturalized.—*a* Flowers rotate, deeply 5-lobed....................No. 1
—*a* Flowers campanulate, few, or scattered.........Nos. 2—4
—*a* Flowers funnel-form, crowded above...............No. 5
§ Exotic.—*b* Sepals appendaged at base. Stig. 3 or 5. Corolla bell-shaped..Nos. 6, 7
—*b* Sepals not appendaged. Stig. 3.—*c* Corolla bowl-shaped.......Nos. 8, 9
—*c* Corolla bell-shaped... ...Nos. 10, 11
—*c* Cor. rotate-spreading....Nos. 12, 13

1 **C. Americàna** L. Tall, erect; lvs. ovate-lanceolate, acuminate, uncinately serrate, contracted to a winged petiole, veins often ciliate; fls. axillary, sessile; style exserted, decurved. ♃ Dry copses: common. 2—4f. Fls. 1' broad, spreading, flat. Aug. †
2 **C. rotundifòlia** L. *Hare-bell.* St. weak, slender; radical lvs. ovate or reniform-cordate, cauline linear, entire; flowers few, nodding, bell-shaped and blue. ♃ Damp rocks, N. States. 1f. Very delicate. June, July. Rt. lvs. seldom found with the fls.
3 **C. aparinoìdes** Ph. Stem weak, slender, branching above, triangular, the angles inversely aculeate; lvs. lance-linear, subentire; fls. terminal, 4" long, white. In wet meadows. 1—1½f, leaning on the grass like a Galium. June—Aug.
4 **C. divaricàta** Mx. Glabrous, erect, with slender, divaricate, paniculate branches; lvs. narrow-lanceolate, pointed at each end, sharply dentate; fls. campanulate, pendulous on the slender branchlets. Rocky woods. Va., W. and S. 2f. July.
5 **C. glomeràta** L. St. angular, simple, smooth; lvs. lance-oblong, cordate, the lower petiolate; fls. crowded above, cor. funnel-form, violet-blue. Fields, Mass. 2f. § † Eu.
β. AGGREGÀTA. Flowers pale blue, in a dense head, and other var. are cultivated.
6 **C. MÈDIUM**. *Canterbury-bells.* Erect, hispid; lvs. lanceolate; fls. 1½'; stig. 5. ⓑ Eu. 3f.
7 **C.** SPECIÒSA. Erect; lvs. lance-linear; fls. racemed, nodding; stig. 3. ♃ Eur. 2f.
8 **C.** PYRAMIDÀLIS. Smooth, branched; lvs. lance-ovate; fls. broad, racemed. ♃ Eu. 6f.
9 **C.** PERSICIFÒLIA. Smooth; lvs. lance-linear, thick; fls. broad, axillary. ♃ Eur. 3f.
10 **C.** TRACHÈLIUM. St. angular, hairy; lvs. ovate, cord. dentate; ped. 1–3-flwd. ♃ Eu. 4f.
11 **C.** RAPUNCULOÌDES. Rough; lvs. ovate, pointed; rac. spicate; fls. nodding. ♃ Eu. 2f.
12 **C.** LÒREYI. St. erect. ang.; lvs. obov. to lance-ovate; cal. hairy; cor. 2' broad. ⓘ Eu.
13 **C.** GARGÀNICA. St. diffuse; lvs. cord.-reniform to ovate; fls. small, star-shaped. ♃ Eu.

2. SPECULÀRIA, Heist. Calyx 5-lobed, tube elongated. Cor. rotate, 5-lobed. Fil. hairy, shorter than the anthers. Sty. included, hairy. Stig. 3. Caps. prismatic, 3-celled, opening laterally in the upper part. ⓘ Fls. axillary and terminal, sessile, erect.

1 S. perfollàta Lam. St. mostly simple, erect; lvs. reniform-ovate, cordate-clasping, crenate; fls. sessile, aggregate, axillary. Fields, copses. 1f. Fls. deep blue. Jn., Jl.
2 S. Ludovicàna Torr. St. branched, branches slender; lvs. ovate, acute, subentire, sess. or slightly clasping; ovaries slender, fls. smaller (5'' broad). S. Car. to La. 1—2f.
3 S. Spéculum. *Venus' Looking-glass.* Stem diffusely branching; lvs. oblong, crenate; fls. solitary, with shallow lobes, blue varying to white, all Summer. S. Eur. Hardy.

3. PLATYCÒDON, A. DC. Cor. large, bowl-shaped. Stig. 5, thick, spreading. Caps. ovoid, opening at the top by 5 acute valves. ♃ Siberia. Smooth and glaucous.

P. grandiflòrum. Lvs. lance-ovate, serrate; fls. 2', blue var. to wh., few, terminal. 18'.

Order LXXIII. ERICACEÆ. Heathworts.

Plants shrubby or suffruticous, sometimes herbaceous, with *Leaves* simple, alternate or opposite, mostly evergreen, without stipules. *Corolla* regular or somewhat irregular, 4–5-cleft, the petals rarely distinct. *Stamens* as many or twice as many as the petals, free, hypogynous. *Anthers* 2-celled, generally open by pores, often appendaged. *Pollen* (except in Monotropeæ) compounded of 4 united grains. *Embryo* straight, lying in the axis of, or in the end of fleshy albumen. Figs. 64, 89, 90, 99, 114, 248, 255, 311, 438.

§ Ovary adherent, in fruit a berry crowned by the calyx teeth. Shrubs...(Suborder I.)
§ Ovary free.—*x* Shrubs, trees. Capsule or berry with the cells ∞-seeded...(Suborder II.)
 —*x* Shrubs. Fruit a capsule with the cells one-seeded...(Suborder III.)
 —*x* Herbs half-woody, low.—*y* Leaves evergreen. Stamens distinct...(Suborder IV.)
   —*y* Leaves evergreen. Filaments united...(Suborder V.)
   —*y* Leaves none. Plants without verdure...(Suborder VI.)

I. VACCINEÆ.—*a* Fls. 5-parted. Berries 10-seeded. Shrubs often resinous-dotted...Gaylussacia. 1
   —*a* Flowers 5-parted. Berries ∞-seeded. Shrubs dotless..........Vaccinium. 2
   —*a* Flowers 4-parted.—*b* Petals narrow, reflexed. Berries red........Oxycoccus. 3
     —*b* Petals short, spreading. Berries white......Chiogenes. 4
II. ERICINEÆ.—*c* Flowers 4-parted. Sepals colored, larger than the corolla........Calluna. 5
   —*c* Flowers 4-parted. Sepals small...(Gen. 11, or)..................Erica. 6
   —*c* Flowers 5-parted.—*d* Petals distinct, or very nearly polypetalous...(*m*)
       —*d* Petals united,—monopetalous...(*e*)
     e Corolla funnel- or bell-form, with spreading lobes...(*k*)
     e Corolla urceolate (ovoid, cylindric or globular), lobes small...(*f*)
     e Corolla saucer-form, holding the anthers in 10 pits............Kalmia. 7
     e Corolla salver-form, very fragrant. Trailing shrublet..........Epigæa. 8
       f Fruit fleshy, the matured ovary 5-seeded................Arctostaphylos. 9
       f Fruit fleshy, the matured calyx ∞-seeded.................Gaultheria. 10
       f Fr. dry, capsular,—*g* septicidal. Lvs. linear, heath-like....Menziesia. 11
           —*g* loculicidal.—*h* Lvs. linear, moss-like..Cassiope. 12
               —*h* Lvs. ample. Shrubs...Andromeda. 13
               —*h* Lvs. ample. Trees....Oxydendrum. 14
         k Stamens 5, included. Plant and leaves very small.........Loiseleuria. 15
         k Stamens 5 (rarely more), long-exserted. Cor. funnel-form...Azalea. 16
         k Stamens 10 (rarely fewer), exserted. Cor. bell-form.....Rhododendron. 17
       m Corolla very irregular, open before the leaves appear............Rhodora. 18
       m Cor. regular,—*n* 7-petalled. Stamens 14.....................Befaria. 19
         —*n* 5-petalled.—*o* Capsule 5-celled....................... 20
           —*o* Caps. 3-celled.—*p* Fls. umbelled..Leiophyllum. 21
             —*p* Fls. racemed...Clethra. 22
III. CYRILLEÆ.—*r* Flowers 4-parted, with 8 stamens and a 2-celled capsule.......Elliottia. 23
   —*r* Flowers 5-parted,—*s* with 5 stamens and a 2-celled capsule......Cyrilla. 24
       —*s* with 10 stamens. Caps. 3-celled, 2 winged..Mylocarium. 25

ORDER 73.—ERICACEÆ.

IV. PYROLEÆ.—*s* Flowers racemed, many. Herbs nearly acaulescent..............PYROLA. 26
—*s* Flowers umbelled, few. Stems ascending. Style very short......CHIMAPHILA. 27
—*s* Flowers solitary (one only).—*t* Capsule 5-celled................. MONESES. 28
—*t* Capsule 3-celled...................SHORTIA. 29
V. ? GALACINEÆ. Anthers 5, one-celled. Capsule 3-celled. Scape spicate.........GALAX. 30
VI. MONOTROPEÆ.—*u* Corolla polypetalous. Plant white, reddish or tawny.......MONOTROPA. 31
—*u* Corolla monopetalous,—*v* campanulate, in a short spike....SCHWEINITZIA. 32
—*v* ovoid, in a loose raceme..........PTEROSPORA. 33

1. **GAYLUSSÀCIA**, H. B. K. HUCKLEBERRY. Calyx adherent, 5-toothed. Cor. urceolate or campanulate, 5-cleft or toothed. Sta. 10. Anth. awnless, the cells produced upward into tubular beaks opening at the apex. Berry drupe-like, globular, 10-celled, 10-seeded. ♄ ♄ Leaves alternate. Flowers in lateral, bracted racemes, white or reddish, small. Fruit black or dark blue, sweet. May, June.

§ Leaves evergreen, very smooth, with no resinous dots, crenulate................No. 1
§ Leaves deciduous, sprinkled with resinous dots beneath, entire..........Nos. 2—4

1 **G. brachýcera** (Michx). *Box H.* Lvs. oval to ovate, thick and firm; rac. dense, ped. very short; cor. short-ovoid; berries light blue. Rocky hills, Pa. to Va.; rare. 1f.
2 **G. dumòsa** T. & G. Minutely hairy and glandular; lvs. obovate-oblong, *mucronate*; bracts persistent; cor. short-bell-form; ber. black, large, insipid. Me. to Fla. 1—2f.
3 **G. resinòsa** T. & G. *Black H.* Branches ashy; lvs. oval to lance-obl.; rac. 1-sided, deciduous bracts, ped. short as the fls.; cor. 5-angled, contracted at mouth; sty. exserted; fr. black, round, sweet and eatable, ripe in Aug. Thickets, Can. to Va., and W. 2f.
4 **G. frondòsa** T. & G. *Blue Dangles. High Blueberry.* Lvs. oblong-obovate, pale-glaucous beneath; rac. loose, bracts deciduous, shorter than the ped.; cor. egg-bell-form; berries large, blue, sweet and eatable, in Aug. Thickets, N. Eng. to La. 3—6f.

2. **VACCÍNIUM**, L. BLUEBERRY. Calyx adherent, 5-toothed. Cor. urceolate, campanulate or cylindric, limb 4- or 5-cleft, reflexed. Sta. 8 or 10, included. Anth. with 2 awns on the back, or awnless, the 2 cells prolonged into a tube opening at apex. Berry 4 or 5 (or partly 8–10)-celled, cells ∞-seeded. ♄ ♄ Leaves alternate. Flowers solitary or racemous, white or reddish, small. Fruit generally eatable. Fig. 90.

§ Anthers 2-awned back of the 2 horns. Leaves deciduous...(*a*)
a Filaments smooth. Fruit 4–5-celled, blue. Shrubs 1f or less........ ..Nos. 1, 2
a Filaments hairy. Fruit partly 10-celled. Taller (2—20f high).........Nos. 3, 4
§ Anthers 2-horned, without the awns. Filaments 10, hairy...(*b*)
b Leaves evergreen. Flowers 4-parted. Fruit 4-celled......................No. 5
b Leaves evergreen. Flowers 5-parted. Fruit partly 10-celled...........Nos. 6, 7
b Lvs. deciduous. Fr. partly 10-celled. Fls. in short, close racemes...(*c*)
c Corolla bell-shaped. Leaves hairy both sides, entire..................No 8
c Corolla cylindrical. Leaves smooth or nearly so.............. ....Nos. 9—11
c Corolla ovoid, evidently contracted at the mouth................Nos. 11—13

1 **V. uliginòsum** L. *Bilberry.* Procumbent; lvs. obovate, obtuse, dull, glaucous beneath; fls. solitary, axillary; cor. ovoid-globous, 4-cleft; stam. 8. White Mts. Jn., Jl.
2 **V. cæspitòsum** Mx. *Bilberry.* Dwarf, cæspitous; lvs. obovate, attenuate at the base, thin, serrate, reticulate with veins, shining; flowers subsolitary; corolla oblong, 5-toothed; stamens 10. White Mountains. 2—3'. July.
3 **V. stamíneum** L. *Deerberry.* Lvs. oval-lanceolate, acute, dull, glaucous beneath; pedicels solitary, axillary, nodding; cor. bell-spreading, seg. acute, oblong; anth. 10, with the long tubes exserted. Dry woods. 2—3f. Fruit greenish-white. May, June.
4 **V. arbòreum** Mx. Lvs. obovate, acute at base, mucronate, veiny, shining above

ORDER 73.—ERICACEÆ.

pale green and subpubescent beneath ; pedicels second, in leafy racemes ; cor. cylindric-bell-shaped, rose-white ; anth. 10, included. Woods, S. 8–20f. Fr. black. May, Jn.

5 **V. Vitis-Idæa** L. Decumbent, much branched, smooth, evergreen ; lvs. 4–7″, oval, obtuse, thick, margin revolute, pale beneath ; fls. solitary or in short clusters, 4-parted ; corolla campanulate. Hills and mts., N. Eng.: rare. June, July.

6 **V. Myrsinites** Mx. Erect, much branched ; lvs. small, elliptical, acute at each end, glabrous, serrulate ; fls. in small lateral clusters of 2—5 ; cor. ovoid, urceolate ; style slightly exserted. Woods, S. 1f. Whole plant often purplish. March, April.

7 **V. myrtifòllum** Mx. St. simple, decumbent at base, from long, creeping roots ; lvs. 1—2′, cuneate-obovate or oval, pale beneath ; fls. in dense, sessile, lateral clusters of 6—12 ; cor. oblong-cylindric ; fr. round, black. Woods, S. 1f. Mar., April.

8 **V. Canadénse** Rich. Branches reddish-green, pubescent, leafy ; lvs. elliptic-lanceolate, acute at each end ; rac. fasciculate, sessile, subterminal ; cor. campanulate ; cal. lobes acute. Rocky thickets, N. Eng., and W. 8—12′. Berries blue, sweet. May.

9 **V. Pennsylvánicum** Lam. *Common Low Blueberry*. Branches green, with 2 pubescent lines ; lvs. 1′, crowded, elliptic-oblong, acute at each end, bristly-serrulate, shining ; fls. in short, bracteate, dense rac. Hard soils, Can. to Pa. Ber. blue, sweet.

β. **nìgrum.** Dark green ; berries black and shining, without bloom.

γ. **alpìnum.** Dwarf, decumbent ; lvs. small (3—4″), narrow-oblanceolate. Mts.

10 **V. vacillans** Soland. Low, bushy ; lvs. oval to ovate, acute or mucronate, pale green, dull, glaucous beneath, minutely serrulate ; rac. dense-flowered, preceding the full-grown lvs. Hilly woods, N. Eng. to Tenn. 1—2½f. Fr. blue-black, sweet. May, Jn.

11 **V. corymbòsum** L. *Common High Blueberry*. Tall ; flowering branches nearly leafless ; leaves oval to lanceolate, acute or acuminate at each end, *entire*, pubescent when young, often glaucous beneath ; rac. short, sessile ; cor. cylindrical to ovoid. Low woods. 5—10f. March—June.—Varies exceedingly.

β. **virgàtum.** Branchlets leafless, covered with rose-colored rac. Sts. virgate. 5f. S.

γ. **amœnum.** Lvs. oblong ; fls. cylindric, large, roseate ; sty. included ; fr. blk. 8f.

δ. **fuscàtum.** Lvs. serrulate ; ped. elongated ; sty. exserted ; fls. striped with red. 3f.

ε. **glabrum.** Plant glabrous throughout, the leaves entire. Rare.

12 **V. galèzans** Mx. Flowering branches leafy ; lvs. sessile, cuneate-lanceolate, subserrate, veiny, glabrous when old ; flowers in small, sessile fascicles ; corolla small, yellowish ; style exserted ; fruit small, black. Swamps, S. 1f. April +.

13 **V. hirsùtum** Buckley. Whole plant, with fls. and fr., densely hirsute ; lvs. ovate, entire ; corolla oblong, nearly closed at mouth ; berry round. Mts. of N. Car. 1f.

3. **OXYCÓCCUS**, Pers. CRANBERRY. Calyx adherent, 4-cleft. Cor. 4-parted, with long, narrow, reflexed segments. Sta. 8. Anth. tubular, 2-parted, opening by oblique pores. Berries globous, 4-celled, many-seeded. ♄. Delicate, with alternate lvs., red and purple berries on slender ped.

* Stem erect, with membranous, deciduous leaves. Berries sweetish............No. 1
* Stem prostrate, slender. Leaves evergreen, small. Berries acid............Nos. 2, 3

1 **O. erythrocárpus** Ell. Lvs. oval, acuminate, thin, ciliate-serrulate ; fls. axillary, solitary, the long segments at length reflexed. Mts. of Va. and Car. 1—2f. June.

2 **O. palústris** Pers. Sts. filiform, purple ; lvs. ovate, entire, revolute on the margin ; pedicels terminal, 1-flowered ; corolla pink, segments ovate. Alpine bogs, N.

3 **O. macrocárpus** Pers. St. filiform ; lvs. oblong, obtuse at each end, edges revolute, glaucous beneath ; pedicels axillary, elongated, 1-flowered ; corolla segm. linear-lanceolate. Sphagnous swamps, Va., and N. Fruit large, valuable. June.

4. **CHIÓGENES**, Salisb. Calyx 4-cleft, persistent. Cor. broadly campanulate, limb deeply 4-cleft. Stam. 8. Anth. cells distinct, awnless on the back, bicuspidate at apex, opening longitudinally. Ov. adherent. Fr.

white, 4-celled, many-seeded. ♃ Delicate. Lvs. very small, alternate, with the flavor of the Checkerberry. Cor. small, wh., axillary, solitary. Fig. 248.

C. hispídula T. & G.—In old woods, N. Eng., N. and W. Stems creeping, slender, 1—3f. Leaves oval, 4—6″. Berries very small. May, June.

5. CALLÙNA, Salisb. HEATHER.
Cal. of 4 scarious, colored sepals. Cor. campanulate, 4-parted, shorter than the calyx. Stam. 8. Anth. 2-crested on the back, cells opening laterally. Stig. 4-lobed. Caps. 4-celled, 8-seeded, 4-valved. ♄ Lvs. opposite, minute, crowded. Fls. axillary, or crowded in 1-sided racemes, scarious, roseate, with 4—6 scarious bracts.

C. vulgàris Salisb.—Low grounds, Tewksbury! Mass., Me., and N. 2f. Lvs. ¼″.

6. ERÌCA, L. HEATH.
Cal. 4-parted. Cor. tubular, bell-, cup-, urn-, globe-, egg-, or salver-form, the limb in 4 short lobes. Stam. 8. Sty. filiform. Caps. 4-celled, opening by 4 loculicidal valves. Sds. 2—∞ in each cell. ♄ Very delicate, chiefly S. African, branching and brittle. Leaves whorled, rarely alternate, linear or acerous. Flowers nodding, cyanic.

1 E. cinèrea L. *Scotch Heath.* Stems clustered; branchlets and linear lvs. (1″) in 3's, crowded : fls. racemous-clustered on the upper branchlets; cal. colored, with few or no bractlets, ¼″; cor. purple, oval, 2″; anth. included, awned beneath. Sandy "moors," Nantucket Is.! Found by Mrs. E. E. Atwater, June, 1868. Apparently indigenous.

2 E. CARNEA. Very slender, 6—10′; leaves in 3's or 4's, 2—3″ long, obtuse; flowers axillary; corolla 2″, and calyx 1″, flesh-color; anthers dark-purple, exserted. Alps. April.—Of the 400 known species, only this is yet common in cultivation.

7. KÁLMIA, L. AMERICAN LAUREL.
Cal. 5-parted. Cor. with 10 prominences beneath and 10 corresponding cavities within, including the 10 anthers. Border 5-lobed. Fil. elastic. Caps. 5-celled, many-seeded. ♄ ♄ Beautiful, N. American. Leaves entire, evergreen, coriaceous. Flowers in racemous corymbs, white and red, in May—July.

* Flowers in terminal corymbs. Leaves thick, mostly acute .... Nos. 1, 2
* Flowers in lateral corymbs. Leaves obtuse .... Nos. 3, 4
* Flowers solitary, axillary. Sepals nearly as long as the corolla .... No. 5

1 K. latifòlia L. *Calico Bush. Spoon-wood.* Lvs. alternate and ternate, oval lanceolate, acute at each end, smooth and green on both sides; corymbs terminal, viscidly pubescent. Woods, Me. to O., Ky., and Fla. 5-20f. Profusely and splendidly flowering.

2 K. glauca Ait. *Swamp Laurel.* Branches ancipitous; lvs. opposite, subsessile, lanceolate, polished, glaucous beneath, revolute at the margin; corymbs terminal, the peduncles and bracts smooth. Bogs, Pa., and N. 2-3f. Lvs. 1′. Corymbs 8-10-flowered.
  β. **rosmarinifòlia.** Leaves linear, more revolute, green beneath.

3 K. angustifòlia L. *Sheep-poison.* Lvs. ternate and opposite, elliptical-lanceolate, petiolate, obtuse at each end, smooth; corymbs lateral; bracts linear-lanceolate. Hills and copses, Can. to Ky. and Car. 2—4f. Flowers deep purple, few in each cluster.

4 K. cuneàta Mx. Lvs. scattered, sessile, cuneate-oblong, obtuse, mucronate, glandular-pubescent beneath; flowers white, in sessile clusters. Swamps, Car.: rare. 3f.

5 K. hirsùta Walt. Slender, branched, hairy; leaves scattered or opposite, ovate to linear-oblong, as long as the pedicels (4—6″). Barrens, S.: common. 1f. Fls. 7″.

8. EPIGÆA, L. TRAILING ARBUTUS. MAY-FLOWER.
Cal. large, 5-parted, with 3 bracts at base. Cor. salver-form, tube villous within, limb

5-parted, spreading. Stam. 10. Anth. dehiscent by 2 longitudinal openings. Caps. 5-celled, 5-valved. ↳ Trailing, with cordate, ovate, entire, alternate leaves, and axillary clusters.

E. repens L.—Rocky woods, N. Eng. to Pa., Ky., and N. Stems half-shrubby, hairy, 10—15' long. Lvs. evergreen, 2'. Fls. rose-colored, delightfully fragrant. Apr., May.

9. ARCTOSTÁPHYLOS, Adans. BEAR-BERRY. Cal. 5-parted, persistent. Cor. ovoid, diaphanous at the base, limb with 5 small recurved segments. Anth. 10, with 2 long, reflexed awns, and opening by pores. Drupe or berry 5–10-celled, the cells 1-seeded. ♄ Trailing. Leaves alternate. Racemes terminal.

1 A. Uva-ursi Spr. Lvs. entire, thick, evergreen, shining above, obovate; flowers drooping; drupe red, as large as a currant, the nut 5-seeded. Rocky hills, N. May.
2 A. alpina Spr. Lvs. thin, serrate, deciduous, obovate, acute, strongly netted; ped. hardly longer than the bractlets; drupes black. High mts., Me., N. H., and N.

10. GAULTHÉRIA, Kalm. CHECKERBERRY. WINTERGREEN. Cal. 5-cleft, with 2 bracts at the base. Cor. ovoid-tubular, limb with 5 small, revolute lobes. Fil. 10, hirsute. Caps. 5-celled, invested by the calyx, which becomes a berry. ♄ Leaves alternate. Pedicels bibracteolate.

G. procúmbens L. St. procumbent, with the branches erect or ascending; lvs. obovate, mucronate, denticulate, crowded at the top; fls. few, drooping, terminal. Woods and pastures, Can. to Penn. and Ky. 3'. Red berries and leaves spicy. June—Sept.

11. MENZIÉSIA, Smith. Cal. deeply 4- or 5-cleft. Cor. urceolate or campanulate, 4- or 5-lobed. Sta. 8 or 10, anth. opening by terminal pores. Caps. 4- or 5-celled, opening septicidally. Seeds ∞. Low, shrubby plants, of various habits. Flowers in terminal clusters.

§ PHYLLODOCE, Salisb. Lvs. evergreen, heath-like. Fls. 5-parted, bell-form....No. 1
§ MENZIESIA *proper*. Leaves deciduous. Flowers 4-parted, urceolate..........No. 2

1 M. taxifòlia Robbins. *Mountain Heath*. St. prostrate at base; lvs. linear, obtuse; pedicels erect, slender, terminal, aggregate, 1-flowered. Alpine bogs, N. H., Me., and N. 6—12'. Leaves 6—7''. Flowers purple, the ped. 18''. June.
2 M. ferrugínea Smith. β. *globulàris* Sims. Shrub low, straggling, pubescent; leaves lance-oval, ciliate; flowers small, nodding, on slender pedicels, greenish-purple. Mts., Penn. to Car. 3—4f. June.

12. CASSÍOPE, Don. MOSS-PLANT. Sep. bractless, imbricated, ovate. Cor. globular-campanulate, 4- or 5-lobed. Anth. 8 or 10, pendulous, cells opening by a terminal pore, with a long reflexed awn behind. Caps. 4- or 5-celled, valves 2-parted. Placentæ pendulous, ∞-seeded. ♄ Small, alpine, moss-like or heath-like shrubs. Flowers solitary, pedicellate.

C. hypnoides Don. Stem filiform, tufted; leaves evergreen, subulate, smooth, crowded; flowers 5-parted, purple, nodding. High mts., N. H., N. Y., Me. 2—3'. Ju.

13. ANDRÓMEDA, L. Cal. 5-parted, persistent, not becoming fleshy in fruit. Cor. urceolate, the mouth more or less contracted, 5-toothed. Anth 10, cells 2, opening by a terminal pore. Caps. 5-celled, 5 valved, often re-enforced with 5 external valvelets. Seeds ∞. ♄ ♄ with entire, or serrulate, alternate leaves. Figs. 64, 638.

Order 73.—ERICACEÆ.

§ Sepals valvate in the early bud. Fls. in clusters. Caps. globular...(c)
§ Sepals imbricate in the bud. Capsule depressed...(a)
   a Fls. solitary, axillary. Pericarp double. Anth. awnless. (Cassandra)..Nos. 1, 2
   a Flowers in axillary racemes. Pericarp simple, with 5 entire valves...(b)
     b Anth. awnless. Bractlets at the base of the pedicels. (Leucothoe)..Nos. 3—5
     b Anth. 2-awned. Bractlets at the base of the calyx. (Eubotrys)....Nos. 6, 7
   c Flowers in a terminal nodding umbel. Cor. globular. (Euandromeda).....No. 8
   c Flowers in racemes, panicled or axillary...(d)
     d Capsule with 5 narrow valvelets applied to the sutures...(e)
     d Capsule naked. Corolla ovoid. Anthers 2-awned. (Portunia)...Nos. 9, 10
       e Corolla oblong. Filaments or anthers 2-awned. (Pieris)....Nos. 11—13
       e Corolla globular. Filaments and anth. awnless. (Lyonia)...Nos. 14—16

1 A. calyculàta L. *Leather-leaf.* Lvs. oblong, obtuse, flat, acute at base, rusty beneath; fls. white, each with a leaf, in leafy racemes: cal. 2-bracteted at base, sep. acute; inner pericarp 10-valved, thin. Bogs, Can. to Car. and Wis. 3f. April +.

2 A. angustifòlia Ph. Leaves linear-lanceolate, acute, the margins revolute; calyx segments acuminate, 2-bracteolate. Otherwise as No. 1. Swamps, S. Car., Ga.

3 A. axillàris Lam. Leaves oblong, acute, denticulate, petiolate; rac. dense, short, sepals roundish, obtuse. Banks, low country, Va. to Fla. 2—4f. Evergreen. Mar.

4 A. Catesbæi Walt. Lvs. lance-ovate, conspicuously pointed, petiolate, finely serrulate; rac. dense, nodding, nearly as long as the leaves; sep. ovate, acute. Banks, up-country, Penn. to Ga. 2—5f. Racemes 2—3′, white. Evergreen. May.

5 A. acuminàta L. *Pipe-wood.* Leaves very smooth, rigid, lance-ovate, gradually pointed, entire; rac. loose, short; branches hollow. Swamps, S. 3—10f. April.

6 A. racemòsa L. Lvs. lance-oval, slightly pointed, serrulate, deciduous; rac. strict, ascending, terminal, naked, long and 1-sided; sep. ovate, acuminate; anth. cells each 2-awned at apex; seeds wingless. Wet woods. 2—6f. Rac. 2—3′, white. Jn., July.

7 A. recúrva Buckley. Lvs. deciduous, lance-ovate, acuminate; anth. cells each 1-awned; pod 5-lobed; sds. winged, flat; branches recurved-spreading. Mts.,Va., N. Car.

8 A. polifòlia L. *Wild Rosemary.* Erect, smooth, glaucous; lvs. oblong-linear, with margins revolute, white beneath (2—3′); umb. 5-9-flwd., roseate. Bogs, N. 1f. Jn.

9 A. floribúnda Lyon (Ph.) Lvs. thick, evergreen, lance-oblong, acute or pointed, bristly-serrulate; rac. paniculate, crowded; bractlets minute; cor. white; anth. awns 2, reflexed, white. Mts., Va. to Ga. 2—10f. Flowers numerous and handsome. Apr.

10 A. phillyræfòlia Hook. Lvs. thick, shining, evergreen, elliptic-oblong, obtuse, serrulate above; rac. subterminal, loose; sep. lanceolate; cor. oval; anth. each with 2 long reflexed black awns. Woods, Quincy, Fla. 1—3f. (A. Croomii, C-B.)

11 A. nitida Bartram. *Fetter-bush.* Lvs. thick, evergreen, shining, elliptical, acuminate at each end, margins veined and revolute; umbels axillary, nodding, roseate; branches sharply angled. Low pine-barrens, S. 3—6f. March, April. Elegant.

   β? **rhombifòlia.** Leaves broad-oval; sepals ½ as long as the ovoid corolla. Fla.

12 A. Mariána L. *Stagger-bush.* Lvs. thin, deciduous, oval, entire, acutish; flowering branches leafless; fls. large (4—5″), white or reddish, in lateral crowded fascicles; sepals linear, ½ as long as the cylindric corolla. Sands, N. J. to Fla. 3f. June, July.

13 A. speciòsa Mx. Lvs. oval, obtuse, serrate, veiny, deciduous; flowering stems mostly leafless, branched; sepals ½ as long as the large bell-shaped white corolla. Swamps, S. June.—Varies with the leaves broad, crenate, whitish beneath.

14 A. ligustrìna Muhl. Pubescent; lvs. deciduous, lance-obovate to obovate, short-acuminate, serrulate; rac. panicled on the leafless flowering branches. Wet soils, Ct. to Fla. 6f. June.—Var. with small lvs. scattered among the small (1″) downy fls. S.

15 A. ferrugínea Walt. Lvs. thick, rigid, evergreen, obovate to oblanceolate, rusty beneath, revolute-edged; umb. axillary; fls. small (1″); valvelets nearly as broad as the valves. Pine-barrens, S. 3—20f. Shrub or small tree. Apr., May. (A. rigida Ph.)

16 A. montàna Buckley. Lvs. evergreen, lance-ovate, ciliate-serrulate; fls. in large panicles; pedicels pubescent, with 3 linear bractlets. Mts., N. Car. 4—6f.

ORDER 73.—ERICACEÆ.   203

14. **OXYDÉNDRUM**, DC. SORREL-TREE. Sep. bractless, valvate in the early bud. Cor. urceolate, ovoid, 5-toothed. Anth. 10, linear, erect, awnless, cells opening lengthwise. Capsule oblong, truncate, 5-celled, 5-valved. Seeds ∞. ♄ Lvs. petiolate, oblong-lanceolate, acuminate, serrulate. Flowers white, in terminal panicles of slender, spicate racemes.

O. arbòreum DC.—Ohio, Penn., and S. along the Alleghany Mts. Tree 40–50ft. Jn., Jl.

15. **LOISELEÙRIA**, Desv. ALPINE AZALEA. Calyx 5-parted, lobes equal. Cor. subcampanulate, 5-parted, regular. Sta. 5, equal, erect, shorter than the corolla, anth. dehiscing laterally. Style straight, included. Caps. 2- or 3-celled, 2- or 3-valved, ∞-seeded. ♄ Delicate, procumbent, tufted, with opposite, petiolate, entire leaves. Pedicels terminal, solitary, 1-flowered. Corolla rose-color.

L. procúmbens Desv.— Summit of the White Mts., N. H. A tiny shrub. 3—6′. Lvs. elliptical, 3″, margins revolute. Flowers nearly sessile. June, July.

16. **AZÀLEA**, L. SWAMP PINK. Cal. small, 5-parted. Cor. funnelform, somewhat irregular, with 5 spreading lobes. Sta. 5. Fil. and style long, exserted, declined, anth. opening by pores. Caps. 5-celled, 5-valved, ∞-seeded. ♄ Erect. Lvs. alternate, deciduous, oblong or obovate, entire. Flowers in umbelled clusters, terminal, large and showy. Fig. 114.

§ Calyx lobes all (or rarely one excepted) very short or minute..............Nos. 1, 2
§ Calyx lobes all oblong and of conspicuous length.—a Native................Nos. 3, 4
                                                              —a Exotic....................Nos. 5, 6

1 **A. viscòsa** L. Branchlets hispid; leaves obovate-oblong, the edges, midvein, and petiole bristly; fls. appearing after the lvs., very viscid, the tube much longer than the segments; stamens exserted; style much longer. Swamps. 4—10f. May—July.
  β. **nítida.** Lvs. smooth, green, shining, oblanceolate. Dry woods, N. 1—2f.
  γ. **híspida.** Lvs. very hispid above, smooth and glaucous beneath. Mts., Pa.

2 **A. nudiflòra** L. *Pinxter-bloom.* Young branchlets and lvs. beneath pubescent; clusters naked, appearing with or before the young lvs.; corolla slightly viscid, tube downy, scarcely longer than the segm. Woods; more common S. 3—7f. Apr.+.—Varies with the flowers *pink, deep purple, white-variegated, white* with a buff centre, and *buff* all over; the latter two fragrant. Also, with 10—20 stamens.

3 **A. calendulàcea** Mx. *Flaming Pinxter.* Young branchlets pubescent; lvs. attenuated to the base, corymbs nearly or quite leafless; tube of the cor. hirsute, not viscid, shorter than the ample lobes. Upland woods, O., Pa., and S. 3—10f. May, Jn.
—The splendid flowers vary to *yellow-scarlet, flame-color, brick-red, saffron-yellow,* &c.

4 **A. arboréscens** Ph. Branches smooth; lvs. obovate, glabrous, glaucous beneath, margins ciliate; corymbs leafy with full-grown leaves; corolla tube not viscid, longer than the lobes. Mts., Penn., and S. 10—20f. May—July.

5 **A. INDICA.** Strigous, but not glandular; lvs. wedge-lanceolate, acuminate, ciliate; fls. terminal, 1—3 together. Japan. Fls. scarlet, crimson, white, &c. Splendid.

6 **A. PÒNTICA.** Lvs. oblong, acute, margin ciliate; fls. viscid, corymbed, after the leaves; tube equalling the limb, yellow, very fragrant. Asia Minor.

17. **RHODODÉNDRON**, L. ROSE BAY. Calyx small, deeply 5-parted, persistent. Cor. campanulate, often slightly unequal, 5-lobed. Stam. 10 (rarely fewer), mostly declinate, anthers opening by 2 terminal pores. Caps. 5-celled, 5-valved, many-seeded. ♄ ♄ With alternate, entire leaves. Flowers in dense, terminal umbels from large, scaly buds. Figs. 99, 311

Order 73.—ERICACEÆ.

* Leaves obtuse at each end. Flowers purple or lilac, not spotted............Nos. 1, 2
* Leaves acute or acuminate, dotted or discolored beneath. Fls. spotted...Nos. 3, 4, 5
* Leaves acuminate, scarcely paler beneath. Flowers very broad, purple.........No. 6

1 **R. Lappónicum** Wahl. *Lapland Rose Bay.* Dwarf; lvs. elliptical, very small, roughened with concave rusty scales both sides; fls. small (7''), lobes equal, purple; sta. 5, 7, or 10, exserted. High mts., N. Eng., N. Y. 8—10', very bushy. June, July.
2 **R. Catawbiénse** Mx. *Catawba Rose Bay.* Lvs. oval, rounded-obtuse at each end, paler beneath. smooth; cal. lobes oblong, elongated; cor. broad-campanulate, lilac-purple, large (14''); stam. 10. High mts., Va., N. Car. 3—6f. Lvs. 3—5'. Jn. †
3 **R. punctàtum** Andr. Lvs. elliptical, acute or acuminate, glabrous, the lower surface and dense corymbs covered with resinous dots; fls. bell-funnel-form, pink-red. green-spotted within, the lobes wavy. Uplands, Car., Ga. 4—6f. Lvs. 2—3'. Jn., Jl. †
  β. *Chapmánii.* Lvs. oval-obovate, obtuse, small (1—2'); sepals minute. W. Fla.
4 **R. máximum** L. Lvs. obovate-oblong, acute, smooth, coriaceous, rusty beneath, revolute on the margin; cal. lobes oval, obtuse; cor. white to roseate, spotted within; lobes unequal, roundish. Along streams, N. Eng. to Ga. 6—20f. Splendid. †
5 **R.** ARBÒREUM. Lvs. lanceolate, silvery-spotted beneath; cor. lobes crenulate and curled, white, buff, red, crimson, &c.; calyx downy. Himmaleh Mts. 5—20f.
6 **R.** PÓNTICUM. Lvs. lance-oblong, attenuated to each end, smooth, green both sides; corolla bell-rotate; calyx smooth. Asia Minor. Low bush, flowers broad (2'), purple.

18. **RHODÒRA**, Dunham. Cor. adnate to the 5-toothed calyx, deeply divided into 3 segments, upper one much the broadest, 2–3-lobed at the apex, in bud enfolding the 2 lower. Sta. 10, declinate, fil. unequal, anthers opening by 2 pores. Caps. 5-celled, 5-valved. Cells many-seeded. ♄ With alternate leaves, and pale-purple flowers. April, May.

R. Canadénsis L.—Woods or swamps, N. Eng. to Penn. 2—3f. Fls. in terminal clusters, 1', appearing before the oblong leaves, which are downy-canescent beneath.

19. **BEJÀRIA**, Mut. Fls. heptamerous. Calyx 7-toothed, campanulate. Corolla of 7 distinct petals. Sta. 14. Caps. 7-celled, 7-valved, many-seeded. ♄ With alternate, entire lvs., and fls. in dense, racemous panicles.

B. racemòsa Vent. Branches hispid and glutinous; lvs. ovate-lanceolate, glabrous; racemes terminal, white. Sandy soils, Ga., Fla. 3—4f. June, July.

20. **LÈDUM**, L. LABRADOR TEA. Calyx minute, 4-toothed. Cor. 5-petalled, spreading. Sta. 5—10, exserted, anthers opening by 2 terminal pores. Caps. 5-celled, opening at the base. ♄ Lvs. alternate, entire, ferruginous-tomentous beneath, coriaceous. Fls. in terminal corymbs, white.

L. latifòlium Ait. Lvs. elliptic-oblong, strongly revolute at edge; sta. 5—7, scarcely exserted. Mountains, Penn., to Greenland. 2—4f. May—July.

21. **LEIOPHÝLLUM**, Pers. SAND MYRTLE. Calyx 5-parted. Pet. 5, ovate-oblong, spreading. Sta. 10, exserted, anthers dehiscing by lateral clefts. Caps. 3-celled, 3-valved, many-seeded. ♄ Glabrous, with erect branches. Lvs. alternate, entire, oval, coriaceous, revolute-edged. Corymbs terminal. Flowers white.

L. buxifòlium Ell.—Pine-barrens, N. J. to Car. 8—12f. Leaves shining. May.

22. **CLÈTHRA**, Gært. SWEET PEPPER-BUSH. Cal. 5-parted, persistent. Pet. 5, distinct, obovate. Sta. 10, exserted, anth. inverted in the bud, at length erect. Style persistent, stigma 3-cleft. Caps. 3-celled, 3-valved

ORDER 73.—ERICACEÆ.

∞ -seeded, enclosed by the calyx. ♄♄ Lvs. alternate, petiolate. Flowers white, in downy-canescent racemes. Bracts deciduous.

1 C. alnifòlia L. Lvs. cuneiform-obovate, acute, acuminately serrate, green on both sides, smooth or slightly pubescent beneath; racemes terminal, elongated, simple or branched; bracts subulate. Swamps, N. Eng. to Ga. 3—8f. Fragrant. July, Aug.
β. *tomentòsa.* Lvs. tomentous beneath; spikes subpanicled; fls. 3''. S. Apr.–Ju.
γ. *scabra.* Lvs. coarsely serrate, rough-downy both sides. Ga. (Bainbridge). Pet. 2''.

2 C. acumínàta Mx. Arborescent; lvs. glabrous, glaucous beneath, oval, acuminate, abruptly acute at base, finely serrate, on slender petioles; rac. terminal, solitary; bracts long, caducous. Mts., Va., Ky., to Car. 10—18f. Lvs. 4—6'. July, August.

23. **ELLIÓTTIA**, Muhl. Calyx small, 4-toothed. Corolla of 4 petals slightly cohering at base. Stamens 8, anth. sagittate. Style slender, with a capitate, undivided stig. Caps. 3-celled, 3-seeded. ♄ With virgate-branched, alternate, lanceolate, entire leaves, and terminal racemes of white flowers.

E. racemòsa Muhl.—Dry, rich soils, S. Ga. 4—8f. Racemes bractless. June.

24. **CYRÍLLA**, L. Cal. 5-parted, minute. Pet. 5, distinct, spreading. Sta. 5, anth. opening lengthwise. Style short, with 2 stig. Caps. 2-celled, 2-seeded, indehiscent. Seeds suspended. ♄ Branches irregularly whorled, with entire, elliptic-oblanc. lvs., and the white fls. in slender clustered rac.

C. racemiflòra Walt.—Sandy swamps, S. 12—18f. Lvs. 2—3'. Rac. 4—6'. June.

25. **MYLOCÁRIUM**, Willd. BUCKWHEAT TREE. Calyx 5-toothed, minute. Pet. 5, obovate, obtuse. Sta. 10, very short, fil. thickened below. Caps. corky, 2- or 3-winged, 3-celled, with 3 subulate seeds. ♄ Very smooth, with branches irregularly whorled, elliptical leaves, and terminal racemes of white, fragrant flowers.

M. ligustrinum Willd.—Borders of swamps, Ga. and Fla. 4—8f. April, May.

26. **PÝROLA**, Salisb. WINTERGREEN. Cal. 5-parted. Pet. 5, equal. Sta. 10, anth. large, pendulous, fixed by the apex, 2-horned at base, opening by 2 pores at top. Style thick, as if sheathed. Stig. 5, appearing as rays or tubercles. Caps. 5-celled, opening at the angles, many-seeded. ♃ Low, scarcely shrubby, evergreen herbs. Lvs. radical or nearly so, entire. Scapes mostly racemous, from a decumbent stem or rhizome. Fig. 99.

§ Stamens and style straight. Stigmas peltate, 5-rayed. June, July............Nos. 1, 2
§ Stamens ascending. Style declined and curved. Stigma 5-tubercled...(a)
    a Leaves dull (not shining). Petals greenish-white....................Nos. 3, 4
    a Leaves thick and shining. Flowers white or rose-colored.............Nos. 5, 6

1 P. mìnor L. Lvs. round-ovate, repand-crenulate, longer than their petioles; rac. spike-like; corolla globular, including the short style. Woods, N. H., and N. July

2 P. secúnda L. Lvs. broadly ovate, acute, subserrate, longer than the petiole; rac. secund; cor. oblong; style exserted. Woods, N. States. 5—8'. Lvs. near the base
β. *pumila* (Paine). Lvs. nearly orbicular, thin; scape 3- 6-flowered 4 8' N Y.

3 P. chloràntha Swartz. Lvs. orbicular, crenulate, shorter (1') than the petiole; scape tall (6—12'), few-flowered; segm. of the cal. very short, obtuse; pet. half-open, oval, greenish; anth. conspicuously tubular. Woods, N. States and Can. June, July.

4 P. ellíptica N. Leaves oval or elliptical, thin, longer than their petioles; scape naked, 6-10-flowered; sep. very short and obtuse; anth. pores blunt; fls. nodding, fragrant. Woods, N. States and Can. 3—9'. Petioles white. June, July.

ORDER 73.—ERICACEÆ.

5 P. rotundifòlia L. Lvs. round-ovate, shorter than the petiole, thick; scape 3 angled, bracted below, ∞-flowered; sepals ovate, obtuse; anther pores distinctly tubular. Woods, Can. to Car., and W. 8—14′. Flowers large. June, July.
β. *uliginòsa*. Lvs. dull, 1½′, the stalk much longer; sep. acute; fls. smaller.
6 P. asarifòlia Mx. Lvs. round-reniform, thick, shining, shorter than the petiole; scape angular; rac. lax, ∞-flowered; sepals lanceolate, acute; anther pores blunt. Old woods, N. States and Can. 6—12′. Flowers purple. June.

27. CHIMÁPHILA, Ph. PIPSISSIWA. Cal. 5-parted. Pet. 5, spreading. Stamens 10, fil. dilated in the middle, anth. cells produced into tubes, opening by a 2-lipped pore at apex. Style very short, thick. Capsule 5-celled, opening from the summit. ♄ Small, glabrous. Leaves cauline, serrate, thick. Ped. scape-like. Flowers terminal, nodding, roseate. Fig. 255.

1 C. umbellàta Nutt. *Prince's Pine.* Lvs. cuneate-lanceolate, shining, 1-colored, serrate, in 4's—6's; umbel 4-7-flowered. Dry woods. 8—12′. July.
2 C. maculàta Pursh. Lvs. lanceolate, acuminate, rounded at base, remotely serrate, discolored, opposite or in 3's; ped. 2-3-flowered. Sandy woods. 6—8′. Jn., Jl.

28. MONÈSES, Salisb. Calyx 5-parted. Cor. 5-parted, rotate. Sta. 10, regular, 2-spurred at base, opening by 2 tubular pores at apex. Style straight. Stig. 5-lobed. Caps. 5-valved, 5-celled, ∞-seeded. ♃ Low, simple, smooth. Lvs. at top of the stem, roundish, serrulate, petiolate, veiny. Peduncle terminal, longer than the stem.

M. grandiflòra Salisb.—Mossy woods, N. Eng., N. Y.: rare (com. in Oreg.) 3′. Scape with a bract in the midst, and a single, terminal nodding white flower, 6″ broad. Jn.

29. SHÓRTIA, Gray. (This genus was founded upon an imperfect specimen in the Herbarium of Michaux, labelled, "High mountains of Carolina." It has never been seen in this country, but grows in Japan.)

30? GALAX, L. BEETLE-WEED. Cal. of 5 distinct, persistent sepals. Cor. of 5 oblong-obovate, distinct petals. Fil. 10, united into a tube with as many teeth, those opposite the petals sterile. Anth. 5, 1-celled, opening across the top. Caps. 3-celled. Seeds ∞, enclosed in a loose, cellular testa. ♃ Roots tufted, creeping, deep red, sending up roundish-cordate, long-stalked, glabrous leaves and a scape bearing a dense raceme of white flowers. (Shortia and Galax have been lately referred to Diapensiaceæ.)

C. aphýlla L.—Damp woods, Md. to Tenn., and S. Lvs. 2—3′. Scape 1—2f. Jl., Aug.

31. MONÓTROPA, L. INDIAN PIPE. PINE SAP. Sep. 1—5, bract-like. Pet. 4—5, connivent in a bell-shaped corolla, gibbous at base. Sta. 8—10, anthers opening transversely at apex. Stig. 5-rayed. Caps. 4-5-celled, 4—5-valved. Seeds ∞, minute.—Low, parasitic herbs, destitute of green herbage, furnished with scale-like bracts instead of leaves.

§ Sepals (or bracts) 1—3. Flowers solitary, scentless. Style very short.......No. 1
§ Sepals 4 or 5. Flowers in a secund raceme, fragrant. Style long.............No. 2

1 M. uniflòra L. *Indian Pipe. Bird's-nest.* St. short; scales approximate; fl. nodding; fr. erect. Common in woods. 6—8′. Plant whitish. June—Sept.
2 M. Hypópitys L. *Pine Sap. Bird's-nest.* More or less downy; pedicels as long as the flower; caps. subglobous. Woods: com. 6—10′. Plant tawny. June—Aug.

ORDER 73.—ERICACEÆ.

32. SCHWEINÍTZIA, Ell. CAROLINA BEECH-DROPS. Calyx persistent, of 5 erect, ovate-acuminate sepals. Corolla persistent, campanulate, limb 5-lobed. Sta. 10, anthers awnless, opening by pores at apex. Style thick, stig. large, 5-angled, caps. 5-celled, 5-valved. Seeds numerous, minute. Plant leafless, brownish. Flowers subsessile, capitate, reddish-white, with the odor of the violet.

S. odoráta Ell.—Woods, Md. to Car. 3—5'. Habit of Monotropa. February, March.

33. PTERÓSPORA, Nutt. ALBANY BEECH-DROPS. Calyx 5-parted. Cor. urceolate, roundish-ovoid, the limb 5-toothed and reflexed. Sta. 10, anthers peltate, 2-celled, 2-awned, opening lengthwise. Caps. 5-celled, 5-valved. Seeds very numerous, minute, winged at the apex. ♃ Leafless, brownish-red, simple, viscid-woolly. Fls. racemed, white.

P. Andromedèa Nutt.—Near Albany, N. Y. (A. Stores), N. and W.: rare. 12—30'. Rac. erect, loose, with 40 or more drooping fls. resembling those of Andromeda. Jl.

ORDER LXXIV. AQUIFOLIACEÆ. HOLLYWORTS.

Shrubs or *trees*, with simple, coriaceous, exstipulate leaves. *Flowers* small, axillary, sometimes diœcious. *Sepals* 4—6, imbricate in bud, very minute. *Corolla* regular, 4-6-cleft or parted, hypogynous, imbricate in æstivation. *Stamens* inserted into the very short tube of the corolla and alternate with its segments. *Anthers* adnate. *Ovaries* free from the calyx, 2-6-celled, with a solitary, suspended ovule in each cell. *Fruit* drupaceous, with 2—6 stones or nucules. *Albumen* large, fleshy.

§ Habitually tetramerous.	Drupe with 4, bony, sulcate nutlets............................ILEX.	1
§ Habitually tetramerous.	Drupe with 4, horny, smooth nutlets........................NEMOPANTHES	.
§ Habitually hexamerous.	Berry with 6 (7, 8) smooth, cartilaginous seeds............PRINOS.	3

1. ILEX, L. HOLLY. Fls. 4- (rarely 5-) parted, mostly perfect, but many abortive. Calyx 4-toothed, persistent. Pet. 4, distinct or scarcely united at base. Sta. 4. Stig. 4, or united into one. Drupe red, with 4 bony nutlets, ribbed and furrowed on the convex back. ♄ ♄ ♄ Leaves alternate. Flowers small, white, lateral, single or clustered.

* Trees evergreen. Leaves armed with spinous teeth..........................No. 1
* Shrubs evergreen. Leaves unarmed, serrate or entire ....Nos. 2—4
* Shrubs deciduous. Lvs. thin.—α Pedicels short as the petioles..........Nos. 5, 6
      —α Ped. (the sterile) longer than petioles.....No. 7

1 I. opàca Ait. Lvs. thick, smooth, oval, spinescent at apex, and with remote, repand, spinescent teeth; drupe ovoid, nutlets 5-ribbed on the back. Woods, Mass to Ga. and La. 15—30f. A beautiful evergreen. June.
 β. *integra.* Lvs. entire, only a few of them 1-3-toothed. Tree, S.

2 I. Dahoon Walt. Downy, more or less; lvs. 2—3', oblong to oblanceolate, thick, shining above, pale beneath, entire, acute or obtuse; sterile ped. ⚥-flowered, fertile few-flowered; nutlets 3-ribbed. Swamps, Va., and S. 5—12f. May.
 β. *ligustrina* has narrow, wedge-lanceolate, acute, subserrate leaves. South.

3 I. myrtifòlia Walt. Nearly smooth; lvs. very small (5—9''), oblong-linear, thick, serrulate when young, subsessile; pedicels 1-9-flowered. Pine-barren ponds, Md. to Fla. 12—20f. Stems straggling, light gray. Very unlike No. 2. May.

4 I. Cassine Walt. *Cassena Tea.* Smooth; lvs. small (10—18″), elliptical, obtuse, crenate, thick, shining; ped. about 3-flowered. Coastward, S.: common. 6—15f, bushy. March, April. Was used as a tea by the Creek Indians.
5 I. decídua Walt. Nearly smooth; lvs. thin, 1—2′, lance-ova¹, pointed, blunt-serrate; ped. short as the petioles, the ♂ clustered; seeds obtusely ribbed. S. 6—9f.
  β. *urbana.* Lvs. 2—3′, oval, obtuse, tapering to the base. Ill., and S. May.
6 I. Amelánchier Curt. Leaves (variable) ovate, oblong to lanceolate, acute or pointed, serrulate, thin, downy beneath; ped. short as the petioles, ♂ clustered, ♀ solitary; drupe red. Hills and mts., N. Y. to S. Car. (Prinos ambiguus Ph.)
  β. *monticola.* Lvs. large (3—5′), glabrous, the short ped. and cal. some downy.
7 I. ambígua Chapm. Lvs. oval or elliptical, acute (scarcely pointed), serrulate or nearly entire, smoothish; ♂ ped. much longer than the pet., clustered, ♀ short, solitary. Wet grounds, S. 4—8f. March, April. (Prinos ambiguus Mx.)

2. NEMOPÁNTHES, Raf. Parts of the flower in 4's or 5's. Calyx very small. Petals linear-oblong, shorter than the stamens. Stig. sessile Drupe globular, red, with 4, rarely 5, smooth, horny nutlets (seeds). ♄ Lvs. entire, smooth, thin. Fls. white, small, on slender pedicels, ♂ ♀ ☿.

N. Canadénsis DC.—N. Eng. to Mich. Shrub 4—6f. Lvs. 2′. Ped. 9—12′. May, Jn.

3. PRINOS, L. Winter-berry. Fls. small, habitually 6-parted and perfect, but often fruitless. Calyx 6-cleft. Cor. monopetalous, subrotate, 6-parted. Sta. 6 (in the sterile flowers rarely fewer, in the fertile rarely more). Berry 6-seeded, seeds with a smooth, cartilaginous testa. ♄ ♄ With alternate lvs., small white fls., and red or black berries. (See *Addenda.*)

§ Leaves deciduous, thin. Berries red. (No. 3a, p. 446, and)............ Nos. 1—3
§ Leaves evergreen, thick, shining. Berries black.........................Nos. 4, 5

1 P. verticillàtus L. *Black Alder.* Lvs. oblanceolate or elliptical, acuminate, mucronate-serrate, small; pedicels shorter than the petioles; berries scarlet, in close bunches as if verticillate, all Winter. Low woods. 8f. Leaves 1—1½′. July.
2 P. lanceolàtus Ph. Lvs. lanceolate, long-acuminate, sharp-serrate, glab., 1—3′; fls. subsessile, the sterile 3-androus; berries large, red. Swamps, S. (Dr. J. Hale.)
3 P. lævigàtus Ph. Leaves lanceolate, appressed-serrulate, glabrous, shining above, short-acuminate; ped. longer than the pet., in 2's or 3's. Swamps, Can. to Va. 7f. Jn.
4 P. glaber L. *Ink Berry.* Lvs. coriaceous, cuneate-lanceolate, glabrous, serrate at the end; ped. longer than the pet., 1–3-flowered. Swamps, Ms. to La. 3—4f. Jn., Jl.
5 P. coriàceus Ph. Lvs. thick, obovate, serrate at the end, glabrous, shining; fls. all solitary, on very short peduncles, 6-8-parted. Woods, S. 4—6f. Lvs. 2′. May.

Order LXXVI. STYRACACEÆ.

Trees or *shrubs* with alternate, simple leaves, destitute of stipules. *Flowers* or *racemes* solitary, axillary, bracteate. *Calyx* 5-, rarely 4-lobed. *Corolla* 5-, rarely 4- or 6-lobed, imbricated in bud. *Stamens* definite or ∞, unequal in length, usually cohering. *Anthers* innate, 2-celled. *Ovaries* adherent, 2–5-celled, the partitions sometimes hardly reaching the centre. *Fruit* drupaceous, generally with but one fertile cell. *Seeds* 5—1.

Tribe I. SYMPLOCINEÆ. Calyx 5-cleft. Anth. ∞, innate, globular. Fls. yellow.....Symplocos. 1
Tribe II. STYRACEÆ. Calyx mostly truncate. Anthers 8—12, linear-oblong, adnate.
  Flowers white,—a 5-parted. Fruit wingless, 1-seeded....................Styrax. 2
      —a 4-parted. Fruit winged, 2- or 3-seeded.................Halesia. 3

1. SÝMPLOCOS, Jacq. Cal. 5-cleft. Cor. 5-parted, spreading. Sta. ∞, in 5 clusters, one attached to the base of each petal. Fil. slender. Anth. globular. Ovary 3-celled, half-adherent. Drupe dry, with a 3-celled, mostly 1-seeded nut. ♄ ♄ With clusters or racemes of small yellow flowers.

S. tinctòria L'Her. Lvs. oval or elliptical, acuminate, acute at base, thick; fls. sessile, in axillary, dense clusters of 6—12; calyx lobes ovate, obtuse. Va., and S. 10—20f. Drupe ovoid, 6″. The dried leaves dye yellow. March, April.

2. STYRAX, Tourn. Cor. deeply 5-parted, much longer than the campanulate calyx. Sta. 10, joined to the base of the corolla, fil. united into a short tube at base. Anth. linear, erect. Ov. adherent at base. Fr. coriaceous, 1-celled, mostly 1-seeded. ♄ With alternate leaves and axillary racemes of white, drooping, showy flowers. March—May.

1 S. pulverulénta Mx. Pulverulent-downy; lvs. broadly oval, obtuse, glandular-serrulate; fls. axillary and terminal. Va. to Fla. 2—3f. Petals 6″.
2 S. Americàna Lam. Plant glabrous; lvs. oblong or elliptical, acute at each end; rac. leafy, few-flowered, cor. often downy. Swamps, Va., and S. 4—8f.
3 S. grandifòlia Ait. Lvs. ample, broadly obovate, acute or short-acuminate, hoary tomentous beneath; racemes leafless, longer than the leaves. Va. to Fla. 6—12f.

3. HALÈSIA, Ellis. SNOWDROP TREE. Cal. obconic, briefly 4-lobed. Cor. inserted into the calyx, campanulate with a narrow base, 4-parted. Sta. 8—12, connate into a tube below. Sty. filiform. Fr. dry, 2—4-winged. Sds. 1—3. ♄ ♄ Lvs. alternate, abruptly acuminate, finely denticulate or entire. Flowers in advance of the leaves, pendulous, in lateral clusters of 3—5, white, showy.

1 H. tetráptera L. Lvs. oblong-ovate; fls. 6″ long; pet. half-united; stam. 12; fr. equally 4-winged. Woods, Va. to Ky., and S. Shrub 10—20f. April.
2 H. díptera L. Lvs. oblong-ovate; fls. 1′ long; pet. slightly united; stam. 8; fruit 2-winged. Woods, S. Tree 15—30f, often 50f. Lvs. 6′. Pods near 2′. April, May.

ORDER LXXVII. EBENACEÆ. EBONADS.

Trees or *shrubs* without milky juice and with a heavy wood. *Leaves* alternate, exstipulate, coriaceous, entire. Inflorescence axillary. *Flowers* by abortion diœcious, seldom perfect. *Calyx* free, 3–6-cleft, divisions nearly equal, persistent. *Corolla* regular, 3–6-cleft, often pubescent, imbricate in æstivation. *Stamens* twice or 4 times as many as the lobes of the corolla. *Fruit* a fleshy, oval, or globous berry. *Seeds* large, suspended, albuminous.

DIOSPYROS, Dalesch. PERSIMMON. Fls. ♂ ♀. Cor. tubular or campanulate, convolute in bud. ♂ Sta. mostly 16. Fil. shorter than the anthers. Style 0. ♀ Sta. mostly 8, without anthers. Style 2–4-cleft. Berry ovoid or globous, 4-12-, mostly 8-celled, cells 1-seeded. ♄ ♄ A large genus, mostly tropical.

D. Virginiàna L. Lvs. elliptic, abruptly acuminate, entire; racemes axillary, 3-1-flowered, pedicels shorter than the flowers; calyx 4-parted; stamens 8. Woods, lat 42°, and S. 10—30f. Berry large as a plum, sweet after frost.

Order LXXVIII. SAPOTACEÆ. Soapworts.

Trees or *shrubs*, mostly with a milky juice, and simple, entire leaves. *Flowers* small, regular, perfect, mostly in axillary clusters. *Calyx* free, persistent. *Corolla* hypogynous, short, *stamens* usually as many as its lobes and opposite to them, inserted into its tube along with one or more rows or appendages. *Anthers* extrorse. *Ovary* 4–12-celled, with a single anatropous ovule in each cell. *Seeds* large. (Included Theophrastaceæ.)

* Corolla 6–8-cleft, with a pair of appendages at each sinus. S. Fla............MIMUSOPS *Sieberi* DC.
* Corolla 5-cleft,—*a* with a single appendage at each sinus. S. Fla.........SIDEROXYLON *pallidum* Jq.
 —*a* with a pair of, &c.—*b* Sterile stamens fringed. S. Fla...DIPHOLIS *salicifolia* A. DC.
 —*b* Sterile stamens entire...................................BUMELIA. 1

BUMELIA, Swartz. Cal. 5-parted. Cor. 5-cleft, with a pair of appendages between the lobes. Sta. 5, opposite the lobes, alternate with 5 petaloid, sterile stamens. Ov. 5-celled. Sty. filiform. Drupe ellipsoid, 1-seeded, exalbuminous. ♄ ♃ Wood hard and firm. Lvs. entire, of a firm texture. Fls. aggregated, white or greenish. Our species are all more or less spiny, and with very tough twigs.

* Leaves hairy beneath......Nos. 1, 2. ** Leaves glabrous both sides....Nos. 3, 4

1 **B. tenax** Willd. Silky-ferruginous; lvs. wedge-oblong to obovate, obtuse; clusters 20–35-flwd., with slender pedicels; drupe oval, corrugated. Sands, S. 20–30f. Jn., Jl.
2 **B. lanuginòsa** Pers. Woolly-ferruginous; lvs. oval, acutish, thin; fascicles 6–12-flwd., with short pedicels; drupe globular. Damp. S. Ill. and S. 8–12f. June, Jl.
3 **B. lycioìdes** Gært. Lvs. wedge-elliptical, rather acute; clusters densely 20–30-flwd., ped. shorter than petioles (2–3″). Damp, Ky., and S. 15–25f. Branches virgate. May.
4 **B. reclinàta** Vent. Lvs. obovate, obtuse, small (9–12″); clusters 15–20-flwd.; ped. slender, half as long as the leaf. River banks, S. Car. to Fla. A straggling shrub. Jn.,Jl.

Order LXXXI. PRIMULACEÆ. Primworts.

Herbs low, with the leaves mostly radical or mostly opposite. *Flowers* 5- (rarely 4–6-) parted, regular and monopetalous. *Stamens* 5, inserted on the corolla tube and opposite to its lobes. *Ovary* 1-celled, with a free central placenta. *Style* 1. *Stigma* 1. *Capsule* 1-celled, ∞-seeded. *Seeds* with fleshy albumen. Figs. 22, 133, 249.

§ Ovary half-inferior. Capsule opening by valves. Leaves undivided. (Tribe IV.)
§ Ovary superior.—* Capsule opening by valves. Leaves pectinate. (Tribe I.)
  —* Capsule opening by valves. Leaves undivided. (Tribe II.)
  —* Capsule opening by a lid. Leaves undivided. (Tribe III.)
  I. HOTTONIEÆ. Corolla salver-form. Plants floating. Leaves verticillate....HOTTONIA. 1
  II. PRIMULEÆ.—*a* Acaulescent.—*b* Corolla limb spreading, tube cylindrical....PRIMULA. 2
      —*b* Corolla limb spreading, tube ovoid........ANDROSACE. 3
      —*b* Corolla lobes reflexed.—*c* Stam. exserted...DODECATHEON. 4
          —*c* Stam. included...CYCLAMEN. 5
      —*a* Caulescent.—*d* Corolla wanting. Leaves opposite..........GLAUX. 6
          —*d* Corolla 7-parted. Leaves in one whorl......TRIENTALIS. 7
          —*d* Cor. 5- or 6-parted. Lvs. opp. or whorled....LYSIMACHIA. 8
  III. ANAGALLIDEÆ.—*e* Flowers 5-parted, scarlet. Leaves opposite..........ANAGALLIS. 9
          —*e* Flowers 4-parted, white? Leaves scattered..........CENTUNCULUS. 10
  IV. SAMOLEÆ. Flowers 5-parted. Leaves alternate................. ....SAMOLUS. 11

Order 81.—PRIMULACEÆ.

1. HOTTÒNIA, L. WATER-FEATHER. Calyx 5-parted. Cor. sa.ver-form, with a short tube, and a flat, 5-lobed limb. Sta. inserted in the tube of the corolla, included. Stig. globous. Caps. globous-acuminate. ♃ Fleshy, with pectinate-pinnatifid, submersed, radical leaves.

H. inflàta Ell. St. immersed, with a whorl of lvs. (1–2′) at or near the surface; scapes clustered, jointed, hollow, 3—10′, bearing several whorls of small white fls. Pools, N. and S. April—June. Curious.

2. PRÍMULA, L. PRIMROSE. AURICULA. Cal. angular, 5-cleft. Cor. salver-shaped or often rather funnel-shaped, with 5 entire or notched or bifid lobes. Sta. included, fil. very short. Caps. ovoid, 5-valved, valves often bifid, opening at the top, ∞-seeded.—Herbs with the leaves all radical and flowers in an involucrate umbel, often showy.

```
 * Native, wild species. Corolla salver-form, the lobes abruptly spreading...Nos. 1, 2
 * Exotic.—a Corolla salver-form, the lobes abruptly spreading............Nos. 3, 4
    —a Corolla funnel-form.—b Leaves rugous, hairy, toothed..........Nos. 5, 6
                           —b Leaves plain, smooth, often entire.....Nos. 7, 8
```

1 **P. Mistassínica** Mx. Lvs. spatulate, dent-crenate, green both sides; invol. 1-8-flwd., ⅓ as long as pedicels; cor. lobes obcordate, tube much exserted. Lake shores, Vt. (Willoughby) N. Y. (Seneca), and N. 3—7′. Fls. 5″ broad, white. Jn. Delicate.
2 **P. farinòsa** L. *Bird's-eye P.* Lvs. lance-elliptic, obtuse, dentic. at apex, whitish-mealy beneath, as well as the 3-20-flwd. invol.; cor. pale-purple, with a yellow centre, its lobes bifid. Lake shores, Mich., Me. (A. H. Smith), and N. 6—12′. June, July.
3 P. GRANDIFLÒRA. *Common P.* Lvs. obovate-oblong; umb. radical; cor. limb flat, yellow, varying to all shades of orange, and red, to white, single or double. Europe.
4 **P. PURPÙREA.** Lvs. lanceolate, obfuse, yellowish-mealy beneath; scape longer than the leaves; invol. ∞-flwd., as long as the pedicels; lobes entire, dark-purple. Nepal.
5 **P.** OFFICINÀLIS. *Cowslip P.* Lvs. oblong, hairy beneath; fls. all nodding; cal. angular; cor. concave. Endless varieties are raised from the seed. Europe. (P. veris.)
6 **P.** ELÀTIOR. *Ox-lip P.* Lvs. hairy both sides; outer fls. nodding; cor. flat. Eur. 1f. Yel.
7 **P.** AURÍCULA. Lvs. obovate, fleshy; scape ∞-flowered, as long as the leaves; bracts short; calyx powdery. Alps. The varieties are innumerable and beautiful.
8 **P.** CALYCÌNA. Leaves lanceolate, entire, acute, edged with white; invol. 3-5-flwd., as long as the pedicels; cal. tube inflated; corolla lobes emarginate. Austria. Purple.

3. ANDRÓSACE, Tourn. Cal. 5-cleft or toothed. Cor. funnel-form or salver-form, the 5 lobes entire, tube constricted at the throat, ovate, shorter than the calyx. Fil. and style very short. Caps. globous. Minute cæspitous herbs, with radical, rosulate leaves. (Scape bearing an umbel.)

A. occidentàlis Ph. Lvs. oblong-spatulate and ovate, entire, glabrous; scape ∞-flowered; bracts oval, pedicels slender; calyx angular, segments longer than the small white corolla. (1) Gravelly shores, Ill., and W. 1—3′.

4. DODECÁTHEON, L. AMERICAN COWSLIP. PRIDE OF OHIO. Cal. 5-parted, reflexed. Cor. tube very short, limb 5-parted, segm. reflexed. S'a. 5, inserted into the throat of the corolla. Fil. very short. Anth. large, acute, connivent at apex. Style exserted. Caps. oblong-ovoid, 5-valved, ∞-seeded. ♃ Root fibrous, with radical, oblong leaves, an erect, simple scape, and a terminal umbel of nodding white flowers and erect fruit.

D. Meàdia L.—Ohio, Penn. to Cal.! common in prairies. Whole plant glabrous, 1-2f scape 9-20-flowered, usually about 12-flowered. Singularly elegant. May, June.

Order 81.—PRIMULACEÆ.

5. CYCLAMEN, L. Cal. bell-shaped, 5-parted. Corolla tube ovate, short, limb 5-parted, reflexed. Anth. 5, included, sessile. Caps. globous, 5-valved.—Oriental herbs. Root a large tuber. Leaves all radical, ovate or roundish, cordate. Scapes naked, erect, with one nodding flower, but in fruit coiling up and hiding the capsule in the ground.

1 C. Europæum. Lvs. crenate; petals lance-ovate, fragrant, roseate. Europe.
2 C. Coum. Lvs. entire; petals round-ovate, inodorous, purple. Asia Minor.

6. GLAUX, L. Black Saltwort. Calyx campanulate, 5-lobed, colored. Corolla none. Sta. 5. Caps. roundish, surrounded by the calyx, 5-valved, 5-seeded. ♃ Maritime, branching, glabrous, with opposite leaves and small, axillary, solitary flowers.

G. **marítima** L.—Salt marshes, Can. to N. J. Plant fleshy, branching, leafy, 4—6′; lvs. round-ovate, obtuse, entire, darkly glaucous; calyx reddish-white. July.

7. TRIENTÀLIS, L. Chickweed-Wintergreen. Cal. and cor. 7-(6-8-) parted, spreading. Sta. 7 (6—8). Fruit capsular, somewhat fleshy. ∞-seeded. ♃ St. low, simple. Lvs. subverticillate. Pedicels 1-flowered.

T. **Americàna** Ph. St. erect, simple, leafless at base; lvs. glomerate at top of the stem, few, narrow-lanceolate, serrulate, acuminate; sepals linear, acuminate. Rocky woods: com. 3—6′. Pedicels 1—4, filiform; corolla white, starlike, 6″. May, June.

8. LYSIMÁCHIA, L. Loose-strife. Fls. 5-(rarely 6- or 7-) parted. Cor. wheel-shaped, the petals nearly or quite distinct. Sta. 5, on the base of the corolla. Fil. often somewhat connate or with intervening, sterile ones. Capsules globous, 5–10-valved, opening at the apex. Seeds few or many. ♃ With opposite or verticillate entire leaves. (Flowers yellow.)

§ Petals 5—7, distinct, dotted, with 5—7 intervening teeth. (Naumbergia)........No. 1
§ Petals 5, united at base, that is, monopetalous...(a)
  a Sterile filaments 0, the perfect stamens monadelphous...(c)
  a Sterile filaments 5 short teeth alternate with the perfect stamens...(d)
    c Flowers whorled, in a long, terminal, bracted raceme............Nos. 2, 3
    c Flowers not racemed—axillary or paniculate.....................Nos. 4—6
    d Leaves acute at base, tapering to the short petiole. ....Nos. 7, 8
    d Leaves rounded or abrupt at base, long-petioled............Nos. 9, 10

1 L. **thyrsiflòra** L. St. simple; lvs. dotted, linear-elliptical, pointed, sessile; thyrsoid racemes from the middle axils pedunculate, shorter than the leaves; pet. linear, brown-dotted. Meadows, N. Eng. to O., and N, 2f. June. (Naumbergia C-B.)

2 L. **stricta** Ait. Lvs. opposite, rarely in 3's, lanceolate to lance-linear, acute, sessile, dotted; axils producing bulblets after flowering; fls. whorled, in a long, open, terminal raceme, yellow, with purple streaks. Low grounds. 1—2f. July.
    β. *angustifòlia* (Chapm.) Lvs. very narrow, obtuse; petals acute. South.

3 L. **Herbemónti** Ell. St. simple: lvs. whorled in 4's or 5's, ovate to lance-ovate, pointed, sessile, revolute at edge, dotted; fls. racemed, dotted. Carolina: rare. 2f.

4 L. **Fr seri** Duby. Glandular-downy at top; lvs. opposite, ovate or ovate-cordate, pointed, petiolate, dotted; fls. in a terminal panicle; sep. fringed. S. Car. (Fraser).

5 L. **quadrifòlia** L. Erect, simple; lvs. in whorls of 4's (rarely 5's or 3's), lanceolate, pointed, sessile, dotted; ped. slender, solitary in each axil; pet. oval, obtuse. Damp shades, Can. to Car. and Ky. 18′. Corolla yellow, with purple lines. June.

6 L. **nummulària** L. *Moneywort.* Trailing, weak; lvs. roundish, subcordate, on short petioles, opposite, dotless; fls. solitary, large, showy. Fields and gardens. §

Order 82.—PLANTAGINACEÆ.

7 **L. longifòlia** Ph. St. slender, flexuous, 4-angled; lvs. linear, shining, revolute at edge; fls. large, in pairs or 4's, terminal on the stem or short branches; petals broad-ovate. erose-dentate; anthers large. Low prairies. W. and S. 1f—20'. July.

β. *tenuis*. Leaves lance-linear, flat, edges not revolute. Miss. and La.

8 **L. lanceolàta** Walt. St. angular above; leaves lance-oblong, acute at each end, subsessile, veiny, ciliate at base; ped. solitary, axillary. Meadows. 12—18'. July.

β. *heterophylla*. Lower lvs. oval or oblong, petiolate; flowers at the summit.

9 **L. ciliàta** L. St. erect, 4-angled; lvs. opposite, ovate to lance-ovate, rounded at base, petioles distinct, *ciliate ;* flowers nodding, mostly opposite, in the upper axils, large (1'); stamens distinct. Thickets, along streams. 2—3f. Often branched. Jl.

β. *tonsa*. Pet. entire, destitute of ciliæ; lvs. and fls. smaller. Mts., Ky., Tenn.

10 **L. radicans** Hook. St. square, long, trailing, *rooting* at the joints; br. slender; lvs. lance-ovate, acute, on long pet.; fls. small (4''). Swamps, Va., and S. 2—4f. Jl.

9. ANAGÁLLIS, L. Scarlet Pimpernel.
Calyx 5-parted. Cor. rotate, deeply 5-parted, tube 0. Sta. 5, hairy, anth. introrse. Caps. globular, thin, opening all around (pyxis).—Herbs with square stems and opposite or whorled entire leaves. Pedicels axillary, solitary. Fig. 249.

A. arvénsis L. Procumbent; lvs. broad-ovate, sessile, shorter (6—10'') than the curved ped.; sepals lance-linear, as long as the roundish crenate-glandular, red petals. ⑴ Fields, waysides. The flowers (sometimes blue, Dr. Buel) close at 2 p. m., or on the approach of foul weather; hence called the *Poor Man's Weather-glass*.

10. CENTÚNCULUS, L. False Pimpernel.
Cal. 4-parted. Cor. urceolate-rotate, 4-cleft, shorter than the calyx. Sta. 4, beardless, united at base. Capsules globous, circumscissile. Seeds very minute. ⑴ Very diminutive, with alternate lvs. Fls. axillary, solitary, subsessile, white?

C. mínimus L. St. ascending, branched; leaves subsessile, oval, obtuse, entire, the lower opposite; sep. linear-subulate. Wet, Ill., and S. 1—6'. April—July.

11. SÁMOLUS, L. Water Pimpernel.
Calyx partly adherent, 5-cleft. Corolla salver-form, 5-cleft. Sta. 5, alternating with 5 scales (sterile filaments). Caps. dehiscent at top by 5 valves, many-seeded.—Herbs with alternate lvs. Flowers corymbous or racemous. May—Aug. Figs. 22, 133.

1 **S. Valerándi** L. (S. floribundus K.) St. simple or branched; lvs. obtuse, wedge-oval, the lower petiolate; fls. in a raceme or panicle of racemes, pedicels with a minute bract near the middle; petals longer than the sepals. Wet gravels. 6—12'.

2 **S. ebracteàtus** Kunth. Erect, leafy below; lvs. obovate-spatulate; fls. racemed, ped. bractless; cor. white, 3 times longer than the calyx (3''). Marshes. Fla., and W.

Order LXXXII. PLANTAGINACEÆ. Ribworts.

Herbs rarely shrubby, with radical leaves and the flowers in spikes on scapes. *Flowers* regular, tetramerous. *Stamens* 4—2, alternate with the lobes of the corolla, and inserted on its tube. *Anthers* versatile, *filaments* usually slender and exserted. *Fruit* a membranous pyxis, with 1, 2, or many albuminous seeds.

PLANTÀGO, L. Plantain. Ribwort. Sep. 4, membranous, persistent. Cor. limb 4-toothed, spreading, persistent on the fruit. Stamens 4 (rarely 2), the long, slender filaments exserted, or in some of the fls. in cluded. Ovary 2-(4-) celled. Pyxis membranous, opening below the middle by a lid, when the loose dissepiment falls out with the seeds.—Herbs acaulescent. Fls. small, whitish, in a slender spike raised on a scape.

§ Flowers uniform ; stamens exserted in all of them...(a)
§ Flowers dimorphous, the anthers included in most of them...(b)
   a Seeds 7—16. Leaves broadly ovate, 7-veined. Spike dense............No. 1
   a Seeds 4 only. Leaves oblong or cordate, 3-7-veined...................Nos. 2, 3
   a Seeds 2 only. Leaves lanceolate. Scape tall. May—October........Nos. 4, 5
   a Seeds 2 or 4. Leaves linear, fleshy......................................No. 6
   b Corolla lobes permanently spreading. Seeds 2, concave..........Nos. 7, 8
   b Corolla lobes closing, and erect on the fruit. Summer..........Nos. 9—11

1 **P. major** L. *Common P.* Leaves ovate, some toothed, smoothish, palmately 7-veined, ample ; spikes 1—2f high. ♃ Door-yards: common. Long white elastic fibres are drawn from the veins when the leaf is plucked.

2 **P. Kamptschatica** Cham. Leaves elliptic-oblong, obtuse, 3-5-veined ; spikes loose-flowered ; bracts acute, shorter than the sepals. Ala. (P. Rugelii C-B.)

3 **P. cordàta** Lam. Lvs. ovate, cordate or very abrupt at base, obscurely toothed, subpinnately 5-7-veined ; fls. loosely spicate, larger than in No. 1; the bracts ovate, obtuse. ♃ Along streams, Can. Wis., and S. As large as P. major. June, July.

4 **P. lanceolàta** L. Lvs. lanceolate, pointed at each end ; scape angular, longer than the leaves ; spike dense, ovate or cylindric, brown. ♃ Meadows. 1—2f.

5 **P. sparsiflòra** Mx. Leaves lanceolate or oblong, pointed each way ; scape terete, longer than the leaves ; spike long, loose, interrupted. S. and S-W. 6—18′.

6 **P. marítima** L. β. *juncoides.* Leaves linear, glabrous, fleshy, nearly as long as the slender scape ; spike loose, bracts roundish. Coast, N. J., and N. 4—12′.

7 **P. aristàta** Mx. Lvs. linear, woolly at base, smoothish above ; scape longer; spike dense ; bracts long, rigid, awn-like (5″) ; petals round-cordate, spreading, conspicuous ; seeds 2, boat-shaped. Prairies, Ill. 6—10′. June, July. (P. Patagonica, β. (Gray.))

8 **P. gnaphaloìdes** L. White-woolly ; lvs. oblong to linear ; spike dense, exceeding the lvs. ; bracts deltoid, not exceeding the calyx. Wis. to Tex. 3—6′. June, Jl.

9 **P. Virgínica** L. Hoary pubescent ; lvs. elliptical, 3-5-veined ; scapes and spikes elongated, dense-flowered ; cor. closed on the pod, erect ; seeds rarely more than 2 ; bracts shorter than the cal. ④ Dry hills and rocks, Conn., W. and S. 5—10′. May—Sept.

10 **P. heterophýlla** N. Lvs. linear, entire, or some of them with a few slender teeth ; ped. many, as long as the leaves ; spikes loose ; pod conoid, twice longer than the calyx, crowned with the closed cor., 10-24-seeded. ② Wet, Penn., and S. 4—8′.

11 **P. pusílla** N. Thinly pubescent ; lvs. filiform-linear, shorter than the capillary, few-flowered scapes ; pod crested, longer than the calyx, 4-seeded. ① Conn. (Mr. Bowles), W. and S. 1—3′. Seeds oblong. May—July.

ORDER LXXXIII. PLUMBAGINACEÆ. LEADWORTS.

Herbs or undershrubs with the leaves alternate or all clustered at the root. *Flowers* regular. *Calyx* tubular, 5-toothed, plaited, persistent. *Corolla* hypocrateriform, of 5 petals united at base, or sometimes almost distinct. *Stamens* 5, hypogynous and opposite the petals, or inserted on their claws. *Ovary* 1-celled, free from the calyx. *Styles* 5 (seldom 3 or 4). *Fruit* a utricle, or dehiscent by valves, containing 1 anatropous seed.

I. STATICEÆ. Styles distinct, at least above. Utricle not valvate. Leaves radical...(a)
II PLUMBAGINEÆ. Style 1, with 5 stigmas. Pod subvalvate. Leaves cauline...(b). PLUMBAGO 3
   a Stigmas filiform. Styles glabrous. Scape branching..................STATICE. 1
   a Stigmas filiform. Styles plumose. Scape capitate.....................ARMERIA. 2

Order 84.—LENTIBULACEÆ.

1. STÁTICE, L. Marsh Rosemary. Calyx funnel-form, limb scarious, 5-nerved, 5-parted. Pet. scarcely united at base. Fil. 5, adnate to the very base of the corolla. Ovary crowned with the 5 glabrous, filiform styles, utricle opening crosswise. ♃ Herbs with the scape branching, the flowers 3-bracted, sessile on the 3-bracted branchlet.

S. Limònium L. Very smooth. Leaves oblong to oblanceolate, acute, tipped with a bristle, long-stalked; scapes terete, corymbous-paniculate; fls. separate or in pairs, on the upper side of the branchlets, blue-purple. Marshes. 6—12′. July—October.

2. ARMÈRIA, Willd. Thrift. Flowers collected in a dense head. Invol. 3- to many-leaved. Cal. tubular-campanulate, 5-angled, with 5 shallow lobes, scarious and plaited. Pet., sta., etc., as in Statice. ♃ Lvs. radical, mostly linear. Scape simple, appendaged above with a sheath.

1 **A. vulgàris.** Scape terete, smooth; lvs. linear, flat, obtuse; outc. bracts of the invol. ovate-acute; fls. rose-colored. Sea-coast, Oreg., &c. 1f. June—August.
2 **A. latifòlia.** Scape solitary, tall; lvs. broad-oblong, 5-7-veined; flowers rose-red, bracts cusp-pointed, scarious. Portugal. 1—2f. June—August.

3. PLUMBÀGO, Tourn. Leadwort. Cal. 5-lobed. Corolla salver-form, tube longer than calyx, limb twisted in æstivation. Anth. 5, linear Stig. 5, filiform. Utricle membranous, mucronate with the persistent style. ♄ ♃ Flowers cyanic, numerous through the season.

1 **P. capénsis.** Shrubby; lvs. oblong, entire, white-scaly beneath; fls. in short terminal spikes, pale blue, the tube 1′ or more in length. S. Africa. 2—4f. Hardy S.
2 **P. cœrùlea.** Herbaceous; lvs. acuminate; fls. in loose spikes, blue. 6″. ♃ S. Am.
3 **P. coccínea.** Herb tall; lvs. oblong, large; spikes long, loose; fls. scar. 1—2′. India

Order LXXXIV. LENTIBULACEÆ. Butterworts.

Herbs small, growing in water or wet places, with showy, bilabiate fls. on scapes. *Calyx* inferior, of 2 or 3 sepals. *Corolla* irregular, bilabiate, personate, spurred. *Stamens* 2, included within the corolla and inserted on its upper lip. *Anthers* 1-celled. *Ovary* 1-celled, with a free, central placenta. *Style* 1. *Stigma* cleft. *Fruit*, capsule many-seeded. *Seeds* minute. *Embryo* straight, with no albumen. Fig. 399.

§ In wet, rocky places. Leaves broad, entire. Corolla throat open..................Pinguicula. 1
§ In water, floating. Leaves dissected. Corolla throat closed. .... Utricularia. 2

1. PINGUÍCULA, L. Butterwort. Cal. 5-parted, somewhat bilabiate. Cor. bilabiate, ringent, upper lip bifid, lower trifid, spurred at base beneath. Sta. 2, very short. Stig. sessile, 2-lobed. Caps. erect. Sds. ∞. ♃ Lvs. radical, rosulate, entire, greasy to the touch. Scapes 1-flowered, nodding. March—May.

* Corollas blue, purple, or white, lobes very unequal......................Nos. 1—3
* Corollas yellow, the lobes nearly equal................. ....No. 4

1 **P. vulgàris** L. Scape and calyx a little downy; cor. lips very unequal, lobes obtuse, entire; spur cylindrical, straightish. N. Y. (rare), and N. 6—8′. Cor. 1 long.
2 **P. elàtior** Mx. Lvs ovate to spatulate; scapes villous near the base; cal. glandu-

lar; corolla lobes obtuse, 2-lobulate; spur half as long as the tube, blunt. 8. Car. to Fla. Scape very slender, 8—12' high. Lvs. 1' or less. Fls. 1'. (P. australis N.)
3 **P. pùmila** Mx. Lvs. glabrous, roundish-ovate; corolla tube oblong, lobes emarginate; spur acute, nearly as long as tube. Ga., Fla. 2—4'. Fls. 4—5'' long.
4 **P. lùtea** Walt. Lvs. elliptic to obovate; cor. bell-shaped, nearly regular, the lobes sinuate-dentate; spur slender, ⅓ as long as corolla. S. 5—8'. Fls. 9'' broad.

2. UTRICULÀRIA, L. BLADDERWORT.

Cal. 2-parted, lips subequal. Cor. irregularly bilabiate, personate, spurred. Stamens 2. Stig. bilabiate. Caps. globular, 1-celled. ⁓ Loosely floating, or fixed in the mud. Lvs. radical, multifid or linear and entire, mostly furnished with little inflated *utricles* (whence the name) as buoys. Scape erect. June—Sept. Fig. 399.

§ Floating. Scape involucrate with a whorl of large inflated petioles...............No. 1
§ Floating. Scape naked, branches bearing bulblets and bladders...(a)
§ Stems creeping and rooting in mud, with few or no air-bladders...(b)
    a Flowers purple. Branches whorled, submersed..............................No. 2
    a Flowers yellow.—*c* Bladders borne on the capillaceous leaves...(d)
        —*c* Bladders and leaves borne on separate branches........Nos. 3, 4
    d Spur acute or retuse, about as long as the lips........................Nos. 5—7
    d Spur obtuse, short.—*e* Fls. of 2 kinds, the *lipless* down on the stems......No. 8
        —*e* Fls. of 1 kind only, all on the scapes ....Nos. 9—11
    b Spur appressed to and scarcely equalling the lower lip of the corolla.....Nos. 12, 13
    b Spur remote from the corolla, slender, acute.........................Nos. 14, 15

1 **U. inflàta** Walt. Upper lvs. in a whorl of 5 or 6 at the surface of the water; pet. and midvein inflated, lower lvs. capillaceous, dissected, submerged; scape 4-5-flwd. ♃ In ponds and ditches. Rhizome or stem long. Scape 8'. Fls. 8'' broad, yellow, upper lip rounded, entire, lower lip 3-lobed. August.
2 **U. purpùrea** Walt. Leaves all submersed, fibrillous, whorled on the long stem; scape assurgent, 2-3-flowered; lower lip 3-lobed, bisaccate, longer than the conical spur beneath it. ① Ponds. Scape 3—5'. Flowers 6'' broad, violet-purple.
3 **U. intermèdia** Hayne. Lvs. 2-ranked, crowded, 4—5 times forked, divisions linear-subulate, ciliate-denticulate, rigid, 2—3'' long; bladders all on leafless branches; scape 2-3-flowered; spur conical, acute; corolla 6—8''. ① Pools, Pa., and N. 6—8'.
4 **U. Robbinsii** Wood. Leaves alternate, 3—4 times forked, divisions flaccid, linear-capillary, entire, 8—12'' long; bladders all on leafless branches; scape tall (8—13'), 4-7-flowered; spur fusiform; corolla 4—5''. ① Mass. (Dr. Robbins.)
5 **U. strìata** Le Conte. Lvs. 3-4-furcate, divisions capillary; scape 2-6-flowered, 8—12'; fls. 6'', on slender pedicels, lips subequal, 3-lobed, the upper *striate* with red, concave, the lower as long as the obtuse, notched spur. ① L. I. to Fla.
6 **U. longiròstris** Ell. Lvs. 2-3-furcate, with setaceous segments; scape 1-3-flowered (3—4'); lower lip entire, shorter than the subulate spur. South.
7 **U. biflòra** Lam. Lvs. capillary, root-like, bearing numerous bladders; scape 2—5', 2-flowered; spur obtuse, notched, equalling the lower lips. W. and S.
8 **U. clandestìna** N. Lvs. capillaceous-multifid, scattered, bladder-bearing; scape slender, 3—4', 2-3-flwd., seldom seen; cor. 5'', spur shorter than the 6-lobed lower lip; ped. down on the stems 1', with 1 apetalous flower. ♃ Ponds, Mass. to N. J. and Pa.
9 **U. gibba** L. Minute, with hair-like leaves and few utricles; scape 1-2-flwd., naked (2—3'); corolla spur blunt (*gibbous*) and short, lips many-lobed. ♃ R. I. to Car.
10 **U. vulgàris** L. Lvs. capillaceous-multifid, fibrillous; sc. scaly, 5-12-flwd., 6—12'; spur conical, shorter than the closed lips (3—4''), divergent; fr. nodding. ♃ Ponds.
11 **U. minor** L. Lvs. short, several times forked; sc. 3-6-flwd., 4—7'; cor. ringent, spur blunt, deflexed, much shorter than the obovate, flat lower lip; fr. nodding. ♃.
12 **U. bipartìta** Ell. Lvs. fibrillous-multifid; sc. 1-3-flwd., 2—3'; cal. lower lip 2-*parted ;* spur obtuse, half as long as the entire lower lip. Soft mud, South.

ORDER 85.—OROBANCHACEÆ.   217

13 **U. subulàta** L. Minute, creeping; lvs. few, linear, entire, obtuse; sc. few, 1-5-flwd., 3′, with ovate bracts; spur acute, appressed to the lower 3-lobed lip. Springs.
14 **U. resupinàta** Green. Rooting; lvs. linear-capillaceous, erect, undivided (1′); scapes ∞, simple, 1-flwd., 1-bracted (3—6′); spur ascending, remote from and shorter than the erect lips of the *light-purple* corolla (which is 4″). Muddy shores, N. Eng.
15 **U. cornùta** Mx. Scape rooting, tall (9—12′), scaly, 2-5-flwd.; lvs. fugacious or 0; flowers subsessile, palate very prominent; spur subulate, decurved away from the erect tube and limb. Mud or shallow pools. Flowers large, yellow.

ORDER LXXXV.   OROBANCHACEÆ.   BROOM-RAPES.

Herbs fleshy, leafless, growing parasitically upon the roots of other plants. *Calyx* 4–5-toothed, inferior, persistent. *Corolla* irregular, persistent, imbricate in æstivation. *Stamens* 4, didynamous. *Anthers* 2-celled, cells distinct, parallel, often bearded, at base. *Ovary* 1-celled, free from the calyx, with 2 or 4 parietal placentæ. *Capsule* enclosed within the withered corolla, 1-celled, 2-valved. *Seeds* very numerous and minute, with albumen.

* Flowers polygamous, on spicate branches, sterile above, fertile below..................EPIPHEGUS. 1
* Flowers perfect,—*a* in one dense spike. Calyx split in front.........................CONOPHOLIS. 2
    —*a* in one dense spike. Calyx 5-toothed.................. ....PHELIPÆA. 3
    —*a* solitary on long peduncles or scapes............................APHYLLON. 4

1. **EPIPHÉGUS**, Nutt. BEECHDROPS. ☉ ☿ ♀ Upper fls. complete, but sterile, with a tubular, curved, 2-lipped cor. barely including the stamens. Lower fls. ♀, with a short, 4-toothed cor. and imperfect stamens. Caps. 2-valved, with 2 placentæ on each valve.—A smooth, dull-red, leafless, branching plant, with sessile flowers all along the branches.

E. **Virginiàna** Bart.—In beech-woods: common. 1f. Fls. brownish, 5″. Aug., Sept.

2. **CONÓPHOLIS**, Wallroth. SQUAW-ROOT. Fls. ☿, crowded in a thick, scaly spike. Cal. with 2 bractlets at base, 4-toothed, split down in front. Cor. ringent, upper lip arched, notched, lower 3-lobed. Sta. exserted. Caps. 1-celled, 2-valved, with 2 placentæ on each valve.—Stem simple, thick, short, covered with scales, the flowers in the upper axils.

C. **Americàna** Wal.—In old woods: com. 4–7′ high, and 1′ thick, pale-yellowish. Jl.

3. **PHELIPÆA**, Tourn. BROOM-RAPE. Fls. ☿, spiked or racemed. Cal. 2-bracted at base, 4–5-cleft. Cor. 2-lipped, including the stam. Caps. 1-celled, 2-valved, with 2 placentæ on each valve.—Stem thick, scaly.

P. **Ludoviciàna** Don. Glandular-pubescent; stem thick, short; spike dense; cal. 5-cleft; cor. funnel-form, lips subequal; bracts ovate, obtuse. Alluvion, Ill.

4. **APHÝLLON**, Mitchell. NAKED BROOM-RAPE. Fls. ☿, solitary, on long, bractless ped. or scapes. Cal. 5-cleft. Cor. tube elongated, curved, limb spreading, subequally 5-lobed. Anthers included. Capsule with 4 placentæ.—Plants glandular-pubescent. Stem nearly subterraneous.

1 A. **unifiòrn** T. & G. Ped. *in pairs*, simple, naked, each 1-flwd. Woods and thickets. Ped. 4–5′, scape-like, purplish-yellow, like the nodding flowers. June.
2 A. **fasciculàta** T. & G. Stem 2–3′ high, bearing *many* peduncles from near the summit, each with few scales and 1 purple flower. Mich., and W. 4–6′. May.

Order LXXXVI. BIGNONIACEÆ. Trumpet-flowers.

Trees, shrubs, or *herbs,* often climbing, with opposite, exstipulate leaves, and large, showy, monopetalous, irregular, 5-parted flowers. Stamens 2 or 4, often with 1 or 3 sterile rudiments. *Anthers* 2-celled. *Ovary* 2-carpelled. *Style* 1. *Stigma* divided. *Capsule* woody, 2-valved, with few or many large seeds. Figs. 30, 31, 95, 199, 445.

§ Plants woody, with the leaves mostly opposite, and the flat seeds winged...(I.)
§ Plants herbaceous, leaves all simple, some alternate. Seeds wingless...(II.)
  I. BIGNONIADS.—Trees, with simple leaves, and long, cylindric pods..............CATALPA. 1
    —Shrubs climbing. Leaves compound (binate). Calyx truncate...BIGNONIA. 2
    —Shrubs climbing. Leaves pinnate. Calyx 5-toothed. ....TECOMA. 3
    —Half-shrubby climbers (exotic). Lvs. compd. (bipinnate)...ECCREMOCARPUS. 4
  II. SESAMEÆ.—Coarse, clammy herbs, the fleshy pods 2-horned..............MARTYNIA. 5
    —Smoothish, erect. Pods dry, 4-celled, not beaked................SESAMUM. 6

1. CATALPA, Scop. CATALPA. Cal. 2-parted. Cor. campanulate, 4- or 5-cleft, the tube inflated. Sta. 2 fertile, 2 or 3 sterile. Stig. 2-lipped. Caps. 2-celled, long, cylindric. ♄ Lvs. opposite or in 3's, simple, petiolate. Flowers in large, showy, terminal panicles, May—July. Figs. 30-1, 445.

1 **C. bignonioìdes** Walt. Lvs. ample, thin, cordate-ovate, lustrous above, downy beneath, long-petioled; fls. in erect, pyramidal panicles, large, irregularly bell-shaped, white, with yellow and violet spots. A beautiful tree 30—50f. Native and cultivated.

2 **C.** Kæmpferi. Lvs. smaller, entire or lobed, glabrous both sides; fls. smaller. Japan.

2. BIGNÓNIA, Tourn. Cal. margin nearly entire. Cor. somewhat bilabiate, 5-cleft, bell-funnel-shaped. Sta. didynamous, 4 fertile, 1 a sterile filament. Caps. long and narrow, valves flat or scarcely convex, parallel with the partition. ♄ ♄ ♄ Often with tendrils.

1 **B. capreolàta** L. Climbing, smooth; leaves binate, consisting of a pair of ever green, cordate-lanceolate leaflets and a branching tendril between them; fls. axillary, near 2', red-yellow; pod 6—7' long. Woods, S. 50f. Very slender. March—May.

2 **B.** Tweediàna. With yellow fls. 2', in panicles; cal. bilabiate. From Buenos Ayres.

3. TECÓMA, Juss. TRUMPET-FLOWER. Cal. campanulate, 5-toothed. Cor. tube short, throat dilated, limb 5-lobed, subequal. Sta. 4, didynamous, with the rudiment of a fifth, anther-cells 2, diverging. Caps. 2-celled, 2-valved, the valves contrary to the partition. Seeds winged. ♄ ♄ ♄ Lvs. opposite, odd-pinnate in the following.

1 **T. radicans** Juss. Climbing by *radicating* tendrils; lfts. 4 or 5 pairs, ovate, dentate-serrate, pointed; corolla thrice longer than the calyx; stam. included. Woods, thickets, Penn., S. and W. 20—80f. Fls. red, 2' long. June—Aug. Very showy.

2 **T.** Capénsis. Climbing; lfts. broad-ovate, crenate-serrate; cor. long, trumpet-shaped, incurved, stam. and style exserted. S. Afr. Flowers corymbed, 2' long, orange.

3 **T.** grandiflòra. Climbing; lfts. lance-ovate, pointed, dent-serrate; cor. scarcely longer than the 5-toothed calyx (3'), scarlet. China and Japan.

4 **T.** jasminoìdes. Climbing; lfts. ovate, shining, entire; pan. terminal; cor. trumpet-shaped, white, roseate in the throat. Australia. Common in greenhouses.

4. ECCREMOCÁRPUS, R. & P. Calyx acutely 5-cleft, broader and much shorter than the tubular corolla, whose lobes are 5, rounded, reflexed.

Sta. 4, included. Caps. 1-celled, 2-valved, valves placentiferous in the middle. Half-shrubby climbers, from S. Am. Tender. (Calampelis, Don.)

1 E. scaber. Lvs. bipinnate; cor. tube inflated above the calyx, scarlet, drooping, 1'.
2 E. longiflora. Lvs. tripinnate; cor. tube cylindric, curved, yellow, 3', drooping.

5. MARTÝNIA, L. Unicorn Plant. Cal. 5-cleft, bracteolate at base. Cor. campanulate, tube gibbous at base, limb 5-lobed, unequal. Sta. 5, one rudimentary and sterile, four didynamous. Caps. coriaceous, ligneous, 4-celled, 2-valved, each valve terminating in a long, hooked beak. ① Chiefly southern, branching, viscid-hairy, strong-scented. Flowers large.

1 M. proboscídea Glox. Branches mostly decumbent; lvs. cordate, entire, roundish, villous, upper ones alternate; fls. on long, axillary peduncles; beaks 2 (when the valves separate), hooked; corolla dull yellowish. Fields, thickets, S. and W. 2f. Jn.
2 M. frágrans. Lvs. roundish-3-lobed, sinuate-dentate; raceme few-flowered; corolla purple, yellow inside, fragrant; beaks shorter than the pod. Mexico.
3 M. lútea, with large yellow funnel-form corollas, is from Brazil.

6. SÉSAMUM, L. Oil-seed. Cal. 5-parted. Cor. campanulate, 3-cleft, the lower lobes the longest. Sta. 4, didynamous. Stig. lanceolate. Caps. 2-celled, the cells divided by the inflexed edges of the valves. ① E. India. Leaves petiolate, the lower opposite, upper alternate.

S. Indicum DC. Lvs. lance-ovate, lower ones 3-lobed, upper ones undivided, serrate; flowers axillary, sessile, pale purple. Fields and gardens. Seeds rich in oil. §

Order LXXXVII. GESNERIACEÆ. Gesnerworts.

Tropical plants, somewhat fleshy, with opposite or radical leaves, no stipules, and showy, somewhat irregular flowers. *Calyx* half adherent to the ovary (in the following genera), 5-parted. *Corolla* tubular, 5-lobed, imbricated in bud. *Stamens* 2 or 4, didynamous, with a rudiment. *Style* 1. *Fruit* a capsule nearly free, 1-celled, with 2 double, many-seeded placentæ.

Corolla tube bell-form, equally tumid at base, limb oblique ....Gesneria, 1
Corolla bell-funnel-form, gibbous at base, limb short.. ....Gloxinia, 2
Corolla salver-form, subequal, limb flat-spreading..Achimenes, 3

1. GESNÉRIA, L. ♃ With tuberous roots and toothed leaves. Sta. 4, with a rudiment, anthers cohering at first. Brazil.

1 G. Líndleyi. Lvs. opposite, ovate-oblong, rugous; flowers in a terminal raceme; corolla 18'', scarlet or red, the limb very short. Brazil.
2 G. Douglasii. Leaves whorled, ovate, pubescent, with the numerous red-yellow flowers in their axils.—The species are many and much mixed.

2. GLOXÍNIA, L'Her. Has often radical leaves (or with very short stems), crenate, and large axillary or radical flowers. Stamens 4, with a fifth rudiment, anthers cohering. Brazil.

G. speciósa. Leaves oval-oblong, on long radical petioles; ped. subradical, 1-flowered; corolla bell-shaped, 1½', violet, varying to white.

3. ACHÍMENES, Br. Erect, downy herbs, with scaly buds. Anth. 4, separate, the rudiment on the base of the corolla.

220   ORDER 88.—SCROPHULARIACEÆ.

1 **A. LONGIFLORA.** Leaves oblong, pointed at both ends, serrate; corolla violet-purple 15″; calyx 4—5″, pedicel still shorter, 1-flowered, axillary. Mexico.
2 **A. COCCÍNEA.** Leaves ovate, acuminate; corolla scarlet, 10″, calyx 5″, the pedicel longer, axillary, erect, with the flower nodding. Jamaica.

ORDER LXXXVIII. SCROPHULARIACEÆ. FIGWORTS.

Herbs chiefly, without fragrance, the leaves and inflorescence various. *Fls.* irreg., 5-(rarely 4-)parted, didynamous or diandrous (rarely pentandrous). *Calyx* free from the ovary, persistent. *Corolla* monopetalous, imbricated in bud. *Stamens* inserted in the tube of the corolla, 1 or 3 of them usually rudimentary. *Ovary* free, 2-celled, with 1 style, a 2-lobed stigma, and becoming in fruit a 2-celled, ∞-seeded capsule, with axile placentæ and albuminous seeds. Figs. 70, 106, 134, 167, 434, 502.

```
1 Leaves alternate (or opposite, and the corolla spurred or saccate behind)...(2)
1 Leaves opposite, and the corolla lower lip an inflated sac. (Tribe 2.)
1 Leaves opposite, and the corolla not spurred nor saccate...(5)
    2 Inflorescence compound, centrifugal or terminal. Exotics. Tribe 1...(x)
    2 Inflorescence simple, centripetal or axillary...(3)
        3 Stamens 5. Corolla large, rotate, more or less irregular. Tribe 3...(a)
        3 Stamens 4 or 2. Corolla minute, 4- or 5-lobed. Little herbs. Tribe 7...(k)
        3 Stamens 4. Corolla large, upper lip exterior in the bud. Tribe 4...(b)
        3 Stamens 4 or 2. Corolla lower lip exterior in the bud...(4)
            4 Corolla bell- or thimble-shaped, oblique, lobes spreading. Tribe 8...(m)
            4 Corolla bilabiate, upper lip vaulted and arched. Tribe 12...(p)
    5 Stamens 2, exserted. Corolla rotate or salver-form. (Tribe 9.)
    5 Stamens 2 (rarely 3), included. Corolla tubular, labiate, rotate, &c. Tribe 6...(f)
    5 Stamens 4, perfect,—* the 5th a large, conspicuous rudiment. Tribe 5...(c)
                —* the 5th a minute rudiment, or none...(8)
        8 Inflorescence compound, in cymes or panicles. Tribe 5...(d)
        8 Inflorescence simple.—† Corolla wheel-shaped, largest lobe upward. Tribe 3...(a)
                    —† Corolla salver-form, lobes about equal. (Tribe 10.)
                    —† Corolla bell-shaped, not helmeted. Tribe 11...(n)
                    —† Corolla bilabiate, not helmeted. Tribe 6...(e)
                    —† Corolla bilabiate and helmeted. Tribe 12...(q)
```

I. SALPIGLOSSIDEÆ. (Corolla in bud plicate at the clefts. Inflorescence cymous.)

   TRIBE 1. SALPIGLOSSIEÆ.—*x* Stamens 2. Corolla deeply many-cleft............SCHIZANTHUS. 1
        —*x* Stamens 4.—*y* Corolla tubular-funnel-form............SALPIGLOSSIS. 2
           —*y* Cor. salver-form. Anth. unlike....BROWALLIA. 3
           —*y* Cor. salver-form. Anth. all alike..BRUNFELSIA. 4

II. ANTIRRHINIDEÆ. (Corolla in bud imbricate, the upper lip covering the lower.)

   TRIBE 2. CALCEOLARIEÆ. Flowers in cymes, very showy, cultivated............CALCEOLARIA. 5
   TRIBE 3. VERBASCEÆ.—*a* Stamens 5, corolla not inverted, subregular............VERBASCUM. 6
       —*a* Stamens 4. Cor. inverted on the twisted pedicels......ALONSOA. 7
   TRIBE 4. ANTIRRHINEÆ.—*b* Corolla spurred. Pod opens by valves............NEMESIA. 8
       —*b* Corolla spurred. Pod opens by pores............LINARIA. 9
       —*b* Corolla saccate at base, throat closed............ANTIRRHINUM. 10
       —*b* Corolla throat open, naked inside. Climbers......MAURANDIA. 11
       —*b* Corolla throat open, with 2 hairy lines. Climbers..LOPHOSPERMUM 12
   TRIBE 5. CHELONEÆ.—*c* Sterile filament a scale. Flowers small, lurid............SCROPHULARIA. 13
       —*c* Sterile filament shorter than the rest. Seeds winged....CHELONE. 14
       —*c* Sterile filament equalling the rest. Seeds wingless....PENTSTEMON. 15
          —*d* Herbs. Corolla labiate, blue and white............COLLINSIA. 16
          —*d* Shrubs slender. Corolla tube straight............RUSSELIA. 17
          —*d* Shrubs erect. Corolla tube incurved............PHYGELIUS. 18
          —*d* Trees. Corolla blue, tubular-bell-form............PAULOWNIA. 19
   TRIBE 6. GRATIOLEÆ.—*e* Calyx 5-angled. Corolla 2-lipped, 5-lobed, large.......MIMULUS. 20
       —*e* Calyx 5-angled. Corolla oblique, 4-lobed, large......TORENIA. 21

Order 88.—SCROPHULARIACEÆ.

```
                —e Calyx 5-parted, equal.   Leaves many-cleft............CONOBEA.      22
                —e Calyx 5-parted, unequal.   Leaves undivided............HERPESTIS.   23
                —f Calyx 5-parted.   Sterile filaments short, or 0........GRATIOLA.    24
                —f Calyx 5-parted.   Sterile filaments exserted...........ILYSANTHES.  25
                —f Calyx 4-lobed.  Stamens 2.   Flowers minute.....MICRANTHEMUM.       26
                —f Calyx 4-lobed.  Stamens 3.   Flowers small.  S...HYDRANTHELIUM.     27
III. RHINANTHIDEÆ.  (Corolla in bud imbricate, the lower or lateral lobes exterior.)
     TRIBE 7.  SIBTHORPEÆ.—k Stamens 2.   Corolla 4-cleft......................AMPHIANTHUS.   28
                          —k Stamens 4.   Corolla 5-cleft......................LIMOSELLA.     29
     TRIBE 8.  DIGITALEÆ.—m Stameus 2.   Calyx 4-parted.  Flowers small........SYNTHIRIS.     30
                         —m Stamens 4.   Calyx 5-parted.  Flowers large........DIGIATLIS.     31
     TRIBE 9.  VERONICEÆ.—Stamens divergent.   Upper leaves often alternate......VERONICA.    32
     TRIBE 10. BUCHNEREÆ.—Stamens approximate by pairs.   Upper lvs. altern....BUCHNERA.      33
     TRIBE 11. GERARDIEÆ.—n Stamens long-exserted.   Corolla tubular............MACRANTHERA.  34
                         —n Stamens short.—o Cor. yellow, tube short as limb...SEYMERIA.      35
                                          —o Corolla yellow, tube elongated....DASYSTOMA.     36
                                          —o Cor. purple.  Lvs. very slender...GERARDIA.      37
     TRIBE 12. EUPHRASIEÆ.—p Anther-cells unequal, separated..  ................CASTILLEJA.   38
                          —p Anther-cells equal.—r Calyx 10-ribbed............SCHWALBEA.      39
                                                —r Calyx not ribbed............PEDICULARIS.   40
                      —q Calyx inflated.  Seeds many, winged..........RHINANTHUS.              41
                      —q Calyx not inflated.—s Seeds many, wingless...EUPHRASIA.               42
                                            —s Seeds 1—4, oblong.......MELAMPYRUM.            43
```

1. **SCHIZÁNTHUS**, R. & P. Cut-flower. Cor. irregular, the upper lip 5-cleft, external in æstivation, lower much smaller, 3-parted. Fil. 4, 2 of them sterile. Capsules 2-celled. ① Chili. Leaves pinnatifid, alternate. Cymes supra-axillary.

S. PINNÀTUS. Lvs. once or twice pinnatisected; cor. segm. longer than tube, the middle segm. of the posterior lip 2-lobed and hood-like; stam. exserted. 1—2f. Fls. delicate and handsome, 1' broad, purple and yellow, with a dark spot in the midst. Aug.—Oct.

2. **SALPIGLÓSSIS**, R. & P. Trumpet-tongue. Corolla obliquely tubular-funnel-form, with an ample throat, lobes all emarginate. Sta. 4, fertile, with a short rudiment. Style trumpet-shaped at apex and incurved. Capsules oblong, valves bifid. ♃ Chili. Resembles Petunia.

S. SINUÀTA. Annual in our gardens, 1—2f, weak, viscid-downy. Leaves elliptic-oblong, sinuate-toothed or pinnatifid. Fls. 1½' long, very showy, dark-purple, striped, &c.

3. **BROWÁLLIA**, L. Cor. salver-form, with a long tube, and oblique, 5-lobed limb. Anth. of the two posterior stamens halved, sub-1-celled. Lobes of the stigma broad, divaricate. Caps. membranous, valves bifid.—S. American herbs, with alternate, entire leaves and cyanic flowers.

1 B. DEMÍSSA (also elata). Leaves petiolate, ovate; lower fls. axillary, upper racemed; calyx hairy; cor. tube 6'', limb 1', blue or violet, varying to wh. ① 1—2f. Summer.

4. **BRUNFÉLSIA**, Sw. Corolla salver-form, with a long tube, and a broad 5-lobed limb. Sta. 4, all equal. Style incurved at apex, stig. of 2 broad lobes. Caps. coriaceous, valves entire.—S. American shrubs, with alternate, entire leaves and large blue flowers. (Franciscea, Pohl.)

1 B. HOPEÀNA. Lvs. obovate to ovate; fls. solitary; cor. tube little exceeding the cal., lobes rounded, subequal, violet, blue, or white, 1' broad. 3f. Much branched.
2 B. LATIFÒLIA. Leaves elliptic to oblong; fls. in loose cymes; cor. tube thrice longer than the calyx, and longer than the limb (1'). Leaves 3—5' long, shining above.

5. CALCEOLÀRIA, L. SLIPPER-FLOWER. Calyx 4-parted, valvate in bud. Cor. tube very short, limb 2-lobed, lobes entire, concave or spur-like, the lower inflated. Sta. 2, lateral, with no rudiments. Caps. ovoid conical, valves bifid.—S. American and New-Zealand herbs or shrubs, with opposite or whorled leaves and very curious flowers, of all colors, endlessly varied in cultivation.

§ Leaves pinnatisect. Anther cells separated, one empty. Annual............ No. 1
§ Leaves ovate to lanceolate. Fls. corymbous. Anth. cells contiguous.....Nos. 2—4

1 **C. PINNÀTA.** Rough-downy, weak, 1f, the lower lip orbicular, pale-yellow.
2 **C. CORYMBÒSA.** Erect; lower lip broad-ovate, obtuse, open beyond the middle, ylw.
3 **C. CRENATIFLÒRA.** Villous; lower lip hanging, large, obovate, 3-furrowed, spotted, ylw.
4 **C. INTEGRIFÒLIA.** Viscid; lower lip orbicular, little longer than the upper, scarcely contracted at the base; upper lip twice longer than the calyx. Shrub. 2—3f.

6. VERBÁSCUM, L. MULLEIN. Cor. rotate, 5-lobed, unequal. Sta. 5, declinate, all perfect. Caps. ovoid-globous, 2-valved. ① Rarely ♃ or suffruticous. Leaves alternate. Flowers in spikes or paniculate racemes. June—August. Fig. 434.

§ Leaves decurrent on the stem. Flowers in a long, thick spike, yellow......No. 1
§ Leaves not decurrent.—a Flowers in racemes, white, yellow or purple.....Nos. 2, 3
—a Flowers paniculate, white or yellow............Nos. 4, 5

1 **V. Thápsus** L. *Common Mullein.* Leaves decurrent, densely tomentous on both sides; rac. spiked, dense; 3 of the sta. downy, 2 of them smooth. ② Fields, waysides. 3—5f. Almost never branched, woolly all over. Flowers numerous. §
2 **V. Blatt ̀ria** L. *Moth Mullein.* Lvs. clasping, oblong, smooth, serrate; ped. 1-flwd., solitary, racemous; filaments all bearing violet wool. ① Waste grounds, waysides. 3f. Flowers 1', white or yellow. Stem often branched.
3 **V. PHŒNÍCEUM.** Leaves mostly radical, ovate to oblong, petiolate, smooth above, downy beneath; racemes rarely branched; flowers violet to red. ② Eur. 3f.
4 **V. Lýchnitis** L. *White Mullein.* Whitish tomentous; st. angular; leaves green above, the lower petiolate; fls. in loose fascicles, forming a pyramidal panicle; fll. all white-woolly. ② Sandy fields, N. Y. to Ga.: rare. Flowers pale yellow. § Eur.
5 **V. PULVERULÉNTUM.** Clothed in cottony, deciduous tomentum; lvs. tomentous both sides, ovate-oblong; fls. numerous, yellow, in a large panicle. ② Eur.

7. ALÓNSOA, R. & P. Cor. resupinate by the twisted pedicel, rotate, 5-cleft, lobes very obtuse, unequal. Sta. 4, short, declinate. Caps. obtuse, flattened, septicidal.—S. American, very branching herbs, with opposite leaves, square branches, and terminal racemes of scarlet flowers.

1 **A. INCISÆFÒLIA.** Leaves lance-ovate, incisely serrate, petiolate; cor. 1' or less wide, 3—4 times longer than the calyx. ① All Summer. From Chili.

8. NEMÈSIA, Vent. Calyx 5-parted. Corolla personate, saccate or spurred behind, upper lip 4-lobed, lower entire. Sta. 4, lower pair circumflexed at base. Caps. compressed, with 2 keeled valves, and winged seeds. ① S. Africa. Lvs. opposite. Fls. solitary and axillary, or racemed.

1 **N. VERSÍCOLOR.** Lvs. ovate to lanceolate and linear, entire or toothed; cor. lobes ob long, all subequal (4—5''), spur 4'', incurved, acute. 3f. Blue-white.
2 **N. FLORIBÚNDA,** has ovate leaves, an obtuse spur, and white-yellow flowers.

9. LINÀRIA, Juss. TOAD-FLAX. Calyx 5-parted. Corolla personate,

Order 88.—SCROPHULARIACEÆ.

upper lip bifid, reflexed, lower 3-cleft, throat closed by the prominent palate, tube inflated, with a spur behind. Caps. 2-celled, bursting below the summit.—Herbs. Lower leaves generally opposite, upper alternate. Fls. solitary, axillary, often forming terminal, leafy racemes. Fig. 70.

* Stems prostrate, creeping. Leaves broad, reniform or hastate. Eur.. ...Nos. 1, 2
* Stems erect, with narrow leaves, mostly scattered.................Nos. 3—5
* Stems erect, with broad lanceolate leaves, all verticillate.................No. 6

1 **L. Cymbalària.** Lvs. palmate-veined, reniform, 5–7-lobed, mostly alternate; fls. axillary, small, yellow, spur shorter than tube. ♃ Smooth, delicate.

2 **L. Elátine** L. Hairy; lvs. feather-veined, hastate, entire, alternate; ped. solitary, long; cor. yellow and purple. ① Fields. 1—2f. Very slender. § Eur. July.

3 **L. Canadénsis** Dumont. Lvs. scattered, erect, linear, obtuse; fls. racemed; st. simple; scions procumbent; fls. blue. ① Fields, waysides. 6—12'. Very slender. Flowers small, in a loose raceme. Spur filiform, long, short, or 0. June—Sept.

4 **L. vulgàris** Mill. *Common Toad-flax.* Leaves linear-lanceolate, crowded; spikes terminal; fls. dense, imbricate; cal. smooth, shorter than the spur. ♃ Meadows, waysides. 1—2f. Very leafy, with showy rac. of yellow and orange fls. Jl., Aug. § Eur.
β. **Pelòria.** Corolla with 3—5 spurs, and a regular border of 3—5 lobes, with 5 stamens. Penn. (Dr. Darlington). Poughkeepsie, N. Y. (Mr. W. R. Gerard).

5 **L. bipartìta.** Erect; lvs. linear, alternate; ped. much longer than the lance-linear, scarious-edged sepals; cor. 8—10'', violet, the palate orange.

6 **L. triornithóphorum.** *Three Birds.* Smooth, glaucous; leaves in 3's and 4's; fls. whorled, each resembling 3 little birds. ♃ Eur. 2—3f. Curious.

10. ANTIRRHÌNUM, L. Snap-dragon.

Calyx 5-sepalled. Corolla gibbous (not spurred) at base of tube, throat closed (personate) by the prominent palate, upper lip bifid, reflexed, lower trifid. Sta. 4. Capsules opening by 2 or 3 pores, as in Linaria.—Herbs, European, &c., with the lower leaves opposite, the upper alternate. Flowers axillary, large, racemed above. Fig. 502.

1 **A. majus.** Erect; leaves lanceolate; fls. evidently racemed; sep. hairy, shorter than the cor. tube; cor. pink, purple, or scarlet, mouth yellow. ♃ 18'. Fls. 1'. Summer.

2 **A. Oróntium.** Low, spreading; lvs. oblong-lanceolate; fls. smaller than in A. majus (6''), the sepals equalling the cor., which is rose or white, with purp. spots. ① Sum.

11. MAURÁNDIA, Ort.

Calyx 5-parted. Cor. bilabiate, tube scarcely gibbous at base, throat open, with 2 prominent glabrous folds, upper lip of 2 rounded lobes, lower of 3. Sta. 4. Caps. oblique, opening by chinks below the apex. ♃ Mexican, climbing and twining, with large purple flowers all Summer.

1 **M. antirrhiniflòra.** Leaves mostly triangular-hastate; fls. glabrous, 1', tube some gibbous at base, throat partly closed by the prominent hairy palate. 10f.

2 **M. semperflòrens.** Lvs. cordate-hastate, angular; calyx glabrous; cor. bell-form, not gibbous (throat open), 1½' long, pale violet or rose-colored. 10f.

3 **M. Barclayàna.** Leaves broadly triangular-cordate or hastate; calyx clothed with long glandular hairs; cor. near 2' long, very oblique, purple, throat open. 10f.

12. LOPHÓSPERMUM, Don.

Corolla tubular-campanulate, limb 5-lobed, subregular, throat open, between two hairy lines. Caps. globular. Seeds winged. Otherwise as in Maurandia. Fig. 106.

1 L. erubéscens. Lvs. triangular-cordate, dentate-lobed, pubescent; cal. segm. ovate, hirsute; cor. downy, 2½—3' long, red, with an ample border. 10—20f.

2 L. scandens. Lvs. cordate-ovate, pointed, coarse-toothed, smoothish; calyx segm. lance-ovate; cor. glabrous, 2', scarlet, limb erect-spreading. 10f.

13. SCROPHULÀRIA, L. Figwort.
Calyx in 5 acute segments. Cor. subglobous, limb contracted, sub-bilabiate, lip with an internal, intermediate scale (sterile filament). Capsules 2-celled. Valves with 2 inflated margins.—Herbs or suffruticous, often fœtid. Leaves opposite. Cymes in simple or compound, terminal, thyrsoid panicles. Fig. 167.

S. nodosa L. Glabrous, tall, branching; leaves ovate, oblong, or lanceolate; fls. in loose pedunculate cymes, combined into an oblong panicle; sterile anther a roundish green scale on the dull, olive-colored corolla. ♃ Thickets. 4—6f. July—Oct.

14. CHELÒNE, L. Turtle-head. Snake-head.
Calyx deeply 5-parted, with 3 bracts at base. Cor. inflated, bilabiate. Sta. 4, woolly, the sterile filament shorter than the rest. Caps. valves entire. Seeds broadly winged. ♃ With opposite leaves and sessile flowers in the upper axils.

1 C. glabra L. Smooth; lvs. subsessile, oblong-lanceolate, acuminate, serrate, acute at base; flowers densely spiked. By brooks and in wet places. 2f. Stems simple, in clumps. Flowers 1' long, white or roseate, with short gaping lips. Aug., Sept.
  β. *purpùrea.* Lvs. distinctly petiolate, acuminate; flowers rose-purple. West.

2 C. Lyòni Ph. Smooth; lvs. ovate, acuminate, petiolate, serrate, the lower cordate; fls. in a dense spike. Mts. of Car. and Ga. 1—2f. Corolla purple, 1¼'. July—Sept.

15. PENTSTÉMON, L. Beard-tongue.
Calyx deeply 5-cleft. Cor. elongated, often ventricous, lower lip 3-lobed, spreading. The fifth filament (tongue) sterile, bearded, longer than the rest or about as long; anth. smooth. Seeds ∞, angular, not margined. ♃ N. American, branching, paniculate. Leaves opposite, the lower petiolate, upper sessile or clasping. Flowers showy, red, violet, blue, or white, in Summer.

  * Native E. of the Mississippi River, sometimes cultivated...(*a*)
  a Leaves dissected. Corolla bell-shaped, lobes rounded, subequal..........No. 1
  a Leaves undivided, serrulate. Sterile filament (tongue) bearded.......Nos. 2, 3
  a Leaves entire. Tongue puberulent, widened and incurved at the apex....No. 4
  * Native W. of the Mississippi, cultivated for ornament...(*b*)
  b Leaves incisely pinnatifid. Corolla lobes subequal. Tongue smoothish...No. 5
  b Leaves serrate, with pale purple or blue flowers. Tongue bearded....Nos. 6—8
  b Leaves entire.—*c* Cor. strongly bilabiate, scarlet. Tongue bearded........No. 9
        —*c* Cor. scarcely bilabiate,—*d* scarlet or crimson......Nos. 10—12
                    —*d* blue or violet..........Nos. 13—15

1 P. disséctus Ell. Lvs. pinnately divided into linear segm.; fls. in a loose panicle; cor. with a curved tube, 9—10″, purple; tongue bearded at apex. Dry. Ga. 2f. Jn., Jl.

2 P. pubéscens Sol. Pubescent or glabrous; lvs. ovate-oblong to lanceolate; fls. in a loose panicle; cor. tube 7—9″, gradually enlarged upward, pale purple, lower lip with two bearded folds inside, some longer than the upper. Hills and bluffs. 1—2f. †

3 P. Digitàlis N. Glabrous; lvs. elliptic to lanceolate, the upper clasping; fls. many, large, corolla tube abruptly enlarged to bell-form, pale blue or purplish, 12—15″ long, throat widely open, beardless. Rich soils, Pa., W. and S. 2f. Leaves 3—6'.

4 P. grandiflòrus Fras. Glabrous and glaucous; lvs. oblong-obovate to roundish-ovate, upper clasping, all entire; panicle long, slender; corolla bell-shaped, 15″, limb nearly regular, bluish purple. Ill., Wis., and W. 3f. Handsome. †

ORDER 88.—SCROPHULARIACEÆ.

5 P. RICHARDSÒNI. Smoothish, branching; fls. 1', violet, in leafy panicles. Oreg. 2f.
6 P. OVÀTUS. Puberulent; lvs. cordate-clasping; fls. 9", numerous, light blue. Oreg. 2f.
7 P. COBÆA. Puber., tall; lvs. lance-ovate, clasping; fls. 2', broad-campanulate. Tex.
8 P. CAMPANULÀTUS. Glabrous; lvs. lance-linear to lance-ovate, long-pointed; panicle long, loose, 1-sided; corolla tube inflated, large, bell-shaped. Mexico.
9 P. BARBÀTUS. Smooth and glaucous; lvs. oblong to lance-linear; cor. tube long (13"), scarcely dilated upward, lower lip and tongue densely bearded. Mexico. 2—4f.
10 P. MURRAYÀNUS. Glaucous; lvs. connate-clasping, upper roundish; cor. 18", bright red, dilated upward, in a long virgate panicle; tongue smooth. Texas. 3f.
11 P. HARTWEGI. Upper lvs. clasping; cor. tubular, 2', crimson; tongue glab. Mex. 3f.
12 P. GLABER. Smooth and glaucous; sts. in bunches, simple; lvs. lanceolate to ovate, entire; flowers 18", in slender panicles, blue-crimson. Nebraska, and W. 2f.
13 P. SPECIÒSUS. Tall; st. lvs. lanceolate, sessile; cor. blue, 18", mouth ample, tongue filiform, the panicle long, virgate, secund, each cyme with 5—9 fls., very showy. Oreg.
14 P. GENTIANOÌDES. Tall; st. lvs. broad-clasping; cor. 16", violet, mouth ample, tongue glabrous, dilated and retuse at apex, the panicle long, some leafy. Mexico. 3—4f.
15 P. CŒRÙLEUS. Low, leafy; lvs. lance., sessile; cor. blue, 8"; tongue bearded. Neb.

16. COLLÍNSIA, Nutt. INNOCENCE.
Calyx 5-cleft. Cor. bilabiate, orifice closed, upper lip bifid, lower trifid, with the middle segment carinately saccate and closed over the declinate style and stamens. Caps. with 2 bifid valves. Seeds large, concavo-convex. ① With verticillate or opposite leaves, axillary and terminal flowers, very pretty.

1 C. verna N. Lvs. ovate to lanceolate, the cauline cordate-clasping, dentate; verticils 4-6-flwd.; cor. blue and white, twice longer than the calyx, 2 or 3 times shorter than the pedicel. Banks of streams, N. Y., and W. 8—18', branching. May, June.
2 C. parviflòra Doug. Lvs. ovate to lanceolate; verticils 2-6-flwd; cor. blue, little longer than the calyx and little shorter than the pedicels. L. Sup., and W. 6-10'. Jn.
3 C. bícolor. Stem lvs. ovate, crenate, sessile; verticils 6-10-flwd. : calyx hairy, longer than the ped.; cor. 9", rose-violet and white. California. 2f. Hardy and handsome.
4 C. GRANDIFLÒRA has lvs. thickish and all entire, with ∞ large blue-purple fls. Oreg.

17. RUSSÉLIA, Jacq.
Cal. 5-parted. Cor. tubular, limb sub-bilabiate, of 5 short rounded lobes, the 2 upper twin. Sta. 4, the fifth a small rudiment. Caps. subglobous, septicidal, valves bifid. Sds. ∞, mixed with hairs. ♃ Mexican. Lvs. opposite or whorled, often minute or scale-like.

R. JÚNCEA. Very smooth, with long, drooping, rush-like branches; lvs. lanceolate to linear, or scale-like on the branches. Flowers scarlet, 1', remote in drooping racemes.

18. PHYGÈLIUS, Mey.
Cal. 5-parted. Cor. tube long, enlarged above, limb oblique, lobes rounded. Fifth stamen a minute rudiment. Caps. very oblique, with unequal cells. ♃ Caffraria. Leaves opposite. Flowers in a loose panicle of cymes.

P. CAPÉNSIS.—Shrub 2f, smooth and beautiful. Leaves lance-ovate, crenate, petiolate. Flowers pendulous, 1½', crimson, yellow within.

19. PAULÓWNIA, Siebold.
Calyx deeply 5-cleft, fleshy. Cor. tube long, declinate, enlarged above, limb oblique, with rounded segments. Sta. 4, arched downward, with no rudiment. Caps. acuminate, valves septiferous in the middle. Seeds ∞, winged. ♃ From Japan, with very large cordate, ovate leaves and large blue-purple fragrant panicles.

P. imperiàlis.—In parks, 40f high. Flower-buds formed in Autumn, opening in the following Spring. Corolla near 2'. Tree of rapid growth and kingly port.

20. **MÍMULUS**, L. MONKEY-FLOWER. Calyx tubular, 5-angled, 5-toothed. Corolla ringent, the upper lip reflected at the sides, palate of the lower lip prominent. Stig. thick, bifid. Caps. ∞-seeded.—Herbs prostrate or erect, with square stems and opposite lvs. Ped. axillary, solitary, 1-flwd.

§ Leaves pinnate-veined. Flowers blue (wild) or yellow (cultivated)......Nos. 1, 2, 6
§ Leaves palmate-veined. Flowers yellow or scarlet............ ....Nos. 3, 4, 5

1 **M. ríngens** L. Lvs. sessile, smooth, lanceolate, acuminate; ped. axillary, longer than the flowers. ♃ A common inhabitant of ditches and mud soils. 2f. Flowers large. (1'), pale blue, yellow-mouthed, appearing in July and August.

2 **M. alátus** Ait. Leaves petiolate, smooth, ovate, acuminate; ped. shorter than the fls.; st. winged at the 4 corners. ♃ N. Y., W. and S., in muddy places. 2f. Aug.

3 **M. Jamèsii** Torr. Stems diffuse, rooting; leaves subentire, round-reniform, 5-7-veined, the upper as long as the peduncles of the small yellow fls. L. Sup., and W.

4 **M. lùteus**. Lvs. round-ovate, the cauline sessile or clasping, shorter than the peduncles; calyx ovoid, half as long as the broad, large, yellow, spotted flowers. Cal.

5 **M. cardinàlis**. Branching, villous-clammy; leaves ovate, narrowed to the clasping base, shorter than the long ped.; cal. large, inflated; cor. ample, rose-orange. Cal.

6 **M. moschàtus**. *Musk Plant.* Decumbent, hairy-viscid; leaves ovate, dentate; cor. tube exceeding the calyx, yellow. Oregon. Smells strongly of musk.

21. **TORÈNIA**, L. Calyx tubular, with prominent angles, oblique. Cor. ringent, upper lip notched, lower larger, trifid. Sta. 4, arched beneath the upper lip, the longer pair appendaged at base. Stigma double. Capsules included.—Herbs tropical, diffuse, with opp. leaves and racemed fls.

T. Asiática. Lvs. petiolate, lance-ovate, crenate-dentate; calyx acute at base, ½'; cor. twice longer, ample, pale purple tipped with violet. 2f+, trailing.

22. **CONÓBEA**, Aublet. Calyx 5-parted, equal. Upper lip of the corolla 2-lobed, lower lip 3-parted. Fertile sta. 4, anth. approximating by pairs, cells parallel. Caps. round-ovoid, ∞-seeded.—Herbs, with opposite leaves. Peduncles axillary, solitary or in pairs, 1-flowered.

C. multífida Benth. Low, diffusely-branched, puberulent; leaves petiolate, pinnately dissected; segments linear or cuneate, lobed or entire, obtuse; cor. greenish, scarcely exserted (2''), lobes entire. ① Sandy banks of rivers, O. to La. 6—12'. July.

23. **HERPÉSTIS**, Gært. Calyx unequally 5-parted. Corolla subbilabiate, upper lip emarginate or 2-lobed, lower 3-lobed. Sta. 4, fertile. Caps. 2-furrowed, valves parallel with the dissepiment. Seeds ∞, small. ♃ Obscure weeds with opposite leaves. Peduncles 1-flowered, axillary, or subracemous, often with two bractlets near the calyx.

§ Leaves feather-veined, or obscurely 1-3-veined. Cor. yellow, or bluish.....Nos. 1, 2
§ Leaves palmately many-(7-9-)veined, subentire. Corolla blue..............Nos. 3, 4

1 **H. nigréscens** Benth Erect; st. square, branched; leaves oblanceolate, crenate-serrate above; ped. equalling or exceeding the leaves; corolla yellowish, upper lip rounded, entire. Wet pl., S. 1—2f. Cor. rather longer (5'') than cal. Blackens in drying.

2 **H. Monnièra** Humb. Prostrate, fleshy; lvs. wedge-obovate, subentire; ped. as long (9'') as the lvs.; fls. few, bluish; cor. 4'' wide, nearly regular. Wet banks, Pa., & S.

3 **H. amplexicaùlis** Ph. Stem submersed, woolly; leaves ovate, cordate-clasping,

obscurely crenate, obtuse; ped. shorter than the calyx, cor. ‡ longer, the upper lip emarginate; disk 10-toothed. Swamps, N. J., and S. 6—12'. August,

4 **H. rotundifòlia** Ph. Creeping, smooth; lvs. round-obovate, entire; ped. 2 or 3 times longer than cal.; cor. upper lip notched. Pools, Ill. to La. 1f. Fls. 5''. Aug.

24. GRATÌOLA. HEDGE HYSSOP. Calyx 5-parted, subequal. Cor. upper lip entire or slightly bifid, lower trifid, the palate not prominent. Sta. 2, fertile, mostly with 3 sterile filaments. Capsules 2-celled, 4-valved, valves inflexed at margin. ♃ Low, with opposite leaves. Peduncles axillary, 1-flowered, usually bibracteolate near the calyx.

§ Flowers sessile. Cells of anthers vertical. Plants rigid, bristly-hairy.....Nos. 7, 8
§ Flowers pedunculate. Anther cells transverse. Plants smooth or viscid...(a)
  a Sterile filaments none, or very minute and pointed....................Nos. 1—3
  a Sterile filaments thread-like, tipped with a small head..............Nos. 4—6

1 **G. Virginiàna** L. St. ascending, branched; leaves lanceolate, sparingly toothed; ped. as long or longer than the lvs.; cor. twice longer than the cal.; sterile fil. none. ♃ Common. 4—8'. St. terete, branching, with white or pale-yellow flowers. July.

2 **G. Floridàna** Nutt. St. erect, branched; lvs. lanceolate, few-toothed; ped. longer than the leaves; cor. 4 times longer than the calyx (7''), yellow. ③ Fields, S. 6—9'.

3 **G. sphærocárpa** Ell. Ascending, branched; leaves lance-ovate, attenuate to the base, sparingly toothed; ped. scarcely longer than the cal. Damp. 3—7'. W. and S.

4 **G. aùrea** Muhl. Smooth; lvs. oblong-lanceolate, subentire, clasping; ped. as long as or longer than the leaves; cor. golden yellow. Muddy soils. 6—8'. August.

5 **G. viscòsa** Schw. Viscid-downy; leaves lance-ovate, sharp-serrate, clasping; ped. longer than the leaves; corolla white, twice longer than calyx, which is 2 or 3 times longer than the capsule. Wet places, Ky. to N. Car., and S. 9—12'. (G. Drummondii.)

6 **G. ramòsa** Walt. St. terete, creeping at base; leaves linear, acute, with few teeth near the apex; bractlets nearly 0 · sep. linear; cor. white. Muddy shores, S. May–Jl.

7 **G. pilòsa** Mx. Erect, hispid; vs. ovate, few-toothed, clasping, rugons; cor. tube scarcely longer than the calyx, white. Wet, Md., and S. 9—12'. July—September.

8 **G. subulàta** Baldw. Erect, hispid; lvs. linear or lance-linear, margins revolute, entire; cor. tube slender, thrice longer than the calyx. Wet sands, Ga., Fla. Sept.

25. ILYSÁNTHES, Raf. Cal. 5-parted. Cor. upper lip short, erect, bifid, lower lip larger, spreading, trifid. Sta. 2, fertile; 2 sterile fil. forked, one of the divisions tipped with an obtuse gland, the other acute, or rarely with half an anther. Caps. ovate or oblong, about equalling the calyx. ① With opp. lvs. and axillary, 1-flwd. ped., resembling Gratiola in habit.

1 **I. gratioloides** Benth. Branching, ascending 3—8'; lvs. oblong, obtuse, subsessile, obscurely dentate; cor. twice longer than the calyx, bluish-white, 4''. A small weed-like herb, in wet places; common. Peduncles 3—6''. July, August.

2 **I. grandiflòra** Benth. Diffusely creeping; lvs. thick, roundish, entire, subclasping; ped. hirsute, 1', corolla 6'' long, violet-blue. Sandy swamps, Ga. (Nuttall.)

3 **I. refrácta** Benth. Erect, slender; lvs. clustered below, obovate to oblong, entire the cauline remote, bract-like, linear-subulate; ped. filiform, refracted in fruit; cor. light-blue, 4 times longer than the calyx (5''). Damp pine woods, S. 6—10. June.

4 **I. saxícola** (Curtis). Stems leafy, clustered; leaves oblong, obtuse, entire, sessile; ped. 3—4 times longer than the leaves (7—9''), refracted in fr.; cor. blue, 4''. S. Aug.

26. MICRÁNTHEMUM, Rich. Cal. 4-toothed or cleft. Cor. upper lip shorter, entire, lower trifid. Sta. 2, fertile, a glandular scale at the base of each, sterile filament none. Style short, apex clavate or spatulate. Caps. 2-valved. ※ (i) Slender, glabrous, with opposite lvs. and minute fls.

M. orbiculàtum Mx. Sts. creeping and rooting, branches ascending 1—2′; lvs. orbicular to obovate, 3-veined, entire, subsessile; fls. ¼″ long, lower lip of cor. longer than the calyx. Brackish mud, Del., and S. (M. micranthum, &c.)

27. **HYDRANTHÈLIUM**, H. B. K. Calyx 4-cleft. Cor. 3-cleft, the upper lobe broader, emarginate. Sta. 3, on the corolla, anth. cells parallel, distinct. Style with two short lobes. Caps. ∞-seeded. ☞ Tropical, with opposite leaves and minute, axillary flowers. Habit of Callitriche.

H. crenàtum Wood. Submersed stems flaccid, bearing the lvs. above; lvs. roundish, glabrous, *crenate*, abrupt at base, 7–9-veined, on flat, veiny petioles; pedicels 3″, reflexed; corolla little exserted, white. Pools, Miss., La. (Dr. Hale).

28. **AMPHIÁNTHUS**, Torr. Calyx 5-parted. Corolla small, funnelform, limb 4-lobed, lower lobe larger. Sta. 2, included, style lightly bifid, lobes acute. Capsule obcordate, compressed, ∞-seeded. ① Minute, with flowers both axillary, and on terminal, 2-bracted peduncles 1′ long.

A. pusíllus Torr.—On wet rocks, Newton Co., Ga. Leaves nearly radical, linear, obtuse; 1—2″ long; flowers minute, white. March, April.

29. **LIMOSÉLLA**, L. Mudwort. Calyx 5-cleft. Cor. shortly campanulate, 5-cleft, equal. Sta. approximating in pairs. Capsule partly 2-celled, 2-valved, many-seeded. ☞ ① Minute. Scape 1-flowered.

L. tenuifòlia Nutt. Lvs. linear, scarcely distinct from the petiole; scape as long as the leaves; cor. segments oval-oblong, shorter than the cal. Mud, Penn., and N. 1′.

30. **SÝNTHYRIS**, Benth. Calyx 4-parted. Corolla subcampanulate, segments 4, erect-spreading or 0. Sta. 2 (rarely 4), on the cor., exserted, anth. cells parallel, distinct. Caps. compressed, obtuse or emarginate. ♃ N. American, with a thick root. Radical leaves petiolate, cauline bract-like, on the scape-like stem, alternate. Fls. racemed or spicate. May.

S. Houghtoniàna Benth. Hairy; lvs. ovate, subcordate, crenulate, obtuse; stem or scape dense-flwd. above; cor. greenish, as long as the cal. Hills, Mich., and W. 1f.

31. **DIGITÀLIS**, L. Fox-glove. Calyx 5-parted. Cor. campanulate, ventricous, upper lip reflexed, spreading, middle segment of the lower lip broadest. Caps. ovate, 2-celled, 2-valved, with a double dissepiment. ♃ Europe, Asia. Lower leaves crowded, petiolate, upper alternate. Flowers in showy racemes. Poisonous and medicinal. July, August.

§ Corolla light-yellow, tube twice longer than the lower lip.................Nos. 1, 2
§ Corolla purple, white, brown, often spotted, tube inflated and short......Nos. 3—5

1 D. grandiflòra (or ochroleuca). *Great Yellow F.* Leaves ovate, veiny, serrulate, clasping; racemes downy, loose; corolla 1¼′ long, segments very broad. 4f.

2 D. lùtea. Plant very smooth, with lance-oblong leaves; raceme smooth, with many flowers, all on one side; corolla 8—10″ long, tube not inflated. 2f.

3 D. purpùrea. *Purple F.* Lvs. oblong, rugous, petiolate, crenate, large; flowers in a long, 1-sided raceme, thimble-shaped, purple or white, spotted. 2—3f.

4 D. ferrugínea. Leaves very smooth, lance-oblong; corolla rusty-brown, the lower lip densely bearded, its middle segment ovate. 4f.

5 D. lanàta. Leaves lance-oblong, often woolly; flowers downy or woolly, white or brown; lower segment of the corolla obovate. 2f.

ORDER 88.—SCROPHULARIACEÆ.   229

32. VERÓNICA, L. SPEEDWELL. Calyx 4-parted. Cor. subrotate, deeply 4-cleft, lower segments mostly narrow. Sta. 2, inserted into the tube, exserted. Caps. flattened, often obcordate, 2-celled, few-seeded.—Our species are herbs. Leaves opposite. Flowers solitary, axillary or in racemes, blue, flesh-colored, or white.

§ Tender shrubs (Australian) with axillary racemes of blue flowers........Nos. 16, 17
§ Herbs tall (European) with opposite lvs. and terminal rac. of blue fls....Nos. 14, 15
§ Herbs tall, with whorled leaves, terminal racemes, and *tubular* flowers.....Nos. 1, 2
§ Herbs low, weak (3—12'). Leaves opposite (at base). Corolla rotate...(*a*)
   a Racemes opposite, axillary. Capsule roundish, emarginate...........Nos. 3, 4
   a Racemes alternate, axillary. Capsule not rounded, very flat...........Nos. 5, 6
   a Racemes terminal, or the flowers axillary and not racemed...(*b*)
     b Floral lvs. like the rest, not longer than the recurved peduncles...Nos. 7—9
     b Floral leaves bract-like, longer than the erect peduncles...(*c*)
        c Perennial. Peduncles equalling or exceeding the calyx.....Nos. 10—1.
        c Annual. Peduncles shorter than the calyx or none..........Nos. 12—13

1 V. Virgínica L. *Culver's Physic.* Erect, tall, glabrous or downy; lvs. whorled in 4's–6's, lance-ovate to lance-linear; spikes mostly several, paniculate. ♃ In thickets, Vt., W. and S. 2—5f. Corolla white, with exserted style and stamens. July.

2 V. Sibírica. Hardly different from No. 1, but it has blue flowers. Siberia. 3f.

3 V. Anagállis L. Glabrous, erect; lvs. sessile, clasping and subcordate, lanceolate, acutish, entire or serrulate; rac. in opposite axils; caps. orbicular, slightly notched. ♃ Brooks and pools. Plant fleshy, 1f. Flowers small, blue-purple. June, July.

4 V. Americàna Schw. *Brooklime.* Glabrous, decumbent at base, erect above; lvs. ovate or ovate-oblong, serrate, petiolate, abrupt at base; rac. loose; caps. roundish, turgid, emarginate. ♃ In clear streams. 12—18', fleshy. Fls. blue. June, July.

5 V. scutellàta L. Glabrous, ascending, weak; lvs. linear or lance-linear, sessile, acute, remotely denticulate; rac. very loose; capsule flat, broader than long, cordate at both ends. ♃ Swamps, N. and W. 1f. Fls. flesh-color, rather large. June—Aug.

6 V. officinàlis L. Roughish-downy, prostrate, branching; lvs. wedge-oblong, obtuse, serrate, short-petioled; racemes dense, with pale-blue flowers; capsule downy, triangular-obcordate. ♃ Dry fields. 6—12'. May—July. § Europe.

7 V. Buxbàumii Tenore. Prostrate, hairy; lvs. roundish-ovate, coarsely crenate-serrate, the floral similar, all on short petioles; ped. longer than the lvs.; caps. triangular-obcordate, broader than long. ⓐ Waste grounds, E.: rare. 7–12'. Cor. blue. § Eu.

8 V. agréstis L. *Neckweed.* Hairy, procumbent, diffuse; lvs. cordate-ovate, deeply crenate-serrate, floral similar, all petiolate; ped. as long as the lvs.; caps. roundish, acutely notched, ∞-seeded. ⓐ Fields, E.: rare. 2—8'. Light blue. May—Sept. § Eu.

9 V. hederæfòlia L. Prostrate, pilous; lvs. petiolate, cordate, roundish, coarsely 3-5-toothed or lobed, shorter than the ped.; sep. triangular, subcordate, acute, closed in fruit; caps. turgid, 4-seeded. ⓐ Hard soils, E.: rare. Cor. blue. Mar.—May. § Eu.

10 V. alpìna L. Branched at base, ascending 1—5'; lvs. roundish-oval to elliptical, very obtuse, toothed or entire, subsessile; *racemes hairy*, densely few-flwd.; capsule obovate, notched. ♃ Summits of White Mts., N. H., and R. Mts. Fls. small, blue.

11 V. serpyllifòlia L. Branched below, ascending 3—12'; lvs. oval, obtuse, subcrenate, the lower roundled and petiolate, upper bract-like, oblong, entire; *rac. smoothish*; loose; caps. obcordate, broader than long. ♃ Pastures: com. Cor. blue-wh. May-Aug. §

12 V. peregrìna L. Smoothish, ascending; lvs. petiolate, oblong, few-toothed, obtuse, upper obl.-lin., entire; fls. subsessile, whitish; caps. roundish, slightly notched, ∞ -seeded. ⓐ Clay soils, fields: com. 4—10'. Plant rather fleshy. May, June.

13 V. arvénsis L. *Corn S.* Hairy, branched; lvs. below round-ovate, subcordate, petiolate, crenate, the upper lanceolate; corolla pale blue, pencilled, shorter than the calyx (as in No. 12); caps. obcordate. ⓐ Dry fields: com. 2–6'. May, June. § Eur

ORDER 88.—SCROPHULARIACEÆ.

14 V. SPICĀTA. Erect, 1—2f; leaves opposite, lanceolate, petiolate, serrate; racemes mostly solitary; pedicels shorter than the calyx; corollas blue, showy. ♃ Europe.

15 V. PANICULĀTA. Erect, bushy, 1—3f; lvs. opposite and in 3's, lanceolate, acute at base, petiolate; rac. panicled; ped. longer than the calyx. ♃ Many garden varieties, hybrids between this and No. 14, all with handsome blue racemes. Europe.

16 V. SPECIŌSA. Very smooth, shrubby, with oblong-obovate entire lvs., dense short (2′) racemes in the upper axils, and violet-blue flowers, very beautiful. 1—3f.

17 V. SALICIFŌLIA. Smooth (tree-like at home), with lanceolate, acute, entire leaves, dense glandular-downy racemes (3′), and innumerable blue flowers. 2—5f.

33. **BUCHNĒRA**, L. BLUE-HEARTS. Calyx 5-toothed. Cor. salverform, with a slender tube, and flat limb in 5 subequal lobes. Stam. 4, included, with halved (1-celled) anthers. Caps. 2-valved. ♃ Turns blackish in drying. Leaves opposite. Flowers in a terminal spike. June—Aug.

B. **Americāna** L. Rough-hispid, slender; leaves oblong to linear, few-toothed, obtuse, 3-veined; spike long-stalked, 6-12-flowered; cor. tube 6—7″ long, limb half as long, deep blue. Woods, N. Y., and S. 2—3f, nearly leafless above.

34. **MACRANTHĒRA**, Torr. Calyx lobes 5, long and narrow. Cor. tubular, with an oblique limb, short entire segments, and 4 long exserted subequal stamens. Style long, filiform. Caps. ovate, acuminate. ♃ Tall, with opposite pinnatifid leaves and yellow fls. on long decurved peduncles.

M. **fuchsioìdes** Torr.—Pine-barrens, Ga., Fla., and W. 2—3f. Lvs. lanceolate in outline, with lanceolate segments. Rac. long, loose, 1-sided. Cal. seg. denticulate, shorter than the corolla (or *entire* and still shorter in β. LECONTII). Sept., Oct.

35. **SEYMĒRIA**, Ph. Calyx deeply 5-cleft. Cor. tube short, dilated, lobes 5, ovate or oblong, entire. Stam. 4, subequal, valves of the capsule loculicidal, entire. Seeds ∞.—Herbs erect, branching. Cauline leaves mostly opposite and incised. Flowers yellow.

§ Tube of the corolla woolly within, incurved, as long as the limb................No. 1
§ Tube of the corolla much shorter than the subrotate limb. Leaves small..Nos. 2, 3

1 S. **macrophýlla** N. Tall, smoothish; lvs. large, pinnatifid, with lance-oblong incised segments, upper serrate or entire. ♃? Woods, W. 4—6f. Cor. 6″. July.

2 S. **pectināta** Ph. Viscid-downy, profusely branched; lvs. small (1′ and less), pinnatifid, seg. few, narrow and entire; caps. acute at base. Dry, S. 3f. Aug.—Oct.

3 S. **tenuifōlia** Ph. Smoothish, much branched; lvs. bipinnatifid, 6″ long, segments and rachis filiform; capsule obtuse at base. Wet, S. 2—3f. Cor. 4″. Aug., Sept.

36. **DASÝSTOMA**, Raf. WOOL-MOUTH. WILD FOXGLOVE. Cal. campanulate, 5-cleft. Cor. tube dilated, longer than the 5 entire lobes, woolly within. Stam. didynamous, scarcely included, woolly, anthers all equal, awned at base. Caps ovate, acute, 2 valves bearing a septum in the middle. Seeds ∞. ♃ Tall, erect. Lower leaves opposite. Corolla large, yellow. July—Sept. All blacken in drying. (Gerardia, L.)

* Calyx segments entire.—a Plants pubescent................................Nos. 1, 2
               —a Plants glabrous ....Nos. 3 4
* Calyx segments toothed or pinnatifid. Plants downy....................Nos. 5, 1

1 D. **flava** Wood. Plant pubescent, subsimple; lvs. lance-oblong, entire, or toothed, the lower pinnatifid or incised; cal. lobes oblong, obtuse, shorter than the tu e; ped very short. Woods. 2—4f. A showy herb. Corollas 18″. (G. flava L.)

ORDER 88.—SCROPHULARIACEÆ.

2 D. grandiflòra Wood. Minutely pubescent, branched; lvs. petiolate, lance-ovate, pinnatifid, toothed, or entire; ped. as long as the calyx; cal. tube as long as the lobes (½'), corolla 2' long. Wis., Ill. (J. Wolf), and S. (G. grandiflora Benth.)
3 D. quercifòlia Benth. Glabrous and glaucous, branched; lvs. petiolate, the lower bipinnatifid, upper lance-oblong; cal. lobes longer than the tube, both as long as the pedicels; corolla 2'. Thickets. 3—5f. Common.
4 D. integrifòlia Wood. Glabrous, subsimple; lvs. lanceolate, acute, entire or nearly so; pedicels shorter than the calyx. Woods, Pa., and W. 1—2f. August.
5 D. pediculària Benth. Smoothish or downy; lvs. lance-ovate, pinnatifid with toothed segments; pedicels longer than the hairy calyx, whose toothed segments are about as long as its top-shaped tube. Dry woods. 2—3f. Corolla 15''.
6 D. pectinàta (Torr.) Very hairy; lvs. lanceolate, pectinate-pinnatifid, seg. toothed; calyx longer than the pedicels, segm. longer than tube. Woods, S. 3f. Corolla 18''.

37. GERÁRDIA, L.
Cal. 5-toothed or cleft. Cor. tubular, ventricous or subcampanulate, tube longer than the 5 broad, entire, unequal lobes. Sta. didynamous, in pairs, shorter than the corolla. Caps. obtuse or pointed, ∞-seeded.—American herbs, rarely shrubby. Leaves opposite (except No. 4). Flowers axillary, solitary, purple or rose-color. July—Sept.

§ OTOPHÝLLA. Calyx segments longer than its tube. Two anthers smaller......No. 1
§ GERÁRDIA *proper*. Calyx segments short, equal. Anthers all equal...(*a*)
  a Cor. 2-lipped, upper lip very short, straight. Peduncles slender. S...Nos. 2, 3
  a Corolla lobes subequal, all spreading, throat often hairy...(*b*)
    b Leaves *all* alternate, filiform. Flowers large, on long peduncles. S....No. 4
    b Leaves opposite, rarely the upper alternate and bract-like...(*c*)
      c Peduncles equalling or exceeding the small (½-inch) flowers............Nos. 5—7
      c Peduncles much shorter than the flowers.—*d* Lvs. setaceous or none....Nos. 8, 9
        —*d* Lvs. linear, 1—2' long...Nos. 10—12

1 G. auriculàta Mx. Erect, subsimple, rough-hairy; lvs. lance-ovate, the upper auriculate at base; fls. nearly sessile, 7'' long. (i) Low grounds, Pa. to Car., and W. 2f.
2 G. Mettaùeri Wood. Smooth, slender, diffusely branched; lvs. linear-filiform; ped. filiform, many times longer than the calyx; cor. 8'', upper lip vaulted, notched, lower of 3 rounded lobes, tube with spots and 2 yellow stripes within. (i) Wet sandy places, Mid. Fla. (Dr. Mettauer, 1855). 1—2f. Lvs. 1'—1''. (G. divaricata Chapm.)
  β.? *clausa*. Cor. tube flattened on the back, throat closed by the inflected lip. Fla.
3 G. nùda Wood. Smooth, filiform, branched; lvs. (except a few at the base) all reduced to minute bracts scarcely 1'' long; fls. all terminal, small (5''); caps. globous, exceeding the calyx. Middle Fla. (Dr. Mettauer, 1855). (G. filicaulis Chapm.)
4 G. filifòlia N. St. terete, much branched; leaves filiform, alternate and fascicled; ped. 1', much longer than the leaves; cor. smooth, 9''. (⚥ Barrens, Ga., Fla. 2—3f.
5 G. linifòlia N. ♃ Stems virgate, clustered at root, smooth; lvs. opposite, erect, linear, 3—1'; ped. 8—12'', cal. 2'', truncate; cor. 1', spotted. Wet barrens, S. 2—3f.
6 G. tenuifòlia Vahl. Smooth, paniculately branched; leaves linear to filiform, 1', often coiled; ped. as long as the leaves, longer than the flowers, which are 9'' long; calyx teeth very short, acute; capsule globular. Fields and woods. 1f.
7 G. Skinneriàna Wood (1848). Roughish; st. virgate, angular, few-branched; lvs. linear, rather obtuse, 1'; ped. axillary, very long (1—2'); cal. 1'', teeth obtuse; cor. small (5''), rose-color, not fringed. Low grounds, W. and S. 1 2f. Unlike all the rest, this species does not blacken in drying. (G. parvifolia, Cham.)
8 G. setàcea Walt. (not Benth.) Glabrous, widely branched; lvs. bristle-form, 1' and less; fls. mostly terminal on the filiform, bracted branchlets, large; ped. 2—4''; cal. 1'', teeth very acute, short; cor. 10'', densely fringed. (i) Barrens, Pa., S. and W. 2f.
9 G. aphýlla N. Slender, angular, branched above; lvs. minute, setaceous, 1'', or 0; ped. lateral and term., 1—3''; calyx 1'', teeth obtuse; corolla 8''. (i) Wet, S. 2 3f

10 G. marítima Raf. St. angular, with short branches; lvs. linear, fleshy 6—8″; cor. 7″, some of the lobes fringed; ped. very short; cal. trunc. ①Salt marshes. 4—10′.

11 G. purpùrea L. St. angular, branched; leaves linear, acute, rough-edged, 1—2′; ped. shorter than the calyx, tube truncate with setaceously acute teeth; corolla large (1′), smooth or downy. ①Low grounds. 1—2—4f. Variable.

12 G. àspera Doug. St. roughish, branched; lvs. narrowly linear, rough-hispid, 1′; ped. 1—2 times as long as the cal. (3—6″), teeth lance-acute; cor. 1′. ①W. 1—2f.

38. CASTILLÈJA, L. Painted Cup.
Calyx tubular, 2-4-cleft. Cor. upper lip linear, very long, arched and keeled, enfolding the didynamous stamens, anth. oblong-linear, with unequal lobes, the exterior fixed by the middle, interior pendulous. ♃ ♭ Leaves alternate, the floral often colored at the apex. Flowers subsessile, in terminal, leafy bracts.

1 C. coccínea Spreng. Lvs. sessile, pinnatifid with linear segments; bracts about 3-cleft, *scarlet* (sometimes *yellow*), exceeding the corolla; cal. 2-cleft, nearly equalling the cor., segments notched. ♃ Wet meadows, E. (rare) and W. 8—12′. May, June.

2 C. sessiliflòra Ph. Hairy-downy; lvs. sessile, clasping, oblong-linear, mostly trifid, *not colored;* calyx sessile, elongated; spikes dense; corolla long, exserted, arched, segments of the lower lip acuminate. ♃ Prairies, N-W. 1f. May.

3 C. pállida Kunth. Lvs. linear, undivided, 3-veined, the upper lanceolate, the floral subovate, subdentate at the end, *whitish;* calyx with acute teeth. shorter than the corolla. ♃ ? White Mountains, Green Mountains, and N-W. 1f. August.

39. SCHWÁLBEA, L. Chaff-seed.
Calyx tube 10-ribbed, inflated, obliquely 4-cleft, upper division small, lower large, emarginate or 2-toothed. Cor. ringent, upper lip entire, arched, lower 3-lobed. Caps. oblong. Sds. ∞, chaffy. ♃ With alternate leaves and flowers in a terminal spike.

S. Americàna L.—Sandy marshes, N. Y. to Fla. 1—2f, stout, simple, downy. Lvs lance-ovate, 3-veined, diminishing upward; corolla brown, 1—1½′ long. June.

40. PEDICULÀRIS, L. Lousewort.
Calyx inflated, 2-5-cleft, the segments leafy, or sometimes obliquely truncate. Cor. vaulted, upper lip compressed, emarginate, lower lip spreading, 3-lobed. Capsule 2-celled, oblique, mucronate. Seeds angular.—Herbs. Leaves often pinnatifid. Flowers spicate, yellowish.

1 P. Canadénsis L. Hairy, simple; lvs. alternate, petiolate, lance-oblong, pinnatifid with toothed segments; spike short, dense, leafy; cor. abruptly incurved, with 2 setaceous teeth; capsule ensiform-beaked. ♃ Pastures, copses. 1f. May—July.

2 P. lanceolàta Mx. Smoothish, branching; lvs. subopposite, lance-oblong, doubly cut-crenate; spike elongated, loose at base; corolla upper lip larger and covering the lower; capsule short, ovoid. ♃ Shady banks, N. Y. to Va. and Wis. 1—2f. Sept.

41. RHINÁNTHUS, L. Yellow Rattle.
Calyx 4-toothed, ventricous. Cor. tube cylindrical, as long as the calyx, limb ringent, galea appendaged, compressed, lip broader, deeply divided into 3 obtuse segments. Caps. 2-valved, compressed, obtuse. ① Erect, with opposite leaves.

R. Crista-galli L. Mostly glabrous; lvs. oblong or lanceolate; cor. ½ longer than the calyx; appendages of the galea (upper lip) transversely ovate, broader than long; seeds winged, rattling when ripe. Plymouth, Mass., Lake Superior, and N. 1f.

42. EUPHRÀSIA, L. Eyebright.
Calyx 4-cleft. Upper lip of the

cor. galeate, concave, apex 2-lobed, the lobes broad and spreading, lower lip spreading, trifid, palate not folded. Sta. unequal, ascending beneath the galea. Capsule oblong, compressed, ∞-seeded.—Herbs with opposite leaves and the flowers in spikes.

E. officinàlis L. Lvs. ovate or oblong, the cauline obtuse, crenate, bracts acute, cut-serrate with cuspidate teeth; calyx lobes subequal; corolla light-blue, lower lobes deeply notched. ⓛ White Mountains, Lake Superior. 2–6'. Leaves 1–3".

43. MELAMPŸRUM, L. Cow Wheat. Calyx 4-cleft. Upper lip of the corolla compressed, the margin folded back, lower lip grooved, trifid. Caps. 2-celled, oblique, opening laterally. Seeds 1—4, cylindric-oblong, smooth.—Herbs with opposite lvs. Fls. solitary in the upper axils.

M. praténse, β. *Americànum* (Benth.) Leaves linear-lanceolate, petiolate, glabrous, the upper broader and toothed at base; fls. axillary, distinct; cal. teeth slender, half as long as the yellowish corolla. ⓛ Woods: common. 6—10', branched. Jn.—Sept.

Order LXXXIX. ACANTHACEÆ. Acanthads.

Herbs or *shrubs* with opposite, simple leaves and regular, bracted floweis. *Calyx* 5-parted, equal or unequal, imbricated in the bud. *Corolla* 5-merous, tubular below, limb more or less bilabiate, convolute in bud. *Stamens* didynamous or diandrous, inserted on the tube of the corolla. *Fruit* a 2-celled, 4-12-seeded capsule. *Seeds* supported by hooks or cup-shaped processes of the placentæ, exalbuminous.

§ Seeds destitute of *hooked* supports...(a)
  a Corolla regular. Seeds few, resting on little cups. Vines........................Thunbergia. 1
  a Corolla bilabiate. Seeds many, with no supports,................................Elytraria. 2
§ Seeds resting on hooks proceeding from the placentæ....(b)
  b Corolla funnel-form, subregular. Stamens 4, unequal..... ....Ruellia. 3
  b Corolla bilabiate, ringent. Stamens 4. Pod terete.........................Hygrophila. 4
  b Corolla labiate, the upper lip wanting. Stamens 4. † Rare..................Acanthus *mollis*
  b Corolla bilabiate. Stamens 2.—c Corolla inverted, upper lip 3-toothed... ....Dicliptera. 5
                          —c Corolla straight, lower lip 3-lobed.............Dianthera. 6
                          —c Corolla straight, lower lip 3-parted. ....Cyrtandra. 7

1. THUNBÉRGIA, L. Calyx short, toothed or truncate, with 2 large bractlets at base. Cor. funnel-bell-form, limb 5-lobed, nearly regular. Sta. 4, unequal, included. Caps. beaked, 3–4-seeded. ♃ ♭ Fls. showy, axillary.

T. aláta. A climbing vine, silky-hairy, with cordate-sagittate lvs. on winged pet.; fls. 1½' deep and broad, purple, with a yellow, buff, orange or white border. E. Africa.

2. ELYTRÀRIA, Vahl. Calyx with 4 or 5 unequal segments. Cor. bilabiate, lower lip of 3 bifid segments. Sta. 2 fertile, 2 sterile, included. Caps. 8-seeded.—Herbs acaulescent, with (oblong) leaves at base and clasping bracts on the scapes, and the small flowers in a terminal spike.

E. virgàta Mx. Scapes several, glabrous, covered with the bracts, which are ovate, cuspidate, ciliate, the upper subtending the white flowers; calyx with 2 linear bractlets at base, ciliate. ♃ Wet plains, S. Car. to Fla. 1f. August.

3. RUÉLLIA, L. Calyx 5-parted into slender segments. Cor. funnel-form, limb spreading, subequally 5-lobed. Sta. 4, included, didynamous

Caps. narrow. Seeds 4—16, resting on hooks. ♃ Low, with tumid joints, opposite leaves, and showy axillary blue, purple, or white flowers.

§ DIPTERACÁNTHUS. Anthers pointless. Style bifid. Seeds 8—12.........Nos. 1—2
§ CALÓPHANES. Anthers pointed at base. Style simple. Seeds 4. South...(a)
   a Stems erect from a creeping base, with obtuse lvs............Nos. 4, 5
   a Stems creeping, diffuse, with the leaves entire.. . ...Nos. 6, 7

1 **R. strepens** L. Erect, smoothish, with obovate to oblong,.petiolate 'vs.; ped. very short, 1–4-flowered; bractlets as long as the narrow sepals, little shorter. " In the slender corolla tube. Dry soils, W. and S. 9—16'. Leaves 2—5'. June, July.

2 **R. ciliòsa** Ph. Erect, white-hairy, with lvs. obovate to oblong, abrupt at base and subsessile; bractlets and sepals not half as long as the tube of the corolla. Rich soils, W. and S. 1f or more. Leaves 1—2'. Flowers 2—2½'. June—September.
   β. *hýbridus.* Low, decumbent, and very hairy. Georgia (Dr. Feay).

3 **R. tubiflòra** Le Conte. Downy; leaves oblong to lanceolate, sessile; fls. solitary; *sep. lance-linear,* ½ as long as the long tube of the white cor. Ga., Fla. June—Aug.

4 **R. oblongifòlia** Mx. Very downy; lvs. obovate to oval, subsessile; fls. 1—3 together, bractlets and sepals as long as tube of the spotted corolla. Dry, Ga., Fla.

5 **R. ripària** (Chapm.) Smoothish, simple; lvs. oblong, petiolate; flowers clustered, small (6''), white, bractlets, sepals, and corolla tube equal. Mid. Fla. 12—18'.

6 **R. humistràta** Mx. Smooth; lvs. oblong-oval, petiolate; flowers 1—3 together, bractlets shorter than the setaceous sepals. Rich soils, S. Car. to Fla.

7 **R. lineàris** T. & G. Small, rough-downy; leaves linear-oblong, imbricated, the bractlets similar; capsule 4-angled, with 2—4 seeds. S. Fla.

4. **HYGRÓPHILA**, R. Br. Calyx half-5-cleft, with narrow segments Cor. ringent, lower lip trifid. Sta. 4, unequal, cells of the anth. divergent-sagittate, violet. Stig. subulate. Caps. terete, ∞-seeded. ♃ ≈ Stoloniferous, 4-angled. Flowers clustered in the axils. (See Addenda.)

H. lacústris Nees. Erect, simple, smooth; leaves lance-oblong, sessile (3—4'); fls. sessile, appearing whorled, white. Borders of lakes. New Orleans. 1—2f.

5. **DICLÍPTERA**, Juss. Fls. in bracted heads. Cal. 5-parted. Cor. bilabiate, inverted, upper lip 3-toothed, sta. 2, anth. cells equal, one placed above the other. Caps. 4-seeded, the partitions and valves separating. ♃

1 **D. brachiàta** Spr. Smooth; st. 6-angled, brachiately branched: leaves lanceolate, long-petiolate, acuminate; heads few-flowered, the upper approximate, sessile, lower often pedunculate; flowers purple, 5—6''. River banks, S. 1—2f. June—Aug.

2 **D. Hàlei** Riddell. St. downy,·mostly simple; leaves lance-ovate, petiolate; bractlets and sepals fringed with long hairs; corolla 5'' long. Fla. to La. 1—2f. Jn.—Aug.

3 **D. assúrgens** Juss., with scarlet (1') corollas in 1-sided spikes, grows in S. Fla.

6. **DIANTHÈRA**, Gron. (Rhytiglóssa, Nees, and C-B.) Cal. 5-parted. Cor. bilabiate, upper lip notched, lower 3-lobed. Stamens 2, anth. cells unequal, one placed above the other. Capsule flattened, 4-seeded above the middle. ♃ ≈ Lvs. smooth, entire. Flowers in bracted spikes or heads.

1 **D. Americàna** L. Erect, angular, tall; leaves long-lanceolate, wavy, as long (3—4') as the peduncles; bracts and sepals lanceolate, 3'', the ringent corolla 6'', violet-purple. Banks, N. Y., W. and S. 2—3f. June, July.

2 **D. ensifórmis** Wood. Erect from a creeping base, slender; leaves linear, oblique or cusiform, thick, shorter (3—4') than the peduncles; flowers spicate; calyx 6''; corolla purple, 1'. Fla. April. (D. crassifolia Chapm.)

ORDER 90.—VERBENACEÆ.

3 D. ovàta Walt. St. square, ascending, 4—8'; leaves lance-ovate, acute, longer than the 3-4-flowered peduncles; corolla pale-purple, 3—4''. S. Car. to Fla.

4 D. hùmilis Wood. Erect, square, 1—1½f; leaves lance-elliptical, shorter than the ∞-flowered, 1-sided spikes; corolla 5'', purple. Fla. to La. (Justicia Mx.)

7. CYRTANTHÈRA, Nees   Corolla ringent, upper lip falcate, lower in 3 narrow segments. Sta. 2, anth. nodding. Caps. 4-seeded? ♃ Brazil

C. CARNEA. Stem tall, stout, with ample ovate to oblong leaves, and large, showy, terminal spikes of many flesh-colored flowers. In the greenhouse.

ORDER XC. VERBENACEÆ. VERVAINS.

Herbs (or generally shrubs and trees) with opposite, exstipulate leaves. *Flowers* with a bilabiate or more or less irregular monopetalous corolla. *Stamens* 4, didynamous, rarely equal, sometimes only 2. *Style* 1. *Fruit* dry or drupaceous, 2-4-celled (1-celled in Phryma), forming as many 1-seeded nutlets. *Seeds* erect or pendulous, with little or no albumen.

§ Herbs. Fruit dry, consisting of—
   a 4 one-seeded nutlets. Stamens 4. Corolla 5-lobed.........................VERBENA. 1
   a 2 two-seeded nutlets. Stamens 4. Corolla 5-cleft, minute, spicate. S. Fla......PRIVA echinata.
   a 2 one-seeded nutlets. Stamens 2. Flowers spicate, imbedded. S. Fla..........STACHYTARPHA.
   a 2 one-seeded nutlets. Stamens 4. Corolla 4-parted............................LIPPIA. 2
   a 1 one-seeded nutlet, reflexed. Stamens 4. Corolla bilabiate..................PHRYMA. 3
§ Shrubs. Fruit fleshy, berry-like (or a 2-celled capsule in No. 7)...(*b*)
   b Leaves compound, digitate. Flowers 5-parted. Seed 1.........................VITEX. 6
   b Leaves simple, toothed.—*c* Cymes axillary. Drupes 4-seeded................CALLICARPA. 4
      —*c* Heads axillary. Drupes 2-seeded............................LANTANA. 5
   b Leaves simple, entire.—*c* Drupe 2-seeded. Spikes terminal................(ALOYSIA.) 2
      —*d* Drupe 4-seeded. S. Fla..........................CITHAREXYLUM villosum.
      —*d* Drupe 8-seeded. S. Fla.........................DURANTA Plumieri.
      —*d* Capsule 4-seeded. Flowers in heads..............AVICENNIA tomentosa.

1. VERBÈNA, L. VERVAIN. Calyx 5-toothed, with one of the teeth often shorter. Cor. funnel-form, limb somewhat unequally 5-lobed, lobes emarginate. Sta. 4, included, the upper pair sometimes abortive. Drupe splitting into 4, 1-seeded, indehiscent carpels.—Herbs or undershrubs. Leaves opposite. Flowers sessile, mostly in spikes or heads.

§ Corymbed; the open corollas of the spike forming a corymb. Stems weak...(*a*)
   a Leaves 3-cleft or pinnatifid, the lobes cut-serrate or toothed..........Nos. 1—3
   a Leaves merely serrate or toothed, somewhat incised...................Nos. 4—7
§ Spicate; the open corollas lateral, in slender spikes...(*b*)
   b Stem simple (mostly), bearing a single spike. Leaves oblong........Nos. 8, 9
   b Stem branched, with many spikes.—*c* Leaves mostly simple........Nos. 10—12
      —*c* Leaves much divided........Nos. 13—15

1 V. Alublètla L. Lvs. ovate-oblong in outline, 3-parted, cut, acute and petiolate at base; spikes pedunculate; bracts half as long as the cylindrical calyx. Dry soils, Va. to Ill., and S. If. Flowers lilac, varying in the gardens to purple. April, May.

2 V. INCÌSA. Leaves oblong to deltoid, rugous, cut-lobed and serrate, abrupt at base, petiolate; bracts ovate, a fourth as long as the glandular calyx; corolla rose-purple. ♃ Brazil. Stems some shrubby, ascending.

3 V. MULTÍFIDA. Small, creeping, branched; leaves multifid into narrow, acute segments; bracts subulate, shorter than calyx. ♃ Brazil. Red to white.

ORDER 90.—VERBENACEÆ.

4 V. venòsa. Nearly simple, with rigid, oblong-sessile, cut-serrate leaves; bracts subulate, longer than the calyx, both colored; corolla lilac to blue. ♃ Brazil.
5 V. chamædrifòlia. Leaves oblong-ovate, short-petiolate; bracts subulate, not half the length of the long calyx; corollas scarlet to crimson. ♃ Buenos Ayres.
6 V. phlogiflòra. With many erect branches, and long-petioled, lance-deltoid eaves; bracts lanceolate, half as long as the calyx. Flowers large, red to blue.
7 V. teucrioìdes has very hairy, wrinkled, ovate-triangular, crenate leaves on short stalks, with large white to roseate sweet-scented flowers. ♃ Brazil.
8 V. angustifòlia Mx. Leaves oblong-linear, tapering to base, serrate, with furrowed veins; spikes 1 or few, slender; corolla deep-blue, bracts as long as the calyx (1″). ♃ Rocks and hills. N. Y. to Va., and W. 1f. Leaves 2—3′. July.
9 V. Caroliniàna L. Leaves oblong-obovate to oblong, crenate-toothed, sessile; spike loose; corolla large, roseate, bracts minute. ♃ Dry soils, S. 1—2f. June.
10 V. hastàta L. *Common Vervain.* Lvs. lanceolate, acuminate, cut-serrate, petiolate, the lower lobed or *hastate;* spikes panicled, dense, slender, erect and parallel; flowers blue. ♃ Waysides: common. 3—6f. § Europe. July—September. Hy, brids occur, with cleft leaves and loose-flowered spikes.
11 V. urticæfòlia L. Leaves ovate to lance-ovate, serrate, acute, petiolate; spikes axillary and terminal, filiform, lax; bracts shorter than the calyx. ♃ A homely weed, in waste grounds. 3f. Flowers minute, white. § Europe. July, August.
12 V. stricta Vent. *Mullein V.* Hairy and hoary, rigidly erect; leaves oval to obovate, unequally dentate, sessile, rugous; spikes dense. ♃ Dry fields, W. 1—3f. Very leafy, rather handsome. Corolla blue, 4″ broad. July.
13 V. bracteòsa Mx. Hairy, divaricately branched, leaves laciniate; bracts lance-linear, squarrous on the peduncle and spikes, longer than the small blue flowers. ♃ Dry fields, roadsides, N. Y., W. and S. 8—16′. June—September. (V. canescens.)
14 V. officinàlis L. Smoothish, erect; leaves lanceolate to oblong, pinnately lobed or toothed, subsessile; spikes slender, panicled; bracts not longer than the calyx; flowers purple, small. ♃ Waysides, Conn. to Ga. 2—3f. (V. spuria L.)
15 V. strigòsa Hook. Hoary, rough-downy, rigid; leaves oblong, 3-parted, incised, sessile; spikes strict, lax-flowered bracts long as calyx; corolla large. N. Orl. 2—3f.

2. **LÍPPIA,** L. FOG-FRUIT. Cal. 2-parted. Cor. funnel-shaped, limb sublabiate, upper limb entire or emarginate, lower 3-lobed. Sta. didynamous, included. Drupe dry, thin, enclosed in the calyx, 2-seeded. ♄ ♃ Leaves opposite or whorled. Flowers small, whitish, in heads or spikes.

1 L. nodiflòra Mx. St. 4-angled, geniculate, simple, creeping; lvs. lanceolate to oblanceolate, cuneate at base, petiolate, shorter than the ped. Banks, Pa. to Ill., and S.
2 L. (ALOYSIA) **citriodòra.** *Lemon Verbena.* Shrub smooth; leaves in 3's, lance-linear, punctate beneath, straight-veined, delightfully fragrant. 3f.

3. **PHRYMA,** L. LOP-SEED. Cal. cylindric, bilabiate, upper lip longer, 3-cleft, lower lip 2-toothed. Corolla bilabiate, upper lip emarginate, much smaller than the 3-lobed lower one. Stamens included. Fruit dry, oblong, striate, 1-celled, 1-seeded. ♃ With opposite leaves. Flowers opposite, spicate, deflexed in fruit.

P. leptostáchya L.—Rocky woods. 2—3f. Leaves large (3—6′), thin, coarsely-toothed; flowers small, light-purple, in very slender spikes. July.

4. **CALLICÁRPA,** L. FRENCH MULBERRY. Calyx 4-toothed, bell-shaped. Corolla short-bell-shaped, limb of 4 obtuse segments. Sta. 4, unequal, exserted. Stig. capitate, 2-lobed. Drupe juicy, enclosing 4 nutlets. ♄ With opposite leaves and axillary cymes.

ORDER 91.—LABIATÆ. 237

C. Americàna L. Pubescent; lvs. ovate, acuminate at each end, crenate-dentate, smooth above; clusters shorter than the petioles; fruit forming dense verticils Light soils, S. Shrub much branched, 3—6f, with purple flowers and fruit.

5. LANTÀNA, L. Cal. minute, obsoletely 4-toothed. Corolla funnelform, the tube long-exserted, limb oblique, upper lip bifid or entire, lower trifid. Sta. 4, didynamous, included. Drupe fleshy, double, the parts separable, 1-seeded. ♄ 3—6f. Tropical, with square stems, opposite petiolate leaves, and capitate, handsome flowers, often fragrant.

* Corollas white or lilac, not becoming yellow or scarlet .... Nos. 1—3
* Corollas white or yellow, changing to saffron, scarlet, crimson, &c .... Nos. 4, 5

1 L. NÍVEA. Branches with reversed prickles; lvs. ovate to elliptic, crenate-serrate, as long as the peduncles; no involucre; flowers white, turning to blue. Brazil.
2 L. SELLOWIÀNA. Branches strigous; lvs. rhombic to oblong, coarse-serrate, shorter than the peduncles; heads some involucrate; flowers reddish lilac. Brazil.
3 L. involucràta L. Whitish-downy; lvs. obovate to roundish, crenulate, as long as the peduncles; heads involucrate with the outer ovate bracts, lilac. S. Fla.
4 L. MIXTA. Prickles reversed or 0; lvs. ovate, crenate, abrupt at base, shorter than the peduncles; bracts as long as the corollas, which are white at first, then changing to yellow, then orange, and lastly red. Brazil.
5 L. Còmara L. Often prickly; lvs. as in No. 4, but equalling the peduncles; bracts half as long as the corollas, which are successively yellow, orange, red. Ga., Fla.

6. VITEX, L. CHASTE-TREE. Calyx 5-toothed. Cor. cup-shaped, 5-lobed, somewhat 2-lipped. Stamens 4, unequal, exserted. Drupe entire, 4-celled, 4-seeded. ♄ With opposite, digitate leaves and paniculate cymes.

1 V. AGNUS-CÁSTUS. Leaflets 5 or 7. lanceolate, entire, pointed both ways; panicles white-tomentous, terminal, interrupted; corolla purplish. Hardy. S.
2 V. NEGÚNDO. Leaflets 3 or 5, oblong, serrate, acuminate. Mauritius.
3 V. INCÌSA. Leaflets 5 or 7, incisely pinnatifid, acuminate. China.

ORDER XCI. LABIATÆ. LABIATE PLANTS.

Herbs with square stems, and opposite, aromatic, exstipulate leaves. *Flowers* axillary, in verticillasters, sometimes as if spiked or in heads. *Corolla* labiate (rarely regular), upper lip external in the bud. *Stamens* 4, didynamous, or only 2. *Ovary* free, deeply 4-lobed, the single style arising from between the lobes. *Fruit* composed of 4 (or by abortion fewer) separable 1-seeded nuts or achenia. Figs. 23, 69, 96, 281, 292, 318, 384.

§ Stamens 2, perfect,—*p* ascending beneath the galea; anthers 1-celled. (Tribe IV.)
    —*p* ascending through a cleft in the galea; anthers 2-celled...(*b*)
    —*p* exserted, distant; anthers 2-celled...(*a*)
§ Stamens 4, perfect,—*q* all declined toward the lower lip. (Tribe I.)
    —*q* erect, or ascending toward the upper lip...(2)
  2 Stamens of equal length, corolla almost regular, 4-5-lobed...(*c*)
  2 Stamens, the upper pair longer than the lower (outer), and calyx 13-15-veined. (Tribe VJ
    Stamens, the lower pair longer than the upper (interior) pair...(3)
    3 Stamens divergent, apart, mostly straight and exserted...(*e*)
    3 Stamens parallel, ascending and long-exserted from the upper side...(*b*)
    3 Stamens parallel, ascending in pairs beneath the upper lip...(4)
      4 Calyx 13-veined, 5-toothed, and somewhat 2-lipped...(*f*)
      4 Calyx 5-10-veined, or irregularly notted...(5)

ORDER 91.—LABIATÆ.

§ Calyx strongly 2-lipped, upper lip truncate, closed in fruit...(*h*)
§ Calyx not 2-lipped, 3 or 4-lobed, open in fruit...(*k*)
§ Calyx subequally 5-toothed, teeth not spinescent...(*m*)
§ Calyx subequally 5-toothed, teeth spinescent...(*n*)
§ Calyx unequally 8–10-toothed...(*o*)

I. OCIMOIDEÆ.—*r* Corolla upper lip 4-lobed, lower entire, flattish..............OCIMUM 1
—*r* Corolla upper lip 4-lobed, lower saccate, deflexed...........HYPTIS. 2
—*r* Corolla upper lip 3–4-fid, lower boat-form, involving the sta..COLEUS. 3
—*r* Corolla upper lip 2-lobed, lower 3-lobed....... ....LAVANDULA. 4
II. AJUGOIDEÆ.—*b* Stamens 2, exserted through a fissure in the tube......... ..AMETHYSTEA. 5
—*b* Stamens 4, exserted through a fissure in the tube..........TEUCRIUM. 6
—*b* Stamens very long, involute, arching the corolla... .......TRICHOSTEMMA. 7
—*c* Corolla limb equally 5-lobed. Stamens short..............ISANTHUS. 8
III. SATUREJEÆ.—(Stamens diverging or ascending, 2-celled. Corolla lobes flattish, spreading.)
—*c* Corolla limb obliquely 5-lobed. Leaves purple..........PERILLA. 9
—*c* Corolla limb 4-lobed, upper lobe broadest.... ....MENTHA. 10
d Corolla nearly regular, 4-lobed. Calyx naked in the throat......LYCOPUS. 11
d Corolla bilabiate,—*s* cyanic, throat naked. Stamens straight..........CUNILA. 12
—*s* cyanic, throat naked. Stamens ascending........HEDEOMA. 13
—*s* yellow, throat with a hairy ring inside.............COLLINSONIA. 14
e Calyx 15-veined. Stamens exserted, divergent.........................HYSSOPUS. 15
e Calyx 10-veined, the veins obscured by hairs. Corolla yellow, fringed.......COLLINSONIA. 14
e Calyx 10–13-veined,—*t* throat naked.—*u* Stamens straight, divergent.. .....PYCNANTHEMUM. 16
—*u* Sta. ascending, anth. spurless.....SATUREJA. 17
—*u* Sta. ascending, anth. spurred.......DICERANDRA. 18
—*t* throat hairy.—*v* Bracts roundish, large..............ORIGANUM. 19
—*v* Bracts narrow, minute....... ....THYMUS. 20
f Tube of the corolla straight. Leaves small, subcrenate or entire......CALAMINTHA. 21
f Tube of the corolla curved upward. Leaves large, coarse-crenate...........MELISSA. 22
IV. MONARDEÆ.—Connectile long, transverse, distancing the anther cells......SALVIA. 23
—Connectile continuous with fil. toothed at the juncture.......ROSMARINUS. 24
—Connectile inconspicuous.—*w* Calyx subequally toothed.....MONARDA. 25
—*w* Calyx bilabiate, aristate.......BLEPHYLLIA. 26
V. NEPETEÆ.—Stamens distant, exserted. Flowers in terminal spikes. ....LOPHANTHUS. 27
—Stamens all ascending.—*x* Anther cells divergent, much........NEPETA. 28
—*x* Anther cells divergent, little .......DRACOCEPHALUM. 29
—*x* Anther cells parallel. Fls. large......CEDRONELLA. 30
VI. STACHYDEÆ.—(Stamens parallel, ascending. Cor. upper lip galeate. Cal. 5–10-veined.)
h Calyx lips toothed, upper 3 teeth minute, lower 2 large.....................BRUNELLA. 31
h Calyx lips entire, upper with an appendage on the back..................SCUTELLARIA. 32
k Calyx 3-lobed. Anthers all distinct. Flowers purple streaked..... .....MACBRIDEA. 33
k Calyx 4-lobed. Anthers, the highest pair connate........................SYNANDRA. 34
m Corolla tube inflated in the midst, whitish. Lips small....................PHYSOSTEGIA. 35
m Corolla tube inflated at the throat, purple. Lower lip long..............LAMIUM. 36
m Corolla inflated in the broad, concave upper lip, purple or yellow..........PHLOMIS. 37
m Corolla not inflated, short.—*y* Calyx salver-form, 10-ribbed................BALLOTA. 38
—*y* Calyx broad-bell-form, netted.............MOLLUCELLA. 39
n Anthers opening transversely, ciliate-fringed. Leaves notched..........GALEOPSIS. 40
n Anthers opening lengthwise.—*z* Achenia rounded at the top. Native...STACHYS. 41
—*z* Achenia rounded at the top. Cultivated..BETONICA. 42
—*z* Achenia truncate, 3-angled at top..... ..LEONURUS. 43
o Corolla white, upper lip flattish. Style equally bifid........ ...MARRUBIUM. 44
o Corolla white, upper lip concave. Style unequally bifid. South..LEUCAS.
o Corolla scarlet, exserted. Calyx upper tooth longest...........LEONOTIS. 45

1. ÓCIMUM, L. SWEET BASIL. Upper lip of calyx orbicular, lower 4-fid. Cor. resupinate, one lip 4-cleft, the other undivided. Stam. 4, declined, the lower pair longer, the upper often with a process at their base. Verticils 6-flowered, in terminal, interrupted racemes.

O. basilicum L. Lvs. smooth, ovate-oblong, subdentate, petiolate; cal. ciliate. ①
Plant 6–12′, in the kitchen-gardens. Very fragrant.

2. COLEUS, Lour. Cal. deflexed in fruit, unequally 5-toothed. Cor. decurved, upper lip obtusely 3–4-cleft, lower longer, entire, concave, involving the 4 stamens. ① Verticils 6–∞-flowered. Asia.

C. Blùmei. Leaves large. ovate, bluntly serrate; verticillasters distinct, ∞-flowered. 2f. Tender, cultivated for its splendid leaves, which are marked with crimson, green, and bronze. Flowers inconspicuous.

3. HYPTIS, L. Calyx 5-toothed, teeth acute or subulate. Cor. tube cylindric, limb 5-lobed, the lower abruptly deflexed, contracted at its base, the 4 others flat, erect or spreading. Stam. 4, declinate. Ach. ovoid or oblong.—In our species the flowers are in involucrate heads. Summer.

H. radiàta Willd. Erect, glabrous; leaves lance-ovate to lance-linear, unequally and bluntly serrate, tapering to the petiole; heads opposite, pedunculate, at length globular, bracts seeming radiate. ⚘ Damp, S. 2—3f.

4. LAVÁNDULA, L. Lavender. Cal. ovoid-cylindric, with 5 short teeth, the upper one often largest. Cor. upper lip 2-lobed, lower 3-lobed, lobes all nearly equal, tube exserted, stamens included. ♄

L. spica. Leaves hoary, linear-oblanceolate to linear-lanceolate, rolled at edge, sessile, in the interrupted spike bract-like; flowers small, lilac. Very fragrant, and yielding the well-known *Oil-of-Lavender*. 12—18'. July.

5. AMETHÝSTEA, L. Flowers as in Teucrium, but the stamens are only 2. ① From Siberia.

A. cœrùlea.—A branching, smooth herb, 1f high, with the leaves 3-parted and incised, and blue (to white) corollas little exceeding the calyx. July—Oct.

6. TEÙCRIUM, L. Germander. Cal. subcampanulate and subregular, in 5 acute segments. Cor. with the 4 upper lobes nearly equal, the lowest largest, roundish. Stam. 4, exserted from the deep cleft in the upper side of the tube.

T. Canadénse L. Plant erect, hoary-pubescent; lvs. ovate-lanceolate, acute, serrate, petiolate; bracts linear-lanceolate, about as long as the calyx; spike long, of many crowded verticils of odd-looking purplish flowers. ⚘ Damp grounds. 2f. July.

7. TRICHOSTÈMA, Lin. Blue Curls. Calyx very oblique, veiny, lower lip of 2 short teeth, upper twice as long, of 3, all acute. Cor. tube slender, limb obliquely 5-lobed. Filam. 4, very long, exserted and curved. ① Cymes loose, panicled. Corolla blue.

1 **T. dichótoma** L. Lvs. oblong-lanceolate, attenuate at base, obtuse, entire pubescent, as well as the stem and branches. Dry soils, Mass., and S. 1f. August.

2 **T. lineàre** N. Leaves linear, nearly smooth; stem and branches puberulent. Dry soils, N. Y. (at Salem), and S. 1f. Flowers as in the other, 4″. July, Aug.

8. ISÁNTHUS, Mx. False Pennyroyal. Calyx equally 5-toothed, throat naked. Cor. 5-parted, tube straight and narrow, segm. ovate and equal. Stam. subequal, incurved, ascending, longer than the corolla. ⚘ Viscid, pubescent, with entire leaves acute at each end. Flowers axillary.

I. cœrùleus Mx.—Dry fields, N. and W. 1—1½f. Branching and leafy, resembling Pennyroyal. Leaves lance-elliptic, 3-veined. Flowers 1—2 in each axil, blue. July.

ORDER 91.—LABIATÆ.

9. PERÍLLA, L. Calyx subequally 5-toothed, in fruit becoming gibbous and 2-lipped. Cor. bell-form, 5-cleft, lower lobe a little longer. Sta 4, erect, distant, included.—Asia.

P. ocimoïdes, β. Nankinénsis, is the PURPLE PERÍLLA, a fine leaf-plant, 2f high, with large bronze-purple, ovate, cut-fringed leaves. (β. crispa Benth.) ①

10. MENTHA, L. MINT. Cal. equally 5-toothed. Cor. nearly regular, tube scarcely exserted, border 4-cleft, the broadest segment emarginate. Stam. 4, straight, distant, anth. cells parallel, fil. naked. ♃ Strong-scented herbs. Flowers in dense verticils, pale purple. Summer.

* Leaves sessile. Verticils in a slender, terminal spike.....................Nos. 1.–3
* Leaves petiolate.—*x* Verticils in dense oblong spikes.....................Nos. 4, 5
  —*x* Verticils axillary, not in spikes........ ....Nos. 6—8

1 **M. víridis** L. *Spearmint.* Smoothish; lvs. lance-oblong, acute, cut-serrate; spikes interrupted, attenuate above. Damp soils. 1—2f. § Europe.
2 **M. rotundifòlia** L. Whitish-downy; lvs. roundish to broad-ovate, sharp-serrate; spikes cylindric, nearly continuous. N. J., Pa. (at Easton, Prof. Porter). Ascending 2—3f. Spikes 2—3'. § Europe.
3 **M. sylvéstris** L. Woolly-tomentous; lvs. lance-ovate, canescent, finely serrate; spikes conic-cylindric. Delaware Co., Pa. (A. H. Smith).
4 **M. piperìta** L. *Peppermint.* Smooth; lvs. ovate to lanceolate, serrate; spikes 1', oblong to cylindric ; calyx smooth. Wet. 2f. § Europe.
5 **M. aquática** L. Stem reflex-hairy; leaves ovate, serrate, hairy or smoothish ; spike globular or oblong, calyx villous. Muddy. §
6 **M. sativa** L. Stem reflex-hairy, erect, branched; leaves ovate, canescent beneath ; calyx teeth subulate-awned. Lancaster, Pa. (Porter). § Europe.
7 **M. arvénsis** L. Smoothish, ascending; leaves ovate, serrate above, entire and acute at base; calyx teeth acute. Fields, M. and W.: rare. §
8 **M. Canadénsis** L. *Horsemint.* Upright, hoary-pubescent with spreading hairs; leaves lanceolate, very acute both ways; cymes shorter than the petioles; stamens exserted. Damp. Can. to Pa. and Ky.
  β. *boreàlis.* Plant nearly smooth, with narrower leaves.

11. LÝCOPUS, L. WATER HOARHOUND. Cal. tubular, 4–5-cleft. Cor. subregular, 4-cleft, the tube as long as the calyx, upper segment broadest, emarginate. Stam. 2, distant, diverging, simple. ♃ Bog herbs, with the very small flowers in axillary, dense clusters.

1 **L. Virgínicus** L. *Bugle Weed.* Lvs. broad lanceolate, serrate, tapering and entire at both ends; calyx teeth 4, obtuse, spineless, shorter than the achenia. Common. 1—1½f. Plant often purple, and often with long slender runners. July, August.
2 **L. Européus** L. Lvs. lance-ovate to lance-oblong, petiolate, acute, sinuate-toothed or lobed, the lower incised; calyx teeth 5, acuminate-spinescent, longer than the smooth achenia. Common, and very variable. 1—2f. August.
  β. *rubéllus,* with creeping stolons, and downy toothed lvs. (L. rubellus Mœnch.)
  γ. *sinuàtus,* with smooth sinuate-dentate leaves—no runners. (L. sinuatus Ell.)
  δ. *exaltàtus.* Tall, with smooth leaves cut into linear teeth. (L. exaltatus Ell.)
  ε. *angustifòlius.* Leaves narrow, slightly toothed or subentire. (L. angust. N.)
  ζ. *sessilifòlius.* Lvs. oblong, sessile or clasping, remotely toothed. N. J. (Porter).

12. CUNÌLA, L. DITTANY. Cal. 10-ribbed, equally 5-toothed, throat densely villous; upper lip of corolla flat, emarginate. Stam. 2, erect, exserted, distant.—Flowers numerous, pale red.

ORDER 91.—LABIATÆ. 241

C. Mariàna L. Lvs. ovate, serrate, subsessile, 1′; cymes pedunculate, corymbous, axillary and terminal. ♃ Rocky woods, N. Y. to Ga. and Ark. 1—2f. July, Aug.

13. **HEDEÒMA**, Pers. AMERICAN PENNYROYAL. Calyx 13-striate, gibbous at base, bilabiate, throat hairy, upper lip 3-toothed, lower 2-cleft. Cor. bilabiate, upper lip erect, flat, emarginate, lower spreading, 3-lobed. Stam. 2, fertile, ascending.—Low, fragrant herbs.

1 **H. pulegioìdes** Pers. Lvs. oblong, few-toothed, petiolate, narrowed to each end; verticils axillary, 6-flowered; corolla equalling the calyx. ① Dry pastures. 6—12′. June—Aug. A small herb of pungent fragrance and taste, common and much used.

2 **H. hispida** Ph. Hairy, branching, with sessile, linear, obtuse leaves and verticils 6-flowered; corolla scarcely exceeding the calyx. ① Banks, W. 2—5′. July.

14. **COLLINSÒNIA**, L. HORSE BALM. Cal. ovoid, 10-striate, upper lip truncate, 3-toothed, lower 2-cleft. Cor. exserted, bell-ringent, upper lip in 4 subequal lobes, lower longer, declined, fringed. Stam. 2 or 4, much exserted, divergent. ♃ Coarse, strong scented, with large, ovate, serrate, petiolate lvs. and yellowish fls., in a terminal, leafless panicle or raceme.

* Stamens 4, perfect, long exserted. Leaves very large. South............Nos. 1, 2
* Stamens 2, perfect,—*a* the upper pair of filaments mere points............Nos. 3, 4
        —*a* the upper pair of filaments capitate. South...........No. 5

1 **C. verticillàta** Baldw. Viscid-downy above; lvs. broad-oval, 6—8′, acute, petioles 1—2′; racemes long, naked; flowers in whorls, 9″; lower lip strongly fringed. Lookout Mt., Tenn., and Middle Ga. 1—2f. Raceme 1f. May, June.

2 **C. anisàta** Ph. Viscid-downy; lvs. cordate, acuminate, crenate, 5—7′, petiole 1′; panicle 3—6′, bracts ovate, flowers 5—6″. Ga., Fla., Ala. 1—2f. July—Sept.

3 **C. Canadénsis** L. Sparsely downy; leaves mucronate-serrate, acuminate, abrupt at base, 4—7′; lower petioles slender; panicle 5—8′, loose, bracts ovate; flowers 5—6″. Damp shades, Can. to La. (Hale). 3—4f. Summer.

4 **C. scabriúscula** Ait. Leaves scabrous above, small (1½—2′), acuminate, acute at base, petioles slender, 1′; panicle leafy, fls. 4—5″, calyx 1″. Woods, S. 2f. Sept.

5 **C. punctàta** Ell. Pubescent; lvs. 4—7′, lance-ovate, pointed, acute at base, resinous-punctate beneath; panicle leafy below, flowers 5″. Woods, S. 2—6f. Sept.+

15. **HYSSÒPUS**, L. HYSSOP. Calyx tubular, 15-striate, equally 5-toothed. Upper lip of the corolla erect, flat, emarginate, lower 3-parted, the middle segment largest, tube about as long as the calyx. Stamens 4.

H. officinàlis L.—Native of Europe and Asia, occasionally cultivated for its medicinal properties. A bushy herb, 2f, with oblong-lanceolate leaves, and bright blue fls. in 1-sided verticils approximate in a terminal spike. St. exserted, diverging. §

16. **PYCNÁNTHEMUM**, Mx. BASIL. Calyx tubular, 10–13-striate, 5-toothed, teeth equal or subbilabiate, throat naked within. Upper lip of corolla nearly entire, lower trifid, middle lobe longest, all ovate, obtuse, stam. 4, distant, subequal, anth. with parallel cells. ♃ Erect, rigid branching herbs, all N. American. Verticils dense, many-flowered. Aug., Sept.

§ Calyx 2-lipped, in flat or loose cymes. Leaves petiolate, subserrate...(*a*)
§ Calyx subregular, in roundish dense heads...(*b*)
  a Teeth of the calyx ovate, acute, awnless........................No. 1
  a Teeth of the calyx tipped with bearded awns ....Nos. 2—4
    b Calyx teeth and bracts with naked awns as long as the corolla.....Nos 5, 6
    b Calyx teeth awnless, shorter than the corolla...(*c*)

c Heads panicled. Leaves subpetiolate, subentire..............Nos. 7– 9
c Heads corymbed. Leaves sessile, entire...................Nos. 10–12
c Heads solitary, involucrate. Leaves serrate.....................No. 13

1 **P. albéscens** T. & G. Leaves lance-ovate, acute, whitened beneath, the upper whitened both sides; flowers in little secund racemes. Ala. to La. 2–3f.

2 **P. Túllia** Benth. Villous-pubescent; leaves ovate to lanceolate, acute or pointed; the floral whitened; inflorescence as in No. 1. Mountains, S.

3 **P. incànum** Mx. *Wild Basil.* Whitish, with a soft down; leaves ovate, rounded at base, slightly acuminate; the floral whitened both sides; cymes 1' and less broad, not racemed; corolla pale red, dotted. Rocky woods, N. and W. 2–4f.

4 **P. clinopodioides** T. & G. Villous-canescent; leaves lanceolate, acute both ways; cymes small, dense, terminal and subterminal. Dry soils, N. Y., N. J., and W. 2–3f. Plant not whitened. Heads 6'' diameter.

5 **P. aristàtum** Mx. Smoothish; leaves ovate-oblong, acute, subserrate, rounded at base, petiolate; bracts rigid; heads few, 6—9'' diam. Barrens, N. J., and S. 1—2f.

6 **P. hyssopifòlium** Benth. Smoothish; leaves linear-oblong, obtuse, nearly sessile and entire; heads few, large, 1' diameter. Barrens, Va. to Fla. 1—2f.

7 **P. Tórreyi** Benth. Slightly pubescent; lvs. lin.-lanceolate, acute, subentire; bracts and subulate calyx teeth white-pubescent. Dry hills, New York Island, N. J.: rare.

8 **P. pilòsum** N. Hoary with soft, spreading hairs; leaves lanceolate, acute at each end, subentire, subsessile; calyx teeth ovate-lanceolate, and with the bracts white-tomentous. Prairies, W. States, to Ga. 2f. Cymes dense, 6—9''.

9 **P. mùticum** Pers. Minutely white-downy at top; leaves ovate to lance-ovate, acute, rounded or subcordate at base; calyx teeth short, merely acute. In dry woods. 2—3f. Heads roundish, dense, 4—6''.

10 **P. lanceolàtum** Ph. Leaves linear-lanceolate, entire, acute, rigid, abrupt at base, sessile; calyx teeth short, hairy; heads small (3—5''). Dry woods, Mass. to Car., and W. 1—2f. Handsome, fragrant, nearly smooth.

11 **P. linifòlium** Ph. Glabrous; leaves linear, attenuated both ways; heads compact, corymbed; calyx teeth pungently awn-pointed. Dry soils. 1—2f.

12 **P. nudum** N. Glabrous, pale, subsimple; leaves few and small, ovate-oblong, obtuse, entire, sessile; calyx teeth acute, pubescent. Mts., N. Car. to Ga. 1—2f.

13 **P. montànum** Mx. Glabrous except the villous-ciliate ovate and linear bracts; leaves lanceolate, serrate, acute; heads involucrate. Mountains, Va. and Car. 1—2f. Resembles a Monarda. Fragrant.

17. **SATURÈJA,** L. SUMMER SAVORY. Calyx tubular, 10-ribbed, throat not hairy. Segments of the bilabiate corolla not equal. Stamens diverging, scarcely exserted.—Herbs with small leaves and purplish fls.

S. horténsis L. St. branching; lvs. linear-oblong, entire, acute at the end; ped. axillary, cymous. ① River banks, W., escaped from gardens: rare. §

18. **DICERÁNDRA,** Benth. Calyx 13-striate, tubular, upper lip subentire, lower bifid, throat hairy. Cor. tube exserted, straight, strongly 2-lipped, the upper erect, emarginate, the lower spreading. Sta. 4, exserted, distant, anther cells divaricate, each with a little horn. ① Branching, smooth, with loose cymes.

1 **D. linearifòlia** B. Stem and branches strict; lvs. linear, or linear-oblong; cymes stalked, of 1—5 showy pink flowers, forming slender panicles. Dry woods, Prince Edward County, Va. (Dr. Mettauer), to Fla. (Miss Keen). 1f. October.

2 **D. densiflòra** B. Leaves lance-oblong; cymes sessile, 5-10-flowered. E. Fla.

19. **ORÍGANUM,** L. MARJORAM. Calyx tube 10-striate, 5-toothed,

hairy in throat. Corolla tube scarcely exserted, upper lip erect, flat, emarginate, lower with 3 nearly equal segments. Stamens 4, ascending, distant. ♃ Leaves subentire. Fls. in dense oblong spikes, with imbricated bracts.

1 O. vulgàre L. *Wild M.* Leaves ovate, petiolate, hairy; spikes corymbed; bracts ovate, purplish; calyx teeth equal. Fields : rare. 1f. June, July.

2 O. Marjoràna. *Sweet M.* Leaves oval or obovate, obtuse, petiolate, hoary-pubescent; bracts roundish; calyx tube split below. 1f. A kitchen vegetable.

20. THỲMUS, L. Thyme. Cal. 2-lipped, ovoid, 13-veined, upper lip of 3, the lower of 2 subulate teeth, throat hairy. Cor. moderately 2-lipped. Sta. straight, exserted, distant. ♄ Leaves small, entire, strongly veined. Bracts minute. Flowers purple. European culinary herbs.

1 T. Serpýllum L. *Wild T.* Stems creeping and ascending, leafy, each terminated with a small, dense, oblong head of flowers grateful to bees. † §. June.

2 T. vulgàris. Stems erect from the decumbent base; lvs. oblong-ovate to lanceolate, the sides revolute; fls. in term., leafy spikes. Much branched. 6—10′ high. Jn., Jl.

21. CALAMÍNTHA, Mœnch. Calaminth. Cal. tubular, 13-nerved, throat mostly hairy, upper lip 3-cleft, lower 2-cleft. Corolla tube straight, exserted, throat inflated, limb bilabiate, upper lip erect, entire or emarginate, lower spreading, its middle lobe largest. Stamens 4, the lower pair longer, usually ascending. ♃

§ Herbs hairy. Cymes dense, capitate, bracted. Calyx tube curved, 2-lipped..No. 1
§ Herbs hairy. Cymes loose, pedunculate. Calyx tube straight, 2-lipped.......No. 2
§ Herbs smooth. Cy. loose, sessile, bracted. Cal. straight, teeth subregular...No. 3
§ Shrubs low, slender, nearly smooth. Cymes few-flowered. Fls. large....Nos. 4—7

1 C. Clinopòdium Benth. *Wild Basil.* Plant clothed with whitish hairs; leaves ovate, subserrate; fls. purplish, in dense verticils or heads, with many subulate bracts. Low woods, N. and W. 1—2f. Heads near 1′ wide. June—August.

2 C. Népeta Link. Branched below, soft-villous; leaves small, broad-ovate, obtuse; cymes few-flowered, becoming some racemed; corolla white, 3—4″; calyx 1″. Va., Tenn., to Ga. Roadsides, &c. 2f. Strongly aromatic. July, August. § Europe.

3 C. glabélla B. Smooth, decumbent at base, diffusely branched; leaves narrowly oblong, tapering to base; verticils 6-10-flowered. Rocks, O. to Ark. 18′. Cor. 4—5″, pale violet. Fragrant like Pennyroyal. Often produces runners, and runs into
β. *diversifolia.* Flowering stems nearly erect, the barren prostrate like runners, bearing small ovate leaves (3—4″). Rocks, Niagara, and N-W. 10′.

4 C. Caroliniàna Sw. Smooth, simple; lvs. ovate, obtuse, crenate-serrate; bracts similar; cy. few-flwd., on short stalks; cor. rose-purp., 7—9″. Dry woods, S. 15′. Jl.

5 C. coccínea B. Shrub with virgate branches; lvs. narrowly ovate-oblong; verticils of 2—6 ample scar. fls.; cor. 15—18″, gland.-pubescent. Sandy shores, Fla. 2f.

6 C. canéscens T. & G. Low shrub, minutely canescent-downy; leaves linear, with rolled edges, obtuse, crowded; fls. sol., opp., 8″, rose-red. Sandy shores, Fla. 8—12′.

7 C. dentàta Chapm. Tomentous; lvs. wedge-obovate, 2-4-toothed at apex. Fla. 2f.

22. MELÍSSA, Tourn. Balm. Calyx 13 ribbed, the upper lip 3-toothed, flattened and dilated, lower bifid. Cor. tube recurved-ascending, upper lip erect, flattish, lower spreading, 3-lobed, the middle lobe mostly broadest. Stamens ascending.

M officinàlis L. Pubescent; st. erect, branching; fls. in loose, axillary cymes leaves ovate, crenate-serrate, petiolate; bracts similar; corolla 7″, yellowish. Gardens, whence it has escaped into the fields and woods. 1—3f. July, August.

23. SÁLVIA, L. SAGE. Cal. striate, bilabiate, upper lip 3-toothed or entire, lower bifid, throat naked. Corolla ringent, tube equal, upper lip straight or falcate, lower spreading or pendent, 3-lobed. Stamens 2, connectile transverse on the filament, supporting at each end a cell of the halved anther. ♄ ♃ Figs. 96, 281.

* Native species.—§ Calyx limb 3-lobed. Lower anther cell wanting.......Nos. 1—3
  —§ Calyx deeply 2-lipped, 5-toothed. Both cells present..Nos. 4—6
* Species cultivated (No. 7 spontaneous).—*a* Flowers blue............. ....Nos. 7—9
  —*a* Flowers white...............Nos. 10, 11
  —*a* Flowers yellow..................No. 12
  —*a* Flowers red.—*b* Herbaceous..Nos. 13, 14
  —*b* Shrubby.....Nos. 15, 16

1 S. azùrea Lam. Smoothish, branching; lvs. linear-oblong and linear, subentire, acutish; racemes slender; verticils 2–6-flowered; corolla pubescent, tube barely exserted; limb azure blue. ♃ S. Car. to Fla. and La. 1—3f. Summer.

2 S. longifòlia N.? Tall, branched, puberulent; leaves oblong-lanceolate, serrate; racemes slender; corolla 8—9″, tube twice longer than calyx. ♃ Ga. to Ark. 3—6f.

3 S. urticifòlia L. Thinly pubescent; leaves rhomb-ovate, acute, serrate, decurrent on the petiole; verticils 4–10-flowered, distant in the raceme; corolla smooth, tube little longer than the calyx. ♃ Hilly woods, Va., and S. 18′. May.

4 S. lyràta L. Lvs. radical, lyrate, erose-dentate, many, stem lvs. about 1 pair, linear spatulate, bract-like; fls. in whorls, racemed at top of the square scape. ♃ In woods 6—15′. Flowers 1′, violet-purple. April—June.

5 S. obovàta Ell. Lvs. broad-obovate, entire, the floral ovate; verticils remote in the raceme; corolla blue, 8″, calyx 3″. ♃ Ga. to La. 1—2f. June, July.

6 S. Claytòni Ell. Lvs. cordate- to lance-ovate, sinuate-pinnatifid, and toothed, rugous, bracts ovate, pointed. ♃ Sandy fields, S. Car.

7 S. Sclàrea L. Lvs. ample, rugous, broad-cordate, doubly crenate; bracts colored; corolla pale purple, upper lip high-arched. ② Gardens, § in Penn.

8 S. OFFICINÀLIS. *Garden Sage.* Shrubby; leaves lance-oblong, crenulate, rugous; corolla upper lip vaulted, equalling the lower. From S. Europe. 1f. July.

9 S. PATENS. Hispid and hairy; leaves ovate-deltoid or ovate-hastate, crenate; flowers very large; calyx bell-form, 7″; corolla blue, 2′ long; stamen exserted. Mexico. 3f.

10 S. ARGÉNTEA. Leaves white with wool, large, ovate, sinuate-lobed, the floral concave; flowers 18″, racemed, the upper lip long-falcate. S. Europe.

11 S. CHIONÁNTHA, with large white-woolly, linear-lanceolate leaves and very large (2′) white flowers with arched galea, is from Asia Minor.

12 S. AÙREA. Shrub 3—4f, with roundish ovate whitened leaves, the splendid yellow flowers 2½′, calyx 1′, in dense racemes. From Africa.

13 S. COCCÍNEA. Stem and ovate-cordate leaves beneath hoary-downy; verticils of 6—10 red smooth flowers (8″) in a raceme; cal. 2-colored, 4″. ♃ Cuba, § in Ga., &c. 1—2f.

14 S. PSEUDO-COCCÍNEA, 3f high, is hispid with long spreading hairs, has ovate leaves rounded at base. Otherwise like No. 14. ♃

15 S. FULGENS. Plant branching, weak-stemmed, pubescent, with lance-ovate, subcordate leaves, the corollas 2′, bright red, opposite, in terminal racemes. Mexico.

16 S. SPLENDENS. Plant erect, smooth, with ovate lvs. and opposite pubescent flowers; calyx 1′, scarlet as well as the 2′ corollas. The commonest species. Mexico. 3f.

24. ROSMARÌNUS, L. ROSEMARY. Calyx upper lip entire, lower bifid. Cor. upper lip 2-parted, lower lip reflexed, in 3 divisions, of which the middle is the largest. Fil. 2, fertile, elongated, ascending toward the upper lip, having a tooth on the side. ♄ S. Europe.

B. officinàlis. Shrub evergreen with opposite, linear-oblong, obtuse, shining leaves. Flowers axillary and terminal, bright blue, fragrant of camphor. 4f.

25. MONÁRDA, L. MOUNTAIN MINT.
Calyx elongated, cylindric, striate, subequally 5-toothed. Cor. ringent, tubular, upper lip linear, lower lip reflexed, 3-lobed, the middle lobe narrowest. Sta. 2, fertile, ascending beneath the upper lip, and mostly exserted, anth. cells divaricate at base, connate at apex. ♃ Verticils few, dense, many-flwd., bracted. Jl.—Sept.

* Calyx densely hairy in the throat. Corolla purple or whitish..............Nos. 1, 2
* Calyx naked in the throat. Corolla scarlet or yellow.....................Nos. 3, 4

1 **M. fistulòsa** L. *Horsemint. Wild Bergamot.* Lvs. ovate to lanceolate, pointed, serrate or subentire, petiolate; flowers in large terminal heads; corollas 1', exserted, greenish white, pale lilac, or blue. Thickets, W. Vt., W. and S. 2—4f. Variable.
2 **M. Bradburiàna** N. Lvs. ovate to lanceolate, acute, rounded at base, subsessile; cal. curved, teeth spinescent (as in No. 1); bracts and corolla purple. Prairies, W. 3f.
3 **M. punctàta** L. Lvs. lance-oblong, tapering to the petiole; bracts leafy, colored, longer than the pale yellow, brown-spotted corollas. Barrens, N. J., S. and W. 2—3f.
4 **M. didyma** L. St. branching, acutely 4-angled; lvs. broad-ovate, pointed, short-petiolate; heads terminal and subterminal, with large (15'') showy crimson corollas, and bracts stained with the same hue. Swamps: rare. Often cultivated. 2f.

26. BLEPHÍLIA, Raf.
Calyx 13-veined, upper lip 3-toothed, lower lip shorter, 2-toothed, the teeth setaceous. Cor. upper lip short, erect, oblong, obtuse, entire, lower lip of 3 unequal, spreading lobes, the lateral ones orbicular. Stam. 2, fertile, ascending, exserted. ♃ Verticils dense, approximate in a spike.

1 **B. hirsùta** Benth. Hirsute all over, wide-branched; lvs. ov.-lanceolate, pointed, serrate, petiolate; bracts oblong, acuminate, colored, shorter than the pale, purple-spotted flowers; cor. 5''. Damp woods, N. Eng., W. Pa., and W. 1—2f. June, July.
2 **B. ciliàta** Raf. Thinly hirsute, simple; lvs. lance-oblong, distant, subsessile; verticils 3—5, the ovate bracts long as the calyx. Barrens, Pa., S. and W. 2—4f. Jl.-Sept.

27. LOPHÁNTHUS, Benth. HEDGE HYSSOP.
Cal. 15-ribbed, oblique, 5-cleft, upper segments longer. Cor. upper 1 ρ bifidly emarginate, lower lip 3-lobed, the middle lobe broader and crenate. Stam. diverging. ♃ Tall, erect. Verticils spicate.

1 **L. nepetoìdes** B. Stem smooth, stout, angles sharp; lvs. ovate, pointed, serrate; calyx teeth ovate, obtusish, green, in spikes 2—3' long; corollas inconspicuous, greenish white. Fence-rows, &c., M. and W. 3—6f. July, Aug.
2 **L. scrophularifòlius** B. Stem pubescent, angles obtuse; leaves ovate, crenate-serrate; calyx teeth lanceolate, acute, colored; corolla pale purple. Borders of fields, M. and W. 3—4f. July, Aug. Closely resembles No. 1.
3 **L. anisàtus** B. Smooth; leaves ovate, &c., whitened beneath; calyx teeth as in No. 2; corolla azure-blue, fragrant of anise. Wis. to Dak. (Dr. Matthews.)

28. NÉPETA, L. CATMINT. GROUND IVY.
Cal. tubular, 5-toothed; Cor. tube slender below, dilated and naked in the throat, upper lip emarginate, lower 3-lobed, the middle lobe largest and crenate, margin of the orifice reflected. Sta. ascending, anther cells divergent. Figs. 318, 384.

246   Order 91.—LABIATÆ.

§ Tall. Verticils in a terminal raceme. Calyx nearly regular..................No. 1
§ Glechòma. Low, diffuse. Flowers axillary. Calyx curved, oblique ....No. 2
1 **N. catària** L. *Catnip.* Erect, hoary-tomentous; lvs. petiolate, cordate, deltoid-ovate, coarsely crenate-serrate; flowers spiked, the whorls slightly peduncled. ♃ About old buildings, &c. 2—3f. July. §. The delight of cats.
2 **N. Glechòma** B. *Gill-over-the-Ground.* Leaves reniform, crenate; corolla 3 times longer than the calyx (10''), bluish purple, anthers forming 2 little crosses. ♃ Creeping in grass, about walls, &c. 3'—1f. May. § Europe.

29. **DRACOCÉPHALUM, L.** Dragon-head. Calyx subequal, oblique, 5-cleft, upper segment larger. Cor. bilabiate, upper lip vaulted, emarginate, throat inflated, lower lip spreading, 3-cleft, middle lobe much larger, rounded or subdivided. Sta. 4, distinct, ascending, the upper pair longer. ②

D. parviflòrum N. Plant some downy, slender; leaves petiolate, lanceolate, deeply serrate; flowers small, bluish, spicate. N. New York, and W.: rare. 18'. July.

30. **CEDRONÉLLA, Mœnch.** Cal. subcampanulate, 5-toothed. Cor tube exserted, throat dilated, upper lip straight, flattish, emarginate or cleft lower 3-fid, middle lobe largest. Stam. 4, ascending, the upper longer, anther-cells parallel. Flowers spicate, bracted. Summer.

1 **C. cordàta** N. Pubescent, producing runners; leaves cordate, petiolate, bluntly crenate; spikes unilateral, corolla pale blue, 1'. ♃ Rocks, O., Va.: rare. 1f. June.
2 **C.** Mexicàna. Erect, with cordate-lanceolate, dentate leaves; flowers in a spike of close whorls, purple, large. Mexico. 2—3f. (Gardoquia (Lind.))

31. **BRUNÉLLA, Tourn.** Self-heal. Blue-curls. Cal. about 10-ribbed, upper lip dilated, truncate, with 3 short teeth, lower lip with 2 lanceolate teeth. Filam. forked, one point of the fork bearing the anther. ♃

B. vulgàris L. Stem simple; leaves oblong-ovate, toothed, petiolate; flowers blue, in a large oblong-ovoid spike of dense verticils with reniform bracts. Low grounds, very common, varying from 8' to 2f. All Summer.

32. **SCUTELLÀRIA, L.** Skull-cap. Cal. campanulate, lips entire, upper one appendaged on the back and closed after flowering. Cor. upper lip vaulted, lower dilated, convex, tube much exserted, ascending, throat dilated. Stam. ascending beneath the upper lip, anthers approximate in pairs, achenia tubercular. ♃

§ Flowers large (7 to 13'' long), racemed at top of the stem, with bracts...(*a*)
  a Bracts ovate, abrupt at base. Lips of corolla short. Petioles long...Nos. 1, 2, 3
  a Bracts lance-oblong, acute at base. Leaves notched, petiolate...(*b*)
    b Galea of the corolla longer than the lower lip.....................Nos. 4, 5
    b Galea of the corolla not longer than the lip...........................Nos. 6, 7
  a Bracts leaf-like, longer than the calyx. Leaves entire, subsessile .. Nos. 8—10
Flowers large or small, opposite, solitary, in the axils of the leaves.....Nos. 11—13
§ Flowers small (3'' long), in slender, axillary, one-sided racemes..............No. 14

1 **S. versícolor** N. Glandular-hairy, erect, branched; leaves broad-ovate, cordate, crenate, veiny; racemes long, many-flowered; bracts ovate, entire, subsessile; corolla 6—7''', lips blue, subequal, lateral lobes distinct. Pa., and W. States. 1½—4f.
2 **S. rugòsa** Wood. Hairs and leaves as in No. 1, but the stem is weak, ascending, bracts petiolate, and the lower lip of the (8'') corolla pendent and twice longer than the upper. Rocky shores, Harper's Ferry, Va., and S-W. 18'.

3 S. saxátilis Rid. Smoothish and not glandular, weak, ascending; leaves as in Nos. 1, 2; bracts as in No. 2; corolla 8″, lips equal, upper 3-lobed. Del., Va., and W. 2f.
4 S. canéscens N. Erect, pubescent; leaves ovate to oblong, lower cordate; rac. terminal and axillary; bracts lance-linear; corolla 8″, *canescent*, upper lip arched, remote from the lower. Dry soils, M. and W.: common. 1—3f. (S. argṹta Bkly.)
5 S. villòsa Ell.? Slender, erect, stem finely tomentous; leaves lanceolate, acute both ways, serrate; flowers paniculate, bracts lance-linear; corolla 9″, tube slender, galea strongly arched, 5 times longer than lip. Ga. (Dr. Feay). 2—3f.
6 S. serràta Andr. Erect, subsimple, green, smoothish; leaves ovate, pointed both ways, serrate; rac. few-flwd.; cor. 13″, lips subequal. Woods, E. Pa., Ill., and S. 2—3f.
7 S. pilòsa Mx. Erect, subsimple, pubescent; leaves rhomb-ovate or oval, obtuse, remote, crenate-serr.; racemes terminal; corolla 9—12″, lips distant. Pa. to Ga. 2f.
8 S. integrifòlia L. Erect, subsimple, tomentous or downy; leaves ovate to lance-linear, obtuse, entire, or the lower crenate; flowers 9″, much enlarged above, the lips subequal, in a terminal raceme. Dry soils, M. and S. 9′—2f.
9 S. Floridàna Chapm. Slender, branching; leaves all linear, obtuse, entire, with rolled edges, lowest minute; corolla 1′, enlarged above, lips subequal. W. Fla. 1f.
10 S. macrántha (or Japonica). In gardens, 1f, smooth (except the hairy calyx); lvs. clasping, lance-linear; flowers 1′, blue, with subequal lips, handsome. China.
11 S. nervósa Ph. Slender, erect, producing runners; leaves sessile, broad-cordate, crenate-serr., 3-5-veined; fls. few, 4″, with small floral lvs. Rocks, M. and W. 8—15′.
12 S. párvula Mx. Root a string of tubers, stem erect, 3—9′; lvs. ovate to oblong, obtuse, small (6″), sessile, entire; fls. 3″, exceeding the lvs., blue. Fields, M. and W.
13 S. galericulàta L. *Common S.* Erect, branched, smoothish or downy; leaves nearly sessile, cordate-oblong or lanceolate, obscurely crenate, acute; flowers few, large (9—12″), sessile, axillary. Low meadows, Can. to Penn. 12—18′.
14 S. lateriflòra L. *Mad-dog S.* Smoothish, subsimple; lvs. petiolate, lanceolate, serrate; fls. 4″; rac. axillary, secund, equalling the lvs. Ditches, N., W., M. 1—2f.

33. MACBRÌDEA, Ell.
Calyx 3-lobed, upper lobe oblong, narrow, lower rounded. Cor. tube long-exserted, throat inflated, upper lip erect, concave, lower short, spreading, the middle lobe rounded, broadest. Sta. ascending under the upper lip, anthers approximate by pairs. ♃ Erect, subsimple, with large purple-white flowers in heads.

1 M. púlchra Ell. Lvs. wedge-lanceolate, acute, serrulate, the floral ovate; corolla 18″, streaked with purple and white. Wet pine-barrens, S. 12—18′. Aug., Sept.
2 M. alba Chapm. Lvs. wedge-oblong, obtuse, dentate; the floral roundish; corolla white; lower lobes of the calyx notched. Pine-barrens, W. Fla. 12—18′. July, Aug.

34. SYNÁNDRA, N.
Cal. 4-cleft, segm. unequal, subulate, converging to one side. Upper lip of corolla entire, vaulted, the lower obtusely and unequally 3-lobed, throat inflated. Sta. ascending beneath the galea upper pair of anth. cohering, having the contiguous cells empty. ② Flowers solitary, axillary, somewhat spicate above. Figs. 69, 292.

S. grandiflòra N.—Woods, O. to Tenn. 6—18′. Stem simple. Lvs. cordate-ovate acuminate, petiolate. Cor. white, 1′, with large lobes, purple-striate. June.

35. PHYSOSTÈGIA, Benth.
LION-HEART. Cal. bell-form, 5-toothed Cor. much exserted, throat inflated, upper lip concave, entire, lower of 5 broad-spreading lobes. Sta. 4, separate, ascending beneath the upper lip. ♃ Smooth, with lanc., serrate lvs. and term. spikes of showy rose-white fls.

P. Virginiàna B. Stems mostly simple; lvs. oblong to narrow-lanceolate, sessile thick; spikes 4-rowed, ∞-flowered; corolla 8—15", spotted inside. Wet banks, M., W., and S. Often cultivated. 1—4f. August, September.—Variable.

36. LÀMIUM, L. HENBIT. Cal. 5-veined, with 5 subequal, subulate teeth. Cor. dilated at throat, upper lip vaulted, galeate, lower lip broad, emarginate, lateral lobes truncate, often toothed on each side near the margin of the dilated throat. Stamens 4, ascending. May—November.

* Weeds in waste grounds, with roundish leaves and small purple flowers...Nos. 1, 2
* Lvs. cordate, ovate. Fls. larger (1'), hairy in throat, side-lobes toothed...Nos. 3, 4

1 **L. amplexicaùle** L. Leaves cut-crenate, petiolate, the floral sessile-clasping; corolla slender (6"), galea entire, side-lobes not toothed, throat spotted. ① 6—10'.
2 **L. purpùreum** L. Lvs. roundish to ovate, crenate, all petiolate; corolla slender, 6", hairy within, side-lobes with a subulate tooth, galea entire. ① Penn., &c.: rare.
3 **L. alba** L. Lvs. hairy, petiolate; cor. white, curved, a hairy ring within, and the side-lobes with a tooth. Waysides: rare. Flowers in whorls. Cultivated.
4 **L. MACULÀTUM** (or rugòsum). Leaves hairy, rugous, petiolate, marked with a white oblong spot along the midvein. Flowers as in No. 3, but purplish. Cultivated.

37. PHLÒMIS, L. JERUSALEM SAGE. Calyx truncately 5-toothed. Cor. galea broad, keeled, lower lip spreading, 3-fid. Stamens ascending beneath the galea, the upper pair appendaged at base. Leaves rugous. Verticils showy, axillary.

P. tuberòsa. Tall, smoothish, with large ovate-cordate, crenate leaves; fls. 30—40 in a whorl, purple, hairy inside. Scarce in gardens and waste grounds.

38. BALLÒTA, L. BLACK HOARHOUND. Cal. funnel-form, 10-veined, 5-toothed. Cor. tube cylindrical, as long as the calyx, upper lip concave, crenate, lower lip 3-cleft, middle segment largest, emarginate. Sta. 4, ascending, exserted. Achenia ovoid-triangular. ♃

B. nìgra L. Lvs. ovate, subcordate, serrate; bracts linear-subulate; cal. throat dilated, teeth spreading, acuminate. Waste places, N. Eng.: rare. July. § Europe.

39. MOLUCCÉLLA, L. MOLUCCA BALM. SHELL FLOWER. Calyx campanulate, very large, the margin expanding, often repand-spinous. Corolla tube included, limb bilabiate. Stamens 4, ascending. ①

M. LÆVIS. A curious plant, noted for its ample calyx, much larger than its small, yellowish corolla. Stem smooth, 2f; leaves round-ovate, cut-crenate. Syria.

40. GALEÓPSIS, L. HEMP NETTLE. Cal. 5-cleft, spinescent. Upper lip of the corolla vaulted, subcrenate, lower lip with 3 unequal lobes, having 2 teeth on its upper side, middle lobe largest, cleft and crenate. Sta. exserted, anth. cells transverse. ① Vert. distant, many-flwd. § Eur.

1 **G. Tetràhit** L. St. hispid, the internodes thickened upward; leaves ovate, hispid, serrate; cor. twice as long as the calyx, upper lip nearly straight, concave; corolla white-purple. A common weed in fields and waste grounds, N. States. 1—3f. Jn., Jl.
2 **G. Làdanum** L. Internodes equal; lvs. lanceolate, subserrate, pubescent; upper lip of the corolla slightly crenate; corolla roseate. Waste soils: rare. 1f. August.

41. STACHYS, L. HEDGE NETTLE. Cal. tube angular, bell-form, 5- or 10-ribbed, with 5 acute or pungent teeth. Cor. upper lip erect-spreading or some vaulted, lower spreading, 3-lobed, mid-lobe largest. Stamens ascending, lower pair longer, anthers approximating by pairs. Verticils 2-10-flowered, approximating in a terminal raceme.

Our species are much alike, yet easily distinguished. They have stems mostly hispid, leaves elliptic-lanceolate, crenate-serrate, narrowed to an abrupt base, and corolla pale-purple with deeper spots. Summer.

* Plants ♃, leaves smooth. Calyx teeth divaricately spreading.........Nos. 1, 2
* Plants hispid or hairy.—a ♃ Calyx teeth spinescent. Lvs. subsessile....Nos. 3, 4
  —a Calyx teeth acute. Leaves petiolate...........Nos. 5, 6

1 **S. hyssopifòlia** Mx. Leaves sessile, linear-lanceolate, serrulate, small (1—2′); calyx teeth half as long as the 7″ corolla. Mass. to Mo., and S. 6—12′.
2 **S. glàbra** Rid. Leaves all petiolate, serrate; calyx teeth much spreading, as long as the corolla tube. Woods, N. Y. to Mich., and S. 15′—3f. Racemes 3—7′.
3 **S. palústris** L. Stout, hirsute; leaves some pointed, large, hoary beneath; corolla twice longer (7—8″) than the calyx teeth. Moist shades, Can. to Car. 1—4f.
4 **S. áspera** Mx. Slender, hispid; leaves pointed, sharp-serrate; calyx glabrous. teeth hispid, equalling the corolla tube. Damp soils: common. 2f. Not leafy.
5 **S. cordàta** Rid. Stout, with large, pointed leaves, crenate-dentate; calyx teeth triangular, much shorter than the corolla. ♃ Shady banks, W. 2—5f.
6 **S. arvénsis** L. Weak, diffuse; lvs. ovate-cordate, obtuse; bracts very short; cal. teeth lanceolate; corolla tube included, lips short. ① Waste grounds, N.: rare. §

42. BETÓNICA, Tourn. BETONY. Calyx tubular-bell-form, with 5 awn-like teeth. Cor. as in Stachys, but beardless inside. Stam. ascending parallel beneath the galea. Style bifid. Lower leaves long petioled, cordate, all crenate. Verticils large, dense, in a terminal spike.

1 **B. officinàlis** L. *Wood B.* Spike interrupted at base; flowers purple, cor. twice longer than calyx (7″), galea entire. Gardens, and escaped. 1f. Rare. § Europe.
2 **B.** GRANDIFLÒRA. Villous; floral leaves clasping; verticils separate; corolla violet, large (15″), handsome, galea obcordate, glabrous. Gardens. 2f. Siberian.

43. LEONÙRUS, L. MOTHER-WORT. Calyx 5-10-striate, 5-toothed, teeth subspinescent. Upper lip of the corolla entire, hairy, concave, erect, lower lip 3-lobed, the middle lobe obcordate. Stam. 4, ascending beneath the upper lip. Mostly ♃. Verticils axillary. Flowers purplish. Summer.

1 **L. Cardìaca** L. Lvs. palmate-lobed, 3-fid, to lanceolate; corolla longer than the calyx, a hairy ring within. About dwellings. 3—5f. § Asia.
2 **L. marrublástrum** L. Leaves oblong-ovate, coarsely cut-serrate; cor. shorter than the calyx teeth, naked within. Waste grounds. 2—4f. § Europe.

44. MARRÙBIUM, L. HOARHOUND. Cal. tubular, 5-10-striate, with 5 or 10 subequal teeth. Cor. upper lip erect, flattish or concave, entire or bifid, lower lip spreading, 3-lobed, middle lobe broadest, emarginate, tube included. Stam. included in the tube. ♃ Fls. in dense verticils, white.

M. vulgàre L. Hoary-pubescent; lvs. roundish, ovate, crenate-dentate, downy canescent beneath; cal. of 10 setaceous, hooked teeth. Fields, &c. 1—2f. Ju., Jl. § Eur.

45. LEONÒTIS, Br. LION'S-EARS. Calyx 10-veined, apex incurved, throat oblique, sub-10-toothed, upper tooth largest. Cor. tube exserted, upper lip concave, erect, entire, lower short, spreading, trifid. Sta. 4, under the galea, anth. in pairs.—Vert. dense, with numerous lin.-subulate bracts.

L. nepetæfòlia Br. Erect, stout; lvs. thin, ovate, crenate, on slender petioles; cal. teeth 8, spinescent; whorls very large; cor. scarlet, 10″. ① Fields, S. 4—7f. § Afr.

ORDER XCII. BORRAGINACEÆ. BORRAGEWORTS.

Herbs (shrubs or trees), with round stems and branches, not aromatic. *Leaves* alternate, generally rough, with stiff hairs. *Stipules* none. *Flowers* seldom yellow, generally in a coiled (scorpoid) inflorescence. *Sepals* 5. *Petals* 5, united below, regular, very rarely irregular. *Stamens* 5, inserted in the tube. *Ovary* 4-lobed, or entire, forming in *fruit* 4 separate, 1-seeded achenia in the bottom of the persistent calyx. Figs. 141, 455.

I. EHRETIEÆ. Ovary entire, style terminal. Fruit 4-seeded, fleshy. Shrubs...(a)
    a Calyx 4-5-toothed, in heads. Corolla funnel-form, white. Fla. and †..........CORDIA *bullata.*
    a Calyx 4-5-toothed, in corymbs. Corolla funnel-form, white. Fla..............EHRETIA *Beurreria.*
    a Calyx 5-parted, in secund spikes. Corolla salver-form, pale..................TOURNEFORTIA. 1
II. HELIOTROPEÆ. Ov. entire, style terminal. Fr. dry, separating into parts...(b)
    b Corolla tube cylindrical, throat open. Fruit separating into 4 parts.......HELIOTROPIUM. 2
    b Corolla tube conical, throat constricted. Fruit separating into 2 parts.........HELIOPHYTUM. 3
III. BORRAGEÆ. Ovary deeply 4-lobed, style basilar. Fruit 4 achenia...(c)
    c Corolla irregular, blue,—*d* having the border obliquely lobed......... ....ECHIUM. 4
        —*d* having the slender tube bent.......................LYCOPSIS. 5
    c Corolla regular in both tube and border...(e)
        e Achenia armed with barbed prickles.—*f* Corolla salver-form...............ECHINOSPERMUM. 6
        —*f* Corolla funnel-form..... ....CYNOGLOSSUM. 7
    e Achenia unarmed. Corolla throat closed by scales...(g)
        g Corolla wheel-form, no tube. Anthers exserted................BORRAGO. 8
        g Corolla wheel-form, a very short tube. Anthers included........ ..OMPHALODES. 9
        g Corolla tubular-bell-form, white. Style exserted......... ....SYMPHYTUM. 10
        g Corolla funnel-form, blue. Stamens included.................ANCHUSA. 11
    e Achenia unarmed. Corolla throat not closed with scales...(h)
        h Corolla tubular, with erect, acute lobes, white...........................ONOSMODIUM. 12
        h Corolla lobes rounded, convolute in the bud.. ....MYOSOTIS. 13
        h Corolla lobes rounded, imbricate in bud,—*k* white or yellow.....……LITHOSPERMUM. 14
        —*k* purple-blue.................MERTENSIA. 15

1. **TOURNEFÓRTIA,** L. SUMMER HELIOTROPE. Cal. 5-parted. Cor. salver-form, throat naked. Sta. 5, included. Sty. short. Fr. 2-carpelled, 4-celled and 4-seeded. ♭ ♄ With entire leaves and secund spikes.

1 T. HELIOTROPOÏDES Hook. Shrubby at base, erect, hairy, with oval obtuse wavy-edged leaves; ped. terminal, 2 or 3 times forked, with numerous small inodorous, pale-lilac, pretty flowers. Buenos Ayres. ·
2 T. **gnaphaloïdes** all white-silky, and T. **volùbilis,** climbing; in S. Fla.

2. **HELIOTRÒPIUM,** Tournef. HELIOTROPE. Calyx 5-parted. Cor. salver-form, throat open, folded between the lobes. Anth. sessile. Sty. short, stigma conical, the achenia cohering at base, at length separable. ♃ ♭ Fls. white or purple, in 1-sided, scorpoid spikes. Summer.

§ Flowers white, in forked terminal spikes, or single lateral ones........... Nos. 1, 2
§ Flowers white-purple, in a cluster of terminal spikes. Cultivated.........Nos. 3, 4

1 H. **Europæum** L. Erect, pubescent; lvs. oval, veiny, obtuse, petiolate; calyx spreading in fruit, hairy. ① Rocky banks, moist fields, Va., and N.: rare. 8—12'. §
2 H. **Curassávicum** L. Glabrous, ascending; leaves linear-oblong to spatulate, obtuse, tapering to base, veinless and glaucous. ① Shores, W. and S. 1f.
3 H. PERUVIÀNUM. Shrubby, erect, pubescent; leaves rugose, lance-ovate, short-petiolate; corolla twice longer than the calyx, peculiarly fragrant. Peru.
4 H. CORYMBÒSUM. Pubescent, with lance-oblong leaves tapering both ways: flowers deep purple, less fragrant, but larger than in No. 3.

ORDER 92.—BORRAGINACEÆ.

3. **HELIÓPHYTUM**, DC. Calyx 5-parted. Cor. salver-form, throat constricted, 5-rayed. Anth. included. Sty. very short. Nuts 2, each 2-celled (sometimes with 2 additional empty cells).—Herbs with habit of Heliotrope.

H. Indicum DC. Erect, branching, hairy; lvs. ovate, erose-serrulate, acute, veiny, rugous, abrupt or subcordate at base; spike terminal, single (rarely forked); corolla much exserted; fruit with four empty cells. ① Fields, W. and S. 1—2f. §

4. **ÉCHIUM**, Tourn. VIPER'S BUGLOSS. Calyx 5-parted, segm. subulate, erect. Cor. campanulate, obliquely and unequally lobed, with a short tube and naked throat. Stigma cleft. Achenia tuberculate, base flat. Flowers irregular, in spicate, panicled racemes. Summer.

E. vulgàre L. Plant rough with bristles and tubercles; lvs. lanceolate; fls. large, handsome, violet-blue, many and crowded. ② Fields, Pa. to Va. 1½f.

5. **LYCÓPSIS**, L. WILD BUGLOSS. Calyx 5-cleft. Cor. funnel-form, tube incurved, throat closed with ovate, converging scales. Ach. perforated at base, ovoid, angular. ① Distinguished mainly by the curved cor. tube.

L. arvénsis L. Plant hispid, erect, branched above, with lanceolate, repand-denticulate leaves; flowers small, sky-blue with white scales, the bent tube longer than the calyx, in leafy racemes. Fields and waysides. 1f. § S. Europe.

6. **ECHINOSPÉRMUM**, Swartz. BURR-SEED. Calyx 5-parted. Cor. hypocrateriform, throat closed with concave scales. Ach. erect, bearing 1—3 rows of echinate prickles, smooth between, compressed or angular, fixed to a central column.—Herbs with bracted racemes and small blue fls.

E. Láppula Lehm. Branched above; lvs. hairy, lanceolate to linear; corolla longer than calyx, border concave; ach. with prickles in two rows. ① Dry soils. 1f. July.

7. **CYNOGLÓSSUM**, Tourn. HOUND'S TONGUE. Cal. 5-parted. Cor. short, funnel-form, concave, throat closed by 5 converging, convex scales. Ach. covered with echinate prickles, depressed, forming a broad pyramidal fruit, each fixed laterally to the style. Lvs. large. Cor. blue, purple or white.

§ Racemes without bracts, or nearly so..Nos. 1, 2
§ Racemes bracted at base, but the pedicels always extra-axillary...............No. 3

1 **C. officinàlis** L. *Common H.* Silky-pubescent, leafy to the top; leaves oblong-lanceolate, the upper sessile; naked racemes panicled; corolla dull purple. ♃ Pastures, &c. 1—2f. Plant dull green, ill-scented. July. § Europe.

2 **C. Virgínicum** L. Plant hairy, leafless above, with oblong-oval lvs. below, and a terminal cluster of short spikes of pale-purple flowers. ♃ Woods, Va., N. and W.

3 **C. Morrisòni** DC. *Beggar-ticks.* Rough-pubescent, widely-branched; leaves acuminate; racemes forked; flowers very small, white; fruit with doubly barbed prickles adhering to all that pass. ① Rocky places. 2—3f. July.

8. **BORRÀGO**, Tourn. BOURAGE. Cal. 5-parted. Cor. rotate, with acute segments, a scale at base of each. Sta. converging. Ach. ovoid, muricate, excavated at base, inserted lengthwise into an excavated recep.—Eur.

B. officinàlis. Rough-haired, branching; leaves ovate; flowers sky-blue, showy, in terminal, loose racemes. ② In old gardens, sowing itself. 1 2f. All Summer.

9. **OMPHALÒDES**, Tourn. NAVELWORT. Calyx deeply 5-parted.

Cor. rotate, tube shorter than the calyx tube, throat closed. Sta. included Achenia cup-form, toothed at the edges.—Oriental herbs.

1 O. linifòlia. Erect, smooth, glaucous; leaves obovate to linear-lanceolate; corolla white, twice longer than calyx. ① Spain. 1f. June—August.

2 O. verna. Runners creeping; leaves cordate to ovate, puberulent; racemes in pairs, few-flowered; flowers bright blue. ♃ S. Europe. 6′. April, May.

10. SÝMPHYTUM, Tourn. Comfrey. Cal. 5-parted. Cor. tubular-campanulate, orifice closed with 5, subulate scales, converging into a cone. Ach. smooth, ovoid, fixed by an excavated base. ♃ Oriental herbs.

S. officinàle L. Stem hairy, winged with the decurrent, lance-ovate leaves; fls. white or pink, in revolute racemes. Gardens and fields. 2—4f. Summer.

11. ANCHÙSA, L. Bugloss. Cal. 5-parted. Cor. funnel-form, throat closed with 5 scales. Sta. included. Achenia excavated at base.—Europe.

A. Itálica. Plant bristly-hispid, with lanceolate leaves and panicled racemes of numerous bright-blue, small mellifluous flowers. A hardy biennial. Summer.

12. ONOSMÒDIUM, Mx. Cal. deeply 5-parted, with linear segments. Cor. cylindrical, having a ventricous, half 5-cleft limb, with the segments converging and the throat open. Anth. sessile, included. Style much exserted. Achenia whitish, shining. ♃ North American. Racemes terminal, subspicate, one-sided. Flowers white. Summer.

1 O. Virginiànum A. DC. Very rough with appressed, stiff bristles; lvs. oblong, sessile, 5-veined; cor. hispid, ‡ longer than the lance-linear sepals, the segm. lance-subulate; anthers arrow-shaped. Dry soils. 15—30′. Corolla 4—5″.

2 O. Caroliniànum DC. Shaggy with long, spreading, rusty-white bristles; leaves lance-oblong, 7-veined; flowers shaggy-bristly; corolla near twice longer than sepals, the segments ovate, obtuse. By streams, M., W., S. 2—4f.

3 O. molle Mx. Hoary with soft appressed hairs; lvs. oblong-ovate; corolla hirsute, lobes triangular, pointed. Dry soils, W. 2—3f.

13. MYOSÒTIS, Dill. Forget-me-not. Cal. 5-cleft. Cor. salver- or funnel-form, tube about equalling the calyx, the 5 lobes convolute in bud, throat closed with short, concave scales. Ach. ovate, smooth, with a small cavity at base.—Herbs slightly villous. Racemes bractless, or with a few small leaves at the base. Flowers never axillary. May—Aug. Fig. 455.

§ Racemes one-sided. Calyx clothed with minute, appressed hairs, if any......No. 1
§ Rac. two-sided. Calyx beset with spreading, minutely-hooked bristles....Nos. 2, 4

1 M. palùstris Roth. Roughish-downy, or nearly smooth, branching; leaves lance-oblong, obtuse; ped. spreading, longer (2—3″) than the equal cal.; cor. 2—3″ broad, blue, with a yellow centre. ♃ Gardens; from Europe, also escaped in fields, &c.

β. *laxa,* taller (1f), very slender; lvs. lin.-obl.; ped. 4—6″ long. Swamps, ditches.

2 M. arvénsis L. Rough with tubercled hairs, branched; leaves oblong-lanceolate, acute; rac. loose, naked; ped. twice as long as the open, equal cal. ② Fields. 6—15′.

3 M. verna N. (stricta Link.) Rough-bristly, with spatulate to lin.-oblong lvs.; ped. ascending, as long as the closed, bilabiate calyx; racemes leafy at base. ① Dry hills.

4 M. versícolor Pers. Stem very slender, hispid-villous; leaves oblong; racemes leafless; pedicels shorter than the deeply and equally 5-cleft calyx; flowers yellow, varying to blue. Del. (Canby, Porter). § Europe. The true Forget-me-not.

14. LITHOSPÉRMUM, L. Gromwell. Puccoon. Cor. funnel- or

ORDER 93.—HYDROPHYLLACEÆ.

salver-form, limb 5-lobed, orifice open, with or without appendages, anth. included. Stig. obtuse, bifid. Ach. bony, rugous or smooth, flat at base. —Herbaceous or suffruticous, generally with a thick, reddish root. Flowers spiked or racemed, bracted, white or yellow. (See Addenda.)

§ Achenia rugous-tubercled. Corolla throat open, not appendaged, white.......No. 1
§ Achenia smooth and white. Corolla throat appendaged.—*a* Fls. white...Nos. 2—4
—*a* Fls. yellow..Nos. 5—7

1 **L. arvénse** L. *Wheat-thief.* Leaves linear-lanceolate, obtuse, hairy; calyx nearly equal to the corolla, with spreading segments. ① A rough weed in fields. 1f—18'. Root reddish. Fls. small, solitary in the upper axils. May, June. § Europe.
2 **L. officinàle** L. Erect, very branching above; lvs. lanceolate, acute, veiny; calyx nearly equal to the tube of the corolla. ♃ Dry soils, N. and M. 1—2f. Flowers small, pedicellate, in recurved, leafy racemes. July. § Europe.
3 **L. latifòlium** Mx. Rough, erect, subsimple; leaves ovate, sessile, pointed both ways; racemes leafy, sepals lance-linear. ♃ Thickets, N. Y. to Va., and W. 2f.
4 **L. angustifòlium** Mx. Ascending, much branched; leaves linear, rigid; flowers scattered; corolla hardly exserted. ♃ Sandy banks, W. 6—15'. Leaves 1'.
5 **L. canéscens** Lehm. *Puccoon.* Erect, subsimple, soft-villous; leaves oblong or linear-oblong, obtuse; stem revolute at top, with the showy orange-yellow flowers axillary. ♃ Fields, prairies, N. Y., W. and S. 8—12'. June, July.
6 **L. hírtum** Lehm. Erect, simple, rough-haired; lvs. lance-linear, the floral lance-ovate; corolla twice longer than the linear sepals. ♃ Pa., W. and S. 8—15'. May.
7 **L. longiflòrum** Spr. Slender, simple, cinereous-strigous; leaves linear; corolla tube 4 times longer than the calyx (9—12''). Plains, W. 10—15'. July.

15. **MERTÉNSIA**, Roth. SMOOTH LUNGWORT. Calyx short, 5-cleft. Cor. tube cylindric, limb subcampanulate, 5-cleft, throat open, often with 5 folds or ridges between the insertion of the stamens. Sta. inserted at top of the tube. Ach. smooth or reticulated. ♃ St. and lvs. usually glabrous, pellucid-punctate, the radical many-veined, cauline sessile. Rac. terminal.

1 **M. Virgínica** DC. Ascending, very smooth; root leaves large, obovate to ovate, stem leaves sessile, lance-oblong, all entire, obtuse; fls. somewhat trumpet-shaped pendent, 10'', blue to lilac, very handsome. Rich soils, N. Y., S. & W. 1—1½f. May. †
2 **M. marítima** Don. Glabrous, weak; lvs. ovate, obtuse, fleshy, glaucous; corolla twice longer than calyx, blue-purple. Sea-shore, N. H., and N.: rare.
3 **M. paniculàta** Don. Scabrous, erect; lvs. acuminate, cordate-ovate to oblong · corolla thrice longer than calyx, blue to white. Lake Superior, and N. †

ORDER XCIII. HYDROPHYLLACEÆ. HYDROPHYLLS.

Herbs mostly, with alternate-lobed leaves and regular bluish flowers. *Calyx* 5-cleft, usually with appendages at the clefts, persistent, free. *Corolla* 5-lobed, often with 10 honey scales or furrows near the base. *Stamens* 5, inserted into the corolla, with a deeply bifid style. *Ovary* entire, ovoid, free, 1-celled, with 2 parietal, several-seeded placentæ. *Fruit* 2-valved, filled by the placentæ. *Seeds* reticulated, albuminous.

§ HYDROPHYLLEÆ. Ovary and pod 1-celled. Style bifid. Leaves cleft...(a)
§ HYDROLEÆ. Ovary and pod 2-celled, ∞-seeded. Styles 2. Leaves entire...(a)
    a Lobes of the corolla convolute in the bud...(b)
    a Lobes of the corolla imbricate (quincuncial) in the bud...(c)

ORDER 93.—HYDROPHYLLACEÆ.

   b Stamens exserted. Flowers in forked, revolute cymes....................HYDROPHYLLUM 1
   b Stamens included. Flowers solitary, opposite the leaves..................NEMOPHILA. 2
  c Flowers solitary. Calyx enlarged in fruit...ELLISIA. 3
  c Flowers racemed.—*d* Lobes of the corolla fringe-toothed.....................COSMANTHUS. 4
                   —*d* Lobes of the wheel-bell-form corolla entire.............PHACELIA. 5
                   —*d* Lobes of the tube-bell-form corolla entire.............WHITLAVIA. 6
  e Corolla wheel-bell-form. Leaves ordinary, with soft hairs....................HYDROLEA. 7
  e Corolla funnel-form. Leaves large, with stinging hairs......................WIGANDIA. 8

1. **HYDROPHYLLUM**, Tourn. WATER-LEAF. BURR-FLOWER. Sepals slightly united at base. Corolla bell-form, convolute in bud, with 5 double folds (nectaries) inside. Sta. exserted. Caps. globous, 1-celled, 2-valved, 4-seeded, 3 of the seeds mostly abortive. Placentæ 2, fleshy, free except at the base and apex. ♃ Leaves large, long-stalked, pinnately or palmately veined, cauline alternate. Cymes scorpoid, bractless.

   § Calyx appendaged between the sepals at base. Stamens as long as the cor...No. 1
   § Calyx not appendaged. Filaments much exserted........................Nos. 2—4

1 **H. appendiculàtum** Mx. Hairy; lvs. palmately 5-lobed, the lower pinnately divided, lobes pointed and toothed; sta. often included; appendages deflexed, much shorter (1″) than sep. (4—5″); cor. blue. Woods, N. Y. to Wis., & Va. 1—1½f. May.

2 **H. Virgínicum** L. Nearly smooth; leaves pinnatifid; segments oval-lanceolate, pointed, incised, the upper 3 confluent; petioles long; ped. still longer, bearing a roundish tuft of pale flowers with hirsute calyxes. Moist woods. 1f. June.

3 **H. Canadénse** L. Lvs. smoothish, palmate, roundish, with 5—7 shallow lobes, unequally dentate, teeth obtuse-mucronate; fls. in crowded fascicles; ped. shorter than the forked petioles: cor. white or purplish. Alpine woods. 1—1½f. June, Jl.

4 **H. macrophýllum** N. Whitish, with reversed hairs; leaves oblong-oval in outline, pinnatifid, and cut into blunt-mucronate teeth; cymes dense, globous, on long peduncles; corolla white, 6″; stamens 10″. Rocky woods, W. and S. 1f. June.

2. **NEMÓPHILA**, N. Cal. 5-parted, the sinuses with reflexed appendages. Cor. wheel-bell-form, lobes rounded, convolute in bud, tube with 5 pairs of folds within. Sta. included. Ov. and caps. as in Hydrophyllum, the placentæ each 2–12-ovuled. ① Tender and fragile, with pinnately-parted leaves and solitary, showy flowers.

   * Leaves all or the lower alternate. Flowers not spotted................Nos. 1, 2
   * Leaves all opposite. Flowers spotted with blue or brown................Nos. 3, 4

1 **N. microcàlyx** F. & M. Smooth; leaves triangular, 5-3-cleft, with rounded, mucronate teeth; ped. and petioles slender; corolla 1—2″, white, calyx still smaller; seeds 1 or 2. Damp woods, S. 3—12′, very weak. April.

2 **N. insígnis**. Lvs. oblong, with 7—9 ovate, acute lobes, shorter than peduncles; fls. 1′ or more broad, the border pure blue with a white centre. California.

3 **N. maculàta**. Leaves 3-7-lobed, tapering and entire at base; flowers on long ped., 1½′ broad, white, with a violet spot on the apex of each lobe. California.

4 **N. atomària**. Leaves and peduncles nearly as in the last; flowers white, 10—12″, sprinkled all over with small brown spots. Sierra Mountains.

3. **ELLÍSIA**, L. Cal. 5-parted, equalling the tubular-bell-form corolla, enlarged in fruit. Cor. tube minutely appendaged within. Sta. included. Caps. 2-valved, 4-2-seeded. Leaves pinnatifid, flowers white, May—July.

E. Nyctelæa L. Weak, slender; lvs. petiolate, the upper alternate, lobes 9—11, lin. oblong; ped. 1-flowered, with calyx larger than corolla. Woods, Pa., W and S. 1f

ORDER 93.—HYDROPHYLLACEÆ. 255

4. **COSMÁNTHUS**, Nolte. MIAMI MIST. Cal. 5-parted. Cor. wheel-bell-form, tube not appendaged, lobes delicately fringe-toothed, as long as the stamens. Ovary hairy. Capsule 2-valved, 4-seeded. ① Delicate, with alternate leaves and small pale flowers in long, bractless racemes.

1 **C. Púrshii** Wood. Nearly smooth, erect; lvs. pinnatifid, the upper sessile, lobes 5—7, oblong, acute; rac. 9-15-flowered; pedicels longer than the lance-linear, ciliate sepals; fls. light blue, 5—6″. River bottoms, Ill., Ky., to Ga. 8—12′. May, June.
2 **C. fimbriàtus** Mx. Pubescent; stems clustered, assurgent; leaves pinnate, with 5—7 roundish or oblong-obtuse lobes; pedicels as long as the oblong-spatulate, obtuse sepals; corolla white, 4—5″. Mountains, Tenn., Va., to Ga. May.

5. **PHACÈLIA**, L. Cal. not appendaged. Corolla tubular-bell-form, lobes entire, imbricate in bud, tube appendaged within. Sta. 5, generally exserted. Ov. and caps. hispid, ovoid, 4 – ∞-seeded.—Herbs hispid, with alternate leaves and 1-sided racemes. May, June.

§ Capsule 4-seeded. Corolla tube evidently appendaged within.............Nos. 1—3
§ EUTÒCA. Caps. (or ovary) 8 - ∞-seeded. Cor. obscurely appendaged...(a)
　a Seeds or ovules 6—8. Racemes simple. Native South..................Nos. 4—6
　a Seeds or ovules 20 or more. Rac. forked or corymbed. Gardens. ①..Nos. 7—9

1 **P. bipinnatífida** Mx. Stem hairy, suberect, much branched; lvs. cut-pinnatifid, long-petioled, segm. again incised; rac. forked or simple, loose; corolla twice longer than calyx, 6″, blue. ② Hilly woods, Ill. to N. C. and Ala. 1—2f.
2 **P.** TANACETIFÒLIA. Hispid or hairy, tall, with pinnatisect leaves, long, dense racemes, corollas blue, and long, exserted stamens. California. 1—2f.
3 **P.** CONGÉSTA. Hoary-pubescent; lvs. pinnate with very unequal alternate-cut lfts. racemes loose, spicate; flowers small, blue; stamens little exserted. California. 1f.
4 **P. parviflòra** Ph. Stems smoothish, weak; lvs. all petiolate, pinnatifid or 3-fld. lobes distant, small; fls. 4″, pale; sep. smoothish. ② Shady banks, Pa., and S. 9′.
5 **P. maculàta** Wood. Erect, branched, sparingly hirsute; lvs. pinnatifid, 5-7-lobed, lower petiolate, upper sessile; fls. 7″, violet-blue, 10-spotted around the yellow throat; sepals bristly-ciliate, linear-oblong. ② Stone Mountain, Ga., and W. 6—12′.
6 **P. pusilla** Buckley. Pubescent; leaves sessile, pinnatifid, lobes abruptly pointed; fls. pale-blue or white; sepals linear-oblong; stamens exserted. Prairies, Ala.
7 **P. Franklinii** Gray. Soft-hairy, erect; lvs. bipinnatifid with crowded lobes; racemes short, dense, crowded, with blue fls. Isl. Royal (Porter) to Oreg.! Cultivated.
8 **P.** VÍSCIDA. Viscid with glandular hairs, ovate, coarsely-toothed leaves, and long, revolute racemes, uncoiling as the large (9″) purple-blue flowers expand. Cal. 1f.
9 **P.** MENZIÈSII. Lvs. linear, entire, or the lower with few linear-oblong lobes; flowers sessile, light-blue, in short spikes. Oregon.

6. **WHITLÀVIA**, Harvey. Cal. 5-parted. Cor. tubular-campanulate, the 5 lobes abruptly spreading, throat slightly contracted. Sta. exserted. Capsule ∞-seeded. ① From Texas and California.

W. GRANDIFLÒRA. Some viscid, with broad, ovate, petiolate, coarsely-toothed leaves, loose racemes of large (1′) deep-blue (or white) bell-shaped flowers. June—October.

7. **HYDRÒLEA**, L. Sep. 5. Cor. rotate-campanulate, 5-lobed, bearing the 5 stamens. Styles 2, distinct. Capsule 2-celled, 2-valved, the placentæ large, with ∞ minute seeds.—Herbs with entire leaves and cymes of blue flowers. July—September.

1 **H. corymbòsa** Macbride. Not spiny, some hairy above; lvs. lance-ovate, sessile; branchlets corymbed, each with a terminal, showy, azure flower. Ponds, S. 1—2f.

2 H. quadriválvis Walt. Spiny, hispid; leaves lanceolate, petiolate; cymes 4-6-flowered; cor. azure-blue, 5—6" broad; sep. ovate. Slow waters, S. C., and W. ♃

8. WIGÁNDIA, H. B. K. Cor. funnel-form.—Herbs with large leaves.

W. CARACASÀNA. Half-shrubby, with ovate-cordate, doubly-crenate, variegated, ample leaves, stinging hairs, and revolute spikes of small flowers. S. Am. Greenhouse.

ORDER XCIV. POLEMONIACEÆ. PHLOXWORTS.

Herbs with alternate or opposite leaves and 5-parted, regular, showy flowers. *Corolla* monopetalous, the lobes convolute, rarely imbricate in æstivation. *Stamens* 5, adherent to the corolla tube, and alternate with its lobes. *Ovary* 3-celled. *Stigma* 3-cleft. *Capsule* 3-celled, 3-valved, loculicidal. *Seeds* few or many, albuminous, attached to a permanent columella. Fig. 46.

```
I. POLEMONIEÆ.  Sepals united at base. Lobes of the corolla convolute in bud...(a)
II. DIAPENSIEÆ.  Sepals distinct, oval. Lobes of the corolla imbricated in bud....DIAPENSIA.  7
  a Stamens unequal, included in the tube of the salver-form corolla..................PHLOX.  1
  a Stamens unequal, in the tube of the funnel-form (scarlet) corolla................COLLOMIA.  2
  a Stamens equal and protruded from the corolla tube.  Seeds ∞...(b)
    b Leaves undivided, opposite. Corolla wheel-funnel-form, dentate.............FENZLIA.  3
    b Leaves variously divided.  Ovary and pod ∞-seeded...(c)
      c Stamens equal and straight.  Corolla of various forms..............  .....GILIA.  4
      c Stamens declined in the bell-form corolla.—d Low herbs.......  .....POLEMONIUM.  5
                                                 —d Climbing shrubs............COBÆA.  6
```

1. PHLOX, L. PHLOX. LYCHNIDEA. Calyx prismatic, deeply 5-cleft. Corolla salver-form, the tube more or less curved. Sta. very unequally inserted, and included in the tube. Caps. 3-celled, cells each 1-seeded.—A highly ornamental North American genus. Lvs. mostly opposite, sessile, simple, entire. Fls. in terminal cymes, corymbed or panicled. Fig. 46.

```
* Lobes of the corolla rounded and entire at the end...(1)
  1 Panicle of cymes oblong or pyramidal, many-flowered.................Nos. 1, 2
  1 Panicle of cymes corymbed, level-topped, flowers fewer...(2)
    2 Plants glabrous.  Calyx teeth shorter than its tube...  ...........Nos. 3, 4
    2 Plants hairy.  Calyx teeth attenuated, longer than the tube...(3)
      3 Leaves narrow, linear, or nearly so............................Nos. 5, 6
      3 Leaves broad, ovate or lanceolate, &c......................Nos. 7, 8 β, 9
* Lobes of corolla notched or bifid at the end.—4 Leaves distant..........Nos. 8, 10
                                              —4 Leaves imbricated....  .....No. 11
```

1 P. paniculàta L. Smooth, erect; leaves oblong- or ovate-lanceolate, pointed at each end; fls. numerous, in a terminal panicle, pink-purple, varying to white; calyx teeth setaceous-pointed. ♃ Shady banks, Penn., W. and S. 2—3f. July—Sept. †

β. acumìnàta. Lvs. ovate-acuminate, downy beneath; stem hairy.

2 P. maculàta L. Stem roughish, purple-spotted, upright; leaves thickish, lanceolate, the upper ovate-cordate; fls. many, purple, in an oblong panicle; calyx teeth lanceolate, acute. ♃ Moist fields, Penn. to Car., and W. 2—3f. June—August.

β. gracìlior. Tall, slender, rough; leaves lance-linear and linear. Ga. (Feay.)

γ. suaveolens. Smooth; flowers white, sweet-scented. Gardens.

3 P. Carolìna L. Ascending, often branched; leaves lanceolate, rounded at base, pointed; fls. rose-purple, in small, dense cymes. ♃ Prairies, woods, Pa., W. and S. 9'—2f. May—July.—*β. ovàta* has roughish stems and ovate leaves.

ORDER 94.—POLEMONIACEÆ.

4 P. glabérrima L. Slender, erect; leaves oblong- to lance-linear, taper-pointed, thick, with rolled margins; calyx teeth sharp-pointed; corollas pale-pink, few. ♃ Prairies and barrens, Wis. to Ga. 1—3f. June, July.

5 P. pilòsa L. Ascending, slender, glandular-hairy above; lvs. lanceolate to linear, attenuate to an acute apex; corymbs loose; calyx teeth bristle-pointed, much longer than the tube; corolla small. ♃ Wis. to N. J., and S. May, June.
  β. *Floridàna.* Leaves oblong-lanceolate; calyx teeth lance-setaceous. Fla.

6 P. involucràta Wood. Hoary-pubescent, branched and ascending at base; lvs. linear-oblong, rather obtuse, clasping, flat, the floral similar and closely subtending the dense corymbs as if *involucrate;* calyx teeth linear or subulate-spatulate; flowers purple to carmine. ♃ Dry soils, S. 6—12'. May, June.

7 P. reptans Mx. Assurgent, with creeping stolons; lvs. obovate to ovate, obtuse, fls. few; sep. linear-subulate; cor. blue-purple. ♃ Hills, Ind. to Pa., and S. 9'. Jn.

8 P. divaricàta L. Low, diffuse, downy; lvs. ovate to lance-oblong, acute; flowers grayish-blue, lobes notched; sep. lin.-subulate. ♃ N. Y. to Wis., and S. 1f. Apr., May.
  β. *Laphamii.* Leaves ovate; corolla lobes obtuse, entire. Wis. (Lapham).

9 P. Drummóndii Hook. Upright, forking, glandular-hairy; lvs. lanceolate to oblong, mostly alternate; sepals lance-setaceous, revolute; flowers in dense corymbs, all shades in the gardens, white to purple, with a star. ① Ga.! to Texas.

10 P. bífida Beck. Low, assurgent, diffuse; lvs. lance-ovate to lance-linear; fls. few, sepals linear, petals deeply bifid, purple. ♃ Ill. to Mo.: rare. 6'. April.

11 P. subulàta L. *Moss Pink.* Procumbent, much branched and very leafy, in tufts; leaves rigid, linear to subulate, fascicled; flowers pink to white, covering the tufts in May. 5—8'. Penn., S. and W., and in gardens.

2. COLLÒMIA coccínea. ① From Chili, has bright carmine-red fls in heads subtended by broad bracts. Leaves ovate-lanceolate, often 3-cleft at apex, alternate. Pods 3-seeded. 10—15'. June, July.

3. FÉNZLIA diantiioìdes. ① California. A small pink-like herb, 3—6', with exquisitely beautiful flowers, 1'. solitary, pink with 5 purple dots around a yellow eye, and the 5 lobes evenly notched at the end. Leaves linear, opposite.

4. GÍLIA, R. & P. Cal. teeth acute. Cor. funnel-form, the tube short or long, bearing the equal sta. more or less exserted and not declined. Pet. entire. Pod ∞-seeded.—Herbs with elegant, showy flowers.

§ Ipomópsis. Corolla tube long exserted, in thyrse-like racemes. Tall.........No. 1
§ Leptosìphon. Corolla tube long, slender, in involucrate heads. Low.........No. 2
§ Eugílìa. Corolla tube included in the calyx, scattered or capitate.........Nos. 3, 4

1 G. coronopifòlia Pers. *Standing Cypress.* A splendid herb 2—4f, plume-like in form, closely beset with delicate pinnatifid lvs. and bearing at top a long (1f) thyrse of bright red flowers (15''). ② Sandy banks, S. C. to Fla., and W. July—Sept. †

2 G. androsàcea. Strict, simple, downy; lvs. opp., digitately 5-9-cleft into very narrow segments; cor. 1' or more long, lilac, purple or white. ① Cal. 6—12'. May, Jn.

3 G. trícolor. Diffusely branched; lvs. 2-3-pinnatifid; flowers many, 3-colored, limb lilac, throat purple, tube yellow. A great favorite, from California.

4 G. capitàta, with the blue 6'' flowers at length in round dense heads. Cal. and Oreg.

5. POLEMÒNIUM, L. Greek Valerian. Calyx and corolla bell-form, with suberect segments. Stamens equally inserted, declined, hairy at base. Capsules 3-valved, 3-celled.—Herbs weak, with alternate pinnately-divided leaves and terminal cymes, blue to white.

1 P. reptans L. Diffusely branched; leaves 7-11-foliate, leaflets acute; fls. nodding, pod cells 2- or 3-seeded. ♃ Damp uplands, N. Y. to Wis., and S. 1—1½f. May.

2 P. cœrùleum. Tall, with erect branches; leaflets 11—17, pointed; fls. erect; seeds ∞. Swamps, Vt., N. Y., N. J. (Dr. Howe, Prof. Porter). 2—3f. Often cultivated.

6. COBÆA SCANDENS. Calyx large and leaf-like. Cor. large, throat ample, limb spreading, dull purple. Leaves pinnatisect, ending in a tendril. Coarse climbers, from Mexico. The lower leaf-segments resemble stipules.

7. DIAPÉNSIA, L. Cal. of 5 oval sepals, closely subtended by bracts. Corolla bell-form, imbricated in the bud. Fil. flat, arising from the sinuses of the corolla, anth. cells diverging at base and the dehiscence transverse. Caps. 3-celled, ∞-seeded. ↳. Prostrate, with densely imbricated, entire leaves and solitary terminal flowers.

§ DIAPÉNSIA *proper.* Anthers without awns. Flowers pedicellate..............No. 1
§ PYXIDÁNTHERA. Anthers with the lower valve awned. Flowers sessile......No. 2

1 D. Lappónica L. A little tufted shrublet, with fleshy, evergreen, obtuse leaves, and the tiny white fls. raised on pedicels 1' long. White Mountains. 2—3'. July.

2 D. barbulàta Ell. Prostrate, creeping, forming dense beds, with short branches; flowers terminal, sessile; anth. short-awned at base. Barrens, N. J., and S. 3—6'. J?

ORDER XCV. CONVOLVULACEÆ. BINDWEEDS.

Chiefly twining or trailing *herbs*, sometimes parasitic, sometimes shrubby. *Leaves* (or scales when leafless) alternate. *Flowers* regular, pentamerous and 5-androus. *Sepals* imbricated. *Corolla* monopetalous, 5-plaited or lobed, convolute in bud. *Ovary* free, 2-(rarely 3-)celled or falsely 4-celled, or of 2 distinct, 1-ovuled pistils. *Capsule* 2-6-seeded. *Embryo* large, coiled in mucilaginous albumen. Figs. 48, 65, 81, 82, 209–10, 262.

III. CUSCUTINEÆ. Leafless, twining, orange-yellow parasites. ....CUSCUTA. 11
II. DICHONDREÆ. Leafy. 2 distinct ovaries with 2 distinct styles............ ......DICHONDRA. 10
I. CONVOLVULEÆ. Leafy. Ovary 1. Capsule dehiscent. Seed-lobes leafy...(a)
  a Styles united into one...(b)
  a Styles 2 or 3, distinct or nearly so. Stamens included...(z)
    b Ovary and pod 4-celled.—*c* Stamens exserted. Flowers small........... ......QUAMOCLIT. 1
      —*c* Stamens included. Flowers large...............BATATAS. 2
    b Ovary and pod 3-celled. Stigma capitate, granulate.......................PHARBITIS. 3
    b Ovary and pod 2-celled...(d)
      d Stigma 1, capitate.—*e* Stamens included.................................IPOMŒA. 4
        —*e* Stamens exserted..................................CALONYCTION. 5
      d Stigmas 2,—*x* ovate, flattened. S. Fla........................JACQUEMONTIA *violacea*.
        —*x* linear-terete. Calyx not bracted........................ .... CONVOLVULUS. 6
        —*x* oblong-terete. Calyx in 2 large bracts................... .CALYSTEGIA. 7
  z Styles each bifid. Peduncle very short..EVCLYULUS. 8
  z Styles each simple. Peduncles longer than the leaves... .... STYLISMA. 9

1. QUÁMOCLIT, Tourn. CYPRESS-VINE. Sep. 5, mostly mucronate. Cor. tubular-cylindric, with a salver-form border. Sta. exserted. Style 1, stigma capitate, 2-lobed. Ov. 4-celled, cells 1-seeded. ♄ From Tropical Am.

1 Q. vulgàris Choisy. *Cypress-vine.* Lvs. pinnatifid to the midvein, segm. linear, parallel, acute; ped. 1-flwd.; sep. ovate-lanceolate; cor. scarlet. ⓛ An exceedingly delicate vine, in gardens, and often escaped S. July, Aug. §

2 Q. coccínea Mœnch. Leaves cordate, acuminate, entire or angular at base; ped. elongated, about 5-flowered; calyx awned; flowers light scarlet, limb nearly entire ½'' broad. ⓛ Along rivers. S. and W. June—Aug. § †

ORDER 95.—CONVOLVULACEÆ.

2. BATÀTAS, Rumph. SWEET POTATO. Cal. of 5 sepals. Cor. campanulate, with a spreading limb. Stam. 5, included. Style simple, stigma capitate, 2-lobed. Capsule 4-celled, 4-valved, with 4 erect seeds. ♃ Herbs, or shrubby, with milky juice.

1 B. littorális Chois. Creeping, sending out runners; lvs. smooth, thick, sinuate with 3–5 rounded lobes and cordate at base; ped. 1-flowered, as long as the leaf; sep. abrupt-pointed; seeds tomentous; corolla white. ♃ Coast sands, S. Aug.—Oct.
2 B. macrorhiza Wood. Creeping or twining; lvs. cordate, lobed or entire, softdowny beneath; ped. 1-5-flowered, shorter than the leaves; cor. purple; seeds villous. ♃ Sands, S. C. to Fla. Root very large. (Ipomœa Michauxii Swt.)
3 B. édulis. *Sweet Potato.* Lvs. 3-5-lobed or angled, lobes acute; ped. 3-5-flowered. as long as the petioles. ♃ W. Indies. Extensively cult. for its sweet tubers. Purple.

3. PHÁRBITIS, Chois. MORNING GLORY. Calyx 5-sepalled. Cor. bellfunnel-form. Sty. single, stig. capitate, granulate. Ov. 3-(rarely 4-)celled, cells 2-seeded. ♃ Beautiful, cultivated and spontaneous.

1 P. purpùrea Wood. Twining stem clothed with reversed hairs; lvs. cordate, entire; ped. 2-5-flowered; corolla large, dark purple, varying to blue, flesh-color, &c., appearing in long succession, in fields and gardens. June, July. §
2 P. Nil Chois. Stems hairy; leaves cordate, 3-lobed; ped. 1-3-flowered, shorter than the petioles; sepals ovate, long-pointed, corolla tube white, border indigo (*nil*) blue. Gardens, and in fields. July, Aug. §
3 P. HEDERÀCEA, from S. Am., differs from P. Nil in the middle lobe of its lvs., which is ovate, and contracted at base; ped. 1-flwd.; cor. 2' or more broad, varying in purple and blue, blue and white, pink and white, &c.—The hybrid P. LIMBÀTA has a purple star with a white border and leaves scarcely lobed. ①
4 P. LEÀRII, from Mexico, has ped. longer than the cordate, velvet-silky leaves, each bearing a cluster of magenta-blue-red flowers. Greenhouse. ♃. 10—15f.

4. IPOMŒA, L. Cal. 5-sepalled. Cor. bell-funnel-form. Sta. included. Style 1, stigma capitate. Ov. and capsule 2-celled, cells 2-seeded.—Herbs, shrubs, or trees. Our species are herbs creeping or climbing.

* Flowers capitate, involucrate, small, blue. Sepals hairy..................No. 1
* Flowers separate.—*a* Sepals bristly ciliate, capsules somewhat hairy......Nos. 2, 3
   —*a* Sepals glabrous.—*b* Flowers purple. Maritime......Nos. 4, 5
   —*b* Flowers white, rarely yellow...Nos. 6—8

1 I. tamnifòlia L. Hairy; leaves ovate, cordate, acuminate, large, equalling the peduncles; fls. crowded, 9″, with linear bracts and sepals. ① Ga. to La. Jl.—Sept.
2 I. commutàta R. & S. Smoothish; lvs. cordate, entire or 3-lobed; ped. as long as the petioles; flowers 2—5, purple to pink, 18″; sep. 5″. ① Fields, S. July—Oct.
3 I. lacunòsa L. Puberulent; lvs. cordate, entire or angular-lobed; ped. ♀ as long as the petioles; flowers 1—3, white, with a purplish rim, 1', sepals ♀ as long. ① Dry fields and hills, Penn. to Ill., and S. 2—6f. August, September.
4 I. Pes-Capræ Sw. Roughish; leaves roundish, emarginate or 2-lobed, thick; ped. as long as the petioles; fls. 1—5, purple, 3' long. Coasts of Ga. and Fla. June +.
5 I. sagittifòlia (Mx.) Glabrous; lvs. cordate-sagittate; ped. as long as the petiole, much shorter than the one large (3') purple flower. ♃ Marshes, S. June +.
6 I. sinuàta Ort. Lvs. palmately 7-cleft, varying to sinuate-lobed; segments pinnatifid; ped. 1- or 2-flowered; corolla white, 1'. ♃ Ga., Fla. 20f. July—October.
7 I. elliolàta Pers. Leaves cordate, entire, acuminate; ped. 1-flowered, 2-bracted above; corolla large, yellow; sepals 8″ long. ♃ N. Car. and Tenn.
8 I. pandurata Meyer. *Wild Potato.* Leaves broad-cordate to panduriform; ped.

1-5-flowered, longer than the petioles; sepals ⅓ as long as the corolla; corolla 3', white with a purple centre. ♃ N. Y. to Ill., and S. July, August.

5. CALONÝCTION SPECIÓSUM (or Ipomœa Bona-nox), GOOD-NIGHT, is a tall climber of the W. Indies and S. Fla., often cultivated in the greenhouse. Flowers 4—7 on each long peduncle, very large, funnel-form, white.

6. CONVÓLVULUS, L. BINDWEED. Sep. 5. Cor. bell-form. Style 1. Stigmas 2, thread-form, often revolute. Ovary and capsule 2-celled, 4-seeded.—Herbs or shrubs, twining or erect.

1 **C. arvénsis** L. Prostrate or climbing; leaves arrow-shaped to ear-shaped; ped. bearing 1 small rose-white flower and 2 bracts. ♃ Fields: rare. June. §
2 **C. TRÍCOLOR.** Stem weak, 1—3f high; leaves lance-obovate, sessile, shorter than the 1-flowered ped.; corolla yellow in centre, white next, border blue. ① Europe.

7. CALYSTÈGIA, Br. Calyx 5-parted, included in 2 leaf-like bracts. Cor. bell-form, 5-plicate. Style 1. Stigmas 2, obtuse. Capsule 1-celled, 4-seeded.—Herbs, with the flowers solitary.

1 **C. spithamæa** Br. Erect or assurgent, 6—8' (a *span*) high; leaves lance-oblong, as long as the peduncles; flowers white. ♃ Can. to Penn., and W. June.
2 **C. Sèpium** Br. *Rutland Beauty.* Glabrous, twining; lvs. cordate-sagittate, lobes truncate; bracts cordate; flowers many, large, white with a reddish tinge. ♃ Hedges, thickets, Can. to Fla. 6—10f. May—July.
   β. *Catesbeiàna.* Pubescent, with small leaves and short peduncles. S.
   γ.? *paradóxa.* Tomentous; bracts linear, remote from the flowers. (Pursh.)

8. EVÓLVULUS, L. Sep. 5. Cor. bell-, funnel-, or wheel-form. Sty. 2, each bifid. Ovary and capsule 2-celled, 4-seeded.—Herbs diffuse.

E. seríceus Swtz. Stem dividing at base into simple, filiform, procumbent branches; leaves lance-linear, sessile, 3-veined, silky beneath, 9''; ped. 1—2'', 1-flowered; corolla wheel-form, 5'', white. ♃ Prairies, Ga., Fla., to La. 1f.

9. STYLÍSMA, Raf. Sepals 5, equal. Corolla bell-form. Stamens included. Styles 2, rarely 3. Stig. capitate. ♃ Slender creepers.

1 **S. humistràta** (and aquática) Walt. Hairy or smoothish; leaves oval, oblong, or linear, obtuse or retuse both ways, on short petioles; ped. longer than the leaves, 3-(1—5-)flowered; bracts minute; styles less than ⅓ united; corolla 6—9'', white. Sandy soils, Va. to O., and S. 2—5f. Lvs. 12—18''. (S. evolvuloides Choisy.) Jn.-Sept.
2 **S. Pickeríngii** (Torr.) Leaves linear, narrowed to subsessile base; bracts leafy, equalling the flower; styles more than ⅓ united, otherwise as No. 1. N. J. to N. C.

10. DICHÓNDRA, Forst. Sep. 5, obtuse. Corolla bell-form, 5-cleft. Pistils 2, distinct. Capsules 2, utricular, 1-seeded. ♃ Prostrate.

D. repens Forst. Lvs. round-cordate or reniform, the petiole longer than the blade or the 1-flowered peduncles; calyx villous, larger (3'') than the whitish corolla (2''). Wet grounds, S. 3—12'. March—May.

11. CUSCÙTA, Tourn. DODDER. Fls. 5-(rarely 4-)parted. Corolla globular-bell-form. Sta. appendaged with scales or fringes at base. Styles 2. Caps. 2-celled, 4-seeded. ① Stems yellow to orange, thread-form, with minute scales for leaves, twining against the sun and living on other plants.

§ Stigmas filiform as well as the styles. Capsule regularly circumscissile..... No 1
§ Stigmas capitate. Capsule indehiscent or bursting irregularly...(*)

ORDER 96.—SOLANACEÆ. 261

* Sepals distinct. with imbricated bracts added. Flowers sessile.. Nos. 2, 3
* Sepals united, bracts few and scattered. Flowers pedicellate...(a)
 a Corolla cylindrical, withering on the top of the capsule..........Nos. 4—6
 a Corolla bell-shaped, persistent at the base of the capsule...(b)
  b Lobes of the corolla acute or acuminate........................Nos. 7, 8
  b Lobes of the corolla obtuse................... ....Nos. 9—11

1 **C. Epilinum** Weih. *Flax D.* Fls. sessile in small, dense, remote heads; calyx 5-parted, scarcely shorter than the globular corolla or capsule. Flax fields. Jn. § Eur.
2 **C. glomeràta** Choisy. Fls. in compact masses surrounding the foster stem while its own filiform stems decay; sepals 1″, with many squarrous bracts; corolla white, 2″, tube-bell-form, 5-lobed. On the Compositæ, &c., W. and S.
3 **C. compácta** Juss. Fls. in large (1—2′) masses, with thick stems; sep. and 3—5 bracts minute (⅓″); cor. slender, with 5 oblong lobes. N. Y., W. and S., on shrubs.
4 **C. tenuiflòra** Eng. Pale, much branched, on high plants; fls. short-pedicelled; cor. tube slender. twice longer than the calyx or its own short obtuse lobes; capsule often but 1- or 2-seeded. Wet grounds, N. J., Pa., to Ill., and W.
5 **C. infléxa** Eng. Fls. pedicelled, mostly 4-parted; cor. fleshy, its lobes erect and inflexed, margins crenulate; capsule brown, capped with the dead corolla. Prairies and open woods, Ill. to Va. and Ga. On Hazel, Rhus, &c.
6 **C. decòra** Chois. Fls. pedicellate, 5-parted, large (1¼″), fleshy, white; cor. broad-bell-form, lobes acute; capsule enveloped by the dead corolla. Wet, Ill. to Fla.
7 **C. chlorocárpa** Eng. Low, branching, orange; fls. 4-parted, short-pedicelled, 1″, bell-form, the lobes of cal. and cor. acute; caps. large, greenish. Wis. to Del., & S.
8 **C. arvénsis** Beyr. On low plants; flowers small (⅓″), 5-parted, pedicellate; corolla tube shorter than its pointed lobes, or the rounded sepals. N. Y. to Ill., and S. Ju., Jl.
9 **C. obtusiflòra** H. B. K. Low, bright orange; fls. pedicell., dotted with red glands (β. glandulosa); sep. round-obtuse; caps. 1½″. Mostly on Polygonum. Ga., S. and W.
10 **C. Gronòvii** Willd. Stems thick, often high-climbing; fls. mostly 5-parted, a: length densely paniclcd; corolla tube bell-form, longer than the calyx, its lobes obtuse, entire, spreading. Common in all the country. Flowers 1½″.
11 **C. rostràta** Shutt. Fls. large (2—3″), in loose cymes; corolla deeply bell-form lobes obtuse; capsule 2—3″, with a 2-pointed beak. Mountains, Md. to S. Car.

ORDER XCVI. SOLANACEÆ. NIGHTSHADES.

Plants herbaceous, rarely shrubby, with a colorless juice and alternate leaves often in pairs. *Flowers* mostly regular, often extra-axillary, 5-parted, on bractless pedicels. *Corolla* valvate or plicate in the bud, and often convolute. *Calyx* persistent. *Stamens* 5, adherent to the corolla tube, alternate with its lobes; anthers 2-celled. *Fruit* a 2-(rarely 3- or more)celled capsule or berry. *Seeds* ∞, with a curved embryo in fleshy albumen. Figs. 66, 113, 168, 260, 483–4.

§ NOLANEÆ. Ovaries few or ∞, distinct, simple. Corolla funnel-bell-form........NOLANA. 1
§ SOLANEÆ. Ovary 1, compound, 2-(or more)celled...(*)
  * Corolla wheel-form, the tube very short. Anthers convergent...(b)
  * Corolla bell-form, the broad tube including the erect anthers...(c)
  * Corolla funnel-form, tube long and—α the limb somewhat unequal...(d)
                                       —α the limb quite regular...(e)
    b Stamens connate, opening by slits inside. Berry torous.................LYCOPERSICUM. 2
    b Stamens connivent, opening by terminal pores. Berry round..........SOLANUM.   3
    b Stamens connivent, opening by slits. Berry dryish, angular..........CAPSICUM.   4
    c Corolla bluish. Berry dry, enclosed in the enlarged calyx...........NICANDRA.   5
    c Corolla yellowish. Berry juicy, enclosed in the enlarged calyx. . . . . . PHYSALIS. 6
    c Corolla purplish. Berry blackish, sitting on the open calyx...... ..ATROPA.   7

d Stamens exserted, declinate. Capsule opening by a lid	HYOSCYAMUS.	8
d Stamens included, unequal. Capsule opening by valves	PETUNIA.	9
e Stamens exserted, growing to the summit of the tube	NEIREMBERGIA.	10
e Stamens exserted, growing to the bottom of the tube	LYCIUM.	11
e Stamens included.—*x* Flowers 3′—12′ long. Calyx prismatic	DATURA.	12
—*x* Flowers 1′—4′ long. Calyx terete	NICOTIANA.	13
—*x* Flowers 6—10″ long. Calyx terete, short.	CESTRUM.	14
—*x* Flowers 5″ long. Leaves very small	FABIANA.	15

1. **NOLÀNA,** L. Calyx 5-parted. Cor. showy, funnel-bell-form. Ovaries 3—40, distinct, 1–6-celled, becoming as many drupes around the base of the style. ♄ ♄ From S. America, with blue flowers.

1 **N.** ATRIPLICIFÒLIA. Stems procumbent; leaves thick, entire, ovate to spatulate, obtuse; flowers solitary, supra-axillary, with a yellow tube, azure-blue border, and white zone, numerous all Summer.

2 **N.** PROSTRÀTA. Leaves ovate-oblong, tapering both ways; calyx segments triangular-arrow-shaped; corolla blue with dark-purple streaks. Otherwise as No. 1.

2. **LYCOPÉRSICUM,** Tourn. TOMATO. Calyx 5–6–∞ - parted. Cor. rotate, with a short tube and a plicate-valvate limb. Stamens 5–6 – ∞, exserted, anth. connate at apex, longitudinally dehiscent on the inner face. Berry fleshy, 2–3–∞ - celled. Ped. extra-axillary, ∞-flowered.

L. ESCULÉNTUM Mill. Hairy; st. herbaceous, weak; lvs. unequally pinnatifid, segments cut: corolla many-lobed; fruit torulous, furrowed, smooth. ① A coarse, strong-scented herb with yellowish flowers and splendid fruit.

3. **SOLÀNUM,** L. POTATO. Calyx 5-parted, persistent. Cor. rotate, subcampanulate, tube very short, limb plicate, 5-cleft, lobed or angular. Anth. erect, connivent, distinct, opening at the top by 2 pores. Berry 2-celled, subglobous or depressed. Seeds ∞.—Herbs or shrubs. Peduncles terminal, becoming lateral by the extension of the axis. Figs. 260, 483–4.

§ Prickles none. Anthers obtuse...(*a*)

a Herbs, with the flowers and fruit in clusters	Nos. 1, 2
a Shrubby climbers, with clustered flowers and fruit	Nos. 3, 4
a Shrubs erect, with orange or scarlet berries	Nos. 5, 6

§ Plants armed with prickles. Anthers linear-oblong, pointed...(*b*)

b Flowers 5-parted. Calyx open in fruit. Anthers equal	Nos. 7—9
b Flowers 5-parted. Calyx closed on the fruit. Anthers unequal	Nos. 10, 11
b Flowers 6-9-parted. Calyx open with the large fruit	Nos. 12, 13

1 **S. tuberòsum** L. *Common Potato.* Subterranean branches bearing tubers; leaves pinnatifid unequally and interruptedly; corolla 5-angled, ped. jointed. S. America. Cultivated since the 17th century. Many varieties.

2 **S. nigrum** L. *Nightshade.* Smoothish; leaves ovate, toothed, wavy, or entire; umbels lateral, drooping, flowers small (2—3″). whitish; berries black, as large as a peppercorn. Weed in old fields. 2—3f. Summer. § Europe.

3 **S. Dulcamàra** L. *Bittersweet.* Stems shrubby, slender, climbing; leaves cordate, entire or with 1 or 2 pairs of lobes at base; clusters terminal and lateral, corolla purple, with 5 green spots; fruit red. July. § Europe.

4 **S.** JASMINOÌDES. Climbing high, smooth, lvs. ovate, entire; clusters blue-wh. Brazil.

5 **S.** PSEUDO-CÁPSICUM. *Jerusalem Cherry.* Erect, like a dwarf tree; leaves oblong-lanceolate, smooth, shining; flowers solitary, white, berries scarlet, as large as cherries. Mauritius. 2—4f. Handsome.

6 **S.** LACINIÀTUM. Shrub erect, smooth: lvs. pinnatifid; fls. blue; fr. orange. Australia.

ORDER 96.—SOLANACEÆ. 263

7 S. **Carolinénse** L. *Horse Nettle.* Prickles large, yellow, scattered on the stem, petioles, and veins; leaves angular-lobed, acute; flowers white, 10—15'', racemed; berries yellow. Roadsides, N. Y., S. and W. 1—2f. June.
8 S. **Virginiànum** L. Hairy and prickly; leaves deeply pinnatifid with angular sinuate lobes; flowers pale-violet, 15'', in leafy racemes. Va., and S. July.
9 S. **mammòsum** L. *Apple-of-Sodom.* Villous and with scattered spines; leaves roundish-ovate, subcordate, lobed; berries inversely pear-shaped. ① Waste grounds. Ga., Fla., and W. Flowers violet, 15''. Fruit yellow.
10 S. ROSTRÀTUM. Hoary-tomentous and very prickly; leaves doubly sinuate-lobed · flowers yellow, 12—15''; fruit closed in the burr-like calyx. ① Kansas.
11 S. HETERODÓXUM. Very hairy and prickly; leaves doubly pinnatifid, lobes runcinate; flowers violet-blue. ① From Texas. Fruit black.
12 S. MELÓNGENA (or esculentum). *Egg Plant.* Prickly; lvs. ovate, wavy or sinuate; flowers violet; fruit very large, glossy-purple, prized as a great delicacy. E. India.- A variety has *white* fruit exactly imitating a goose-egg.
13 S. TEXÀNUM. With scarlet fruit depressed-globous and lobed. From Tex. Mex.

4. **CÁPSICUM**, Tourn. PEPPER. Calyx erect, 5-cleft. Cor. rotate, tube very short, limb plaited, 5-lobed. Anth. connivent. Fr. capsular, dry, inflated, 2–3-celled. Seeds flat, very acrid.—Herbs or shrubs, with hot and acrid taste. Leaves often in pairs. Ped. axillary, solitary.

C. ÁNNUUM. *Red* or *Cayenne P.* Herb with angular, branching stem, smooth ovate entire leaves and large roundish or lance-form red fruit. ① Many varieties.

5. **NICÁNDRA**, Adans. APPLE OF PERU. Cal. 5-cleft, 5-angled, the angles compressed, sepals sagittate. Cor. campanulate. Sta. 5, incurved. Berry enveloped in the persistent calyx. ① Peruvian. Summer.

N. **physaloìdes** Adans. Herb smooth, with ample ovate-oblong, sinuate-angled lvs.; flowers solitary, axillary, white, with blue spots. Gardens and fields. 2—5f. §

6. **PHÝSALIS**, L. GROUND CHERRY. Calyx 5-cleft, persistent, at length inflated. Cor. bell-rotate, tube very short, limb obscurely 5-lobed. Sta. 5, connivent. Berry globous, enclosed within the 5-angled calyx.— Herbs (rarely shrubs) with angular branches. Leaves alternate or unequally twin. Flowers solitary, nodding, extra-axillary, all Summer.

§ Anthers yellow. Ped. elongated. Fruit edible, not filling the calyx...(a)
  a Corolla yellow with brown-purple in the centre........................Nos. 1—3
  a Corolla yellow in centre as well as border............................Nos. 4, 5
§ Anthers blue or violet. Ped. shorter than the petioles...(b)
  b Peduncles near 1' long. Berry not filling the closed calyx..........Nos. 6—8
  b Peduncles 2—3'' long. Berry filling the open calyx....................No. 9

1 P. **viscòsa** L. Viscid-pubescent, diffuse; leaves ovate to oblong, mostly abrupt at base and bluntly toothed; corolla 8—10''; fruiting-calyx 1½'. ♃ Dry soils. 1f.
2 P. **Pennsylvánica** L. Puberulent, decumbent; leaves ovate to lanceolate, repand-toothed or entire, base obtuse or acute; corolla slightly spotted, 6—8''; fruit-calyx rounded, 1'. ♃ Dry soils, Penn., S. and W. 6—15'.
  β. *lanceolata.* Pubescent; leaves tapering and acute both ways. S.
3 P. **angustifòlia** N. Glabrous; leaves lance-linear, entire, thickish; fruit-calyx wing-angled, 1'; corolla 10--12''. ♃ Wet sands, Fla. 6—12'.
4 P. **nyctaginea** Dun. Pubescent; leaves small, elliptic-ovate, blunt-toothed; calyx hairy; corolla small (5—6''), wholly yellow. South. 6—12'.
5 P. **Alkekéngi** L. *Strawberry Tomato.* Pubescent, erect; leaves deltoid-ovate, acuminate, repand; calyx reddening in fruit. ♃ Gardens and fields. 1—2f.

264   ORDER 96.—SOLANACEÆ.

6 P. pubéscens L. Viscid-tomentous, decumbent; leaves ovate or cordate, base unequal, repand; corolla spotted, 6''; fruit-calyx 5-angled. ① Damp. S. and W. 9—18'.
7 P. angulàta L. Smooth, erect; lvs. ovate to oblong, acutely toothed; cor. small (3—6''); fruit-calyx ovoid-conic, longer than its stalk. ① Dry fields.
8 P. Linkiàna Nees. Smooth, diffuse, 2f or more; leaves lance-oblong, *attenuate* both ways, *subulate*-toothed; corolla 6''; fruit-calyx 1½'. ① S. C., Ga. (Dr. Feay).
9 P. Philadélphica Lam. Smoothish, erect; lvs. obliquely ovate, pointed, angular-repand; corolla 9'', spotted and striped; berry large, red. ① M. and W. †

7. **ÁTROPA, L.** DEADLY NIGHTSHADE. Calyx 5-parted. Cor. campanulate, limb 5-cleft, valvate-plicate in bud. Stam. 5, distant, included Berry globous, 2-celled, sitting on the enlarged calyx. ♃ Herbs of lurid colors. Leaves often twin.

A. BELLADÓNNA.—Europe. Leaves ovate, entire, large. Berries dark-purple, handsome but poisonous, like the whole plant. Medicinal.

8. **HYOSCYAMUS,** Tourn. HENBANE. Calyx tubular, 5-cleft. Cor. funnel-form, one of the 5 obtuse lobes larger. Sta. 5, declinate. Stigma capitate. Capsule ovoid, 2-celled, opening with a lid near the summit.—Coarse herbs, native in Eastern countries.

H. niger L. Branched, very leafy, viscid-hairy and fœtid; leaves sinuate-lobed, clasping; corolla straw-color, netted with purple, in one-sided spikes. ② In old fields, and rubbish. 2f. Poisonous—medicinal. July.

9. **PETÚNIA,** Juss. Cal. segments oblong-spatulate. Cor. funnel- or salver-form, tube cylindric, limb spreading, slightly unequal. Sta. 5, inserted in the middle of the tube, unequal, included. Caps. 2-celled. Seeds minute. South American herbs. Leaves alternate, entire, the floral twin. Flowers solitary, large, all Summer. Fig. 66.

1 P. NYCTAGINIFLÒRA. Erect, diffusely branched, viscid-hairy; flowers white, tube slender, thrice longer than the calyx, limb spreading 1½—3'. ♃
2 P. VIOLÀCEA. Prostrate at base, then erect, viscid-hairy; flowers violet-purple, tube inflated, twice longer than the calyx. By admixture numerous varieties, single, double, striped, &c., are raised.

10. **NIEREMBÉRGIA,** Ruiz & Pav. Cal. curved, 5-cleft. Cor. funnel-form, tube long and slender, limb ample, spreading, plicate, slightly unequal. Sta. 5, inserted in the throat, unequal, connivent, anth. hid beneath the stigma. Capsule 2-celled, ∞-seeded.—South American, chiefly herbs, creeping, with elegant, solitary, extra-axillary flowers.

N. GRÁCILIS. Stems very slender and much branched; lvs. linear to spatulate; flowers 1' or more, white, lilac, purple, with a yellow eye.

11. **LÝCIUM,** L. MATRIMONY VINE. Cal. 2–5-cleft. Cor. tubular, bell- or funnel-form, 4- or 5-lobed. Sta. 4 or 5, exserted. Berry 2-celled, seeds several. ♄ ♄ Often spiny. Leaves alternate, entire, often clustered. Flowers small, solitary or in pairs.

L. Bárbarum L. Branches spiny, slender, pendulous or climbing; leaves lanceolate; corolla greenish-purple, 5-parted; calyx 3- or 4-toothed; berries small, orange-red. From Barbary. Planted for arbors walls, &c.

Order 96.—SOLANACEÆ.

2 L. Caroliniànum Mx. Branches rigid, spiny, upright; lvs. fleshy, club-shaped, clustered; flowers small, 4-parted, purple. Salt marshes, S. 3f.

12. DATÙRA, L. Thorn Apple. Calyx large, tubular, inflated, deciduous, or spathe-form. Cor. funnel-form, limb plicate in bud, with 5 or 10 cuspidate angles. Sta. 5. Caps. 2-celled, 4-valved, cells 2-parted. ①♄ Coarse, fœtid, poisonous, with large, often handsome flowers. Fig. 168.

§ Calyx deciduous, its base persistent. Flowers suberect. ①...(*a*)
   a Limb of the corolla 5-toothed. Pods erect............................Nos. 1—3
   a Limb of the corolla 10-toothed. Pods drooping....................Nos. 4, 5
§ Calyx persistent, splitting and spathaceous. Flowers erect. ①.............No. 6
§ Calyx persistent, often splitting. Flowers pendulous. Tree-like ....Nos. 7—9

1 **D. Stramònium** L. *Jimson Weed.* Stem forked; lvs. large, ovate, with unequal sides and angular teeth; corolla cream-white, 2′ long. Waste grounds. 3f. §
  β. *Tátula.* Stem purple; flowers bluish-white; stem 3—4f. S. and W. §
2 **D.** quercifòlia. Leaves sinuate-pinnatifid; flowers white, 5′ broad. Mexico. 2f.
3 **D.** fastuòsa. Stem dark purple, with whitish, shining dots; lvs. lance-ovate; cor violet without, white within, single or double, 7′ long. ① Egypt. Splendid.
4 **D.** Metel. Villous-pubescent; lvs. ovate; flowers white, 4′ broad. Mexico. 3—4f.
5 **D.** meteloìdes. Smoothish, slender; leaves ovate-oblong; flowers pure white or tinged with blue, 5′ broad. Very fine. From Mexico.
6 **D.** ceratocaùla. Stem terete, thick, purple; leaves lance-ovate; corolla thrice longer (5—7′) than the calyx, tube incurved, limb 10-toothed. Cuba.
7 **D.** arbòrea. Leaves lance-ovate, downy; calyx spathaceous, entire; corolla 8—10′ long, white, green-veined; anthers distinct. Peru. Flowers often double.
8 **D.** suavèolens. Leaves ovate-oblong, entire; calyx 5-toothed; corolla 9—12′ long, sweet-scented, white; anthers cohering. Mexico.
9 **D.** sanguínea, has flowers 8′ long, limb red, tube yellow, with purple veins. Peru.

13. NICOTIÀNA, Tourn. Tobacco. Calyx urn-shaped, 5-toothed. Cor. funnel-form, 5-lobed. Sta. 5. Caps. 2-celled, 2–4-valved. ① Coarse narcotics, with large, entire leaves and terminal fls. Jn.—Aug. Fig. 113.

1 **N. rústica** L. Viscid-pubescent; lvs. petiolate, ovate; corolla tube cylindric, lobes round-obtuse, greenish-yellow. Weed in N. Y., &c. 1—1½f. §
2 **N.** Tabàcum. *Virginia T.* Viscid-pubescent; leaves lanceolate, sessile and decurrent; corolla tube inflated in throat, lobes acute, rose-color. 4—6f.
3 **N.** longiflòra. Branches spreading; upper leaves sessile, cordate-lanceolate; flowers racemed, white-purple-yellow, tube slender, 4′. Hardy South.

14. CESTRUM, L. Calyx often colored, 5-cleft. Cor. tubular-funnel-form, tube clavate, limb 5-cleft or 5-parted, plicate in bud. Sta. 5, included, adnate to cor. below. Style 1. Berry few-seeded. ♄ S. American, with entire leaves and brilliant flowers in clusters, fragrant.

§ Habrothámnus. Corolla clavate, red or purple, limb suberect............Nos. 1, 2
§ Eucéstrum. Corolla club-funnel-form, yellow-orange, limb spreading... Nos. 3, 4

1 **C. élegans.** Lvs. lance-ovate; corolla purple, shining, 9′′; calyx purple, 3′′. 5–6f.
2 **C.** fasciculàtum. Lvs. broad-ovate; corolla scarlet, 9′′; calyx reddened, 3′′. 5–6f.
3 **C.** aurantìacum. Leaves lance-ovate; corolla tube inflated, orange-colored, 5′′. 4f.
4 **C.** Parqui. Leaves narrow-lanceolate; corolla dull yellow, 6′′, tube terete.

15. FABIÁNA imbricàta, Ruiz & Pav., is a fine little shrub resembling a Tamarix, with small (6′′ long) ovate leaves covering the numerous branches, and small violet-white flowers. † Chili.

Order XCVII. GENTIANACEÆ. Gentianworts.

Herbs smooth, with a colorless, bitter juice, with entire, exstipulate leaves. *Flowers* regular, mostly centrifugal in inflorescence and convolute in the bud. *Calyx* persistent. *Corolla* withering, its lobes alternate with the stamens. *Ovary* free, 1-celled, with 2 more or less projecting parieta. placentæ. *Fruit* a 2-valved, septicidal, ∞-seeded capsule, rarely baccate. *Seeds* with a minute, straight embryo in the axis of fleshy albumen. Fig. 140.

I. GENTIANEÆ. Corolla convolute (in No. 8 imbricate) in the bud. Leaves opposite...(b)
II. MENYANTHEÆ. Corolla valvate-induplicate in the bud. Leaves alternate or radical...(a)
    a Petals beardless or nearly so. Leaves simple, floating..................LIMNANTHEMUM. 10
    a Petals bearded inside. Leaves trifoliate, erect.......................MENYANTHES. 9
  b Sepals only 2. Corolla 4-parted, tubular-campanulate.......................OBOLARIA. 8
  b Sepals as many as the petals, more or less united...(c)
    c Corolla lobes furnished each with a spur in the middle of the back........HALENIA. 7
    c Corolla lobes furnished each with a large central gland..................FRASERA. 6
    c Corolla lobes plain, without spurs or glands...(d)
      d Leaves reduced to scales. Corolla deeply 4-parted.....................BARTONIA. 5
      d Leafy.—*e* Style none, stig. sessile. Corolla tubular.........GENTIANA. 4
        —*e* Style present.—*x* Corolla tube longer than the limb........ERYTHRÆA. 3
          —*x* Corolla tube shorter than the limb.......EUSTOMA. 2
          —*x* Corolla wheel-form, tube none............SABBATIA. 1

1. **SABBÀTIA**, Adams. AMERICAN CENTAURY. Calyx 5–12-parted. Cor. rotate, 5–12-parted. Sta. 5—12, anth. soon recurved. Style 2-parted. Caps. 1-celled. ①② Slender, with very beautiful flowers, in Summer.

§ LAPITHÆA. Corolla 7-12-(mostly 9-)parted, rose-red......................Nos. 1, 2
§ SABBÀTIA *proper*. Corolla 5-(rarely 6-)parted...(*a*)
  a Flowers white but } —*x* paniculate or scattered....................Nos. 3, 4
    drying yellowish } —*x* in a level-topped cyme. Branches opposite.......Nos. 5, 6
  a Flowers rose-red.—*b* Branches opposite......................Nos. 7, 8
    —*b* Branches alternate...........................Nos. 9, 10

1 S. chloroìdes Ph. Simple or forked; flowers 1—5, *pedunculate*, 20″; petals oblanceolate, 10″; sepals linear-spatulate, 6″; leaves lanceolate to oblong. Wet grounds, Plymouth, Mass., R. I., and S. 1—2f. †

2 S. gentianoìdes Ell. Strict, subsimple; leaves linear, exceeding the internodes; flowers *sessile*, 2-bracted, solitary, or several together; petals obovate, 10″; sepals lance-subulate, 4″. Wet barrens, Ga., Fla., and W. 1—2f.

β. *Boykìnii* (Gray). Leaves lance-oblong, at least the lower. Ga.

3 S. calycòsa Ph. Rigid, divaricately-forked; flowers few, distant; sepals oblanceolate (5—8″), as long as the petals; leaves oblong, 3-veined. Va., and S. 1f.

4 S. paniculàta Ph. Stem much branched, terete, with 4 thread-like ridges · branches mostly opposite; leaves small, oval, oblong to linear; panicle diffuse; sepals subulate, 3″; petals 6″. Low grounds, Va., and S. 1—2f.

β. *Elliòttii.* Branches alternate; leaves mostly linear; petals 7 or 8″.

5 S. lanceolàta (Walt.) Corymbously-branched and 4-angled above; leaves ovate to lanceolate, 3-5-veined; flowers 6-parted, 1′ broad. Barrens, N. J. to Fla. 2f.

6 S. macrophýlla Hook. Stem terete throughout, corymbed at top; leaves erect, thick, ovate, acuminate, 3-5-veined; flowers small (½′ broad). Fla., La.

7 S. angulàris Ph. Stem with 4 winged angles, corymbous-panicled; leaves ovate, 5-veined, clasping; flowers 15—18′ broad, with a greenish star. Wet meadows, N. Y. to Ill., and S. 10—18′.

8 S. brachiàta Ell. Stem obtusely 4-angled, panicled; leaves lance-linear to linear

Order 97.—GENTIANACEÆ.

lowest ovate; flowers 15″, the star purple, bordered with green; petals oblong-obovate, obtuse. Prairies, Ind. to Va., and S. 1f.

9 S. grácilis Salisb. Very slender, diffuse; leaves oblong to linear-filiform; flowers distant; pet. elliptic, obtuse, 5″; sep. filiform, 4″. Wet, Mass. to Fla., and La. 2f.
  β. **stelláris.** Suberect, the flowers larger (13″ broad), the star yellow.
10 S. campéstris. Low (6—10′), erect; lvs. ovate to oblong; fls. few, 15″ broad, the star yellow; calyx tube 5-winged; sepals as long as the broad petals. La.

2. EÚSTOMA, Don. Calyx 5- or 6-parted, with subulate segments. Cor. wheel-funnel-form, 5–6-parted. Sta. shorter than the style.—Herbs glaucous, with few large splendid blue flowers.

1 E. Russelliánum. Stem 1—2f, forked; lvs. ovate, cuspidate, subconnate; fls. long-stalked, expanding 3—4′, petals oval. ① Ark. (Mr. Robertson).
2 E. exàltàtum, taller, with flowers 2′ broad, grows in S. Fla. (Chapman).

3. ERYTHRǼA, Renealm. Calyx 5-4-parted, angular. Cor. funnel-form, 5–4-parted, tube slender. Anth. 5-4, exserted, spirally twisted. Style slender. ① Stem squarish, 3—10′. Leaves connate at base.

1 E. ramosíssima, β. *Muhlenbérgii* (Griseb.) Stem 1-3-times-forked into a loose cyme; leaves ovate-oblong; flowers *pedicellate*, bright purple, 4″. L. Is. to Va.; rare.
2 E. spicàta Pers. Stem forking, erect; leaves oval to lanceolate; fls. sessile, 8″, spicate on the long branches, rose-white. Nantucket to Md. § Europe.
3 E. Centaùrium Pers. Erect; lvs. oblong, acutish at each end; flowers subsessile in the loosely corymbed cymes, rose-purple, 6″. Oswego, N. Y. August. §

4. GENTIÀNA, Tourn. Gentian. Calyx 5- or 4-parted or entire. Cor. tubular, limb 5- or 4-cleft, closed or open. Sta. 5 or 4. Stig. 2, style 0 or very short. Capsule oblong, 1-celled, seeds numerous and minute.— Herbs with showy flowers in August to October.

§ Fls. 4-parted, fringed, sky-blue; no crown or folds. ① ....Nos. 1, 2
§ Fls. 5-parted, blue, pedicellate, clustered; no fringe or folds. ① ....No. 3
§ Fls. 5-parted, corolla with folded appendages between the lobes. ⚄...(a)
  a Flower solitary, terminal, somewhat stalked. Leaves linear............No. 4
  a Flowers clustered, sessile,—b ochroleucous or whitish....... ....Nos. 5, 6
      —b blue; the corolla always closed............No. 7
      —b blue; the corolla open or expanding...Nos. 8—10

1 G. crinìta Frœl. *Fringed G.* Stem and branches erect; leaves lanceolate, acute; petals obovate, finely fringed at margin. ① Moist soils, Can. to Ga., and W. 1f. A beautiful and interesting plant.
2 G. detónsa L. Stem and few branches strict; leaves lance-linear; flowers solitary, long-stalked, petals crenate-ciliate. ① N. Y. to Wis. 1f.
3 G. quinquefloèra L. St. 4-angled; lvs. ovate to lanceolate, acute; fls. 7—8″, pedicellate, clustered; sepals subulate, very short, or (in β. *parviflora*) lance-linear, 4″; corolla segments bristle-pointed. ② Fields and woods. 1f.
4 G. angustifòlia Mx. Slender, erect; fl. 18—20″ long; lvs. linear; sepals linear, 7—10″; corolla blue, lobes ovate, the cleft folds much shorter. N. J. to Fla. 1f.
  β. *viridiflora.* Flower nearly sessile, 15″, greenish white, folds very short. S.
5 G. ochroleùca Frœl. Lvs. smoothish, oval to elliptical, acutish both ends; calyx segments lance-linear, nearly equalling the 20″ corolla. Pa. (Prof. Porter) to Fls. 1f.
6 G. alba Muhl. Very smooth, stout; lvs. lanceolate, the broad base clasping; fls. 2′ long, calyx segments ovate, very short. Woods, prairies, M. and W. 1½—2f.
7 G. Andréwsii Griseb. *Closed Blue G.* Simple, smooth; leaves oval-lanceolate;

cluster dense, terminal; calyx segments ovate-oblong, 3—4''; corolla 18'', inflated, *never opening*, folds as long as segments. Woods, N. Eng. to Fla. 2f.

8 G. Saponaria L. Subsimple, stout, smooth; leaves oblanceolate to lance-oblong, 3-veined; calyx segments linear, 6—8''; corolla 2', folds much shorter than the open erect lobes. N. J., Pa., to Ill., and S. 2f. Leaves 2—3'.

9 G. linearis Wood. Simple, slender; lvs. lance-linear to linear, 1-(rarely 3-)veined; calyx segments subulate, 4—7''; corolla folds subentire, much shorter than the erect or spreading lobes. N. Eng. (rare) to Iowa and Ky. 1—1½f. July—Sept.

10 G. puberula Mx. Slender, rough or puberulent; leaves 1', oval to ovate, very rough-edged, clasping, acute; calyx segm. lanceolate, 5''; corolla subcampanulate, 15'', lobes very acute, folds short, cleft. Prairies, W. and S. 9—18'.

5. BARTONIA, Muhl. SCREW-STEM. Fls. 4-parted, persistent. Cor. subcampanulate, pet. slightly united. Stig. thick, some bifid. Sds. very ∞ and minute. ♃ Slender, erect, with scale-like lvs. and small white fls.

1 B. verna Muhl. Low, simple, 3—5', clustered; ped. 1-flowered, petals 3'', oblong, obtuse, sepals 1'', acute. Bogs and barrens, Va. to Fla. March.

2 B. tenella Muhl. Branched above, very slender, 5—12'; ped. opposite, erect, subequal, 4''; petals pointed, 1'', sepals nearly as long. Wet. Mass. to Fla. August

β. *brachiata*. Pedicels bent outward and upward, some alternate. S.

6. FRASERA, Walt. COLUMBO. Fls. mostly 4-parted. Pet. united at base, oval, spreading, each with 1 or 2 bearded glands in the middle. Sty. 1, stig. 2, distinct. Caps. compressed, 1-celled. Seeds few, large, elliptic, margined. ♃ Showy and tall, with opposite or verticillate leaves.

F. Carolinensis Walt. Smooth, 4—9f high! paniculate above; lvs. oblong, sessile, in 4's—6's; petals greenish with blue dots, and a large purple gland. Rich soils, N. Y., S. and W. A stately plant, and a good tonic. June, July.

7. HALENIA, Borkh. FELWORT. Flowers 4-parted, broad bell-form. Each petal prolonged at base into a spur, which is glandular at the end. Stigmas 2, sessile.—Flowers panicled.

H. deflexa Griseb. Erect, branched, lower leaves oblanceolate, upper lance-ovate, 3-5-veined; spurs slender, curved outward, half as long as the 4'' greenish-yellow petals. ③ N. Eng. (rare) to Wis. 18'. August.

8. OBOLARIA, L. PENNYWORT. Calyx of 2 wedge-oblong sepals. Corolla tube-bell-form, 4-cleft. Sta. on the corolla. Stigma sessile, bifid. Seeds ∞, very minute. ♃ Flowers sessile, pale.

O Virginica L.—Woods, N. J., W. and S. Stem 4—8', subsimple. Leaves roundish, sessile, thick, crowded above, sepals similar. April, May.

9. MENYANTHES, Tourn. BUCK BEAN. Cal. 5-parted. Cor. rotate or funnel-form, limb spreading, 5-lobed, villous within, no glands at the base. Stamens 5. Style 1, stigma bifid. Capsule 1-celled.—Bitter herbs, actively medicinal. Leaves trifoliate, nearly radical.

M. trifoliata L.—In muddy places, Penn. to Cal., and N. 8—12'. Petioles long and round. Scapes bearing racemes of handsome, flesh-colored flowers. May.

10. LIMNANTHEMUM, Gmel. FLOATING HEART. Cal. 5-parted. Cor. rotate, each seg. with a glandular scale at base. Sty. short or 0, stig.

2-lobed. Caps. opening by decay. ~ Stagnant water. Pet. long, bearing an umbel of small white fls. below the roundish leaf-blade, also oblong tubers.

1 L. lacunòsum Griseb. Leaves small (1—2'), smooth, round-reniform; seeds smooth and shining. N. Eng. to Fla. (Villarsia lacunosa Vent.)
2 L. trachyspérmum Gray. Lvs. large (3—5'), dotted and pitted beneath; seeds muricate about the margins. Md. to Fla. and La. (Menyanthes, Mx.)

ORDER XCVIII. LOGANIACEÆ.

Herbs or *shrubs* with opposite leaves, stipules between the petioles or at least a ridge, and with 4- or 5-parted regular gamopetalous flowers. *Ovary* superior, *stigmas* as many as the cells. *Fruit* a 2-celled capsule, or a 1–2-seeded drupe. *Seeds* winged or peltate, with albumen. Fig. 47.

* Delicate, twining shrubs, with large yellow flowers. S. ....GELSEMIUM. 1
* Low herbs.—∞ Flowers scarlet, tubular, with one style...............................SPIGELIA. 2
  —∞ Flowers small, white, 5-parted, in 1-sided racemes....................MITREOLA. 3
  —∞ Flowers small, white, 4-parted, in axillary cymes..... ....POLYPREMUM. 4

1. GELSÈMIUM, Juss. YELLOW JESSAMINE. Cor. bell-funnel-form with 5 short rounded lobes. Sta. 5, now longer and now shorter than the style (*dimorphous*). Caps. flattened, twin, cells each with 4—6 winged sds. ♄ Very slender, with numerous flowers. The stipules a mere ridge.

G. sempérvirens Ait.—Woods and banks, Va., and S., overrunning bushes and low trees. Leaves thick, shining, lanceolate. Flowers 1'. March—May.

2. SPIGÈLIA, L. PINK-ROOT. Calyx seg. linear-subulate. Cor. narrowly funnel-form, limb 5-cleft. Anth. 5, convergent. Caps. twin-lobed, few-seeded.—Herbs, with the flowers sessile in terminal spikes. Fig. 47.

S. Marilándica L. Stem square, erect, simple; leaves sessile, ovate-lanceolate; spike scorpoid, uncoiling as the 3—8 handsome flowers expand; corolla 1½—2' long. ♃ Thickets, Pa. to Ill., and S. June. Medicinal.

3. MITRÈOLA, L. Corolla tubular, short, 5-cleft, hairy in the throat. Sta. 5, included. Ovary 2-celled, styles 2, united only at top with 1 stigma. Capsule 2-horned, ∞-seeded. ① Flowers in several scorpoid spikes at top of a long terminal peduncle. June—August.

1 M. petiolàta T. & G. Branching; leaves ovate to lanceolate, tapering at base to a petiole; raceme loose-flowered. Va., and S. 1—2f.
2 M. sessilifòlia T. & G. Nearly simple; leaves oval to elliptical, sessile, shorter than the internodes; raceme close-flowered. S. C. to Fla. 10—18'.

4. POLYPRÈMUM, L. Calyx seg. 4, subulate. Corolla broad bell-form, lobes a little unequal, obtuse, throat bearded. Stamens 4, included. Stigma subsessile. Capsule ovoid. ① Smooth, diffusely branched from base, with linear-subulate leaves. Flowers sessile.

P. procúmbens L.—Dry fields, Va., and S. 6—12'. In dense patches. May—Sept.

ORDER XCIX. APOCYNACEÆ. DOG-BANES.

Plant with an acrid, milky juice, entire, exstipulate, mostly opposite lvs.

Order 99.—APOCYNACEÆ.

Flowers 5-parted, regular, the calyx persistent, the corolla twisted in æstivation. *Stamens* 5, with distinct filaments, anthers filled with granular pollen. *Ovaries* 2, distinct, but their stigmas blended into a head-shaped mass. *Fruit* 1—2 follicles, or capsular or baccate, with albuminous seeds.

§ Herbs erect, native.—*a* Corolla bell-form, whitish.   Leaves opposite..................APOCYNUM.   1
   —*a* Corolla salver-form, blue.   Leaves alternate..................AMSONIA.   2
§ Half-shrubby, cultivated, trailing or erect.   Corolla wide-spread ....VINCA.   3
§ Shrubs twining.—*b* Native.   Flowers small, yellowish.....................FORSTERONIA.   4
  —*b* Cultivated.   Flowers large, white..............................ECHITES.   5
§ Shrubs erect.—*c* Leaves opposite or in 4's.   Corolla yellow........................ALLAMANDA.   6
 —*c* Leaves opposite or in 3's.   Corolla roseate............................NERIUM.   7
 —*c* Leaves alternate.   Flowers 3″.   Fruit a drupe.   S. Fla............VALLESIA.

1. **APÓCYNUM**, Tourn. Dog's-BANE. Cor. bell-form with short lobes. Sta. included, alternating with 5 glandular teeth on the base of the corolla. Ovaries 2. Stigma connate. Follicles slender, distinct. Seeds comous. ♃ Leaves entire, mucronate, opposite. Flowers pale, in cymes, June—Aug.

1 **A. androsæmifòlium** L. Leaves ovate; cymes terminal and lateral; cor. 3″, with red stripes, tube longer than the calyx, lobes spreading. Hedges and fields. 3f. A handsome plant, smooth or downy.

2 **A. cannabìnum** L. Leaves oval to lance-oblong, often downy beneath; cymes terminal; corolla 1″, tube not longer than the calyx, lobes erect. In shades. 2—4f Pods 3′ long. (A. hypericifolium Ait.)

2. **AMSÒNIA**, Walt. Calyx segment pointed. Cor. tube hispid, funnel-form, limb in 5 linear segments twisted in bud. Style 1. Ovaries 2, connate at base, follicles 2, erect, slender. Seeds not comous. ♃ Leaves alternate, entire. Clusters terminal, blue.

1 **A. Tabernæmontàna** Walt. Leaves ovate-lanceolate, acuminate; sepals lance-acuminate; corolla 8″, livid blue. Damp grounds, W. and S. 2f. May, June.—Varies with leaves lance-elliptic, and sepals acute.

2 **A. ciliàta** Walt. Leaves more or less crowded, linear or filiform, the margins ciliate; cluster long-stalked, corymbed, or soon panicled; corolla glabrous outside. Sands, S.: common. 1—2f. April, May.

3. **FORSTERÒNIA**, Meyer. Corolla funnel-form, deeply 5-cleft, twisted in bud. Anthers adherent to the stigma. Stigma 2-lobed. Follicles 2, spreading, seeds comous. ♄ Leaves opposite.

F. diffórmis DC. Climbing; leaves round-oval to lance-oval, cuspidate-pointed; cymes axillary and terminal, stalked; calyx segments ovate, long-pointed; corolla 3—4″, pale yellow. Swamps, Va., and S. May—August.

4. **VINCA**, L. PERIWINKLE. Cor. funnel- or salver-form, convolute, with the 5 lobes oblique, orifice 5-angled. Two glands at base of the ovary. Follicles 2, erect, slender. ♃ ♄ Lvs. opposite. Flowers solitary, axillary.

1 **V. MINOR.** Procumbent; leaves elliptic-lanceolate, not ciliate; sepals lanceolate; flowers scentless, violet, purple, or white. May, June. Europe.

2 **V. MAJOR.** Decumbent; leaves ovate, ciliate at edges; sepals long, bristle-pointed. In shades, forming loose masses, leaves often silver-edged. Europe.

3 **V. ROSEA.** Erect, soft-downy; leaves oval, obtuse; flowers large, roseate, often white or white-edged, perpetual. From Madagascar.

ORDER 100.—ASCLEPIADACEÆ. 271

5. ECHÌTES, Br. Cor. funnel- or salver-form, not appendaged, lobes convolute, bearing the subsessile anthers in the throat; 5 glands at base of ovaries. Foll. 2, slender. Sds. comous. ♄ ♃ Lvs. opp. (Mandevilla, Lindl.)

E. suavèolens. Climbing; leaves cordate-ovate, acuminate, shorter than the axillary or terminal racemes; flowers fragrant, 2'. S. America.
E. umbellàta Jacq. and E. Andréwsii Chapm. are indigenous in S. Fla.

6. ALLAMÁNDA cathártica. Shrub from Guyana, with slender branches, oblong thin-pointed leaves, and bright-yellow flowers 2½—3'. Cor. funnel-bell-form, lobes 5, rounded, throat appendaged. Ova. 1, becoming a prickly, 1-celled capsule.

7. NÈRIUM, L. Oleander. Corolla salver-form, convolute, throat crowned with 5 cleft scales. Anth. arrow-shaped, tipped with a long hairy bristle. ♄ Lvs. lanceolate, acute both ways, thick and leathery, in 2's or 3's.

1 N. Oleánder. Leaves lanceolate; scales of the crown each of 3 or 4 pointed unequl teeth; fls. clustered, inodorous, often double, 2'. Palestine. 5—10f, very handsome.
2 N. odòrum. Leaves linear-lanceolate; scales of the crown each 4–7-cleft; appendages of the anthers exserted; flowers fragrant. India.

ORDER C. ASCLEPIADACEÆ. Asclepiads.

Plants (chiefly herbs in the United States) with a milky juice, often twining. *Leaves* opposite (rarely whorled or scattered), without stipules, entire. *Flowers* generally umbellate, 5-parted, regular, the *sepals* and also the *petals* united at base, both valvate in æstivation. *Stamens* united, adherent to and covering the fleshy mass of the two united stigmas. *Pollen* cohering in masses. *Ovaries* 2, forming follicles in fruit.

FIG. 530.—1. Asclepias cornuti. 2. A flower, the petals and sepals reflexed, and the corona erect. 3. One of the segments of the corona with the horn bent inwardly. 4. A pair of pollen masses suspended from the glands. 5. A mature follicle. 6. Vertical section of P. phytolaccoides showing the two ovaries. 7. Lobe and horn of the corona.

§ Stems erect, leafy, herbaceous...(*)
§ Stems climbing, often shrubby...(*)
§ Stems low, leaves fleshy, all radical...Stapelia. 12
a A little horn in each *hood* of the crown. Petals reflexed............Asclepias. 1
a No horns in the crown.—*b* Petals reflexed or spreading. ....Acerates. 2
—*b* Petals erect. ....Podostigma. 3
c Corolla salver-form, white, the crown in the bottom of the tube .....Stylandra. 10
e Corolla wheel-form, flattish, the lobes spreading...(*)
e Corolla segments erect, crown 5-leaved,—*d* each leaflet 2-awned. ....Enslenia. 4
—*d* leaflets awnless. ....Metastelma. 5

ORDER 100.—ASCLEPIADACEÆ.

n Crown double, the outer a ring, the inner 5-leaved. S. Fla..................SARCOSTEMMA.
n Crown simple,—x deeply 5-parted. Leaves linear. ....SEUTERA. 1
—x of 5 awned scales. Leaves ovate................ ....PERIPLOCA. 2
—x a ring 5-10-lobed, or merely wavy...(y)
y Anther slits vertical, pollinia pendulous. Leaves thin ............VINCETOXICUM. 7
y Anther slits horizontal, pollinia spreading. Leaves cordate..............GONOLOBUS. 8
y Anther slits vertical, pollinia erect. Leaves thick..........................HOYA. 11

1. ASCLĒPIAS, L. MILK-WEED. SILK-WEED. Calyx and cor. segm. soon reflexed. Staminal crown of 5 distinct *hoods* (cucullate leaflets), each with a little curved horn from within. Anth. consolidated with the stig., forming a 5-angled truncate mass (antheridium), opening by 5 chinks. Pollen masses (pollinia) 5 pairs, hanging vertically by a pedicel from a cleft gland. Follicles 2, lance-shaped, seeds comous. ♃ Erect, with the flowers in simple umbels which are between the petioles or terminal. Jn.—Aug.

* Flowers whitish, greenish, or purple in various shades...(a)
* Flowers orange-colored or scarlet. Leaves narrowly lanceolate........Nos. 15—17
  a Leaves ovate to lanceolate, narrowed to a petiole...(b)
  a Leaves ovate-oblong to cordate, sessile or clasping................Nos. 12—14
  a Leaves linear, very narrow...(x)
  b Both crown and corolla greenish-purple. Pods woolly-spiny.............Nos. 1, 2
  b Both crown and corolla pure purple. Pods smooth.....................Nos. 3, 4
  b Crown white; corolla white tinged with pink. Flowers small............Nos. 5—7
  b Crown white; corolla greenish-white.—c Umbels pedunculate...........Nos. 8, 9
                        —c Umbels subsessile. S.........Nos. 10, 11
  x Leaves all opposite, or rarely the highest alternate................Nos. 18, 19
  x Leaves mostly verticillate or scattered. Flowers greenish..........Nos. 20, 21

1 A. Cornùti Desn. Leaves oblong-ovate, downy beneath, acutish at base and short-stalked, longer than the many-flowered umbels; hoods ovate; horns acute. Road sides and hedges. 2—4f. Leaves 5—8'. Flowers 6''.
2 A. Sullivántii Eng. Leaves ovate-oblong, smooth both sides, nearly sessile; hoods obovate; horns blunt; flowers 9''. Ohio to Ill. July.
3 A. purpuráscens L. Simple; leaves ovate to *elliptical*, acute mucronate; umbels subsolitary, terminal; peduncle 1—2'; pedicels 1'; horns horizontal. N. Eng. to N. Car., and W. 3—4f. Flowers large (6''), dark purple. Hoods lance-ovate.
4 A. incarnàta L. Branching above; leaves lanceolate; umbels many or few, somewhat panicled; flowers small (3''); ped. ¼—2'. Wet places. 3—5f: common.
   β. *pulchra*. Hairy; leaves lance-oblong or -ovate. Very handsome. †
5 A. ovalifòlia Desn. Low, downy; lvs. ovate, acutish; umbels subsessile, 10-15-flwd.; pet. oval; hoods yellowish, obtuse, longer than the horns. W. (A. Vaseyi C-B.)
6 A. perénnis Walt. Branched at base, half-shrubby, smooth; leaves thin, lanceolate, pointed both ways, long-stalked, exceeding the small white umbels; hoods shorter than the horns. Low grounds, W. and S. 2f. (A. parviflora C-B.)
7 A. quadrifòlia Ph. Simple, smooth; leaves ovate, acuminate, some of them in whorls of 4; umbels few, loose-flowered, long-stalked. Dry woods. 2f.
8 A. variegàta L. Simple, smoothish; leaves oval to lance-oval, short-pointed, acute at base; umbels densely ∞-flowered, small (1'—18'' diam.); hoods orbicular.
   β. *nivea*. Lvs. elliptical, pointed both ways; umb. 10-15-flwd. N. J., W. & S. 1—3f.
9 A. phytolaccoides Ph. Tall, simple; leaves broadly ovate, pointed both ways, glaucous; umbels lateral, with about 20 drooping fls.; peduncles and pedicels 1—3' long; hoods truncate, with 4 unequal teeth; horns exserted. Damp shades. 4—5f.
10 A. tomentòsa Ell. Woolly, stout; leaves lance-oblong, wavy, cuspidate; umbels lateral, with many large flowers; hoods obovate, truncate. Barrens, S.

11 **A. obováta** Ell. Tomentous; leaves obovate, obtuse, mucronate; umbels 10-14-flowered, lateral; fls. large, yellowish-green; hoods elongated. Gravels, Ga., Fla.
12 **A. rubra** L. Simple, glabrous; lvs. ovate, long and acutely pointed, subsessile; umbels panicled above, few; flowers red-purple; hoods acute, some longer than the slender exserted horns. Barrens, N. J., and S. 2—3f. Leaves 3—5'.
13 **A. obtusifòlia** Mx. Simple, smooth; leaves oblong to oblong-ovate, subcordate, obtuse-mucronate; umbels 1—3, terminal, pedunculate, 15-25-flowered; hoods truncate, shorter than the sickle-shaped horn; flowers 6'', red-green. M., W., S. 3f.
14 **A. amplexicaùlis** Mx. Simple, flexuous, glaucous; lvs. ovate, cordate-clasping, obtuse, *not mucronate;* ped. lateral and terminal, with ∞ dull-purplish flowers; pedicels slender; hoods ovate, including the horns. Copses, S. 1—2f.
15 **A. tuberòsa** L. *Butterfly-weed.* Stem ascending, hairy, umbellate branched; leaves sessile, *alternate*, lance-oblong; umbels many, erect; flowers bright orange-red; hoods oblong; horns suberect. Dry fields. Root tuberous. Stem 2f. †
16 **A. paupércula** Mx. Smooth and virgate; leaves linear and oblong-linear, 4—6' long; umbels with few large yellow-red flowers at the naked summit. N. J., and S.
17 **A. Curassávica** L. Half-shrubby and branching at base; branches terete, leafy to the top; leaves lance-linear; umbels with few large scarlet flowers. S. Fla. Cult.
18 **A. cinèrea** Walt. Stem wiry, simple, naked above; leaves linear-filiform, 1—3', erect; umbels terminal, several, bracteolate, 3-5-flowered; peduncles 4—6''; pedicels 6—8''; corolla ashy-purple, 3—4''. Damp barrens, S. C. to Fla. 2—3f.
19 **A. virídula** Chapm. Stem and leaves as in No. 18; umbels 6-12-flowered, yellowish green, shorter than the leaves. Fla.
20 **A. Michaùxii** Desn. Stems diffuse; leaves linear, 3—4', scattered; umbels ∞-flowered, often panicled, mostly shorter than the lvs.; fls. 3'', *fragrant.* Sands, S. 1f.
21 **A. verticillàta** Ell. Simple, slender, erect; leaves linear, very narrow, generally *verticillate;* umbels small, many, lateral, 1' diameter, pedunculate. Swamps. 2f.

2. **ACERÀTES**, Ell. Hoods of the crown destitute of a horn. Otherwise nearly as in Asclepias. ♃ Flowers greenish. June—August.

§ ACERÀTES *proper.* Umb. lateral; pet. reflexed; crown adnate to anth...Nos. 1—3
§ ANÁNTHERIX. Umbels terminal; pet. spreading; crown free from anth..Nos. 4, 5

1 **A. viridiflòra** Ell. Stout, whitish-downy; leaves thick, oval, obtuse, petiolate, varying to elliptic-lanceolate, or even to orbicular (Ga., Prof. Pond); umbels small, dense, subsessile. Sands. 2f. Leaves exceedingly variable.
2 **A. longifòlia** Ell. Rough-puberulent, simple; leaves alternate, lance-linear to linear; umbels lateral, pedunculate, densely many-flowered; flowers small, 3'', crown stipitate. Prairies, W. 2—3f. Peduncles 1'.
3 **A. lanuginòsa** Desn. Low, stout, hairy; leaves lanceolate; umbel 1, on the naked summit of the stem, dense; crown sessile. Prairies, Wis. 1f.
4 **A. connivens** Desn. Strict, half-shrubby; leaves oval-oblong; umbels 7-12-flwd., along the naked summit of the stem; pet. 5'', oval, with a short cusp; hoods *connivent* over the anthers. Barrens, Ga., Fla. 2f. Leaves 20—30''.
5 **A. paniculàta** Desn. St. angular; lvs. lance-oblong, obtuse; umbels clustered at the leafy top, 5-9-flowered; pet. large, half-erect, 7''; pods glabrous, seeds with long silky tufts. Ga. to Ill. and Kan. (Rev. J. H. Carruth.)

3. **PODOSTÍGMA**, Ell. Cor. seg. 5, erect, oblong. Crown *stipitate*, hoods without horns. Follicles 2, long, slender, smooth. ♃ Low and simple, with opposite leaves and supra-axillary few-flowered umbels.

P. pubéscens Ell.—Wet grounds, S. A curious plant, with linear-oblong leaves and 3—5 umbels of yellowish-green flowers, in May, June. 1f.

4. **ENSLÈNIA**, Nutt. Cor. 5-parted, segments erect; hoods or scales

of the crown 5, free, each terminated by 2 filiform, flexuous lobes. Pollinia oblong, pendulous. Stig. 5-angled, conical. Follicles cylindraceous, smooth. ♄ A twining herb, with opposite, cordate leaves, and cream-white flowers in small lateral corymbs.

E. álbida N.—W. and S.: common. 6—10f. Clusters 5-8-flwd., fragrant. July, Aug.

5. METASTÉLMA, Br. Cor. somewhat bell-form, segments incurved at apex. Crown of 5 distinct scales. Stigma flat. Pods smooth, slender, seeds comous. ♄ Lvs. cuspidate, smooth. Umb. of few small flowers.

M. Fráseri Desn. Leaves oval; umbels sessile; pet. ovate, ciliate, as long as the linear crown-scales. In Carolina (Fraser, in DC.).
M. Schlectendahlii and other species grow in S. Fla. (Dr. Chapman.)

6. SEUTÉRA, Reich. Sepals 5, lanceolate. Cor. rotate, segm. acute. Crown on the base of the sessile anthers, of 5 retuse segments. Pollinia ovoid, pendulous. Stigma bifid. Pods smooth, seeds comous. ♄ Leaves linear, fleshy. Umbels few-flowered.

S. marítima Desn.—Salt marshes, S., twining on the rushes, &c. Leaves opposite, 1'. Umbels 7-10-flowered. Pet. greenish, crown short, white. June—October.

7. VINCETÓXICUM, Mœnch. Calyx and cor. 5-parted, wheel-form. Crown a fleshy, 5–10-lobed disk. Anth. tipped with a membrane. Pollinia and fruit as in Asclepias. ♃ ♄ Flowers small, in dense clusters.

1 V. nigrum Mœnch. Herb somewhat twining, with lance-ovate, attenuately-acute leaves and small blackish clusters in the axils. Gardens and fields: rare.
2 V. scopàrium (N.) Shrubby at base, much branched; leaves thin, linear, 1'; clusters short-stalked, downy, with few green flowers; pods slender, 1'. Fla.

8. GONÓLOBUS, Mx. Corolla subrotate, 5-parted, convolute in bud. Crown a small, fleshy, undulate-lobed ring, attached to the throat of the corolla. Anth. opening *transversely* beneath the stigma. Pollinia 5 pairs, horizontal. Pods turgid, seeds comous. ♄ Leaves cordate. Umbels few flowered, short, extra-axillary. Flowers brownish.

* GONÓLOBUS *proper*. Cor. rotate, flat, lobes linear to oblong, smoothish...Nos. 1—3
* CHTHAMÀLIA. Corolla bell-form, small (woolly), lobes ovate, 1" long ....No. 4

1 G. macrophýllus (and lævis) Mx. Smooth, or with minute down and scattered hairs; leaves short-pointed, base-lobes open; umbels 5-flowered, buds conic-pointed; pet. linear-subulate, 4"; pod smooth, ribbed. Shady banks, Va. to Ky., and S. 3—5f.
2 G. oblìquus Br. Hirsute with spreading, unequal hairs; leaves acuminate, base-lobes closed and some oblique; umbels 2-5-flowered, buds oblong, pet. linear-oblong 6"; pod muricate, ribless. Banks, O. to Pa. and Ga. 3—5f.
3 G. hirsùtus Mx. Hirsute; leaves acuminate; umbels 5-8-flowered, buds : void. petals oblong, 3", yellow, downy; pod muricate. Woods, South. 4—8f.
4 G. prostràtus Ell. Branches from base, prostrate, 6—12'; leaves small (1'), reniform-cordate; umbels sessile, 3-5-flowered; corolla segments ovate, 1", very woolly inside, dark purple. Sands, Ga. (Dr. Feay). (Chthamalia pubera Desn.)

9. PERÍPLOCA, L. Cor. rotate, flat, 5-parted. Crown 5-cleft, tipped with 5 filiform awns. Filaments distinct, anthers cohering. Pollinia 5, each 4-lobed, single. Follicles 2, smooth, divaricate. Seeds comous. ♄

P. Græca L. Leaves ovate, acuminate, 3—4'; flowers panicled on a long peduncle; petals very hairy, linear, obtuse, purple. Gardens, &c. 10—15f. August. §

10. STEPHANÒTIS, Pet.-Th. Sepals distinct. Cor. salver-form, limb 5-lobed, convolute in bud, tube including the 5-leaved crown in its enlarged base. ♃ Leaves thick, very smooth.

S. FLORIBÚNDA. Leaves oval; flowers 5—8 on each peduncle, white and fragrant, tube 1', limb 1½' broad. Greenhouse plant, from Madagascar.

11. HOYA, Br. WAX-PLANT. Sepals 5. Corolla rotate, flat, valvate in bud. Crown of 5 depressed, spreading segm. Pollinia fixed by the base, connivent. Pods smooth, seeds comous. ♃ Smooth, fleshy.

H. CARNÒSA. Branchlets puberulent; leaves oval-oblong; flowers in dense umbels, pink-colored, wax-like. Greenhouse plant, from E. India.

12. STAPÈLIA, L. CARRION-FLOWER. Calyx 5-parted. Cor. rotate, fleshy, 5-cleft. Crown double, of 2 rings entire or lobed. Pollinia erect. Pods erect, smooth.—Fleshy, leafless, cactus-like plants, from S. Africa, with large, dark-red *fœtid* flowers, in the greenhouse.

S. HIRSÙTA, with erect, dull-green 4-sided branches, toothed on the angles, and flowers 3—4' broad, with purple, ciliate, lance-ovate petals.

ORDER CI. OLEACEÆ. OLIVEWORTS.

Trees and *shrubs*, with opposite, simple or compound leaves, and regular 4–8-parted *diandrous* flowers. *Corolla* rarely wanting, its divisions more in number than the stamens. *Ovary* free, 2-celled, with 2 (rarely 1 or ∞) ovules in each cell. Fig. 16.

```
I. JASMINEÆ.  Corolla 5-8-parted.  Ovary cells each with 1 erect ovule..........JASMINUM.   1
II OLEACEÆ proper. Corolla valvate, 4-parted or 0. Ovary cells 2- or ∞-ovuled...(*)
    * Flowers perfect, corolla present. Leaves simple...(a)
    * Flowers imperfect, inconspicuous, often apetalous...(c)
        a Flowers yellow. Ovary with many ovules in each cell................FORSYTHIA.  2
        a Flowers white, or lilac. Ovary cells 2-ovuled...(b)
        b Stamens exserted.  Fruit a fleshy drupe or berry....................OLEA.      3
        b Stamens included.—x Corolla salver-form, tube longer than lobes....SYRINGA.    4
                —x Corolla funnel-form, tube shorter than lobes...............LIGUSTRUM.  5
                —x Corolla lobes long, linear, drooping......................CHIONANTHUS. 6
        c Leaves simple. Corolla 0. Fruit a fleshy drupe......................FORESTIERA.  7
        c Leaves pinnate. Corolla 0, or present. Fruit a winged samara........FRAXINUS.   8
```

1. JASMÍNUM, L. JESSAMINE. Calyx 5–8-lobed. Cor. salver-form, limb 5–8-cleft, convolute in bud. Sta. included. Berry double, 2-seeded. ♃ ♄ Petioles jointed.

§ Leaves opposite, unifoliate. Flowers white, 8-10-parted............ Nos. 1, 2
§ Leaves opposite, 3-9-foliate. Flowers white, 5-parted............. Nos. 3– 5
§ Leaves alternate, 3-7-foliate. Flowers yellow, 5-parted............ Nos. 6, 7

1 **J. SAMBAC.** Scarcely climbing; leaves ovate; petals 8, rounded, fragrant. India.
2 **J. LAURIFÒLIUM.** Climbing; leaves lanceolate; pet. 9 or 10, linear, fragrant. India.
3 **J. AZÓRICUM.** Diffuse; leaflets 3, ovate, shining; flowers very fragrant. Azores.
4 **J. OFFICINÀLE.** Climbing; lfts. 7, lanceolate; sep. linear, equaling cor tube. Asia.

Order 101.—OLEACEÆ.

5 **J.** GRANDIFLÒRUM. Climbing; leaflets 9, oval, some confluent, the odd one pointed; sepals thrice shorter than the corolla tube; petals oval. India.
6 **J.** REVOLÙTUM. Not climbing; lfts. ovate, pointed; pet. roundish, recurved. Asia.
7 **J.** ODORATÍSSIMUM. Climbing; lfts. oval, obtuse; fls. less fragrant than No. 6. Azores.

2. **FORSYTHIA,** Vahl. Calyx very short, deciduous. Cor. subcampanulate, lobes long, twisted in bud. Sta. inserted in the base of the tube, included. Seeds ∞ in the 2-celled pod. ♄ Leaves opposite or in 3's, appearing after the yellow flowers.

1 **F.** VIRIDÍSSIMA. Branches erect, strict, covered with flowers in early Spring, each flower separate, pedicellate, lateral; leaves lanceolate. China.
2 **F.** SUSPÉNSA. Branches weak, pendulous; leaves ovate; flowers scattered. Japan.

3. **SYRÍNGA,** L. LILAC. Calyx small, persistent, many times shorter than the tube of the salver-form corolla. Sta. included. Pod 2-celled, valves bearing the septum in the middle, seeds 4. ♄ Leaves opposite.

1 **S.** VULGÀRIS. *Common L.* Leaves cordate-ovate, entire, glabrous; flowers *lilac* to lilac-purple, in a dense thyrse, very fragrant. A beautiful shrub, from Hungary: varying with flowers *bluish,* or *white.* April—June.
2 **S.** PÉRSICA. *Persian L.* Leaves lanceolate, acute, smooth, often pinnately cleft; thyrse loose, smaller, white, or lilac-blue. Persia.
3 **S.** VILLÒSA. *Chinese L.* Leaves elliptic, acute, hairy beneath. N. China.

4. **ÒLEA,** Tourn. OLIVE. Calyx short. Corolla tube short, limb 4-parted, spreading. Stamens 2, inserted in the base of the tube, exserted. Ovary with 4 suspended ovules, ripening only 1 or 2 seeds. Drupe fleshy, oily. ♄ ♄ Leaves opposite. Flowers white.

* Racemes axillary, shorter than the coriaceous leaves.....................Nos. 1—3
* Racemes in a large terminal panicle. (Visiania paniculata C-B)............No. 4

1 **O.** AMERICÀNA L. Leaves oblanceolate to elliptic, entire, smooth, shining, attenuated to a petiole; raceme compound, scarce longer than the petiole; flowers diœcious; drupes globular. Swamps, N. J. to Fla. 15—20f.
2 **O.** EUROPÆA. Leaves lanceolate, mucronate; racemes longer than the petioles; drupes oval. Europe. Cultivated in California, rarely far South. 20—40f.
3 **O.** FRÀGRANS. Shrub; leaves lance-oblong, *serrate;* flowers small, white, very fragrant, in axillary corymbs, white-red; *styles* 2. China. (Osmanthus.)
4 **O.** CLAVÀTA. Shrub with ovate entire leaves and many small flowers in large panicles; style 1, club-shaped, exserted like the stamens. China. Hardy S.

5. **LIGÙSTRUM,** L. PRIVET. PRIM. Cal. minutely toothed. Cor. funnel-form, 4-lobed. Sta. subincluded. Sty. very short. Berry 2-celled, 2-4-seeded. Sds. angular. ♄ With simple lvs. and term. panicles of white fls.

L. VULGÀRE L. Leaves lanceolate to obovate, 1—2', obtuse or acute, thick but deciduous; flowers small, in small thyrses; anthers partly exserted, but shorter than the ovate corolla lobes. Planted in hedges. May, June. § Europe.

6. **CHIONÁNTHUS,** L. FRINGE TREE. Cal. short, 4-parted. Cor. tube very short, including the 2 stamens, the limb of 4 linear lobes. Style very short. Drupe fleshy, with a bony 1-seeded nut. ♄ ♄ With opposite leaves and white flowers in panicles.

C. VIRGÍNICUS L. Leaves oval to oblong; panicle with filiform branches and pedicels;

petals very narrow, drooping, 10″. A highly ornamental shrub or small tree, in woods. S. Penn., and S. April—June.

7. FORESTIÈRA, Poir. Dioecious, apetalous; buds ∞-flowered. ♂ Flowers sessile, crowded, each flower a pair of stamens surrounded by a calyx of 4 sepals. ♀ Flowers pedicellate, umbellate, no calyx, an ovary tipped with a slender style and capitate stigma, cells 2, ovules 4. Drupe 1-seeded. ♄ ♄ Leaves opposite, simple. Flowers minute.

1 **F. acuminàta** Poir. Glabrous; leaves lance-elliptic, pointed both ways, serrulate, petiolate; drupe linear-oblong, pointed. Streams, Ill. to Ga. 15f.
2 **F. ligustrìna** Poir. Some downy; leaves ovate to oblong, obtuse, attenuate to a petiole, serrulate; drupe oval-oblong. Banks, Ga., Fla.
3 **F. porulòsa** Poir. Smooth; leaves lance-oblong, obtuse, sessile, dotted and rusty beneath; drupe round-ovoid. Coast of E. Ga. and Fla.

8. FRÁXINUS, Tourn. ASH. Fls. ♂ ☿ ♀ or ♂ ♀. Cal. 4-toothed, rarely 0. Cor. of 2 or 4 oblong or linear petals, or 0. Sta. 2. Stig. bifid. Samara 2-celled, flattened, winged at apex, 4-ovuled, but 2-seeded. ♄ ♄ Leaves opposite, odd-pinnate, petiolate. Flowers racemed or panicled. Wood valuable for timber. April, May. Fig. 16.

§ Native species, all dioecious and apetalous, in woods, &c...(a)
§ European species, polygamous, planted for shade, &c..................Nos. 1, 2
  a Calyx persistent at the *terete* base of the samara........................No. 3
  a Calyx persistent at the *narrow, flattened base* of the samara..........Nos. 4—6
  a Calyx none, the samara naked at the *broad* base........................Nos. 7, 8

1 **F.** ORNUS. *Flowering Ash.* Lfts. 7—9, lanceolate, serrate above; buds pubescent; panicles dense; petals 2 or 4, linear-oblong, white; fruit lance-linear. Parks.
2 **F.** EXCÉLSIOR. *European Ash.* Leaflets 11—13, lance-oblong, serrate; racemes short, dense; fruit linear-oblong, notched at end; pet. and calyx 0. A tall tree, in parks, &c. β. PÉNDULA, the *Weeping Ash*, is one of its varieties.
3 **F. Americàna** L. *White Ash.* Leaflets 7—9, ovate, acuminate, subentire, shining; panicles loose; fruit calyculate, the seed portion terete, half as long as the oblong wing. A forest tree 40—80f. Timber excellent.
4 **F. pubéscens** Walt. *Red Ash.* Leaflets 7—9, lance-ovate, acuminate, subserrate, petioles and branchlets velvety-pubescent; fruit calyculate at the acute base, gradually widened into the oblanceolate wing. Wet woods. 30—60f.
5 **F. víridis** Mx.*f.* *Green Ash.* Lfts. 7—9, lance-ovate, serrate, long-pointed, bright green, and, with the petioles and branchlets, *glabrous;* fruit calyculate, spatulate, obtuse, the seed portion as long as the wing. Woods, W. and S. 15—25f.
6 **F. platycárpa** Mx. Leaflets 5—7, elliptical, acute, obscurely serrate, some downy, fruits broadly-spatulate, attenuate to the calyculate base, some of them (especially in β. *triptera*) with 3 angles winged! Va., and S.
7 **F. quadrangulàta** Mx. *Blue Ash.* Leaflets 7—9, short-petiolulate, lance-ovate, acuminate, sharply serrate; branchlets square or acutely 4-angled; buds velvety; fruit oblong, winged to the base. Woods, W. 60—80f.
8 **F. sambucifòlia** Lam. *Black Ash.* Leaflets 7—11, lance-ovate, *sessile,* serrulate, pointed; fruit oblong with equal ends, notched at apex. Swamps, Can. to Pa. and Ky. 40—70f. Wood used for hoops, baskets, &c.

Cohort 3. APETALÆ,

Or Monochlamydeous Exogens. Plants with no corolla, the calyx or perianth green or colored, consisting of a single series of similar organs, or often wholly wanting.

Order CII. ARISTOLOCHIACEÆ. Birthworts.

Low herbs or *climbing shrubs*, with alternate leaves and perfect flowers. *Perianth* tube adherent to the ovary, brown or dull, valvate in the bud. *Stamens* 6 to 12, epigynous and adherent to the base of the styles. *Ovary* 6-celled, becoming a 6-celled, many-seeded capsule or berry. *Seed* albuminous, embryo minute. Figs. 24, 333.

1. **ASÀRUM**, Tourn. Wild Ginger. Calyx bell-form, regular, 3-cleft. Sta. 12, placed upon the ovary, anth. adnate to the middle or summit of the filaments. Style very short, stigma 6-rayed. Fruit fleshy, 6-celled, crowned with the calyx. ♃ Acaulescent, with creeping rhizomes and 1 or 2 leaves on each branch. Flowers solitary.

§ Leaves in pairs. Calyx lobes pointed, reflexed. Ovary wholly adherent......No. 1
§ Leaves solitary. Calyx lobes obtuse, suberect. Ovary partly free........ Nos. 2, 3

1 A. **Canadénse** L. Lvs. 2, broad-reniform, on long, opposite, radical petioles with the flower between; sepals greenish-purple, pointed, reflexed; filaments extended above the anthers. Rich shades. The root is a popular remedy. May, June.
2 A. **Virginicum** L. Leaf orbicular-ovate, glabrous, coriaceous, deeply cordate, entire, obtuse; flowers subsessile; calyx short, smooth outside; segments obtuse, dull purple. Rocky soils, Va., Ky., and S. April.
3 A. **arifòlium** Mx. Leaf broadly hastate with a deep sinus; fl. 7—9″, tubular, soon urceolate, lobes short and obtuse. Rich soils, Va., and S. March—May.

2. **ARISTOLÒCHIA**, Tourn. Birthwort. Calyx tubular, tube variously bent and inflected above the ovary, limb irregular. Anth. 6, subsessile on the style. Stig. 6-lobed. Caps. 6-celled, ∞-seeded. ♃ Caulescent, with alternate leaves and lateral lurid purple flowers.

§ Stem erect. Calyx tube sigmoid (*i. e.*, twice bent like the letter S)........Nos. 1, 2
§ Stem climbing, woody. Calyx tube recurved, once bent upward. May, Jn..Nos. 3, 4

1 A. **serpentària** L. *Virginia Snake-root.* Stem flexuous; lvs. petiolate, oblong or ovate, thin, cordate, acuminate; ped. radical, many bracted; cal. tube smoothish, contracted in the midst. Thickets, Pa., S. and W. 8—13′. June, July.
  β. **hastàta.** Leaves narrowly oblong, auricled at base, short-stalked. S.
2 A. **reticulàta** N. St. very flexuous; lvs. oval, cordate-clasping, with decussating lobes, strongly reticulated; flowers radical, small (5″). La. 1f.
3 A. **Sipho** L'Her. *Dutchman's Pipe.* Lvs. glabrous, ample, round-reniform; ped. 1-flowered, with 1 clasping bract; flowers 1½′, bent like a *siphon* or tobacco-pipe, limb spreading. A vigorous climber, 30—40f, in hilly woods, Pa. to Ky., and S. †
4 A. **tomentòsa** Sims. Leaves downy or hairy beneath, round-cordate, very veiny; ped. solitary, 1-flowered, bractless; flowers 20″, tube yellowish, limb purple, reflexed. throat nearly closed. Banks, Ill., and S. 30—40f. May.

Order CIII. NYCTAGINACEÆ. Marvelworts.

Herbs (shrubs or trees) with tumid joints, entire and opposite leaves *Flowers* generally surrounded with an involucre (calyx-like when the flower is solitary). *Calyx* a delicate, colored, funnel-form or tubular perianth, deciduous above the 1-celled, 1-seeded ovary, leaving its persistent base to harden and envelop the fruit (achenium) as a kind of pericarp. *Stamens* 1 to several, definite, slender, hypogynous, exserted, unequal. *Embryo* coiled around the copious white albumen. Figs. 143, 207.

§ Involucre just like a calyx, including one flower..MIRABILIS. 1
§ Involucre 5-leaved, including many flowers in an umbel-like head..................ABRONIA. 2
§ Involucre 5-lobed, including 3—5 flowers..OXYBAPHUS. 3
§ Involucre 0.—*x* Herbs, with minute flowers in little clusters.........................BOERHAAVIA. 4
—*x* Shrubs. Flowers diœcious, cymous. S. Fla.......................................PISONIA.

1. MIRÁBILIS, L. Marvel of Peru. Four-o'clock. Involucre calyx-like, 5-lobed, 1-flowered, lobes acuminate. Perianth (calyx) tubular funnel-form, limb spreading. Sta. 5, and style more or less exserted. Fruit (as in all the genera) an achenium invested in the permanent base of the calyx. ♃ Cultivated. Leaves ovate, more or less cordate, acuminate.

1 **M. Jalàpa.** Erect, glabrous; flowers 3—6 in each terminal fascicle, short-stalked opening at about 4 o'clock P. M., and remaining in bloom all night, infinitely various in color. Peru. 2f. Summer.
2 **M. Dichótoma.** Erect, glabrous; flowers sessile, mostly yellow, smaller than in M. Jalapa; limb 6″. Mexico. 2f. Summer.
3 **M. Longiflòra.** Weak, diffuse, viscid-pubescent; lower leaves long-petioled; flowers sessile, tube 6′ long, hairy, border 1′, white. Mexico.

2. ABRÒNIA, Juss. Involucre 5-leaved, surrounding an umbel-like head of many small flowers on a long peduncle. Perianth salver-form, limb 5-lobed, corolla-like, deciduous. Sta. 5, and style included. ♃ Fleshy.

1 **A. Umbellàta.** St. prostrate; lvs. ovate, long-petioled; umbellate heads compact; fls. rosy-lilac or pink, the lobes obcordate. Sandy sea-coasts, California. 1—2f.
2 **A. Fragrans.** Stem ascending; leaves lance-ovate, long-stalked; umbels loose, fls. and involucre white, tubes near 1′. Dalles, Oregon.

3. OXÝBAPHUS, Vahl. Invol. 5-cleft, containing 3—5 fls., persistent. Perianth tube very short, limb bell-form, plicate, deciduous. Sta. 3, and style exserted. Fruit obovoid, ribbed. ♃ Flowers small, purple.

1 **O. nyctagineus** Sweet. Smoothish, erect, forked; lvs. broad-ovate to lanceolate, subcordate, acute; ped. solitary; involucre 3-5-flowered. Banks, W. June—Aug.
2 **O. angustifòlius** Sweet. Bushy, with alternate branches; lvs. lanceolate, acute both ways, subsessile, 1—2′; ped. ½—¼′, axillary; involucre cup-shaped, hispid, 3 flowered; ovary hispid. Dry soils, S. 2—3f. June—July.
3 **O. àlbidus** Sweet. Stem with strict slender branches, or simple; leaves linear oblong, petiolate, the upper often bract-like; ped. half as long (6″—1′) as the leaves, involucre hairy, 3-flowered. S. 1—2f. May.

4. BOERHAÀVIA, L. Involucre 0, bractlets deciduous. Perianth funnel- or bell-form, colored, 5-lobed, upper half deciduous, lower persistent. Sta. 1—4. Fruit 5-ribbed, truncate at apex, 1-seeded. ① Leaves petiolate. Flowers very small.

B. erécta L. Glabrous; lvs. ovate, wavy, pale beneath; clusters 3–6-flwd., distant in a strict panicle with filiform branchlets. Sands, S. 2—4f. June—Sept.

B. hirsùta, and **B. viscòsa**, grow in S. Fla., according to Dr. Chapman.

ORDER CIV. POLYGONACEÆ. SORRELWORTS.

Herbs (rarely shrubs) with alternate leaves and mostly sheathing stipules (*ochreæ*) surrounding the stem above each tumid joint. *Flowers* mostly perfect. *Perianth* (or calyx) 3–6-cleft, mostly colored, imbricated in bud and persistent. *Stamens* 4—15. *Ovary* 1-celled, free, with a single, erect ovule. *Styles* or stigmas 2 or 3. *Fruit* a 3-angled achenium enclosed in the calyx. *Seed* erect, albuminous, with a curved embryo. Figs. 147, 151–4, 286, 304, 313, 337, 521.

§ Ochreæ, or sheathing stipules, present at each joint...(b)
§ Ochreæ none.—a Flowers in involucrate umbels, 6-sepalled........................ERIOGONUM. 1
    —a Flowers in bracted racemes, 5-sepalled. Stems with tendrils........BRUNNICHIA. 2
  b Sepals 4, equal by pairs. Stamens 6. White Mountains: rare....................OXYRIA. 3
  b Sepals 6, all similar. Stamens 9. In gardens: common..........................RHEUM. 4
  b Sepals 6, the 3 inner increasing, tuberculate.................................RUMEX. 5
  b Sepals 5 (in one Polygonum 4 irregular)...(c)
    c Sepals all or the 3 inner fringed. Pedicels solitary..........................THYSANELLA. 6
    c Sepals all entire,—x open, or 3 closed on the fruit. Pedicels solitary......POLYGONELLA. 7
      —x open at base of fruit. Pedicels fascicled................FAGOPYRUM. 8
      —x closed on the angular fruit. Herbs ....POLYGONUM. 9
      —x combined with the round fruit. Trees. Fla............COCCOLOBUS.

1. ERIÓGONUM, Mx. Fls. many in each common 5-toothed involucre. Cal. deeply 5-cleft. Sta. 9, sty. 3. Ach. 3-angled or 3-lobed.—Herbs clothed with down or wool. Lvs. alternate, exstipulate, mostly at the base of the stem, the upper bract-like, often whorled at the forks of the umbel late inflorescence. Very abundant in the Pacific States. June—Aug.

1 **E. tomentòsum** Mx. Lower lvs. crowded, oblong-obovate, rusty-white beneath, the upper whorled in 3's; involucre sessile; calyx colored. ♃ Dry soils, S. 2—3f.

2 **E. longifòlium** N. Lower lvs. crowded, oblong-linear, white beneath, the upper scattered; involucre pedunculate; calyx green, woolly. Fla., and W. 2—4f.

2. BRUNNÍCHIA, Banks. Calyx colored, 5-parted, lobes oblong, at length increased and closed on the obscurely 3-angled achenium. Fil. 8, capillary, styles 3, slender, stigmas entire. ♄ Tendrils from the ends of the branches. Flowers racemed, greenish.

B. cirrhòsa Banks.—A smooth, shrubby vine, 10—20f, on river banks, Car. to Fla., and W. Leaves cordate to ovate, entire. Sheaths obsolete. May.

3. OXÝRIA, R. Br. MOUNTAIN SORREL. Cal. herbaceous, 4-sepalled, the 2 inner sepals erect, larger, the 2 outer reflexed. Ach. lens-shaped, thin, girt with a broad, membranous wing. Sta. 6, equal. Stig. 2, sessile, penicillate. ♃ Low, nearly acaulescent, alpine plants.

O. renifòrmis Hook (or digyna Camp.) Root leaves on long stalks, reniform; outer sepals ⅓ as long as the inner; fruit orbicular. White Mountains, and N. 3—4'. June.

ORDER 104.—POLYGONACEÆ.

4. RHEUM, L. RHUBARB. Calyx colored, 6-sepalled, persistent. Sta 9. Sty. 3, very short, spreading, stig. multifid, reflexed. Ach. 3-angled, the angles margined. ♃ Flowers fasciculate in racemous panicles.

R. RHAPÓNTICUM L. *Pie-plant.* Leaves smooth, cordate-ovate, very large (1—2f), the petioles juicy and pleasantly acid, of equal length; stems hollow, 3—4f, panicles bursting from large white bracts. Siberia.

5. RUMEX, L. DOCK. SORREL. Calyx of 6 sepals nearly distinct, the 3 inner (valves) larger, petaloid, connivent over the achenium, 1 or more of them usually bearing a tubercle or grain on the back, the 3 outer green. Sta. 6. Styles 3, short, stigmas penicillate-fringed. Ach. and seed 3-angled, embryo lateral.—Weed-like herbs with small, greenish flowers often whorled, in racemes or panicles. May—July. (See *Addenda*.)

§ Docks. Flowers all or mostly perfect. Valves bearing grains on the back...(*)
§ Sorrels. Flowers diœcious. Valves grainless. Leaves acid (hastate)....Nos. 11, 12
* Valves entire, or merely angular...(a)
* Valves conspicuously toothed on each side near the base.........Nos, 8—10
   a Pedicels in fruit 2—5 times longer than the *subcordate* valves......Nos. 1—3
   a Pedicels in fruit shorter or not longer than the valves...(b)
      b Leaves flat, all tapering to both ends............Nos. 4, 5
      b Leaves wavy, the lower cordate or subcordate............Nos. 6, 7

1 **R. crispus** L. *Yellow D.* Root fusiform, yellow; lvs. lanceolate, wavy, acute, the lower oblong, subcordate; ped. twice longer than calyx; valves broad-ovate, cordate, each bearing a grain; rac. long, some leafy. ♃ Fields. 2—3f. § Europe.

2 **R. verticillàtus** L. *Water D.* Leaves acute at each end, lance-oblong; rac. leafless, dense; ped. 7—9″ long, deflexed; valves broad-ovate, each bearing a large grain. ♃ In muddy places. 2f. Whorls 10-30-flowered.

3 **R. Hydrolápathum** Huds. *Great Water D. β. orbiculàtus.* Tall (3—5f); lvs. lance-obl., acute both ways, crose-crenulate, the lower very long; pan. naked. dense ped. 5—6″; valves round-ovate, obtuse, all grain-bearing. ♃ Pools, M. and N.
γ. **Floridànus.** Valves deltoid-ovate, obtusely-pointed. Fla.

4 **R. altíssimus** Wood. *Peach-leaved D.* Tall (3—6f); leaves entire, lance-elliptical, acute both ways; rac. leafless, panicled, slender; valves broadly subcordate, one of them grain-bearing, one obscurely so, and one naked. ♃ Wet, M. and W. (R. Britannicus Meisn. nec Linn. who says "valves all grain-bearing.")

5 **R. salicifòlius** Weinm. *Pale D.* Lvs. lin.-lanceolate, attenuate-acute both ways; pan. leafy at base; ped. very short; valves all grain-bearing. ♃ Coast, N-E. 3f.

6 **R. conglomeràtus** Murr. Lvs. oblong to lanceolate, lower subcordate; whorls mostly axillary; valves oblong-ovate, all grain-bearing. ♃ Wet. N. 2—3f. §

7 **R. sanguineus** L. Lvs. as in No. 6, mostly with red veins; pan. leafy at base, whorls distant; valves oblong-obovate, one or two grain-bearing. ♃ Fields. §

8 **R. obtusifòlius** L. Lower leaves ovate-cordate, obtuse, upper narrow, acute; panicle leafy, whorls distant; valves hastate-ovate, one chiefly grain-bearing, all with some bristle-shaped lateral teeth. ♃ Fields, &c. 2—3f. § Europe.

9 **R. maritimus** L. *Golden D.* Low (1f); leaves lance-linear, the lowest cordate, wavy; whorls crowded; valves rhomb-ovate, pointed, each with 4 lateral awns and a large grain, yellowish. (①) Brackish waters, Mass. to Car.

10 **R. pulcher** L. Lower lvs. cordate, some fiddle-shaped, upper lanceolate; whorls distant, leafy; valves strongly toothed, unequally grain-bearing. S. §

11 **R. Acetosélla** L. *Sheep Sorrel.* Leaves oblanceolate, the base lobes conspicuous; valves not increasing in fruit. A common weed. 6′- 1f

12 **R. hastulàtus** Baldw. Leaves with small auricles or none, glaucous; valves increasing to round-cordate in fruit; ped. jointed. Mo. to Ga., rare.

Order 104.—POLYGONACEÆ.

6. THYSANÉLLA, Gray. Fls. ♂ ⚥ ♀. Cal. colored, 5-parted, lobes all erect, the 2 outer cordate, the 3 inner smaller, pectinate-fringed. Sta. 8. Styles 3. Achenia 3-angled, acuminate.—A smooth, erect herb, with the habit of Polygonella. (Polygonum, Ell.)

T. **fimbriàta** Gr.—Pine-barrens, Ga., Fla. Stem branched, 2—3f. Sheaths bristle-fringed. Lvs. linear, 1—2′. Fls. rose-white, in crowded, panicled spikes. July—Oct.

7. POLYGONÉLLA, Mx. Calyx colored, 5-sepalled, persistent. Sta. 8, included. Styles 3 or almost 0. Ach. 3-cornered, naked or enclosed in the 3 inner sepals enlarged and become scarious valves. Embryo straight. —Herbs or delicate shrubs, with very narrow leaves and the small flowers solitary in each ochrea.

§ Fls. diœcious. Pedicel 1″. Filaments all filiform. Stig. nearly sessile...Nos. 1—3
§ Fls. all ⚥. Pedicel 2″. The 3 inner filaments dilated. Styles manifest...Nos. 4, 5

1 P. **parvifòlia** Mx. Shrubby, branches strict, leafless above; lvs. linear-cuneate; panicle oblong; inner sepals equalling the acute achenia. S. 1—2f.
2 P. **grácile** N. Annual, glaucous; branches filiform; leaves spatulate; 3 inner sepals exceeding the pointed achenia. Dry sands, S. 2—3f.
3 P. **Croomia** Chapm. Shrubby; branches slender; leaves linear (2—3″); 3 valves unequal, 2 roundish, 1 oblong, exceeding the achenia. Uplands, S.
4 P. **Meisneriàna** Shutt. Shrubby, very leafy, leaves linear, filiform, 6—10″, evergreen, ochrea tipped with a white membrane; 2 outer sepals reflexed. Uplands, Ga., Ala., Fla. 1—2f. A delicate bushy shrub.
5 P. **articulàta** Meisn. Annual, strict, with erect branches, which are soon nearly naked; leaves linear, caducous from the tops of the truncate sheaths; sepals flesh-colored, expanding. Dry. N. J., and W.: rare.

8. POLÝGONUM, L. KNOT-GRASS. Calyx of 5 sepals, rarely fewer, colored or greenish, similar, imbricated in bud, at length all connivent, persistent. Sta. 8, rarely fewer. Sty. 2 or 3, mostly 3, short filiform. Ach. 3-cornered or lens-shaped, enclosed in the dry, withered calyx. Embryo curved, lateral, lying in a groove at one angle of the albumen. Herbs with ochreate-jointed stems and small, white, red, or greenish fls. June—Sept.

§ Stems armed with retrorse prickles. Lvs. cordate-sagit. ECHINOCAULON..Nos. 21, 22
§ Stems unarmed, twining. Leaves cordate-hastate. TINIARIA............Nos. 18—20
§ Stems erect or decumbent, unarmed. Leaves hardly ever cordate...(*)
* Calyx unequally 4-cleft. Styles 2, long deflexed. TOVARIA................No. 17
* Calyx equally 5-parted. Styles erect...(a)
  a Sheaths salver-form. Stamens 7. Style 2-parted. Tall. AMBLYOGONUM...No. 16
  a Sheaths subcylindrical. Stamens 5, 6, 8. Styles 2 or 3...(b)
    b Flowers in leafless, terminal, spike-like racemes. PERSICARIA...(c)
    b Flowers axillary, or seldom forming a leafy raceme...(e)
      c Raceme 1, dense. Stem at base or rhizome decumbent....Nos. 14, 15
      c Racemes several. Sheaths naked, not fringed..............Nos. 12, 13
      c Racemes several. Sheaths bristly, fringe-ciliate...(d)
        d Style 2-(or 3-)cleft. Achenia flat or lens-shaped.... ....Nos. 9—11
        d Style 3-cleft. Achenia sharply 3-cornered...................Nos. 5—8
          e Achenium protruding beyond the calyx, 3-angled ....Nos. 3, 4
          e Achenium included in the calyx, 3-angled....................Nos. 1, 2

1 P. **aviculàre** L. *Bird's K. Doorweed.* Procumbent, diffuse; leaves lance-ellip-

ORDER 104.—POLYGONACEÆ.

tic, acutish, 1'; flowers 2 or 3 together, subsessile, reddish; achenia striate, dull, enclosed; stamens 5—8. ① A common weed, 6—16'. In rich shady soils it arises to
β. **eréctum**, with larger oval leaves and pedicel ate flowers.

2 P. ténue Mx. Slender, rigid, erect, with long simple-angular branches; lvs. linear, erect; sheaths bristle-fringed; flowers solitary; achenia shining. Dry. ↓—1f.

3 P. marítimum L. Prostrate, diffuse, glaucous, with very short joints and swelling torn sheaths; lvs. fleshy, oblong, 1—6''; fls. sessile, at length spicate; fruit little exserted, smooth and shining. ① Sandy coasts, Mass. to Ga. ↓—1f.

4 P. ramosíssimum Mx. Erect or ascending, *much branched*, striate; lvs. linear-oblong, 1—2'; flowers greenish, pedicellate; fruit ↓ exserted, olive-green, shining, 1¼''. ① Sandy shores, R. I. to Mich. and Md. 2—3f.

5 P. hirsùtum Walt. Densely hirsute with spreading tawny hairs, erect; lvs. lanceolate; sheaths fringed; flowers white, in 2 or 3 slender spikes. ② S. 2—3f.

6 P. hydropiperoides Mx. *Mild Water-pepper*. Stem smooth, slender, sheaths long, close, fringed and hispid; lvs. linear-lanceolate, not acrid; spikes erect, slender, loose at base; calyx glandless, achenia shining.
β. **setàcea** the leaves and stem above are more or less hispid. ⚄ Wet. 1—3f.

7 P. acre H. B. K. *Water Smartweed*. Glabrous, virgate, slender; sheath loose, bristle-fringed; lvs. lanceolate, *acrid;* spikes filiform, erect; flowers reddish-green, dotted like the leaves; fruit shining. ① Wet places. S. and W. 2—5f.

8 P. tinctòrium. *Madder*. Lvs. oval; spikes oblong, dense, roseate. China. 1—2f.

9 P. Hydropìper L. *Water Pepper*. Glabrous; sheaths bristly-ciliate; lvs. lanceolate, very acrid, finely punctate; spikes nodding, loose, slender, greenish; calyx punctate; stamens mostly 6; achenia roughened, black. ① Damp. 1—2f. §

10 P. Càreyi Olney. Stem erect, 3—5f, bristly and much branched; leaves lanceolate, some hispid; stipules tubular-truncate, ciliate; spikes dense, purplish, nodding on long hairy peduncles. ① Swamps, N. Eng. to Penn. (See p. 447.)

11 P. Persicària L. *Smart-weed*. Glabrous, erect; leaves lanceolate, usually marked with a brown spot; sheaths fringed; spikes dense, erect, oblong; stamens 6; style 2-cleft; achenia shining. ① Waste grounds: common. 1—2f. §

12 P. Pennsylvánicum L. Branches above and pedicels glandular-hispid; leaves lanceolate; spikes erect, oblong, crowded, rose-colored, showy; achenia lens-shaped, with flat sides. ① Margins of waters. 2—4f.
β. **densiflòrum.** Smooth; racemes slender; achenia truly lens-shaped. South.

13 P. incarnàtum Ell. Smoothish; leaves lanceolate; branches and ped. glandular-dotted; spikes linear, nodding, becoming long; achenia lens-shaped, with concave sides. ① Ditches and pools, W. and S. 2—3f.

14 P. amphíbium L. Stem prostrate and rooting below, ascending; leaves thick, smooth, lance-oblong, variable; spikes oblong, ovoid or dense; stamens 5; style 2-cleft. Pools and swamps. 3—1f. Spike 1' or more.
β. **terrestre.** Plant more or less hirsute; spikes elongated.

15 P. vivíparum L. Low, simple, erect from a creeping rhizome; leaves lance-linear, with rolled edges; spike 1, linear. ⚄ White Mountains, and N.

16 P. orientàle L. *Prince's Feather*. Tall, erect, branched; leaves large, with hairy salver-form sheaths; stamens 7; styles 2; spikes large, red, nodding, showy. ① Fields and gardens. 3—8f. §

17 P. Virginiànum L. Stem simple; leaves lance-ovate, acuminate; flowers remote, 1 from each sheath, in a slender raceme, greenish. ⚄ Shades. 3 1f.

18 P. convólvulus L. *Knot Bindweed*. Prostrate or climbing, roughish, sheaths naked; leaves hastate, pointed; flowers in axillary fascicles or in interrupted racemes; fruit exserted, dull, blackish. ① Fields. 2—1f. §

19 P. cilinode Mx. Climbing; sheaths *ciliate* at base; leaves deeply cordate, pointed; racemes paniculate, loose; achenia shining. Hedges. 3—8f.

20 P. dumetòrum L. *Hedge Bindweed*. Climbing high; joints not ciliate; leaves cordate-hastate, with acute lobes ; outer sepal keeled and winged on the back ; fruit smooth, black. Thickets. 3—12f. §.—A native form,

β. **scandens,** has the raceme panicled and the sepals with very broad wings.

21 P. sagittàtum L. *Scratch-grass.* Climbing, 3—5f, rough backwards ; leaves lance-sagittate ; flowers in small heads, whitish ; stamens 8 ; style 3. ① Wet.

22 P. arifòlium L. Rough with reversed prickles, 3—5f ; leaves hastate, apex and lobes pointed ; flowers racemed ; stamens 6 ; styles 2. Wet.

9. FAGOPÝRUM, Tourn. BUCKWHEAT.

Calyx colored, equally 5-parted, persistent, unchanged. Stamens 8, alternate with 8 honey-glands. Styles 3, with capitate stigmas. Ach. 3-angled, much exceeding the calyx. ① Leaves cordate-hastate. Flowers rose-white, in panicled racemes.

1 F. esculéntum Mœnch. Smoothish ; leaves with obtuse lobes ; flowers showy, numerous, sought by bees ; achenia ovoid-triangular, wingless, black. Fields. 2—4f. §

2 F. TARTÁRICUM. *India Wheat.* Glabrous ; leaves broader than long, lobes acutish ; racemes axillary and terminal, scarcely panicled ; achenia lance-triangular, angles sinuate-dentate, rather obtuse ; calyx minute. Tartary. Cultivated.

ORDER CV. PHYTOLACCACEÆ. POKEWORTS

Herbs with alternate, entire leaves and perfect, 5-parted flowers. *Calyx* free. *Stamens* 5—30, alternate with the sepals when of the same number. *Ovary* of 1 to several carpels, each 1-ovuled. *Styles* and *stigmas* as many as carpels. *Fruit* baccate or acheniate. *Seeds* erect, with the embryo coiled around the albumen.

§ Styles and carpels 5—12. Fruit baccate. Leaves exstipulate...................PHYTOLACCA. 1
§§ Style and carpel 1. Leaves with stipules.—*a* Berry globular, smooth...........RIVINA. 2
—*a* Achenium with 2 hooks..................PETIVERIA. 3

1. PHYTOLÁCCA, Tourn. POKE. GARGET-WEED. Calyx 5-parted. Stamens 5—25. Styles 5—12. Berry depressed-globular, with as many seeds as styles.—Herbaceous. Racemes terminal, soon opposite the leaves.

P. decándra L. Stem stout, purplish, tall ; leaves ovate ; flowers with 10 stamens and 10 styles ; berries black, full of crimson juice. Hedges. 5—8f. July +.

2. RIVÌNA, Plum. Calyx 4-parted, 3-bracted. Sta. 4 or 8. Berry at last dry, 1-seeded, embryo a vertical ring. Shrubby, with racemes terminal, soon lateral.

R. lævis L. Branching, smooth, 6—8f ; lvs. ovate ; fls. rose-white, in long racemes ; stamens 4. Fla., and W. Herbage bright-green.

3. PETIVÈRIA ALLIÀCEA L. Half-shrubby, 2—3f, with obovate-obtuse leaves and spicate flowers. Grows in S. Car. (Michaux), and S. to the tropics.

ORDER CVI. CHENOPODIACEÆ. CHENOPODS OR GOOSE-FOOTS.

Herbs chiefly weed-like and homely, more or less fleshy, with alternate exstipulate *leaves*. *Bracts* not scarious. *Flowers* greenish, regular. *Calyx* imbricated in bud. *Stamens* as many as, and opposite to the calyx lobes, or fewer. *Ovary* 2-styled, 1-celled, becoming a 1-seeded, thin utricle, or caryopsis. *Embryo* coiled or spiral.

ORDER 106.—CHENOPODIACEÆ. 285

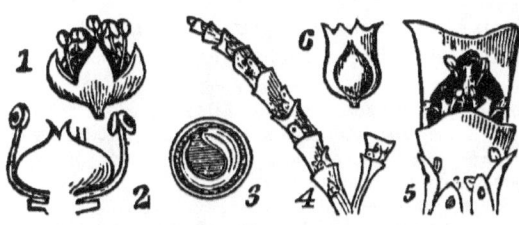

1. Flower of Chenopodium album. 2. Calyx, &c., removed, showing the ovary and 2 stamens. 3. Seed cut across, showing the coiled embryo. 4. Branch of Salicornia herbacea. 5. Two joints magnified. 6. Ovary of a flower.

§ Leaves flat, neither fleshy nor spiny. Embryo a ring around copious albumen...(a)
§ Leaves none, or linear and fleshy or spinescent. Embryo a spiral or folio. Albumen 0...(y)
   a Stems twining and climbing. Flowers white, in racemes............. .......BOUSSINGAULTIA. 1
   a Stems erect. Flowers greenish, all similar and perfect...(b)
   a Stems erect. Flowers greenish, of two sorts, monœcious or diœcious...(x)
      b Seed horizontal.—c Pericarp rough and corky. Calyx ribbed............BETA. 2
          —c Pericarp thin, in a calyx bordered all around........CYCLOLOMA. 3
          —c Pericarp thin, in a plain, unbordered calyx..........CHENOPODIUM. 4
      h Seed vertical.—d Fruit smooth, sepals distinct, mostly fleshy.............BLITUM. 6
          —d Fruit in a veiny, wrinkled calyx. Leaves pinnatifid....ROUBIEVA. 5
          —d Fruit axillary to a bract, no calyx. Leaves linear......CORISPERMUM. 10
   x Fruit enclosed in a hardened calyx without bracts. Cultivated..........SPINACIA. 9
   x Fruit naked (no calyx) between two bracts. Leaves oval or triangular...ATRIPLEX. 7
   y Embryo folded, not spiral. Stems jointed, leafless..................SALICORNIA. 11
   y Embryo a flat spiral, horizontal. Leaves acute................CHENOPODINA. 12
   y Embryo a conic spiral. Sepals appendaged. Leaves spinescent......SALSOLA. 13

1. **BOUSSINGAÚLTIA,** Kunth. MEXICAN VINE. Cal. corolla-like, open, 5- or 6-parted, with several imbricated bracts. Stig. 3, club-shaped. Pericarp thin. ♭ Twining to the right. Leaves thick, petiolate. Flowers in many spike-like racemes. S. America.

   B. BASELLOÌDES. Leaves broadly cordate-ovate, acuminate. ♃ Arbors. 15f.

2. **BETA,** Tourn. BEET. Cal. 5-cleft, persistent. Sta. 5. Ovary half-adherent. Stig. 2. Utricle depressed, corky, enclosed in and consolidated with the ribbed calyx.—Herbs with fleshy roots, furrowed stems, alternate leaves, and greenish, spicate flowers.

   B. **vulgàris.** Glabrous; leaves large, wavy, acute both ways; spikes in a large panicle the second year. ⊕ S. Eur. Cultivated for its root, which is commonly red.
      β. *Cicla.* Scarcity. Leaves roughish ; root slender, whitish ; flowers in 3's.
      γ. *Rapa.* Turnip Beet. Root napiform, white or red, very sweet.
      δ. *Mangel-wurtzel.* Root very large, mostly white. Cultivated for stock.

3. **CYCLOLÒMA,** Moquin. Calyx 5-cleft, lobes strongly keeled, at length appendaged outside with a circular membranous border or crown. Sta. 5, styles 3. Utricle depressed, enclosed. ① With furrowed stems, alternate lobed leaves, and small sessile flowers.

   C. **platyphýllum** Moq.—Banks of the Mississippi, Ill., and W. 1—1½f, white-downy above. Leaves lance-oblong, sinuate-toothed or lobed, 2'. Flowers at length in small panicles. July—Sept.

4. **CHENOPÒDIUM,** Tourn. PIGWEED. GOOSEFOOT. Calyx bractless, 5-cleft, lobes often keeled, never appendaged, more or less enclosing the fruit. Sta. 5, styles 2. Utricle depressed, membranous, seed mostly

286   Order 106.—CHENOPODIACEÆ.

horizontal, lenticular. Weeds often glaucous or glandular, with alternate, often rhombic lvs., and the minute fls. in panicled spikes. June—Aug.

§ Plants ill-scented, smooth, never glandular. Embryo a complete ring...(*)
§ Plants glandular-puberulent, green, aromatic. Embryo a half ring...(b)
    * Herbage glaucous or whitish, covered with mealiness..................Nos. 1—3
    * Herbage green, rarely purplish, not glaucous or mealy...(a)
       a Leaves entire, ovate-oblong, on slender petioles ....No. 4
       a Leaves toothed or lobed, petiolate.............................Nos. 5—7
         b Flowers glomerate, axillary, in spike-like racemes............Nos. 8, 9
         b Flowers cymous, innumerable, in long raceme-like panicles......No. 10

1 C. glaùcum L. Prostrate or ascending, branched; leaves ovate to oblong, obtuse, sinuate-angled or -dentate; racemes simple; seed partly enclosed. ① Mass. to Pa.: rare. 1f. Leaves 1—2', whitish beneath. § Europe.

2 C. album L. *Common P.* Erect, loosely branched, striate; lvs. rhombic ovate, sinuate-toothed to subentire; racemes some panicled; seed wholly enclosed. ① The commonest of weeds, 2—7f, often striped with purple.

3 C. Boscianum Moq. Erect, branched; lvs. small, lance-linear, entire, canescent beneath; seed partly enclosed. ① Shades, Pa. (Prof. Porter), and S. 2f.

4 C. polyspérmum L. Ascending, branched from base; lvs. ovate to oblong, entire, bright green; racemes spike-like, strict; fruit partly enclosed. Rare. § Eur.

5 C. hýbridum L. Leaves ample, subcordate, deeply sinuate-angled, with pointed lobes; racemes leafless; seed rugous, dull. ① Common, 2—4f. § Europe.

6 C. muràle L. Ascending; leaves ovate-rhombic, acute at base, unequally and acutely toothed; seed acute-edged, dull-rugous. ① Rare. 12—18'. §

7 C. úrbicum L. Erect; leaves as in No. 6, but slightly mealy; racemes strict, dense, in an erect narrow panicle; seed blunt-edged, shining. ① 2—4f. §

8 C. ambrosioides L. *Mexican Tea.* Branched; leaves oblong to lance-linear, attenuate both ways, sinuate-toothed to entire; spikes dense, leafy, seed shining, obtuse-edged; fruit wholly enclosed. ① 1—2f. § Mexico.

9 C. anthelmínticum L. *Worm-seed.* Subsimple; leaves ovate-oblong, deeply sinuate-serrate or pinnatifid; racemes spike-like, long; styles mostly 3; fruit as in No. 8. ♃ Waste grounds. 1—3f. § Mexico.

10 C. Botrys L. *Oak-of-Jerusalem.* Leaves oblong, obtuse, sinuate-subpinnatifid; branches strict, panicles slender, spirally twisted. ① Sands, &c. 1—2f. Plants strongly aromatic of turpentine.

5. **ROUBIÈVA**, Moq. Calyx 5-toothed. Sta. 5. Styles and stig. 3. Seed lens-shaped, quite vertical, enclosed in the veiny rugous calyx. ♃ Pubescent, much branched. Leaves pinnatifid.

R. multífida Moq.—Roadsides about New York. Prostrate and ascending. 1—2f. Flowers minute, in numerous panicled racemes. Leaves 1'. § S. America.

6. **BLITUM**, Tourn. BLITE. Calyx 3–5-sepalled, mostly becoming juicy and berry-like in fruit, enclosing the utricle. Sta. 1—5. Styles 2. ① Leaves petiolate. Flowers glomerate.

§ Heads forming a dense terminal spike. Calyx dry.....................No. 1
§ Heads axillary, some spicate above. Calyx thickened..................Nos. 2, 3

1 B. Bonus-Henrícus Reich. *Good King Henry.* Subsimple, ascending, mealy; leaves triangular-hastate; stamens 5. Waysides, N.: rare. § Europe.

2 B. maritimum N. Much branched; leaves lanceolate, attenuate to both ends; stamen 1; seed shining. Marshes, R. I., N. Y., and N. J. 1—2f. August.

3 B. capitàtum L. *Strawberry B.* Branched; leaves triangular-hastate glomerate fruit reddened like strawberries, insipid. Va., and N. 1—2f. June.

7. ÁTRIPLEX, Gært. ♂ Bractless. Calyx 3–5-sepalled. Sta. 3—5. ♀ Ovary 2-styled, with no stam., enclosed between 2 leaf-like bracts, with or without a calyx.—Herbs or shrubs, often mealy or scurfy, with opposite or alternate hastate leaves and glomerate-spiked green flowers.

§ Leaves partly opposite. Bracts triangular-ovate..........................Nos. 1, 2
§ Leaves all alternate.—*x* Bracts rhombic, canescent, toothed ....No. 3
—*x* Bracts round-ovate or -cuneate....................Nos. 4, 5

1 **A. hastàta** L. Slender, weak, green; leaves petiolate, hastate, remotely-toothed; flowers single in the slender spikes, bracts triangular-ovate, denticulate. N. Eng. to S. Car., coastward. 1—3f.—β. *Purshiàna* is scurfy.

2 **A. littoràlis** L. Erect with many strict branches; leaves short-stalked, lanceolate to linear, subentire; flowers glomerate, forming interrupted spikes; bracts hispid, triangular-hastate, denticulate. Lake shores, N-W.

3 **A. ròsea** L. Canescent, ascending, branched; lvs. ovate to oblong, sinuate-toothed; glomerules axillary, bracts rhombic, toothed. Albany, N.Y. (Prof. Porter). 2f. § Eur.

4 **A. HORTÉNSIS.** *Garden Orache.* Erect, branched; lvs. triangular-hastate or -oblong, subentire, bright green; bracts roundish, entire. Asia.

5 **A. arenària** N. *Sand Orache.* Mealy-canescent, branched; leaves oval to oblong, entire, short-petioled; bracts broad-cuneate, united, denticulate. ① Sea-beaches, Mass. to Fla. (Obione, C-B.) 6—12′. July—Sept.

9. SPINÀCIA, Tourn. SPINAGE. Diœcious, bractless. ♂ Cal. 3–5-sepalled. Sta. 4 or 5, exserted. ♀ Calyx tubular, 2- or 4-toothed, soon hardening and enclosing the compressed achenium. Styles 4. ① Leaves petiolate. Flowers green, axillary. June, July.

S. OLERÀCEA. Leaves hastate-lanceolate to arrow-shaped; fruit-calyx solitary, 3-angled, armed with 2 or 4 slender prickles, or unarmed. ① Gardens. 1—2f.

10. CORISPÉRMUM, Juss. Calyx 1–2-sepalled or 0. Stam. 1—5. Styles 2, short. Pericarp oval, flat and thin, adnate to the seed, vertical. ① With narrow, sessile leaves, and sessile, solitary, axillary flowers.

C. hyssopifòlium L. Hairy or glabrous, much branched; flowers in many bracted spikes, bracts ovate, subulate-pointed; leaves 1′ and less; fruit a pellucid disk Sandy lake-shores, Buffalo, and W. §

11. SALICÓRNIA, Tourn. SALTWORT. SAMPHIRE. Flowers 2 or 3 together, sunk in the cavities of the jointed stem. Calyx bladder-like, denticulate, enclosing the compressed vertical fruit. Stamens 1 or 2. Styles 2. Embryo folded.—Seaside, jointed, fleshy herbs almost leafless, with opposite branches.

1 **S. herbàcea** L. Suberect; spikes elongated, green; joints truncate and bractless; middle flower largest. ① Salt marshes. 8—12′. August.

2 **S. Virgínica** L. Erect; spikes short, soon red; joints short, tipped with 2 acute bracts: flowers all alike. ♃ Salt marshes. 6—9′. Sept. (S. mucronata C-B.)

3 **S. fruticòsa** L. Prostrate, with ascending branches; spikes slender, joints tipped with 2 obtuse bracts. ♃ Sandy beaches. (S. ambigua C-B.)

12. CHENOPODÌNA, Moq. GLASSWORT. Calyx bracteolate, cup-shaped, 5-parted, fleshy in fruit with the seed horizontal. Sta. 5. Stigma sessile. Embryo a flat spiral.—Smooth seaside fleshy plants, with alternate sessile leaves and axillary flowers. (Suæda, Forsk.)

C. maritima Moq. Diffusely branched; lvs. linear, 2' and less, sem¿terete; flowers minute, green, clustered, sessile; seed black, shining. ① Marshes. August.

13. SÁLSOLA, Gært. SALTWORT. Fls. ☿ sessile. Sep. 5, transversely-winged on the back. Wings enlarged and scarious in fruit. Sta. 5. Styles 2. Utricle depressed, horizontal. Embryo cochleate.—Seaside fleshy plants, with terete leaves and axillary, whitish flowers.

S. Kali L. Branches diffuse on the sand, rigid, with crowded subulate leaves, each tipped with a spine; flowers solitary, wings purplish; seed with a thin testa and green embryo coiled like a snail-shell. ①

ORDER CVII. AMARANTACEÆ. AMARANTHS.

Herbs similar to the last Order, but with an imbricated involucre of 3 dry, scarious bracts added to the flowers. *Sepals* 3—5 (rarely but 1), persistent and often colored, unchanged in fruit. *Stamens* 3—5. *Ovary* compressed, 1-celled, 1–∞-ovuled. *Style* 1. *Fruit* a utricle, caryopsis or berry. *Seed* vertical, albuminous. *Embryo* annular.

§ Anthers 2-celled. Ovary with many ovules. Cultivated............................CELOSIA. 1
§ Anthers 2-celled. Ovary 1-ovuled. Leaves alternate...(*)
§ Anthers 1-celled. Ovary 1-ovuled. Leaves opposite...(α)
   * Flowers monœcious or polygamous, all with a calyx and stamens................AMARANTUS. 2
   * Flowers diœcious, the pistillate with neither calyx nor stamens.................ACNIDA. 3
     α Sterile stamens none.—(Flowers white, paniculate).....................IRESINE. 4
     α Sterile stamens none.—(Flowers crimson, &c. Capitate. Cultivated)......GOMPHRENA. 5
     α Sterile stamens 5, the 5 fertile in a tube.—*x* Heads axillary...............TELANTHERA. 6
                       —*x* Spikes terminal and axillary....FRŒLICHIA. 7

1. CELÒSIA, L. COCKSCOMB. Fls. perfect, 3-bracted. Calyx of 5 sepals. Sta. 5, anth. 2-celled. Stig. 2 or 3, recurved. Utricle circumscissile, many-seeded, more or less enclosed in the calyx.—Herbs or shrubs, smooth, erect, with alternate leaves and brilliant, scarious flowers.

1 C. CRISTÀTA. Leaves lance-ovate; spikes ovoid-pyramidal, varying in cultivation to fantastic shapes, crimson or even white. ① E. India. 2—4f.

2. AMARÁNTUS, Tourn. AMARANTH. Fls. ♂ ☿ ♀ or ♂, 3-bracted. Cal. of 5 or 3 sepals. Stamens 3—5, rarely 2, anth. 2-celled. Stig. 2 or 3. Fruit a 1-seeded utricle, circumscissile, or tearing, or not opening. ① Coarse weeds, with alternate petioled lvs. and minute fls. in clusters. Aug.

§ AMARÁNTUS *proper*. Utricle thin, regularly circumscissile. Not spiny...(a)
§ EUXÒLUS. Utricle somewhat fleshy, indehiscent, or tearing open...(c)
  a Flowers 5-parted, in long panicled spikes,—b crimson-tinged.........Nos. 1, 2
                     —b green..............Nos. 3, 4
  a Flowers 3-parted, in separate, axillary, dense glomerules... ... .Nos. 5, 6
    c Spines 2 in each axil. Bracts not longer than the 5 sepals.. ....No. 7
    c Spines none.—*x* Bracts longer than the 3–5-sepalled calyx.........Nos. 8, 9
           —*x* Bracts shorter than the 5-sepalled calyx........Nos. 10, 11

1 **A. hypochondrìacus** L. *Prince's Feather*. Smoothish; leaves lance-oblong, on long stalks. some reddened; spikes very obtuse, the terminal one much the largest; flowers deep purple. Fields and gardens. 3—6f. § Mexico.

2 **A. paniculàtus** Mcq. *Prince's F.* Pubescent, pale-green; leaves lance-ovate

ORDER 107.—AMARANTACEÆ.

spikes slender, acutish, crowded, all nearly equal, reddish-green, or in β. *sanguíneus*, crimson; bracts short-awned. Fields and gardens. 2—3f. § Mexico.

3 A. retrofléxus L. Pubescent, erect, stout; leaves ovate or subrhombic, obtuse-pointed; panicle of thick, crowded, dense spikes; bracts awned, longer than calyx. A common weed in gardens and fields. 2—4f. Plant green or glaucous. §

4 A. hýbridus L. Erect, glabrous, green; leaves ovate, bright green; panicle loose; spikes terete, obtuse; calyx shorter than the awned bracts. § Mexico.

5 A. albus L. *White Pigweed.* Whitish, diffusely spreading; leaves long-petioled, rhomb-ovate, very obtuse; glomerules remote, in pairs, 4- or 5-flowered : common. §

6 A. melanchólicus. *Love-lies-bleeding.* Erect, usually dark-purple; leaves lance-oblong, obtuse, emarginate; glomerules dark-purple. Asia. 2—4f.
   β. *tricolor.* Leaves variegated with purple, green, and yellow.

7 A. spinòsus L. Much branched; leaves rhomb-ovate, obtuse, with 2 spines in each axil; spikes panicled, erect, acute; bracts equalling the sepals; utricle falling without opening. Waysides, Penn. to Fla., and W. §

8 A. lívidus Moq. Erect, smooth, livid-purplish; lvs. elliptic, obtuse, emarginate; spikes slender, rigid, acute; sepals thrice longer than bracts; fruit rugous. §

9 A. defléxus L. Ascending, ashy-green, branches deflexed; leaves rhomb-lanceolate, obtuse; spikes thick, obtuse; sepals longer than bracts; fruit smooth. §

10 A. víridis L. Erect; livid-purple; leaves long-petioled, ovate; spikes panicled, rather loose and long; sepals twice longer than the bracts. Waste grounds, S.

11 A. pùmilus Raf. Diffuse or prostrate; leaves subsessile, obovate; flowers in axillary, sessile glomerules; fruit twice longer than the calyx. Sandy sea-coasts.

3. ACNÌDA, L. WATER HEMP. Fls. ♂ ♀, 3-bracted. ♂ Calyx of 5 equal, erect sepals. Stamens 5, anth. 2-celled. ♀ Cal. 0. Ovary 1-ovuled, with 3—5 stig. Utricle 1-seeded, naked. ① Glabrous, tall, branched, with long-stalked, entire leaves and fls. small, green, in slender spikes. Jl.—Oct.

  § ACNÌDA *proper*. Utricle indehiscent, longer than its stigmas..................No. 1
  § MONTÈLIA. Utricle circumscissile, shorter than its stigmas..................No. 2

1 A. cannabìna L. Leaves lanceolate to linear, *pointed*, 2—8'; ♂ spikes numerous, rather dense, 2—4'; ♀ spikes interrupted; panicle leafy; fr. 1¼', obovoid, bracts ½ as long. Salt marshes. 3—8f. The two sorts quite dissimilar.

2 A. tamariscina. Leaves lance-oval, 1—3'; spikes interrupted and leafy at base, or throughout; ♀ bracts longer than the ovary. Wet shores, E. and W. 1—6f. The ♂ plant scarcely differs from ♂ No. 1.

4. IRESÌNE, Br. Fls. ♂ ♀ or ☿, 3-bracted. Calyx of 5 erect sepals. Sta. 5, anth. 1-celled. Stigmas 2 or 3. Utricle valveless, included in the calyx.—Leaves opposite, petiolate. Flowers minute, scarious, white, in dense spikes or heads. September, October.

  I. celosioìdes L. Branches opposite, strict; leaves ovate-lanceolate; flowers in numerous delicate panicled spikes. ① Banks, W. and S-W. 2—4f.

5. GOMPHRÈNA, L. GLOBE AMARANTH. Fls. 3-bracted. Cal. 5-sepalled, erect. Fil. 5, 3-cleft at apex, middle tooth bearing the 1-celled anth. Stig. capitate. Fr. as in Iresine. Tropical plants. Lvs. opposite. Flowers in heads.

  G. GLOBÒSA. Trichotomously much branched; leaves oblong, entire; flowers fadeless bright purple, in heads 1' diameter. ① E. India. 1—2f.

6. TELANTHÈRA, Br. Fls. 3-bracted. Cal. of 5 sepals. Stamens 5, with 5 intervening sterile filaments, anth. 1-celled. Stig. capitate. Fr. as in Iresine. Leaves opposite. Heads axillary and terminal.

T. polygonoìdes Moq. Procumbent, diffuse, hairy; leaves oval, obtuse, attenuate to a winged petiole; flowers silvery whitish. ♃ Waste grounds, S.

7. **FRŒLÍCHIA,** Mœnch. Fls. 3-bracted. Calyx tubular, 5-cleft at apex. Sta. 5, connate into a tube, with 5 sterile filaments. Anth. 1-celled Stigmas capitate or tufted. Utricle enclosed in the hardened calyx. ① Hairy or woolly stems, long-jointed.

F. Floridàna Moq. Nearly simple, strictly erect; leaves linear; flowers in short dense, cottony spikes. River banks, W. and S. 1—3f. July, August.

ORDER CVIII. LAURACEÆ. LAURELS.

Trees and *shrubs* aromatic, mostly with alternate, simple, punctate *leaves*. *Flowers* with a colored perianth of 4—6 slightly united, strongly imbricated sepals. *Anthers* 2- or 4-celled, opening upward by as many recurved, lid-like valves. *Ovary* 1-celled, 1-ovuled, free, in fruit a berry or a drupe. *Seed* without albumen.

§ Flowers perfect. Stamens 12, the 3 inner sterile.—a Evergreen trees................PERSEA. 1
—a Leafless vines. S. Fla.........CASSYTA.
§ Flowers diœcious. Stamens 9, all fertile. Leaves deciduous...(*)
* Involucre none. Anthers 4-celled, 4-valved. Leaves lobed........................SASSAFRAS. 2
* Involucre 4-leaved. Anthers 2-celled, 2-valved. Leaves entire..............BENZOIN. 3
* Involucre 4-leaved. Anthers 4-celled, 4-valved. Leaves entire..................TETRANTHERA. 4

1. **PÉRSEA,** Gært. RED BAY. BAY GALLS. Fls. ☿, umbellate, with no involucre. Cal. of 6 sepals. Sta. 12, the 3 inner sterile, reduced to mere glands, anth. 4-celled (2 cells above and 2 below). Drupe oval, seated on the persistent calyx, containing 1 large seed. ♄

P. Carolinénsis Mx. Tree 30—40f, often but a shrub, with lance-oblong, entire, firm leaves, 6′; umbels small, on ped. 1—2′; drupe oval, blue. Swamps, Va. to Fla. Bark deep-furrowed; wood fine, rose-colored. April, May.

2. **SÁSSAFRAS,** Nees. SASSAFRAS. Fls. ♂ ♀. Calyx 6-parted, deciduous. ♂ Sta. 9, the 3 inner with a pair of glands at base, anth. 4-celled. ♀ Sta. 6, all sterile. Ov., style, and stig. 1. Drupes ovoid, blue, on thick red pedicels. ♄ Flowers yellow, appearing before the leaves in Mar.—Jn.

S. officinàle Nees. Shrub or small tree, 10—20f; leaves of two forms—ovate and entire, or 3-lobed, cuneate at base; flowers handsome, in racemes or corymbs. Fields and woods. Bark pleasantly aromatic.

3. **BÉNZOIN,** Nees. SPICE WOOD. Flowers ♂ ♀, with 4 involucrate scales. Cal. 5- or 6-parted. ♂ Sta. 9, the inner 3 glandular at base, anth. 2-celled. ♀ Sta. 15—18 rudiments. Drupe obovoid, red. ♄ ♄ Lvs. entire. Fls. yellow, in small lateral clusters before the leaves. (Lindera, Thunb.)

1 **B. odoríferum** Nees. Shrub 6—12f; leaves lance-obovate, acute at base; buds and pedicels smooth. Moist woods: common. May.

2 **B. melissæfòlium** Nees. Shrub 2—3f; leaves lance-oblong, abrupt or cordate at base; buds and pedicels villous. Swamps, S. February, March

4. TETRANTHÈRA, Jacq. POND SPICE. Fls. as in Benzoin, but the anthers are 4-celled and 4-valved as in Sassafras. Drupe globular (red). ♄ Flowers yellow, precocious. February, March.

T. **geniculàta** Nees. Shrub 8—15f, with branches and branchlets very crooked and divaricate. Leaves small, oval to oblong. Swamps, S.

Order CIX. LORANTHACEÆ. Loranths.

Shrubby plants, parasitic on trees, with thick, opposite, exstipulate leaves. *Flowers* mostly diclinous, an adherent *calyx* of 2—8 lobes, with *stamens* of the same number, opposite the calyx lobes. *Ovary* 1-celled, becoming a fleshy fruit with one albuminous seed. (See Addenda.)

PHORODÉNDRON, N. MISTLETOE. Fls. ♂ ♀, in jointed spikes, mostly 3-lobed. ♂ Anth. sessile on the base of each lobe, the 2 cells divergent. ♀ Stig. sessile (no stamens). Fr. a pulpy, viscous berry.—Herbage yellowish-green. Stems brittle, woody, firmly engrafted on the limbs of oaks, elms, &c.

P. **flavéscens** N. Stems much branched, 1—1½f; leaves wedge-obovate, thick, entire, as long as the spikes; berry white, pellucid, sticking to the limb which it touches until it takes root.

Order CX. SANTALACEÆ. Sandalworts.

Trees, shrubs, and *herbs*, with alternate, undivided leaves, with the *calyx* tube adherent to the ovary, limb 4–5-cleft, valvate. *Stamens* as many as the sepals, and opposite to them. *Ovary* 1-celled, with a free central placenta bearing at top 2—4 suspended ovules, but in *fruit* drupaceous, 1-seeded, crowned with the persistent calyx.

§ Half-shrubby. Anthers connected to the sepals by a tuft of hairs.................COMANDRA. 1
§ Shrubs. Anthers free.—* Flowers 4-parted, with 4 petals in the ♀ flowers............BUCKLEYA. 2
—* Flowers 5-parted, all apetalous. Leaves alternate..........PYRULARIA. 3

1. COMÁNDRA, N. BASTARD TOADFLAX. Calyx tube adherent, limb 4- or 5-parted. Anth. 4 or 5, connected as above mentioned. Fil. on a 5-lobed perigynous disk.—Smooth plants, with herbaceous branches and whitish flowers in small umbels.

1 C. **umbellàta** N. Flowers perfect; branches strict, corymbed above; leaves oblanceolate, subsessile; umbels 3-flowered, exceeding the leaves; connecting hairs yellow. Rocky woods. 1f. Leaves scattered, 9″. June.
2 C. **Darbya** A. DC. Flowers diœcious; branches short, leafy; leaves elliptical, mostly opposite; umbels 5-flowered, shorter than the leaves; connecting hairs white. Woods, S.; rare. 1—2f. The fertile plant unknown.

2. BUCKLÉYA, Torr. Fls. ♂ ♀, the ♀ with a double calyx, the inner (corolla) caducous, and without stamens. Sty. 4-lobed. ♂ Calyx single, 4-lobed, with 4 stam. Fruit oblong, 10-furrowed, 1-seeded. ♄ Leaves subsessile, entire. Sterile flowers clustered, fertile solitary.

B. distychophýlla Torr.—Mountains of E. Tenn. Shrub 10—20f; leaves ovate, acuminate; fruit 8—9" long, resembling that of Forestiera.

3. PYRULÀRIA, Mx. OIL-NUT. Fls. ♂ ♀. Calyx 5-cleft, half-adherent by the 5-toothed disk. Style 1, stigmas 2 or 3. Drupe pear-shaped, 1-seeded, with the albumen very oily. ♄

P. púbera Mx. Shrub 4—6f, spineless, with oval-oblong leaves and small greenish flowers in terminal racemes; drupe 7—9". Mountain streams, Pa., and S. May.

ORDER CXI. THYMELACEÆ. DAPHNADS.

Shrubs with a very tough, acrid bark, entire leaves and perfect flowers, with the calyx tubular, colored, the limb 4-(4- or 5-)parted, regular, the tube bearing the *stamens*, as many or usually twice as many as its lobes, and free from the *ovary*, which is 1-celled, 1-ovuled, the suspended seed with little or no albumen.

1. DIRCA, L. LEATHERWOOD. Cal. colored, tubular, limb obscurely 4-toothed. Sta. 8, exserted. Style 1. Berry 1-seeded. ♄ Fls. opening before the oblong-obovate, alternate leaves, 3 from each bud.

D. palústris L. Shrub 3—5f, along streams, with very tough bark; flowers 4″, yellowish, in April, May; berry oval, small, red.

2. DAPHNE, L. Cal. colored, funnel-form, limb spreading, 4-parted. Anthers 8, subincluded. Stigmas capitate. Berry fleshy, 1-seeded. ♄ Native of the Old World.

1 D. MEZÈREUM. Shrub 1—3f, with very smooth lanceolate leaves appearing later than the lateral clusters of rose-purple, sweet-scented flowers.

2 D. ÒDORA. Shrub 2—3f; leaves lance-oblong, evergreen; clusters terminal, roseate, very fragrant. Greenhouse.

3 D. LAURÈOLA. Shrub 1—5f, hardy, with large oblanceolate, shining, evergreen leaves and axillary clusters of greenish flowers.

ORDER CXII. ELÆAGNACEÆ. OLEASTERS.

Shrubs or *trees* usually with the leaves covered with a silvery scurf, entire. *Flowers* mostly diœcious, the calyx free, entire, persistent, becoming in *fruit* pulpy and berry-like, enclosing the 1-celled, 1-seeded achenium. Embryo straight, with little albumen.

* Flowers perfect. Stamens 4. Leaves alternate, petiolate, entire..................ELÆAGNUS. 1
* Flowers diœcious. Stamens 8. Leaves opposite, after the flowers..................SHEPHERDIA. 2
* Flowers diœcious. Stamens 4. Leaves alternate, after the flowers..................HIPPOPHÆ. 3

1. ELÆÁGNUS, L. OLEASTER. Cal. 4-cleft, colored within. Sta. 4, alternate with the sepals. Achenium enclosed in the mealy, 8-furrowed calyx tube. ♄ ♄ With silvery foliage.

1 E. ARGÉNTEA Ph. *Silverberry.* Shrub 8—12f; leaves broadly or narrowly elliptical, acute, 1—2′; flowers axillary, deflexed, canescent. Dakota, and W.

2 E. HORTÉNSIS. Tree with narrow-lanceolate, acute leaves; flowers axillary, erect.— Also, E. LATIFÒLIA, with evergreen leaves, is cultivated.

ORDER 113.—EUPHORBIACEÆ. 293

2. SHEPHÉRDIA, N. Fls. ♂ ♀. Cal. 4-cleft. Sta. 8, with 8 glands. ♀ Calyx tube closely investing the ovary, limb 4-lobed. Sty. and stig. 1. Berry globular, fleshy. ♄ Spinescent.

1 S. **Canadénsis** N. Shrub 6—8f; leaves elliptic-ovate, clothed beneath with stellate hairs and rusty scales, nearly smooth above. Banks of streams, N. Clusters subsessile. Berry sweetish.
2 S. **argéntea** N. *Buffalo Berry*. Tree 12—18f; leaves oblong-ovate, obtuse, both surfaces smooth and covered with silvery scales. Fruit the size of a currant, scarlet, well-flavored. Missouri.

3. HIPPÓPHÆ RHAMNOÍDES. Shrub with lance-linear leaves, silvery white beneath, and a crowd of yellow, acid drupes. Europe.

ORDER CXIII. EUPHORBIACEÆ. SPURGEWORTS.

Herbs, shrubs, or *trees*, usually with a milky, acrid juice. *Flowers* dioclinous, sometimes enclosed in a cup-shaped involucre. *Calyx* inferior, sometimes wanting. *Corolla* scale-like or colored, often wanting. *Ovary* free, sessile or stipitate, 2-, 3-(or more)-carpelled; styles distinct or united. *Fruit* of 2, 3 (or more) 1-2-seeded carpels (rarely of 1 carpel) united to a common axis, at length separating. *Embryo* in fleshy albumen. Fig. 142.

547. Head or capitulum of Euphorbia corollata. 8. The involucre tube cut open, showing the monandrous, staminate flowers surrounding the pistillate. 9. One of the ♂ flowers, with a toothed bract at base. 50 Cross-section of the ovary, showing the 3 one-seeded cells or carpels.

ɣ Cells of the ovary 1-ovuled; fruit cells or carpels 1-seeded...(*)
ɣ Cells of the ovary 2-ovuled; fruit cells or carpels each 2-seeded...(x)
* Flowers in a cup-shaped involucre, the ♂ many, each merely a stamen, the ♀ only 1,—an ovary exserted on a pedicel.................................EUPHORBIA. 1
* Flowers not in an involucre, ♂, all apetalous, with a calyx only...(a)
  a Stigmas and carpels 6—9. Fruit fleshy, apple-like. Trees. S. Fla....HIPPOMANE *Mancinella*.
  a Stigmas and carpels 3. Fruit dry, capsular...(b)
    b Stamens erect in the bud, 2—4 in number...(c)
    b Stamens erect in the bud, 8—∞ is number...(d)
    b Stamens inflexed in the bud. ♂ Flowers usually with small petals...(e)
      c Staminate calyx imbricated in bud. Anthers pendulous. Tree. S. Fla....SEBASTIANIA *lucida*.
      c Stam. calyx imbricated in bud. Anthers erect. Flowers in spikes.........STILLINGIA. 2
      c Stam. calyx valvate in bud. Flowers in racemes. Plant downy......... TRAGIA. 3
      d Flowers in cymes, with white, imbricated sepals. Stinging............JATROPHA. 4
      d Flowers in small spikes with large bracts. Sepals valvate....... ACALYPHA. 5
      d Flowers in long interrupted spikes. Sepals 3, valvate in bud........MERCURIALIS. 6
      d Flowers in panicles. Leaves palmately lobed, glabrous.............RICINUS. 7
      e Ovary 3- or 2-celled -seeded. Plants hairy, downy, &c........ ....CROTON. 8
      e Ovary 1-celled and -seeded. Plants silvery-scurfy....................CROTONOPSIS. 9
    ӕ Calyx 5–6-parted; stamens 3, united. Flowers axillary, small.............PHYLLANTHUS. 10
    x Calyx 4-parted; stamens 4, distinct, large. Flowers in bracted spikes......PACHYSANDRA. 11
    ӕ Calyx 4-parted; stamens 4, distinct. Fls. axillary. Shrub. Lvs. opposite....BUXUS. 12

1. EUPHÓRBIA, L. SPURGE. Fls. monoecious and achlamydeous, several in an involucrate cluster, simulating one flower (see figures). In-

ORDER 113.—EUPHORBIACEÆ.

volucre calyx-like, 4- or 5-lobed, often with 4 or 5 large glands. ♂ Fls. 9 or more, each a stamen with a bract. ♀ Flower central, a 3-celled, 3-ovuled ovary on a pedicel. Styles 3, 2-cleft. Caps. 3-lobed, separating into 3 nutlets.—Plants with a milky juice.

§ Shrubs of the greenhouse, with scarlet bracts or involucres............Nos. 33—35
§ Herbs, erect, without stipules. Leaves alternate or opposite...(a)
§ Herbs, mostly prostrate, diffuse. Leaves all opposite, oblique at base, small, furnished with small stipules at base. Glands of the involucre 4, usually white-margined. ① May—Nov....(x)
  a Glands of the involucre 5, bordered with white petaloid appendages...(b)
  a Glands of the involucre 4 or 5, crescent-shaped or 2-horned...(c)
  a Glands of the involucre 1—5, neither white nor horned...(d)
    b Heads pedunculate. Branches regular. Leaves oblong to linear.......Nos. 1, 2
    b Heads pedunculate. Branches irregular. Leaves oval or ovate........Nos. 3, 4
    b Heads nearly sessile. Leaves with broad white margins................No. 5
  c Umbel of many rays. Stem leaves narrow, alternate. Seeds smooth. ♃...Nos. 6, 7
  c Umbel of 3 rays, and forked. Stem leaves alternate, thin ....Nos. 8, 9, 10
  c Umbel of 3 or 4 rays, and forked. Stem leaves opposite, thick........ ...No. 11
    d Inflorescence a simple terminal cluster. Leaves toothed or lobed....Nos. 12, 13
    d Inflorescence a forked cyme, peduncles in the forks. Lvs. entire....Nos. 14, 15
    d Inflorescence a compound umbel. Heads terminal...(e)
      e Seeds reticulated or wrinkled. Leaves serrulate..................Nos. 16, 17
      e Seeds smooth and even,—k in a rough, warty fruit..............Nos. 18—20
        —k in a smooth and even fruit..........Nos. 21—23
 x Leaves serrulate or serrate. Seeds roughened with wrinkles or pits...(y)
  y Stems ascending or erect. Plants smooth or smoothish.........Nos. 24—26
  y Stems flat on the ground, spreading, mostly hairy.................Nos. 27, 28
 x Leaves entire. Seeds smooth and even. Plant glabrous...........Nos. 29—32

1 E. corollàta L. *Flowering S.* Erect, glabrous, or subglabrous; umbel 3–7-rayed, rays 3- and 2-forked; lvs. oblong to oblong-linear, obtuse, those of the umbel whorled or opposite; involucre glands obovate, petaloid. ♃ Dry fields, 1—2f. July, Aug.
β. *angustifolia.* Leaves oblong-linear; umbel becoming irregular. S.

2 E. Curtísii Eng. Smooth, slender, branched from base, divisions about 3-forked, then 2-forked; leaves opposite or in 3's, linear-oblong or linear; heads minute; involucre glands narrowly white-bordered. ♃ Barrens, S. 1f. (E. discoidalis Chapm.)

3 E. pubentíssima Mx. Hairy, 2 or 3 times forked; leaves oval or ovate-oblong, petiolate or subsessile, scattered, the floral much smaller; heads minute; involucre glands minutely white-margined, entire. Dry. S. 1f. (E. paniculata Ell.)

4 E. mercurialina Mx. Stem naked below, leafy, and 3- or 2-forked above, pubescent; leaves oval or ovate, petiolate, mostly opposite; involucre *lobes crenulate*, white. Tenn.: rare. 8—10′. Too near to the preceding.

5 E. marginàta Ph. Leaves lance-oblong, sessile, the floral crowded, with a broad white margin; umbel 3-rayed, capitate. ① Ky., and W. 1f. †

6 E. Cyparíssias L. Lvs. linear, much crowded, the floral broad-cordate, all sessile; umbel of many simple rays; glands lunate. ♃ Fields and gardens. 1f. §

7 E. Esula L. Lvs. lance-linear, the floral broadly cordate; umbel of many forked rays, and scattered branches below; glands 2-horned. Fields: rare. §

8 E. Peplus L. Leaves round-cuneate, the floral ovate; umbel of 3 (rarely 5) forked rays; carpels doubly wing-keeled on the back. Fields, N. Eng.: rare. § Europe.

9 E. Ohiótica Steud. Smooth, erect from a decumbent branching base; lvs. mostly floral, reniform, sessile, the pairs appearing orbicular; carpels not winged; glands 2-horned. ♃ Woods, Ohio, W. and S. 1f. (E. commutata Eng.)

ORDER 113.—EUPHORBIACEÆ.

10 E. tetrápora Eng. Leaves linear-spatulate, the floral larger, transversely ovate; umbel 3-rayed; *seeds* 4-*pitted* on the inner face. ① Ga. to La. 10'.

11 E. Lathỳris L. *Caper S.* Stout, 2 or 3f high; leaves sessile, lance-linear, all opposite; umbel 4-rayed, then forked; glands horned. Gardens, and §.

12 E. heterophýlla Mx. Stem with scattered branches, 1—3f; leaves ovate, or sinuate-lobed, or panduriform, all petiolate and scattered, the upper stained red on the margins; gland 1, sessile. Iowa to Ga. June, July.

13 E. dentàta Mx. Stem 8'—2f, hairy, with opposite branches; leaves opposite, ovate, *dentate*, petiolate; heads subsessile; seed tubercled, round and black; gland 1 or more, stalked. ① Shades, Penn. to Iowa and La.

14 E. Ipecacuánhæ L. Root long, stems clustered, slender, diffusely forked; lvs. opposite, all oblong to linear, obtuse, sessile; heads on filiform pedicels; seed white, compressed, pitted. ♃ Sands, coastward. 8—12'. (E. gracilis Ell.)

15 E. nudicaùlis Chapm. Slender, forking above; leaves minute (¼''), obovate, the upper opposite; heads minute, glands margined, greenish. ♃ Fla.

16 E. Helioscòpia L. Stout; umbel 5-rayed, rays trifid, and forked; lvs. cuneate to obovate, whorled above; glands round, stalked. ① Waysides, N. §

17 E. dictyospérma F. & M. Slender; umb. once or twice 3-forked, then 2-forked; floral leaves roundish-ovate, subcordate, cauline oblong-spatulate to obovate; fruit warty, *seeds reticulated.* ① Ky., and S-W. (E. Arkansana C-B.)

18 E. Darlingtònii Gray. Tall (2—3f); umbel 5-8-rayed, rays forked or trifid; leaves entire, oblanceolate, the floral oval. ♃ Woods, Penn., and S.

19 E. platyphýlla L. Erect, 8—16'; umbel 5-rayed; leaves lance-oblong, subcordate, serrulate, the floral triangular-ovate. ① Lake shores, N. §

20 E. obtusàta Ph. Erect, 1—2f; umbel 3-rayed, rays trifid or forked; leaves all sessile, serrulate, obtuse, the floral roundish-cordate, the lower oblanceolate; fruit very warty. ① Woods, Va., and W.

21 E. inundàta Torr. Smooth, erect; umbel 3-rayed, and forked; leaves entire, sessile, lanceolate to oblong-ovate; glands round, entire; seeds globous. ♃ Wet barrens, Fla. 6—12'. Heads on slender peduncles. Root woody.

22 E. sphærospérma Shutt. (E. Floridana Chapm.) Lvs. lance-linear to cordate-ovate; heads green, glands crenate. Otherwise like No. 21. ♃ Dry. Fla. 1—2f.

23 E. telephioìdes Chapm. Plant some fleshy, 2—5' high; lvs. ovate, large on the stem, small on the umbel. Otherwise like No. 22. ♃ West Fla. May, June.

24 E. hypericifòlia St. 1—2f; lvs. 6—12'', oval-oblong, serrate all around; sds. oval, obtusely 4-angled, wrinkled and tubercled, black. ① Fields: common.

25 E. glyptospérma Eng. St. 5—10'; lvs. 4—6'', linear-oblong, serrulate toward the apex; stip. fringed; sds. ovoid, obtuse-angled, wrinkled, amber-color. Wis., and S-W.

26 E. maculàta L. Hairy; leaves oblong, serrulate, often with a brown spot; stip. minute; seeds sharply angled, obscurely wrinkled, reddish. Sandy fields: common.

27 E. humistràta Eng. Hairy; lvs. elliptic-obovate, serrulate at apex, rarely spotted; stipules fringed; seeds obtuse-angled, oval, roughened, brownish. Banks, W.

28 E. serpyllifòlia Pers. Smooth; lvs. obovate-oblong, serrulate at apex, seldom spotted; stipules fringed; seeds acutely 4-angled, cross-wrinkled. Banks, W.

29 E. polygonifòlia L. Lvs. oblong-linear; glands of invol. not appendaged; seeds large (1'' long), ovoid, smooth and whitish. Sandy sea and lake coasts.

30 E. Geyerl Eng. Leaves oblong-obovate; glands with narrow appendages; seeds small (¼''), ovoid, acute, obtusely 3-angled, ash-colored. Sandy soils, N-W.

31 E. serpens H. B. K. Lvs. round-ovate, very small (1—2''); stip. triangular; glands scarcely appendaged; pod acutely keeled, seeds ovoid-3-angled. Ill. to La.

32 E. cordifòlia Ell. Lvs. 4—6'', cordate-oval; glands conspicuously white-appendaged; pods and seeds as in No. 31. Fields, South. Spreading 1f.

33 E. splendens. Shrubby and fleshy, thorny; lvs. ovate, acute both ways; ped. axillary; floral leaves in pairs, broader than long, scarlet. Madagascar.

ORDER 113.—EUPHORBIACEÆ.

34 E. fulgens. Not spiny; lvs. lanceolate, pointed both ways, floral lvs. soon falling; lobes and appendages of the involucre red and purple. Mexico.

35 E. pulchérrima (or Poinsettia). Floral leaves lanceolate, of a brilliant red, lower leaves wedge-oblong, often fiddle-shaped, all pointed. Mexico.

2. STILLÍNGIA, Gard.

Fls. ☿, in a terminal, dense spike, apetalous. ♂ Calyx cup-form, lobed and crenulate. Sta. 2 or 3. Fil. exserted, with short, 2-lobed anthers. ♀ Calyx 3-lobed. Style trifid, with 3 diverging, simple stigmas. Capsule 3-lobed, 3-celled, 3-seeded.—Plants smooth, erect, with alternate leaves. Fertile flowers at the base of the sterile spike. Bracts of the spike biglandular at base. May—Sept.

1 **S. sylvática** L. Herbaceous; stems clustered; leaves subsessile, lance-linear to lance-oblong, and obtuse to acuminate, crenate-serrulate; spikes yellowish, longer than the leaves; glands cup-shaped. ♃ S. 1—3f.

2 **S. aquática** Chapm. Shrubby; stem single; lvs. short-stalked, lanceolate, acute, sharply serrulate; spikes shorter than the leaves; glands peltate. Fla. 3—6f.

3 **S. ligustrìna** Mx. Shrubby; leaves lance-ovate, petiolate, entire; stipules ovate; spikes shorter than the leaves; sta. 3. Swamps, S. 6—12f. (Sebastiania, Muller.)

4 **S. sebífera** L. *Tallow Tree.* Tree 30—40f; lvs. long-petioled, rhomboidal, acuminate, entire; fruit rough, blackish, seeds white. S. §. (Excœcaria, Mul.)

3. TRÀGIA, Plum.

Fls. ☿. Cor. 0. ♂ Calyx 3-parted. Sta. 2 or 3, distinct. ♀ Calyx 5- to 6- to 8-parted, persistent. Style 3-cleft. Stig. 3. Fruit 3-lobed, 3-celled, separating into 3 bivalve, 1-seeded nutlets. ♃ ♭ Homely weeds. Lvs. mostly alternate, pubescent, stipulate. Fls. small, racemed. May—August.

1 **T. macrocárpa** Willd. Slender summits of the branches twining; lvs. cordate-ovate, acuminate, serrate; rac. long (3—4′); fr. 5—6″. Copses, Ky., and S. 2—4f.

2 **T. urticæfòlia** Mx. Erect, hairy, sparingly branched; leaves deltoid-lanceolate, truncate at base, sharp-serrate; fruit very hairy. Dry. S. 1—2f.

3 **T. innócua** Walt. Erect, branched, puberulent; leaves ovate-oblong, varying to linear, coarsely few-toothed or entire. Dry. S. 1f. (T. urens L., but it does not sting as Linnæus supposed.)

4. JÁTROPHA, L. Spurge Nettle.

Fls. ☿, in forked cymes; the fertile generally in the forks. Calyx colored, imbricate in bud. Corolla present or not. Sta. 10—30, monadelphous. Styles 3, forked. Pod 3-carpelled. ♃ Leaves palmi-veined, stipulate.

J. urens, β. *stimulosa* Mul. Low, hispid with bristly stings; leaves half 3-5-lobed, cordate, lobes lanceolate, serrate; sepals white, oval, spreading; corolla 0. Sandy woods, S.: common. Stings white, ½′ long. March—July. (Cnidoscolus, Pohl.)

5. ACALÝPHA, L. Three-seeded Mercury.

Fls. ☿, in short clusters or little spikes, surrounded by a large cut-toothed bract. Cor. 0. ♂ Calyx 4-parted. Sta. 8—12, monadelphous, with halved anthers. ♀ Calyx 3-parted. Styles 3, each 2-∞-cleft. Fr. 3 nutlets. ① Weeds resembling Nettles, with stalked alternate leaves (and ♄ tropical). Summer.

1 **A. Virgínica** L. Leaves lance-ovate, obtusely pointed, obscurely serrate, equalling their petioles (1—2′); sterile spikes hardly exserted. Dry. 10—20′.

β. *gracilénta.* Leaves narrower, on shorter stalks; ♂ spikes exserted.

2 **A. Caroliniàna** Walt. Lvs. ovate, cordate, closely and strongly serrate; ♂ spikes axillary, ♀ terminal, fruit soft-echinate, bracts with linear lobes. W. and S.

6. MERCURIÀLIS, Tourn. Fls. ☿, apetalous, axillary, in bractless spikes or fascicles. Calyx 3-parted. Sta. 10—20, anth. 2-celled, extrorse. Fruit 2-carpelled, 2-seeded.—Herbs with opposite, petiolate leaves.

M. ánnua Willd. Lvs. lanceolate, &c., thrice longer than the stalks; branches opposite; ♂ spikes long, interrupted, seeds oval, pitted. ① Waysides, S.: rare. §

7. RÍCINUS, Tourn. CASTOR OIL PLANT. Fls. ☿, apetalous. Calyx 3–5-parted, valvate in the bud. ♂ Sta. ∞, with irregularly united filaments. ♀ Style short, stigmas 3, 2-parted, plumous, colored. Capsule echinate, 3-lobed, 3-celled, 3-seeded.—Herbs or shrubs.

R. COMMÙNIS L.—A stout ① herb with peltate, palmi-lobed leaves, 4—12′, divided into lance-shaped lobes. Southward it becomes a shrub, or tree 10—20f. Cult. for its seeds, yielding the *castor oil*, or for the ornament of its splendid foliage. E. India.

8. CROTON, L. Fls. ☿. Calyx 4–8-parted. Petals hypogynous, 4—8, mostly minute, often (especially in the ♀) wanting. ♂ Disk with 4—6 lobes. Sta. 5 or more, anthers inflexed in the bud. ♀ Ovary 3-celled, styles 3, 1–3-times forked. Fruit 3-carpelled, 3-seeded.—Plants glandular, clothed with scurf or stellate hairs. Leaves alternate.

§ Downy. Fertile calyx 5-parted, with 2 styles, and pendulous..................No. 1
§ Hairy or scaly. Fertile calyx 5-parted, with 3 styles, each 2- or 3-cleft ....Nos. 2—4
§ Densely woolly. Fertile calyx 8-parted. Styles 3, each twice 2-cleft.......Nos. 5, 6

1 **C. monanthógynus** Mx. Stellate-downy, di- and tri-chotomously branched; lvs. ovate or subcordate, silvery beneath; fls. in the forks. ① Prairies, Ill., and S. 1f.

2 **C. glandulòsus** L. Hispid, tri-(or 4-)chotomously branched; lvs. clustered at the forks, lance- to linear-oblong, serrate, with 2 concave glands at base; fls. in clusters, the sterile 4-parted, 8-androus. ① A straggling weed, W. and S. 1—2f.

3 **C. argyránthemus** Mx. Clothed with silvery glandular scales, branched at base; lvs. oval to oblong; fls. in a hd. or spike, silvery all over, all 5-parted. ♃ Ga., Fla. 1f.

4 **C. marítimus** Walt. Half-shrubby, bushy, trichotomously branched, tomentous; lvs. broad-oval, silvery beneath; flowers in dense heads on long stalks; stam. about 10; stigmas 18—20. Drifting sands, sea-coast, S. 2—3f. July—Oct.

5 **C. capitàtus** Mx. Lvs. ovate to oblong, long-petioled, obtuse; ♀ cal. large (7″), 7-8-cleft; styles 3, each 4-parted to base; seed double-convex. W. and S.

6 **C. Ellióttii** Chapm. Lvs. lance-oblong, short-petiolate, acutish; ♀ cal. 6″ diam., 5-8-cleft; styles 3, each 4-cleft to the middle; seeds plano-convex. ① S. 2—3f.

9. CROTONÓPSIS, Mx. Fls. ☿, minute, in spikes. Calyx 5-parted. ♂ Petals 5, spatulate. Sta. 5, distinct. ♀ Petals 0, 5 scales instead. Stig. 3, each bifid. Ovary and pod 1-celled, 1-seeded. ① Slender, silvery-scurfy, with small, alternate leaves. Upper flowers sterile.

C. lineàris Mx.—Sandy swamps, N. J. to Ill., and S. Stems as slender as Flax, repeatedly trifid and forked, 1—2f. Leaves linear-oblong, 6—10″. June—Sept.

10. PHYLLÁNTHUS, L. Flowers ☿, axillary. Calyx in 5 or 6 segments. Petals 0. Stam. 3, very short. Styles 3, bifid. Ovules and seeds 2 in each 2-valved carpel.—Leaves alternate, in 2 ranks.

P. Caroliniénsis Walt. St. slender, with alternate branches; lvs. oval, 6—10″, the ramial much smaller; flowers subsolitary. ① Pa. to Ill., and S. 6—18′. June—Aug

11. PACHYSÁNDRA, Mx. Flowers ♂♀, apetalous, in bracted spikes. Calyx 4-parted. ♂ Filaments 4, long-exserted, flat. ♀ Styles 3, recurved. Capsule 3-horned, 3-celled, cells 2-seeded. ♃ Procumbent stems from long creeping root-stocks. Leaves alternate.

P. procúmbens Mx. Lvs. ovate to obovate, coarsely toothed, clustered above the spikes, which are all near the base of the stem. Va. to Ky., and S. March—May.

12. BUXUS, L. BOXWOOD. Flowers ♂♀, axillary. ♂ Calyx 3-leaved, petals 2. Sta. 4. ♀ Cal. 4-sepalled. Pet. 3. Sty. 3. Caps. with 3 beaks and 3 cells. Seeds 6. ♄ ♄ Leaves opposite, ovate, entire, smooth.

B. SEMPÉRVIRENS. A tree of slow growth, fine-grained wood, in Europe. The dwarfed varieties are planted in gardens for edgings.

ORDER CXIV. URTICACEÆ. NETTLEWORTS.

Plants of various habit, with stipules (which are often early deciduous) and with small inconspicuous, mostly diclinous flowers. *Calyx* regular, free from the 1-celled ovary. *Stamens* as many as the calyx lobes and opposite to them. *Fruit* a 1-seeded samara, drupe or achenium, separate or aggregated. The following groups have usually been regarded as Orders.

§ ULMACEÆ. *Trees* with colorless innoxious juice. Flowers habitually perfect, not in aments. Fruits separate. No albumen. (Figs. 181, 256, 295, 316, 382. 509)...(*a*)

§§ ARTOCARPEÆ. *Trees* with milky poisonous juice. Flowers diclinous, in aments or heads. Fruits aggregated. Seed albuminous. (Figs. 195–6, 298, 349)...(*b*)

§§§ URTICEÆ. *Herbs*. Flowers diclinous, not in aments. Filaments crenulate. Fertile calyx 3–5-parted. Embryo straight. (Fig. 503)...(*c*)

§§§§ CANNABINEÆ. *Herbs*. Flowers diclinous. Filaments straight. Fertile calyx of 1 sepal, spathe-like. Embryo curved or coiled. (Fig. 213)...(*d*)

a Flowers appearing before the leaves. Fruit a samara winged all around.. .....ULMUS. 1
a Flowers *with* the leaves. Fruit wingless,—*x* a dry nut from a 1-celled ovary....PLANERA. 2
—*x* a sweet, fleshy drupe..............CELTIS. 3
b Flowers enclosed within a hollow receptacle, both kinds together...............FICUS. 4
b Flowers external.—*y* Fertile aments globular. Branches thorny................MACLURA. 5
—*y* Fertile aments globular. Plants thornless................BROUSSONETIA. 6
—*y* Fertile aments oblong, fruit sweet, juicy..................MORUS. 7
c Herbs with stinging hairs.—*z* Stamens 4. Leaves opposite......................URTICA. 8
—*z* Stamens 5. Leaves alternate ....LAPORTEA. 9
c Herbs stingless.—*n* Stamens 3. Fertile calyx 3-sepalled.......................PILEA. 10
—*n* Stamens 4.—*o* Flowers in slender spikes... ....BŒHMERIA. 11
—*o* Flowers in involucrate clusters.... ....PARIETARIA. 12
d Herbs twining. Fruit in an imbricate strobile-like ament........................HUMULUS. 13
d Herbs erect. Fruit a 2-valved caryopsis in axillary pairs.......................CANNABINUM. 14

1. ULMUS, L. ELM. Fls. ☿. Calyx 4–9-cleft. Stam. 4—9, fil. long and slender. Styles 2. Ovary 2-celled. Samara flat, 1-seeded. ♄ Fls. yellowish, or reddish, in precocious clusters. Figs. 181, 256, 295.

* Samara fringed with hairs, hanging on slender ped., 2-beaked ....Nos. 1—3
* Samara not fringed, nearly sessile,—*x* slightly notched at apex............Nos. 4, 5
—*x* cleft down to the seed..............Nos. 6, 7

1 U Americàna L. *White Elm*. Lvs. oval, acuminate, doubly serrate; flowers in loose, umbel-like clusters; fruit oval, 6″, its 2 beaks with points incurved and meeting. A majestic tree, with ascending branches and often long pendulous "weeping" branchlets. Native, and everywhere cultivated.

2 **U. racemòsa** Thomas. *Cork Elm.* Smaller tree, with rigid branches; branchlets downy, often with wing-like corky ridges; *flowers* 2—4 in each fascicle, which are arranged in racemes. N. H. to Wis., and S. 20—30f.

3 **U. Floridàna** Chapm. Tree 30—40f, with brittle branches, smooth; lvs. thick, acute; fruit orbicular, 2—3″, its teeth broad and erect. W. Fla.

4 **U. alàta** Mx. *Winged Elm. Whahoo.* Tree, with its branchlets here and there winged with 2 corky ridges; leaves lance-oblong, acute, 1—2½′; flowers racemed; fruit downy all over, with its 2 beaks slender. Ill. to Va., and S.

5 **U. fulva** L. *Red Elm. Slippery Elm.* Tree 20—40f; buds covered with fulvous down; leaves oblong-ovate, acuminate; flowers reddish, 7-parted, sessile; fruit orbicular. Low grounds. Valued for its very mucilaginous *liber.*

6 **U.** CAMPÉSTRIS. *English Elm.* A stately tree, 50—70f, with rigid branches and dense foliage; leaves small, ovate; stamens 5; fruit nearly orbicular. Europe.

β. SUBERÒSA. Branchlets with thick corky wings; stamens mostly 4. Europe.

7 **U.** MONTÀNA. *Scotch Elm. Witch Elm.* Large tree, with ample obovate, cuspidate leaves, rough above, downy beneath; flowers 5-parted; fruit oblong, 1′. Europe.

2. **PLÁNERA,** Gmel. Fls. ♂ ☿ ♀. Cal. lobes and sta. 4 or 5. Stig. 2, oblong, diverging; ova. 1-celled, fruit 1-seeded, wingless, indehiscent. ♄

P. aquática Gm. Tree 30—40f, elm-like, with small smooth, ovate, acute, serrate leaves and axillary flowers in clusters of 2—5; nut roughened. Swamps, S.

3. **CELTIS,** Tourn. NETTLE TREE. SUGAR-BERRY. Fls. ♂ ☿ ♀, the ♂ 6-parted and the ☿ 5-parted. Sty. 2, elongated, spreading. Drupe globular. ♄ ♄ Leaves mostly oblique at base. Flowers subsolitary. Fig. 316.

1 **C. occidentàlis** L. Tree 30—70f, with wide-spread branches; lvs. ovate, subcordate, acuminate, serrate, rough-hairy beneath; ped. longer than the petiole; sepals triangular-ovate, erect, white; drupe 3″, dark purple. Woods, &c.

β. *crassifolia.* Leaves cordate, thick, mottled with dark and light green.

γ. *integrifolia.* Leaves smooth, subentire; bark smooth. W. and S.

2 **C. pùmila** Ph. A straggling shrub, 3—10f, with broad-ovate, acute, smooth, serrate leaves; calyx of 6 oblong-linear spreading segments, 2″. Woods, S.

4. **FICUS,** Tourn. FIG. BANIAN. Fls. ☿, minute, fixed upon the inner surface of a hollow receptacle. ♂ Calyx 3-parted, sta. 3. ♀ Calyx 5-parted, ovary 1, seed 1. Fruit (syconus) composed of the enlarged, fleshy receptacle enclosing the numerous dry, imbedded achenia. Fig. 195.

1 **F.** CÁRICA. *Common Fig.* Leaves cordate, 3-5-lobed, repand-dentate, rough and downy; fig pear-shaped. From Asia. A shrub in our conservatories, a small tree S.

2 **F.** ELÁSTICA. *India-rubber Tree,* in the greenhouse, with a straight, simple trunk, and very large (8—10′), shining, thick, oblong leaves. E. India.

3 **F.** REPENS. Creeping on walls, &c., with ovate, cordate, acute, serrate lvs. E. India.

4 **F.** INDICA, the *Banian* (§ 207), with many trunks, may grow South.

5. **MACLÙRA,** N. OSAGE ORANGE. Flowers ♂ ♀, the ♂ racemous, calyx 4 parted. ♀ Flowers in a dense globular head. Calyx 4-sepalled, fleshy, finally embracing the obconic achenium, all ripening into a globular *sorosis,* resembling an orange. Style terminal. ♄ Juice milky. Leaves alternate, entire. Branches with sharp spines. Fig. 298.

M. aurantiaca. Lvs. shining, ovate-oblong, thickish, pointed; fruit yellow when ripe, lactescent, pendulous. Arkansas. Planted for hedges. May, June.

6. **BROUSSONÈTIA,** L'Her. PAPER MULBERRY. Fls. ♂ ♀, in aments

the ♂ cylindric, the ♀ globular, style lateral, ovary becoming a fleshy club-shaped 1-seeded fr. protruding from the tubular, 3- or 4-toothed calyx. ♄

B. papyrífera. Tree with a low bushy head, of rapid growth, with rough and downy leaves, ovate or variously lobed; fruit dark red, hispid. Japan. Fig. 349.

7. MORUS, Tourn. MULBERRY. Fls. ☿, in aments, the ♂ loose, the ♀ dense and spike-like. Cal. 4-parted, sta. 4, sty. 2. Achenium compressed, enclosed in the fleshy calyx, the whole spike thus constituting a compound berry (sorosis). ♄ Leaves alternate, broad, often palmately lobed. Fig. 196.

1 **M. rubra** L. Tree or shrub, 15—60f; roots yellow; leaves rough and downy, subcordate, serrate; fertile spikes cylindric; fruit dark red, very sweet.

2 **M. alba.** *Chinese M.* Shrubs (here), with smooth and shining, cordate, unequally serrate leaves; fruit whitish. Introduced for silkworms.

3 **M. nigra.** Tree for ornament and shade, from Persia, with rough, ovate or lobed leaves; fertile spikes oval; fruit reddish-black, acid.

8. URTÌCA, Tourn. NETTLE. Fls. ☿, sometimes ♂ ♀. ♂ Calyx 4-sepalled. Ovary a cup-shaped rudiment. Sta. 4. ♀ Sepals 4, the outer pair minute, the inner at length surrounding the shining, compressed achenium. Stig. 1, sessile.—Herbs with stinging hairs. Leaves opposite. Fls. green, in axillary or subterminal clusters or racemes. Summer. Fig. 503.

§ Clusters compound, longer than the petioles. Perennials................ Nos. 1, 2
§ Clusters simple, shorter, or not longer than the petioles. Annuals........Nos. 3, 4

1 **U. pròcera** Willd. Stem tall (3—6f), slightly hispid, with few stings; leaves lance-ovate, 5-veined, uncinate-serrate; spikes panicled above. Waste places: common.

2 **U. dioìca** L. St. 1—3f, very hispid and stinging; leaves ovate, deeply serrate, the slender point entire; spikes clustered in the axils. Wastes: common. §

3 **U. urens** L. Low (1f), hairy; lvs. broadly ovate, coarsely serrate, 5-veined; clusters pedunculate, loose, by pairs in each axil. Waste grounds, E. §

4 **U. chamædrioìdes** Ph. St. 1—2f, with scattered bristles; leaves ovate, crenate-serrate; clusters capitate, 1 or 2 in each axil, spiked above. Ky., and S.

9. LAPÓRTEA, Gaudich. WOOD NETTLE. Fls. in axillary panicles, the ♂ calyx 5-parted, the ♀ of 4 sepals, the 2 inner larger. Sta. 5. Stig. subulate. Achenium flat, ovate, very oblique. ♃ Hairs stinging. Lvs. ample, ovate, petiolate.

L. Canadénsis Gaud. Leaves 3—5′, acuminate, serrate; flowers minute, green, in panicles, 1—2′, the lower sterile. Damp woods. 2—6f.

10. PÌLEA, Lindl. RICHWEED. Fls. in dense axillary clusters, the ♂ with 3 or 4 sep. and sta. ♀ Sepals 3, unequal, oblong. Sta. 3 rudiments. Achenia roughened, erect, ovate. ① Smooth, stingless. Stipules united.

P. pùmila Gray. Stem succulent, weak; leaves rhomb-ovate, crenate-serrate, long-stalked; flowers green, in short clusters. Moist shades. 3—18′. July, Aug.

11. BŒHMÈRIA, Jacc FALSE NETTLE. ♂ Calyx 4-parted, with lanceolate, acute segments. Stamens 4. ♀ Calyx tubular, truncate, or 4-toothed, persistent and closely investing the ovate, pointed achenium.—Herbs or shrubs, stingless. Flowers minute.

B. cylíndrica Willd. Erect, simple; leaves generally opposite, on long petioles ovate, acuminate, dentate; upper spikes interrupted, leafy at top, sterile, lower dense, fertile. ♃ A coarse weed in swamps. 2—3f. Spikes 1—6'. July, August.

β. *lateriflora* has narrower leaves, shorter stalks, all alternate.

12. PARIETÀRIA, Tourn. PELLITORY. Fls. polygamous, in clusters, surrounded by a many-bracted involucre. ♂ Cal. 4-sepalled. Sta. 4, at first incurved, elastically expanding. ♀ Stigma tufted. Ach. polished, enclosed within the persistent, 4-lobed calyx.—Herbs weed-like, with alternate leaves. Clusters of green flowers axillary.

1 **P. Pennsylvánica** Muhl. Lvs. oblong-lanceolate, veiny, tapering to an obtuse point, entire; involucre longer than the flowers. ① Rocky shades. 6—12'.
2 **P. Floridàna** N. Leaves round-ovate, obtuse, entire, on long petioles; flowers as long as the involucre. ① Damp sands, S. 10'. (P. debilis Forst. ?)

13. HÙMULUS, L. HOP. Fls. ♂ ♀, the ♂ panicled, with 5 sep. and sta. Anth. with 2 terminal pores. ♀ Aments with large imbricated, entire, 1-flowered bracts. Cal. of 1 sepal, investing the achenium. Styles 2. Embryo coiled. ♭ Twining with the sun. Leaves opposite. Fig. 213.

H. lùpulus L.—Rich alluvion, along streams, and extensively cultivated. Stems 10—20f. Leaves cordate, 3-5-lobed, rough, on long stalks. Bitter, narcotic. July.

14. CÁNNABIS, Tourn. HEMP. Flowers ♂ ♀, the ♂ with 5 sep. and sta., in panicles. ♀ In spikes. Cal. a single spathe-like sepal enfolding the 2-valved cariopsis. Embryo curved. ① Leaves opposite, digitate.

C. satìva L.—Fields, waste grounds. Tall, erect, 4—8f. Leaves petiolate, regularly formed of 5—7 lanceolate-serrate leaflets. Cultivated S-W. June. §

ORDER CXV. SAURURACEÆ. SAURURADS.

Herbs with jointed stems, alternate, entire *leaves* furnished with stipules. *Flowers* in spikes, perfect, naked, having neither corolla nor calyx. *Stamens* definite. *Ovaries* 3—5, more or less united. Fig. 15.

SAURÙRUS, L. LIZARD-TAIL. Inflorescence a terminal spike of 1-flowered scales. Sta. 6, 7, 8 or more. Ovaries 4. Berries 4, 1-seeded. ♃ Stem angular. Leaves cordate, acuminate, petiolate.

S. cérnuus Willd.—Common in marshes, 1—2f. Leaves 4—6'. Spikes slender, recurved at the more slender top, its flowers whitish. July, August.

ORDER CXVI. CALLITRICHACEÆ. STARWORTS.

Herbs aquatic, small, with opposite, simple, entire leaves. *Flowers* axillary, solitary, very minute, polygamous, achlamydeous, with 2 colored bracts. *Stamen* 1, rarely 2; *filament* slender; *anther* 1-celled, 2-valved, reniform. *Ovary* 4-celled, 4-lobed; *ovules* solitary. *Styles* 2; *stigmas* simple points. *Fruit* 1-celled, 4-seeded, indehiscent. *Seeds* albuminous.

CALLÍTRICHE, L. Character the same as that of the order. ⚥ Very delicate.

* Stems short (6''- 2'), spreading on moist grounds. Leaves reniform......Nos. 1, 2
* Stems (3—16') growing in water. Fruit sessile.—*x* Leaves of two kinds...Nos. 3, 4
　　　　　　　　　　　　　　　　　　　　—*x* Leaves all linear..........No. 5

1 **C. Austínii** Eng. Lvs. obovate, 1—2''; fruit depressed, 4-lobed all around, its pedicel and stig. nearly as long, lobes narrowly winged. N. J. (Porter), N. Y., and W.
2 **C. peploides** N. Lvs. elliptical, 1''; fruit roundish, 4-lobed above, *sessile*, its stigmas twice as long, lobes not winged. Tenn. to La. (Hale). 1—2'.
3 **C. verna** L. Floating lvs. 3'', rosulate, obovate, narrowed below, the submersed leaves 6'', oblong-linear; fruit oval, emarginate, longer than its stigmas. Pools.
4 **C. heterophýlla** Ph. Floating leaves spatulate, attenuate below, 4—6'', the submersed linear, 6—9''; fruit globous, obcordate, its stigmas rather longer. Pools.
5 **C. autumnàlis** L. Leaves all submersed, 3—5''. linear, obtuse at both ends; fruit rounded, its lobes slightly united, winged; styles slender. Lakes and rivers.

Order CXVII. PODOSTEMIACEÆ. Threadfoots.

Herbs aquatic, with the habit of seaweeds, with alternate, dissected *leaves*, with *flowers* minute, perfect, naked or with 3 sepals. *Stamens* 1 or many, hypogynous. *Ovary* compound, 2–3-celled, with as many stigmas, and numerous ovules. *Fruit* a many-seeded capsule, ribbed and somewhat pedicelled. *Albumen* none.

PODOSTÈMUM, L. C. Rich. Threadfoot. River Weed. Fls. axillary, solitary. Sta. 2, fil. united below. Ovary oblong-ovoid. Stig. 2, sessile, recurved. Caps. 2-celled. Seeds minute.—Small, submersed ≈, adhering to stones and pebbles.

P. ceratophýllum Mx. Leaves alternate, repeatedly forking into linear, threadform segments; stem a few inches long, in running water: common.

Order CXVIII. CERATOPHYLLACEÆ. Hornworts.

Herbs aquatic, with whorled, dichotomously dissected leaves. *Flowers* ♂, sessile, axillary, minute, with neither calyx nor corolla. *Involucre* 8–12-cleft. *Anthers* (12—24) sessile. *Fertile flower* a simple 1-celled ovary with one ovule. *Cotyledons* 4.

CERATOPHÝLLUM, L. Hornwort. Character that of the order. ≈
C. demérsum L. Stem floating or prostrate, 8—16', with numerous branches and whorls; leaf-segment filiform, sharply toothed. Pools.

Order CXIX. EMPETRACEÆ. Crowberries.

Heath-like *shrubs*, with evergreen, linear, exstipulate *leaves*, and small, imperfect *flowers*. *Calyx* of 4—6 hypogynous, imbricated scales, the inner often colored and petal-like. *Stamens* 2—4, with compound pollen. *Ovary* free, 2-9-celled, 2-9-ovuled. *Fruit* fleshy, with as many seeds. In Batis the *drupes* are consolidated.

* Stamens 3. Stigmas 6-9-rayed. Berry 6-9-seeded.....................................Empetrum. 1
* Stamens 3. Stigmas 3 or 4. Style slender. Drupe 3- or 4-seeded............ ......Corema. 2
* Stamens 2. Stigmas 4. Berry 2-seeded. Shrub erect................ ....Ceratiola. 3
* Stamens 4. Stigma 1. Berry 4-seeded. Prostrate... ....Batis. 4

1. **ÉMPETRUM**, Tourn. CROWBERRY. Fls. ♂ ♀. Perianth consisting of 2 series of scales, the 3 inner petaloid. ♂ Sta. 3, anth. pendulous on long filaments. ♀ Stig. subsessile, 6–9-rayed. Drupe globular, with 6—9 seed-like nutlets. ♄ Alpine.

E. nigrum L. A small prostrate shrub, 1—4f; branches closely beset with oblong-linear leaves with rolled edges, 2—3″; berries black, eatable. High mountains of N. Eng., N. Y. May, June.

2. **COREMA**, Don. Perianth of 5 or 6 bractlets, the 3 inner sepaloid. ♂ Sta. 3, exserted. ♀ Ovary 3- or 4-celled. Style filiform, 3- or 4-cleft, with narrow stigmas. Drupe globular, minute, with 3 or 4 seeds. ♄

C. Conradii Torr. Shrublets diffusely branched, 6–12′, with narrowly linear leaves, 2—3″; flowers in terminal clusters, with brownish scales and purple stamens Saudy barrens, N. J. and N-E., forming blackish tufts. April.

3. **CERATÌOLA**, Mx. SAND-HILL ROSEMARY. Fls. ☿, of 6—8 imbricated, concave, fimbriate scales, the 2 or 4 inner membranous. ♂ Sta. 2, exserted, anth. 2-celled, roundish. ♀ Ovary 2-celled. Style short. Stig. 4 or 6, spreading, toothed. Drupe 2-seeded. ♄ Branches whorled, erect.

C. ericoìdes Mx.—Sandy places, Ga., Fla. 3—6f. Leaves whorled, crowded, linear-terete, 5—6″. Flowers reddish, followed by yellowish drupes. March, April.

4. **BATIS**, P. Br. Fls. ♂ ♀, in cone-like spikes. ♂ Calyx of 2 unequal, united sepals. Pet. 4, clawed. Sta. 4, anthers introrse, exserted. ♀ A mass of 4-celled ovaries only, becoming a sorosis of 4-seeded drupes. ♄

B. marítima L.—Salt marshes, Fla. Stems prostrate, 2—3f; leaves club-shaped fleshy, 1′. Spikes 5″, fleshy. Petals white. June—September.

ORDER CXX. PLATANACEÆ. SYCAMORES.

Trees with a watery juice, alternate, palmate *leaves*, and sheathing, scarious *stipules*. *Flowers* monœcious, in globular aments, destitute of both calyx and corolla. *Sterile.—Stamens* single, with only small scales intermixed. *Anthers* 2-celled, linear. *Fertile.—Ovary* terminated by a thick style with one side stigmatic. *Nut* clavate, tipped with the persistent, recurved style. *Seed* solitary, albuminous. Fig. 288.

PLÁTANUS, L. PLANE TREE. BUTTON WOOD. SYCAMORE. Character of the genus the same as that of the order. The ♂ and ♀ flowers in separate aments.

P. occidentàlis L. Tree in hard, gravelly soil, 50—80f. The trunk grows to great size, and hollow; bark whitish; leaves large, angularly lobed and toothed; stipules oblique; balls pendulous, solitary. May.

ORDER CXXI. JUGLANDACEÆ. WALNUTS.

Trees with alternate, pinnate, exstipulate *leaves* and monœcious *flowers*. *Sterile flowers* in aments, with an irregular perianth. *Fertile*, solitary or clustered. ♀ *Calyx* regular, 3–5-lobed, tube adherent to the partly 2–4-celled ovary. *Fruit* a tryma (§ 157), with a fibrous epicarp (*shuck*) and a bony endocarp (*shell*). *Seed* large, orthotropous, exalbuminous, with lobed, often sinuous, oily cotyledons.

* Sterile aments solitary, simple. Epicarp persistent on the *tryma*.................JUGLANS. 1
* Sterile aments clustered, lateral. Epicarp 4-valved and separating ....CARYA. 2

1. JUGLANS, L. WALNUT. ♂ Fl. a calyx, scale-like, 5- or 6-parted, with about 20 stamens. ♀ Fls. terminal, 4-parted, with 4 greenish petals and 2 fringed stigmas. Tryma with a spongy epicarp closely investing the very rough endocarp. ♄ Leaflets many. Pith in transverse plates.

1 **J. cinèrea** L. *White W. Butternut.* Tree 40—50f, with a large but short trunk, and wide-spread branches; leaflets 15—17, lanceolate; fruit oblong-ovate, viscid-hairy. Good for its fruit and handsome wood. April, May.

2 **J. nigra** L. *Black W.* Tree 60—90f, with a long, straight trunk; leaflets 15—21, lance-ovate, subcordate; fruit globous, glabrous, uneven, the kernel edible. The wood is dark-purple, used in cabinet-work. April, May.

3 **J. règia**, from Persia, but called *English* walnut, has 7—11 leaflets, and a smoothish endocarp (shell) with a rich kernel. Rarely cultivated.

2. CÁRYA, N. HICKORY. ♂ Calyx scale-like, 3-parted, with 4—6 stamens. ♀ Calyx 4-cleft, no petals. Stig. 2-lobed, lobes bifid. Epicarp 4-valved, disclosing a smooth, even nut. ♄ Timber very strong. Leaves and both kinds of flowers from same bud, in March—May.

§ Leaflets 13—15, scythe-shaped. Nut oblong, thin-shelled, very sweet.........No. 1
§ Leaflets 7—11. Nut with a tender shell and very bitter kernel.... ....Nos. 2, 3
§ Leaflets 5—9. Nut roundish, hard-shelled, sweet and eatable...(*)
  * Valves of the epicarp distinct to the base. Bark with loose plates....Nos. 4, 5
  * Valves of the epicarp united below. Bark continuous, firm...........Nos. 6—8

1 **C. olivæfórmis** N. *Pecan Nut.* Tree 60—90f; leaflets falcate, 5—6'; ♂ aments separate to base; nut with its kernel loose in the thin, oblong shell. River bottoms, Ind., Ill., and S. Bark at length shaggy.

2 **C. amara** N. *Bitter Nut.* Tree 20—40f; leaflets about 9, ovate-oblong, sharply serrate; fruit roundish, valves half-united; nut white. Moist.

3 **C. aquática** N. Tree 30—40f; leaflets about 11, lanceolate, oblique, subentire; fruit pedunculate, ovate, with a thin, reddish shell. Swamps, S.

4 **C. alba** N. *Shagbark.* Tree 40—50f, with a rough, shaggy bark; leaflets 5, the two lower much smaller; fruit and nut roundish, squarish, with a thin shell and very sweet meat: common. Fruit and timber excellent.

5 **C. sulcàta** N. *Thick-shellbark.* Tree 40—80f, with shaggy bark; leaflets 7 or 9, the odd one subsessile; fruit large, oval, 4-furrowed; nut pointed at each end, 1¼—2' long, with thick shell. Common West.

6 **C. tomentòsa** N. *Mocker Nut.* Tree 40—60f; bark rugged, but not shaggy; leaflets 7—9, odd one stalked, all and the petiole rough-downy; aments hairy; nut with a very thick shell and small kernel.

7 **C. porcìna** N. *Pignut.* Tree 60—100f; leaflets 5 or 7, nearly glabrous; fruit ovate to pyriform, with a bitterish kernel: common. (*C. glabra* Torr.)

8 **C. microcárpa** N. Tree 60—80f; leaflets 5 or 7, glabrous; aments glabrous; fruit roundish-ovoid, as small as a nutmeg. Woods, N. Y., and S.

ORDER CXXII. CUPULIFERÆ. MASTWORTS.

Trees or *shrubs. Leaves* alternate, simple, straight-veined, with deciduous stipules. *Flowers* ♂, the sterile in aments which are racemed or capi-

Order 122.—CUPULIFERÆ.

tate. ♂ *Calyx* scale-like or regular, with 5—20 stamens inserted at its base. ♀ *Calyx* adherent to the 2-3-celled, 2-6-ovuled ovary. *Fruit* a 1-celled, 1-seeded nut, solitary or several together, invested by an *involucre* which forms a scaly or echinate cupule. *Seed* destitute of albumen, filled by the embryo with its large cotyledons. Figs. 1–4, 182, 256, 277, 218–22, 338–40, 381, 386, 435, 507, 511.

§ Sterile flowers in aments, fertile, solitary, or few together...(*)
* Involucre of many scales, valveless, cup-like, partly enclosing the 1 nut............QUERCUS. 1
* Involucre of prickly scales, 4-valved, enclosing 2 or 3 nuts................. ....CASTANEA. 2
* Involucre of soft, prickly scales, 4-valved, enclosing 2 nuts........ ....FAGUS. 3
* Involucre of 2 or 3 large, lacerated, united scales, valveless, with 1—2 nuts..........CORYLUS. 4
§ Sterile flowers and fertile, both kinds in pendulous aments...(*)
* Involucre scales in pairs, with their edges united, inflated.......................OSTRYA. 5
* Involucre scales in pairs, distinct, 3-lobed, becoming leaf-like... ....CARPINUS. 6

1. QUERCUS, L. Oak. ♂ Fls. in loose aments. Calyx mostly 5-cleft. Sta. 5—10. ♀ Fls. in clusters or scattered. Ov. 3-celled, 6-ovuled (Fig. 162), with 3 stig., but in fruit a 1-seeded nut (acorn) seated in a scaly cup or involucre. ♃♄ A noble genus. In many oaks the fruit is ②, that is, 2 years in ripening, known by its occupying the old wood below the leaves of the season.

§ Leaves mostly entire, the ends subequal, petioles very short...(*)
* Peduncle longer than the oblong acorn. Leaves evergreen. Fruit ①....No 1
* Peduncle shorter than the acorn. Fruit ②.—*x* Lvs. downy beneath...Nos. 2, 3
   —*x* Lvs. smooth both sides...No. 4
§ Leaves 3-lobed and dilated above, *awnless* when mature. Fruit ②........Nos. 5, 6
§ Leaves 3-9-lobed or pinnatifid, broad, lobes *setaceously awned*. Fruit ②...(*)
* Lvs. at base cuneate, short-pet., 3- or 5-lobed. Shrubs or small trees..Nos. 7—9
* Leaves at base abrupt or truncate, mostly long-petioled, 7-9-lobed...(*a*)
   a Nut one-third immersed in the saucer-shaped, fine-scaled cup...Nos. 10, 11
   a Nut near half immersed in the hemispherical, coarse-scaled cup...(*b*)
      b Leaves cinereous-downy beneath, acorn also downy....................No. 12
      b Leaves (except when young) glabrous both sides... ....Nos. 13, 14
§ Lvs. 5-9-lobed, divisions obtuse, never bristle-awned. Fr. ①, sessile...Nos. 15—18
§ Lvs. 9-25-toothed, downy beneath, awnless. Acorn ①, sweet, eatable..Nos. 19, 20

1 **Q. virens** Ait. *Live Oak.* Tree 40—50f, often much smaller, very valuable for timber; leaves small, firm, elliptic-oblong, obtuse, downy and pale beneath, rarely a few sharp teeth; nut oblong-obovoid; ped. 1'. Va., and S.

2 **Q. cinerea** Ph. *Upland Willow O.* Shrub 4—20f; lvs. as in No. 1, but more downy beneath; nut roundish, in a saucer-shaped cup. Barrens, Va. to Fla.

3 **Q. Imbricaria** Mx. *Laurel O. Shingle O.* (Fig. 338.) Tree beautiful, 40—50f, with dense dark-green foliage; lvs. 3—5', lance-oblong, wavy, shining above; nut roundish, in a shallow cup. Common W. and S. Makes poor shingles.

4 **Q. Phellos** L. *Willow O.* Tree 30—60f, with poor timber; lvs. linear-lanceolate, entire, 3—4', glabrous; acorn roundish, 6'', in a shallow cup. Borders of swamps. N. J. to Ky. and Fla. Young shoots with toothed leaves.

β. **laurifolia.** A large handsome tree; lvs. 3—5', often with a few teeth. S. †

5 **Q. aquatica** Mx. *Water O.* Tree 20—40f, of rounded form and dense, shining foliage; leaves wedge-obovate, entire or obscurely 3-lobed above, attenuate to base, short-petioled; nut round-ovoid. Swamps, Md. to Fla., and cultivated.

6 **Q. nigra** L. *Black-Jack. Barren O. Iron O.* Tree small and gnarled, with dark mossy foliage; leaves short-petioled, wedge-form, mostly with 3 subequal rounded lobes at apex, subcordate at base, rust-downy beneath. N. Y., W. and S.

Order 122.—CUPULIFERÆ.

7 Q. tríloba Mx. *Downy Black-Jack.* Tree of rapid growth, 20—30f; leaves oblong-cuneiform, acute at base, rusty-tomentous beneath; lobes at apex often toothed. bristle-pointed; nut depressed. Barrens, N. J. to Fla.

8 Q. Catesbǽi Mx. *Turkey O.* Tree 20—25f; leaves large, very irregular, glabrous, cuneate at base, lobes deep, narrow, with bristle-pointed, divaricate teeth; cup large, half covering the ovoid, mealy nut. Barrens, South.

9 Q. Ilicifòlia Wang. *Scrub O. Bear O.* Shrub 3—7f, straggling; lvs. *petiolate*, obovate, angularly 5-(3–7-)lobed, 3—4′, whitish-downy beneath; acorn small (5—6″), cup very shallow. Barren tracts: common. Animals feed on the acorns.

β. *Georgiàna.* Leaves smaller and smoother, of the same form, on Stone Mt.!

10 Q. rubra L. *Red O.* Tree 50—70f, wide and high; leaves long-stalked, glabrous, sinuses rounded, shallow, lobes 7—9, with bristle-pointed teeth; acorn 1′, ellipsoid. ‡ immersed in the shallow cup. Wood reddish, coarse: common.

11 Q. palùstris Mx. *Pin O.* (Figs. 1-4.) Sinuses deep and broad, lobes oftener 5. petioles long (1—2′), toothed as in Q. rubra; acorns 7—8″; nut ‡ immersed in the cup. Tree 60—80f, with a light open foliage, in wet, cool soils.

12 Q. falcàta L. *Spanish O.* Tree 60—70f; lvs. long-stalked, obtuse at base, ashy-tomentous beneath, lobes 5—7, narrow, simple or toothed, more or less *falcate* ; acorn globular, 4—5″, in a shallow subsessile cup. Va. to Fla.

13 Q. coccínea Wang. *Scarlet O.* Trees very large (80f); lvs. much like Q. rubra, but changing to *scarlet* in Autumn, while that becomes red-brown; acorn 7—8″, nut ‡—‡ immersed in the cup. In young shoots the leaves almost lose their lobes and teeth, but keep their bristles. Not rare.

β. *tinctòria. Black O.* Leaves oftener obovate in outline; bark black and bitter.

14 Q. Leàna N. *Lea's O.* Leaves oblong, blunt at base, margin with a few angular, very irregular lobes: acorn roundish, in a hemispherical cup. Rare. Ohio (Clark), Ill. (Wolf). A hybrid? but very constant.

15 Q. alba L. *White O.* (Fig. 339.) Lvs. short-petioled, acute at base, oblong, sinuate-pinnatifid, lobes subequal, obtuse; acorn sessile; nut oblong-ovoid, ‡ immersed in the tubercled cup. Timber very useful.

16 Q. obtusíloba Mx. *Iron O. Post O.* Tree middle size, wide-spreading; leaves cuneiform at base, downy beneath, deeply sinuate, the 3 upper lobes dilated, each 2-lobed; nut oval, half immersed, sweet. Timber good.

17 Q. macrocárpa Mx. *Moss-cup O.* (Figs. 340, 435.) Leaves deeply and lyrately sinuate-lobed (most deeply in the middle); *cup* very deep, *fringed* with the pointed scales, nut ‡ or more immersed, 1′. Common. W. and S.

18 Q. lyràta Walt. *Over-cup O.* Tree large; leaves acute at base, whitish beneath, with 7—9 triangular acute lobes; *cup rugged* with the scales, nearly or quite including the round nut. Swamps, S.

19 Q. bícolor Willd. *Swamp White O.* Tree handsome, 70f; leaves obovate, acute and entire at base, white-downy beneath, with 9 or more obtuse teeth or lobes; acorns in pairs on long (1—2′) peduncles. Low woods.

20 Q. Prinus L. *Swamp Chestnut O.* Tree 50—70f, with large (1′) sweet acorns; leaves 4—7′, obovate, crenate-undulate, downy beneath, with straight, strong veins; fruit ped. shorter than the petioles; nut ‡ immersed. (Q. monticola.)

β. *acuminàta.* Leaves oblanceolate, pointed, teeth sharp; fruit subsessile.

γ. *prinoìdes.* Shrub 3—4f; fruit crowded, sessile; leaves small.

2. CASTÀNEA, Tourn. CHESTNUT.

Sterile flowers in long, slender aments, fertile fls. few, 3 together, in an involucre. Cal. 6-lobed or parted. Sta. 8—20. ♀ Ovary 3–6-celled, with as many stigmas. Fr. a prickly involucre (burr), 4-valved, enclosing 1—3 coriaceous 1-seeded nuts. ♄ ♄ Leaves acuminate, expanding before the flowers. Fig. 381.

1 C. vesca L. Tree 50—80f, with a large straight trunk. Lvs. 6—9′ long, lance-oblong, serrate, smooth; nuts mostly 2 or 3 together; aments 6—9′, yellowish, in July, the brown nuts ripe in October. In woods.

2 **C. pùmila** Mx. *Chinquapin.* Shrub 6—12f, much branched; leaves obovate to oblong-ovate, downy beneath; nut solitary. N. J., W. and S.

3. **FAGUS**, Tourn. BEECH. Sterile flowers in capitate aments, suspended by a slender peduncle, fertile 2 within an involucre. Calyx 5- or 6 cleft or lobed. Stam. 5—12. ♀ Ovary 3-celled with 3 stigmas. Fruit a pair of 1-seeded, sharply 3-angled nuts in a prickly involucre. ♄ Leaves plicate in bud. May. Figs. 182, 256, B.

1 **F. ferrugínea** Ait. Tree 50—80f, with a smoothish ash-colored bark; lvs. ovate to oval, short-petioled, pointed, regularly and remotely toothed, hairy when young. Timber fine-grained. Hardly distinct from

2 F. SYLVÁTICA, the European Beech, which has broader leaves, and is occasionally cultivated, especially the variety with *purple leaves.*

4. **CÓRYLUS,** Tourn. HAZEL-NUT. Sterile flowers in a cylindrical ament, fertile flowers in a capitate one. Calyx represented by 2 scales in the axil of a bract. Stam. 8, with half-anthers. ♀ Ovary adherent, 2-ovuled, 2-styled. Nut bony, roundish, 1-seeded, enclosed in a many-cleft involucre. ♄ Leaves acuminate, expanding after the flowers. May.

1 **C. Americàna** Walt. Shrub 5—10f; leaves roundish, cordate; involucre bell-form, much wider than the nut, coarsely toothed. Thickets: common.

2 **C. rostràta** Ait. Shrub 3—6f; leaves ovate to oval; involucre bottle-shaped, longer than the nut, 2-parted, with toothed segments. Thickets.

3 C. AVELLÀNA. *Filbert.* Shrub 3—10f; leaves as in No. 1; involucre not larger than the large rounded nut. From Europe, rarely cultivated.

5. **ÓSTRYA,** Michl. LEVER-WOOD. HOP HORNBEAM. ♂ Aments cylindrical, hairy. Calyx a scale, with 8 1-celled bearded anthers. ♀ Aments loose, flowers in pairs under each deciduous scale; ovary with 2 stigmas, enclosed in a sac (involucre), which in the *hop-like* fruit is inflated, ovoid, and much larger than the nut. ♄ Wood very hard and strong.

O. Virgínica Willd. Small tree 20—30f; leaves elliptical, acuminate, serrate; buds acute; fertile ament oblong, pendulous, 2′. Woods. April, May.

6. **CARPÌNUS,** L. HORNBEAM. IRON-WOOD. ♂ Aments long, cylindric. Calyx a roundish ciliate scale, with 8—14 stamens, slightly bearded ♀ Aments loose, with large oblong *3-lobed bracts*, each 1-3-flowered. Calyx 6-toothed. Stigmas 2. Nut ribbed. ♄ April, May.

C. Americàna L. Tree small, 12—20f; leaves ovate-oblong, acuminate, serrate; bracts of the fertile aments becoming leaf-like, 1′ long. In woods.

ORDER CXXIII. BETULACEÆ. BIRCHWORTS.

Trees or *shrubs* with *bark* in thin layers, *leaves* alternate, simple, straight-veined, and with deciduous *stipules*. *Flowers* ☿, 3 together, in the axil of each 3-lobed bract of the ament. *Calyx* 0. ♂ *Stamens* distinct, definite. *Anthers* 2-celled. ♀ *Ovary* 2-celled, 2-ovuled, becoming in fruit a thin, 1-celled, 1-seeded nut. Figs. 163-4, 283, 296, 307, 312, 437

ORDER 124.—MYRICACEÆ.

1. BÉTULA, Tourn. BIRCH. ♂ Fls. in clustered, drooping, slender aments, bracts peltate, deeply 3-parted. *Calyx a scale,* sta. 4. ♀ Aments oblong-ovoid, bracts 3-lobed, 3-flowered. *Calyx* 0. Ovary tipped with 2 styles. Nut flattened, winged. ♄♄ Buds sessile. Flowers yellow, precocious, in Spring. Figs. 163–4, 437.

* Trees with a yellowish bark, smoothish leaves, and short, erect, ♀ aments......No. 1
* Trees with a reddish-brown bark and ovate-oblong, suberect, ♀ aments.....Nos. 2, 3
* Trees with a white bark, long-stalked leaves, and drooping ♀ aments........Nos. 4, 5
* Shrubs with brownish bark, roundish leaves, and short, erect, ♀ aments....Nos. 6, 7

1 **B. lùtea** Mx. *f.* *Yellow B.* A forest tree 40—80f, known at sight by its silver-yellow bark; leaves ovate, deeply and doubly serrate; ♂ aments 2—4', drooping, the ♀ ovoid-oblong, 1', erect. Can. to N. Car. (B. excelsa C-B. not of Ait. ?)

2 **B. lenta** L. *Black, Sweet,* or *Cherry B.* A noble tree, about 60f; lvs. cordate-oval, acuminate, sharply serrulate; ♂ aments 3—4', ♀ aments erect, pedunculate, much shorter. Woods, Can. to Ga. Timber rose-colored. Cambium (§ 418) sweet and spicy.

3 **B. nìgra** Ait. *Red B.* Tree 30—50f, the bark loose and torn; leaves rhomb-ovate, acute both ends, repand and serrulate, small, petioles hairy; ♂ aments 2—3', drooping, ♀ oval, *sessile,* erect, 6''. Swamps, Mass. to Fla. Twigs very slender.

4 **B. populifòlia** Ait. *White B.* Tree 30—40f, trunk white, twigs brown; leaves *deltoid* (Fig. 307), lobed and serrulate, acuminate. Thickets, Me. to Pa.

5 **B. papyràcea** Ait. *Paper,* or *Canoe B.* Tree 50—70f, trunk white, branches brown; lvs. ovate, acuminate, doubly serrate; ♀ aments 1' long. Mt. woods, Can. to Pa.

   β. *minor.* Shrub 6—9f, with smaller and merely *acute* leaves. White Mountains.

6 **B. pùmila** L. *Dwarf B.* Shrub 2—7f, branches (not glandular) and young leaves downy; lvs. rounded to obovate, serrate, 6—16''. Swamps, Ct. to Pa. (Prof. Porter).

7 **B. glandulòsa** Mx. Shrub 1—4f, upright, branches glabrous, *dotted with wart like glands;* leaves round-obovate, glabrous, crenate, 9''. Mts., N. and N-W.

   β. *rotundifòlia.* Shrublet prostrate, 6—12'; lvs. orbic. White Mts. (B. nana C-B.)

2. ALNUS, Tourn. ALDER. ♂ Flowers in cylindric, drooping aments, bracts peltate, with 5 scales and several flowers beneath. *Calyx 4-parted,* sta. 4, anth. 2-celled. ♀ Aments ovoid, bracts cuneate, truncate, thick, 2-flowered. *Calyx of 4 scales,* persistent. ♄♄ Buds peduncled.

* Fls. developed before the lvs. in early Spring. Fruit almost wingless.....Nos. 1, 2
* Fls. developed with or after the leaves. Fruit winged, No. 3,......wingless, No. 4

1 **A. incàna** Willd. *Speckled,* or *Black A.* Stems 8—20f; leaves obtuse at base, broad oval or ovate, sharp-serrate and some lobed, glaucous-downy beneath; stipules lance-oblong. Thickets by streams, N. Eng. to Wis. and Can.

2 **A. serrulàta** Ait. *Smooth A.* Stems in clumps, straightish, 10—15f; lvs. obovate, pointed, doubly serrulate, smooth; stipules elliptical, obtuse. Swamps.

3 **A. víridis** DC. *Mountain A.* Shrub 3—4f; lvs. oval, acute, clammy; stip. broad-ovate; fertile aments on long stalks, oval. Streams in mountains, northward.

4 **A. marítima** Muhl. Tree 20f; leaves glabrous, ovate to obovate, cuneate, serrulate; fertile aments ovoid-oblong, 1'. River banks, Del., and S.

ORDER CXXIV. MYRICACEÆ. GALEWORTS.

Shrubs with alternate, resinous-dotted, often fragrant leaves, with the *flowers* monœcious or diœcious, both kinds in scaly *aments,* and destitute of corolla or calyx. ♂ *Stamens* 2—8. ♀ *Ovary* 1-celled, with 1 erect ovule. *Stig.* filiform. *Fr.* dry or drupaceous, indehiscent. *Seed* with no albumen.

ORDER 125.—SALICACEÆ.

1. MYRÌCA, L. CANDLEBERRY MYRTLE. Fls. ♂ ♀, the ♂ in cylindrical aments; anth. 4—10 in each scale, large, 2-celled. ♀ Aments ovoid, ovary 1 to each bract, in a cup of 3—5 scales, stigmas 1—4, spreading. Drupes covered with wax or resinous dots. ♄ *Leaves undivided.*

* Stigmas 2 or 4. Fruit small (1—3″), ovoid..................................Nos. 1—3
* * Stigma solitary. Fruit large (6″), oblong. (Leitneria, Chapm.).............No. 4
1 **M. cerífera** L. *Bayberry.* Shrub 3—4f; lvs. 1—2′, oblong to oblanceolate, entire or a few remote teeth above; stam. about 6; aments 6—9″; *drupe* oval, 2″, covered *with white wax* (bayberry tallow). Coasts, Can. to Fla.
  β. **Carolinénsis.** Lvs. large (3—5′), evergreen, tapering to the petiole. M. and S.
  γ. **pùmila.** Leaves linear-oblanceolate, acute at each end. 1—3f. S.
2 **M. Gale** L. *Sweet Gale.* Shrub 3—4f; leaves wedge-oblong, obtuse and serrulate at apex, 1—1½′; aments 4—8″; *nuts crowded,* 1″, *reddish.* Shores.
3 **M. inodòra** Bartr. Shrub 6—16f, with whitish bark; lvs. thick, evergreen, 1—2′, oblong, obtuse, entire, with rolled edges; *drupe* 3″, *ovoid, black.* Fla.
4 **M. Floridàna** (Chapm.) Shrub 2—6f, with brown bark; lvs. oblanceolate, acute, entire, long-stalked, deciduous; *drupe oblong, greenish,* 6″. Mid. Fla.!

2. COMPTÒNIA, Sol. SWEET FERN. Fls. ♂, the ♂ in cylindric aments, with reniform pointed bracts and 3—6 stamens. ♀ Aments globular. Ovary surrounded by 6 linear scales longer than the bracts. Nut ovoid. ♄ *Leaves pinnatifid,* narrow, fern-like, stipulate.

C. **asplenifòlia** Ait.—Dry hills, Can. to Va. Shrub 2f, with brown twigs, the very fragrant leaves 3—5′ long, with 20—30 wing-like lobes. Stipules pointed.

ORDER CXXV. SALICACEÆ. WILLOW-WORTS.

Trees or *shrubs* with alternate, simple *leaves* and deciduous or persistent *stipules. Flowers* ♂ ♀, both kinds in *aments,* one under each bract of the ament. *Calyx* none or cup-form and entire. *Ovary* 1-2-celled, with 2 short *styles. Fruit* a capsule, 2-valved, ∞-seeded. *Seeds* with a tuft of hairs (coma) and no albumen. Figs. 17-20, 200, 287.

1. SALIX, Tourn. WILLOW. OSIER. Aments cylindric, *bracts* imbricated, *entire,* 1-flowered, no calyx, but a little nectariferous gland instead. ♂ Sta. 2—7. ♀ Ovary ovoid-acuminate, stigmas 2, short. Caps. 1-celled, the valves revolute when open. Seeds ∞. ♄ ♄ ♄ Branches mostly long and slender. Leaves mostly narrow and pointed, and with stipules. Nos. 4, 10, and 21 are used in basket-making.

§ Stamens 3—10. Aments *with* the leaves, scales green-yellow, caducous ....Nos. 1—3
§ Stamens 2, the filaments united. Aments precocious, scales black................No. 4
§ Stamens 2, rarely 3 (1 in No. 13), the filaments distinct...(*)
  * Scales yellow-green. Am. *with* the lvs.—*a* Ov. subsessile, glabrous. Trees....5—7
      - *a* Ovaries stalked. Shrubs......Nos. 8, 9
  * Scales of the ♀ aments brownish or blackish, persistent...(*b*)
    b Ovaries and pods sessile. Shrubs..................................Nos. 10, 11
    b Ovaries and pods stalked, and glabrous. Aments *with* the lvs.....Nos. 12, 13
    b Ovaries and pods stalked, and downy or silky...(*c*)
      c Aments appearing with the leaves. Shrubs.................Nos. 14—16
      c Aments appearing before the subentire hairy leaves........Nos. 17—19
      c Am. before the serrate, smooth or downy long petioled lvs....Nos. 20, 21

ORDER 125.—SALICACEÆ.

1 S. lúcida Muhl. *Shining W.* Tree small, handsome, 5—15f; branches green; lvs. smooth and shining, lance-ovate, acuminate with a long point; stip. serrate; stam. mostly 5. Along streams, especially northward and northwest. Often cultivated.

2 S. pentándra. *Bay W.* Tree 20—40f, very elegant, in shrubberies; lvs. lance-ovate, cuspidate-pointed, shining; twigs reddened; aments yellow; sta. 5+. Europe.

3 S. nigra Marshall. *Black W.* Shrub 10—20f; leaves linear-lanceolate, attenuate to both ends; stip. small, caducous; branches pale yellow; stamens 3—5. Common.

4 S. purpùrea L. Shrub 6—10f, with long, slender, olive-colored twigs; leaves very smooth, oblanceolate; 1 filament with 2 anthers. Low grounds. †

5 S. frágilis L. *Crack W. Bedford W.* Trees tall (60—80f), of quick growth, with greenish divergent twigs brittle at base (like many other species); leaves lanceolate; stipules caducous; stamens 2, rarely 3. Often planted in parks. § Europe.

   β. *decípiens.* A smaller tree, with red polished twigs and upper leaves obovate.

   γ. *Russelliana,* has long-pointed, serrate, bright lvs. with conspicuous stipules.

6 S. alba L. *White W. Yellow W.* Large trees, with straight branches and yellowish tough twigs; lvs. lanceolate with a straight point, and silky-whitish, especially beneath; stigmas subsessile, 2-lobed. Common, of rapid growth. §

   β. *vitellina,* has shining, yellow branches, with narrower leaves.

   γ. *cœrùlea,* leaves bluish, nearly or quite smooth beneath. By rivers.

7 S. Babylónica L. *Weeping W.* Tree of large size, with long, slender, pendent branches; lvs. linear-lanceolate, acuminate; stipules roundish; ♀ aments 1—2′ long, the ♂ unknown in U. S.—β. annuláris, leaves curled into a ring. Not drooping.

8 S. longifòlia Muhl. Shrub diffuse, 2—10f, with whitish twigs; lvs. long, linear, pointed both ways, remotely toothed, hairy. River banks, N. Eng. and W.

9 S. myrtilloides L. Shrub low, erect, glabrous; lvs. elliptic-oblong, entire, acute or obtuse. Mountain bogs, N. and N-W. (S. pedicellaris Ph.)

10 S. viminàlis L. *Basket Osier.* Stems long, straight, slender, 10—12f; lvs. lance-linear, long, pointed, silky-canescent beneath; aments precocious. Wet.

11 S. herbàcea L. *Arctic W.* Low, creeping, 1—2′ high; lvs. round-oval, cordate, serrate, glabrous; aments few-flowered, terminal. Summits of White Mountains.

12 S. cordata Muhl. Shrub 6—8f, with smooth, green branches; lvs. lance-oblong, cordate, acuminate, smooth; stipules large, serrate. Wet grounds.

   β. *myricoides.* Leaves not cordate, with 2 glands at base, glaucous beneath.

   γ. *angustàta.* Leaves lanceolate, acute at base; stipules much smaller.

13 S. Cútleri Tuckm. Low, prostrate; lvs. elliptic to obovate, shining above; stamen single; aments pedunculate, dense. White Mountains. (S. uva-ursi C-B.)

14 S. vagans, β. *rostráta* (Andersson). Shrub 3—12f, with straight, erect, yellowish branches; leaves lance-ovate to lance-obovate, acute, subentire, glaucous-downy beneath; stip. toothed; fertile aments becoming long and loose; ovaries long-pointed (*rostrate*). Dry grounds, Penn., N. and W. (S. livida Wahl.)

15 S. argyrocárpa And. Shrub low, creeping; leaves lance-oblong or -linear, glaucous beneath with appressed silvery hairs; pod short-conical, silvery-silky, style slender. White Mountains. Young plants all silvery. (S. repens C-B.)

16 S. chlorophýlla And. Shrub low, spreading; lvs. glabrous, glaucous beneath; lanceolate to oblanceolate, subentire; fruit very short-stalked; style very long, stigma entire; stipules 0. White Mountains, and N. (S. phylicifolia C-B.)

17 S. trístis Ait. *Sage W.* Small downy shrub with a profusion of small naked aments; leaves lance-linear to oblanceolate; stipules minute, caducous. Dry fields.

18 S. hùmilis Marsh. Shrub 4—8f, with brown twigs; lvs. oblanceolate; stip. lunate, subdentate, shorter than the distinct petioles. Dry. (S. Muhlenberghiana Barr.)

19 S. cándida Willd. Shrub 4—6f, handsome, all whitish; leaves linear-lanceolate, very long; stipules lanceolate, as long as the petioles. In damp woods. Common.

20 S. díscolor Muhl. Shrub 7—15f; branches greenish-brown; leaves lance-oblong, remotely toothed, glaucous beneath; stipules lunate, toothed or entire; ov. conical, densely silky; *stigmas long, linear.* Swamps. (S. eriocephalus Mx.)

21 S. petiolàris Sm. Shrub 4—15f, twigs long, slender, tough, purplish or yellowish; lvs. linear-lanceolate, smooth, glaucous beneath; stipules lunate, dentate; ovaries ovoid, densely silky, *stigmas very short*. Sandy banks of streams.

β. *sericea*. Lvs. grayish-silky beneath; stigma sessile; stipules deciduous.

2. PÓPULUS, Tourn. POPLAR. ASPEN. Aments cylindric, scales lacerate-fringed. Cal. an oblique, disk-like cup, its margin entire. ♂ Sta. 8—30. ♀ Ova. free, stig. very large, 2-lobed. Caps. 2-valved, 2-celled. ♄ Large trees, with soft, light wood. Leaves broad, on long, often compressed petioles. Aments lateral, before the leaves.

§ Buds not viscid. Leaves lobed, always white-downy beneath..............No. 1
§ Buds not viscid. Leaves round-ovate, soon glabrous and green..........Nos. 2—4
§ Buds viscid with a resinous varnish. Leaves always glabrous...(*x*)
   x Leaves ovate, whitened beneath. Stamens 20—30 ....Nos. 5, 6
   x Leaves deltoid or deltoid-ovate Stam. 6—30. Petioles compressed..Nos. 7—9

1 P. ALBA. *Abele P. Silver-leaf P.* Tree rapidly growing, and spreading by the roots; leaves cordate, lobed, dark green above, very white beneath. Europe.

2 P. tremuloìdes Mx. *American Aspen.* Tree 25—40f; bark smooth, greenish; lvs. roundish-cordate, abruptly pointed, dent-serrate; petioles compressed, rendering the leaves *tremulous* in the slightest breeze. Woods: common.

3 P. heterophýlla L. *Cotton-wood.* Tree 40—60f, with smooth greenish bark; lvs. roundish, cordate or ovate, serrate, white-downy when young; buds very downy, short, obtuse; stamens very many; seed with much *cotton*. Wet woods.

4 P. grandidentàta Mx. Tree some 40f. with smoothish gray bark; lvs. round-ovate, acute, with large unequal sinuate teeth, villous when young; buds subpubescent; petioles compressed. Woods. Common northward.

5 P. balsamìfera L. *Balsam P. Tacmahac.* Tree 40—80f, with rough bark; lvs. ovate, acuminate, with appressed serratures; buds very fragrant. Wet. N.

6 P. candicans Ait. *Balm-of-Gilead.* Tree 30—50f; lvs. ovate, cordate, acuminate, serrate; petiole hairy; buds full of fragrant resin. Woods, and cultivated.

7 P. angulàta Ait. *Western Cotton-wood.* Tree 40—80f, branches acutely angular or winged; leaves deltoid-ovate, or broad-cordate; buds little viscid. S. and W.

8 P. Canadénsis De f. *Necklace Cotton-wood.* Tree 40—80f; young branchlets angular; lvs. deltoid to oval, acuminate, crose-denticulate, subcordate; ament scales not hairy. By rivers and lakes. N. and W. (P. monilifera Ait.)

9 P. nìgra, β. *betulifolia*. *Black P.* Tree with an ovoid form, 30—40f; young branches a d lvs. pubes.; lvs. deltoid-rhombic, pointed, crenate-serrate. N. Y.; rare. †
  γ. *dilatàta*. *Lombardy P.* Tree very tall, pyramidal in form; lvs. deltoid. Com.

Class II. GYMNOSPERMÆ.

Pistils none, or represented by open scales, with ovules in their axils. Stigma none, but the pollen applied directly to the ovules, which become naked seeds, destitute of a true pericarp. Flowers always diclinous. Cotyledons often more than 2. (§ 510.)

COHORT 4. CONOIDEÆ. Equivalent to the Class. (§ 515.)

ORDER CXXVI. CYCADACEÆ. CYCADS.

Trees of low stature, simple trunks with their internodes undeveloped and the surface scarred with the fallen leaves. *Leaves* pinnate, parallel-veined, circinnate. *Flowers* diœcious, naked, in cones, ♂ *anther* covering the under surface of the connectile. ♀ *Scales* peltate, bearing naked ovules dorsal or marginal.

1. CYCAS REVOLÙTA, from Japan, hardy South, has a short thick trunk, crowned with numerous pinnate leaves, 4—5f long, with innumerable linear 1-veined leaflets rolled at the edges. Fruit in an oblong spadix.

2. ZÀMIA INTEGRIFÒLIA. COONTIE. FLORIDA ARROW-ROOT. Stem corm-like, abounding in starch. Leaves 3—5f long, leaflets 3—5', lance-oblong. *jointed* to the rachis, entire, ∞-veined; fruit in a large oblong cone. S. Fla.

ORDER CXXVII. CONIFERÆ. CONIFERS.

Trees or *shrubs* mostly evergreen, abounding with a resinous juice. *Leaves* scattered or fascicled, mostly linear, parallel or fork-veined. *Flowers* ♂ ♀ or ☿, destitute of corolla or calyx, in aments and cones. ♂ *Stamen* 1, or several united. ♀ *Ovary, style,* and *stigma* wanting. *Ovules* 1—∞ at the base of the carpellary scale. *Fruit* a strobile (cone) with the scales woody and distinct, or baccate with the scales fleshy and coherent. Figs. 107, 166, 194, 216, 256, 293, 299, 352–3, 473–4, 491. See Hoopes' *Book of Evergreens.*

§ ABIETINEÆ. Scales of the cone each with a bract beneath it. Seeds 2, winged...(*)
§§ CUPRESSINEÆ. Scales bractless. Seeds 1—9, mostly with 2 wings...(**)

```
* Leaves evergreen, linear, 2—5 together in each fascicle ................PINUS.       1
* Leaves evergreen, linear, solitary, scattered..........................ABIES.       2
* Leaves in fascicles of many together,—a evergreen......................CEDRUS.      3
                              —a deciduous..............................LARIX.       4
   ** Cones baccate, consisting of the fleshy coherent scales............JUNIPERUS.   5
   ** Cones dry, scales imbricated.—x Leaves lance-linear................CUNNINGHAMIA. 6
                    —x Leaves scale-form, opposite, 4-rowed.............THUYA.        7
   ** Cones dry, scales valvately closed.—y lvs. scale-form, opposite, 4-rowed...CUPRESSUS. 8
                    —y Lvs. linear, alternate, deciduous................TAXODIUM.    9
                    —y Lvs. alternate, evergreen. †....................SEQUOYA.     10
```

1. PINUS, L. PINE. Fls. ☿, the ♂ in clustered aments. Stamen 1, with a 2-celled anther. ♀ Aments of many open imbricated carpellary scales, each with a bract at the back and 2 inverted ovules at base within. Cone woody, persistent two years, the scales often thickened and awned at the tip. Seeds nut-like, winged. Cotyledons 3—12. ♄ Fascicles of 2—5 linear-filiform leaves, sheathed at base.

§ Leaves in 5's.—*x* Scales spineless, hardly thickened at end............Nos. 1—3
                  —*x* Scales ending with a cap and a spine.............No 4
§ Leaves in 3's.—*y* Cones oblong, with small recurved spines...........Nos. 5, 6
                  —*y* Cones ovoid, with weak or strong spines..........Nos. 7—9
§ Leaves in 2's.—*z* Scales tipped with spines or prickles.............Nos. 10—12
                  —*z* Scales spineless.—*a* Trees native..............Nos. 13, 14
                                         —*a* Trees European..........Nos. 15, 16

1 P. stròbus L. *White Pine.* A majestic tree, 100—170f. in the forests; lvs. needle-shaped, 4—5', not rigid; cones oblong, 5—7', pendulous. Woods, Penn., N. and N-W. Timber of great value in architecture.

2 P. EXCÉLSA. *Bhotan P.* Lvs. glaucous, 5—7'; cones cylindric, 6—9'; nuts winged. Asia

Order 127.—CONIFERÆ.

3 P. cembra. *Stone P.* Leaves 2—3′; cones ovate, erect; seeds hard, wingless. Alps.
4 P. aristàta. *Colorado P.* Leaves 1—1½′, crowded; cones oval, 2½′. Tree 40—50f.
5 P. austràlis Mx. *Long-leaved P.* Tree 60—100f, very resinous; leaves 10—15′, crowded; cones lance-oblong, nearly as long as the leaves. Stands in extensive forests, South. Very valuable for turpentine, timber, or fuel.
6 P. Tæda L. *Loblolly P.* Tree 50—90f; leaves 6—10′, with long sheaths; cones deflexed, half as long as the leaves, with small but strong spines. S.
7 P. scrótina Mx. *Pond P.* Tree 30—50f; leaves 5—8′, rigid; cones broadly ovoid, polished, nearly spineless, as large as a goose-egg. Wet lands, S.
8 P. rígida Mill. *Pitch P.* Tree 30—70f, with very rough bark; leaves *rigid*, 4—6′, with short sheaths; cones clustered, ovoid-conic, 2—3′. Sandy barrens.
9 P. ponderòsa. Tree 50—100f in California, with sturdy trunk, smoothish bark, heavy wood; leaves 9—12′; cones 3½′, conical, with short strong spines.
10 P. mitis Mx. *Yellow P. Spruce P.* Tree of slow growth, 30—60f; lvs. covering the branchlets, some of them in 3's, mostly in pairs, 3—5′, slender; cones 1½—2′, ovoid-conic, clustered. In dry lands. Timber very valuable.
11 P. pungens Mx. Tree with crooked branches, 20—30f; leaves stout, crowded, about 2′; cones ovoid, 3′, with stout spines 3″ long. Mountains, Penn., and S.
12 P. inops Ait. *Jersey P. Scrub P.* Tree 15—25f, rough and crooked; lvs. rigid, obtuse, 2—3′; cones ovoid-oblong, 2—3′, with straightish prickles. Barrens.
13 P. resinòsa Ait. *Norway P. Red P.* Tree 60f, bark smoothish; lvs. slender, 5—6′, sheaths 6—12″; cones conic with a rounded base, half as long as the leaves. Dry woods, Penn. to Wis., and N. Timber compact, moderately resinous.
14 P. Hudsònica Poir. (P. Banksiana Lamb.) A straggling pine 5—25f; lvs. rigid, curved, 1′, the cones longer (1½—2′), recurved, smooth. Rocks, Me., W. and N.
15 P. Làrico. *Corsican Pine.* A large tree of rapid growth, very handsome in parks; leaves slender, bright green, wavy, 4—6′; cones 2—3′. Branches whorled.
  β. **austrìaca.** *Austrian P.* Leaves more rigid, of a rich dark-green color.
16 P. sylvéstris. *Scotch P.* Tree of rapid growth, perfectly hardy; lvs. 2—4′, twisted, rigid, bluish green; cones ovoid-conic, 2—3′. Common in Europe.

2. **ÁBIES**, Tourn. Spruce. Fir. Hemlock. ♂ Aments clustered with the old lvs. ♀ Am. solitary, cones with thin, flat, spineless scales, persistent one year. Seeds winged. Cotyledons 3—9. ♅ Lvs. solitary, not sheathed, scattered over the branches, linear, short, mostly petioled.

§ *Fir.* Cones erect, the scales deciduous. Lvs. flat, spreading two ways...(x)
§ *Spruce.* Cones nodding. Lvs. 4-sided or ensiform, pointing all around...(a)
§ *Hemlock.* Cones hanging. Leaves flat, mostly spreading two ways.........Nos. 1—3
  a Cones oval, 1—2′ long, the scales nearly entire. Native. †.............Nos. 4, 5
  a Cones oblong, 3—8′ long, the scales erose-dentate. Cultivated.........Nos. 6, 7
    x Bracts conspicuously exserted, much longer than the scales........Nos. 8—10
    x Bracts shorter than the scales or rarely a little exserted...........Nos. 11—13

1 A. Canadénsis Mx. *Common H.* Tree 50—80f, very beautiful when young; lvs. short-linear (6—8″), glaucous beneath; cones ovoid, terminal, as long as the leaves, scales concealing the bracts. Rocky woods: common N.
2 A. Williausònii (or Pattoniana). Large tree in Oregon, very fine and hardly here, but rare; leaves yellowish, 6—8″, the cones three times longer, bracts concealed.
3 A. Douglássii. A huge tree in Oregon, handsome; cones with long, 3-forked bracts.
4 A. nigra Mx. *Double S.* Tree pyramidal, 60—80f; leaves 6—7″, dark green; cones ovoid, 1—2, scales erose-denticulate. Damp mountain woods, northward.
5 A. alba Mx. *Single S.* Tree 30—80f, subpyramidal; leaves 6—9″, glaucous; cones deciduous, cylindrical, 2′, with the scales entire. Rocky woods: common.
6 A. Pícea (or excelsa). *Norway S.* A stately tree with dense dark-green foliage; lvs. 9—12″; cones 5—8′ long, light brown, scales notched. Very common.

7 A. Menziesii. Tree 50—70f in Oregon ; lvs. ½', silvery-glaucous ; cones 3—4', man .
8 A. bracteata. Tree 100f in California ; leaves 2—3', silvery-glaucous beneath ; cones 4', bracts 3-lobed, middle lobe much exceeding the scale, and recurved.
9 A. pectinata. Tree from Europe, 80f; leaves 9'', obtuse, glaucous beneath ; cones 1—7', brown when ripe, bracts fringed, the cuspidate point spreading.
    β. Cephalónica, from Greece, bracts linear-oblong, toothed, reflexed.
    γ. Nordmánnia, from Crimea, bracts with an entire recurved point.
10 A. Fráseri Ph. *Double Balsam F.* Tree small (15—30f); bark smooth, *blistered* as in the next ; leaves 8—10'', seeming 3-veined beneath ; cones 1—2', oblong ; bracts denticulate, long-pointed, reflexed. White Mountains ! and Alleghanies.
11 A. balsámea Marsh. *Balsam F.* Tree 30—50f, with smooth bark filled with *blisters* (reservoirs) of *balsam ;* leaves 8—10'', obtuse, silvery beneath ; cones cylindrical, 3—4' × 1', bracts concealed or slightly exserted. Damp woods. Cultivated.
12 A. Sibírica (or Pichta). Small tree from Asia ; leaves 1' ; cones ovoid-conic, 3—4'.
13 A. grandis. Tree 200f in Oreg.; lvs. 1'—13'', bifid, silvery beneath ; cones oblong, 4'.

    3. CEDRUS, Link. ♂ Am. solitary, terminal. ♀ Cones persistent two or three years ; scales persistent, close-pressed ; bracts concealed adnate to the scales. ♄ Leaves sessile, fascicled as in Larix, rigid, evergreen.

1 C. Libáni. *Cedar of Lebanon.* Tree with wide-spread branches ; leaves 9—15'', dark green, acute ; cones oval, obtuse, brown, 3 × 2', scales very many.
2 C. Deódara. Huge tree in the Himalayas ; lvs. 1—2', light glaucous ; cones ovoid, 4'.

    4. LARIX, Tourn. Larch. Tamarack. ♂ Anthers 2-celled, cells opening lengthwise, with simple pollen grains. ♀ Cones erect, oval or roundish, scales colored, persistent. Seeds with a proper wing. ♄ Leaves deciduous, acerous, soft, scattered, and in axillary, many-leaved fascicles.

1 L. Americàna Mx. ↘A splendid tree 70—100f, with straight axis and horizontal branches ; leaves filiform, very slender, 1—2', in bunches of 12—20 ; cones deep purple, 6—10'', scales few, with inflexed edges. Woods northward. Common in cult.
    β. *péndula.* Branchlets slender and drooping. Exquisitely beautiful.
2 L. Europæa. Large tree ; lvs. flattened, linear-spatulate ; cones 1—1½' long.

    5. JUNIPERUS, L. Juniper. Fls. ♂ ♀, aments very small, roundish. ♂ Scales peltate, each with 4—7 anther-cells beneath. ♀ Scales few, united at base, 1-2-ovuled, forming a sort of berry in fruit. Cotyledons 2. ♄ ♄ Leaves subulate or scale-like, pungent, opposite or whorled.

    § Lvs. scale-form, opp., 4-rowed, and subulate in 3's, not jointed, nerveless...Nos. 1—3
    § Lvs. all subulate and in 3's, divaricate, jointed to the stem, 1-nerved .....Nos. 4—7

1 J. Virginiàna L. *Red Cedar.* Tree of middle size, dark green ; early lvs. very slender, 3—4'', little divergent, in 3's, later ones 1—2''. scale-form, 4-rowed, opposite, appressed ; cones or berries small, blue-white, on short branchlets. Rocky soils.
2 J. sabìna, β. *procúmbens* Ph. Shrub trailing ; lvs. opposite, obtuse, a gland in the middle, imbricated in 4 rows ; fruit larger (3''), nodding, dark purple. Rocks, N.
3 J. Bermudiàna L. Late branchlets very slender, covered with scale-form pungent lvs. in 4 rows, divergent, 1'' ; fr. brown, no bloom, 2'', subsessile. Fla. 15—20f.
4 J. commùnis L. *Common J.* (Fig. 353.) Tree or shrub ; leaves in 3's, crowded, pungent-acuminate, 6–8'', fruit small (2''), subsessile, dark-purple, sweetish. Woods.
    β. *alpìna.* Shrub trailing ; leaves more crowded, less spreading, curved. N.
    γ. oblónga. Branchlets drooping ; leaves lance-linear, glaucous ; fruit clustered.
5 J. rígida. *Weeping J.* Branchlets drooping ; lvs. channelled on the upper side. Japan.
6 J. Oxycèdrus. Shrub 10-12f, from Eur., is known by its red-brown berries 3—4'' long.
7 J. drupácea. Shrub from Syria, 8—12f, with berries dark-purple, as large as a plum

ORDER 128.—TAXACEÆ. 315

6. CUNNINGHÀMIA Sinénsis. Tree from China, 30—40f, very unique. Leaves 1—1½′, lance-linear, stiff and pungent, in 2 rows. Cones ovoid, 1½′, with toothed and pointed scales (or bracts ?) each 3-seeded.

7. THÚYA, Tourn. ARBOR VITÆ. Fls. ♂, on different branches, terminal. ♂ Anther-cells 4 on each imbricated scale. ♀ Scales few, in pairs, opposite, imbricated, each 2–6-ovuled. Seeds winged. ♄ Leaves scale-form, opposite, imbricated in 4 rows.

1 T. occidentàlis L. Tree branched from base to summit; leaves rhombic-ovate, tubercled on the back; cones oblong, scales not reflexed, each 2-seeded. On rocky banks, common N., now very frequent in cultivation. Many varieties.
2 T. (THUYOPSIS) DOLABRÀTA. Tree from Japan, 40—60f, with ovate scale-form lvs., not appressed; cones small, roundish, each scale 5-seeded. Rare.
3 T. (BIOTA) ORIENTÀLIS. Shrub light green, or yellowish; ramifications vertical; cones broad, with thick scales and horn-like reflexed points. China.

8. CUPRÉSSUS, Tourn. Aments ♂, small, roundish. ♀ Scales each with 2—∞ erect ovules. Cone globular, the scales angular, peltate, valvately closed until ripe. ♄ Leaves scale-form, flat, imbricated as in Thuya, often with a tubercle on the back. CYPRESS.

1 C. sempérvirens. Cone large, oval, 1′, scales ∞-seeded; lvs. minute, ovate, obtuse, very closely imbricated. Cultivated South. Tree strict, conical, 20—40f.
2 C. thyoìdes L. *White Cedar.* Tree pyramidal, filiform branchlets square; leaves minute, lance-ovate, close, the tubercle manifest. Swamps. Cones small as peas.
3 C. LAWSÒNII. Splendid tree from Oregon; branchlets flattened, feather-like, bluish-green; leaves lance-ovate, tubercled; cones 1½″. Becoming common.

9. TAXÒDIUM, Rich. BALD CYPRESS. Fls. ♂, sessile, small, roundish, the ♂ in spikes, ♀ in pairs below. Cone globular, the scales peltate, angular, thick, firmly closed till ripe, with 2 angular seeds at base. Cotyledons 6—9. ♄ With deciduous, linear, 2-rowed leaves.

T. dístichum Rich. Tree 100—125f, trunk 6—9f diam.; large conical excrescences grow up from the roots; lvs. light-green, scattered, in 2 rows on the slender branchlets. Swamps, Va., and S. Timber valuable.

10. SEQUÓYA, Endl. RED-WOOD. Cones roundish, with peltate trapezoid, 5-seeded scales, valvately closed. Seeds winged both sides. ♄ Immense, Californian. Leaves linear or subulate, alternate.

1 S. sempérvirens, Tree 200f, with a diam. of 10f; bark blackish, with rose-purple wood almost imperishable; cones globular, 1′; leaves of 2 kinds.
2 S. gigántea. Tree 300f, with a diam. of 20f (often larger!); bark cinnamon color, wood dull red, cones oval, near 2′; leaves mostly subulate. Rarely planted.

ORDER CXXVIII. TAXACEÆ. YEWS.

Trees or *shrubs*, with the general habit of the Pines, but with no cones, nor even the carpellary scale. *Flowers* consisting simply of anthers or an ovule involucrate with bracts. *Fruit* a nut-like seed, naked, or in a cup form dry or pulpy disk. *Cotyledons* 2. Fig. 166.

ORDER 129.—PALMACEÆ.

* Leaves linear. Anthers 5—8 on each scale. Seed sitting in a fleshy cup..............TAXUS. 1
* Leaves lance-linear. Anthers 4. Seed fleshy-coated or dry, not in a cup..............TORREYA. 2
* Leaves linear to ovate, 1-veined. Anthers 2. Seed inverted, in a shallow cup... ....PODOCARPUS. 3
* Leaves flabelliform, fork-veined. Anthers 2. Seed erect, in a deep cup....... ....SALISBURIA. 4

1. TAXUS, Tourn. YEW. Flowers axillary, the ♂ in aments. Stam. or bracts peltate, 5—8-lobed, with 5—8 anther-cells. ♀ Flower solitary. Ovule erect, becoming a nut-like seed, sitting in a deep fleshy cup-shaped disk. ♄♄ Leaves rigid, alternate, in 2 rows.

1 **T. Canadénsis** L. *Dwarf Y.* (Fig. 166.) Shrub low or prostrate, branches ascending; lvs. mucronate, revolute-edged, 9—12''; stam. with 5 anther-cells; fruit depressed-globous, a black seed in an amber-colored cup. Rocky soils, northward.

2 **T.** BACCÀTA. *English Y.* Tree of low stature, widely spreading; lvs. falcate, acute, flat, 10—12''; stam. with 6—8 anther-cells; fruit oblong-bell-form. Europe.

3 **T. brevifòlia** N. Tree 15—50f, branches ascending; lvs. 7—10'', very narrow; sta. with 6 anther-cells; fruit oval. Fla.? and Oreg. The species are all closely related.

2. TORRÈYA, Arn. Flowers axillary, the ♂ many in the ament, bracts in 4 rows. Stamens with 4 anther-cells. ♀ Ovule with few bracts, becoming drupe-like, at length a dry ovoid bony nut or seed. ♄♄ Leaves rigid, alternate, 2-rowed, pungent, lance-linear.

T. taxifòlia Arn. Tree 15—30f, with erect strict form, dark green; lvs. 1—1½' long, 2-ranked as well as the branchlets; fruit smooth, glaucous, ovoid, 9—11''. Fla. †

3. PODOCÁRPUS, L'Her., contains some rare evergreens with remarkably large leaves (2—3' long). As yet very sparingly cultivated.

4. SALISBÙRIA ADIANTIFÒLIA (or Ginkgo biloba). Tree 40—80f, from Japan, strict and pyramidal. Lvs. fan-shaped, 2-lobed, fork-veined and petiolate, in structure much like the Maidenhair Fern. The flowers and fruit are seldom seen.

PROVINCE, ENDOGENS,

THE MONOCOTYLEDONOUS PLANTS. Stems without the distinction of bark, wood, and pith, endogenous in growth (§ 421). Leaves mostly parallel-veined and alternate. Flowers 3-parted (rarely ⚥). Embryo with one cotyledon. (Prov. Acrogens, 360.)

CLASS III. **PETALIFERÆ.** Endogenous plants having flowers either with a whorled perianth or without one, but *never glumaceous.* (Class IV. GLUMIFERÆ. Page 355.)

COHORT 5. **SPADICIFLORÆ.** Flowers crowded on a thickened or club-shaped rachis (spadix), mostly naked, rarely with a scale-like perianth. (Cohort 6, p. 322.)

ORDER CXXIX. PALMACEÆ. PALMS.

Trees or *shrubs*, chiefly with unbranched trunks growing by the terminal bud. *Leaves* large, plaited, on sheathing petioles, collected in one terminal cluster. *Flowers* perfect or polygamous, on a branching spadix bursting from a spathe. *Perianth* double, 3-merous, hexandrous, *ovaries* (and *styles*) 3, distinct or commonly united into 1, each 1-ovuled. *Fruit* fleshy, 1-3-seeded. Fig. 508.

* Flowers all perfect. Ovaries and styles united into 1. Berry single..................SABAL. 1
* Flowers perfect and staminate. Ovaries and styles distinct. Drupes 3...............CHAMÆROPS. 2

1. SABAL, Adans. PALMETTO. Fls. ☿, sessile, complete. Sepals 3 united, petals 3, subdistinct. Sta. 6, fil. distinct. Ovaries 3, soon united, Sty. 1. Fr. a dryish 3-seeded berry. ♄ ♄ Caudex (§ 227) procumbent or erect, beset with the persistent bases of the petioles. Lvs. palmately fan-shaped, many-cleft. Flowers small, greenish. June—Aug.

1 S. Palmétto Loddig. Caudex erect, 20—50f, usually enlarged above; the majestic lvs. are 6—10f long, all from one terminal bud; spadix much shorter than the leaves, spathe double; berry globular. Along the coast, Fla. to S. C.

2 S. Adansòni Guern. Caudex prostrate; lvs. rigid, longer than the even-edged petioles; spadix slender, much branched, as high (3—4f) as the leaves; style thick, obtuse; berry depressed. Along the coast, in low grounds, S.

3 S. serrulàta R. & S. Caudex creeping; petioles *aculeate-serrate;* spadix thick, 2—3f; style subulate; berry oblong-ovoid. Barrens, S. C. to Fla.

β. *minima.* Every way smaller; leaves about 7-cleft. E. Fla.

2. CHAMÆROPS, L. BLUE PALMETTO. Fls. ☿ and ♂. Perianth as in Sabal. Sta. 6 or 9, connate at base. Ovaries 3, distinct, stig. sessile. Berries 3, 1-seeded. Palms acaulescent. Petioles aculeate. Spadix dense-flowered, flowers yellowish. June—Aug.

C. Hystrix Fraser. Caudex low, making offsets at base; leaves 3—4f, the petioles spiny in the axils; drupes ovoid, hairy, in masses. Clay soils, Ga., Fla.

ORDER CXXX. ARACEÆ. AROIDS.

Herbs with a creeping rhizome or corm, an acrid or pungent juice, *leaves* often veiny, and the *flowers* mostly diclinous and naked. *Inflorescence* a spadix, dense-flowered, naked or mostly surrounded with a large spathe. *Perianth* none, or of 4—6 scales. *Anthers* extrorse. *Ovary* free, *stigma* sessile. *Fruit* baccate or dry, *seeds* albuminous. Figs. 432, 436.

* House, or greenhouse plants, usually with very large leaves...(y)
* Wild native plants, growing in water or damp places...(α)
   a Spadix growing to the spathe. ♀ Flower solitary. Floating................PISTIA. 1
   α Spadix free, enveloped in the spathe...(c)
    c Spadix naked, destitute of a spathe.—*h* Leaves ensiform......................ACORUS. 7
                  —*h* Leaves oval, &c..................ORONTIUM. 6
    z Flowers covering only the base of the spadix. Perianth 0.............ARISÆMA. 2
    c Flowers covering the whole spadix, or all but the base, and...(f)
      d Monœcious. Spathe involute. Stamen around a shield..............PELTANDRA. 3
      d All perfect.—*x* Perianth 0. Spathe open, white. Swamps ......CALLA. 4
            —*x* Perianth regular. Spathe shell-form.............SYMPLOCARPUS. 5
   y Spadix naked at the top. Spathe yellowish. Leaves peltate...... ......COLOCASIA. 8
   y Spadix naked at the top. Spathe yellowish. Leaves not peltate........ PHILODENDRON. 9
   y Spadix covered with flowers. Spathe white.—*z* Leaves green only RICHARDIA. 10
                —*z* Leaves variegated .... CALADIUM. 11

1. PÍSTIA, L. Spathe tubular at base, spreading above. Fls. ⚥, few, the upper ♂ in an involucre, of 3—8 anther-cells. ♀ Fl. solitary, of a 1-celled ovary and thick style. Berry several-seeded. ⚘

P. **spathulàta** Mx. Floating free in still water; leaves 1—2', obovate-spatulate, rosulate, the veins lamellated beneath; spathe white. E. Fla.

2. ARISÆMA, Mart. DRAGON-ROOT. INDIAN TURNIP. Spathe convolute at base. Spadix with a long naked summit, flower-bearing at base. ♂ Fls. above the fertile, each merely a cluster of 4 or more stamens. ♀ Ovary 1-celled, stig. flat. Berry red, 1- or few-seeded. ♃ Root tuberous. Scape sheathed with the petioles.

1 A. **triphýllum** Torr. *Jack-in-the-pulpit.* Stem a large corm fiercely acrid; scape round, thick, 8—12'; leaves 2, trifoliate; leaflets oval, pointed, sessile; spathe striped, inflected over the club-shaped spadix. Rocky woods. April+.

2 A. **quinàtum** Wood. Leaves 1 or 2, with very long sheaths, one or both *quinate;* leaflets oval to lance-oval, acute, or obtuse, cuspidate, narrowed to a petiolule. Ga. to Car., in hilly woods. 1—2f. (A. polymorphum Buckley.)

3 A. **Dracóntium** Schott. *Green Dragon.* Leaf mostly 1, pedate, with 7—11 lance-oblong leaflets; spadix subulate, longer than the spathe. Bogs. 2f.

3. PELTÁNDRA, Raf. Spathe convolute. Spadix staminate above, pistillate below. Anth.-cells 8—12, opening at top, adnate to a thickened peltate connectile. Berries 1-∞-seeded. ♃ Leaves sagittate, the long petioles sheathing the scape. May, June.

1 P. **Virgínica** Raf. Leaves sagittate-hastate, the base lobes long and turned outward; spathe green, 4—6' long; berries green, 1-3-seeded. Marshes. 9—18'.

2 P. **glauca** Feay. Leaves sagittate-cordate, lobes rounded; spathe white and open at the top, 3'; berries ∞-seeded, red. Coastward, S. (Xanthosoma, Sch.)

4. CALLA, L. Spathe ovate, spreading, white. Spadix covered with the naked fls. Perianth 0. Fil. 6, slender, with 2-celled anthers. Berry red, depressed, 3–6-seeded. ♃ ⚘ Rhizome creeping. Leaves cordate.

C. **palústris** L.—Shallow waters, Pa., and N. Scape 4—6'. Leaves 2—3'. July.

5. SYMPLOCÁRPUS, Salisb. SKUNK CABBAGE. Spathe shell-form, thick, close to the ground in early Spring, preceding the leaves, incurved at base and apex. Spadix oval, covered with the dull purple, perfect fls. Perianth 4-parted. Berries 1-seeded. ♃ ⚘ Leaves all radical, very large.

S. **fœtidus** Salisb.—Swamps, meadows: common. Leaves cordate-oval, 12—20'.

6. ORÓNTIUM, L. GOLDEN CLUB. Spathe 0. Spadix cylindrical, yellow, crowning the naked scape. Perianth 4–6-sepalled. Sta. 4—6. Fr. a dry utricle, 1-seeded. ♃ ⚘ Leaves lanceolate, all radical.

O. **aquáticum** L.—Pools and brooks. 1f. Very smooth. Scape thickened upward, green at base, white above, the summit (flowers) golden yellow. June.

7. ÁCORUS, L. SWEET FLAG. Spathe 0. Spadix cylindric, sessile, issuing from the side of a leaf-like scape. Perianth 6-sepalled. Sta. 6. Ova and fruit 3-celled, capsular, ∞-seeded. ♃ Rhizome thick, aromatic. Lvs all radical, linear-ensiform like the scape.

ORDER 132.—TYPHACEÆ. 319

A. Cálamus L. Scape ensiform, continued long and leaf-like above the green, dense-flowered spadix. In wet soils. 2—3f. Root tastes warmly pungent. June, July.

8. COLOCÀSIA ANTIQUÒRUM, from Egypt, &c., has large (2—3f) ovate-sagittate, peltate, repand leaves, on petioles longer than the scape. Spathe erect, much longer than the spadix. Cultivated for food, and for ornament.

9. PHYLLODÉNDRON GRANDIFÒLIUM. Stems rooting, running or climbing. Leaves very large (2—4f), opaque, strongly veined, cordate-sagittate, acute, entire. Petioles terete, red-spotted. Spathe yellowish. S. America.

10. RICHÁRDIA AFRICÀNA (Kunth, Calla Æthiopica L.). Known everywhere as the Ægyptian Calla, but native of the Cape of Good Hope: is a grand house-plant, 2—4f, with large hastate-cordate leaves, round scapes, a large milk-white spathe rolled in at base and back at apex, surrounding a yellow cylindric spadix.

11. CALÀDIUM BÍCOLOR. Roots tuberous. Lvs. radical, peltate, hastate-cordate, short-pointed, variegated with crimson or purple at the centre, or pellucid at base, or white-spotted. A splendid leaf-plant. Panama!

ORDER CXXXI. LEMNACEÆ. DUCKMEATS.

Herbs minute, stemless, floating free upon the water, and consisting of a leaf-like frond, or a tuft of leaves, with one or more fibrous roots. *Flowers* bursting from the substance of the frond, or axillary, enclosed in a spathe, *the sterile* consisting of 1 or 2 stamens, *the fertile* of a 1-celled ovary. *Fruit* a utricle, with 1 or more seeds. *Emb.* straight, in fleshy albumen. Fig. 516.

1. LEMNA, L. DUCKMEAT. Fls. from a chink in the edge of the frond, 2 sterile, each a single recurved stamen, with 1 fertile,—an ovary with style and stigma. Ovules and seeds 1—7. ① ♃ Fronds 1—7" long Roots hair-like. Flowers rarely seen.

§ Ovule solitary. Frond with a single root. (LEMNA *proper*) ....Nos. 1—3
§ Ovules 2. Frond many-rooted. (Spirodela, Schleiden)........................No. 4

1 **L. trisúlca** L. Fronds oblong, as long (2—3") as their stalks, proliferous from their sides, thin, obtuse. Pools of clear water, in patches.
2 **L. perpusilla** Torr. Fronds thin, 3-veined, round-obovate, 1—2", in groups of 3—7; style slender; seed round-oblong, erect. Ponds, N. Y., W. and S. August.
3 **L. minor** L. Fronds thick, veinless, obovate or roundish, 1—2", single or in groups of 2—4; style short; seed ovoid, half-erect. Stagnant waters: common.
4 **L. polyrrhiza** L. Fronds oval, 2—3", thickish, 5-7-veined, purplish beneath, each with a bundle of black roots beneath. Stagnant waters: rare.

2. WÓLFFIA, Horkel. Fls. from the centre of the minute frond, 2 only; ♂ flower a stamen with a 1-celled anther. ♀ Ovary with a very short style, ovule and seed 1. ① Fronds ½—¼", rootless, separate.

W. Columbiàna Karsten. Frond round-oval. Floating, with Lemna, seeming mere specks of green—the least of all flowering plants. Not rare.

ORDER CXXXII. TYPHACEÆ. TYPHADS.

Herbs growing in marshes and ditches, with rigid, ensiform, sessile leaves. *Flowers* monœcious, arranged on a spadix or in heads, with no spathe

Perianth of a few scales, or a tuft of hairs, or 0. *Stamens* 1—4, with long, slender filaments. *Ovary* with 1 pendulous ovule. *Seed* albuminous, with an axial embryo. Fig. 211.

1. TYPHA, L. CAT-TAIL. REED-MACE. Spadix long, cylindric, dense, sterile above. ♂ Sta. 3 together, united into a common filament. ♀ Ova. pedicellate, surrounded at base by a hair-like pappus or calyx. ♃ Fls. very numerous, packed solid in the large brown terminal spadix.

- **1 T. latifòlia** L. Leaves linear, flat, exceeding the stem; spadix cylindric, the sterile and fertile contiguous. Tall and smooth, 3—5f, in swamps.
- **2 T. angustifòlia** L. Leaves linear, channelled, exceeding the stem; spadix cylindric, the sterile some remote from the fertile. Swamps. 2—4f.

2. SPARGÀNIUM, L. BURR REED. Spadices or globular heads many, the lower fertile, consisting of sessile pistils, each with 3—6 sepals, and forming 1-seeded nuts. Sterile heads a mass of stamens with scales intermixed. ♃ ≈ August.

- * Stigmas mostly 2. Stems of the inflorescence branching, erect..............No, 1
- * Stigma always single. Stem subsimple, erect or floating..................Nos. 2, 3
- **1 S. eurycárpum** Eng. Stout, 1—3f; lvs. very long, carinate beneath; fruit heads 1', nuts large, obpyramidal, truncate, sessile; sterile heads numerous. Borders of rivers and ponds, N. Eng. to Pa., and W. (S. ramosum C-B.)
- **2 S. simplex** Huds. Erect, slender, 1—2f; leaves triangular at base, long and narrow; sepals spatulate, denticulate; nuts beaked and stiped; heads 6—8'' broad, the ♂ more than the ♀. Ponds and bogs, N. and W.
  β. *natans*. Leaves floating, flat; stigma shorter than the style; heads few.
- **3 S. mínimum** Bauhin. Slender, weak, simple, erect or floating; leaves narrow, flat; heads few, axillary, small (3—4''); fruit scarcely beaked, sessile. Streams, N. Eng., and W. (S. angustifolium C-B.)

ORDER CXXXIII. NAIADACEÆ. NAIADS.

Water plants with jointed stems, and sheathing stipules, or sheathing petioles. *Flowers* perfect or diclinous, naked or with a 2–4-parted perianth. *Stamens* definite. *Ovaries* free, sessile, 1-ovuled. *Stigma* simple, often sessile. *Fruit* indehiscent. *Seed* without albumen, with a straight or curved embryo.

- * Flowers axillary, sessile, the staminate reduced to a single stamen...(a)
  - a Fertile flowers reduced to a single pistil, with 2 or 3 stigmas. Leaves opposite...NAJAS. 1
  - a Fertile flowers with about 4 pistils in a cup, with as many stigmas..............ZANNICHELLIA. 2
- * Flowers spadaceous, or 2—20, sessile on a spadix or spike...(b)
  - b Flowers monœcious, seated in 2 rows on the side of a linear, flat spadix ....ZOSTERA. 3
  - b Flowers perfect, naked, 2—5, 4-merous. Fruit raised on slender stipes..........RUPPIA. 4
  - b Flowers perfect. Perianth 4-sepalled. Stamens 4. Pistils and achenia 4.......POTAMOGRTON. 5

1. NAJAS, L. WATER NYMPH. ♂ Fl. a solitary stamen, in a little hooded spathe. ♀ Fl. a naked pistil with 2—4 subulate stigmas. Fr. a little 1-seeded, drupe-like nutlet. ≈ Entirely submersed. Lvs. opposite, linear, broader at base, toothed. Flowers axillary.

- **1 N. major** All. Stem frail and slender, 1—3f; leaves 1' and less, crowded above with conspicuous spinulous teeth; nutlets ovoid, 1½'' long. N. Y. (Clinton).

ORDER 133.—NAIADACEÆ. 321

2 N. Indica Cham., β. *gracillima.* Stems filiform, forking; leaves opposite and in 3's, very narrowly linear, remotely spinulous-serrate. N. Y. and Pa. (Porter).

3 N. fléxilis Rostk. Leaves narrowly linear, in 3's, 4's, and 6's, minutely serrulate, as well as their abruptly-widened sheathing base, 3—12″. Ponds: common.

2. **ZANNICHÉLLIA,** Micheli. HORN PONDWEED. Fls. ☿, both kinds together in the same axil. ♂ Sta. 1, with a slender fil. ♀ Cal. of 1 sepal, cor. 0. Ova. 4 or more, each with a style and stig. Fr. 4 or more oblique achenia. ≈ Submersed, with filiform branches, and linear, entire leaves.

Z. palústris L. Stems round, leafy, 1—2f; leaves opposite, grass-like, 2—3′; anther 4-celled; achenia 4—6, toothed on the back. Pools and ditches: rare.

3. **ZÓSTERA,** L. SEA WRACK. Spadix linear, leaf-like, bearing the ☿ fls. in 2 rows on one side. Perianth 0. ♂ Anther ovoid, sessile, opening lengthwise, with hair-like pollen. ♀ Ova. as long as the anther, style bifid. Utricle 1-seeded. ♃ ≈ Stipules united into a sheath. Leaves grass-like.

Z. marìna L. Rhizome creeping, sending up long simple stems; lvs. alternate, ribbon-like, 1—5f long; spadix 2′, in a spathe at base of a leaf. Grows in the sea, along shore, Me. to Ga., and is washed up by the waves.

4. **RÚPPIA,** L. DITCH-GRASS. Fls. ☿, 2 together on a spadix arising from the sheath of a leaf. Perianth 0. Anthers 2, large, sessile, 2-celled. Ovaries 4, fruit 2—4 dry drupes on pedicels. ♃ ≈ A grass-like plant, all submersed but the flowers. Flower-stalk at length very long.

R. marítima L. Stems filiform, branched, 2—5f; leaves linear-setaceous, 2—6′, on inflated sheaths; flowers arising to the surface. Seas, and lakes (Hankenson), E.

5. **POTAMOGÈTON,** Tourn. POND-WEED. Fls. ☿ on a spadix arising from a spathe. Cal. 4-sepalled. Anth. 4, alternate with sepals. Ova. 4. Ach. 4, sessile, flattened on one or two sides. Seeds curved or coiled. ≈ Mostly ♃, only the spadix with its 3—10 small green fls. arising to the surface of the water. Lvs. stipulate, the upper often opposite. Fr. July, Aug.

§ Leaves of two kinds, the floating oval-elliptical, coriaceous, petiolate;
   stipules free from the petiole, connate; submersed leaves thin...(*)
    * Submersed leaves linear or reduced to mere petioles............Nos. 1—4
    * Submersed leaves lanceolate, rarely lance-linear.................Nos. 5—8
§ Leaves all similar, submersed, mostly thin and membranous...(*a*)
   a Leaves lanceolate or lance-oblong, petiolate or merely sessile........Nos. 9, 10
   a Leaves oval or oblong, broad and clasping at base..................Nos. 11—13
   a Leaves linear or setaceous.—*x* Stipules 0, or adnate to the leaf......Nos. 14, 15
           — *x* Stipules free.—*y* Stems flat............Nos. 16, 17
                  — *y* Stems filiform......Nos. 18—20

1 P. natans L. Subsimple; floating lvs. 2—3′, lance-oblong, narrowly obtuse, on slender (2—6′) petioles; stipules long, linear; lower lvs. few, linear, 2—6′; spikes 1—2′, on thick peduncles much longer; fruit turgid, 3-keeled. Ponds and ditches.

2 P. Claytòni Tuckm. Simple; floating leaves lance-oblong, about 15-veined, 1—1¼′, longer than their petioles, opposite; lower lvs. linear, 3-veined, 3—6′ × 1″, spikes and their peduncles near 1′; fruit orbicular, 3-keeled. Streams and ponds: common.

β. *heterophýllus.* Petioles and peduncles longer than the leaves (2—3′). Mass.

3 P. hýbridus Mx. Stems branching, filiform; floating lvs. oval, 5-7-veined, 7—10″ their petioles shorter, suboppposite; spikes and their stalks 4—6″; lower lvs. linear-setaceous, 1—3′, many; fruit minute, dentate. Common.

β. *diversifolius.* Leaves nearly all floating, oval, the lower few and short.
4 **P. Spiríllus** Tuckm. Very delicate, branched ; floating lvs. oval to lanceolate, 5–9 veined, 7—10″, on short broad petioles ; lower leaves narrowly linear, obtuse, 1—2 , submersed ped. 1-2-flowered ; embryo a little *spiral.* Streams : rare.
5 **P. gramíneus** L. Stem much branched, terete ; floating lvs. long-stalked, ovate to oblong, acutish, 13-veined ; lower leaves lanceolate to lance-linear, pointed, stip obtuse ; fruit small, obtuse-angled. Common, and very variable.
6 **P. flúitans** Roth. Lvs. long-stalked, the floating thinnish, opposite, elliptic-oblong, the submersed linear-oblong, all acute both ways. 11–21-veined ; fruit acutely 3-keeled on the back. In ponds and rivers. (P. lonchitis Tuckm.)
7 **P. púlcher** Tuckm. Stem simple ; floating leaves ovate, subcordate, 25–35-veined, 5–5′, alternate ; upper submersed lvs. lanceolate, long-acuminate, undulate, the lower oval-oblong ; fruit 3-keeled. Penn., N. J. (Prof. Porter), N. and W. Rare.
8 **P. amplifòlius** Tuckm. Stems simple ; floating leaves oval to elliptical, 2¼—4′, 35–45-veined, on long, opposite stalks ; submersed lvs. larger than the floating, 5—7′, lanceolate, short-stalked, or sessile. Ponds. (P. fluitans C-B.)
9 **P. lucens** L. Leaves large, often *shining,* lance-oval, 3—5′ × 1′, pointed and *mucronate,* on short stalks ; spike 2′ ; fruit roundish, slightly keeled. Rivers and lakes.
10 **P. obrútus** Wood. Stem simple ; leaves *all submersed,* narrow-lanceolate, 3′, ob scurely 7-veined, subsessile, acute ; spike 1′, the stalk 2′ ; fruit inflated, acutely keeled, conspicuously umbilicate both sides. Slow waters. No floating leaves.
11 **P. prælóngus** Wulf. St. *very long,* branched ; lvs. lance-ovate to lanceolate, obtuse, half-clasping, often large ; peduncle very long (3—5′) ; fruit sharp-keeled. Rivers.
12 **P. perfollàtus** L. Stem branched ; lvs. cordate-clasping, roundish to ovate, obtuse ; ped. short, few-flowered ; fruit not keeled. Ponds and slow waters : common.
13 **P. críspus** L. Branched below ; leaves 3-veined, half-clasping, narrow-oblong obtuse. 1—2′, crisp-wavy ; fruit acuminate-beaked. Penn., and E. (Prof. Porter).
14 **P. pectinàtus** L. Stem flexuous, repeatedly forking ; leaves linear-setaceous, 2—3′ ; spike interrupted, on a long filiform peduncle ; fruit large (2″), rough. E. and N.
15 **P. Robbínsii** Oakes. Stem very branching ; leaves lance-linear, crowded, sheathing the stem with their bases ; spikes on short peduncles. N. and W.
16 **P. compréssus** L. St. branching, flattened ; lvs. linear, ∞-veined, 2—4′ × 1—2″ ; stip. obtuse ; spike 12-15-flowered, much shorter than the peduncle. Ponds.
17 **P. obtusifòlius** Mert. and Ktch. St. branching, flattened ; lvs. linear, 3-veined ; stip. obtuse ; spike 6–8-flowered, as long as the peduncle. Pa., and N·W.
18 **P. pauciflòrus** Ph. St. slightly flattened, much forked ; lvs. linear, 1—3″ × ½—1″ ; flowers few (3—12) in the spike ; fruit distinctly crested. Rivers, &c.
19 **P. pusíllus** L. Stem filiform, branched ; leaves linear, varying to capillary, 1-3 veined ; spikes 3-5-flowered, long-stalked ; fruit not keeled. Shallow waters.
20 **P. Tuckermàni** Robbins. Very slender and delicate, forked ; lvs. capillary and confervoid ; spike 6-9-flowered, on a very long peduncle (5′). Ponds, Pa., and N.

Cohort 6. FLORIDEÆ.

ENDOGENOUS PLANTS with the flowers usually perfect and complete, the perianth double, 3-parted, the outer often, and sometimes both, green.

Order CXXXIV. ALISMACEÆ. WATER PLANTAINS.

Marsh *herbs,* with parallel-veined, petiolate leaves and branching peduncles. *Flowers* perfect or monœcious, with a regular double perianth.

Order 134.—ALISMACEÆ.

Sepals 3, green. *Petals* 3, colored or green. *Stamens* hypogynous. *Ovaries* 3 or more, separating into as many distinct fruits.

§ BUTTOMEÆ. Petals colored. Carpels 6—20, each with ∞ ovules..............Hydrocleis. 1
§ ALISMEÆ. Petals colored. Carpels many, 1-2-seeded...(x)
§ JUNCAGINEÆ. Petals green. Carpels 3, each 1-3-seeded...(y)
   x Flowers monœcious. Stamens many..Sagittaria. 4
   x Flowers all perfect.—z Stamens 9—24 ....Echinodorus. 3
        —z Stamens 6. Flowers panicled.. ....Alisma. 2
   y Anthers oval. Carpels 1-seeded. Leaves radical..............................Triglochin. 5
   y Anthers linear. Carpels 2-3-seeded. Leaves cauline...........................Scheuchzeria. 6

1. HYDRÓCLEIS Humbóldtii (or Limnocharis), from Brazil, grows in pools, like Sagittaria, with long-stalked, oval, 7-veined leaves and large (2—3') orange-yellow flowers. Sepals small. Stamens 18—24. Ovaries 6.

2. ALÍSMA, L. Water Plantain. Sepals persistent. Petals involute in the bud. Ovaries and styles arranged in a circle, forming many flattened achenia. ♃ ⚬ Acaulescent.

A. Plantàgo L. β. *Americànum*. Lvs. 5-7-veined, ovate or oval, subcordate, pointed; scape many-flowered, fls. whorled, small, rose-white. Pools. 1—2f. July, Aug.

3. ECHINODÒRUS, Rich. Sepals persistent. Petals imbricate in bud. Sta. 6 — ∞. Ovaries and styles ∞, imbricated, forming many flattened, beaked achenia. ⚬ Scape creeping or erect. Fls. small, white, whorled.

1 E. **radicans** Eng. Leaves large (5—12'), 7-veined, cordate, ovate, on long petioles; scape prostrate, running and rooting; flowers clustered at the nodes, white; stam. 18—24; ovaries very many. ♃ Swamps, Ill. to Ga. June, July.

2 E. **rostràtus** Eng. Leaves 1—3', ovate, cordate, on long petioles; scapes erect, sharply angled; stamens 12; carpels ∞, strongly ribbed and beaked. ☉ West.

3 E. **párvulus** Eng. Leaves lance-elliptic, as long as the petioles (1'); scapes 3-6 flowered; stamens 9; carpels about 20, beakless; flowers about 3''. ☉ E. and W.

4. SAGITTÀRIA, L. Arrowhead. Fls. ♂ or ♂ ♀, in whorls of 3 on the scape, the lower fertile. Petals white, larger than the sepals, imbricated in bud. Sta. ∞. Ovaries very ∞, crowded in a head. Achenia flattened, margined, and beaked. ⚬ Juice milky. Leaves on long radical stalks, sagittate to linear. Summer.

* Leaves mostly arrow-shaped. Filaments slender, elongated............Nos. 1, 2
* Leaves lanceolate to linear, very rarely with narrow, base lobes...(a)
   a Filaments as long as the anthers. Pedicels all subequal..............No. 3
   a Filaments thick, shorter than anthers.—z Fertile pedicels very short.....No 4
        —z Pedicels subequal................. Nos. 5, 6

1 S. **variábilis** Eng. Scape 1—2f, 12 angled; sterile pedicels twice longer than the fertile; filaments much *longer* than the anthers; achenia with a conspicuous averted beak. Waters: common. Flowers about 1' broad. Varies exceedingly.
   a. Leaves lanceolate, with lance-linear lobes of the same length.
   β. *obtùsa*. Leaves ample (6—10'), broad-ovate, obtuse. Fls ♂ ♀. M., W., and S
   γ. *latifòlia*. Leaves ample, ovate, acute, their lobes ovate, pointed.
   δ. *gracilis*. Leaves and their spreading lobes long, linear, acute.
   ε. *pubéscens*. Plant pubescent all over; leaves and lobes ovate.

2 S. **calycìna** Eng. Scape soon procumbent; pedicels all subequal; bracts roundish; calyx closed on the fruit; filaments as long as the anthers. Waters. Leaves as in No. 1, but sometimes all linear and floating.

3 S. lanceolàta L. Leaves lance-oblong, rarely linear, tapering to the long petiole; scape branched; 2—3f; achenia obovate-falcate. Swamps, Va. to Fla.

4 S. heterophýlla Ph. Leaves linear-lanceolate, rarely some of them with 1 or 2 base lobes; scape simple, weak; achenia narrow, long-beaked. Common S. and W.

5 S. gramínea Mx. Scape erect, slender, 5—20′; leaves lance-ovate to linear, rarely sagittate; pedicels all equally slender; achenia beakless; flowers 8—9″ diameter.
β. *platyphýlla.* Leaves lance-ovate; flowers larger, 1′ broad. South.

6 S. pusílla N. Scape shorter than the leaves (2—4′); leaves linear, shorter than the petioles; flowers few, the fertile but one, deflexed; stamens about 7. N. J., and S.

7 S. natans Mx. Scape mostly erect, 3—6′; leaves oval-lanceolate, floating, obtuse, 3-veined; lower pedicels longest; achenia angular, short-beaked. South.

5. TRIGLÒCHIN, L. Arrow-Grass.

Sepals and petals concave, deciduous (green). Sta. 6, very short, anth. large, extrorse. Ova. 1-ovuled, 3—6, united and indehiscent in fruit. ♃ Leaves all radical, grass-like Scape jointless, and bractless. Flowers small. July.

1 T. marítimum L. Fruit ovate-oblong, grooved, of 6 united carpels; scape longer (9—18′) than the leaves. Salt marshes and Lake shores, northward.

2 T. palústre L. Fruit nearly linear, of 3 united carpels; scape scarcely longer than the numerous and very narrow leaves. Marshes, N. Y., and N. 6—12′.

6. SCHEUCHZÈRIA, L.

Sep. and pet. oblong, acute, persistent. Sta. 6, with linear anthers. Ovaries 1-2-ovuled, becoming flattened inflated capsules. ♃ Leaves cauline, sheathing at base, linear.

S. palústris L.—A rush-like plant, in swamps, Vt. to Ill. (J. Wolf). Root-stock horizontal, fleshy. Stem 1f. Leaves semicylindric, 4—8′. Flowers yellowish green, in a bracted raceme. Stamens large, exserted. July.

Order CXXXV. HYDROCHARIDACEÆ. Frogbits.

Aquatic *herbs*, with parallel-veined *leaves* and diclinous *flowers* solitary or spicate. *Perianth* regular, 3-6-parted, the inner segments petaloid. *Stamens* 3—12. *Ovary* adherent, 1-9-celled, with 3, 6, or 9 *stigmas*. *Fruit* dry or juicy, ∞-seeded, indehiscent.

* Leaves all radical, roundish, floating in stagnant waters..................LIMNOBIUM. 1
* Leaves opposite or verticillate in 3's and 4's on the stems, submersed.............ANACHARIS. 2
* Leaves all radical, grass-like, in water.................................VALLISNERIA. 3

1. LIMNÒBIUM, Rich. Frog's-bit. Fls. ⚥. Spathes subsessile, the ♂ 1-leaved, about 3-flwd., the ♀ 2-leaved, 1-flwd. Perianth showy, white. Sta. 6—12 (mere rudiments in ♀). Ov. 6-9-celled, becoming a ∞-seeded berry. ♃ Stoloniferous. Lvs. on long stalks, subcordate. July, Aug.

L. Spóngia Rich.—Lake Ont. (rare), and S. Lvs. 1—1½′, purplish and *spongy* beneath

2. ANÁCHARIS, Rich. Ditch Moss. Fls. ♂ ⚥ ♀, solitary. Spathe axillary, bifid. Perianth 6-parted, colored, small, the fertile excessively produced above the adherent ovary into a capillary tube. Style capillary, with 3 large stigmas. Fruit few-seeded. ♃ Wholly submersed. Aug.

A. Canadénsis Planc. Stems filiform, long, forking; very leafy; leaves linear-oblong, serrulate, 5—10″; tube of the dingy-white fls. 2—10′ long! Streams and bogs.

3. VALLISNÈRIA, Mich. EEL-GRASS. Fls. ♂ ♀. Spathe ovate, 2-4-parted. ♂ Spadix or spike covered with minute naked fls. ♀ Fl. solitary, a slender perianth with linear segm. and 3 bifid stig. Fr. cylindrical, ∞-seeded. ♃ ≈ Fertile flowers on long spiral scapes. July, Aug.

V. **spiràlis** L. Lvs. 1—2f long, obtuse, ¼' wide, scapes of the sterile plants short, of the fertile filiform, tortuous, 2—4f, bearing the single white fl. at or near the surface

ORDER CXXXVI. BURMANNIACEÆ.

Small annual *herbs*, with naked or scaly *stems* and scale-like tufted *leaves*. *Flowers* perfect. *Perianth* tubular, 6-toothed, adherent. *Stamens* 3 or 6. *Capsule* 1- or 3-celled. *Seeds* ∞, minute, in a loose testa.

1. APTÈRIA, N. Perianth tube longer than the slender teeth, which are alternately narrower. Caps. globular, 1-celled. ① Apparently leafless.
A. **setàcea** N. Erect, very slender, 4—6f, with remote subulate scales, and bearing above 1 or 2 racemes; flowers 3—4'', purplish, distant. Woods, Fla., and W.

2. BURMÀNNIA, L. Perianth tube scarcely produced above the ovary, often 3-winged below, limb with the 3 inner teeth much shorter. Capsule prismatic, often 3-winged, 3-celled. ① Leafless.

1 B. **biflòra** L. Stems capillary, simple, 2—3', with scarcely perceptible bracts, and 1 or 2 (rarely more) light-blue flowers, 2—3'' long at top. Swamps, Va., and S. Oct.
2 B. **capitàta** (L). Stem setaceous, 6—8', simple, bearing at top a dense cluster of white flowers, and a few subulate bracts. Uplands, S.; less common. Sept.

ORDER CXXXVII. ORCHIDACEÆ. ORCHIDS.

Herbs perennial with fleshy *roots*, simple, entire, parallel-veined *leaves*. *Flowers* very irregular, with an adherent, ringent *perianth* of 6 parts. *Sepals* 3, usually colored. *Petals* 3, odd one (lowest by the twisting of the ovary), called the lip, diverse in form from the others, sometimes lobed, often spurred. *Stamens* 3, gynandrous (consolidated with the style), some of them abortive, *pollen* powdery or waxy. *Ovary* inferior, 1-celled, *capsule* 3-valved. *Seeds* innumerable. Figs. 71, 105, 240, 247, b. 263, 291, 435.

§ CYPRIPEDIEÆ. Anthers, the 2 lateral fertile, the terminal petaloid...(a)
   a Lip a large, inflated, spurless sac. Petals and sepals spreading................CYPRIPEDIUM. 1
§ OPHRYDEÆ, &c. Anthers, only the upper one fertile, 2-celled...(b)
   b Lip a large inflated sac, 2-spurred under the apex. Leaf 1...............CALYPSO. 2
   b Lip produced behind into a spur, which is free from the ovary...(c)
   b Lip spurless, or the spur adheres to the ovary (except in No. 13)...(f)
     c Anther fixed; pollen-masses 2, club-shaped, in 2 separate cells......ORCHIS. 3
     c Anther lid-like, on the end of the stigma; pollen-masses 4.........TIPULARIA. 4
       d Plants brown and leafless, rarely with radical leaves...(e)
       d Plants green and (except No. 16) furnished with leaves...(m)
        e Lip hooded, i. e., its margins involute. Perianth spreading............BLETIA. 5
        e Lip concave, sessile, often with an adnate spur...............CORALLORHIZA. 6
        e Lip concave, raised on a claw. Plant with 1 late leaf ....APLECTRUM. 7
          m Lip flat. Flowers obscure, in racemes, nearly bractless...(n)
          m Lip flat, expanded and lobed, tubercled at base. Flowers showy.....ONCIDIUM. 8
          n Lip channelled, reflexed. Flowers whitish, in bracted spikes...(o)
          m Lip bearded or 3-lobed. Stamen lid-like. Flowers showy...(p)

Order 137.—ORCHIDACEÆ.

```
n  Lip entire, dilated.  Column minute.  (Leaf 1). ................MICROSTYLIS    9
n  Lip sagittate or cordate.  Column elongated.  Leaves 2........LIPPARIS.       10
n  Lip 2-lobed or cleft at apex.  Leaves 2 cauline, opposite.....LISTERA.        11
o  Lip with 2 lateral callosities, not at all saccate............SPIRANTHES.     12
o  Lip without callosities, saccate, or even spurred at base.....GOODYERA.       13
   x  Flowers greenish.  Lip posterior, and beardless............PONTHIEVA.      14
   x  Flowers purple.  Lip posterior, and bearded.................CALOPOGON.     15
   x  Flowers purplish.  Lip anterior (as in most Orchids)...(y)
y  Column free from the lip.  Calyx spreading...................POGONIA.        16
y  Column adnate to the lip below.  Calyx erect.  Leaves 0......ARETHUSA.       17
y  Column adherent to the lip.  Calyx spreading.  On trees, South..EPIDENDRUM.  18
```

1. CYPRIPÉDIUM, L. Lady's Slipper. The 2 lower sepals united into 1 leaf, or rarely distinct. Pet. spreading. Lip inflated, saccate, obtuse. Column terminated by a petaloid lobe (barren stamen), and bearing a 2-celled anther under each wing. ♃ With large plaited leaves and large showy flowers. May, June. Fig. 71.

§ Sepals 3, the two lower entirely distinct. Stem leafy........ ....No. 1
§ Sepals 2, the lower composed of two united nearly to the tip...(a)
   a Stem a leafless scape, 2-leaved at base. Flower rose-colored.............No. 2
   a Stem leafy.—x Flowers solitary or several, white or rose-colored.....Nos. 3, 4
                 —x Flowers 1–3, mostly 1, yellow. Plant pubescent......Nos. 5, 6

1 C. arietinum Ait. *Ram's Head.* Stems usually clustered, 8–12', each 1- or 2-flwd.; leaves elliptical; upper sep. oblong-ovate, the lateral sep. and pet. lin.-lanceolate, lip obconic, as long as the pet. Damp woods, N. Eng. to Wis., and N. Curious.

2 C. acaùle Ait. Scape 10–14', bearing a single large (2') flower; lvs. elliptic-oblong; pet. lanceolate, shorter than the large boat-shaped lip. In damp woods. Beautiful.

3 C. spectàbile Sw. Stem leafy, 2f, hairy; lvs. lance-ovate, acuminate; sep. broad-ovate, obtuse, the lower (double) one smaller; lip 2', white-purple. Swamps. Superb.

4 C. candìdum Willd. St. leafy, 1f; lvs. oblong-lanceolate, acute; fl. 1; sep. subequal; lip 1', compressed, white, shorter than the (2') pet. Woods and prairies.

5 C. parviflòrum Salisb. St. very leafy, 8–12'; lvs. lanceolate, acuminate; sepals ovate to lance-ovate; lip *depressed*, shorter than the petals. Low woods and prairies.

6 C. pubéscens Sw. *Large Yellow L.* Stems usually clustered, 1f or more; leaves broadly lanceolate, acuminate; sepals lanceolate; lip compressed laterally, *moccasin-shaped*, shorter than the linear, twisted petals. Woods, meadows, and prairies.

2. CALÝPSO, Salisb. Sep. and pet. subequal, ascending. Lip large, inflated, with 2 spurs dependent beneath near the apex. Column petaloid. Pollinia 4. ♃ Scape 1-leafed at base, 1-flwd. above, arising from a corm.

C. boreàlis Salisb.—Old mossy woods, Vt., N. Y., W. to Oregon! Scape 6–8'; leaf broad-ovate, 1–2'; flowers purple and yellow, 1½'. Rare eastward. May.

3. ORCHIS, L. Sepals and pet. similar, some of them ascending and arching over the column. Lip turned downward, produced at base into a spur which is free from the twisted ovary. Sta. 1, anth. 2-celled, a pollen-mass in each cell.—Fls. racemed on the stem or scape. June—August. (Includes Habenaria, Gymnadenia, and Platanthera.)

* Leaves only 2,—a ovate, nearly as long as scape. Flowers rose-white........No. 1
              —a roundish, the scape much longer. Flowers greenish...Nos. 2, 3
* Leaf only 1. Flowers greenish-white. Lip entire or 3-lobed ....Nos. 4, 5
* Leaves several, clothing the stem more or less...(b)
   b Lip undivided,—c entire, white or greenish........................Nos. 6, 7

ORDER 137.—ORCHIDACEÆ. 327

```
         —c crenulate or wavy, white or yellow............... Nos. 8, 9
         —c 3-toothed.  Flowers yellowish or greenish..... Nos. 10—12
         —c fringed.  Flowers bright yellow or white...... Nos. 13—15
b Lip 3-parted,—x segments fringed.  Flowers white or greenish.....Nos. 16, 17
         —x segments fringed.  Flowers purple................Nos. 18, 19
         —x segments merely toothed.  Flowers violet-purple.......No. 20
         —x segments entire, long, linear-setaceous............Nos. 21, 22
```

1 O. spectábilis L. Lvs. rarely more than 2, 3—6′; scape 4—6′, bearing 1 or 2 lanceolate bracts and 3—5 showy flowers above; spur clavate. Rocky thickets. Pretty.

2 O. orbiculàta Ph. Lvs. 2, roundish, 3—6′, fleshy; scape bracted, 1—2f; upper sepals round, the lateral ovate, half as long as the lip (9—12″). Woods, E. and W.

3 O. Hookeri Wood. Lvs. 2, round-oval, fleshy, 4—5′; scape naked, 8—12′; upper sepals ovate, erect, the lateral deflexed and meeting behind; spur 1′. Woods, N.

4 O. obtusàta Ph. Leaf oblong-ovate, obtuse, 2—3′, near the base of the stem; lip linear, entire, with 2 tubercles at base, as long as the spur. In mud, N.

5 O. rotundifòlia Ph. Leaf round-ovate, radical; scape few-flowered; lip 3-lobed, obcordate, side lobes falcate; spur as long as the lip. Penn., and N.

6 O. hyperbòrea Willd. Lvs. very erect, lanceolate; spike long: bracts longer than the greenish flowers; petals and lip linear, subequal. Shades, northward. 1—4f.

7 O. dilatàta Ph. Slender, 8′—2f; lvs. lance-linear and linear; spike virgate; bracts short; flowers white; lip linear, *dilated-rhombic* at base. Swamps, N.

8 O. nívea Baldw. Very slender, 1—2f; lowest leaf linear, 6—8′, the others subulate, bract-like; flowers white, in an oblong spike; lips oblong. South.

9 O. integra N. Stem leafy, flexuous, 12—15′; lvs. narrow-lanceolate; spike dense, oval; flowers orange-yellow; lip ovate, longer than sepals. Swamps, N. J., and S.

10 O. tridentàta Willd. St. slender, 12—18′; lowest leaf linear-oblong, obtuse, 6′, the others few, small and bract-like; fls. few, greenish; lip 3-toothed at end. Woods.

11 O. bracteàta Muhl. St. leafy; lvs. oblong, obtuse or acutish; bracts 2—3 times longer than the small green fls.; lip 3-(or 2-)toothed at end, lin.-cuneate. Shades. 6—9′.

12 O. flava L. St. leafy; lvs. oblong to lanceolate; bracts longer than the yellowish-brown flowers; lip oblong, obtuse, a tooth each side at base, and a tubercle in the palate; spur shorter than the ovary. Alluvial soils. (O. virescens Muhl.)

13 O. cristàta Mx. Slender, 1½—2f; leaves lance-linear to linear; flowers numerous, small, yellow; sep. and pet. roundish, 1—2″; spur ½ as long as ovary. N. J., and S.

14 O. ciliàris L. *Yellow Fringed Orchis*. Stem 2f; leaves lanceolate; flowers large, numerous, orange-colored; lip 4″ long, twice longer than the linear, notched petals, spur 1′. Swamps. Delicately beautiful.

15 O. Blephariglóttis Willd. *White Fringed Orchis*. Stem 1—2f; leaves lanceolate; flowers pure white; lip fringed in the middle, 2″ long, lanceolate; spur much longer (1′). Swamps, N. Y. to Car., and westward.

16 O. lácera Mx. *Ragged O.* St. smooth, slender, 1—2f; leaves oblong to linear, bracts longer than the flowers; sepals retuse; petals emarginate; flowers ∞; lip segments capillaceous-multifid; spur as long as the ovary. Meadows.

17 O. leucophæa N. *White Prairie O.* Lvs. lanceolate, tapering to a narrow obtuse point; bracts shorter than the ovaries; fls. about 12; spur yellowish, curved, twice longer than the ovary; petals white. Wet prairies.

18 O. Psycodes L. *Purple Fringed O.* Leaves lanceolate; lip segments cuneiform, scarcely longer than the ovate, crenulate, slightly fringed petals; spur longer than the ovary. Meadows. 1½—2½f. Flowers light purple.

19 O. grandiflòra Bw. *Large Fringed O.* Tall, 2—3f; lvs. oval, oblong, and linear, obtuse; lip segments dependent, fan-shaped, twice longer than the fringed petals. Wet meadows, Penn., and N. Superb. (O. fimbriata.)

20 O. peramœna (Gr.) Tall, leafy; leaves lanceolate to lance-linear; sepals round ovate; petals denticulate; lip middle segment 2-lobed, all merely toothed; spur longer than the ovary. Pa. to Ind., and S. Flowers 20—50, large.

21 O. Michaúxii (N.) Very leafy; leaves elliptic-oval, the upper reduced; flowers few, white; petals 2-parted, the lower divisions linear-setaceous, like those of the lip; spur twice as long as the ovary; flowers white. South.

22 O. repens (N.) Stem very leafy from a creeping rhizome; leaves all lance-linear, long; flowers greenish-yellow, dense in the spike, much smaller than in No. 21, but otherwise similar. Pine-barrens, S. August, September.

4. TIPULÀRIA, N. Sepals spatulate, spreading. Petals lance-linear. Lip sessile, 3-lobed, middle lobe linear. Spur filiform, very long. Column free. Anth. opening by a lid, with 4 pollen-masses. ♃ Corms several, connected by a thick fibre. Leaf 1. Flowers bractless.

T. díscolor N.—Pine woods, Vt. to Ga. Leaf ovate, petiolate, 2—3′. Scape 10—15′; raceme with many small, greenish, nodding flowers. July.

5. BLÈTIA, R. & P. Pet. and sep. subequal, distinct. Lip hooded at end (spurless in our species). Column free. Pollinia 8, in pairs, waxy, each pair pedicellate. ♃ Flowers racemed, showy.

1 B. aphýlla N. Leafless; scape 15—30′, with few bracts; racemes long and loose; flowers purplish and yellowish-brown; lip 3-lobed. Swamps, S. August.

2 B. verecúnda H. K. Leaves all radical, broad-lanceolate; scape 2—3f; flowers purple, large and showy; lip broad and crisp at the end. Ga., Fla. July.

6. CORALLORHÍZA, Br. CORAL-ROOT. Sepals and petals subequal, converging. Lip produced behind into a spur, which is adnate to the ovary or obsolete. Pollinia 4. ♃ Plants leafless, brown, arising from coralline roots, sheathed with bracts. Flowers racemed. Fig. 240.

* Spur conspicuously prominent, but adnate. Lip 3-lobed............ ....No. 1
* Spur wholly obliterated.—*x* Lip crenulate, wavy, not at all lobed............No. 2
　　　　　　　　　　—*x* Lip entire, slightly toothed near the base.....Nos. 3, 4

1 C. multiflòra N. Scape 10—15′, all brownish-purple, bearing 15—20 fls. in a long rac.; lip 3-lobed, white, spotted, 3—4″; caps. elliptical, pendulous. Woods, M., N. Jl.

2 C. odontorhíza N. Scape 9—14′, all brownish-purple, bearing 10—20 fls. in a long spike; lip undivided, oval, obtuse, spotted? caps. roundish, reflexed. Old woods. Jl.

3 C. innàta Br. Scape 5-10-flwd.; lip oblong, angularly 2-toothed toward the base, spotless, white; caps. elliptic-obovoid, reflexed. Damp woods, N.: rare. 5—8′. Jn.

4 C. Macræi Gr. Scape 15-20-flwd., fls. large; lip oval, obtuse, obscurely auriculate at base; caps. oval, 6″, reflexed; sepals and petals 6″. N. H., N. and W. 10—16′.

7. APLÉCTRUM, N. ADAM-AND-EVE. PUTTY-ROOT. Sepals and petals distinct, subequal, converging. Lip unguiculate, 3-lobed, middle lobe crenulate. Spur 0. Column free, anth. a little below the apex, pollinia 4, lens-shaped. ♃ Root a globous corm. Leaf 1, large, biennial. Scape after the leaf, bracted, racemed, and brown, as in Corallorhiza. Fig. 263.

A. hyemàle N.—Woods: rare. Corm near 1′ diam., a new one each year. Leaf elliptic-ovate, 3—5′, green all Winter. Scape 12--18′, with a dozen brownish flowers.

8. ONCÍDIUM, Sw. Lip expanded, lobed, tubercled at base. Perianth expanding. Sepals sometimes but 2. Column winged. Pollen masses 2, each 2-lobed. ♃ Splendid flowers, tropical, of easy culture in the **greenhouse**. Flowers large, in open racemes, olive, yellow, &c.

ORDER 137.—ORCHIDACEÆ. 329

1 O. flexuòsum. Scape panicled, arising from the base of a bulb; leaves lanceolate; lip 2-lobed, spotted, much longer than the other petals. Brazil.
2 O. lùridum. Scape erect, branched; leaves elliptical; lip reniform, not longer than the wavy, retuse petals; flowers large, olive-colored. From S. America. 2f.
3 O. Papílio, has one spotted ovate leaf and large yellow-red butterfly-shaped flowers.

9. MICROSTYLIS, N.
Sepals spreading, petals filiform or linear, lip concave, sessile. Column minute, with 2 teeth or lobes at tip. Pollinia 4. ♃ Root tuberous, with 1 or 2 leaves and small racemed flowers.

1 M. ophioglossoides N. St. 5—9′, with a single ovate (2′) leaf near the middle, rac. short (1′), ped. much longer than the minute whitish flowers. Woods, N. June.
2 M. monophýllus Lindl. St. 2—6′, 3-angled, with a single ovate leaf; rac. elongated, 20-40-flowered; pedicels about as long as the flowers (2″). Woods, N.: rare. Jl.

10. LIPPARIS, Rich. Tway-blade.
Sep. and pet. very narrow. Lip spreading, flat. Column winged. Pollinia 4, parallel with each other, without pedicels or glands. ♃ Root tuberous, with 2 lvs. and a rac. of greenish fls.

1 L. liliifòlia Rich. Scape about 6′; leaves 2, radical, lance-ovate, 3—4′; petals filiform, reflexed; lip purple, 6″, abruptly cuspidate; pedicels 1′. Damp woods. June.
2 L. Loesélii Rich. Scape 3—5′, about 6-flowered; pedicels 2″; lip 2″, oblong, mucronate, incurved, wavy; sepals and petals linear. Fields, Can. to Penn. June.

11. LISTERA, Br. Tway-blade.
Sep. and pet. subequal, lip pendulous, 2-lobed or 2-cleft. Column wingless, anth. dorsal, pollen powdery. ♃ Root fibrous. Stem (4—9′) with 2 opposite leaves above the middle. Flowers small, racemed. May—July, in damp woods.

1 L. cordàta Br. Lvs. roundish, subcordate, acute; fls. 10—15, in a short raceme; pedicels length of the ovary; lip-segment linear, length of the sepals. Penn., and N.
2 L. austràlis Lindl. Lvs. ovate; fls. in a loose raceme; ped. 3—4 times longer than the ovary; lip-segment linear-setaceous, twice the length of the sepals. N. J., and S.
3 L. convallarioides Hook. Lvs. round-oval; fls. few, loose, on slender pedicels; lip twice the length of the sepals (4″), 2-lobed at the dilated apex. Ga., and N.

12. SPIRÀNTHES, Rich. Ladies' Tresses.
Spike spiral. Perianth ringent, the 3 upper pieces ascending and connivent, lip oblong, recurved, channelled, the base embracing the column, and with 2 callous processes. Stigma ovate, beaked, 2-toothed at tip. Anthers dorsal, pollinia 2, each 2-lobed, powdery. ♃ Stem nearly naked, bearing many white flowers, bent to a horizontal position.

* Spike dense, with the flowers on all sides. Lvs. present with the flowers..Nos. 1—3
* Spike slender, flowers all in 1 straight or spiral row.—* Lvs. permanent....Nos. 4—6
　　　　　　　　　　　　　　　　　　　　　　　—* Lvs. evanescent... Nos. 7, 8

1 S. cérnua Rich. Leaves lance-linear, the upper bract-like; spike oblong to cylindric, 2—4′; lip very obtuse, crenulate-wavy, conduplicate and recurved; sepals and petals not connivent. 4—5″. Wet, 9—20′. Aug.—Oct.
2 S. Romanzoviàna Cham. Lvs. lance-oblong to linear; spike dense, 1—3′; lip much recurved, ovate-oblong, crenulate-wavy; sepals and petals all connivent above into a galea. Bogs, Me. (Miss Towle) to Lake Superior (Prof. Porter). July, Aug.
3 S. latifòlia Torr. Leaves nearly radical, 3-5-veined, lance-oblong; scape bracted, 4—8′; flowers small (2—3′); plant glabrous. Meadows, Penn., and N. June, July.
4 S. odoràta N. St. stout, 1—2f; lvs. lance-oblong; fls. yellowish, fragrant, 6″, in a spiral row, with leafy bracts; lip 2-toothed at base. Muddy streams, S. October.

5 S. gramíne1 Lindl. Lvs. below lance-linear to linear, the cauline mere sheaths; spike dense, much twisted; flowers white, 3—5″, pubescent, scarcely ringent; lip oblong-ovate, crisped, obtuse. Wet meadows. June—Aug. (S. tortilis C-B.)

6 S. brevifòlia Chapm. Lowest leaves elliptical, evanescent, cauline bract-like; flowers 5—15, in a nearly straight row, ringent, 3—4″; lip entire. S.

7 S. grácilis Bigel. Lvs. all radical, ovate to oblong, fugacious; scape very slender, 8—18′, with a few bracts; flowers 3—4″, in a nearly straight row, pure white; root fasciculate; plant glabrous. Woods: common. July, Aug.

8 S. símplex Gr. Lvs. all radical, fugacious; scape 5—9′, flowers very small (1—2″) in a thin 1-sided spike; lip obovate-oblong. Dry, N. J. (Porter), and S.

13. GOODYÉRA, Br. RATTLESNAKE PLANTAIN.
Spike and perianth as in Spiranthes. Lip sessile, concave or sack-like or even spur-like at base, contracted at the end to a reflexed, channelled point. ♃ Root-stock creeping, branching. Leaves ovate, on sheathing petioles.

* Leaves radical, generally netted with white veins. Lip not spurred.......Nos. 1, 2
* Leaves cauline, uniformly green. Lip spurred at the base behind..............No. 3

1 G. Menzièsii Lindl. Lip concave at base, gradually narrowed and folded at apex; leaves elliptic-ovate; scape 9—12′; spike loose-flowered; flowers pubescent (as are Nos. 2 and 3), *suberect*. Woods. N. Y. to Mich. (Dr. Leidy) and Oreg. ! July, Aug.

2 G. repens Br. Lip saccate-inflated at base; leaves ovate, beautifully netted; scape 6—12′; flowers ovoid, *nodding*, in 1 row, which is more or less spiral; perianth greenish, about 2″ long and nearly as wide. Woods. June, July. (G. pubescens Br.)

3 G. quercícola Lindl. Rooting on the bark of Oaks, &c.; stem leafy; lvs. lance-ovate, thin; spike glabrous, dense, 6—20″; sheaths and bracts membranous; lip ovate at apex, the spur pouch-like, half as long as the ovary. F.a. to La. 6—12′.

14. PONTHIÉVA, Br.
Lip on the upper or inner side, ovate, spreading, and with the other petals inserted into the middle of the column. Anthers with 4 pollinia. Otherwise like Spiranthes.

P. glandulòsa Br. Lvs. radical, oblong-oval; root fasciculate; scape 1f, bracted, with a spike of many greenish pubescent fls. Woods, S. Sept., Oct. (Cranichis N.)

15. CALOPÒGON, Br. GRASS PINK.
Sepals and petals similar, distinct. Lip on the upper (inner) side (the ovary not twisted), unguiculate, bearded. Column free, winged at the summit. ♃ Corm bearing a grasslike leaf, and a scape with several showy flowers.

C. pulchéllus Br. Leaf linear, 8—12′ by 6″, veined; fls. 3—8, large, purple; lip spatulate, crested with colored hairs, erect over the column. Wet meadows. June, July.

16. POGÒNIA, Juss.
Perianth irregular, its pieces distinct. Lip sessile or unguiculate, hooded, bearded inside. Column wingless, free. Anth. terminal, lid-form, with 2 pollinia. ♃

§ Sepals about equal, and similar to the petals, light purple. Lip scarcely lobed ..Nos. 1, 2
§ Sepals much longer than, and unlike the petals, dark brown. Lip 3-lobed ..Nos. 3, 4

1 P. ophioglossoìdes N. Root fibrous; stem 9—16′, with an oval-lanceolate leaf near the middle, and a leaf-like bract near the single large pale-purple flower; lip crested and fringed, as long as the sepals and petals. Swamps. June, July.

2 P. péndula Lindl. *Three-birds*. Root tuberous; stem 4—8′, with 4—8 small scattered leaves and 3 (1—4) drooping bird-like flowers 1′ long. Woods: rare. August.

3 P. divaricàta Br. Stem 1—2f, erect, with 2 linear-oblong lvs. and 1 terminal large flower; sepals linear, recurved at apex, 1½′ long; petals lanceolate, pink-colored, acuminate, 1′, lip a little longer. Swamps, Del. to Fla April, May.

4 P. verticillàta N. Stem 8—12′, bracted at base, bearing 4 or 5 oval lvs. in a whorl at the top, with a curious flower; sepals linear, 2 or 3 times longer than the lanceolate, obtuse petals, which are about 9″ long. Swamps. June, July.

17. ARETHÙSA, Gron. Fl. ringent. Sep. and pet. similar, cohering at base and connivent above. Lip adnate to the column at base, recurved and dilated at apex. Anthers terminal, 2-celled, with 4 pollinia. ♃ Stem low, with sheathing bracts. Flowers purple, beautiful.

A. bulbòsa L. Flower single, 1—2′, erect, with 2 small bracts at its base; lip crenulate-wavy, bearded along the middle. Root a corm. Bogs. 6—12′. June.

18. EPIDÉNDRUM, Swtz. TREE ORCHIS. Sep. and pet. spreading. Lip united with the column forming a tube which is sometimes decurrent on the ovary. Anth. terminal, opercular, 4-celled. Pollinia 4. ♃ Grows on the rough bark of trees. Stems many-flowered.

E. conópseum H. K. Stems clustered, 5—8′, each with a pair of opposite, lancelinear, coriaceous leaves below, and 3—7 purplish fls. 6″ broad. Low lands, S. Ang.

ORDER CXXXVIII. SCITAMINEÆ. GINGERWORTS.

Tropical herbs. Leaves parallel-veined, with the veins diverging from the midvein. *Flowers* irregular and unsymmetrical, with *perianth* 3–6-parted and adherent to the 3-celled *ovary. Stamens* 3—6, some of them abortive. *Styles* united. *Fruit* dry or fleshy. *Seeds* albuminous. Here belong the Cinnamons, Gingers, Bananas, and Arrow-roots.

§ MUSACEÆ. Anthers 5, each 2-celled. Fruit many-seeded. Filaments 6...(*x*)
§ ZINGIBEREÆ. Anther 1, 2-celled. Filaments 3, not petaloid. Fruit ∞-seeded...(*y*)
§ MARANTEÆ. Anther 1, with 1 cell. Filaments 3, petaloid. Capsules 1-3-seeded...(*z*)

   x Perianth of 2 unequal leaves or lips, the lower 5-toothed. Berry oblong............MUSA. 1
   x Perianth of 6 very unequal leaves, with large spathes. Fruit capsular............STRELITZIA. 2
     y Perianth tube slender, lower petal lip-like. Stamens and style long-exserted. HEDYCHIUM. 3
     y Perianth short, in spikes, with large bracts. Stamens and style included......ALPINIA. 4
   z Pistil petaloid, stigma 3-sided. Flowers inconspicuous. Leaves colored..........MARANTA. 5
   z Pistil petaloid, stigma flat, linear. Flowers red, showy. Caps. 3-seeded..........CANNA. 6
   z Pistil short, twisted, with a large gaping stigma. Fls. small. Caps. 1-seeded...., THALIA. 7

1. MÚSA SAPIÉNTUM. BANANA. Scape 7—20f, sheathed below by the stalks of the majestic leaves, the summit a nodding spike of pink-colored flowers, becoming a huge cluster of delicious fruits in which the seeds are abortive.

2. STRELÍTZIA REGÌNÆ. Scape 5—8f, with sheathing bracts, upper bract spathe-like, horizontal, with a cluster of splendid flowers. Sepals lanceolate, 3—4′, yellow. Petals hastate, light blue, enclosing the stamens and style. S. Africa.

3. HEDÝCHIUM ANGUSTIFÒLIUM. Stem 5f, very leafy. Leaves linear-lanceolate. Sepals and pet. linear, the .p oblong, all scarlet, in a dense cluster. **H. CARNEUM** has similar leaves, with pink-colored flowers in a loose cluster. E. India.

4. ALPÍNIA MAGNÍFICA, from Mauritius, 10f high, has the flowers in a head with many large rose-colored bracts, which are bordered with a white line. **A. NUTANS**, still taller, from E. India, has a drooping raceme of pink colored bracts and flowers, with curled and curved petals. Very splendid.

5. MARÁNTA BÍCOLOR, from Brazil, is cultivated for the large ovate leaves, which are beautifully feather-marked with light-green above and purple beneath

6. CANNA, L. INDIAN SHOT. Sepals 3, persistent on the tubercled fruit. Petals 6, the innermost 2- or 3-lobed at the end. Stamen petaloid, with a half anther on one edge. Stigma petaloid, flat, obtuse. ♃ Handsome evergreen herbs, with tall stems and large smooth leaves.

§ CORYTHIUM. Corolla tube manifest. Petals dilated. Anther wholly adnate...No. 1
§ CANNA *proper*. Cor. tube short or 0. Petals narrow. Anther free above..Nos. 2—4

1 **C. fláccida** Rosc. Stem 3—4f; lvs. lanceolate, 2f, pointed both ways; sep. erect, not ⅓ the length of the tube of the funnel-form corolla; petals and filaments obovate, thin, *flaccid*, wavy, yellow, spirally arranged; stig. spatulate. Ponds, South.
2 **C.** INDICA. Stem 3—6f, leafy; lvs. ovate, pointed, 1—2f, abrupt at base; sep. green, 6″; 3 outer pet. erect, green-tipped, the 3 inner recurved or reflexed, the 5th double (2-lobed at end), the stamens and style similar (2′), all scarlet. W. Indies.
3 **C.** DÍSCOLOR. Stem 6—10f; lvs. very large, green and purple; fls. in pairs, crimson.
4 **C.** IRIDIFLÒRA. From Peru. Downy; sheaths colored at edge; fls. drooping, 3′, red.

7. THÀLIA, L. Flowers in a 2-leaved spathe. Cal. 3-sepalled, small. Cor. 6-parted, 3 inner pet. very unequal. Sta. 2-parted, the inner segment slender, bearing the ½ anther. Caps. thin. ♃ ≈ Scape sheathed at base by the petioles, tall, paniculate above. Flowers small, purple.

1 **T. dealbàta** Rosc. Plant 4f, covered with a white powder; lvs. cordate-ovate, on long petioles; panicles dense, erect, the branches as short as the lanceolate bracts. S.
2 **T. divaricàta** Chapm. Plant not powdery, 7f; lvs. lance-ovate, rounded at base; panicle open, divaricate, branches zigzag, much longer than the linear bracts. Fla.

ORDER CXXXIX. AMARYLLIDACEÆ. AMARYLLIDS.

Herbs perennial, chiefly bulbous, with linear *leaves* not scurfy nor woolly. *Flowers* showy, mostly regular and on scapes, with an adherent, 6-parted *perianth*. *Stamens* 6, *anthers* introrse. *Ovary* 3-celled, with *styles* united into 1. *Fruit* a 3-celled capsule or berry. *Seeds* 1 to ∞, with fleshy albumen. Figs. 58, 86, 486, 495.

§ Perianth crowned with a firm cup containing the stamens (§§ 78, 79)..............NARCISSUS. 1
§ Perianth crowned with a thin membrane connecting the stamens...................PANCRATIUM. 2
§ Perianth not crowned.—*a* Segments united into a tube above the ovary...(b)
  —*a* Segments distinct down to the ovary...(z)
  b Flowers in umbels or solitary on the naked scape...(d)
  b Flowers in spikes, racemes, or panicles. Scape bracted...(e)
    d Tube long and slender, segments narrow, abruptly spreading...........CRINUM. 3
    d Tube short or long, gradually expanding. Perianth subirregular..........AMARYLLIS. 4
      e Tube of the perianth straight. Stamens exserted..................AGAVE. 5
      e Tube of the perianth curved. Stamens included..................POLYANTHES. 6
    x Perianth irregular. Stems leafy, flowers umbelled.....................ALSTRŒMERIA. 7
    x Perianth irregular. Scape naked, with 1 large flower.................SPREKELIA. 8
    x Perianth regular.—*y* Sepals all white, larger than the petals...........GALANTHUS. 9
      —*y* Sepals green-tipped, as large as the petals.................LEUCOJUM. 10
      —*y* Sepals and petals equal, yellow.............................HYPOXIS. 11

1. NARCÍSSUS, L. Perianth regular, 6-parted, bearing a bell- or cup-form crown on the throat. Sta. 6, inserted in the tube, and concealed within the crown. ♃ Stems bulbous, scapes bearing a long deciduous spathe with 1 or more yellow or white fragrant flowers. Leaves linear.

§ Crown longer than the tube of the perianth. Scape 1-flowered........... Nos. 1, 2

Order 139.—AMARYLLIDACEÆ.

§ Crown shorter than the tube,—x its border crenated. Flowers 1—5 ......Nos. 3—5
—x its border 6-lobed. Flowers 1—3............No. 6
—x its border entire. Flowers 5—20 ....Nos. 7, 8

1 **N. Pseudo-Narcíssus.** *Daffodil.* Scape 2-edged, 1f; lvs. linear, 1f; fl. large, ylw.; crown bell-form, serrate-crenate, as long as the pet. Often double: com. Apr., May.
2 **N. Bulbocòdium.** *Hoop-petticoat.* Fl. ylw.; cr. much larger than perianth. Apr., May.
3 **N. Jonquílla.** *Jonquils.* Fls. 2—5, yellow, frag., small; crown saucer-shaped, much shorter than the petals; scape terete; lvs. half round, 1f. From Spain. May, June.
4 **N. biflòrus.** *Primrose-peerless.* Fls. generally 2, cream-wh., crown cup-shaped, ylw.
5 **N. poéticus.** *Poet's N.* Fl. 1, white, crown flattish, very small, pale-yellow, edged with crimson, throat yellow. Fl. often double. Scape 1f. Lvs. flat. June. S. Eur.
6 **N. odòrus.** *Great Jonquil.* Fl. mostly solitary, yellow, powerfully fragrant, crown bell-form, 6", the lobes entire; limb 1' long, tube slender, 9". S. Europe. 1f. May.
7 **N. Tazétta.** Crown yellow, bell-form, half as long as the white or yellow petals, the border truncate; leaves glaucous, flat. Spain. May, June. Numerous varieties.
8 **N. polyánthus.** Crown white, thrice shorter than the ovate white petals, border nearly entire; leaves green, flat. Spain. Beautiful, but too tender north.

2. PANCRÀTIUM, L. Perianth tube produced above the (sessile) ovary, long and slender, the 6 segm. long and narrow. Stam. 6, adnate to the crown, exserted; anth. versatile. ♃ Bulb coated, scape solid, 2-edged, bearing a bracted umbel of large (white) flowers. (Leaves linear.)

§ Crown adnate below to the dilated throat and segment of the perianth ....Nos. 1, 2
§ Crown free, funnel-form, throat of perianth not dilated. Tube straight....Nos. 3, 4

1 **P. marítimum** L. Plant glaucous; lvs. longer than scape; tube 3—4', longer than the lin.-lanceolate segm.; crown half-adherent, 12-toothed. Marshes, S. July—Sept.
2 **P. nutans** Gawl. Plant green; lvs. very long (2f); fls. nodding, with a green curved tube 2', seg. nearly 3'; sta. incurved; crown slightly adherent. S. Car. (*Herbert.*)
3 **P. rotàtum** Gawl. Plant glaucous, 1—2f; lvs. long, strap-shaped, obtuse; tube 3', green, shorter than the linear segments; crown irregularly toothed. S. April, May.
4 **P. coronàrium** Leconte. Plant green, 2f; lvs. lance-linear, obtuse; tube 3—4', seg. as long; crown funnel-form, 1½', jagged at edge; sta. 2½'. Wet or dry. South.

3. CRINUM, L. Flowers nearly as in Pancratium, but destitute of a crown. ♃ Bulb coated. Leaves in many rows. Scape solid.

1 **C. Americànum** L. Lvs. lin.-oblong; ova. sessile, 3—4 in the umbel; tube green and lance-lin., white segm. about equal (4'); caps. 1-6-seeded. Swamps, Fla., and W.
2 **C. amábile.** Bulb stem-like; lvs. broad-linear; scape flattened, 3—4f, bearing an umbel of 20—30 purple fragrant flowers 9' long; pet. ligulate, recurved. E. India.
3 **C. ornàtum.** Bulb globular; lvs. undulate; scape 3f, 10-20-flowered; fls. white to roseate, very large; segments lance-oblong. E. India. Many varieties.

4. AMARÝLLIS, L. Perianth tube long or short, expanding upward; limb regular or nearly so. Sta. free, anth. versatile. Style long, declinate. ♃ Bulb coated. Leaves narrow. Scape 1-few-flowered.

1 **A. Atamásco** L. *Atamasco Lily.* Scape 1-flwd.; perianth bell-form, erect, 3', pink-white; tube slender below, 1'; filaments included. An attractive flower, in wet clay soils,Va. to Fla. Scape terete, 6—12'. Lvs. linear, 1f. Mar.-May. (*Zephyranthus Herb.*)
2 **A. vittàta.** Per. 3—4', nodding, white, red striped inside, margins crisped. S. Am.
3 **A. Regìnæ.** Per. nodding, scarlet with a green star, throat fringed; fls. 2—4. S. Am.
4 **A. speciòsa** Fls. 2—4, blood-red, erect, 3' long, funnel-form. S. Afr. (Vallota, Hb.)

5. AGAVE, L. American Aloe. Perianth funnel-form, 6-parted. Sta.

6, exserted, anth. soon versatile. Caps. obtusely 3-angled, ∞-seeded. ♃ Monocarpic herbs (§ 42). Crown-root with thick fibres, a dense clump of thick, rigid, often spiny lvs. Scape bracted, with numerous flowers. July

1 A. Virgínica L. Lvs. lin.-lanceolate, spine-pointed, denticulate; scape simple, 4–6f, loosely spicate above; fls. greenish-yellow, 1', sessile, fragrant. Rocks, Va., and S.
2 A. Americàna. *Century Plant.* Lvs. glaucous, striped with cream-color in some varieties, lanceolate, spine-pointed and toothed, very thick and stout, 3–8f; scape produced but once, after 50–100 years, tree-like, with innumerable flowers. Mexico.

6. POLYÁNTHES (or Polianthes), L. TUBE-ROSE.
Perianth funnel-form, with a curved tube. Fil. inserted into the throat, included. Ovary at the bottom of the tube, its summit free. ♃ Root an upright rhizome.

P. tuberòsa. Stem simple, slender, leafy-bracted, 3f, with a spike of rose-white flowers, 1½', subregular, of exquisite fragrance. From Ceylon. Aug., Sept.

7. ALSTRŒMÈRIA, L.
Perianth funnel-form, some irregular, of 6 leaves distinct to the ovary. Sta. diclinate. Stig. 3-cleft. ♃ Root a rhizome, bearing tubers. Stems leafy, umbellate at top.

1 A. psittacìna. Erect, 1–2f, with remote, lanceolate, sessile leaves; fls. 6–8, in a leafy cluster, pedicellate, 1½'; segments spatulate, red, spotted with green. Brazil.
2 A. Pelegrìna. Lvs. sessile, lance-linear, twisted; fls. 2–6, pink-white, purp.-spotted.
3 A. versícolor. Perianth nearly regular, yellow, with purple spots. Chili.

8. SPREKÈLIA, Endl. JACOBÆA LILY.
Perianth bilabiate, segments distinct to the ovary, the upper 3 spreading. Sta. epigynous, unequal, and with the style declinate, the ends incurved. ♃ Bulbous. Scape hollow, 1-flowered. Leaves linear, erect.

S. formosíssima.—A splendid flower from S. America. Scape 1f. Flower dark red.

9. GALÁNTHUS, L. SNOW-DROP.
Petals shorter than the sepals, notched or lobed. Sta. epigynous, erect, included, shorter than the straight style. ♃ Bulb coated, acrid. Scape 2-edged, solid. Flowers white, pendulous. Pods maturing under ground.

G. nivàlis. Scape 6', 2-leaved; flower 1, as white as snow, in early Spring. Europe.

10. LEUCÒJUM, L. SNOW-FLAKE.
Sep. and pet. subequal, often thickened at apex. Sta. epigynous, included, and style erect. Stig. entire, obtuse. ♃ Bulb coated. Scape 2-edged, hollow. Flowers drooping.

1 L. vernum. Lvs. linear; scape 1-2-flwd.; sep. white, tipped with green or yellow, with divergent veins; spathe 1-leaved; seeds straw-color. March, April.
2 L. æstìvum. Lvs. linear; scape 4-8-flwd., umbellate, 6–10'; sepals 6–8'', pure white with green tips; spathe 1-leaved; seeds black. May, June. Europe.

11. HYPÓXIS, L. STAR-GRASS.
Spathe 2-leaved. Perianth regular, rotate. Seeds ∞, black. ♃ Small, bulbous, grass-like, with yellow flowers on filiform scapes. Meadows and copses.

1 H. erécta L. Hairy; scape about 4-flowered, shorter than the linear leaves, which are 3–5'' wide; flowers greenish without, yellow within. June.
2 H. filifòlia Ell. Smoothish; scape 2-flowered, shorter than the filiform leaves, which are not ¼'' wide. Dry soils, S. Flowers rather larger (9–11'').

ORDER CXL. BROMELIACEÆ. BROMELIADS.

Herbs hard, dry, rigid, and often scurfy, with regular double *perianths*, nearly or quite free from the ovary. *Stamens* 6, *anthers* introrse. *Ovary* 3-celled. *Seeds* numerous, with mealy *albumen*. All tropical, and capable of living in air alone.

1. **TILLÁNDSIA**, L. Sepals 3, membranous, convolute. Pet. 3, petaloid, imbricate, spreading above. Sta. hypogynous. Ovary free. Caps. with 3 double cartilaginous valves. Seeds slender, on comous stipes. ♃ Scurfy air plants, with perennial 2-ranked narrow leaves.

* Stems rigidly erect. Lvs. linear-filiform. Fls. in bracted spikes, blue .... Nos. 2—4
*1 **T. usneoides** L. *Long Moss.* Stems filiform, pendulous, branched; lvs. linear-filiform, curled, 1—2'; fls. solitary, green or gray. Low lands, Va., and S. Hangs in gray festoons from the branches of every tree. Used in upholstery.
2 **T. Bartrámii** Ell. Stems slender, 1f; lvs. shorter, smooth; spike branched, 3—4', loose-flowered; pet. spreading at apex, as long as the bracts. Ga., Fla.
3 **T. cæspitòsa** Leconte. Stems in dense clusters, 3—6'; leaves scurfy, much longer, erect; spike 3- or 4-flowered, 1—2'; pet. recurved, longer than the bracts. E. Fla.
4 **T. recurvàta** Willd. Scapes filiform, 2-flowered, 6'; lvs. scurfy, recurved. E. Fla.

2. **ANANÁSSA** SATÌVA. PINEAPPLE. Raised in hothouses for its well-known fruit, which consists of a consolidated abortive flower-spike. From S. Am

ORDER CXLI. HÆMODORACEÆ. BLOODWORTS.

Herbs perennial, with fibrous *roots*, equitant or rosulate *leaves*, and perfect *flowers*. *Perianth* regular, 6-parted, scurfy or woolly outside, more or less adherent. *Stamens* 6 or 3, and opposite the *petals*, *anthers* introrse. *Ovary* 3-celled, 1-styled. *Capsule* covered with the withered perianth. *Seeds* with cartilaginous albumen.

§ Ovary wholly adherent. Stamens 3, exserted. Perianth woolly outside............LACNANTHES. 1
§ Ovary half free. Stamens 6, included.—x Corymbed perianths woolly all over........LOPHIOLA. 2
—x Racemed perianths rugous-scurfy..........ALETRIS. 3

1. **LACNÁNTHES**, Ell. RED-ROOT. Fls. woolly outside, oblong. Sep. linear. Sta. 3, and style filiform, exserted. Caps. ∞-seeded. ♃ Roots fibrous, red. Lvs. ensiform, equitant. Fls. in a dense corymb. July—Sept

L. tinctòria Ell.—Swamps, R. I. to Fla. Stem strictly erect, 1½—2f; leaves mostly radical, 3—4" wide by 9', or more; flowers 4—5", glabrous and yellow inside.

2. **LOPHÌOLA**, Ker. CREST-FLOWER. Fls. woolly outside and inside, oval. Sepals oblong. Sta. 6, glabrous, not exserted. Styles separable, conical with the 1 stigma. Seeds white. ♃ Root creeping. Stem flexuous, corymbous above, densely clothed with soft white wool. Jl., Aug.

L. aùrea Ker.—Sandy swamps, N. J. to Fla. Stem 1—2½f; leaves mostly radical shorter than the stem; flowers yellowish under the white wool, 2". (Conostylis, Ph.)

3. **ALÈTRIS**, L. STAR-GRASS. COLIC-ROOT. Perianths rugous, as if scurfy or mealy, tubular, 6-cleft, arranged in a slender raceme. Styles

scarcely united. Ovary adherent at base only, opening at top, ∞-seeded. ♃ Smooth, intensely bitter. Leaves all radical, lin.-lanceolate. Jl., Aug.

1 A. farinòsa L. Lvs. rosulate, very acute, many-veined, 3—6'; scape 2—3f, simple; rac. about 9'; *fls. white*, 4—5", on very short ped., oblong bell-form. Low grounds.
2 A. aùrea Walt. *Fls. yellow.* Otherwise scarcely diff. Both plants dry, yellowish.

Order CXLII. IRIDACEÆ. Irids.

Herbs with *corms, bulbs,* or *rhizomes,* equitant, 2-ranked *leaves* and spathaceous *bracts. Perianth* tube adherent to the *ovary. Segments* in 2 sets, often unequal and convolute in bud. *Stamens* 3, alternate with the *petals, anthers* extrorse. *Style* 1, *stigmas* 3, often petaloid. *Capsule* 3-valved, 3-celled, loculicidal. *Seeds* many, with hard, fleshy albumen. Figs. 85, 169, 170, 267–8, 282, 351.

§ Flowers irregular, somewhat bilabiate, nodding..GLADIOLUS. 5
§ Flowers regular and equilateral, mostly erect...(*)
  * Sepals similar to the petals in form, size, and position...(a)
    a Stamens monadelphous. Flowers small, blue. Plant grass-like............SISYRINCHIUM. 7
    a Stamens distinct.—x Flowers radical, with a very long tube................CROCUS. 6
        —x Flowers cauline. Style 3-parted at top................PARDANTHUS. 5
        —x Flowers cauline. Style deeply 3-parted...............SCHIZOSTYLIS. 4
  * Sepals larger than the petals, and otherwise dissimilar...(b)
    b Stamens monadelphous. Petals spreading, panduriform...................TIGRIDIA. 3
    b Stamens distinct,—z stigmas slender, on a slender style..................NEMASTYLIS. 2
        —z stigmas petaloid, on a very short style................IRIS. 1

1. IRIS, L. Flower-de-luce. Sepals 3, reflexed, larger than the 3 erect petals. Sta. distinct. Style short or 0. Stig. petaloid, covering the stamens. ♃ Mostly from tuberous, horizontal rhizomes, with ensiform leaves and large, showy flowers.

* Species growing wild, all (except Nos. 6, 7) in wet meadows or swamps. Apr.—Jn. (§)
  § Stems leafy, tall (1—3f). Tube short; sepals beardless and crestless...(a)
    a Leaves linear, grass-like. Ovary and pod 2-grooved on the sides.........No. 1
    a Leaves sword-shaped. Fls. blue. Sepals much larger than the petals...Nos. 2—4
    a Leaves sword-shaped. Fls. tawny or copper-colored. Petals reflexed...No. 5
  § Stems or scapes low (2—6'), nearly leafless. Tube long and slender...(b)
    b Sepals beardless and crestless. In hilly woods, southward...............No. 6
    b Sepals beardless, but crested with 3 longitudinal folds.................Nos. 7, 8
* Species cultivated for ornament, mostly from Europe...(x)
    x Sepals densely bearded.—y Stems very short, 1-flowered......................No. 9
        —y Stems tall, leafy, 1-5-flowered......... ... Nos. 10—13
    x Sepals beardless.—z Root a rhizome...Nos. 14, 15.—z Root bulbous....Nos. 16—18

1 I. Virgínica L. *Boston Iris.* Stem slender, 1—2f, branching; leaves 2—3" wide; fls. 2—6, on slender ped.; sep. narrow, yellow, edged with purple. Mass. to N. J. Jn.
2 I. versícolor L. *Blue Flag.* Stem flexuous, 2—3f; pet. as long as the stigmas; ovary triangular, with concave sides and rounded angles. Common. June.
3 I. hexágona Walt. Lvs. longer than the flexuous stem; tube longer than the 6-sided ovary; sepals larger than the petals, blue-purple, crested. S., coastward.
4 I. tripétala Walt. Lvs. shorter than the slender stem; tube shorter than the 3-sided ovary; sepals many times larger than the petals. S.: rare. Purple.
5 I. cùprea Ph. Tall and flexuous, 2—3f; petals twice longer than the linear stigmas; capsules sharply 6-angled, shorter than the tube. S. and W. April—July

Order 142.—IRIDACEÆ.

6 **I. verna** L. Scape 1-flowered, 3—5′, shorter than the rigid leaves; tube, sep., and pet. subequa. (2′); stigmas deeply 2-cleft; fls. blue, with some yellow. Mar., Apr.
7 **I. cristàta** Ait. Scape compressed, and, with the lvs., 3—5′; tube longer than the sepals (2′), which are distinctly crested along the middle. Barrens, Va. to Ga. April.
8 **I. lacústris** N. Like No. 7, but the sep. are longer than the tube, &c. L. Huron.
9 **I. pùmila.** *Dwarf I.* Fls. large, blue-purple; pet. larger than sepals. In Spring. 3.
10 **I.** Germánica. Flowers many, deep blue, the spathe also colored. Common.
11 **I.** sambucìna. *Fleur-de-lis.* Flowers ∞, blue-white; segments notched. Common.
12 **I.** Suziàna. Flower 1, very large. purple and spotted; petals reflexed.
13 **I.** Florentìna. *Orris-root.* With broad leaves and large white flowers.
14 **I.** gramínea. Linear leaves much longer than the 1f, 2-flowered scape. Blue.
15 **I.** pseud-ácorus. Flowers yellow; petals smaller than the stigmas, 3f. June.
16 **I.** Xíphium. *Spanish I.* Lvs. subulate; 2 fls.; pet. narrow as stig. All colors. 1—2f.
17 **I.** xiphioìdes. *English I.* Leaves subulate; fls. 2; petals broader than the stigmas.
18 **I.** Pérsica. *Persian I.* Lvs. linear; scape very short; petals smaller than the blue sepals.—All the above are hardy, except this, which is a house-plant.

2. **NEMÁSTYLIS,** N. No tube above the ovary. Sepals spreading, larger than the ascending, cucullate petals. Filam. shorter than the anth. Style enlarged above, and parted into 6 radiating, subulate stigmas. ♃ Bulb ovoid. Lvs. lance-linear. St. very slender, with 1 or 2 bright-blue fls.

N. cœlestina N. Leaves very veiny, 1f; stem 15—20′, few-leaved; spathe 2-leaved sepals obovate. 1′, ⅓ larger than the hooded petals. Swamps, Fla. to La.

3. **TIGRÍDIA,** L. Tiger-flower. Spathe 2-leaved. Perianth regular, the 3 sepals larger than the 3 petals. Stamens monadelphous, filaments united into a long tube. ♃ Bulbous.

T. pavònia. St. simple, flexuous; leaves ensiform, veined; fls. inodorous, 5—6′ broad. ephemeral, several in succession, yellow, with crimson spots. Mexico.

4. **SCHIZÓSTYLIS** coccínea. Stem 3f. Leaves channelled, lance-linear. Flowers concave, regular, 2′ broad, in long spikes, crimson to scarlet, the styles slender and nearly distinct. Lately introduced from S. Africa.

5. **PARDÁNTHUS,** Ker. Blackberry Lily. Sepals and pet. subequal, oblanceolate, spreading. Fil. slender. Style clavate, 3-parted, with 3 stigmas. Caps. oblong. Seeds black, attached to the column, and resembling a blackberry after the valves have fallen. ♃ Root a rhizome. Stem branching, leafy. July, August. (Ixia, L.)

P. Chinénsis Ker.—Leaves ensiform, as in Iris; flowers 1½′ broad, many, orange-yellow, crimson-spotted. Stems 3—4f. Escaped from cultivation.

6. **CROCUS,** L. Lvs. radical. Fls. nearly sessile on the bulb. Tube very long and slender, bearing the funnel-form perianth above the ground. Stigmas 3-cleft.

1 **C.** vernus. *Spring C.* Stigmas short, wedge-shaped; leaves linear. The beautiful flowers are white, blue, and variegated,—the earliest in the garden.
2 **C.** Suziànus, is golden yellow, with the 3 sepals revolute. Turkey.
3 **C.** satìvus. *Saffron. Fall C.* Stigmas slender, reflexed; segments purple. Europe.

7. **SISYRÍNCHIUM,** L. Blue-eyed Grass. Spathe 2-leaved. Segments of the perianth flat, equal. Sta. monadelphous Stig. 3-cleft. ♃

Grass-like plants, with compressed, winged or ancipital scapes, from fibrous roots. June, July.

S. Bermudiàna L. In tufts; lvs. linear, erect, about as long as the scapes; spathe 2-5-flowered; flowers small, blue; segments obovate, notched and mucronate; pedicels slender; pods globular, 8—12'.

α. *anceps.* Scapes winged, so as to resemble the leaves.
β. *mucronàtum.* Scapes barely 2-edged, filiform; spathe pointed.

8. GLADIOLUS, L. CORN-FLAG. Spathe 2-leaved. Perianth irregular, 6-parted, somewhat 2-lipped. Stamens 3, distinct, ascending. Stig. 3, broader above. Seeds winged. ♃ A large genus of bulbous plants, chiefly from S. Africa. Fls. large and splendid. The species are badly confused.

1 **G.** PSITTACÌNUS. Spike 8-10-flowered; flowers scarlet and yellow, spotted, the tube as long as the segments. From this is derived many hybrids, as
β. GANDAVÉNSIS, variegated with orange, scarlet, and yellow. Common.
2 **G.** CARDINÀLIS. Spikes few-flowered, the flowers crimson, with a white stripe in the lower 3 segments; stem branched above, 2f. Not hardy.
3 **G.** FLORIBÚNDUS. Flowers very large, nearly erect, upper segments broader, pink varying to white; spike long and crowded. Very delicate.

ORDER CXLIII. DIOSCOREACEÆ. YAM-ROOTS.

Plants shrubby, twining, arising from tuberous rhizomes, with broad, net-veined *leaves. Flowers* diœcious, regular, hexandrous, *tube* adherent, *limb* 6-parted. *Ovary* 3-celled, 3–6-ovuled, 3-styled. ♂ *Stamens* 6, perigynous. *Fruit* a capsule, 3- or (by abortion) 1-celled, or a berry. *Seeds* compressed, albuminous.

DIOSCÒREA, L. YAM-ROOT. Flowers ♂ ♀. Styles of the fertile 3. Cells of the caps. 2-seeded. Sds. membranaceously margined. ♄ Slender, twining with the sun. Lvs. simple, palmately-veined or divided. Flowers green, inconspicuous, in axillary spikes or panicles.

1 **D. villòsa** L. *Wild Yam.* Leaves broadly ovate, cordate, acuminate, 9-11-veined, the lower opposite or in 4's, upper alternate, petioles long, under surface downy, (never *villous*); stem slender, climbing 5—15f, over bushes, &c. June, July.
2 **D.** SATÌVA. *Yam.* Leaves round-ovate, long-cuspidate, sinuate, cordate, all alternate, smooth; stems sometimes prickly. Root large and sweet. S.

ORDER CXLIV. SMILACEÆ. SARSAPARILLAS.

Herbs or *shrubs*, often climbing. *Leaves* reticulate-veined. *Flowers* diœcious. *Perianth* free from the ovary, 6-parted, regular. *Stamens* 6, inserted into the base of the segments. *Anthers* 1-celled (2-lamellate). *Ovary* 3-celled, cells 1- or 2-ovuled. *Style* 1 or none. *Stigmas* 3. *Berry* roundish. *Seeds* orthotropous, albuminous. Fig. 396.

SMÌLAX, L. GREEN-BRIER. SARSAPARILLA. Character nearly as above. ♄ ♭ Lvs. palmately-veined, entire, petiolate, with a pair of stipular (§ 325, Fig. 306) tendrils. Flowers green or yellowish, small, in stalked, axillary umbels.

ORDER 145.—ROXBURGHIACEÆ.

§ Herbs spineless. Lvs. and fœtid umbels long-stalked. Berries bluish..Nos. 12—14
§ Shrubby vines. Leaves short-stalked. Berries 1–3-seeded...(a)
  a Pubescent, prostrate, spineless. Leaves cordate, evergreen. South.....No. 11
  a Glabrous, climbing, and more or less prickly (except Nos. 5, 6)...(*b*)
    b Lvs. acute at the base, 3–5-veined. Ped. shorter than the pet.....Nos. 8—10
    b Leaves abrupt or cordate at base, 5–9-veined...(*c*)
      c Leaves panduriform, or some hastate. Peduncles elongated..............No. 7
      c Lvs. ovate or oblong, deciduous.—*x* Plants spineless.................Nos. 5, 6
        —*x* Prickly.—*z* Leaves glaucous.........No. 4
          —*z* Leaves green........Nos. 1—3

1 **S. rotundifòlia** L. *Common G.* Vine green, strong, and thorny, some 4-angled; leaves round-ovate, 5–7-veined, cusp.-pointed; ped. a little longer (6—7″) than the petioles; berries glaucous-black. Common in thickets. 10—30f. June, July.

2 **S. híspida** Muhl. Vine terete, hispid below, with weak, slender prickles, nearly unarmed above; leaves thin, deciduous, ovate, cuspidate; ped. twice as long (1′) as the petioles; berries black. Thickets, N. J., and N. 8—12f. June.

3 **S. Walteri** Ph. Vine unarmed, or prickly at base; lvs. cordate-ovate, 3–5-veined; ped. as long as the petioles; berries red, 1–3-seeded. N. J., and S. April—June.

4 **S. glauca** Walt. Vine more or less prickly above, angular; lvs. broad-ovate, glaucous at least beneath; ped. twice longer than the petiole; berries black, with a bloom; flowers yellowish white. Thickets, L. Isl. to Ga., W. to Ky. March—June.

5 **S. Pseudo-China** L. Root-stock tuberous; vine terete; leaves cordate-ovate to oblong, 5-veined; ped. flat, nearly as long as the lvs.; fr. black. N. J. to Ky., and S. Jn.

6 **S. sarsaparilla** L. Root-stock creeping, long; branchlets 4-angled; leaves thin, oblong-ovate; ped. flat, a little longer than the petioles; fruit *red*, 1-seeded. S-W.

7 **S. tamnoides** L. Vine terete; branches 4-angular, aculeate; leaves ovate-cordate to fiddle-form, and hastate, cusp.-pointed, rough-edged. N. J., W. and S.

8 **S. auriculàta** Walt. Vine prickly; branchlets angular, unarmed; leaves lance-auriculate-hastate, thick, small, smooth-edged, evergreen; berries finally black; flowers sweet-scented. S., near the coast. June. (S. maritima C-B.)

9 **S. laurifòlia** L. Vine prickly; branchlets unarmed, zigzag; leaves thick, evergreen, lance-oblong, obtuse, mucronate, 3-veined; fr. black, 1-seeded. N. J., and S.

10 **S. lanceolàta** L. Like No. 9, but the lvs. are thin, and berr. 3-seeded. Va., and S.

11 **S. pùmila** Walt. Lvs. shining above, soft-downy beneath; ped. as long as the petiole (6′); berries red, 1–3-seeded. Shady, rich soils, S. 1—3f. October.

12 **S. herbàcea** L. *Carrion-flower.* Stem erect or reclined, terete; leaves pubescent beneath, or nearly glaucous, ovate-oblong, 7-veined, with or without tendrils; ped. longer than the long petioles (3—4′), 8–20-flowered. Low grounds. 2—5f. June.
  β. *pedunculàris.* Ped. very stout and long (6—8′), 30–50-flowered.

13 **S. lasioneùron** Hook. Vine climbing, glabrous; lvs. all with tendrils, cordate, ovate-oblong; ped. little longer than the petioles (3—4′). Thickets, W. 10f. June.

14 **S. tamnifòlia** Mx. Erect or climbing, glabrous; lvs. 5-veined, cordate-hastate, tapering to the obtuse apex; ped. longer than petioles; fr. blue-black. N. J., and S.

ORDER CXLV. ROXBURGHIACEÆ.

Herbs or *shrubby vines*, with many-veined netted *leaves* and perfect *flowers*. *Perianth* 4-parted, petaloid, persistent. *Stamens* 4, hypogynous. *Ovary* free, 1-celled. *Capsule* 2-valved. *Seeds* several, on hairy stalks, albuminous.

CROÒMIA, Torr. Fls. very small and few, axillary. Perianth seg. in pairs (2 sepals and 2 petals), oval. Ovules 4—6, suspended. Seeds 1—3. ♃ Rhizome creeping. Leaves lance-ovate, cordate.

C. pauciflòra Torr.—Woods, Ga., Fla., Ala. Stem simple, 1f. Leaves about 6, thin, glabrous, pedately arranged, 7-9-veined. Ped. 1'. Flowers 2" wide when open. April.

ORDER CXLVI. TRILLIACEÆ. TRILLIADS.

Herbs with simple *stems*, tuberous *roots*, and verticillate, net-veined *leaves*. *Flowers* terminal, 1 or few, perfect, mostly 3-parted. *Calyx* herbaceous, *corolla* more or less colored. *Stamens* 6—10. *Ovary* free, 3-5-celled, bearing in fruit a juicy, ∞-seeded pod. Figs. 115, 259, 294.

§ Leaves in one whorl. Sepals green, petals colored .... TRILLIUM. 1
§ Leaves in two whorls. Sepals and petals alike greenish .... MEDEOLA. 2

1. TRÍLLIUM, L. WAKE-ROBIN. Perianth deeply 6-parted, in 2 distinct series, outer of 3 sepals, inner of 3 colored pet. Sta. 6, anth. longer than the filaments. Stig. sessile. Berry purple, 3-celled, ∞-seeded. ♃ St. simple. Leaves 3, whorled at the top of the stem, palmi-net-veined. Flowers solitary, terminal. In Spring.

§ Flowers sessile. Petals dark purple, erect .... Nos. 1, 2
§ Flowers on a peduncle raised above the leaves...(*)
  * Leaves petiolate, ovate, rounded at the base. Petals thin, delicate .... Nos. 3, 4
  * Leaves sessile, rhomboidal, nearly as broad as long. Petals thickish .. Nos. 5, 6
§ Flowers on a peduncle deflexed beneath the leaves .... Nos. 7, 8

1 **T. séssile** L. Leaves sessile, roundish-ovate to rhomb-ovate, acute, mottled with dark purple; petals sessile, some spreading, dull purple. Pa., W. and S. 6—12'.

2 **T. recurvàtum** Beck. Lvs. ovate to obovate, narrowed to a petiole; sepals reflexed, green; pet. erect, narrowed at base to a claw, purple, 1'. Woods, W. 8—10'.

3 **T. nivàle** Rid. Stem 2—4'; lvs. oval to ovate, distinctly petiolate; fl. erect, 7—8" long; petals ovate-spatulate, white, half longer than the sepals. Penn. to Wis.

4 **T. erythrocárpum** Mx. *Smiling W.* Lvs. ovate, rounded at base, acuminate; petals lance-ovate, recurved, twice longer than the sepals, wavy, white, beautifully pencilled at base with purple. Woods, Can. to Ga. 8—12'.

5 **T. grandiflòrum** Salisb. Lvs. rhomb-obovate, sessile, conspicuously acuminate; petals spatulate-obovate, much longer (1½—2') than the sepals, white, varying to rose-color. Damp, rocky woods, M., S., and W. 8—12'.

6 **T. eréctum** L. *Bath Flower.* Leaves roundish-rhombic, short-pointed, almost petiolate, about as broad as long; ped. scarcely erect; flower nodding; petals oval-ovate, much broader than the sepals, dark purple, ill-scented. Woods.

  β. *album.* Petals white or greenish; ped. inclined. N. Y. (Hankenson), and W.

7 **T. cérnuum** L. Leaves nearly as in No. 6; ped. more than half the length of the leaves, twice that of the flower; petals flat, not reflexed, white, little larger than the sepals; stigmas as long as the anthers. Woods, M., S., and W. 1—1½f.

8 **T. stylòsum** N. Leaves petiolate, ovate, oval, or elliptic; ped. not longer than the flower, decurved; petals recurved, much larger than the sepals, white; styles united, as long as the stigmas, shorter than the recurved anthers. South. 10—20'.

2. MEDÈOLA, Gronov. INDIAN CUCUMBER-ROOT. Perianth deeply parted into 6 petaloid, revolute segments. Sta. 6, with slender filaments. Stigmas 3, divaricate, united at base. Berry 3-celled, cells 3-6-seeded. ♃ Stem simple, arising from a white, tuberous rhizome (which is thought to resemble the cucumber in flavor) bearing 2 whorls of lvs. and 1—3 term. fls.

M. Virgínica L.—Damp woods. Slender, erect, 1—2f, with cottony wool. Lower whorl of 6—8, upper of 3 leaves. Flowers pendulous, yellowish. July. (Fig. 294.)

Order CXLVII. LILIACEÆ. Lilyworts.

Herbs with bulbous or tuberous *stems*, parallel-veined, sessile *leaves*, and perfect, regular *flowers*, with the *perianth* uniformly colored and free from the *ovary*. *Stamens* 6 (4 in Majanthemum), perigynous. *Anthers* introrse (except in Uvularia). *Styles* wholly or partly united. *Fruit* a capsule or berry. *Seeds* albuminous.

§ LILIACEÆ *proper*. Style entire. Fruit a dry capsule. Plants with a scaly or coated bulb...(*)
§ ASPHODELEÆ. Style entire (or 0). Fr. a dry capsule. With a caudex, root-crown, or rhiz...(**)
§ CONVALLARINEÆ. Style entire. Fr. a colored berry. Plants with a rhiz. or fibrous roots...(***)
§ UVULARIEÆ. Style 3-cleft or 3-parted. Fruit a dry capsule. Plants with a rhizome...(****)

```
*   Stem leafy above as well as at the base. Bulbs scaly...(b)
*   Stem (scape) sheathed at base, leafless, many-flowered...(c)
*   Stem (scape) sheathed at base,—a bearing a single nodding flower............ERYTHRONIUM.  1
                                —a bearing a solitary, erect flower............TULIPA.        2
    b Petals equalling the sepals, with a honey-groove at base..................LILIUM.        3
    b Petals equalling the sepals, with a roundish nectary at base..............FRITILLARIA.   4
    b Petals much larger than sepals, nectary in the midst, or 0................CALOCHORTUS.   5
        c Perianth segments united, forming a tubular flower...(e)
        c Perianth segments distinct, not forming a tube...(d)
    d Flowers small, in a panicle of racemes, white............................NOLINA.        6
    d Flowers in a simple raceme, mostly blue..................................SCILLA.         7
    d Flowers in a corymb, white, with bracts..................................ORNITHOGALUM.  8
    d Flowers in an umbel, white or roseate, with 2–4 bracts....................ALLIUM.        9
      e Limb of the perianth revolute, as long as the tube......................HYACINTHUS.   10
      e Limb of the perianth spreading, much shorter than tube..................MUSCARI.      11
**  Perianth segments united more or less into a tube...(m)
**  Perianth segments distinct.—n Flowers racemed, small, yellow...............SCHŒNOLIRION. 12
                                —n Flowers panicled, white....................YUCCA.        13
    m Stamens straight, longer than the tubular, flame-colored perianth........TRITOMA.      14
    m Stamens all curved upward.—o Flowers in an umbel.........................AGAPANTHUS.   15
                                —o Flowers cyanic, racemed....................FUNKIA.       16
                                —o Flowers xanthic, terminal..................HEMEROCALLIS. 17
*** Perianth segments separate, not forming a tube...(s)
*** Perianth segments united.—v Flowers greenish, axillary....................POLYGONATUM.  18
                             —v Flowers pure white, on a scape................CONVALLARIA.  19
    s Scape leafless, bearing an umbel. Berry blue, 2-celled...................CLINTONIA.    20
    s Stem leafy, bearing the flowers solitary or in pairs. Berries red...(y) (See p. 447.)
    s Stem leafy, bearing a white cluster.—x Flowers 6-parted..................SMILACINA.    21
                                        —x Flowers 4-parted..................MAJANTHEMUM.  22
    y Stems much branched, with filiform branchlets for leaves................ASPARAGUS.    23
    y Stem forking, with oval leaves.—z Fls. axillary. Berry ∞-seeded..........STREPTOPUS.   24
                                    —z Fls. terminal. Berry 3-6-seeded........PROSARTES.    25
**** Stem leafy. Flowers solitary, long, yellowish, drooping..................UVULARIA.    26
```

1. ERYTHRÒNIUM, L. Perianth campanulate. Seg. recurved, the 3 inner ones (petals) usually with a callous tooth attached to each side at base, and a groove in the middle. Style long. Caps. somewhat stipitate, seeds ovate. ♃ Lvs. 2, subradical. Scape 1–∞-flwd. Flowers nodding.

1 **E. Americànum** Sm. *Yellow E.* Bulb deep in the ground, sending up a scape which bears 2 unequal, lanceolate, mottled leaves at the surface of the ground, and a handsome drooping yellow flower at top. Woods. 3–5′. April, May.
   β. *bracteàtum.* Leaves very unequal; scape with a bract near the flower. Vt.

2 **E. álbidum** N. *White E.* Scape naked, bearing a white drooping flower; petals without teeth, narrowed to the base. Wet meadows, N. Y. to Wis. May, June.

2. TÙLIPA, Tourn. TULIP. Perianth campanulate. Sta. short, subulate, anth. broad-linear, deeply emarginate at base. Style very short, stig. thick. Caps. oblong, triangular. ♃ Herbs acaulescent, with coated bulbs, sessile leaves, and a simple scape bearing a solitary, erect flower.

T. Gesneriàna. Plant smooth; leaves ovate-lanceolate, near the ground; segments very obtuse, endlessly variegated with red, yellow, and white. Persia. May, June.

3. LÍLIUM, L. Lily. Perianth bell-form, colored. Sep. 6, gradually spreading or recurved, each with a longitudinal honey-groove within from middle to base. Sta. shorter than the style, anth. versatile. Style clavate, stig. 3-lobed. Caps. subtriangular. Seeds 2-rowed in each cell. ♃ Bulbs scaly. Stems leafy. Flowers large, showy. June—August.

* Native wild Lilies, with yellow, orange, or red, spotted,—*x* nodding fls..Nos. 1—3
   —*x* erect fls......Nos. 4, 5
* Exotic Lilies, cultivated, mostly hardy. Fls. nodding (except Nos. 6, 14)...(*a*)
   a Stems bearing bulblets in the axils. Flowers orange-colored.........Nos. 6, 7
   a Stems never bulbiferous.—*y* Fls. white. Lvs. lanceolate, scattered...Nos. 8—10
      —*y* Fls. wh., varieg. and spotted, sweet....Nos. 11—13
      —*y* Fls. yellow or straw-colored...........Nos. 14—16
      —*y* Fls. red or purple.....................Nos. 17—19

1 L. Canadense L. *Yellow L.* Leaves mostly in whorls, lanceolate, the veins beneath hairy; ped. terminal, mostly in 3's; sepals gradually spreading, yellow to orange, with purple spots inside. Meadows, mostly N. 2—5f.

2 L. supérbum L. *Turk's-cap.* Leaves linear-lanceolate, acuminate, the lower whorled, upper scattered; flowers often numerous, orange to red, spotted, the sepals revolute. Wet soils. 4—6f. Flowers 3—30. Plant splendid.

3 L. Caroliniànum Mx. Lvs. 1-veined, oblanceolate, acuminate, tapering to the base, the upper whorled, the lower scattered; sepals lance-linear, recurved (not revolute), deep yellow spotted with purple. Swamps, S. 1½—3f. Flowers 1—3.

4 L. Philadélphicum L. Lvs. lance-linear, the upper whorled, lower scattered; fls. 1—3; sepals erect-spreading, lance-ovate, obtuse or barely acute, clawed, orange-red, spotted at base, 2½' long. Dry pastures and copses. 15—20'.

5 L. Catesbæi Walt. Lvs. all scattered, lance-oblong to linear; flower solitary; sepals lanceolate, wavy, 3—4', the long claws yellow, lamina and long, thickened acumination scarlet, spotted with purple. Damp barrens, Md., and S. 2—3f.

6 L. bulbíferum. Fls. erect, rough inside, 2½'; sep. sessile; lvs. 3-veined. 4f. Italy.
7 L. tigrìnum. Fls. nodding, spotted; sep. sessile, 3½', rev.; lvs. 5-veined. 6f. China
8 L. cándidum. Fls. campanulate, several, smooth inside. From Persia. 3—4f.
9 L. Japónicum. Fl. solitary, campanulate; sep. revolute at apex. Japan. 2—3f.
10 L. longiflòrum. Fls. solitary, tubular-bell-form; sep. 5—6'. From Japan. 1f
11 L. gigánteum. Tall (8f); fls. spicate, trumpet-form, white, with carmine lines
12 L. speciòsum. Stem 2—3f; leaves lance-ovate, scattered; fls. 1—3, fragrant; sepals 5', revolute, white to roseate, with purple warty spots inside. Japan. Splendid.
13 L. auràtum. Stem 1—2f; leaves lanceolate, scattered; fls. 1—3, fragrant; sepals 6—7', spreading, white, with a yellow band and purple spots. Japan. "Glorious."
14 L. cróceum. Lvs. some in 3's, lin.-falcate; fls. erect, often umbellate, rough inside.
15 L. testàceum. Lvs. whorled? lanceolate, many; fls. several, large, straw-col. 6f.
16 L. cólchicum. Lvs. crowded, lance-lin.; fls. sev., funnel-form; sep. recurved. 2f.
17 L. Pompònium. Lvs. lin. to subulate, crowded; fls. small, scarlet; sep. rough, revol.
18 L. Mártagon. Lvs. lance-oblong, whorled; fls. panicled, purple to roseate, revolute, spotted. From Europe. 5f. [not spotted; sepals reflexed. Palestine. 3f
19 L. Chalcedónicum. Lvs. lance-linear, crowded, erect, rough-edged; fls. bright red,

4. FRITILLÀRIA, Tourn. Chequered Lily. Perianth campanulate, with a broad base and nectariferous cavity above the claw of each segment. Stamens as long as the petals. Stig. trifid. Caps. coriaceous, 3-celled, septifragal. ♃ With coated bulbs, simple, leafy stems, bearing 1 or more nodding flowers in Spring.

1 F. IMPERIÀLIS. *Crown Imperial.* Stem 3f, at base invested with long, narrow lvs., the middle naked, the summit bearing a raceme of large drooping red flowers beneath a crown of bracts. Var. FLAVA has yellow flowers. Persia.

2 F. MELEÀGRIS. *Chequered L.* Stem 1-flowered, with alternate, linear, channelled leaves; flower large, nodding, chequered with purple and yellow. Europe. 1f.

3 F. PÉRSICA. Fls. brownish-purple, in a pyramidal, naked raceme. Persia. 3f.

5. CALOCHÓRTUS, Ph.
Perianth twisted in æstivation. Sepals 3, smaller than the 3 petals, which are bearded within except a central glabrous spot. Style very short, anth. recurved. Seeds 1-rowed in each cell of the capsule. ♃ Californian, bulbous. Leaves narrow. Stem erect.

C. SPLENDENS. Stem with 3—5 large, open, lilac flowers; pet. each with a brown-yellow eye in the middle. 1—2f. June.—A splendid flower, yet rare in cultivation.

C. PULCHÉLLUS and **C. ALBUS**, with the petals connivent into pendent globes, the one golden yellow, the other satin white, are very beautiful.

6. NOLÌNA, Rich.
Perianth small, of 6 equal ovate spreading parts, longer than the 6 stamens. Stigmas 3, recurved, with a very short style. Caps. 3-winged, 3-(or 1-3-)seeded. ♃ Bulb coated. Scape widely branched. Flowers racemed, white, nearly bractless.

N. Georgiàna Mx.—Sand hills, S. Car. to Fla. Scape 2—3f, from a large bulb. Leaves long, narrow, all radical, recurved and channelled, rough-edged.

7. SCILLA, L. SQUILL.
Sepals and petals similar, spreading (blue or purple). Filaments 6, slender, style thread-club-shaped. Caps. 3-angled, 3-celled, cells with 1 or several black seeds. ♃ Bulb coated, bearing several linear leaves and a scape with a raceme.

1 S. esculénta Ker. *Quamash.* Lvs. keeled, flaccid, shorter than the scape; bracts subulate, longer than the pedicels; filaments filiform; stigmas 3-toothed; sepals widely spreading, pale blue. Bottoms, W. 1—2f. May. (Camassia, Lindl.)

2 S. PERUVIÀNA. Leaves ciliate on the edges, longer than the scape; flowers stellate, in a dense conical corymb, violet-blue, rarely white. Spain.

8. ORNITHÓGALUM, L. STAR OF BETHLEHEM.
Stem a coated bulb. Sep. and pet. similar, white, spreading, 3–7-veined. Fil. 6, subulate. Style slender, stigma 3-angled. Caps. roundish, 3-angled. Sds. few, black. ♃ Scape with a corymb of bracted flowers, and linear leaves.

O. umbellàtum L. Leaves channelled, as long as the scape (1f); flowers few, on long pedicels, the white sepals each with a green band outside. June. § Europe.

9. ÀLLIUM, L. GARLIC. ONION.
Flowers in a dense umbel, with a membranous 2-(1–4-)leaved spathe. Perianth deeply 6-parted. Seg. mostly spreading, ovate, the 3 inner somewhat smaller. Ovary angular, stigma acute. Caps. 3-lobed. Seeds few, black. Strong-scented, bulbous plants. Leaves mostly radical.

ORDER 147.—LILIACEÆ.

§ Leaves (none at flowering-time) flat, lanceolate. Ovary only 3-ovuled........No. 1
§ Leaves present, flat.—a Ovary 6-ovuled, often with a 6-toothed crest...(y)
   —a Ovary ∞-ovuled, not crested. Leaves linear........No. 5
§ Leaves terete and hollow.—x Scape or stem slender, not inflated..........Nos. 8, 9
   —x Scape inflated in the midst. Cultivated....Nos. 10, 11
   y Wild native species. Leaves linear and very narrow..................Nos. 2—4
   y Exotics cultivated. Leaves lance-linear or broadly linear..............Nos. 6, 7

1 A. tricóccum Ait. Lvs. 5—8', fugacious, mostly gone in June, when the scape, with its rounded umbel of 10—12 white fls., appears. Woods, N. C., Eng. to N. C., and W. 1f.

2 A. cérnuum Roth. Lvs. very long; umbel *cernuous*, with 12—20 bright roseate fls.; sepals oblong-obovate, acute; filam. filiform, exserted. N. Y., W. and S. 1½—2f. Jl.
   β. *stellàtum*. Umbel mostly erect; stam. not exserted. Dry, Ill., and W. 1—1½f.

3 A. Canadénse Kalm. Scape terete; leaves shorter than the scape; umbel erect, capitate, consisting of both (whitish) fls. and bulblets mixed. Shades. 1f. June.

4 A. mutàbile Mx. Lvs. lin.-filiform, thin, shorter than the terete scape; umb. 20-40-flwd., erect; spathe 3-leaved, purplish; sep. ovate-lanceolate, longer than the sta., white or roseate; capsule 3-lobed, 3-seeded. Woods, S. 1—1½f. March—May.

5 A. striàtum Jacq. Lvs. linear, nearly equalling the terotish scape; spathe 2-lvd.; fls. 3—7, sep. lance-ovate, green-striped outside; not garlic-scented. W. and S. 8—12'.

6 A. sativum. *Common Garlic.* Bulb consisting of many small ones in a common sheath; stem leafy to the middle; umbel bulb-bearing; flowers white. Sicily. July.

7 A. porrum. *Leek.* St. compressed, sheathed at base by the channelled leaves; umb. globous, white; stamens a little longer than the rough-keeled sepals. Europe. July.

8 A. vineàle L. *Crow Garlic.* Stem and few fistulous lvs. very slender; umb. bulb-bearing; stamens alternately 3-cuspidate. Fields, June. It spoils the cows' milk.

9 A. schænoprásum L. *Cives.* Scape equalling the terete, filiform, fistulous lvs.; umb. capitate; sep. longer than the simple stamens, rose-purple. Lake shores, N. ‡

10 A. fistulòsum. *Welsh Onion.* Scape inflated in the midst, not taller than the fistulous leaves; umbel dense, globular; stamens exserted. Asia. 18'. ‡

11 A. cepa. *Common O.* Scape inflated near the base, much taller than the fistulous leaves. ⓐ Universally cultivated, and of many varieties.
   β. proliferum. *Top O.* Umbel producing bulblets instead of flowers.

10. HYACÍNTHUS, L. Hyacinth. Perianth tubular-bell-form, segment spreading-recurved. Stam. straight, perigynous. Ovary free. Seeds few. ♃ Bulb coated. Scape racemous.

H. orientàlis. Lvs. thick, lance-linear, half as long as the scape; flowers many, half 6-cleft, tumid at the base, blue, varying to purple, red, white, &c.; stamens deeply included. Levant. March, April. Fine for the bulb-glass.

11. MUSCÀRI, Tourn. Grape Hyacinth. Perianth-tube ventricous, ovoid, globular or urceolate, limb of 6 very short blunt teeth. Otherwise as in Hyacinthus.

1 M. botryoìdes L. Fls. scentless, globular, nodding, blue (&c.), 2"; lvs. broad-lin., obtuse, longer than the scapes (10'). Gardens and fields. May. § Europe.

2 M. moschàtum. Fls. musk-scented, oval, nodding, 3", greenish-blue, or livid, with a little 6-toothed crown in the throat; leaves lance-linear, *erect*. Europe. April.

3 M. racemòsum. Flowers fragrant, nodding, dense, ovoid-cylindric, blue with a white limb; leaves linear, flaccid, channelled, *recurved*. Rare in gardens.

4 M. comòsum occurs in gardens as a monstrosity, with the tall (1f) raceme changed to a sterile, diffuse, feathery panicle of blue filaments. Showy.

12. SCHŒNOLÍRION, Torr. Stem a tuberous rhizome. Perianth yellow, &c. Caps. obovoid, obscurely 3-lobed. Flowers racemed. ♃ Otherwise as in Ornithogalum, and too near it. April, May.

S. cròceum (Mx.) Lvs. narrowly linear, longer than the scape, which is very slender, 15—20′; flowers small, about 15 in the raceme, yellow; sepals ovate, 2″. Damp. S.

13. YUCCA, L. BEAR'S-GRASS. SPANISH DAGGERS. Perianth persistent and withering, of 6 sepals, the 6 stamens shorter. Stigmas 3, sessile. Caps. oblong, 6-sided, the 3 cells partly divided each into 2 by a false partition. Seeds ∞. ♃ Stem subterranean, or arising into a caudex (§ 227), with linear or sword-shaped perennial leaves and a terminal panicle of white, handsome flowers.

1 **Y. filamentòsa** L. *Bear's-thread.* Acaulescent or nearly so; leaves lance-linear, rigid, sharp-pointed, the margin *filamentous,* i. e., bearing thread-like fibres; scape 5—8f; flowers numerous, cup-form, 1½′. Sands, S. June. †
2 **Y. gloriòsa** L. Caulescent; caudex some 3f; leaves clustered at top, lanceolate, stiff, margins very entire; flowers cup-form, very ∞. S. June, July.
3 **Y. aloefòlia** Walt. *Spanish Daggers.* Caudex some 10f, often branched, naked and scarred; leaves clustered at top, stout and sharp, serrulate; flowers white, with violet spots; sepals oblong. Thickets near the coast, S. June—Aug.

14. TRITÒMA, Ker. Perianth *tubular,* regular, 6-toothed. Stamens straight, hypogynous, alternately longer, and with the style exserted. Caps. ∞-seeded. ♃ Leaves linear, keeled. Scape racemed.

T. UvÀRIA. Lvs. in a dense radical crown; scape 3—5f, with a long raceme of innumerable soon-pendent, red, orange, and flame-colored flowers. S. Africa. Aug.—Oct.

15. AGAPÁNTHUS, L'Her. Perianth tubular at base, funnel-form, free from the ovary, regular. Stam. and filiform style upcurved at the end. Caps. 3-angled. Seeds ∞. ♃ Root tuberous. Leaves flat, linear. Scape bearing a 2-leaved umbel. Blue. July.

A. UMBELLÀTUS. Scape 2f, with the thick radical leaves as long; flowers many, large, the pedicels equalling the perianth. S. Africa. A fine parlor plant.

16. FÚNKIA, Spreng. Perianth funnel-form, deciduous. Stam. 6, hypogynous, and with the style declinate-curved. Caps. elongated, 3-angled. Seeds ∞, winged at end. ♃ Root fasciculate. Leaves all radical, ovate or oblong, veined, petiolate. Scape racemed. Japan.

1 **F. SUBCORDÀTA.** *White Day Lily.* Lvs. large, ovate, subcordate, veins strongly impressed; fls. white, fragrant, horizontal, 5′ long, tube longer than the limb. 2½f. Aug.
2 **F. OVÀTA** Spr. *Blue Day Lily.* Lvs. broad-ovate, acuminate; rac. many-flowered; fls. funnel-form, 2′, blue or violet, nodding, tube shorter than the limb. Ohio, §. †
 3 ALBO-MARGINÀTA. Has its leaves irregularly margined with white.

17. HEMEROCÁLLIS, L. DAY LILY. Perianth funnel-shaped, regular, ephemeral, limb spreading. Stam. 6, inserted in the throat, curved upward. Style slender, curved like the stamens and longer. Caps. with 3 few-seeded cells. ♃ Root fasciculate. Scapes branched. Leaves linear. Flowers large, xanthic, solitary, or racemed. July.

1 **H. FULVA.** Lvs. channelled; pet. obtuse, wavy; veins of sep. branched. An old garden plant, with large tawny flowers, lasting but a day. 3f. § Levant.
2 **H. FLAVA.** Lvs. channelled; sep. acute, bright yellow, veins undivided. Siberia. 1f.

18. POLYGONATUM, Tourn. TRUE SOLOMON'S SEAL. Perianth tubular, limb short, 6-lobed, erect. Stamens 6, inserted near and above the middle of the tube, and with the slender style included. Berry globular, black or blue, 3–6-seeded. ♃ Rhizome horizontal, thick. St. leafy above. (Lvs. alternate.) Fls. axillary, pendent, greenish-white. Fig. 258.

   P. biflòrum Ell. Stem recurved, smooth; lvs. lanceolate to elliptic, sessile, obscurely many-veined, glaucous-pale and more or less pubescent beneath; filaments roughened, inserted near the middle of the tube. Woods. 1—3f. April—June.
     β. *gigánteum.* Plant all smooth, tall; lvs. clasping; ped. 2-6-flwd. 3—7f.
     γ. *latifòlium.* Plant pubescent above; leaves ovate, some stalked.

19. CONVALLÀRIA, L. LILY OF THE VALLEY. Perianth campanulate, of 6 united segments, lobes of the limb recurved. Stam. 6, included, perigynous. Ovary 3-celled, 1-styled, cells 4–6-ovuled. Berry (red) few-seeded. ♃ Rhizome creeping, slender. Lvs. radical, and scape very smooth, low, bearing a raceme of white, drooping, sweet-scented flowers.

   C. majàlis L.—Mountain woods, Va. to Ga. Common in gardens. 6—10'. Lvs. ovato-elliptic, 2 or 3 with each scape. Flowers in an open raceme, 3—4''. May, June.

20. CLINTÒNIA, Raf. Perianth campanulate, of 6 equal, distinct segments. Stam. 6, hypogynous, anth. linear-oblong. Ovary oblong, 2-(rarely 3-)celled. Style elongated. Berry (blue) 2-celled, cells 2–10-seeded. ♃ Rhizome creeping. Lvs. few, broad. Scape naked, bearing an umbel.

   1 **C. boreàlis** Raf. Lvs. broad-oval-lanceolate; flowers 2—5 in the bractless umbel, cernuous; berry-cells many-seeded. Mountainous or hilly woods. June. 8—13'. A smooth and elegant plant. (See Fig. No. 715 in the Class-Book.)
   2 **C. umbellàta** Torr. Lvs. lance-oblong; umbel many-(12–30-)flwd., bracted; fls. white, speckled, 4—5''; berry-cells 2-seeded. Woods, W. N-Y., and S. along the mts.

21. SMILACÌNA, Desf. FALSE SOLOMON'S SEAL. Perianth of 6 equal, spreading segm., united at base. Stam. 6, slender, perigynous, anth. short. Ova. globous, 3-celled, with 2 ovules in each cell. Sty. short, thick. Berry globous, pulpy, 1–3-seeded. ♃ Rhizome creeping, thick or slender. Stem leafy, bearing a terminal cluster of white flowers in April—June.

   § Raceme compound. Stamens longer than the perianth. Ovules collateral......No. 1
   § Raceme simple. Stam. shorter than perianth. Ovules one above the other..Nos. 2, 3
   1 **S. racemòsa** Desf. Stem recurved; leaves oval, strongly veined, acuminate, subsessile; raceme compound. Copses: common. Berries red-dotted. 2f.
   2 **S. stellàta** Desf. St. erect; lvs. many, lanceolate, acute, amplexicaul; fls. few, in a simple raceme; berries dark red. Along rivers, N. and W. 10—20'.
   3 **S. trifoliàta** Desf. Erect; lvs. 3 or 4, oval-lanceolate, tapering to both ends, amplexicaul; rac. terminal, simple; berries red. Mountain swamps, N. and W. 3—6'.

22. MAJÁNTHEMUM, Mœnch. TWO-LEAVED SOLOMON'S SEAL. Perianth of 4 ovate, obtuse, spreading segments, united at base. Stam. 4. Ovary 2-celled. Otherwise as in Smilacina.

   M. bifòlium DC.—Common in open woods. Stem with 2 (rarely 3) ovate, subcordate leaves and a simple raceme of small white flowers, 3—6'. May.—In Oregon, the same plant becomes stout, 2f high, with petiolate, strongly cordate leaves!

ORDER 148.—MELANTHACEÆ. 347

23. ASPÁRAGUS, L. Perianth 6-parted, segm. erect, slight-spreading above. Sta. 6, perigynous. Sty. very short, stig. 3. Berry 3-celled, cells 2-seeded. ♃ Rts. fibrous, matted. Stems with filiform branchlets for leaves in the axils of scales.

A. officinàlis L. Stem herbaceous, very branching, erect; lvs. fasciculate; flowers axillary; berries red. Long cultivated, and § in rocky shores.

24. STRÉPTOPUS, Mx. TWIST-FOOT. Perianth bell-form, of 6 distinct, recurved sepals. Anth. longer than the filaments. Style elongated, stigmas 3-lobed. Berry globose, red, ∞-seeded. ♃ Stem fork-branched. Flowers axillary, solitary, on a *geniculate* or curved pedicel. June.

1 S. ròseus Mx. Lvs. oblong-ovate, clasping, margin finely ciliate; pedicels oftener merely recurved; anth. short, 2-horned at apex; stigma trifid. Damp woods, northward. 1f—15′. Flowers reddish, spotted, under the leaves.

2 S. amplexifòlius DC. Leaves oblong-ovate, strongly clasping, margin smooth and entire; pedicels abruptly bent in the middle; anthers and stigmas entire at the apex; sepals long-pointed, reflexed. Woods, Penn., and N. 2f.

25. PROSÁRTES, Don. Perianth as in Uvularia. Fil. 6, perigynous, included, much longer than the linear-oblong anth. Style elongated, trifid. Berry red, ovoid or oblong, 3-6-seeded. ♃ Stem erect, branched. Flowers few, greenish, terminal, drooping. May.

P. lanuginòsa Don. Lvs. ovate-oblong, pointed, clasping, downy beneath; pedicels in pairs; flowers spreading-bell-form; sep. 5—6″ long. Mountains, N. Y. to Car.

26. UVULÀRIA, L. BELLWORT. Perianth of 6 linear-oblong, connivent sepals, each nectariferous at base. Fila. much shorter than the long, linear, included anth. Style trifid. Caps. 3-celled, few-seeded. ♃ Stem forking. Leaves alternate. Flowers yellowish, drooping.

§ Leaves perfoliate near the base. Capsule obovoid-triangular, truncate....Nos. 1—3
§ Leaves sessile or half-clasping. Capsule ovoid or oval-triangular..........Nos. 4—6

1 U. grandiflòra Sm. Sepals acuminate, smooth within and without, greenish yellow, 1½′ long; anthers obtuse (½′). Woods, 1—2f. May.
2 U. perfoliàta L. *Mealy B.* Sepals acute, 1¼′, twisted, covered inside with shining grains, pale yellow; anthers cuspidate. Woods. 10—14′. May.
3 U. flava Sm. Lvs. obtuse; sepals smooth both sides, yellow. 1′. N. J. to Va.
4 U. sessilifòlia L. *Wild Oats.* Lvs. lance-oval, glaucous beneath; capsule stiped; style 3-cleft, nearly as long as the (9″) sepals. Glades: common. 6—10′. May.
5 U. Floridàna Chapm. Leaves oblong, glaucous beneath; style 3-cleft, half as long as the acuminate (8″) sepals. Woods, Fla. 4—6′. March.
6 U. pubérula Mx. Leaves puberulent, oval, green both sides; capsule sessile (no stipe); style 3-parted to near the base, not exceeding the anthers. Mountains, S

ORDER CXLVIII. MELANTHACEÆ. MELANTUS.

Herbs perennial, sometimes bulbous, often poisonous, with parallel-veined *leaves*. *Perianth* double, regular, persistent, of 6 consimilar, green or colored *segments*. *Stamens* 6, with extrorse *anthers*, 3 distinct *styles* or sessile *stigmas*, and a free, 3-celled *ovary*. *Capsule* 3-celled, 3-partible or septicidal, and *seeds* few or many, with a thin seed-coat.—Very near the Lilyworts, but the divided pistils afford a practical distinction.

ORDER 148.—MELANTHACEÆ.

§ Perianth 6-parted, tube very long, radical, like the Crocus....................... COLCHICUM.   1
§ Perianth 6-sepalled, wheel-form, on a scape or stem, with leaves...(*)
  * Anthers 1-celled, extrorse, cordate, becoming peltate by opening...(a)
  * Anthers 2-celled, extrorse. Capsule loculicidal. Flowers racemous...(c)
  * Anthers 2-celled, introrse. Capsule septicidal. Flowers racemous...(d)
    a Inflorescence racemous, with white flowers. Sta. scarce longer than sep...AMIANTHIUM.   2
    a Inflorescence spicate, with green flowers. Sta. twice longer than sepals...SCHŒNOCAULON. 3
    a Inflorescence paniculate, or a raceme somewhat branched at base...(b)
      b Sepals glandular at base inside, clawed. Stamens perigynous..........MELANTHIUM.   4
      b Sepals glandular at base inside, clawed. Stamens hypogynous........ZIGADENUS.   5
      b Sepals not gland-bearing. Stamens perigynous........................VERATRUM.   6
    c Flowers perfect. Filaments dilated at base. Ovary cells 2-ovuled........ZEROPHYLLUM.   7
    c Flowers perfect. Filaments filiform. Ovary cells ∞-ovuled.............HELONIAS.   8
    c Flowers diœcious, white. Stem leafy..................................CHAMÆLIRIUM.   9
      d Stamens 6. Flowers greenish or yellowish, 9—40.....................TOFIELDIA.   10
      d Stamens 9—12. Flowers deep yellow, 6—9, mostly 6..................PLEEA.   11

1. **COLCHICUM** AUTUMNÀLE. A plant of curious habit, from Europe. The 1—3 long-(5–8'-)tubed, lilac-colored, 6-parted flower arises directly from the new tuber in the Autumn, followed in the succeeding Spring by a stem bearing the leaves and fruit.

2. **AMIÁNTHIUM**, Gray. FLY-POISON. Fls. ☿. Sep. sessile, spreading, glandless, shorter than the stamens. Anth. reniform. Caps. 3-horned, 3-partible into 1–4-seeded follicles. ⚃ St. bulbous at base, scape-like. Lvs. grass-like. Fls. on slender pedicels, turning green with age. May—July.

1 A. **muscætóxicum** Gr. Bulb conspicuous; lvs. broad-linear, obtuse, many; rac. dense; sep. oblong; seeds ovate, red and fleshy. Shades, N. J., W. and S. 1—2f.
2 A. **angustifòlium** Gr. Tall, slender, scarcely bulbous; lvs. linear, acute; sepals oval, changing to brown; rac. very dense; seeds linear, dry. Damp woods, S. 2—3f.

3. **SCHŒNOCAÙLON**, Gray. Fls. ☿. Sep. green, linear-oblong, half as long as the hypogynous stam. Ova. 6–8-ovuled, carpels slightly cohering. ⚃ Scape bulbous, rush-like. Lvs. sedge-like. Spike slender. Apr., May.
S. **grácile** Gr.—Sandy soils, Ga., Fla. Scape 2—3f, lvs. half as long. Fruit unknown.

4. **MELÁNTHIUM**, Gronov. Fls. ♂ ☿ ♀. Sep. spreading, unguiculate, with 2 glands at base, the claws bearing the short stamens. Ova. often abortive. Caps. 3-lobed, 3-pointed with the persistent styles. ⚃ St. thickened at base. Racemes panicled. Flowers yellowish. July, Aug.
M. Virgínicum L.—Wet meadows, N. Y., W. and S. Stem 3—4f, leafy. Lvs. lanceolate to linear, 6"—2' wide, subclasping. Flowers 8", in a large panicle.

5. **ZIGADÈNUS**, Mx. ZIGADENE. Segm. colored, spreading, at base united, contracted and 2-glanded. Sta. hypogynous, nearly as long as the segm. Ovary adherent at base or free. Seeds ∞, scarcely winged. ⚃ Smooth and glaucous. Leaves linear. Flowers greenish, panicled.

1 Z. **glabérrimus** Mx. Rhizome creeping; lvs. channelled, recurved; panicle conical; fls. 1' broad; sepals lance-ovate, with 2 round glands. Swamps, S. 2f. June.
2 Z. **glaucus** N. Stem bulbous, nearly naked; lvs. flat, much shorter than the stem; sepals obtuse, 3", each with 1 obcordate gland. Sandy shores, N. Y. to Dakota. 1½f
3 Z. **leimanthoìdes** Gr. Root fibrous; lvs. flat; panicle slender; segm. obovate, the glandular spot obscure. Swamps, N. J., and S. 2—4f. Flowers white.

6. **VERÀTRUM**, Tourn. FALSE HELLEBORE. Fls. ♂ ☿ ♀. Sep. spreading, sessile and without glands. Sta. shorter than the perianth and inserted

ORDER 148.—MELANTHACEÆ. 349

on its base. Ovary 3, united at base, often abortive. Capsule 3-partible. Seeds few, flat, broadly winged. ♃ Flowers in panicles. July.

§ STENÁNTHIUM. Sepals at base united and adherent to base of ovary ....No. 1
§ VERÁTRUM *proper.* Sepals distinct to base and free from the ovary. .....Nos. 2—4

1 **V. angustifòlium** Ph. Lvs. long-linear; stem slender, 2—4f; panicle 1½f, narrow; segm. green-white, subulate, 2''; flowers sessile, the upper fertile. Pa., W. and S.
2 **V. víride** Ait. Stem stout and very leafy, 2—4f; leaves lance-oval, ample, strongly plaited; flowers innumerable, green; sepals lanceolate, 6''. Wet meadows.
3 **V. parviflòrum** Mx. Leaves nearly all radical, oval-elliptic, petiolate, slightly plaited; stem slender, scape-like, long-paniculate; sepals spatulate-unguiculate, 2—3'', half as long as the pedicels, dingy green. S. 2—5f.
4 **V. Woódii** Robbins. Leaves lance-elliptic to lance-linear, the lower long-petioled, plicate; stem rather stout, 4—6f; panicle long and narrow; sepals oblanceolate to obovate, 4'', *almost black*, as long as the pedicels. Ind., and W.

7. **XEROPHYLLUM**, Mx. Fls. ☿. Sep. oval, spreading, sessile, and without glands. Fila. dilated and contiguous at base. Styles linear, revolute. Caps. 3-lobed, cells 2-seeded. ♃ Lvs. numerous, dry, setaceous, the lower longer, rosulately reclined. Rac. simple, with white, showy flowers.

X. asphodeloìdes N.—Sandy plains, N. J. to N. C. 3—5f. Per. 5'' wide. Ped. 1''. Jn.

8. **HELÒNIAS**, L. Fls. ☿. Sep. sessile, spreading, glandless, shorter than the filiform stamens. Anth. blue. Caps. 3-horned, 3-styled. Seeds ∞, linear. ♃ Scape thickish, hollow, with many radical, narrow-oblanceolate leaves, and a short, dense raceme of purple flowers.

H. bullàta L.—N. J. to Va. Rare. 10—18'. Lvs. nearly as long as the scape. May.

9. **CHAMÆLÌRIUM**, Walt. Fls. ♂ ♀. Sepals linear-spatulate, persistent, white, shorter than the filiform stamens. Anthers yellow. Styles club-form. Caps. ovoid, entire. Seeds ∞, winged at each end. ♃ Root premorse. Stem strict. Racemes slender, dense, nodding at top.

C. lùteum (L.) *Blazing Star.*—Damp grounds. Apr.—Jn. 12—30'. Root lvs. lance-obovate, stem lvs. lanceolate, more on the taller ♀ plant. Racemes 3—12'. Spring.

10. **TOFIÈLDIA**, Hudson. Fls. ☿, 3-bracteolate at base. Sep. spreading, sessile, oblong. Caps. 3-lobed, 3-partible. Seeds ∞, oblong. ♃ Lvs. equitant, grass-like, from fibrous roots. Scapes clustered, bearing spikes or narrow, close, greenish racemes. June—August.

* Glabrous. Pedicels separate, very short. Rac. simple, short, spicate.....Nos. 1, 2
* Glandular. Pedicels in 3's (1's—4's), short. Bracteoles united..........Nos. 3, 4

1 **T. glutinòsa** N. Lvs. glabrous, linear-ensiform, ⅓ as long as the rough-*glutinous* stem; rac. short (1—1½'), spicate; sep. oblanc., 2'', pod 4''. Woods, O. to Wis. 15'.
2 **T. pùbens** Dryand. Leaves nearly ⅓ the length of the glandular-puberulent stem; rac. of alternate, remotish fascicles, slender, 6—8' long, 30-40-flowered; pod scarcely longer than the perianth. Barrens, Del. to Fla. Slender. 2—3f.
3 **T. palústris** Huds. Lvs. 3-5-veined, acute; scape filiform; spike ovoid, lengthened in fruit; bractlets only at the base of the pedicels. Shores of L. Sup., and N.
4 **T. glabra** N. Leaves radical, a few on the stem; rac. 2—5' long, dense, 30-30-flowered; bractlets united near the flower, as in Nos. 1 and 2. Barrens, S. 1—2f.

11. **PLEÈA**, L. C. Rich. Sep. wide-spread, lanceolate, sessile, longer

than the 9—12 stamens. Styles subulate. Capsules 3-lobed. Seeds ∞ bristle-pointed. ♃ Rush-like stem and leaves dry and rigid.

P. tenuifòlia Rich.—Bogs, S. 1—2f. Sept., Oct. Leaves perennial, erect, very narrow, 1f, and bracts sheathing. Rac. loose, of few light-yellow, star-like flowers (1').

ORDER CXLIX. PONTEDERIACEÆ. PONTEDERIADS.

Plants aquatic, with the *leaves* parallel-veined, mostly dilated at base. *Flowers* spathaceous. *Perianth* tubular, colored, 6-parted, often irregular. *Stamens* 3 or 6, unequal, perigynous. *Ovary* free, 3-celled. *Style* 1. *Stigma* simple. *Capsule* 3-(sometimes 1-)celled, 3-valved, with loculicidal dehiscence. *Seeds* numerous (sometimes solitary), attached to a central axis. *Albumen* mealy.

* Flowers irregular, blue. Stamens 6. Utricle 1-seeded, (2 cells abortive)............PONTEDERIA. 1
* Flowers regular,—x cyanic. Anthers 3, of 2 forms. Leaves reniform..............HETERANTHERA. 2
  —x yellow. Anthers 3, of 1 form. Leaves linear................SCHOLLERA. 3

1. PONTEDÈRIA, L. PICKEREL WEED. Perianth bilabiate, under side of the tube split with 3 longitudinal clefts (the 2 lower sepals free), circinate after flowering and persistent. Sta. unequally inserted, 3 near the base and 3 at the summit of the tube. Utricle 1-seeded. ♃ ≈ Leaves radical, long-petioled. Stem 1-leaved, bearing a spike of blue flowers. Jl.

1 **P. cordàta** L. Lvs. ovate to oblong-deltoid, cordate, with rounded lobes; petiole shorter than the peduncle; spike cylindrical, pubescent, 2' long. In slow waters: com. A fine, showy plant, its blue spikes and smooth leaves 1—2f above the water.

2 **P. lancifòlia** Muhl. Lvs. lance-oblong to lance-lin.; fls. as above. S. Apr., May.

2. HETERANTHÈRA, R. & P. Tube of the perianth long and slender, limb 6-parted, equal. Stamens 3, lower anther oblong-sagittate, on a longer filament. Capsule 3-celled, ∞-seeded. ♃ ≈ Leaves mostly reniform, long-petioled. July, August.

1 **H. renifórmis** R. &. P. St. prostrate or floating; lvs. roundish, reniform or auriculate at base; spathe acuminate, 3-5-flowered; flowers white. N. Y., Pa., and W.

2 **H. limòsa** Vahl. Leaves ovate-oblong, both ends obtuse; spathe 1-flowered, long-mucronate; flowers blue. S. and W. (Carruth). Lvs. 1—1½', the stalks thrice longer.

3. SCHÓLLERA, Schreber. Tube of the perianth very long and slender, limb 6-parted, equal. Sta. 3, with similar anthers. Caps. 1-celled, ∞ - seeded. ♃ ≈ Leaves sheathing at base, grass-like, submersed. Stem floating, rooting at the lower joints.

S. gramínea Willd.—A grass-like aquatic, in flowing water, N. 1—3f long. Leaves 1—2'' wide. Flower solitary, 2½' long, spathe half as long. July, August.

ORDER CL. JUNCACEÆ. RUSHES.

Grass-like or rush-like *herbs*, with small, dry, greenish *flowers*. *Perianth* liliaceous in form, more or less glume-like, regular, 6-leaved, in 2 series, persistent. *Stamens* 6, rarely 3, hypogynous. *Anthers* 2-celled, introrse. *Style* 1. *Capsule* 3- or 1-celled, 3-valved. *Albumen* fleshy. Figs. 144, 467.

Order 150.—JUNCACEÆ. 351

* Perianth yellow (greenish outside). Stigma 1. Capsule ∞-seeded............. .....NARTHECIUM. 1
* Perianth green or brownish. Stigmas 3.—x Capsule 3-seeded.. ....LUZULA. 2
  —x Capsule ∞-seeded....................JUNCUS. 3

1. NARTHÈCIUM, Mœhr. Sepals spreading, yellowish inside. Fil. hairy. Caps. prismatic, 3-celled, tipped with the single style and stigma. Seeds ∞, bristle-tipped at each end. ♃ Root creeping. Lvs. linear, equitant. Scape bracted, simple, racemous. July, August.

N. **ossífragum** Huds.—Pine-barrens, N. J. Scape terete, 8—12′, the leaves much shorter. Sepals lance-linear, 2″. Pedicels 3—5″, bracteolate. Capsule yellowish, 4″. (N. Americanum Ker.)

2. LÙZULA, DC. Wood Rush. Perianth persistent, with 2 bractlets at base. Stamens 6. Capsule 1-celled, 3-seeded. ♃ Stem jointed, leafy. Lvs. grass-like, on entire sheaths. Fls. terminal, green or brownish.

* Flowers separate, pedicellate, in umbels or paniculate cymes..............Nos. 1, 2
* Flowers aggregate,—x in pedunculate heads forming an umbel or cyme...Nos. 3, 4
  —x in sessile heads forming a nodding black spike ....No. 5

1 L. **pilòsa** Willd. Lvs. lance-linear, fringed with long white hairs; umbel simple, 12-20-flwd.; ped. 5—10″, soon deflexed; fls. 1″, brownish. Groves, Pa., and N. May.
2 L. **parviflòra** Desv. Taller; lvs. lance-linear, glabrous; umb. decompound; fls. nodding, small; sep. ¼″; caps. dark brown, a little longer. Mts., N. 12—18′. Jn., Jl.
3 L. **campéstris** DC. *Field Rush.* Lvs. linear, flat, with cotton-like hairs; fls. in roundish heads, which are umbelled with very unequal peduncles; sep. rust-colored, longer than the obtuse caps. ; seeds appendaged at base. Meadows. 3—12′. May.
  β. **bulbòsa.** Bulbous at base, 3—9′; sep. shorter than the globular caps. Apr.
4 L. **arcuàta** E. Mayer. Lvs. linear, channelled, glabrous; hds. 3-5-flwd., on filiform, often recurved, unequal ped. ; bracts ciliate; seeds not appendaged. White Mts.
5 L. **spicàta** DC. Lvs. linear, hairy at base, very short; spike oblong, 8—12″; sep. bristle-pointed, equalling the roundish, black capsule (¼″). White Mts. 9—12′. Jl.

3. JUNCUS, L. Rush. Stamens 6 or 3. Capsule 3-celled, or (by the dissepiments not reaching the centre) 1-celled. Seeds numerous. ♃ Mostly glabrous. Stems simple, leafless, or with terete or grassy leaves, entire sheaths, and small, 2-bracteolate, green or brown fls. June—Aug.

§ Clusters growing apparently from the side of the simple scape...(*)
§ Clusters terminal on the stem or scape. Leaves never knotted...(**)
§ Clusters terminal. Flowers in heads. Leaves internally knotted...(***)
  * Leaves few, radical, knotless, terete like the scape....................Nos. 1, 2
  * Leaves none. Flowers separate, not in heads.—a Stamens 3.........No. 3
                                                —a Stamens 6.........Nos. 4—6
  ** Flowers separate, not in heads. Stamens 6...(c)
  ** Flowers capitate, few or many in each head.—b Stamens 6.........Nos. 7, 8
                                                —b Stamens 3.........Nos. 9, 10
    c Stems branched. Pod much shorter than the unequal sepals..........No. 11
    c Stems simple.—d Pod globular, not exserted. Flowers green.......Nos. 12, 13
                    —d Pod oblong or ovoid, exserted, brown..............Nos. 14—16
  *** Seeds tailed. Panicle rather erect, longer than its bract.......... Nos. 17—19
  *** Seeds acute, not tailed.—x Stamens 6...(y)
                                —x Stamens 3, bracts shorter than panicle...(z)
    y Heads 2-8-flwd. (or 1-flwd. in No. 20). Bracts shorter than panicle....Nos. 20, 21
    y Heads 5-70-flowered. Leaf or bract overtopping the panicle..... Nos. 22, 23

ORDER 150.—JUNCACEÆ.

z Heads 5–15-flowered, and numerous, in April—June............Nos. 24, 25
z Heads 20–80-flowered, few and large........................Nos. 26, 27

1 **J. setàceus** Rostk. Scape weak, slender, (not *setaceous*), 1–2f; lvs. shorter; panicle small, 20–30-flwd., flowers separate; sepals very acute, pod globous. Sea-coast, S.

2 **J. Rœmeriànus** Scheele. Scape stout. rigid, 2–4f, and leaves pungent; panicle compound; flowers capitate; sep. sharp-pointed; pod turgid, a little shorter; heads 5–8-flowered, dark brown. Marshes, Va. to Fla. (J. maritimus C-B.)

3 **J. effùsus** L. *Soft R.* Scapes straight, not rigid; panicle decompound, often diffuse; flowers green, sep. as long as the obovoid, obtuse pod. Wet: common. 2–3f.

4 **J. filifórmis** L. Scapes very slender, weak, the subsimple panicle near the middle; sepals longer than the obtuse, mucronate pod. Me. to Mich. 1–2f.

5 **J. Smithii** Engelm. Scapes slender, rather rigid, 2–3f; cyme few-flwd.; flowers brown, 1''; pod round-ovoid, mucronate, exserted. Broad Mountain, Pa. (Porter).

6 **J. Bálticus** Dethard. Scapes in dense rows on the rhizome, rigid, pungent; pan. near the top, brown; sep. erect, very acute, equalling the elliptical, mucronate pod (1½''). Sandy shores, Me. to Penn. and Wis. 1–3f.

7 **J. trífidus** L. Stems tufted, 5–8', wiry, sheathed at base, 3-leaved at top, and with a sessile head of 3 blackish flowers; capsule globular. Mountains, N. H., N. Y.

8 **J. Stýgius** L. Stems few-leaved at base, leafless at top, 7–12'; heads 1–3, about 3-flowered; sepals shorter than the elliptic pod; seeds large, tailed. Me., N. Y.

9 **J. repens** Mx. Stems low, tufted, 2–6'; leaves linear, opposite, fascicled; sepals subulate, awn-pointed, 3–4'', the slender pod 2''. ① Md. to Fla. May.

10 **J. marginàtus** Rostk. Stem compressed; leaves linear, flat; cyme compound, heads many, 2–9-flowered, chestnut-brown; pod globular. 1–3f.

β. *biflòrus.* Heads very numerous, 2–3-flowered, nearly black. S.

11 **J. bufònius** L. *Toad R.* Slender, 3–8', tufted; leaves 1–2'; branches 2, flower bearing the whole length; flowers remote, green; the 3 outer sep. longer. Common.

12 **J. ténuis** Willd. Stems wiry, 8–24'; leaves flat-filiform, 3–8'; bracts longer than the loose panicle; sepals green, longer than the roundish pod. Common.

β. *secúndus.* Flowers 1-rowed on the branchlets; bracts shorter than the panicle.

13 **J. dichótomus** Ell. Stem wiry, 1–2f; lvs. terete-filiform, channelled, on long sheaths; panicle forked or dense; pod roundish, long as sepals. S. Too near No. 12.

14 **J. Gerárdi** Loisel. *Black Grass.* Sts. wiry, leafy, 1–2f; lvs. thread-ensiform, 3–8'; pan. longer than the bracts; *style conspicuous;* pod blackish, long as sepals. Marshes.

15 **J. Greénii** Oakes & Tuckm. Wiry scapes and filiform lvs. rigid; bract filiform, twice longer (4') than the small panicle; flowers secund, straw-brown; sepals ovate, shorter than the ovoid pod. Coasts of N. Eng. and Mich. 1–2f.

16 **J. Vaseyi** Engelm. Sepals lanceolate, as long as the oval pod; bract scarcely longer than the panicle. Otherwise like No. 15. Mich. (Prof. Porter).

17 **J. asper** Engelm. Sts. rigid, 2–3f; lvs. rigid and *rough*, 3–10'; hds. scattered, 3–5 flwd., sep. 2¼'', strongly veined, subequal! shorter than the pointed brown pod. N. J

18 **J. caudàtus** Chapm. Sts. rigid, 2–3f; lvs. 3, rigid, erect; panicle large, erect, hds. 2–4-flwd.; sep. 2'', unequal; pod 3'', finally black; sds. with long white tails. S.

19 **J. Canadénsis** Gay. Sts. terete, with 2 or 3 erect, smooth lvs.; fls. in Aug. and Sept., 3–50 in a head, paniculate, brownish; sepals lanceolate, 3 outer shorter, none longer than the oblong-triangular pod; stamens 3. Common and very variable.

α. *coarctàtus.* Heads 2-5-flwd., in a contracted panicle; pod brown, exserted.
β. *brachycéphalus.* Hds. 3-5-flwd., in a spreading panicle; pod brown, exserted.
γ. *subcaudàtus.* Slender; heads 8-20-flwd., remote; seeds with short white tails.
δ. *longicaudàtus.* Stouter; hds. 8-50-flwd., approximate; sds. slender, long-tailed.

20 **J. pelocárpus** Meyr. Sts. slender, 2-3-lvd., 10–20'; panicle much branched; fls. in pairs or solitary, scattered, reddish; pod oblong, pointed with the slender style, longer than the oblong sepals. Wis. to Me. and Fla. (J. Conradi Tuckm.)

21 **J. articulàtus** L. Stems 1f, with 1–2 leaves; heads 3-6-flowered, crowded in a spreading panicle; sepals brownish, oblong; pod deep brown, oblong, exserted. N.

β. *obtusàta.* Heads 5-flowered; sepals and pod green, obtuse, mucronate. Phila
γ. *insignis.* Panicle erect, few-flowered; outer sepals cuspidate, inner obtuse.
22 **J. militàris** Bw. *Bayonet R.* Stem stout, 2—3f, bearing a single terete leaf near the middle, which overtops the panicle; heads 5-15-flowered; sepals brownish, acute, as long as the acuminate capsule. Bogs, coastward, N. Eng. to Del.
23 **J. nodòsus** L. Stem slender, 2- or 3-lvd.; lvs. slender, the upper (bracts) overtopping the cluster; heads few (1—9), approximate, 5-50-flowered; sepals brown, lance-subulate, shorter than the beaked capsule. Wet sands, Can. to Car.
β. *megacéphalus.* Stout, 3f, upper leaf and bract exceeding the simple cluster; heads 50-80-flowered, green; outer sepals subulate-awned, as long as the pod.
24 **J. acuminàtus** Mx. Stems 2- or 3-leaved; hds. 3-15-flowered, in a loose spreading panicle exceeding the bract; sepals lance-subulate, nearly equalling the short-pointed brown pod; seeds minute, acute at both ends. May, June.
β. *debìlis.* Slender or stout; hds. 3-7-flwd.; pod exserted. N. J., Ky., and S. 9'-3f.
γ. *legìtimus.* Heads 8-15-flowered; pods scarcely exserted. (J. Pondii C-B.)
25 **J. Elliòttii** Chapm. Stem, leaves, and panicle very erect, 1—2f; hds. 5-8-flwd., fls. 1''; sepals lanceolate, as long as the turgid-ovoid, blackish pod; seeds acute. April.
26 **J. brachycárpus** Eng. Strict, rigid, 1½—2½f; leaves 2—3; bract short; hds. round, dense, 50-flwd., pale, few (2—10); 3 outer sepals awned, much longer than pod. W.
β. ? *Wolfii.* Pan. spreading; *pod ovoid, blunt,* little shorter than the sep. Ill. (Wolf).
27 **J. scirpoides** Lam. Rigid, 2f; heads and bract as in the last; style usually exserted; sepals pungent-awned, equalling the taper-pointed pod. N. Y. to Ga.
β. *polycéphalus.* Stout, 3f; heads 60-90-flwd., brownish, distant; lvs. flattened.

ORDER CLI. COMMELYNACEÆ. SPIDERWORTS.

Herbs with flat, narrow *leaves,* sheathing at base. *Sepals* 3, green, *petals* 3, colored. *Stamens* 6, some of them usually deformed or abortive. *Styles* and *stigmas* united into one. *Capsule* 2- or 3-valved. *Seeds* 3 or more.

§ Flowers irregular, clustered in a spathe-like, cordate, floral leaf...................COMMELYNA. 1
§ Flowers regular, clustered. Floral leaves like the rest. Stamens 6............TRADESCANTIA. 2
§ Flowers regular, solitary, axillary. Stamens 3. Moss-like herbs..................MAYACA. 3

1. **COMMELỲNA**, Dill. Fls. irregular, 3 of the stamens sterile, with glands for anthers. Caps. 3-celled, one of the cells abortive or 1-seeded.— Leaves contracted to the sheathing base. Floral leaf or spathe erect in flower, recurved before and after. Petals blue, open but a few hours.

1 **C. commùnis** L. Procumbent and much branched; lvs. lance-ovate, rounded at base; spathe lateral, 2-6-flowered; odd petal reniform. Wet soils, S. June—Nov.
2 **C. Cayennénsis** Rich. Procumbent, glabrous, with small (1½—2½') ovate-oblong, obtuse leaves; spathe lateral, 3-4-flowered; odd petal round-ovate. Banks, Ill. to La.
3 **C. Virgínica** L. Stem weak, ascending; lvs. lanceolate to linear; spathe broad-cordate when open; odd petal very small, raised on a claw. Dry. M., S., W. Jl., Aug
4 **C. erécta** L. Erect, pubescent, sheaths hairy; leaves lanceolate; spathe hawk-bill shaped, its base-lobes united; petals nearly equal. Woods, Pa., W. and S. Jl., Aug.

2. **TRADESCÁNTIA**, L. SPIDERWORT. Fls. regular. Sep. persistent, pet. large, roundish, spreading. Fil clothed with jointed hairs, anth. reniform. Caps. 3-celled. ♃ Fls. in terminal, close umbels. Juice viscid.

1 **T. Virgínica** L. Umbels sessile, terminal and axillary, with leafy bracts; ped. soon reflexed; flowers ephemeral, of a rich deep blue; leaves linear, channelled; stem thick, jointed, 2—3f. Damp. M., S., W. Cultivated.

Order 152.—XYRIDACEÆ.

2 **T. pilòsa** Lehm. Umbels sessile, terminal and axillary; leaves lanceolate, hairy both sides; flowers small, bluish purple. Banks, Ill. to O., and S. 2f.

3 **T. ròsea** Mx. Umbels terminal, *pedtrculate*, with subulate bracts; leaves linear; petals rose-colored, twice longer than the smooth calyx. May. 1f.

4 **T. crassifòlia.** From Mexico, a trailing leaf-plant, in vases and baskets, with thick ovate leaves, variegated with purple, green, and white. Flowers roseate.

3. MAYÀCA, Aubl.
Stamens 3, opposite the sepals. Caps. 1-celled. Seeds several, attached to the middle of the valves. ⚹ Moss-like, creeping, branching, beset with narrow, linear leaves. Peduncles solitary, axillary, 1-flowered. Resembles a Sphagnum.

M. Michaùxii Schott. & Endl. Ped. longer than the lvs. (which are 2—3″), reflexed in fruit; pod 9–12-seeded; petals white. Shallow waters, Va. to Fla. July.

Order CLII. XYRIDACEÆ. Xyrids.

Herbs sedge-like, with equitant *leaves* and a scape bearing a head of regular triandrous *flowers*. *Perianth* of 3 glumaceous *sepals* and 3 colored *petals*. *Fertile stamens* on the claws of the petals. *Style* 3-cleft. *Capsule* 3-valved, ∞-seeded.

XYRIS, L. Yellow-eyed Grass. Head of flowers ovoid-cylindrical, invested with an armor of cartilaginous scales. One sepal membranous, involving the yellow corolla in bud, the 2 lateral strongly keeled, persistent. Pet. crenulate, on claws, caducous. 3 sterile sta. alternately with the 3 fertile. ♃ Lvs. radical, linear, sheathing the base of the slender scape. Jn.–Aug.

* Scape 2-edged above (except No. 6). Lvs. long, linear, flat, often twisted...(x)
* Scape teretish, its lvs. shorter than its sheath (No. 9) or longer, and filiform...No. 8
  x Sepals exceeding the bract, and fringed on the winged keel...............Nos. 6, 7
  x Sepals (the 2 lateral) included,—y winged and ciliate on the keel......Nos. 3—5
            —y wingless or very nearly so...........Nos. 1, 2

1 **X. flexuòsa** Muhl. *Common X.* Scape 6—18′, often bulbous at base; lvs. narrowly linear, 3—9′, often twisted; head round-ovoid, 3—4″; sepals minutely bearded at the tip, lance-oblong, quite wingless on the keel. N. Eng. to Ill. and Ga.

2 **X. ambìgua** Beyr. Scape 2—3f; lvs. broad-linear, rough-edged, 6—12′; hd. lance-oblong, 9—15″; sepals lanceolate, slightly winged; petals large (6″). Barrens, S.

3 **X. Caroliniàna** Walt. Scape 1—2½f, the broad-linear lvs. more than half as long; hd. yellowish-brown, 6—9″; sep. obscurely fringed; pet. 4—5″. Swamps, Mass. to Fla.

4 **X. Elliòttii** Chapm. Scape 2-edged throughout, 1—1½f; lvs. narrow-lin., ½ as long; hd. obovoid, 4—5″; sep. cut-fringed on the wing; pet. 3″. Wet barrens, S. Car. to Fla.

5 **X. platýlepis** Chapm. Scape 2—3f, twisted, as well as the broad-linear lvs.; hd. 9—12″, pale; sepals fringed at the apex, wing narrow; petals 2—3″. Sands, S. Car. to Fla.

6 **X. torta** Sm. Bulbous; terete scape and rigid lvs. twisted; hd. oval to oblong, 5—9″; sepal fringe exserted; petals large, roundish, 8″. Sand, N. J. to Fla. (X. bulbosa K.)

7 **X. fimbriàta** Ell. Scape rough, 2—3f, the broad-linear lvs. nearly as long; hd. large, ovoid, 9—12″; sepals much fringed and exserted; petals small (3—4″). N. J. to Fla.

8 **X. Baldwiniàna** R. & S. Scape 6—18′, twice longer than the filiform bristle-pointed leaves; head oval, 2—4″; sep. falcate, keel winged, ciliolate. Fla. (X. filifolia Ch.)

9 **X. brevifòlia** Mx. Scape 4—12′; lvs. linear to subulate, ½—2′, spreading two ways; head oval, 2—3″; sep. wingless; pet. 2″. Wet places, S. (X. flabelliformis Chapm.)

ORDER CLIII. ERIOCAULONACEÆ. PIPEWORTS.

Herbs perennial, aquatic, with linear, cellular, spongy *leaves* sheathing the base of the slender *scapes*, which bear a dense head of minute imperfect *flowers* at top. *Perianth* 2-6-parted or 0. *Stamens* 6, some of them generally abortive. *Ovary* 2- or 3-celled, cells 1-seeded.

* Stamens (4 or 6) twice as many as the petals. (Scape 7-12-ribbed)..................ERIOCAULON. 1
* Stamens 3, as many as the petals. (Scape 5-ribbed, puberulent)....................PÆPALANTHUS. 2
* Stamens 3, and no petals. Scape 5-ribbed, short, hairy............................LACHNOCAULON. 3

1. **ERIOCAÙLON**, L. PIPEWORT. Fls. ☉, in a compact head, with an involucre, the marginal fertile. Sepals 3. ♂ Petals 2 or 3, black-tipped, united, sta. 4 or 6. ♀ Pet. 2 or 3, distinct, sta. 0. Style 1, stigmas 2 or 3. ♃ Lvs. grass-like. Scape fluted. Chaff and fls. white-woolly at tip. Jn.-Aug.

1 **E. decangulàre** L. Scape tall (2—3f), 10-12-ribbed; leaves linear-ensiform, suberect, near ½ as long as the scapes; head 3—5''; chaff pointed. Swamps, Va. to Fla.
2 **E. gnaphalòdes** Mx. Scape tall (1—2½f), 10-ribbed; leaves ensiform-subulate, 2—4'; bracts and chaff obtuse, densely white-fringed. Swamps, N. J. to Fla.
3 **E. septangulàre** Wth. Scape very slender, 7-ribbed, 3—6', or in water several feet according to its depth; leaves linear-setaceous, 1—3'; heads globular. N. J. to Mich.

2. **PÆPALÁNTHUS**, Mart. Flowers 3-parted. Stamens in the sterile flowers 3. Stigmas in the fertile flowers 3. Capsule 3-seeded. Otherwise nearly as in Eriocaulon.

P. flávidus Kunth. In tufts; scapes 5-ribbed, minutely downy, 6-9'; leaves linear setaceous, 1—2'; head finally globular, bracts obtuse, straw-colored. Va. to Fla.

3. **LACHNOCAÙLON**, Kunth. ♂ Calyx 3-sepalled. Cor. 0. Sta. 3, anth. 1-celled, filaments united below. ♀ Cal. 3-sepalled. Cor. reduced to a tuft of hairs surrounding the 3-seeded caps. Otherwise as in Eriocaulon.

L. Michaùxii K. Scapes 1—5', clustered, 5-ribbed, villous, 2—8' (1f, Chapman); lvs ensiform-subulate 1—2'; head globular, 1—2'', brownish. Sands, Va. to Fla

Class IV. GLUMIFERÆ,

Or GLUMACEOUS ENDOGENS. Plants having their flowers invested with one or more alternate imbricated glumes (chaff or husk) instead of petals and sepals, and collected into spikelets, spikes, or heads. The Class is equivalent to

COHORT 7. GRAMINOIDEÆ, the GRAMINOIDS or grass-like plants.

ORDER CLIV. CYPERACEÆ. THE SEDGES.

These are grass-like or rush-like *herbs*, with fibrous *roots* and solid *culms*. *Leaves* generally 3-ranked, linear, channelled, based on entire or tubular *sheaths*. *Flowers* spiked, perfect or imperfect, one in the axil of each glume. *Perianth* none, or represented by a few hypogynous bristles called *setæ*, or a cup-shaped or bottle-shaped *perigynium*. *Stamens* definite, generally 3 (1—12). *Anthers* fixed by their base, 2-celled. *Ovary* 1-celled, 1-ovuled. *Style* 2- or 3-cleft and the *achenium* 2-sided or 3-sided.

The Sedges abound in marshes, meadows, and swamps.

§ CYPEREÆ. Glumes distychous (2-rowed). Flowers all perfect...(*)
§ SCIRPEÆ. Glumes imbricated all around, each (except sometimes the lowest) with a perfect flower Spikes all terminal or all lateral...(**)
§ RHYNCHOSPOREÆ. Glumes imbricated all around or irregularly, the lowest empty. Spikelets both terminal and axillary (except Dichromena and Chætospora)...(***)
§ CARICEÆ. Glumes imbricated all around, or irregularly. Flowers monœcious or diœcious. Achenium enclosed in a bottle-shaped *perigynium*...(****)

* Inflorescence axillary. Perigynium or perianth of 6—10 setæ.....................DULICHIUM. 1
* Inflorescence terminal. Perigynium none.—α Spikes 2 - ∞ - flowered...............CYPERUS. 2
 —α Spikes 1-flowered, capitate............KYLLINGIA. 3
** Perianth of 3 ovate clawed petals and (often) of 3 setæ. Glumes awned............FUIRENA. 4
** Perianth of 2 oblong sessile scales (pales) and no setæ. Spikes ∞ ....LIPOCARPHA. 5
** Perianth of 1 minute double scale and no setæ. Spikes 2, lateral................HEMICARPHA. 6
** Perianth of setæ only, 3 — ∞. No scales or petals...(*b*)
** Perianth none at all...(*d*)
 b Achenium crowned with a tubercle. Spike solitary, terminal .. ..ELEOCHARIS. 7
 b Achenium not tuberculed.—*c* Setæ 3—6, short, or else tawny. (CHÆTOSPORA, 18).....SCIRPUS 8
 —*c* Setæ ∞ (—6), long, cottony, white or reddish....ERIOPHORUM 9
 d Style 2-cleft. Spikes 5—10, terminal (capitate in Gen. 13)..................FIMBRISTYLIS. 10
 l Style 3-cleft. Achenium 3-angled ....TRICHELOSTYLIS. 11
*** Achenia crowned with the persistent style or its bulbous base (a tubercle)...(*g*)
*** Achenia not tuberculate,—*x* brown like the scales. Setæ none...................CLADIUM 16
 —*x* white or whitish, crustaceous. Setæ none............SCLERIA. 17
 g Perianth none (no setæ).—*y* Spikes diffusely cymous......................PSILOCARYA. 12
 —*y* Spikes capitate. Bracts colored..............DICHROMENA. 13
 g Perianth of setæ.—*z* Achenium tuberculate with the base of the style..RHYNCHOSPORA. 14
 —*z* Achenium horned with the entire long style.....CERATOSCHŒNUS. 15
**** Spikes either with ♂ and ♀ flowers, or each wholly ♂ or wholly ♀ ....CAREX. 19

1. DULICHIUM, Rich. Spikes linear-lanceolate, flattened. Glumes sheathing, closely imbricated in two rows. Style long, bifid, the persistent base crowning the flattened achenium. Perianth of 6—9 barbed setæ. ♃ Culm leafy. Racemes of spikes 2-rowed, axillary. August.

D. spathàceum Pers.—A sedge of peculiar and striking aspect, in marshes and by streams; common. Culm erect, 1—2f, leafy to the top, the leaves linear, in 3 ranks. Spikes 1', alternately arranged on the axillary leafless branchlets.

2. CYPÈRUS, L. GALINGALE.

SEDGE. Spikes flattened, distinct, many-flowered. Glumes imbricated in 2 opposite rows, nearly all floriferous. Setæ 0. Stamens 3—2. Style 3-(rarely 2-)cleft, deciduous. ♃ ①
Culms simple, leafy at base, triangular, bearing an involucrate simple or compound head or umbel at top. June to Sept.

§ PYCREUS. Style 2-cleft, nut flattened
   Spikes flattened, 10-30-flowered...(*)
§ CYPERUS. Style 3-cleft, nut 3-angled. Spk
   5-50-flowered...(**)
§ MARISCUS. Style 3-cleft, nut 3-angle..
   Spikes 1-5-flowered, deflexed......(n)

* Stamens 2 (or partly 3 in No 1).......................Nos. 1–?
* Stamens always 3...Nos. 4, ?
** Culm with many joints, teretish, with leafless sheaths at base..No. 6
** Culm jointless, triquetrous, leafy below..(a) (Invol. of 20 lvs. No. 35)
  a A pair of free persistent scales within each glume. Fls. dense. .7
  a Scales adnate to the rachis or wanting...(b)
    b Spikes capitate at the top of the peduncle, flattened...(c)
    b Spikes racemed or clustered, terete or flattened. Stam. 3..(m)
      c Glumes with recurved points. Stamen 1 only..Nos. 8, 9
      c Glumes with erect points or pointless. Stn. 1...Nos. 10, 11
      c Glumes with erect points. Stamens 3...(d)
        d Umbel compound. Spikes flattened, 3—5 in the clusters..Nos. 12—14
        d Umbel simple.—x Spikes flat, 12-30-flowered............Nos. 15, 16
          —x Spikes flat, 5-7-flowered. Head solitary...No. 17
          —x Spikes flattish, 6-12-flwd. Hds. 1—7...Nos. 18—20
        m Spikes flat, 12-24-flowered, 2-rowed in the clusters......Nos. 21—23
        m Spikes flat, 5-12-flwd., many-rowed in the clusters ...Nos. 24, 25, 35
        m Spikes terete.—y few, arranged in 2 rows in the clusters.....No. 26
          —y many, arranged in many rows........Nos. 27—29
  n Spikes 3-5-flowered, with 4—7 glumes......................Nos. 30—32
  n Spikes only 1-flowered, with 3 or 4 glumes....... ....Nos. 33, 34

1 C. dlàndrus Torr. (Fig. 1.) Slender, 4—10'; umbel of 2—5 very short unequal rays; spikes (Fig. 2) flat, oblong, obtusish, 4—8', fascicled; glumes (Fig. 3) 12—24, brown, with a green keel; stamens (Fig. 4) mostly 2; nut dull. (i) August. Pretty.

β. *castàneus.* Glumes numerous, and of a dark chestnut-brown.
γ. *paucifòrus.* Glumes only 5—9, edged with yellow, 2—3', crowded.

2 C. Nuttàllll Torr. Culm erect, 4—12'; rays few and short; spike lance-linear, very acute, ∞-flwd., crowded; glumes acute, yellowish-brown; stamens 2; sch. dull. (i)
β. *minimus.* Very slender, 3—4', hds. few or several, 2-5 flwd.; sta. 1. N. J., Pa.

3 C. microdóntus Torr. Culm and lvs. slender; spk. numerous, crowded, linear. acute; glumes acute, close; stamens 2; achenia oblong, *grey*, dotted. ① South.
  β. *Gatesii.* Culm and leaves filiform; spikes fewer, loose in the umbel. S-W.
4 C. flavéscens L. Culm and leaves 4—10'; rays 2—4, short, the linear obtuse spikes clustered at the end; glumes obtuse, straw-yellow; achenia shining. ① E.
5 C. flavicomus Mx. Culm 1—3f; involucre 3—5-leaved, very long; umbel some compound; spikes numerous, linear, 12—30-flowered, spreading; glumes very obtuse, brownish-yellow, 3-veined, white-edged; achenia obovate, blackish. Va., and South.
6 C. articulàtus L. Culm 2—6f, the joints internal, leaves 0 or mere sheaths; umbel compound, involucre short; spk. subulate; gls. 14—20, scarious. Swamps, S.
7 C. erythrorhizos Muhl. Culm 2—3f; umbel compound, each ray with several sessile clusters; spikes very many, 6″, teretish; glumes 15—30, yellow-brown; *inner* scales very narrow; achenia 3-angled, light colored, minute. ① Pa., S. and W.
8 C. inflèxus Muhl. Culms clustered, 1—3', leaves setaceous; hds. 1—3; spk. very short (1—2″), crowded; gls. 8—10, with a recurved bristle-point. ① Shores. Com.
9 C. acuminàtus Torr. Culm filiform or slender. 3—12'; hds. 1—7, each of ∞ flat obl.-ovate obtuse spikes 2—3″ long; glumes whitish, recurved at tip. ① Ill. to La.
10 C. virens Mx. Culm sharply rough-angled, 1—4f; leaves keeled, 1—3f; heads ∞, of ∞ ovate 15-flwd. spikes; gls. greenish, merely acute; ach. linear. ♃ Va., and S.
  β. *reg'tus*, has smooth culms and spikes *very* densely packed. S.
11 C. Drummóndii Torr. Culm very rough, 6—15', obtuse-angled; hds. ∞, dense, spike oblong-linear, 40—50-flowered, yellowish; glumes ovate, acute. Swamps. Fla.
12 C. Haspan L. β. *leptos.* Culm 1—2f, leaves shorter, involucre 2-leaved, shorter than the compound umbel; spikes linear, acute, 6″, 3—5 in a cluster; glumes minute, 20—40, mucronate, tawny-brown; achenia very minute, white, tumid. Swamps. S.
13 C. dentàtus Torr. Much like C. Haspan, but the involucre is 3- *or 4-leaved*, and *longer* than the umbel; glumes fewer (7—20), larger, the upper often long-pointed.
14 C. Lecóntii Torr. Culm and leaves 1—2f; umbel much compounded, with about 3 oblong, obtuse, flat silvery spikes on each peduncle; glumes 20—40, obtuse, very closely imbricated. ♃ Sandy coasts, Fla. A handsome sedge.
15 C. fuscus L. Culms 3—6', leaves flat; spk. lance-linear, 1—3″, dark-red or brown, densely fascicled in many heads; glumes round-ovate, closely imbricate. Phila. §
16 C. compréssus L. Culm tumid at base, 4—10', lvs. shorter; spikes lance-linear, in loose hds; gls. 12—40, ov.-acuminate, acutely keeled and close-pressed. Pa., and S.
17 C. divérgens Kunth. Tufts 2—3', leaves longer; spikes lance-ovate, flat, acute, 1″, 6-flowered, white, all in a single somewhat compound head. Fla.
18 C. filicúlmis Vahl. Culm tuberous, very slender, 6—12'; leaves very narrow, keeled; spk. lance-lin., in 1—4 *dense* hds.; gls. loose, 3—8, ovate; ach. gray. ♃ Dry.
19 C. Grayii Torr. Differs from No. 18 only in the *looser* heads of 6—8 *linear* spikes, the glumes less scarious and less veiny. ♃ Mass. to N. J.
20 C. Schweinítzii Terr. Culm rough-3-angled, 1—2f; leaves shorter; umbel simple, rays 4—6, erect; fls. large, in little spikes arranged close into cylindric-oblong compound spikes, with setaceous bractlets. ♃ Shores, N. Y. to Ark.
21 C. rotúndus L. β. *Hydra.* Nut Grass. Culm 6'—2f, the leaves shorter; umbel simple, rays 3 or 4, nearly equaling the invol.; spikes in two rows on the rachis; gls. 14—24, veinless, purple-brown. ♃ Va., and S. A rank and troublesome weed.
22 C. esculéntus. Root producing ovoid tubers as large as chestnuts, eatable when roasted (those of No. 23 very small); glumes veiny, yellow-brown. ♃ Eur. Cult.
23 C. phymatòdes Muhl. Culm 1—2f, with long lvs. and invol.; umbel simple or compound; spk. linear, obtuse; gls. veiny, 12—20, yellowish. ♃ Root creeping.
24 C. strigòsus L. Culm 1—3f; leaves broad-linear; umbel dense, large, some compound; rays 1—5'; spikes crowded, flattened, acute; glumes 8—18, tawny, ovate, acute, veined, much longer than the achenia. ♃ Damp. Common.
25 C. stenólepis Torr. Culm 1½—3f, smooth; leaves stiff, rough; rays 3—8; spikes crowded, 6—7″; glumes 5—8, lance-linear, spreading; seed slender, dull. ♃ S.

Order 154—CYPERACEÆ.

26 C. dissitiflòrus Tor. Culm slender, 1—2f, longer than the narrow leaves; invol. 3-leaved; rays 3—5; spike very slender and pointed, 6—9''', separate on the rachis; glumes 5—7, lance-oblong, acute; achenia brown, 3-angled. ♃ Tenn. to La.

27 C. Michauxiànus Schlt. Culm sharply 3-angled, 6—20'; umbel 6-10-rayed, simple or compound; spikes crowded in oblong clusters, 3''', tawny; glumes 5—10, oblong, overlapping, appressed; achenia ovoid, 3-angled. ♃ Swamps, M. and S.

28 C. Engelmánni Steud. Spikes very slender, with the 5—12 glumes remote, and the achenia oblong-linear. Otherwise like No. 27. ♃ Sandy swamps, W. and S.

29 C. tetrágonus Ell. Culm acutely rough-3-angled. leaves rough-edged; spike 4-angled, oblong, 2—3'''; glumes 5—7, ovate, veiny; rays 6—12, slender. ♃ Dry. S.

30 C. echinàtus (Ell.) Culm 10'—2f, the leaves still longer, involucre 5-6-leaved, very long; umbel simple, rays 8—12, each with a globular *cluster;* spikes 3''', about 3-flowered, subulate, radiant; glumes veiny, oblong, acute; ach. obovoid. ♃ Dry. S.

31 C. ovulàris (Vahl.) Culm 6—16', leaves shorter; umbel simple; rays 3'''—3', each with a *dense oval head;* spikes 1½''', 1-3-flowered, very many. ♃ Bogs. M., W., S.

32 C. Lancastriensis Porter. Culm 1—2½f; leaves linear, long: heads 5—9, oval, on as many slender rays; spikes subulate, 4—6''', soon deflexed, glumes about 5, veiny, obtuse, tawny, very acute, with about 3 linear achenia. ♃ Lancaster Co., Pa.

33 C. retrofráctus (Vahl.) Culm 2—3f, leaves shorter, broad; rays 1—6', each with 1 obovate, dense head; spikes 3''', subulate, 1-flowered, soon deflexed. ♃ N. J., and S.

34 C. uniflòrus Torr. & Hook. Has hds. oblong, 1' long, spks. closely deflexed. La.

35 C. ALTERNIFÒLIUS. Greenhouse species from Madagascar. Culm, and leaves, and many-leaved involucre striped with white and green, like *Ribbon Grass*.

3. KYLLÍNGIA, L.
Spikes compressed. Scales about 4, the two lowest short and empty, the third only usually with a fertile flower. Sta. 1—3. Style long, 2-cleft. Achenia lenticular. Culms triangular, leafy at base. Heads sessile, solitary or aggregated, involucrate, odorous. Aug.

1 K. pùmila Mx. In tufts, 2—12' high, very slender; heads solitary, rarely triple, sessile, oval to oblong; invol. 3-lvd., 1—2'; spk. very ∞, 1-flwd., *green*. ⓘ W. and S.

2 K. sesquiflòra Torr. Root creeping; culms 6—12'; heads mostly triple, oval to oblong, the lateral quite small; spk. densely packed, *white;* invol. deflexed. ♃ Fla.

4. FUIRÈNA, Rotboll. CLOT-GRASS.
Glumes imbricated on all sides into a spike, awned below the apex. Petaloid scales 3, corda'e, awned, unguiculate, investing the *stipitate* achenium. ♃ Stems angular, leafy. Spikes solitary or in heads, pedunculate, (brown).

1 F. squarròsa Mx. Culm 1—2f, with several joints and sheathing flat lvs.; spks. ovoid, *squarrous* with the long recurved awns, 4—7 together in each head. Bogs.

    β. *hispida*. Taller, with sheaths and leaves, hispid with white spreading hairs.

2 F. scirpoìdea Mx. Culm slender, 1—2f, leafless but with several sheaths; spikes 1—3, ovoid, 3—5''', not squarrous, the short awns erect. Wet, Ga., Fla.

5. ELEÓCHARIS, R. Br. SPIKED RUSH
Spikes terete. Glumes imbricated all around. Bristles of the perianth (setæ) mostly 6 (3 to 12); rigid, persistent. Style 2-3-cleft, articulated to the ovary. Achenium crowned with a tubercle which is the persistent bulbous base of the style. Mostly ♃, ☼. Stems leafless. Spike solitary, terminal.

§ Spike terete, cylindrical, not thicker than the tall (2—10) culm...(a)
§ Spike terete (glumes spirally imbricated), thicker than the culm...(b)
§ Spikes flat, glumes few, in 2 or 3 rows, often proliferous. Culm capillary (e

ORDER 154.—CYPERACEÆ.

a Glumes many, rounded, coriaceous. Culm stout. Spike 1—2'........Nos. 1—3
a Glumes few, oblong, thin. Culm slender. Spike ¼—1'............Nos. 4, 5
   b Spike white or greenish-white, ovoid, 2—3''. Ach. blackish. S. .Nos. 6, 7
   b Spike brown or the glumes with tawny sides, white-edged...(c)
      c Tubercle nearly as large as the ribbed and dotted achenium......No. 8
      c Tubercle much smaller than the achenium...(d)
d Achenium 3-angled or tumid, style always 3-cleft...(e)
d Achenium flattened, smooth, style 2- ⎰ —*x* Spike lance-shaped......Nos. 9, 10
    cleft (3-2-cleft in No. 11.) ⎱ —*x* Spike globous or ovate.Nos. 11—13
   e Setæ 4—6, retrorsely barbed, longer than—*y* dotted achenium...Nos. 14, 21
                                           —*y* smooth achenium...Nos. 15, 16
   e Setæ 0—2—6, smoothish, shorter than the achenium............Nos. 17—20
     z Culms often proliferous (i. e., bearing young culms at top)....Nos. 21, 22
     z Culms never proliferous, only 2—6' high ....Nos. 23, 24

1 E. equisetoìdes Torr. Culm terete, many jointed, 2—3f, as thick as the spike ; sheath at base obtuse ; spike 1', acute, glumes very obtuse ; setæ 6: style 3-cleft; ach. smooth, brown. Bogs, R. I., W. and S.

2 E. quadrangulàta Br. Culm 2—4f, jointless, acutely 4-angled with the sides unequal ; spike 1—2' ; glumes obtuse ; ach. dull white, obovoid, tipped with the distinct tubercle ; setæ 6. Bogs, N. Y., W. and S. Rare.

3 E. cellulòsa Torr. Culm 2f, obtusely 3-angled below, jointless ; spike 1', glumes round ; setæ 6 ; ach. broad-obovate, deeply pitted. Marshes, Fla. to La.

4 E. Robbínsii Oakes. Culms slender, 9'—2f, sharply 3-angled, many of them abortive and splitting into hair-like fibres in the water : spikes 6—9'', spindle-form, 5-8-flowered ; ach. 1'', half as long as the 6 setæ. Ponds. Rare.

5 E. elongàta Chapm. Culms floating, very long and slender, with many hair-like abortive ones ; spike 12-20-flowered ; ach. and setæ as in No. 4. Ponds, S.

6 E. capitàta Br. Culms tufted, 3—6', striate ; spike ovate, 1—2'' ; glumes 10—15, whitish-scarious, oblong, deciduous ; ach. black, shining ; setæ 6. Ga., Fla.

7 E. álbida Torr. Culm and *whitish* spike much like E. capitata, but the glumes become 10—20, the *style* 3-*cleft* and achenium tumid, *brown*. Ga., Fla., La.

8 E. tuberculòsa Br. Culms angular, wiry, 10—15' ; spike 3—5'', lance-ovate ; gls. ∞, very obtuse ; ach. scarcely larger than its arrow-shaped tubercle. Swamps.

9 E. palústris Br. Rhizome creeping ; culms 9'—2f, with a long sheath ; spike lance-oblong, 3—6—9'' ; glumes reddish-brown, very numerous, oblong-ovate ; with a broad scarious margin ; ach. obovate, yellowish ; setæ 4. Common.
   β. *calva*. Bristles wanting ; culms filiform. Watertown, N. Y.

10 E. compressa Sull. Culms tufted, very erect, narrow-linear, 1—1½f ; spike oblong-ovoid, 3—5'' ; gls. 10—30, ov.-lanceolate, brown ; ach. yellow ; setæ 0. M., W.

11 E. obtùsa Schultes. Culm 6—16' ; spike ovoid, very obtuse, 2—4'' ; gls. ovate, very many and close, red-brown, white-edged ; setæ 6 ; style often 3-cleft. Common.

12 E. olivàcea Torr. Culms 2—4', densely tufted, spreading, flattened and striated ; spike ovate, acutish, 2—3'' ; glumes 20—30, green-brown ; ach. olive. Sands.

13 E. ovàta Br. Culms tufted, 6—10', finely striate ; spike exactly ovoid, 2—3'' ; glumes 20—30, rounded, tawny, with 2 white striæ ; ach. ivory-white. pyriform-compressed, capped with a *brown* tubercle ; setæ 7, long. E. Penn. (*H. Jackson.*)

14 E. simplex Torr. Culm acute-angled, filiform, 12—18' ; spk. 2—3'', ovoid ; glumes ovate, white-edged, few ; ach. olive-green, much larger than its tubercle. Md., and S.

15 E. rostellàta Torr. Culm 12—20', sulcate, rigid, very slender ; spike lance-ovate, acute, 3—4'' ; glumes 12—20 ; ach. olive-brown, tubercle a mere *beak*. E. and N.

16 E. intermèdia Schultes. Wiry setaceous culms 3—8', spreading, in dense tufts ; spk. oblong-ovate, acute, 1—3'' ; gls. oblong, obtuse, 12—25, with 2 brown lines ; ach. smooth, obovoid, light-brown, with a distinct conical brown tubercle. In wet banks.

17 E. melanocárpa Torr. Culm flat, striate, wiry, erect, 12—18'; spike lance-

ORDER 154.—CYPERACEÆ. 361

oblong, 4—6''; glumes 20—40, ovate; ach. blackish when ripe, covered by a broad tubercle which is abruptly-pointed; setæ 3, purple. Sandy bogs, E. and S.
18 E. ténuis Schultes. Culms filiform or wiry, 4-angled, tufted, 8—18'; spk. elliptical or oval, 2—3''; gls. dark-purple, obtuse, 20 +; ach. roughish, the tubercle broad-depressed, setæ 2 or 3, very short. A variety has the culms *capillary*. Wet places: com.
19 E. tricostàta Torr. Culm flattened, slender, 1—2f; spike oblong-cylindrical, 6—9'; glumes obtuse, rusty-brown, crowded; *setæ* 0; ach. sharply 3-angled, roughish, tubercle conical. N. J., and S. A variety has smaller spikes. (*Dr. Feay.*)
20 E. arenícola Torr. Culms flattish, erect, 6—12', wiry; spk. ovate, obtuse; gls. dark-brown, with broad white margins; ach. yellowish, tubercle distinct. Sands, S.
21 E. Baldwínii Torr. Culms 4—14', capillary, 4-angled, densely tufted; spike 1'', ovate, flat, often proliferous; gls. 5—10, in 2 rows; ach. strongly 3-angled. Ga., Fla.
22 E. prolífera Torr. Culms filiform, flattened, erect or diffuse, 10—20'; spike 3'', lance-ov., acute, often proliferous; gls. 10—15, pale; ach. ribbed, tubercle distinct. S.
23 E. aciculàris Br. Culms hair-like, 2—6'; spike elliptic-ovate, 1'', acute; glumes 4—8; ach. ovoid-triangular, longitudinally striate. Muddy places.
24 E. pusillus (Vahl.) Culms bristleform, 1—5', compressed; spk. ovate; gls. 3—6, mostly empty; ach. acutely triangular, smooth. Coasts. (E. pigmæa.)

6. SCÍRPUS, L. CLUB-RUSH. BULLRUSH. Glumes imbricated on all sides. Perianth of 3—6 setæ, persistent. Sty. 2-3-cleft, not tuberculate at base, deciduous. Achenium biconvex or triangular. ♃ Stems mostly triquetrous, simple, rarely leafless. Spikes solitary, conglomerated, or corymbous, usually rust-colored.

§ TRICÓPHORUM. Setæ 6, not barbed, tawny, tortuous, much longer than the achenium and exserted. Culm leafy. Cyme decompound .... Nos. 19, 20
§ SCIRPUS. Setæ downwardly barbellate, about equalling the achenium.... (*)
  * Spike single, terminal,—*a* Involucral bract 0 in No. 1, long (1') in .... No. 5
                          —*a* Involucral bract as short as the spike .... Nos. 2—4
  * Spikes several or many, clustered—*b* laterally on the culm.... (*c*)
                          —*b* terminally, mostly in cymes.... (*x*)
    c Culms terete, jointless, leafless or with a few short lvs at base... Nos. 6—8
    c Culms triangular, jointless.—*d* Spikes in a single cluster .... Nos. 9—11
                          —*d* Spikes in a cyme, bracted .... No. 12
    x Spikes large (6—15''), oblong, with cleft gls. Culm jointed, leafy. Nos. 13,14
    x Spikes small (1''), mostly in globular heads. Culm jointed, leafy. Nos.15—17
    x Spikes small (2—3''), all separate and pendulous. South .... No. 18

1 S. paucifiórus Lightfoot. Culm filiform or capillary, erect, 3—8', leafless; involucre 0; spk. oval, 1—2''; gls. brown, 5—9; ach. 3-angled, netted, beaked but *not tubercled*. Otherwise an Eleocharis. Western N. Y. (*Hankenson*) to Ill. (*Porter*).
2 S. cæspitósus L. Culm round, wiry, 3—10', sheathed below with rudiments of leaves; spike ovate, 2—3'', with an involucral bract same length; setæ 6, longer than the achenium. High Mountains, N. and S. In tufts. Leaves 3—6''.
3 S. Clintonii Gr. Culm acutely 3-angled, 1f, very slender, base sheathed, with short bristle-shaped leaves; bract subulate, shorter than the ovate chestnut-brown spike (3—5''); glumes pointless. N. Y. (*Clinton*). *Porter.*)
4 S. planifólius Muhl. Culms 1f, 3-angled, threadform, with several linear flat leaves; bract as long as the oblong (2'') spikes; gls. pointed. N. Eng., N. Y. to Del.
5 S. subterminàlis Torr. Culm 1—3f, filiform, with several long capillary floating leaves; bract 1—2', exceeding the oblong (3'') spike, continuous with the culm. N.
6 S. débilis Ph. Culm roundish, furrowed, in tufts, 9—16', with a few subulate lvs. at base or 0; spk. 1—7, ovoid, crowded, 3'', tawny, the culm-leaf above them 2—4' at length reflexed; bristles 4—6, inversely barbed; ach. smooth. Muddy shs.Ct.to Car.

7 S. Smithii Gr. Culm slender, 3—12'; sheath often with a short blade; spk. 1—3. ovoid, greenish, 2—3", sessile about halfway up; setæ 0—1; ach. smooth, lenticular; culm-leaf always erect   Shores, Penn. (*Porter*) Sodus Bay (*Hankenson*.)

8 S. válidus Vahl. Culm cylindric, smooth, 5—8f, its sheath with or without a short blade; panicle cymous, overtopping the short pungent culm-leaf; spk. ovoid, brown. 2", numerous; gls. mucronate, ciliate; setæ 3 or 6. Our stoutest Bullrush. Shores

9 S. pungens Vahl. Culm 1—4f, 3-angled, 1-3-leaved; lvs. 3—12', also 3-angled; spk. 1—6, crowded, sessile, ovate, obtuse, 3—5' below the summit; gls. notched and mucronate; anth. ciliolate at apex; style 2-cleft; setæ 2—6. Ponds and marshes.

10 S. Tórreyi Olney. Culm 2—3f, 3-angled; lvs. 1—3 at base, 1—1½f, 3-angled; spk 1—4, oblong, sessile, 2—4' below the summit; gls. ovate; sty. 3-cleft; ach. triq obovate, pointed, shorter than the setæ.   Borders of ponds, N. E. to N. J., and W.

11 S. Olneyi Gr. Culms triquetrous-winged, 2—7f, leafless, or with 1 very short leaf at base; spk. 6—12, in a sessile head an inch or so below the summit; gls. roundovate, mucronate; setæ 6; style 2-cleft.   Salt marshes, E. and S.

12 S. leptólepis Chapm. Culms 3-angled, 2—5f; leaves 1—3, slender, channelled, sheathing at base; spikes loosely umbelled, single, oblong, 4—6", ∞-flowered; invol. of several small bracts besides the long culm-leaf; gls. lance-ovate, acute; style 3-cleft; setæ 6, equalling the 3-sided ach.   Md. (*Porter*), and S. (S. Canbyi Gr.)

13 S. marítimus L. Culm acutely 3-angled, leafy, 1—3f; lvs. broad-linear, channelled, 1—3½f; spk. 3—12", oblong, 6—10 in each cluster; clusters 1—9, sessile and on short rays; invol. of 2 or 3 very long leaves; setæ 1—4, deciduous, short; achenium plano-convex.   Salt marshes.

14 S. fluviátilis Gr. Culm triquetrous-winged, leafy, 2—4f; lvs. as in No. 13; spk 6—10", oblong, 1—5 in a cluster; clusters sessile and on rays; setæ 6; ach. 3-angled Shores, Eastern, Middle, and Western States.

15 S. atrovirens Muhl. Culm obtusely 3-angled, leafy, 2f; invol. of 3 long leaves, spk. ovate, 1½", 10—20 in the round dense heads; hds. 4" in a compound cyme; dark olive-green; setæ 4, as long as the smooth white ach.   Com. in swales.   N., M., & W.

16 S. sylváticus L. Culm 3f, leafy; invol. of 8 leaves, hardly equalling the thrice compounded cyme; spk. 1", olive-green, 1—3—9 in the small heads; hds. on slender pedicels; gls. acute; setæ 6, straight, as long as the pale 3-angled ach.   Mts. N. H., & N

17 S. polyphýllus Vahl. Culm 2—3f, leafy; invol. of 3 leaves; cyme decompound spk. yellow-ferruginous, 1", 3—6 in the clusters; gls. obtuse; ach. yellowish-white. 3-angled, twice shorter than the 4—5 tortuous setæ.   Margins of waters.   Rare.   North

18 S. divaricàtus Ell. Culm 3—4f, very leafy; cyme large, loose, decompound spk. all separate, 2—3", oblong, pendulous, ferruginous; setæ tortuous. Wet barrens.S.

19 S. Eriophorum Mx. Culm teretish, 3—5f, lvs. 2f; invol. 4-5-lvd., longer than the large loose decompound cyme; spk. very numerous, 1—3", pedicellate; setæ 6. hair-like, curled, conspicuous, 5 or 6 times longer than the white ach. Swamps. Com.

20 S. lineàris Mx. Culm 3-angled, 2—3f, very leafy; cymes term. and axillary, de compound, at length nodding; invol. 1-3-bracted, much shorter than the cyme; setæ as long as the glumes, hardly at maturity exserted.   Swamps.   Common.   S.

7. ERIÓPHORUM, L. COTTON GRASS.

Glumes imbricated all around into a spike. Ach. invested with many (rarely but 6) very long, woolly or cottony hairs. ♃ Culms with or without leaves. Spikes showy after the long setæ have grown. June—August.

§ Setæ 6, crisped, woolly. Spike single. Culms scape-like, naked............No. 1
§ Setæ numerous, straight, cottony. Culm jointed, 1-3-leaved....(*a*)
   a Spike single. Culm bearing 2 sheaths instead of leaves..................No 2
   a Spikes several, collected into a subsessile, capitate cluster...............No. 3
   a Spikes several, separate, in umbel-like cymes .... Nos. 4, 5

ORDER 154.—CYPERACEÆ.

1 E. alpìnum L. Culms jointless, slender, 8—16', form a creeping rhizome; lvs. radical, short, subulate; spk. 2'', the white hairs at length 7—9'' long. Bogs, N., M.
2 E. vaginàtum L. Rigid, tufted, 1—2f, culm with 1 or 2 inflated sheaths; leaves radical, filiform; spk. 6—8'', blackish, hairs 1', white, glossy, 30—40 in each flower. N. Eng. to Mich., and N. Pocono Mt. in Penn. (*Prof. Porter.*)
3 E. Virgínicum L. Culm strict, firm, slender, 2—3f, lvs. shorter, narrowly linear; invol. 2-4-lvd.; spk. ovoid, 3'', many, glomerate with very short ped. forming a capitate cluster; setæ 70—200, pale-cinnamon, 6—8'' long. Bogs.
   β. ***confertíssimum.*** Setæ white, in a large and compact tuft. N. H., N. Y., & Can
4 E. polystáchyon L. Culms 1—2f, with 2 or 3 cauline broad linear lvs.; invol. 2-leaved; spk. about 10, on long drooping peduncles; setæ 30—40 to each flower, 6—8'', white. Very conspicuous in meadows and swamps.
5 E. grácile Koch. Culm 1½-2f; lvs. triquetrous, channelled above, scarce 1'' wide; spk. 3—8, on roughish ped. which are 1''—1'—4' long; setæ white, 8-10''..

8. HEMICÁRPHA, Nees.
Spike many-flowered. Glumes imbricated all around. Interior scale 1, embracing the flower and fruit; setæ 0. Sta. 1. Style 2-cleft, not bulbous at base, deciduous. Ach. compressed, oblong, subterete. ① Low, tufted, with setaceous culms and leaves.

H. subsquarròsa Nees. Culms 2—3', curved, the lvs. shorter; spk. 2 or 3, nearly 2'', ovoid, sessile together; invol. 2-lvd., 1 continuing the stem; gls. *subsquarrous.* Sandy shores.—β. ***Drummóndii.*** Sts. 1—2', spk. only 1. Fulton Co. Ill. (*J. Wolf.*)

9. LIPOCÁRPHA, Brown.
Spikes many-flowered; glumes spatulate, imbricated all around; interior scales 2, thin, subequal, involving the flower and coating the fruit. Perianth none. Sta. 1. Style 2- or 3-fid; achenium coated with the scales. ① Culms leafy at base. Spikes numerous, collected into an involucrate, terminal head.

L. maculàta Torr. Culm 3—8', the linear-filiform lvs. shorter; invol. of 2 long lvs. and 1 short; spk. 3—4, ovoid; glumes very ∞, scarious, marked with red dots and a green midvein; ach. oblong. Wet grounds, Phila. (*Leidy*), and S.

10. FIMBRÍSTYLIS, Vahl.
Glumes imbricated on all sides; bristles 0. Style compressed, 2-cleft, bulbous at base, deciduous, *ciliate-fringed* (as the name indicates.)—With the habit of Scirpus. Lvs. mostly radical.

1 F. spadícea Vahl. Culms 1—3f, hard and rigid; lvs. semiterete, rigid, channelled; rays few, exceeding the 2 or 3 invol. bracts; spk. ovate-oblong, 3—6'' by 2'', rust-colored to brown; stn. 2—3; ach. whitish, minutely netted. ♃ Salt marshes.
2 F. laxa Vahl. Culm 3—12', lax, flattened, striate; lvs. flat, linear, glaucous, rough-edged; rays few, shorter than 1 of the invol. bracts; spk. ovoid, 3'', brown; sta. 1; ach. whitish, with 6—8 prominent ribs. ① Clay soils, Pa. to Ill., and S.
3 F. argéntea Vahl. Glaucous, tufted; culms 2—6', setaceous, flattish, like the leaves; spk. straw-colored, 6—9 in a dense head; invol. lvs. 4, *longer than the culm;* gls. lance-ovate, pointed; stn. 1. ① Philad. (*A. H. Smith*), and S. (*F. congesta* Torr.)

11. TRICHELÓSTYLIS, Lestib.
Glumes in 4 to 8 ranks, carinate; bristles none; style 3-cleft, deciduous below the bulb (if any) at the base; achenium triangular. ① ♃. Sts. leafy at the base, tufted. Spikes in a terminal head, or umbel, or solitary.

§ Spikes rusty-brown, in a cymous umbel, the glumes 6—15, in 4 rows......Nos. 1—3
§ Spikes greenish a both capitate and umbellate, with linear lvs. and bracts..No. 4
    a all capitate in a single head; bracts dilated at base...Nos. 5, 6
   —r one only on each culm, or rarely 2 or 3, bracted .... Nos. 7, 8

1 **T. autumnàlis** (L.) (Fig. 5.) Culm flattened, 2-edged, very slender, 3—10'; lvs. narrow-linear, flat, much shorter; spikes (Fig. 6) lance-oblong, very acute, 4-rowed, 2'', 1—3 together, many in the cyme; glumes sharp-pointed, brown; stamens 2; achenium (Fig. 7) white smooth. ① Wet banks, &c.

2 **T. ciliatifòlia** (Ell.) Culm setaceous, angular, 3—12'; leaves setaceous, *with long brown hairs on the sheaths;* cyme 5-9-rayed, often overtopped by 1 bract; spike 1—2'', mostly single; glumes acute, 4-rowed, 6—12; stamens 2; achenium white. ① Dry, S.

β. *coarctata*. Cyme contracted; spks 2—3'', often 2—3 clustered together.

3 **T. capillàris** (L.) Culm capillary, angular, 3—8'; leaves *setaceous*, much shorter, entirely smooth; spk. 2—4 in the simple cyme; gls. 8—12, strongly keeled, 4-rowed; stamens 2; ach. white, equally 3-sided. ① Sandy fields. (Fig. 8, a flower.)

4 **T. boreàlis** Wood. Culm filiform, angular, 2—4'; lvs. linear, flat, ½—2'; bracts similar, as long as the leaves; spikes capitate and in cymes, 1—5 together, ovoid, green, 1''; glumes pointed; sta. 1; ach. white, 3-angled; sty. bulbous at base. ① Ill. Banks of the Miss. R., Ill. (*J. Wolf.*) Shores of Lake Sup., Mich. (*Mr. Perkins.*)

5 **T. stenophýlla** (Ell.) Culm setaceous, grooved, 2—4'; leaves setaceous, 2—3'; bracts many, 3—4 times longer than the dense head; ach. (Fig. 9) blackish. S.

6 **T. Wàrei** (Torr.) Culm filiform, 1f, 3-angled; lvs. and bracts setaceous, silky-fringed at base, the latter twice longer than the head of 8—12 ovate spikes. Fla.

7 **T. carinàta** (Hook. and Arn.) Culm flattened-setaceous, 3—6', with 1 short setaceous leaf at base; spk. ovoid, near the top; gls. 5—8, broad-ovate, acuminate. S-W.

8 **T.** LEPTÀLEA (Schultes?) Culms filiform, bright green, flaccid, 6—12'', sheathed at base, with a short setaceous leaf or 0; spk. ovate, whitish, as long as its bract (3''); sta. 3; ach. 3-angled, shining. Cult. in conservatories. From S. Eur.

12. **PSILOCÁRYA**, Torr. Fls. ☿. Gls. ∞, imbricated all around, all fertile. Setæ 0. Stam. 2, long, persistent. Style 2-cleft, dilated or tuberculate at base. Ach. biconvex, crowned with the persistent style. ① Culms leafy. Spikes lateral and terminal, cymous, brown.

1 **P. scirpoides** Torr. Culm 3-sided, slender, 5—9'; lvs. linear, 3—5', about 2 on the culm, a cyme in each axil; spike ovoid. 2—3''; ach. 20—30, smoothish (slightly rugous), tippid with the long 2-cleft style. Ponds, R. I., and N.

2 **P. nitens** (Vahl.) Culm 1½—2f, flattened, with several long linear leaves; cymes loose, spike lance-ovoid, 2'', all pedicellate; ach. 8—10, conspicuously rugous, tipped with the entire-part of the style, blackish when ripe. S.

13. **DICHRÓMENA**, Rich. Spikes flattened, in a terminal head Gls. imbricated all around, many empty. Perianth 0. Sta. 3. Sty. 2-cleft. Ach. lens-shaped, crowned with the broad tubercular base of the style. Culms leafy. Bracts discolored.

1 **D. leucocéphala** Mx. Culm 3-angled, 1—2f; leaves narrow-linear; invol. of 6—8 narrow leaves, which are whitened at base as well as the spikes; ach. rugulous, truncate, the tubercle not decurrent. Barrens, N. J., and S.

2 D. latifòlia Baldw. Culm teretish, 2—3f; leaves long, linear; bracts 8—10, lance-linear, reddish white, long-pointed; ach. roundish, roughened, dull, the tubercle decurrent on its 2 edges. Ponds, S.

14. RHYNCHÓSPORA, Vahl. Fls. ☿ or ♂ ☿ ♀, few in each spike. Glumes flattish, loosely imbricated, the lowest small and empty. Perianth of 6—12 setæ. Sta. 3 to 12. Style bifid. Achenium lens-shaped or globular, crowned with a tubercle—the distinct, bulbous base of the style. ♃ Stems leafy, 3-sided. Inflor. terminal and axillary, mostly tawny to brown.

§ Setæ densely plumous. Achenium roundish-ovoid
  (not flattened).........................Nos. 1—3
§ Setæ naked, denticulate or hispid. Achenium more
  or less flattened...(*)
  * Ach. transversely wrinkled. Setæ upwardly bearded.(a)
  * Achenium smooth and even...(c)
    a Setæ shorter than the achenium..........Nos. 4—7
    a Setæ equalling or exceeding the ach...(b)
  b Spikes in drooping panicles. Ach. oblong or obovate.Nos.8,9
  b Spikes in erect or spreading panicles. Ach. roundish..10—12
  b Spikes corymbed or fascicled.—x Ach. round-obovate..13, 14
       —x Achenium oval. Nos. 15, 16
    c Setæ retrorsely hispid, or barbed (under a magnifier).(d)
    c Setæ upwardly hispid (or almost none in No. 29)...(e)
    c Setæ none. Culm and leaves setacious or filiform.
       South.........................Nos. 17, 18
      d Culm and leaves very slender, setacious or filiform......Nos. 19—21
      d Culm wiry and firm, leaves linear. Spikes dark-brown..Nos. 22, 23
  e Culms stout, 2—3f. Setæ and stamens 6—12....................Nos. 24, 25
  e Culms wiry and firm, 1—2f. Stamens 3. Setæ 6, 3, or 0........Nos. 26—29
  e Culm and leaves very slender, setaceous or filiform............Nos. 30, 31

1 R. plumòsa Ell. Culm and leaves filiform-wiry, erect, 10—18'; spikelets 1-flwd., 1", in small fascicles forming *a loose spike* at top, often another below it shorter than the bracts; setæ 6, as long as the tumid, rugous ach. Dry, N. J. to Fla.
  β. **minor.** Every way smaller, 5—10'; fascicles 2 or 3; setæ feathery below. S.

2 R. semiplumòsa Gr. Culm and leaves rigid, wiry, erect; spike 1—2", in a capitate corymb at top, often a smaller one below; ach. solitary, tumid, rugous with a broad tubercle; setæ 6, feathery below. Barrens, S. 1—2f.

3 R. oligántha Gr. Culm and leaves filiform-capillary, erect, 8—14'; spikes 1—3 only, fusiform, 3", with 1 long bract; ach. obovoid; setæ 6, densely feathery. S.

4 R. cymòsa N. Culm acutely 3-angled, 1—2f; leaves linear; spike fascicled, in several crowded cymes; ach. broad-obovate, twice longer than the 6 setæ, 4 times longer than the depressed-conical tubercle. N. J., Pa., and S.

5 R. Torreyàna Gr. Culm teretish, 1½—2f; leaves setaceous; cymes small, several, the lateral on capillary peduncles; ach. oblong-obovate, twice longer than the setæ, thrice longer than the broad tubercle. N. J., and S.

6 R. rariflòra Ell. Culms tufted, 6—16', filiform, the setaceous leaves much shorter; spikes 2", scattered in very loose paniculate cymes; ach. round-obovate, strongly rugous, tubercle very short. Barrens, S.

8 R. inexpánsa Vahl. Culm slender, erect 1½—3f; leaves narrow-linear, flat; spikes lanceolate, 2-4-flowered, 3", in several rather large recurved-drooping panicles; ach. oblong, half as long as the setæ; tubercle short. Wet barrens, S.

ORDER 154.—CYPERACEÆ.

9 **R. decúrrens** Chapm. Culm, leaves, and cymes as in the last; spike 1"; ach, obovate, as long as the setæ, the tubercle *decurrent* on its 2 edges. Marshes, Fla.

10 **R. miliàcea** (Lam.) Culm slender, 3-angled, 2—4f; leaves linear, flat, 6—3' by 3—4"; spikes obovate, all pedicellate, in diffusely spreading cymous panicles; ach round-obovate, little shorter than the setæ. Wet barrens, S.

11 **R. cadùca** Ell. Culm acutely 3-angled, 1—3f; leaves linear, 2—3" broad; spikes ovate, large, 4—5", sessile or stalked, in several rather close erect cymous panicles; glumes *caducous;* ach. roundish, ⅓ as long as the setæ. Wet, S.

12 **R. schœnoides** (Ell.) Culm 3-angled, 2—3f; leaves linear, 2" wide; spikes (2") small and numerous, subsessile, clustered, in several paniculate cymes; setæ twice as long as the obovate flat achenium and small tubercle. Bogs, S.

13 **R. pátula** Gr. Culm 3-angled, thick and stout at base, 2—3f; leaves linear, short; spikes ovate, 2", in several spreading loose panicles; ach. strongly rugous, with a large tubercle, some shorter than the setæ. Ga., Fla.

14 **R. Ellióttii** Gr. Culm solitary, 2—3f; leaves shining, rigid; corymbs 3 or 4 few-flowered, subsimple; spikes large; ach. minutely rugous, with a very short tubercle, little shorter than the setæ. Pine barrens, S. (R. distans Ell.)

15 **R. punctàta** Ell. Culm 3-angled, 1—2f; leaves lance-linear; corymbs of fascicles; ach. rugous-netted, with rows of impressed dots. Marshes, Ga., Fla.

16 **R. microcárpa** Baldw. Culm 2f, teretish; leaves narrowly-linear, setaceous at end; spike turgid-ovate, 1—2"; ach. ovate, flat, minute. Wet, S.

17 **R. pusílla** Chapm. Corymbs 2—3, distant, of minute, scattered ovate, 3-flowered spikes; ach. lens-shaped, oblong-ovate, white. Woods, S. Car. to Fla. 1f.

18 **R. Chapmànii** Curtis. Corymb capitate, terminal, dense; spikes with 5 scales and 1 flower; ach. oval, polished; stamens 1 or 2. S. Car. to Fla. 1½f.

19 **R. alba** Vahl. (Fig. 10.) Culm 10—20', very slender; leaves linear-setaceous; spikes (Fig. 11) whitish, lanceolate, in stalked, corymbous fascicles; setæ 9—12, as long as the ach. (Fig. 12) and tubercle. Common in wet shady grounds. July—Sep.

20 **R. Knieskérnii** Carey. In tufts 6—16', filiform; spikes 1", brown, in 3—5 dense, sessile, remote fascicles; setæ 6, as long as the ach. Iron soils, N. J.: rare.

21 **R. capillàcea** Torr. In tufts, 6—10', setaceous, 3-angled; clusters of brown spikes mostly 2, few-flowered; setæ 6, much longer than the ach. Swamps, M., W.

22 **R. glomeràta** Vahl. Culms 1f, leaves linear; fascicles brown, remote, in several pairs; spikes lanceolate, 2"; ach. obovate, as long as its tubercle, which equals the 6 setæ. In bogs, Can. to Fla. July, Aug.

23 **R. cephalántha** Torr. Culms 2—3f, stout; leaves linear; heads globular, dense, remote, sessile, solitary in the axil or terminal, dark-brown; ach. round-ovoid, obtuse, half as long as the 6 setæ. Barrens, N. J.

24 **R. Baldwínii** Gray. Culms slender, 2—3f; leaves linear; spikes ovate, in a dense terminal corymb of fascicles; setæ 12; stamens 6. Pine barrens, Ga.

25 **R. dodecándra** Baldw. Culms rigid, stout, 1—3f; leaves rigid, linear, erect; spikes 4", ovate, in 4 or 5 loose, stalked cymes; stamens 12; setæ 6—12, long as the large (1½"), roundish, smooth achenium. Bogs, S. (R. megalocarpa.)

26 **R. fasciculàris** Nutt. Culm teretish, wiry, 1—2f; leaves short, narrowly linear; spikes small (1½") in several dense fascicles mostly terminal; setæ 4—6, shorter or longer than the obovoid brown ach. Wet, S.

27 **R. distans** N. Like No. 26, but every way smaller; spikes 1" long, in a dense terminal and often a *distant* lateral fascicle; setæ about equalling the ach. S.

28 **R. ciliàta** Vahl. Glaucous, 8'—2f; leaves short, linear, obtuse, *ciliate* on the edges; spikes all in a dense terminal fascicle; setæ 6, half the length of the ach. S.

29 **R. pállida** M. A. Curtis. Culm firmly erect, 1—2f, 3-angled; spikes pale-tawny, (like R. alba) in a dense terminal head with often a lateral head on a long peduncle; ach. roundish, tubercle minute, setæ 0—3, minute. Bogs, N. J. to N. C.

30 **R. fusca** R. & S. Culm (6—12') and leaves setaceous; spikes ovate-oblong, 2"

dark-brown, in 1 or 2 small fascicles; ach. half the length of the setæ which equal the pointed serrulate tubercle. Maine to N. J., and W. Rare. Europe.

21 R. gracilenta Gr. Tufts 1—2f; culm and leaves threadform, curved; spikes 1″, brown, in 2—3 fascicles; ach. oval, as long as its awl-shaped, serrulate tubercle, shorter than the 6 setæ. Low grounds, N. Y. to Fla. (R. filifolia Torr.)

15. CERATOSCHŒNUS, Nees. Spikelets 2–5-flwd., one flower ☿, the rest ♂. Glumes loosely imbricated, somewhat in 2 rows, lower ones empty. Perianth of 5 or 6 rigid, hispid, or scabrous setæ. Stamens 3. Style simple, very long, persistent as a beak on the smooth, compressed achenium. ♃ Stems leafy, 3-angled, 2—4f. Cymes compound, brown.

1 C. longiróstris (Ell.) 3—5f; leaves flat, 4—6″; spikes in loose fascicles, 9″; ach. 2″, beak 7″, setæ 5″; cymes diffuse, terminal and axillary. Penn., W. and S.

2 C. macrostáchyus Torr. Leaves 2—4″ wide; spikes 1″, in dense fascicles; ach. and beak 8″, setæ 2—3″, culm 2—3f. Hardly distinct. Mass., and South.

3 C. capitátus Chapm. Spikes densely clustered in a few heads; beak only 2″, ach. 1″, setæ 2″, culm teretish, 2—3f, leaves 2—4″ wide. W. Fla.

16. CLADIUM, Browne. Flowers ♂ ☿ ♀. Glumes imbricated somewhat in 3 rows, lower ones empty. Setæ 0. Stamens 2. Style 2–3-cleft, deciduous. Achenium subglobous, the pericarp hard, thickened and corky above. ♃ Stem leafy. Cymes terminal and axillary, brown.

1 C. mariscoides (Muhl.) *Bog Rush.* Culm terete, rigid, 20—30′; leaves narrowly linear, much shorter than culm; spikes 3″, in pedunculate or sessile heads, forming small cymes; ach. ovoid, scarcely beaked. Bogs, N. Eng., and West.

2 C. effusum (Swtz.) *Saw Grass.* Culm obtusely 3-angled, 6—10f, leaves 3—10ft sharply serrate-barbed on the edges; cymes diffuse, decompound, forming a large panicle. A coarse, rank Sedge in ponds, N. Car. to La.

17. SCLÈRIA, L. NUT SEDGE. Flowers ♂, staminate spikes intermixed, fertile spikelets 1-flowered, glumes fasciculate. Perianth cup-shaped or 0. Achenium globous, ovoid or triangular, with a thick, bony pericarp. Style 3-cleft, deciduous. ♃ Culms 3-angled, leafy. Spikes in fascicles Nuts white. In bogs. Summer.

§ SCLERIA. Achenium ovoid or globous, base invested with a short perigynium...(*)
* Achenium smooth, ovoid. Perianth annular, subentire. Stamens 3..Nos. 1, 2
* Achenium rugous-warty, globular. Perianth 6- or 3-lobed..................Nos. 3, 4
* Achenium reticulated or hispid-rugous, globular. Perianth 3-lobed....Nos. 5, 6
§ HYPOPORUM. Achenium ovoid-triangular, base fluted. Perigynium none...(a)
  a Fascicles 4 to 7, interruptedly spiked. Achenium smooth or rugous..Nos. 7, 8
  a Fascicles single, terminal. Achenium ribbed or smooth.............Nos. 9, 10

1 S. triglomeráta Mx. *Whip Grass.* Culm erect, rough, 3 lf; leaves broad-linear, rough-edged; fascicles few, composed of triple clusters of green-brown (3″) spikes; ach. white and polished, more than 1″ in diameter. Common.

2 S. leptocúlmis W. Culm very slender, 2f, nearly naked; lvs. smooth, narrowly linear; compound spikes loose, the lateral on a long filiform peduncle; spikes 3—4″; ach. polished, ovoid, minutely corrugated. S. (S. oligantha Ell. ?)

3 S. ciliàta Mx. Culm scabrous above, 2f; leaves 2, pubescent, bracts ciliate-fringed; ach. beset with unequal warts, disk 3-lobed. Pine barrens, S.

4 S. paucillóra Muhl. Smoothish or hairy; leaves and bracts exceeding the culm;

fascicles few-flowered, the lateral, if any, pedunculate; ach. small, rough, the disk 6-lobed. Rare northward, common South. 10—16'.

β. *glabra.* Smoothish, slender, 1f; lateral fascicles 1-flowered, or 0. Ms. to Ohio.
γ. *Caroliniàna.* Scabrous-hirsute, slender; leaves much exceeding the culm. S.
δ. *Elliòttii.* Stout, 2—3f, denticulate-ciliate; lateral spikes pedunculate. S.

5 **S. reticulàris** Mx. Slender, 1f, leaves shorter than culm; fascicles 2—5, distant, subsessile; ach. dead-white, ¼'', conspicuously netted and pitted. R. I. to Fla.

6 **S. laxa** Torr. Slender, weak, diffuse, 1—2f; lvs. flat, 2'' wide; fascicles very remote, spks. distant, in pairs; ach. 1'', with transverse ridges and brown pits. N. J. to Fla.

7 **S. verticillàta** Muhl. Glabrous, 6—12', slender; fascicles 4—6, smooth, purple, sessile, 8''—1' apart; ach. globular, about ¼'', rugous. N. Y. to Ohio, and South.

8 **S. interrúpta** Mx. Sparingly hirsute, 12—30'; leaves 2'' wide; fascicles 5—7, rusty-brown, sessile, ciliate, 4—9'' apart; ach. smooth, ⅓'' diameter. Sonth.

9 **S. grácilis** Ell. Filiform, smooth, 1—2f; spikes few (1—5 pairs), 3'', in a terminal fascicle; bract erect; ach. ovid-triangular, ribbed lengthwise. South.

10 **S. Baldwínii** (Torr.) Culm scape-like, 2—3f, leaves all radical, long; spikes 5'' long, 3—5 pairs in a terminal fascicle, brown-purple, with 3 bracts, middle bract erect; ach. dull-white, 2'' long, even. In Georgia and Florida.

18. **CHAETÓSPORA**, R. Br. Spikes 1-5-flowered, fls. ☿, glumes in two rows, the lower empty. Setæ 3—6. Stam. 3. Style 3-fid, deciduous. Achenium triangular. ♃ Culm leafy only at base. Fls. capitate, chestnut-brown.

C. nígricans K. Culm 1f, erect, teretish, longer than the narrow erect leaves; spikes 4'' long, in one fascicle, bract erect, 1—3'; achenium ⅓'' diameter, white. Fla., Eur.

19. **CAREX**, L. Flowers diclinous. Spks. 1 or more, either with both staminate and pistillate flowers (*androgynous*), or with the two kinds in separate spikes on the same plant (*monœcious*), or rarely on separate plants (*diœcious*). Glumes single, imbricated, each 1-flwd. ♂ Stamens 3. ♀ Stigmas 2 or 3. Nut (*achenium*) 2-edged or 3-angled, enclosed in a sac (*perigynium*) composed of 2 united glumes. ♃ Culms triangular, in tufts, with grass-like leaves and usually with axillary as well as terminal spikes.

The following enumeration of our Carices is reduced from the excellent monograph by the lamented Prof. C. Dewey, contained in the Class-book of Botany, and revised with the assistance of friends before mentioned, and whose names appear below.

Fig. 13, C. flava. 14, One of its perigynia (magnified): 15, a glume. Fig. 16, C. rosea. 17, A perigynium: 18, a glume.

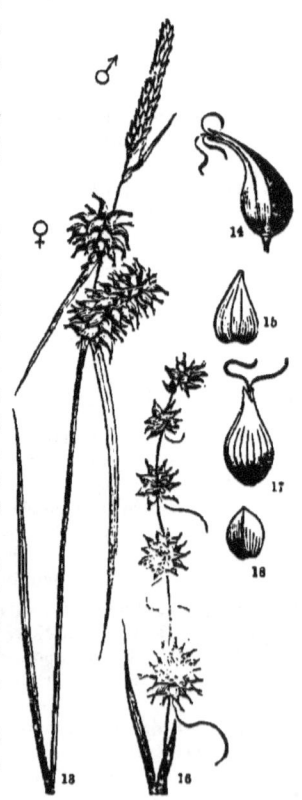

Order 154.—CYPERACEÆ.

§ I. Spike solitary, one (rarely more) borne on each culm...(§)
§ II. Spikes two or more. Stigmas 2. Achenium lens-shaped...(§§)
§ III. Spikes two or more. Stigmas 3. Achenium triangular...(§§§)
    § Stigmas 2. Achenium lens-shaped or flattened...(a)
    § Stigmas 3. Achenium triquetrous or 3-angled...(b)
        a Spike androgynous, staminate at the summit........................No. 1
        a Spike diœcious, or the ♀ spike staminate at the base.............Nos. 2, 3
            b Leaves very narrow, shorter than the culm. Glumes colored...Nos. 4–6
            b Leaves linear, longer than the culms.—Glumes colored......No. 7
                      —Glumes green........Nos. 8–10
            b Leaves very broad, flat, with no midvein. Glumes scarious .....No. 11
§§ Staminate and pistillate flowers in the same (androgynous) spike...(c)
§§ Staminate and pistillate flowers in different spikes—on the same culm...(i)
                                  —on different culms.....No. 12
    c ♂ Flowers variously situated in the approximate spikes....Nos. (12 and) 13—15
    c ♂ Flowers at the summit of the spikes...(d)
    c ♂ Flowers at the base of the spikes...(f)
        d Spikes ∞, paniculate, brown ; perigynia corky, not rostrate....Nos. 16, 17
        d Spikes (or spikelets) 8— ∞, approximate in a compound spike...(e)
            e Perigynium rostrate, scarcely longer than the glume........Nos. 18—21
            e Perigynium long-rostrate, 2 or 3 times longer than the gl....Nos. 22, 23
        d Spikes 3—6, approximate into one—ovoid spike.................Nos. 24—26
                            —cylindric spike a little loose..Nos. 27, 28
        d Spikes 3—8, remote. Perigynia erect in No. 32, radiating in....Nos. 29—31
    f Perigynia radiating in the 3—6 separated spikes. Glumes green....Nos. 33, 34
    f Perig. suberect, few (2—20) in each spikelet. Glumes hyaline white...(g)
    f Perig. suberect, winged, 30—60 in each oblong to obovoid spikelet...(h)
        g Spkl. separate or remote, 2-3-flowered in No. 35, 5-20-flwd. in Nos. 36—39
        g Spikelets closely contiguous, 2-12-flowered.... ....Nos. 40, 41
            h Perigynia lance-linear, long-beaked, 3—1″. Spikelets close. Nos. 42—44
            h Perigynia lanceolate, short-beaked. Spikelets 8—20, club-ovoid.No. 45
            h Perigynia ovate, spreading. Spikelets round-ovoid, close..Nos. 46, 47
            h Perigynia round-obovate, short-beaked, broadly-winged. Five
                nominal species closely related and intermixed..........Nos. 48—52
    i Staminate spike single. Pistillate spikes sessile ....Nos. 53—56
    i Staminate spike single. Pistillate spikes pedunculate..............Nos. 57, 58
    i Staminate spikes 1 or more, and the ♀ spikes often ♂ at the apex...(k)
        k Glumes obtuse, not exceeding the perigynia. Spikes sessile....Nos. 59, 60
        k Gl. acute, little longer or shorter than perig. Lower spikes stalked..61—64
        k Gl. long-awned, much exceeding the perig. Spikes all stalked..Nos. 65—67
§§§ Spikes androgynous, both kinds of fls. in each,— ♂ at the apex........Nos. 68, 69
                                — ♂ at the base.............No. 70
§§§ Spikes—the terminal ♀ at top, the rest all pistillate...(l)
§§§ Spikes—the terminal one wholly ♂, the rest all pistillate...(*)
§§§ Staminate spikes habitually *more than one*...(**)
    l Spikes erect or nearly so, *green*, hairy in Nos. 71, 72, glabrous in....Nos. 73—74
    l Spikes erect, pedunculate, tawny in maturity, glabrous.............Nos. 75, 76
    l Spikes erect (some nodding in No. 79) with black-purple glumes....Nos. 77—79
    l Spikes drooping on filiform stalks, green or some rusty.............Nos. 80—82
    * Pistillate spikes sessile, or solitary on radical peduncles. Perig. with
        a short abrupt beak, not inflated, pubescent. Culm slender...(m)
    * Pistillate spikes with enclosed or nearly enclosed peduncles. Perig.
        inflated, beaked, glabrous, bicuspidate at apex. Spikes turgid,
        often quite large, their leafy bracts longer...(n)
    * Pistillate spikes on exserted peduncles (exserted from the sheaths of

the bracts). Perigynia 3-angled, scarce inflated, not much beaked, and (as well as the glumes) more or less colored...(*p*)
* Pistillate spikes with peduncles (long or short) scarcely sheathed at all, or only the lowest bract on a short sheath...(*x*)
  m Pistillate spikes oblong, brown or hairy, the lowest scarcely sessile.Nos. 84 –87
  m Pistillate spikes ovoid,—all or mostly solitary on radical peduncles.Nos. 88, 89
          —all sessile and crowded on the culm........Nos. 90, 91
          —all sessile and remote on the culm ....Nos. 92, 93
  n ? Spikes small (3—6″), yellowish; perig. with a short recurved beak..94, 95
  n ? Spikes large; perigynia much inflated, with a long straight beak...(*o*)
    o Spikes very short.—Perigynia 3—4″ long....................Nos. 96—98
                —Perigynia 6—8″ long..................Nos. 99—102
    o Spikes oblong-cylindric.—Perigynia ascending...........Nos. 103, 104
                —Perigynia spreading............Nos. 105, 106
p Leaves radical, very broad (6—10″),—triple-veined. ♂ Spikes clavate..107—109
                —one-veined. ♂ Spikes linear..No. 110, β. γ.
p Leaves linear or setaceous, 1—2″, rarely 3—4″ wide...(*r*)
  r Perigynia smooth and not rostrate...(*s*)
  r Perigynia smooth (scabrous in No. 130) and rostrate...(*v*)
  r Perigynia hairy, veined, conical-beaked. South.............Nos. 137—139
    s Bracts leaf-like, exceeding the spikes or culm...(*t*)
    s Bracts shorter than the spikes or culm...(*u*)
      t Perigynia triangular, oblique at the point.............Nos. 110—112
      t Perigynia subterete, straight.— ♂ Spikes pedunculate..Nos. 113, 114
                —♂ Spike sessile........Nos. 115—118
u Fertile spikes white in No. 119, tawny in...............................No. 120
u Fertile spikes green, the sterile pedunculate....................Nos. 121—123
  v Bracts leaf-like, exceeding the spikes or culm. ....Nos. 125—127
  v Bracts not exceeding the spikes or culm...(*w*)
    w Spikes linear, slender, very loose-flowered.................Nos. 128, 129
    w Spikes cylindric, suberect, rather dense....................Nos. 130, 131
    w Spikes oblong,—about 6-flowered, dense...................Nos. 132, 133
              —many-flowered, rather dense............Nos. 134—136
x Perigynia beakless or nearly so.—Spikes suberect, short-ped.....Nos. 140—142
                —Spikes drooping on slender ped.Nos. 143—145
x Perigynia evidently beaked,—diverging in the spike...............Nos. 146—148
                —deflexed in the spike.................Nos. 149, 150
** Perigynium clothed with wool, hairs, or mealiness...(*y*)
** Perigynium glabrous, short-beaked, or evidently longer than its beak...(*z*)
** Perigynium glabrous, long-beaked, or not longer than its beak...(*zz*)
  y Perigynia long-beaked, hispid-pubescent, green........................No. 151
  y Perigynia short-beaked,—mealy-glaucous, chocolate color..............No. 152
              —densely woolly, greenish................Nos. 153, 154
              —hispid-pubescent, brown ....Nos. 155, 156
  z Spikes, or at least the glumes, dark-purple or brown..........Nos. 157—159
  z Spikes green or straw-colored.—Bracts shorter than the culm...... No. 160
                —Bracts exceeding the culm...(*yy*)
    yy ? Spikes long, densely very many(150+)-flowered.............No. 161
    yy ? Spikes not dense. Perigynia much inflated,—30 to 50..Nos. 162—164
                —3 to 12........No. 165
  zz Perigynia 3-nerved or nerveless, in drooping spikes.............No. 166
  zz Perigynia many-nerved,—ascending. Peduncles very short......Nos. 167—170
                —horizontal or deflexed............Nos. 171, 172

1 C. capitàta L. Spike capitate or nearly globous; perigynium roundish-ovate, convex-concave, glabrous, pointed, longer than the ovate obtuse glume. 6—10′. Wht.Mts.

2 **C. gynócrates** Wormesk. ♀ Spike oblong, rather loose-flowered; perigynium oblong, short-beaked, longer than the ovate, acute, colored glume. N. Y., Mich.

3 **C. exílis** Dew. Spk. cylindrical, 1′, dense, ♂ below, or wholly ♂ or ♀; perig. serrulate on the margin, some longer than the ovate-lanceolate glume. Culm and leaves filiform, stiffly erect, 12—20′. Ms. to N. Y. and N. J.

β. **andrógyna.** One or more small ♀ spikes below the terminal. N. Y.

4 **C. scirpoìdea** Mx. Spk. oblong-cylindric (9—12″); perig. oval, pubescent, longer than the ovate dark-purple glume. Leaves flat. 4—10′. N. H. to Mich.

5 **C. polytrichoìdes** Muhl. Spk. oblong, small (3″); perig. 3—8, erect, smooth, twice longer than the ovate obtuse glume. Setaceous, 4—20′. Ms. to Wis.

6 **C. pauciflòra** Ltf. Spk. with about 4 slender *reflexed* ♀ fls. and 1 or 2 ♂ above twice longer than the lanceolate glume. Erect, 3—8′. N. England, New York.

7 **C. Boottiàna** Benth. Culm 6—12′; spk. oblong-cylindric, diœcious; perig. hairy, obovate, smaller than the dark-purple glume. Ala. to La.

8 **C. Willdenòvii** Schk. Peduncles radical, filiform, 2—6′; spk. small, ♂ gls. above, 4—8, ♀ perig. 5—9, scabrous, pointed, the glumes oftener long and bract-like, Leaves 1—2f, grassy. Dry grounds: common.

9 **C. Steudèlii** K. Peduncle radical, 1—6′; spk. with 10—15 ♂ glumes above and 2 or 3 inflated pointed perigynia with long leafy glumes. N. Y., Pa., and W.

10 **C. Bäckii** Boott. Ped. radical, ½—3f, stiff; ♂ fls. about 3, above, ♀ perig. 2—4, glabrous, round-ovate, enclosed in the long leafy glumes. N. Y., O., and N.

11 **C. Fràseri** Sims. Culm 4—10′, lvs. 6—12′ by 1′, flat and thick; spk. oblong, ∞-flowered, perig. ovoid, longer than the hyaline, obtuse glume. Wytheville, Va. (*Shriver*) and Mts. of N. C. A curious and peculiar Carex. Leaves very large.

—— § § ——

12 **C. stérilis** Willd. Culm (and lvs.) slender, erect, 1—2f; oftener diœcious; spks. 3—6, roundish, approximate, ♂ spikes oblong; perig. radiating, ovate, subrostrate, 2-toothed, about equaling the ovate acutish glume. Common in wet places.

13 **C. bromoìdes** Schk. Slender, weak, 1—2f; spikes 4—6, distinct, lanceolate; perig. lanceolate, erect, acuminate, longer than the lanceolate gls. Bogs: common.

14 **C. siccàta** Dew. Erect, 1—2f; spks. 3—7, oval to oblong, ♂ above, or the middle all ♂; perig. lance-ovate, beaked, as long as the brownish gls. Sands, N. Eng. to Ill.

15 **C. distícha** Huds. (C. Sartwellii Dew.) Erect, 2—3f; spks. 12—20, the lower some remote, all ovoid and compact, stam. mostly above; perig. ovate, lanceolate, pointed, equaling the ovate pointed glume. Seneca Co., N. Y. (*Sartwell*), and W.

16 **C. decompósita** Muhl. Culm erect, 18—36′; spikes very many, in a large crowded panicle; perig. round-obovate with a very short beak, biconvex, about equaling the ovate glume. N. Y. to Mich., and S.

17 **C. prairea** Dew. Culm 2—3f; spikes many, in a dense short (3—4′) panicle; perig. erect, lance-ovate, smaller than the glume. N. Eng., and W.

18 **C. teretiúscula** Good. Spikelets roundish, dense, in a cylindrical compound spike 1—2′; perig. brown, corky, ovate, biconvex, short-beaked, diverging; culm 1½—3f; leaves narrowly linear. Common in wet places, northward.

19 **C. vulpinoìdea** Mx. Spikelets very many, dense, ovoid, in a large (2—3′) compound spike; perig. yellowish, very small (½″), ovate, acuminate, diverging, scarcely as long as the pointed glume; culms stout, 2—3f. Common.

β. **setàcea.** Perig. narrower, erect, in a more slender compound spike.

γ. **scàbrior** (Sartwell). Spikes distinct or remote, glume strongly serrulate.

20 **C. conjúncta** Boott. Spikelets in a long (3′) subsimple spike; perig. ovate, subcordate and corky at base, short-beaked; style bulbous at base; nut orbicular; culm weak, 1—2f, flattened. Ohio, and westward. (C. vulpina C-B.)

21 **C. alopecoìdea** Tuckm. Spikelets 8—12, in an oblong 1—2′ spike; perig. ovate nerveless, brown, 1″, subrostrate; culm 3-angled, 2—3f. N. Y., Pa., and W.

22 **C. stipàta** Muhl. Spike often decompound, 2—3′, spikelets ∞, oblong; perig.

ORDER 154.—CYPERACEÆ.

lance-ovate, 1½'', twice longer than the glume ; culm acutely 3-angled with concave sides, leaves nearly as long (2—3f). Marshes : common.

23 **C. Crus-Corvi** Shuttl. Spike decompound or sub-panicled, 3—6'; perig. shortovate, very long-beaked (3''), spreading; glume 1''; culm 2—3f; lvs. linear, flat, many and long. River swamps, Wis. to O., and Fla.

24 **C. cephalóphora** Willd. Head 6—12'' long, dense; perig. broad-ovate, shortbeaked, scarcely longer than the ovate-acuminate glume ; style very short, bulbous at the base; lvs. copious, equaling the slender culm (1f). (C. Leavenworthii Dew.)

25 **C. Muhlenbérgii** Schk. Head ovoid-oblong, 1'; perig. broad-ovate, shortbeaked, strongly nerved, twice larger (1½'') than in No. 24; nut orbicular, style short, bulbous ; culm 1—2f, lvs. shorter, bracts setaceous. In fields, not abundant.

26 **C. chordorhiza** Ehrh. Head ovoid, 9—15'', perig. ovate, nerved, turgid, at length brown, few and large (2''), beakless but minutely pointed ; rhizome creeping; leaves short and narrow, culms 9—15'. Marshes, N. Y. to Wis., and N.

27 **C. cephaloidea** Dew. Spikelets very short, spike 1—1½'; perig. brown (at maturity), acuminate, nerveless, ovate, *shorter than the* thin cuspidate *glume*. Culm 2—4f. Leaves elongated. Fields, hedges, N. Y. (Penn Yan, *Sartwell*), and W.

28 **C. muricàta** L. Spikelets ovoid, often a little remote ; perig. ovate-lanceolate, nerveless, wingless, some longer than the ovate-lanceolate gl. Ms. to N. J., and W. §

29 **C. sparganioides** Muhl. Spikelets 7—10, ovate ; perig. ovate-acuminate, nearly twice longer than the glumes, all green. Culm and leaves 2f. In fields : common.

β. **ràmea,** is a luxuriant form, with the spike large (3—4'), panicled.

γ. **minor,** is a small and delicate form, with the spike 1—2' long.

30 **C. ròsea** Schk. (Fig. 16) Spkl. 5—8, remote, 8-10-flwd.; perig. (Fig. 17) lance-oblong, diverging or reflexed, twice as long as the ovate obtuse glume (Fig. 18). 8—16'. Com.

β. **minor.** Spkl. 4—6, quite remote ; perig. fewer and suberect.

γ. **radiàta.** Spkl. about 3-flwd., perig. oblong, acute. Stem and leaves setaceous.

31 **C. retroflèxa** Muhl. Spkl. 3—5, bracteate, stellate at maturity; perig. 3—6, ovate, acutish, spreading or reflexed, about equaling the acute glume. Woods. 1f.

32 **C. tenélla** Schk. Spkl. 3 or 4, near, erect; perig. 1—3, mostly 2, ovate-obtuse, minutely pointed, brown, smooth, little exceeding the hyaline, ovate, acute gl. In tufts, very slender and flexile, 5—12'. Woods, N. Eng. to Pa., and W. (C.disperma Dew.)

33 **C. stellulàta** L. Culm stiffly erect, 8—24'; spikelets 4—6, ovate, sessile, the spike nearly 2', turning brown ; perig. broad-ovate, short-beaked, a little longer than the ovate, obtuse glume. Wet places, N.

34 **C. scirpoides** Schk. Culm very slender, 6—12'; spkl. 3—4, contiguous, spk. 1', light green; perig. ovate-lanceolate, near twice longer than the ovate-lanceolate, acute glume. Wet. Common. Stam. mostly below the upper spikelet.

35 **C. trispérma** Dew. Very slender, 1f; spikelets 1—3, with long setaceous bracts, about 3-flwd.; perig. oblong, pointed, little longer than the glume. Pa., N. and W.

36 **C. Déweyi** Schk. Slender, leafy, 1—2f; spikelets 3—5, 3-9-flwd., the upper approximate ; perig. oblong-lanceolate, rostrate, 2-toothed, mostly longer than the ovate-lanceolate awned hyaline glume. Woods, N. Eng. to Wis., and Canada,

37 **C. canéscens** L. Erect, 2f, glaucous ; spkl. 5—7, ovate-oblong, remote below, 12-20-flwd.; perig. round-ovate, toothless, eq. the glume. Wet. Com. (C. curta Good.)

38 **C. vítilis** Fries. Slender, flexuous, 1—3f; spkl. 3—5, separate, short-ovoid, 5-10-flwd. ; perig. lance-ovate, pointed, longer than the glume. N. Eng., W. and N.

39 **C. Norvègica** Schk. Yellowish, 6—12', erect; spkl. about 3, 5-12-flwd., the upper often all ♂ ; perig. oval, biconvex, veiny, brown, eq. the obtuse glume. Me. (*Blake.*)

40 **C. Liddòni** Boott. Spike 1—2', of 5—7 oblong spikelets ; perig. and gl. lanceovate, brownish, equal, the latter white-edged ; culm strict, 1—2f. Mich. (*Cooley*), & N.

41 **C. tenuiflòra** Wahl. Spike capitate, ½', of 2 or 3 roundish, about 5-flwd. spkls ; perig. oblong-ovate, plano-convex, acute, equaling the oblong glume. Swamps, N.

42 **C. sychnocéphala** Carey. Spkl. ovoid, in a dense head with long leafy bracts; perig. 2½''', lance-linear, gradually long-beaked, the gl. nearly as long. N. Y.; rare.

43 C. árida Schw. and Torr. Spkl. oblong-oval, large, close and dense, dry and chaff-like in aspect; perig. lance-linear, 4″, clearly bidentate, gl. ♃ as long. W. com.

44 C. scopària Schk. Spkl. 5—8, ovate, approximate, or often crowded in a head perig. 3″, lanceolate, longer than the lanceolate glume; culm 18—24′ high, leafy be low. A very common sedge, in meadows everywhere.

45 C. lagopodioìdes Schk. Spkl. 8—20, ovoid-clavate or globular with a club-shaped base, approximate or crowded; perig. lanceolate, nearly twice as long as the ovate-lanceolate glume. Plant 2f, light green. Common.

46 C. cristàta Schw. Spkl. 6—12, ovoid-globular, crowded into an oblong head; perig. spreading, lance-ovate, pointed both ways, twice longer than the small lanceolate glume. Culm 2—3f, stout. Fields and meadows : common.

47 C. mirábilis Dew. Spkl. as in C. cristata; perig. broadly ovate, rounded at base, acuminate at top, a little longer and broader than the gl. Rigid, 2f. Borders of fields. (C. festucacea β. Carey. C. straminea β. Tuckm. C. cristata Boott.)

48 C. stramínea Schk. Spkl. about 6 (3—12), ovoid to oval or clavate-ovate, remote or contiguous; perig. oval or round-ovate, very flat, broadly winged, abruptly beaked, equaling or exceeding the much narrower glume. Common and variable.

a. týpica. Spkl. 3—6, roundish; perig. spreading, brownish: gl. much smaller
β. ténera. Slender, with 3—6 ovate brownish remote spikes attenuate below.
γ. apérta. Spkl. 4—8, tawny, drooping; perig. long-beaked, thrice longer than gl.
δ. festucàcea. Spkl. 5—8, club-obovate, longer beaked, prominent, brownish.
ε. hyalìna. Spkl. about 6, large, pale ; perig. twice longer than the glume. W.
ζ. monilifórmis. Slender; spkl. about 4, remote, whitish, acute at both ends. E.

49 C. silícea Olney. Spkl. 2—10, pale or silvery-yellow, distant, ovate; perig. orbicular, broadly winged all around, short-beaked, usually longer and broader than the lanceolate glume. Lvs. involute. 8—20′. Sea shore, Maine to Delaware (*Canby*).

50 C. adústa Boott. Spkl. globular with an acute base, large, silvery-green, close or remote; perig. ovate to oval, veined, narrowly winged, acuminate, equaling the glume in length and breadth. N. J., Penn. and N. (C. argyrantha, more delicate.)

51 C. fœna Willd. Spkl. 4—8, pale, oval-oblong, acute, approximate; perig. oval to obovate, appressed, broadly-winged, short-beaked, a little longer than the ovate-lanceolate glume. Plant glaucous, 2—3f. Marshes, R. I. to Pa.

52 C. alàta Torr. Spkl. 4—8, ovate, large, close; perig. roundish or obovate, close, abruptly short-beaked, 3-veined on the back, broad-winged, some longer than the lanceolate white glume. Pale green, 3—4f. N. Y. to Fla.

53 C. Washingtònia Dew. Culm 6—18′; lvs. flat: ♀ spk. 1—4, oblong-cylindric, 6″—1′, the lowest stalked; gls. black, oval, covering the oval apiculate nerved perig.; lower bract often elongated. White Mts., and N. (C. rigida β. ? Bigelovii Gr.)

54 C. rotundàta Wahl. Culm 1f, slender; lvs. channeled; ♀ spk. 1—2, oval or roundish ; perig. ovate, acuminate, equaling the lanceolate brownish gl.; bracts surpassing the culm; ♂ spk. very slender, 1′. Moosehead L., Me. (*Smith*).

55 C. Floridàna Schw. Culms 2—10′, slender, lvs. often longer; ♂ spk. short, sessile, ♀ spk. ovoid, 1—3, crowded; glumes oval, acute, *edged with brown*, covering the obovate, short-beaked perig. Often with solitary ♀ spikes on radical ped. S.

56 C. lenticulàris Mx. Culm 8—18′; lvs. flat; ♂ spk. 1′, ♀ spk. 2—5, ½—1′, with long bracts; perig. ovate-oval, yellowish, nerved, longer than the obtuse glume. Spikes cylindric. Gravelly shores, Me., N. H., N. Y., and northward.

57 C. aùrea Nutt. ♂ Spk. short (6″), ♀ spk. 3 or 4, ½—1′, loose-flowered, spreading; perig. oval, obtuse, yellow-brown, separate, exceeding the hyaline gl. Culm slender, 8—16′; leaves flat, bracts exserted, leafy. Wet. N. Eng., and W.

58 C. Mitchelliàna Curtis. ♂ Spk. often ♀ in the middle ; ♀ spk. 2—3, cylindric, slender, loose; perig. ovate, acute, short-beaked, eq. the gl. 15—20′. Wet. N. Car.

59 C. torta Boott. Spikes cylindric, slender, 2—5′; spikelets 2 or 3, loose below, recurved; perig. lanceolate, the beak recurved or *contorted*, equaling the black banded obtuse lanceolate glume. Very smooth, 2—3f. Wet places.

60 C. vulgàris Fries. ♂ Spikes cylindric, 1—2′, ♀ cylind.-oblong, 1′, ♂ at top; gls black, ovate, obtuse, shorter than the oval, obtuse perig.; culm slender, 6—14′; lvs. flat, bract equaling the culm. Wet, N. Eng., W. and N. (C. cæspitosa C-B.)

61 C. stricta Lam. Spk. cylindric, 1½—2′, erect; glumes lanceolate, acutish, striped, some longer than the ovate-acute perigynia. 2f. Bogs; common.

β. *strictior*. Glumes, especially the upper, a little shorter than the perigynia.

62 C. xerocárpa S. H. Wright. Differs from C. stricta in its extremely r.ender habit; lvs. rolled and rush-like; ♂ spk. almost filiform; gl. shorter than perig. N. Y.

63 C. apérta Boott. Spk. cylindric, erect, 12—15″; perig. brown, round-ovate, shorter than the lance-acuminate glume; culm 1—2f. rough-edged above; lvs. channeled, bracts leafy. Wet meadows, N. Eng., W. and N.

64 C. aquátilis Wahl. Spk. 2—3′, dense, erect, acute, subclavate, the ♂ 2 or 3, ♀ 3—5, with bracts exceeding the culm; gl. lanceolate, usually longer than the roundish, nerveless, reddish, apiculate perigynia. 2—3f. Shores, N.

65 C. crinìta Lam. Spk. pedunculate, long (2—4′), nodding, ♂ mostly but 1, ♀ about 4; perig. round-ovate, apiculate, glume with its long serrulate awn thrice longer—all light brown. Wet meadows: common. 2—3f. Leafy.

β. *gynandra*. Spk. shorter (1—2′), ♀ about 3. perig. inflated, awns spreading, &c.

66 C. marítima Vahl. Spk. 1—2′ long, pendulous or spreading, on peduncles, the ♀ 3—5; perig. orbicular, much shorter than the long-awned green glume; culm 10—20′, erect, with broad, flat, smooth leaves. Salt marshes, Mass., and N.

67 C. salìna Wahl. Spk. cylindric, erect on included stalks, the ♀ 2—4; bracts long; perig. elliptical, apiculate, little shorter than the dark-brown, short-awned glume; culm 8—16′, rough above. Salt marshes, Mass., and N.

———— § § § ————

68 C. pedunculàta Muhl. Spk. 3—7, remote, on filiform stalks; perig. obovate, triquetrous, recurved at tip, few, equaling the brown, oblong, obovate glume. Culm 4—12′. leaves longer, glabrous. Woods. Flowers in early spring.

69 C. Baltzéllii Chapm. Spk. cylindric, 1—2′, ♀ 1—4, ♂ at top, on long canline or subradical peduncles; perig. and gl. oblong-obovate. subequal, the perig. veiny and puberulent. Culm 6—10′, leaves flat, thrice longer. Florida.

70 C. squarròsa L. Spk. 2—4, cylindric-oblong, thick (1′ by 6″), straw-color, stalked, *squarrous* with the long beaks of the globous perig. which conceal the short glumes. Wet places: common. Large and fine, spike showy.

71 C. viréscens Muhl. Spk. 2—4, erect, 6—12″; perig. ovate, *pubescent*, ribbed, longer than the ovate pointed glume or about equal to it. Culm slender, 1—2f, bracts exceeding the culm. Whole plant pubescent and light green. Copses.

72 C. hirsùta Willd. Spk. oval-oblong, 4—9″, erect, near, dense; perig. ovoid-triquetrous, downy, at length only scabrous, longer than the glumes. Culm 1—2f, bracts exceeding it, all pubescent or scabrous. Upland Meadows. (C. Triceps Mx.)

73 C. Smithii Porter. Spikelets 3, oval and oblong, near; perig. globular; achenia broadly obovate with reflexed styles; culm slender; whole plant glabrous, bright green, 2f. Del. Co., Penn. (*A. H. Smith.*) Also in N. J. (See Olney's Carices Am.)

74 C. æstivàlis Curtis. Spk. 3—5, slender, 1—2′, loose, suberect on short stalks; perig. elliptic, pointed both ways, longer than the glume. Tufts 16—24′ high, with flat downy leaves, and bracts exceeding the culm. Mts., Mass. to N. Car.

75 C. Shortiàna Dew. Spk. 4 or 5, cylindric, dense, 1′, erect on naked stalks, tawny in maturity; perig. round-obovate, scarce longer than the ovate glume. Erect, 12—30′. leafy, smooth, handsome. Wet grounds, Penn. to Ill., and S.

76 C. oxýlepis Torr. Spk. 3—6, cylindric, 1—2′, erect on naked ped.; perig. oblong, pointed both ways, little longer than the cuspidate white-edged glume. Fla. to La.

77 C. Buxbaùmii Wahl. Spk. 4, ovoid, sessile, near; lower bract equaling the culm; perig. elliptic, nerveless, rounded on the back, shorter than the pointed blackbanded glume. Culm 10—18′. Common in wet places.

ORDER 154.—CYPERACEÆ. 375

78 C. alpìna Swtz. Spk. 3 or 4, small, oval, close; bract longer than the culm; perig. round-obovate, longer than the black glume. Leaves radical. L. Superior.
79 C. atràta L. Spk. 3—6, oblong-ovate, nodding, the lower stalked; perig. round-ovate, shorter than the dark oval glume. Bract long. White Mountains.
80 C. gracíllima Schw. Spk. 3—4, slender, 12—20″, rather loose, drooping on long filiform remote stalks; bract short; perig. oblong, longer than the oblong short-awned glume. 2f. Meadows.
81 C. formòsa Dew. Spk. 3—4, oblong, 8—12″, on long, distant recurved peduncles; perig. oblong, inflated, twice longer than the ovate acute glume. Culm 2—3f, bract shorter than the culm. Wet meadows.
82 C. glabra Boott. ♀ Spk. short-cylindric (1′), spreading on capillary peduncles; perig. elliptic-oblong, acute at both ends, nerved, twice longer (2″) than the ovate brown-edged glume. Very slender, erect, 18′. N. J., N. Y., Penn.
83 C. Davísii Torr. Spk. 4, 10—15″ long, rather loose, long-stalked, drooping when ripe; bracts much longer; perig. oblong-ovate, nerved, acute, scarce equaling the awned glume. Mass. to Wis., and N.
84 C. præcox Jacq. ♂ Spk. clavate, erect; ♀ spk. about 2, ovate-oblong, 6—9″; perig. 6—12, round-ovate, downy, nearly equal to the ovate colored glume (which is brown, edged with white). Culm 3—6′, leafy at base. Rocky hills, E. Mass.
85 C. Richardsònii R. Br. ♂ Spk. clavate-oblong, erect; ♀ about 2, oblong, near, subsessile; glumes wholly brown; perig. ovoid-triquetrous, obtuse, nearly beakless, shorter than the green-midveined glume. 4—10′. Woods, N. Y. to Ill., and N.
86 C. vestíta Willd. Spk. all sessile, 9″, ♂ cylindric, ♀ 2, ovoid-oblong; perig. ovate, short-beaked, hairy, exceeding the rusty acutish glume. Culm 12—30′, sharp-angled, leafy below. Common in wet places.
87 C. pubéscens Muhl. Spk. oblong, 8—12″, rather loose, the lowest on a short stalk; perig. lance-ovate, beaked, hairy, exceeding the carinate, mucronate glume. Culm 10—20′; leaves downy, flat, 5—10′. Meadows.
88 C. nigro-marginàta Schw. is probably a mere variety of No. 55, having the glumes more extensively colored and the stigmas oftener 3. Hills, Pa., and S.
89 C. umbellàta Schk. Dwarf; ♂ spk. erect, 2—3″, ♀ ovoid, 2—4, each on a sub-radical peduncle, green; perig. 5—8, round-ovate, beaked, nearly equaling the lance-acuminate glume. Leaves 3—5′, far longer than the spike. North.
90 C. Emmónsii Dew. Spikes all sessile, green, ♂ 4—5″, ♀ 2—3, ovoid; perig. about 5, globous, beaked, equal to the pointed glumes. Culm filiform, 6—12′, with very narrow leaves. Fields and hills; common.
91 C. Pennsylvánica Lam. Spikes tawny-red, ♂ 1′ long, pedunculate, the ♀ small, round, sessile, crowded, about 2; perig. round-ovoid, 5—7, downy, short-beaked, equaling the acuminate glume. Culm 4—12′, erect, leaves long. Copses.
92 C. Novæ-Angliæ Schw. Spk. purplish, sessile, ♂ 3—4″, ♀ 2—4, small, near, (except the lowest), with bracts exceeding the culm; perig. 3—7, pyriform, short beaked, larger than the ovate glume. Slender, 4—12′. Open woods.
93 C. vària Muhl. Spikes rusty-green, sessile, oval, 1—3, separated, the ♂ slender, (10″) and stalked, bracts very short; perig. about 7, rouud-oval, abruptly beaked, about equaling the pointed rusty-edged glume. Erect, 8—18, leafy at base. Dry woods.
94 C. flàva L. ♀ Spk. oval, approximate, 2—4; perig. crowded, ovate, ribbed, re-flexed with a long curved beak, longer than the lance-ovate glume. Plant 10—20′, yellowish green. Cold, wet soils; common.
95 C. Œderi Ehrh. ♀ Spk. 3—5, oblong, small (3—5″), close, nearly sessile; perig. globous, diverging with a short abrupt beak; plant yellowish, 8—16′, leaves and bracts erect. Shores, N. Eng., and West. (C. virídula Mx.)
96 C. folliculàta L. ♀ Spk. 2—4, capitate, dense, distant, the lower peduncle exserted; perigynia 4″, lanceolate, nerved, tapering into a long beak, diverging, twice longer than the long-awned glumes; leaves lance linear. Wet.

97 C. rostràta Mx. ♀ Spikes 1—3, capit*te, near; perigynia 3'', suberect, lanceolate, long-rostrate, twice longer than the acutish glume; leaves few, rolled, subulate; culm 1f. Mountain bogs, N. Y., N. H., and North.

98 C. Ellióttii Schw. ♂ Spike slender, 1'; ♀ 2 or 3, globous to oval, distant; perigynia 70—20, ovoid, veined, rostrate, 3''; glume ovate, 1''; culm slender, rigid, 1—2f, the narrow leaves longer. N. Car. to Fla.

99 C. subulàta Mx. ♂ Spike short, subsessile; ♀ spikes 3—5, capitate, distant, 3-7-flowered; perigynia *subulate*, 6'', long-rostrate, divaricate and with 2 divaricate teeth. Slender, smooth, light-green, 1—2f. Can. to N. J.

100 C. turgéscens Torr. ♂ Spike slender, 1½'; ♀ spikes 2 to 3, capitate to oval, loose, the lowest pedunculate, exserted; perigynia 9—12, inflated, striate, conic-rostrate, 6''; glume ovate, acute, 3''. Culm 2—3f, slender; leaves long. Swamps, S.

101 C. intuméscens Rudge. ♂ Spike long-stalked, slender; ♀ 1—3, on very short stalks, capitate; perigynia 5—8, very large (6—7''), acuminate-beaked; glume ovate-cuspidate, 2''; culm 1f; bracts very long. Wet.

102 C. Gràyii Carey. ♀ Spikes 1 or 2, large, capitate, dense; perigynia 15—30, radiating, very large (7—8''), with a long, slender, smooth beak; glume inconspicuous. River bottoms, N. Y., and West.

103 C. lupulìna Muhl. ♀ Spikes 2—4, large, 1—2' by 9—12'', the lower on exserted stalks; perigynia *ascending*, 6½—7'', ovoid and long-beaked, bicuspidate; glume 3'', lance-acuminate. Plant stout, leafy, 2—3f. Wet grounds.

β. *pedunculàta*. Spikes all on long peduncles. ♂ Glumes linear-awned as in α
γ. *andrógyna*. ♀ Spikes staminate at apex. Approaching No. 172.

104 C. lupulifórmis Sartwell. ♀ Spikes 4—5, very large (2—3'); perigynia ascending, 7—8'', the long beak roughish, bicuspidate; glumes long-awned, ovate, 3''; nut as broad as long, the angles knobbed. Swamps: common.

105 C. tentaculàta Muhl. ♀ Spikes 2 or 3, dense, 1½—2' by 7 or 8'', near, on short peduncles; perigynia 4'', ovate, long-beaked, diverging, orifice obliquely 2-toothed; glumes linear-awned, 2''. Stout, leafy, 1—2f. Bogs: common.

β. *altior*. ♀ Spikes 3—4, larger (10'' thick), beak subequally toothed. 2f.

106 C. stenólepis Torr. ♂ Spike small (1') rarely 0; ♀ 1—5, very dense, 1—1½', often ♂ at base; perigynia globous, abruptly beaked, recurved, shorter than the long slender-awned glumes. Related to C. squarròsa. Penn. to Ill., and South.

107 C. plantagínea Lam. ♂ Spike clavate, glumes acute; ♀ spikes 3—5, erect, remote, loose; perigynium 5—10, the point recurved, twice longer than the glume; bracts purple, shorter than the spikes; leaves 6—10'' broad. Woods. March—May.

108 C. Careyàna Torr. ♂ Spike oblong, erect; glumes obtuse; ♀ spikes 2—3, remote, loose; perigynium 3—7, large (2½''), the point oblique, twice longer than the glume; bracts green, much longer than the spikelets; leaves 6—12'' wide. Woods, N. Y., Pa., and W.

109 C. platyphýlla Carey. ♂ Spike clavate, glume acute; ♀ spikes 2—3, very remote, small; perigynia 3—6, small (1½''); glume cuspidate, 1''; bracts as in C. Careyàna; leaves 6—10'' wide, mostly shorter than the culms. Shades, N. States.

110 C. laxiflòra Lam. ♂ Spike linear, glumes lance-oblong, acute; ♀ spikes 3, slender, 1', loose, remote; perigynia 10—15, elliptic-triq., 2'', the point oblique; gl. oblong, mucronate, 1½''; leaves 1-veined, 2—4'' wide, bracts long. Shades: common.

β. *patulifòlia*. Root leaves 6—12'' wide, bracts also wide. Otherwise as in α.
γ. *latifòlia*. Leaves and bracts very broad; perigynia broad, point conspicuous.
δ. *blanda*. Bracts very long, ♂ spike small; ♀ spikes dense; perigynia obovoid.
ε. *intermèdia*. Leaves narrow, ♂ spike on a slender stalk; perigynia as in α.
ζ. *styloflèxa*. Slender, 1—2f, spike small, on long filiform peduncles, 4-6-flowered.

111 C. retrocúrva Dew. Spikes small (5—8''), all on long capillary peduncles, the ♀ 3, loose; perigynia broad-ovate-triquetrous, scarcely oblique-pointed; glumes awned; culms weak, 1f, leaves radical, wide (4''), flat, glaucous. Open woods: rare.

112 C. digitàlis Willd. ♂ Spike slender, 1', stalked; ♀ spikes 3, loose, 6—12'', ro mote, recurved; perigynia 4—10, ovoid-triquetrous, obtuse, longer than the lance-ovate glume; leaves and bracts 1—2" wide, exceeding the 4—12' culm. Open woods.

113 **C. xanthospérma** Dew. ♂ Spike small, sessile; ♀ spikes 4, distant, cylindric, 1', dense, on long slender peduncles; perigynia oval-oblong, obtuse, 2", striate, yellowish when ripe; glumes 1", pointed. Yellowish, 1f. N. J., and South.

114 **C. conoìdea** Schk. Spikes all short-peduncled, ♀ 2 or 3, oblong, dense, erect, 6—10"; perigynia oblong-conic, obtusish; glumes ovate, awned. 1f. Uplands: com.

115 **C. grísea** Wahl. ♂ Spike sessile; ♀ spikes 4, oblong, remote, 6"; perigynia oblong, some longer than the ovate, awned glumes (2½", glumes 2"); leaves light-green, 2—3" broad. Culm 1½f. Woods and meadows.

116 **C. glaucòdea** Tuckm. Spikes short-stalked, 6—12", ♂ clavate, ♀ 3—4, cylindric, dense; perigynia 10—20, ovoid, obtuse, twice longer than the cuspidate glumes. Plant glaucous, 6—10'; leaves 2—4" wide. Mass. to Pa.

117 **C. granulàris** Muhl. ♂ Spike linear, sessile, 1'; ♀ 2—4, cylindric, ½—1½', the lower peduncle long; perigynia close, round-ovate, the point oblique, much longer than the ovate-acuminate glumes. Glaucous, 8—20'. Moist soils: common.

β. *recta*, has the perigynia ovoid, and with a straight point. Ill. to La.

118 **C. júncea** Willd. Spikes slender, on filiform stalks, glumes obtuse; ♂ short; ♀ spikes 2—3, loose; perigynia lanceolate, longer than the glumes; culm 1—1½f, slender, longer than the slender rush-like leaves. Roan Mt., N. C.

119 **C. ebúrnea** Boott. Delicate, erect, 4—10', the setaceous leaves much shorter; spikes 2—3, very small (2—3"), with white, leafless sheaths, the ♀ higher than the ♂; perigynia 3—6, obovoid, beaked, nerveless, ⅓". Rocks, Vt., and West.

120 **C. panícea** L. Spikes 2—4, 1', oblong-cylindric, stalked, tawny; perigynia turgid-ovoid, the very short point oblique, longer than the obtuse glume. Light green, 1f; bracts short. Mass. (*Oakes*). Wis. (*Lapham*). Pa. (*Porter*).

121 **C. lívida** Willd. Spikes 2—4, oblong-cylindric, pale, 8—10", the ♂ and lower ♀ stalked; bracts short; perigynia oval, straight at the obtuse end, longer than the obtuse glumes. Glaucous, 6—16'. Swamps, N. Y., N. J., and North.

122 **C. tetánica** Schk. Spikes 2—4, oblong-cylindric, loose, 1', the ♂ and lower ♀ long-pedunculate; perigynia ovoid to obovoid, apex oblique, longer than the sub-mucronate glumes. Light green, 8—16'; bracts rather short. Wet uplands: rare.

β. *Woodii*, ♀ spikes about 2, very loose; glumes with broad scarious margins.

123 **C. Meadii** Dew. ♂ Spike slender, 1', ♀ oblong-cylindric, loose, 8—10", all pedunculate; perigynia oval, scarce equaling the tawny-edged, ovate-acuminate glumes. Pale, erect, 8—16', the leaves and bracts short. Wet, O. to Ill., and North.

124 **C. Crawei** Dew. Spikes dense, 8—10", erect, ♂ stalked, compound at base, ♀ 2—5, remote, the lowest often long-stalked; perigynia ovoid, acute, twice longer than the ovate glumes. Erect, 6—15'. Spikes dusky green. N. Y., and West. Rare.

125 **C. oligocárpa** Schk. ♂ Spike erect, 9", linear, stalked; ♀ 3, remote, short-stalked, 3- or 4-flowered; perigynium obovoid, *short-beaked*, brown, equaling the awn of the pale glume. Pale, 6—12', bracts long. Open woods and hedges: rare.

126 **C. Hitchcockiàna** Dew. ♂ Spike erect, linear, stalked; ♀ 3, remote, short-stalked, 5-10-flowered; perigynia oval, brown, acute below, *the beak bent back*, scarce equaling the awn of the whitish glume. Subpubescent, 1—2f. N. Eng., and West.

127 **C. exténsa** Good. ♂ Spike subsessile, 6—9"; ♀ 3, oval to oblong, very dense, the lower remote, stalked; perigynia spreading, the short straight beak 2-toothed, gl. much shorter. Rush-like, 1—2f, leaves and bracts rolled. Sands, L. I., Staten I.

128 **C. débilis** Mx. Spikes about 2', very slender; ♀ 3—5, nodding; perigynia 12—20, lance-linear, acuminate-beaked, twice longer than the oblong silvery glumes. Bright green, 1—2f; bracts equal the culm. Moist woods and meadows: common.

β. ? *pubera*. Perig. pubescent, strongly veined, slightly bent. Pa. (*Porter*), and S.

129 **C. arctàta** Boott. Like C. débilis, but with shorter bracts, longer stalks, the perigynium ovoid, taper-beaked, ♂ longer than the ovate-pointed glume. Common.

130 **C. Sullivántii** Boott. Spikes cylindric, 9—15", erect, 4 approximate, or a 5th

if any, remote; perigynium elliptic, rough-hairy, scarcely longer than the ovate-cuspidate glume. Borders of woods, Columbus, Ohio. 2f.

131 C. Kneiskérnii Dew. Spikes rather loose, 1—1½', with recurved peduncles; perigynia ovate-oblong, glabrous, nerved. Otherwise as in C. Sullivántii. Woods, Oriskany and Rome, N. Y., and Cleveland, O.

132 C. vaginàta Tausch. ♂ Spike nodding in flower, stalked; ♀ 2 or 3, remote, loose; bracts short with long sheaths; perig. 5—10, brown-black, globular-ovate, the beak terete, short, bent, exceeding the obtuse gl. Weak, 1—2f. N. Y. (rare), L. Sup.

133 C. capillàris L. Spikes minute, 3—4, oblong, tawny, peduncle *capillary*, perigynia 4—6, oval, nerveless, the short beak exceeding the obtuse rusty glume Pale, delicate, 4—7', leaves long, bracts short. White Mts., N. H.

134 C. fléxilis Rudge. Spikes 3—5, ♂ clavate, ♀ oblong, on *flexile* nodding peduncles; bracts bristle- or scale-form; perigynia ovoid-lanceolate, 2-toothed, scarce longer than the obtuseish rusty glumes. Soft-hairy. 1—1½f. Ct., N. Y.: rare.

135 C. lævigàta Sm. Like C. fléxilis, but with perigynia nerved, bicuspidate, the glumes awn-pointed, and the whole plant smooth. Near Boston. §

136 C. fulva Good. Culm 1f, rough; spikes 3—4, all erect. ♀ ovoid-oblong; perig. ovoid, twice longer than the dark-brown acutish glumes. Near Boston. §

137 C. venústa Dew. Spikes 3 or 4, ♂ linear, 1'—16", rusty, stalked; ♀ loose, 6—16", brown-green; perigynia lance-oblong, 2½", conic-beaked, nerved, rough-hairy, twice longer than the glumes; leaves 1f, culm 2—3f. S. Car. to Fla.

138 C. tenax Chapm. Spikes 2—4, ♂ slender, 1', ♀ oblong, ⊢1', dense, subsessile; bracts longer; perigynia oval, short-beaked, finely-veined, pubescent, twice longer than the ovate glumes; culm 1f; leaves rolled. Ga., Fla.

139 C. dasycárpa Muhl. Spikes 3—4, subsessile, 6—10". ♂ linear, ♀ oblong, hoary, bracts exserted; perigynia oblong-ovate, tomentous, short-beaked, longer than the ovate-acuminate glumes. 1f. Dry fields, South.

140 C. Tórreyi Tuckm. Spikes subsessile, erect, the ♂ oblong, the ♀ ovoid, 2 or 3; perigynia obovoid, very obtuse, scarcely beaked, strongly nerved, longer than the ovate glumes; culm, leaves, and short bracts downy. Penn., and North. Rare.

141 C. Barráttii Schw. & Torr. Spikes cylindric. 6—12", dark-purple, short-pedunculate, the ♀ 2 or 3; perigynium ovoid, little exceeding the ovate glume; culm 1—2f, sharp-angled, leaves much shorter, bracts short. Marshes, N. J. to Car.

142 C. palléscens L. Spikes approximate, 3 or 4, short-stalked, pale. ♂ oblong, 6"; ♀ ovoid, 4—5", bract a little exserted; perigynia ovoid, nerveless, scarce longer than the glumes. Plant pale, 6—15', leaves as long. Dry meadows.

β. *undulàta*. Lower bracts *wavy*-rugous at base; leaves longer.

143 C. limòsa L. Spikes pedunculate, with dark-purple glumes, ♂ linear, erect; ♀ 1—2, oblong, drooping; bracts shorter than the culm; perigynia ovate, scarce equaling the broad, mucronate glumes. Glaucous, 8—16'. Marshes: common.

144 C. rariflòra Sm. Like C. limòsa, but smaller (4—10'), ♀ spikes 1—2. linear, loosely 5-10-flwd.; perig. involved in the glume. Mountains, N. H., Me., and N.

145 C. irrígua Sm. ♀ Spk. 2—4, ovoid-oblong; bract exceeding the culm; perig. oval, much shorter than the long-pointed dark-purple glume, 8—20'. Leaves linear, flat. Spikes drooping as in C. limòsa. Bogs, Pa. to Wis., and N.

146 C. millàcea Muhl. Spikes cylindric, slender, 1½—2', ♂ erect, ♀ nodding, loose below; perig. ovoid-triquetrous, short-beaked, as long as the white-edged awned glume. Culm 1—2f, leaves rather broad. Wet meadows: common.

147 C. scabràta Schw. Spikes 3—6, cylindric, 1½—2', suberect, dense, the lower on long peduncles; bracts long; perig. ovoid-triquetrous, *rough*, the slender beak equaling the acuminate glume. Culm 1—2f, leaves broad. Swamps, Cau. to Car.

148 C. hystricìna Willd. ♂ Spk. linear, stalked, 1', ♀ 3, oblong-cylindric, dense, 12—18", near, nodding; perig. ovoid, inflated, nerved, diverging, the long slender beak bifid, longer than the awned glume. 1—2f, very leafy. Swales: common.

β. *Cooleyi*. Slender; ♀ spikes ovoid, the lowest long-pedunculate.

ORDER 154.—CYPERACEÆ. 379

149 C. pseudo-cypèrus L. ♂ Spk. linear, 1½', ♀ 3—5, cylindric, thick, 1—2', pedunculate, recurved; perig. horizontal or deflexed, lanceolate, with 2 suberect teeth, equaling the lance-aristate glume. Ponds and ditches, Can. to Pa.

150 C. comòsa Boott. ♂ Spike lin.-cylindric, 2—3'; ♀ 3, long (2—3'), cylindric, thick, dense-curved, on recurved ped.; perig. lance-linear, deflexed, the slender beak with 2 long spreading cusps. Stout, 2—3f. Wet.

151 C. trichocárpa Muhl. Spikes erect, ♂ about 3, clustered, ♀ 3, oblong-cylindric, thick but rather loose, 1½—2'; perig. conic-ovoid, 4'', ascending, veined, the beak slender, forked, exceeding the hyaline gl. Puberulent, 15—30'. Marshes: common.
β. *turbinàta*. Spk. ♀ ovoid-oblong, dense; perig. more diverging.

152 C. verrucòsa Ell. ♂ Spk. 2, often 1, erect, ♀ 3—7, remote, all cylindric, dense, heavy, 2—3', bracts long, on long sheaths; perig. ovate-triquetrous, shorter than the awn of the oblong glume. Culm and leaves 2—3f. Wet grounds, S.
β. *glaucéscens*. ♂ Single, ♀ sterile at apex; perig. broader or obovoid. South.

153 C. lanuginòsa Mx. ♂ Spk. 1—3, linear, 1—2', the upper stalked, ♀ mostly 2, nearly sessile, oblong-cylindric, 9—15''; leaves and bracts flat; perig. ovoid, with 2 sharp teeth, equaling the lanceolate awned glume. 1—2f. Wet places: common.

154 C. filifórmis L. Much like the last, but the leaves and bracts are convolute and rush-like, and the ♀ glumes ovate, acute. Pale. Marshes: common.

155 C. striàta Mx. ♂ Spk. 1—4, erect, the lower sessile; ♀ 1—2, remote, cylindric, erect, dense; perigynia ovoid, acuminate, 2-toothed, twice longer than the ovate acute glumes. Stiffly erect, 1—1½f, leaves and bracts rolled at the ends. Pa., and S.

156 C. Houghtònii Torr. ♂ Spikes 1—3, ♀ 2—3, cylindric, thick (12—15'' × 4''), near, subsessile, erect; perigynia ovoid-inflated, bifurcate, much longer than the ovate cuspidate glume. Stout, 2—3f, leaves and bracts flat. Me. to Wis.

157 C. polymórpha Muhl. Spikes oblong, erect; glume obtuse; ♀ 1—2, 1', the lower remote, exsert-pedunculate; bracts and leaves short; perigynia oval-ovate, beak short, purple, exceeding the ovate purplish gl. Erect, 5—20'. Sands, Pa., and N.

158 C. paludòsa Good. Spikes erect, cylindric, 15—20'', dense, near; glume cuspidate; ♀ spikes about 3; bracts long, sheathless; perigynia ovate, short-beaked, equaling the narrow glumes. Erect, 1½—2f; leaves channeled. Marshes, Mass.

159 C. ripària Curtis. Spikes erect, cylindric, 2—3', ♂ 2—5, ♀ 2—3, nearly sessile; bracts and leaves long; perigynia conic-lanceolate, with 2 slender teeth, some longer than the narrow-awned glumes. Stout, 2—4f. Shores. (C. lacústris.)

160 C. Cherokeénsis Schw. ♂ Spikes lance-linear, 6—12'', ♀ cylindric, 1—1½', 2—7, the lower nodding, on exserted peduncles; perigynia lance-ovate, much longer than the ovate glume. Slender, 2f, light green. Ga., Fla., and West.

161 C. ampullàcea Good. ♂ Spikes often bracted, linear; ♀ 3—4, cylindric, thick, 2—3' by ½', very dense, near, suberect; perigynia ovoid, more or less *abruptly* beaked, bifurcate, larger than the pointed glumes. Stout, 2—3f, the flat leaves longer. Swamps, N. Eng. to Pa., and West. (C. utriculàta, Bt.)

162 C. monìle Tuckm. ♂ Spikes slender, 2—4; ♀ 2, rarely 1 or 3, cyl., 1—2', rather loose, suberect, short-ped.; perig. ovoid, polished, 2—3'', the short slender beak bifurcate, twice longer than the lance-oblong glume. Bright green, 2f. N. Eng. to Ill. (C. Vaseyi Dew. is the same plant, as shown by specimens from *Dr. S. H. Wright.*)

163 C. Tuckermàni Boott. ♀ Spikes very remote, short-stalked, cylindric-oblong, thick, 6—15'' by 6—7''; perigynia very large (5'' by 2½''), globous-ovoid, shining; beak short, slender; glumes much shorter. 2f. Wet: common.

164 C. Olneyi Boott. ♂ Spikes 2—3, like those of *C. bullata*; ♀ spk. ofter but 1, 1'—18'' by ½'; ped. short; perig. 50—80, 2½—3'' long, 10-veined, turgid-ovoid, the short *beak and 2 cusps rough-serrulate;* ach. like *C. ampullacea*. Culm 1—1½f; lvs. taller, 1'' wide. Wet grounds, R. I.

165 C. oligospérma Mx. ♂ Spikes 1—2, slender; ♀ 1—2. Globular or oblong, subsessile; perigynia 4—12, turgid-ovoid, 2½'', beak short, 2-lobed, scarce exceeding the ovate glumes. Slender, 2f; leaves and bracts rolled. Pa., and North.

380   Order 154.—CYPERACEÆ.

166 C. longiróstris Torr. ♂ Spikes mostly 3; ♀ mostly 3, cylindric, 1', loose, stalks filiform, recurved; perigynia roundish, the very slender beak 2-toothed, longer than the scarious glumes. 2f. Rocky woods, North.

167 C. aristàta R. Br. ♂ Spikes 2, very slender, remote; ♀ 2—4, cylindric, 1—2', erect; perigynia lanceolate, conspicuously nerved, glabrous, 2-awned; glumes awned, much shorter. 2f. Shores, N. Y., West and North. Akin to No. 151.

168 C. Schweinítzii Dew. ♀ Spikes 2—4, near, ascending, cylindric, 1—2', more or less dense, straw-yellow; perigynia 50—150, ovoid, the long beak 2-toothed, much exceeding the subulate glumes. Very leafy, 1f. N. J., N. Y., and N. Eng.

169 C. bullàta Schk. ♂ Spikes 1—3, linear, with lance-oblong, close glumes; ♀ spikes 1—2, oblong, 1' by 8'', short-stalked; perigynia turgid-ovoid, 5'', beak 2-cuspidate, thrice longer than the obtusish glumes. 1—2f. Swamps, N. E., and S.: com.

170 C. physèma Dew. ? Resembles the last, but has very long leafy bracts, ♂ spk. 3 with loose glumes, and the single large oblong ♀ spike loose-flowered; perigynia radiating, brownish. A variety? Newark, N. Y. (*Hankenson*).

171 C. gigántea Rudge. ♂ Spikes 1—3, glumes pointed; ♀ 2—4, 18—30'', loose, pedunculate, suberect, brownish; perigynium ovoid-acuminate, many(18)-nerved, the very long beak forked, two or three times longer than the lanceolate-awned glume. Stout, 2—3f; leaves 6'' broad. Del. to Ky., and South. Allied to No. 103.

172 C. retrórsa Schw. ♂ Spikes 1—3, often partly fertile; ♀ 4—6, cylindric, thick, near, 1—2' by 7'', spreading; perigynium ovoid, inflated, few(10)-nerved, the long beak forked, deflexed, far exceeding the glume. Bright green, 2f. Pools: common.

β. *Hartii*. ♀ Spikes loose, distant, the lower long-stalked. N. Y. (*S. H. Wright*).

γ.? *lupulus*. ♂ Spikes 2; ♀ very large, short-stalked, straw-yellow; perigynia horizontal, much inflated, 10-nerved; glumes pointed. A fine Carex; 2—3f; allied both to Nos. 103, 171, and 172. N. Y. (*E. L. Hankenson, H. B. Lord*).

Order CLV. GRAMINEÆ. The Grasses.

Herbs (the Canes and Bamboos are woody and tree-like) with culms mostly hollow and jointed. The *leaves* are alternate, 2-ranked, on tubular sheaths split down to the base, and bearing a membranous *ligule* (of the nature of stipules) where the sheath and blade meet. *Flowers* in little spikelets of 1 or several, with the glumes in 2 rows, collected into spikes, racemes, or panicles. *Glumes* (the lower pair of scales in the spikelet) alternate, enclosing the flowers. *Pales* (or palæ, the outer pair of scales of each particular flower) alternate and unequal. *Perianth* 0 or represented by 2 minute hypogynous scales. *Stamens* 1—6, commonly 3, *anthers* versatile, 2-celled, bifid at both ends. *Ovary* simple, 1-ovuled, 1-styled, with 2 feathery *stigmas*. *Fruit* a caryopsis, with mealy *albumen*.

A vast and important Order, contributing largely to the sustenance of man and beast. Both herbage and seed are rich in sweet and nutritious matter. In temperate regions, the Grasses form a *turf*, soft, green, and compact, clothing the hills and plains, pastures and meadows. But in tropical regions this beautiful turf-carpet is unknown, the Grasses becoming larger, even trees (as the stately Bamboo), and stand more isolated, with broader leaves and larger panicles. To this Order belong the Cereal Grains, as the *Indian-Corn, Wheat, Rye, Oats, Barley, Rice*, &c., as well as the Hay-grasses—*Timothy, Red top, Blue-grass, Spear-grass*, &c. Also the *Sugar-Cane*, and various kinds of Sorghum.

§ Spikelet 1-flowered with no apparent rudiment of a second flower...(2)
§ Spikelet 2-flowered, one of the flowers sterile or rudimentary...(7)
§ Spikelet 3-flowered, the two lower (lateral) flowers sterile or rudimentary...(§).................Tribe ƒ
§ Spikelet 2- ᴏ- flowered, two or more of the flowers perfect, or all imperfect (♀ ♂)...(9)

Order 155.—GRAMINEÆ.

2 Inflorescence paniculate...(3)
2 Inflorescence strictly spicate, spikes equilateral...(5)
2 Inflorescence strictly spicate, spikes unilateral...(6)
   3 Glumes none (or minute and the stamens 6)...(a).................................Tribe 1
   3 Glumes present, at least 1 conspicuous...(4)
      4 Pales of the flower thin and soft, often awned...(b)..........................Tribe 2
      4 Pales of the flower coriaceous,—* tipped with awns...(f).....................Tribe 4
          —* awnless...(g)..Tribe 5
 5 Spikes cylindric, the spikelets condensed all around...(e)........................Tribe 3
 5 Spikes prismatic, spikelets sessile in rows...(p)..................................Tribe 9
   6 Spikelets rounded on the back, appressed to the rachis...(q)...................Tribe 5
   6 Spikelets acutely keeled on the back, imbricated on each other...(x) }............Tribe 10
   7 Upper fls. of the spikelet abortive.—* Fls. in unilateral spikes...(z).....
         —* Flowers paniculate...(k)......................................Tribe 7
   7 Lower flower of the spikelet abortive...(8)
      8 Pales coriaceous, firmer in texture than the glumes. Paniculate...(g)...........Tribe 5
      8 Pales membranous, thinner than the glumes. Spicate...(bb).....................Tribe 11
         9 Flowers in 2- or 4-rowed,—* equilateral spikes...(c)..........................Tribe 6
           —* unilateral spikes...(x).................................Tribe 10
      9 Flowers in panicles more or less diffuse...(10)
         10 Pale awned at the tip or awnless...(u).......................................Tribe 8
         10 Pale awned on the back or below the tip...(k)................................Tribe 7

1. ORYZEÆ. (*Spikelets 1-flowered, panicled. Glumes obsolete. Stamens 1—6.*)
  a Flowers perfect, flattened laterally, awnless.—Glumes 0. Stam. 2 or 3. *Cut Grass*....LEERSIA. 1
                           —Glumes minute. Stamens 6. *Rice*.......ORYZA. 2
  a Flowers monœcious, both kinds in the same panicle. Stamens 6. *Indian Rice*.......ZIZANIA. 3
  a Flowers monœcious, each kind in separate panicles. Stamens 5—12. S.........LUZIOLA. 4

2. AGROSTIDEÆ. (*Spikelets 1-flowered, panicled. Glumes and pales thin. Grain free.*)
  b Flowers surrounded at base with a tuft of long, silky hairs..................CALAMAGROSTIS. 10
  b Flowers naked or thinly bearded at base...(c)
    c Glumes both long-awned and longer than the awned pales................POLYPOGON. 9
    c Glumes both awn-pointed (or minute and the pale awned)...............MUHLENBERGIA. 8
    c Glumes awnless, conspicuous...(d)
      d Pale stalked in the glumes, awned on the back, monandrous. *Sweet Reed*......CINNA. 7
      d Pale sess. in the glumes, 3-androus,—acute, awnless. Glumes shorter....SPOROBOLUS. 6
                      —obtuse, often awned on back. *Bent G.*..AGROSTIS. 5

3. PHLEOIDEÆ.—*e* Glumes united at base, awnless. Pale 1, awned................ALOPECURUS. 11
    —*e* Glumes distinct, mucronate. Pales 2, awnless. *Timothy*............PHLEUM. 12
    —*e* Glumes distinct, pointless. Pales 2, awnless........................CRYPSIS. 13

4. STIPACEÆ.—*f* Awn of the flower simple, straight, deciduous.................ORYZOPSIS. 16
  —*f* Awn of the flower simple, twisted, very long........................STIPA. 15
  —*f* Awn of the flower triple or 3-parted. *Poverty Grass*................ARISTIDA. 14

5. PANICEÆ. (*Spikelets 2-flwd., lower flower abortive. Glumes very unequal. g Pale coriaceous.*)
  g Spikelet apparently 1-flowered, the lower glume wanting and the single abortive pale
    supplying its place.—Flowers spicate, unilateral ....PASPALUM. 17
               —Flowers diffusely panicled, all alike. *Millet Grass*........MILIUM. 18
               —Flowers paniculate, 2 sorts, one under ground........AMPHICARPUM. 19
  g Spikelet evidently 2-flowered, both glumes present, abortive flower neutral or ♂ ...(h)
    h Flowers paniculate,—without awns or spines. Pale cartilaginous. *Panic G.*....PANICUM. 20
            —without awns or spines. Pales herbaceous............PENICILLARIA. 21
            —with the glumes and pale coarsely awned. *Cock-spur*..OPLISMENUS. 22
    h Flowers spike-panicled,—each with an invol. of awned pedicels. *Fox-tail*......SETARIA. 23
           —each with a hardened, burr-like invol. *Burr Grass*...CENCHRUS 24

6. PHALARIDEÆ.—*i* Sterile flowers 2 minute rudiments. Panicle spicate............PHALARIS 25
    —*i* Sterile flowers 2 awned pales. Panicle spicate............ANTHOXANTHUM 26
    —*i* Sterile flowers both 2-valved, ♂. Panicle open.........HIEROCHLOA. 27

7. AVENEÆ. (*Spikelets 2 - ∞ -flowered, panicled. Glumes large. Pale awned below the tip.*)
  k Spikelet with 1 perfect flower and 1 awned staminate flower—above. *Soft Grass*...HOLCUS. 28
                  —below. .....§ ARRHENATHERUM. 31
  k Spikelet with definitely 2 perfect flowers. Pale subentire, awn dorsal............AIRA. 29
  k Spikelet with 2 or more perfect flowers. Pale 2 toothed at apex...(m)

382   ORDER 155.—GRAMINEÆ.

 m Awn between the two teeth, twisted ; glumes very large .... DANTHONIA. 30
 m Awn dorsal below the middle (except in the cultivated Oat). *Oat.*................AVENA. 31
 m Awn dorsal above the middle.—Flowers 2—5. Teeth cuspidate.................TRISETUM. 32
    —Flowers 5— ∞. Teeth acutish. *Brome.*........BROMUS. 33
l. **FESTUCACEÆ.** (*Spikelets 2 - ∞-flowered, panicled, awnless, or the lower pale tipped with a straight bristle or awn. Glumes 2.*)
 n Glumes definitely 2, all the lower flowers of the spikelet perfect...(*o*)
 n Glumes several, indefinite, the lower flowers abortive and glume-like...(*p*)
  o Flowers fringe-bearded at the base. Pales 3-cuspidate or entire...(*q*)
  o Flowers beardless. Lower pale mucronate or awn-pointed (except in one Festuca)...(*r*)
  o Flowers beardless. Lower pale obtuse or acute, not at all awned...(*s*)
   q Lower pale 2- or 3-cuspidate and 1-2-awned. Upper pale entire....................TRICUSPIS. 34
   q Lower pale 2-cuspidate and 1-awned. Upper pale entire. 8—12f.....................ARUNDO. 35
   q Lower and upper pale both entire and pointless at apex......................GRAPHEPHORUM. 36
   q Lower pale long-pointed, *white* as well as the glumes and hair. *Pampas Grass*.....GYNERIUM. 37
  r Glumes and pales keeled,—herbaceous, 5-veined. Flowers glomerate...........DACTYLIS. 38
    —membranous, 3-veined. Panicle spicate.............KOELERIA. 39
  r Glumes and pales rounded on the back,—both coriaceous. Grain free........DIARRHENA. 40
    —pale papery, grain adherent. *Fescue.*....FESTUCA. 41
 s Spikelets 2-3-flowered, with some abortive terminal flowers. Pale papery, not keeled...(*t*)
  t Upper glume broad-obovate, shorter than the flower. ....EATONIA. 42
  t Upper glume oblong, 7-9-veined, longer than the flowers. *Melic.*............ ....MELICA. 43
 s Spikelets 2-50-flowered, all perfect. Pales usually thin...(*u*)
  u Lower pale keeled, 3-veined, membranous like the glumes... ....ERAGROSTIS. 44
  u Lower pale keeled, 5-veined, usually cobwebbed at base. *Spear Grass*................POA. 45
  u Lower pale convex-keeled, obscurely 9-veined. Panicle spiked.............BRIZOPYRUM. 46
  u Lower pale convex, 7-(—5)-veined, never webbed at base. *Manna*..............GLYCERIA. 47
  u Lower pale convex-ventricous, cordate, obscurely veined. *Quake*..................BRIZA. 48
  p Herbaceous.—Flowers glabrous, awnless, falcate-pointed.... ....UNIOLA. 49
   —Flowers silky-villous at base. Tall, stout. *Reed.*..........PHRAGMITES. 50
  p Woody, tall (the flowering branches low). Flowers short-awned........ARUNDINARIA. 51
9. **HORDEACE.E.** (*Spikelets 1-10-flowered, sessile, alternate in a spike. Rachis jointed.*)
 v Spikes several. Spikelet solitary at each joint, 1-flowered.............................LEPTURUS. 52
 v Spike single.—Spikelets 1-flowered, 3 at each joint. *Barley*......................HORDEUM. 53
  —Spikelets 2 - ∞- flowered,—several at each joint. *Hedgehog*...........ELYMUS. 54
   —1 at each joint...(*w*)
  w Glume 1, in front of the spikelet which is edgewise to the rachis. *Darnel.*........LOLIUM. 55
  w Glumes 2, opposite.—Spikelet 3 - ∞- flowered. *Witch G. Wheat* ....TRITICUM. 56
   —Spikelet 2-flowered. *Rye.*..................................... ....SECALE. 57
10. **CHLORIDEÆ.** (*Spikelets in 1-sided jointless spikes, 1 - ∞ flowered. Upper flower abortive.*)
 x Spikes very slender, many, in an equilateral raceme...(*y*)
  y Spikes raceme-like. Spikelets with several perfect flowers................. LEPTOCHLOA. 58
  y Spikes with sessile, 2-flowered spikelets, one flower a rudiment............GYMNOPOGON. 59
 x Spikes slender, several, digitately arranged above, or, in No. 60, axillary...(*z*)
  z Spikelets with 1 perfect flower,—awnless, globular, no rudiment.............MANISURUS. 60
   —awnless, oblong, with a rudiment...............CYNODON. 61
   —awned, glume 3-lobed.................EUSTACHYS. 62
   —awned, glume acute....................... ....CHLORIS. 63
  z Spikelets with several perfect flowers.—Flowers awnless.......................ELEUSINE. 64
   —Flowers awned...........DACTYLOCTENIUM. 65
 x Spikes thick and dense, 1 — ∞. Spikelets with 1 perfect flower...(*aa*)
  aa Spikes several or many. Flower with no rudiment...............................SPARTINA. 66
  aa Spikes 1, few, or many. Flower with a terminal rudiment.........BOUTELOUA. 67
  aa Spike solitary, recurved. Awns terminal and dorsal................................CTENIUM. 68
11. **BACCHARIEÆ.** (*Spikelets in pairs or 3's, 2-flowered, the lower flower abortive. Fertile pales thinner than the glumes, except in No. 72.*)
 bb Flowers (the fertile) imbedded in the cavities of glabrous, jointed spikes...(*cc*)
  cc Spikes monoecious, *g* abortive, ♀ below, both naked. *Sesame*...................TRIPSACUM. 69
  cc Spikes monoecious *g* above panicled, ♀ below enveloped in *husks. Maize*..............ZEA. 70
  cc Spikes uniform,—terete. The pedunculate spikelet abortive................ROTTBOELLIA. 71
   —compressed. Both spikelets fertile........... .............STENOTAPHRUM. 72

Order 155.—GRAMINEÆ.

bb Flowers not imbedded, spicate or panicled, mostly long-bearded...(*dd*)
   dd Both spikelets of each pair fertile.—Lower flower awned. *Plume G*..........ERIANTHUS. 73
       —Flowers awnless. *Sugar-cane*............SACCHARUM. 74
   dd Only one spikelet of each pair fertile.—Fls. and rachis hairy. *Beard G*.....ANDROPOGON. 75
       —Flowers and rachis smoothish............SORGHUM. 76
   dd The lower spikelet on each spike fertile, in a bony shell. *Job's-tears*..............COIX. 77

1. LEÉRSIA, Sol. CUT GRASS. FALSE RICE.

Spikelets 1-flwd., flat, fls. ☿. Glumes 0. Pales boat-form, nearly equal, awnless, ciliate, enclosing the free flat grain (caryopsis). ♃ Swampy grasses. Lvs. very rough backward. Fl. in secund panicled racemes. June, Aug.

    * Panicle compound, large, diffuse. Spikelets nearly 3″ long .... Nos. 1, 2
    * Panicle simple or nearly so. Spikelets scarce more than 1″ .... Nos. 3, 4

1 L. oryzoïdes Swtz. (*a*) Spikelets narrowly elliptic, spreading, white, close (*b*); stamens 3; culm 3–5f, retrorsely rough, lvs. broad. By streams. Aug.

2 L. lenticularis Mx. *Catch-fly Grass*. Spkl. round-oval (*c*) when closed, closely imbricated; stam. 2 (*d*); ovary ovate (*e*); plant smoothish. Ponds and low grounds, Ill. to Va., and S.: rare. Fls. said to close on flies.

3 L. Virginica Willd. Spkl. small, closely appressed to the branchlet; stam. 2, pales white, with green veins, slightly ciliate. Wet shades. Aug.

4 L. hexandra Swtz. Panicle erect, narrow, exserted, 2–4′; spkl. loosely imbricated, lance-oblong; stam. 6. Culms branched, 1–5f. Water. Fla.

2. ORŸZA, L. RICE.

Spikelets 1-flwd., ☿ Glumes minute or obsolete, pales compressed-boat-shaped, the lower larger and usually awned. Stamens 6. Grain oblong, smooth, free in the pales. ① Fls. paniculate.

O. SATÍVA. Culm 2–4f, lvs. broadly linear, the ligule 1′ long. A most important cereal, cultivated South in meadows and inundated grounds.

3. ZIZÀNIA, Gron. INDIAN RICE.

Stout water-grasses, with large monœcious panicles. Glumes 0. Pales 2, thin, narrow, the lower one with a straight awn in the ♀. Stam. 6 in the ♂ (*b*).

1 Z. aquatica L. Panicle ample, 1–2f, the lower branches spreading, sterile (*a*), upper fertile; awns (*d*) long (1¼′); grain slender, 6–8″, very caducous, farinaceous. Marshes, Aug. Culm 5–8f. Lvs. broad.

2 Z. miliacea Mx. Sterile and fertile fls. intermixed in the ample panicle; pales with short (1–3″) awns. Culm 6–10f. Leaves narrow. Ohio, and S.

4. LUZÌOLA, Juss.

Spikelets and fls. as in Zizània, but the ♂ and ♀ in separate panicles on the same root. Sta. 5—11, anth. very long. Grain ovoid. ♃ Aquatic, with long narrow leaves.

L. Alabaménsis Chapm. Culms 4–6′, 1-lvd., the leaf 1–2f long, its purple sheath enclosing the bract and peduncle; panicle few-flowered; spikelet lance-ovate, on erect jointed pedicels. Alabama: rare.

ORDER 155.—GRAMINEÆ.

5. AGRÓSTIS, L. BENT GRASS. Spikelets 1-flwd. Glumes 2, subequal, awnless, usually longer than the flower. Paleæ 2, thin, pointless, naked, the lower 3–5-veined, sometimes awned on the back, the upper often minute or wanting. Grain free. Mostly ♃, cœspitous, with slender culms and open panicles.

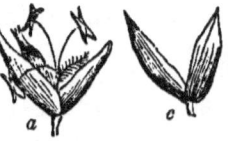

§ AGRÓSTIS. Upper pale conspicuous. Panicle
    rather dense..........................Nos. 1, 2
§ TRICHÒDIUM. Upper pale minute or wanting.
    Panicle thin, capillary...(*)
    * Lower palea with a long exserted awn on the back..................Nos. 3, 4
    * Lower palea awnless, or bearing a very short awn..................Nos. 5, 6

1 **A. vulgàris** With. *Red-top (a).* Culm erect, 1–2f; pan. purple, oblong, with short branches; ligules very short; lower pale (*b*) 3-veined, twice longer than the upper, nearly awnless. A valuable grass: common.

2 **A. alba** L. *Florin G.* Culm decumbent and rooting at the lower joints, then ascending 1–3f, stoloniferous; ligules long (3–4''); pan. greenish-white, or purplish, contracted; pale 5-veined, awned or not. Common.

3 **A. canina** L. *Dog's* or *Brown B.* Decumbent and rooting at base, 1–2f; leaves setaceous-rolled; pan. brownish; lower pale and awn exserted. Wet meadows. E. §
    β. *alpina.* Culms low, in tufts, with wide panicles, and twisted awns. Mts.

4 **A. arachnoides** Ell. Erect, 5–8'. pan. ⅓ its length; lvs. linear-setaceous; lower pale, ''', its awn as fine as a gossamer, twisted, 3–4'' long. S. C., Ga., and W. Apr.

5 **A. scàbra** Willd. *Rough Hair G.* Erect from a decumbent base, 1–2f, very slender, all *scabrous-hispid ;* pan. large, capillary, spkl. purplish, (*c*, glumes, *d*, flower). The thin, airy panicles are at length driven before the wind. Fields and pastures. June–Aug.
    β. *perennans.* Panicle pale-green, the branches shorter. In damp shades.
    γ. *oreóphila.* Pan. less diffuse; lower pale with a short twisted awn. Mts.

6 **A. elàta** Trin. Culms stoutish, simple, erect, 2–3f; lvs. broadly linear; pan. purple, with long suberect whorled branches dense-flowered half their length; gls. 1¼'' long, lower pale 5-veined, 1''. Swamps, N. J. to Ky., and S. Sept., Oct.

6. SPORÓBOLUS, Br. DROP-SEED GRASS. Spikelets 1-flwd. Gls. 2, the lower smaller. Fls. sessile. Paleæ 2, awnless, usually longer than the glumes. Sta. 2 or 3. Grain deciduous, free. ♃ Tough, wiry, with rolled rigid leaves and contracted panicles often half-enclosed in the sheath.

§ VILFA. Grain (caryopsis) linear. Glumes and pales all
    sub-equal. Panicle contracted................Nos 1–3
§ SPORÓBOLUS. Grain oval or globous, its pericarp often
    loose on the seed...(*a*)
    a Glumes very unequal, one of them as long as the purplish pales..........Nos. 4–6
    a Glumes equal or unequal, both shorter than the pales. Sheaths beardless...(*b*)
        b Panicle contracted, spikeform, sheathed or exserted. Lvs. involute...Nos. 7, 8
        b Panicle capillary, open. Often a 2d flower or rudiment. Lvs. flat...Nos. 9, 10

1 **S. vagínæflòrus** Torr. (*a*.) Culms in tufts, simple, ascending, 6–12'; lvs 2–4'; panicles lateral and terminal, mostly concealed in the tumid sheaths; grain ⅓ shorter than the 2'' pales. ① Dry gravel. More common W. and S.

Order 155.—GRAMINEÆ.

2 S. Virgínicus (L.) Like No. 1, but the root is ♃, the culms branched, often decumbent, and the spikelets very small (1″) and many. Coast, S. Sept., Oct.
3 S. cuspidàta (Torr.) Glumes very acute, the lower pale *cuspidate*; pan. terminal, slender, few-flowered; spikelet nearly 2″. ♃ Maine, and Canada.
4 S. cryptándrus (Torr.) Culm 2—3f; sheaths strongly bearded at the throat; terminal panicle pyramidal, exserted, the lateral concealed; pales equaling the upper glume (1″), twice longer than the lower. ♃ Sandy coasts and shores. Aug.
5 S. júnceus (Mx.) Glaucous, erect, 1—2f; leaves erect, 2—6′ by 1″; pan. open, stalked, narrow, loose; glumes ovate, obtuse, the upper 1½″, lower ⅓″, anth. and stig. white. ♃ Common in dry barrens, Penn., W., and S. No lateral pan. Aug.-Oct.
6 S. heterólepis (Gr.) Lowest lvs. as long as the culm, 1—2f; upper gl. 3″, subulate, longer, lower cuspidate, shorter than the pales; panicle very thin, stalked, open; grain globular, 1″. Dry places, Conn. to Wis. Aug.
7 S. asper Kunth. (c) Lowest lvs. very long (1—3f), involute-filiform; culms 1—2f; panicle contracted, partly or wholly enclosed; glumes unequal, white, much shorter than the oblong obtuse pales (3″); grain oval. Sands. Sept.
8 S. Indicus Br. Erect, 2—3f; pan. long (1f), very narrow, its short branches appressed; glumes unequal; grain oval. Dry grounds, S.: common. May—Sept.
9 S. compréssus Kunth. Culm erect, 1—2f, leafy, much *compressed*, branched at base; pan. thin, 6—10′; gl. acute, ⅓″; pales 1″, obtuse. Sandy bogs, N. J. Sept.
10 S. serótinus (Torr.) Culm filiform, compressed, 10—18′, few-lvd.; pan. capillary, diffuse; glumes ⅓″, ovate, obtuse; pales ⅓″. Wet sands, Maine to N. J. Sept.

7. CINNA, L. Sweet Reed-grass.
Spkl. 1-flwd., flat. Gl. 2, subequal, awnless, the upper a little longer than the subequal pales, which are short-stiped. Lower pale with a short awn on the back. Sta. 1. Grain oblong, free. ♃ Erect, tall and simple, with a large panicle, green or slightly purplish. July, Aug.

1 C. péndula Trin. (a) Culm 3—5f; lvs. broad-linear, with conspicuous ligules; pan. pale-green, 1f, nodding, with its drooping branches in whorls of 4's or 5's; awn exserted. A fine grass in damp woods, much sought by cattle.
2 C. arundinàcea Willd. Bright green, 3—6f; pan. erect, green-purple, 10′; lower pale obtuse, its awn not exceeding its obtuse point. Handsomer than No. 1, its spikelets twice larger (¾″). Shady woods.

8. MUHLENBÉRGIA, Schr. Drop-seed Grass.
Spkl. 1-flwd. Glumes persistent, bristle-pointed or acute, rarely obtuse. Pales sessile, usually hairy at base, deciduous with the enclosed grain, green, the lower awned or mucronate at apex. Sta. 2—3. Culms often branched. July—Sept.

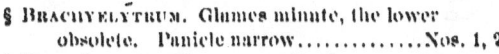

§ Brachyelytrum. Glumes minute, the lower
  obsolete. Panicle narrow..............Nos. 1, 2
§ Trichochloa. Glumes small. Lower pale
  3-veined. Panicle capillary... ....Nos. 3, 4
§ Muhlenbergia. Gl. manifest. Pale 3-veined. Pan. terminal and axillary....(a)
  a Glumes awned and twice longer than the awnless pale............. ....No. 5
  a Glumes pointed, not longer than—the mucronate pale .... .Nos. 6, 7
      — the long-awned pale.............. ..Nos. 8, 9

1 M. aristàta Pers. Erect, simple, 1—2f; lvs. broad-linear; pan. terminal, simple,

ORDER 155.—GRAMINEÆ.

3—4'; spkl. large, few; lower pale 6" (12—18" with its awn), 5-veined; upper pale, with an abortive pedicel in the groove of its back; sta. 2. ⚥ Rocky hills.

2 M. diffūsa Schr. (*d*) Decumbent, diffuse, branching, 8—18', lvs. 2—3'; panicles very slender, terminal and lateral; spikelets 2" (4" with its awn), white with green spots; glumes (*g*) extremely minute, white. Shady places: frequent.

3 M. capillāris Kunth. *Hair G.* Erect, very slender, 1½—3f, simple; pan. purple, large, diffuse, branches 1—4', as fine as hairs; pales long-awned. Dry soils.

4 M. trichópodes (Ell.) Panicle erect, oblong, not diffuse, green; lower pale tipped with a short awn. Culms 3f, leaves flat. Pine barrens, S. (Agrostis, Ell.)

5 M. glomerāta Trin. Glaucous, erect, subsimple, 1½—3f, lvs. 3—5'; pan. spike-like, dense, interrupted, 2—3'; glumes 2", pales 1". Bogs, northward.

6 M. Mexicāna Trin. (*a*) Culms much branched, ascending 2—3—5f; leaves lance linear; pan. many, the lateral half-sheathed, dense, and narrow; glumes and pales subequal (1") or one glume longer. Damp shades: common.

β. *purphrea.* Culms wiry, branched only at base; panicle purple. Ill. *J. Wolf.*

7 M. sobolífera (Muhl.) (*b*) Like the last, but the panicles are more slender, or filiform, and the glumes shorter than the pales. Hardly distinct. Woods.

8 M. sylvática T. & G. (*s*) Culms ascending, branched, diffuse, 2—3f; pan. slender, rather dense; glumes subequal, scarce shorter than the lower pale (1"), whose awn is 2—4". Rocky shades, N. England to N. J., and W. (Agrostis, Muhl.)

β.? *tulpina.* Very glaucous; pan. very dense, raceme-like; glumes abruptly short-awned: *pale about as long as its awn.* N. Y. *H. B. Lord.*

9 M. Willdenōvii Trin. (*w*) Culm and leaves as in the last; pan. very slender, loose-flowered; glume bristle-pointed, ‡ shorter than the pale, whose awn is 3—4 times as long as the spikelet. Rocky woods: com.

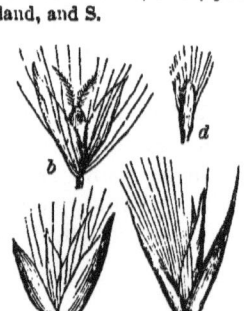

9. POLYPŌGON, Desf. POLYPOG G. Spkl.
1-flwd., densely panicled. Glumes subequal, similarly awned, much longer than the flower (*c*). Lower pale usually awned near the tip. Stam. 3. Grain free.

P. Monspelléŋsis Desf. (*a*) Culm simple, 1f or more; lvs. lance-linear, 2—5'; pan. spike-like, 2—3', pale; gl. (*b*) 1", their awns 2". N. England, and S.

10. CALAMAGRÓSTIS, Adans. Spkl. 1-flwd. Glumes subequal, acute or pointed. Pales bearded at the base, lower one mucronate, mostly awned below the tip, upper often with an abortive rudiment of a second flower. ⚥ Culms simple, tall, paniculate, from creeping rhizomes.

§ CALAMAGRÓSTIS. No rudiment. Panicle expanding, loose. Pales awnless. .....Nos. 1, 2
§ DEYÉUXIA. Rudiment a hairy pedicel. Lower pale awned. Spikelet 2—3"...(*a*)
§ AMMÓPHILA. Rudiment plumous. Panicle spike-form. Spikelet 6"..................No. 10
a Beard nearly equaling the pales. Panicle rather open ....Nos. 3, 4
a Beard nearly equaling the pales. Pan. contracted....................Nos. 5—7
a Beard much shorter than the pales. Awn from near the base.........Nos. 8, 9

1 C. brevipilis (Torr.) Slender, 3—4f; leaves broad-linear, flat; pan. purple, with

Order 155.—GRAMINEÆ.

capillary branches; gl. unequal, shorter than the pales; beard very short, not half the length of the pales. ♃ Sandy swamps, N. J.: rare. Sept.

2 C. longifòlia Hook. Stout, 2–4f; lvs. rigid, involute, long-filiform-pointed; upper glume as long as the pales; ha'rs half as long. Shores of the great lakes. Aug.

3 C. Canadénsis Beauv. (c) *Blue-joint*. Rigidly erect, 3–5f; leaves flat; panicle oblong, its branches in 4's and 5's; gl. longer (1½″) than the pales, purplish; awn from the middle of the pale, as fine as the long beard. A good grass: common N. July.

4 C. Langsdórfii Trin. Spikelets 2½″ long; awn stouter than the soft beard. Otherwise like No. 3. White Mts., N. H., Isle Royal, L. Sup. (*Porter*). August.

5 C. confínis Nutt. (a) Lvs. flat, panicle narrow, dense, reddish; gl. ovate, 2″, equaling the flower (b); beard ⅓ shorter than the pales; awn from below the middle, not exserted. Culm 2–5f. Penn. (*Jackson*), Penn Yan, N. Y. (*Sartwell*). July.

6 C. strícta Trin. Differs from No. 5 only in its rigid leaves rolled at the point, its awn from below the middle, its beard as long as the pales. Lakes, N. Aug.

7 C. Nuttallìàna Steud. Lvs. flat; pan. dense; glumes 3″, long-pointed, ⅓ longer than the pales; awn from near the tip of the pale; beard some shorter than the pale. Swamps, Mass. to N. Car. (*C. coarctàta* Torr.) Aug.

8 C. purpuráscens Br. Culm 1–1½f; pan. spike-like, 3–7′, purplish; gls. rather obtuse, less than 2″; beard scanty, short, ⅓ as long as the rudiment, ⅓ as long as the pales; awn short, straight. White Mountains, N. H., Mt. Marcy, N. Y. (*Peck*.)

9 C. Pórteri Gr. Slender, 2–4f; lvs. flat; pan. very narrow, 4–6′; glumes fully 2″, exceeding the pales; hairs few, short, almost none at the base of the lower pale; awn contorted. Huntingdon Co., Penn. (*Porter*). July.

10 C. arenària Roth. *Sand Reed*. Rhizomes creeping extensively, culms stout, erect, 2–4f; lvs. rolled and rush-like; pan. spike-form, with erect appressed branches 6–10′; spkl. very flat. Sandy beaches, northward. August.

11. ALOPECÙRUS, L. Fox-tail G. Spikelets 1-flwd. Gl. flat-keeled, connate at base, subequal. Upper pale 0, lower flat-keeled, awned on the back below the middle. Sta. 3. Panicle contracted into a cylindric dense spike.

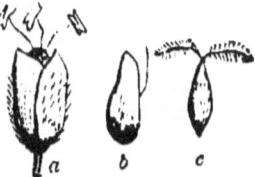

1 A. aristulàtus Mx. *Wild F.* Ascending from a bent base, 1–2f, glaucous; spike slender, 1–2′ by 2½″, grayish; glumes (a) and pale obtuse, equal; awn (b) scarcely exserted (c, ovary and stigmas). In wet places. June—August.

2 A. geniculàtus L. *Bent F.* Ascending from a bent base, 1–2f; spike 2–2½′; upper leaf scarce longer than its sheath; glumes pubescent, obtuse; awn geniculate far surpassing the culm. Wet meadows, East. §

3 A. praténse L. *Meadow F.* Erect, stout, 1½–2½f; spike about 2′; upper leaf shorter than its sheath; gl. ciliate; awn twisted, nearly thrice longer than its pale. Fields and pastures, Northern States. A good grass. §

12. PHLEUM, L. Cat-tail G. Glumes equal, flat-keeled, mucronate or rostrate, longer than the truncate awnless pales. Compound spike cylindric and very dense. June, July.

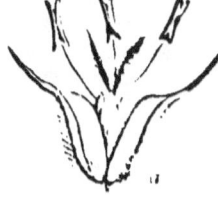

1 P. praténse L. *Timothy. Herd's G.* (a) Erect, rigid, 2–4f; lvs. broad-linear, flat; glumes alike cuspidate, in a long dense terete green spike. A grass of the highest value for hay in the North, but will not flourish South.

2 P. alpìnum L. Erect, 1f; lvs. shorter than the sheaths; spike oblong-ovoid, 4–5′ long; awns as long as their glumes. White Mountains, and Arctic Am.

13. CRŶPSIS, Ait. Compound spk. oblong, many-bracted and sheathed

at base. Glumes and pales awnless, subequal, of similar texture. Grain glabrous, free. Turfy grasses, none native.
C. schenoídes Lam. Tufted, glaucous, 3–12'; lvs. 2–3', long-pntd.; spk. oblong. ① Waste ground, E. Penn., Del., etc. § Eur.

14. **ORYZÓPSIS**, Mx. MOUNTAIN RICE. Spkl. 1-flwd. in a slender spicate panicle. Gl. membranous at edge, subequal, about equaling the oblong, terete, short-stiped flower. Lower pale coriaceous, involute, enclosing the grain, and tipped with a simple, jointed awn. ♃

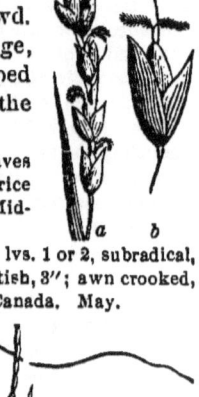

1 **O. melanocárpa** Muhl. Culm leafy to the top, 1–1½f; leaves lance-linear; rachis flexuous; few-flwd.; gl. 5–6''; awn thrice longer (1') than its blackish pale. ♃ Rocky woods and hills, Middle States, and northward. Aug.

2 **O. asperefòlia** Mx. (*a*) Culm 10–20', its sheaths leafless; lvs. 1 or 2, subradical, erect, rigid, pungent, 1f; the simple pan. 2–4' long; gl. (*b*) whitish, 3''; awn crooked, 6'' long, its pale and grain whitish. ♃ Woods, N. States and Canada. May.

3 **O. Canadénsis** (Poir.) Culm slender, 9–18', naked above; lower sheaths bearing rigid, involute-filiform leaves; pan. 1–2'; awn short or 0. Rocks. N. May.

15. **STIPA**, L. FEATHER G. The flower deciduous from the glumes with its sharp and bearded stipe. Pales coriaceous, short, the lower embracing the upper and the slender grain, and bearing a long twisted or bent awn. ♃ Leaves narrow. Pan. loose. (See *Addenda*.)

1 **S. avenàcea** L. *Black Oat-G.* (*c*) Culm naked above, 2–3f; lvs. mostly radical, setaceous; pan. 4–6' long, the capillary branches at length diffuse; gl. (*a*) equaling the blackish fruit; awn (*b*) 2–3' long, twisted below, bent: common. July.

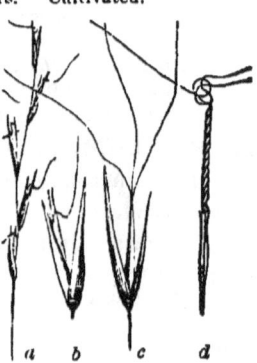

2 **S. júncea** Pursh. *Weather G.* Culm 2–3f; leaves rolled-threadform, long; glume slender-pointed, twice longer than the fruit; awn contorted, bent, 4–6' long. The pungent stipe adheres like tick-seed. Prairies, Ill., Mo., and N. May—July.

3 **S.** PENNÀTA. *Feather G.* From Europe. Culm 2f; lvs. rolled threadform at apex; gl. awn-pointed, 1'; awn 8–16' long, twisted below, softly plumous above, and "worn (says *Gerard*) by sundry ladies instead of feathers." Cultivated.

16. **ARISTÌDA**, L. BEARD G. POVERTY G. Panicle contracted and racemous. The flower stiped in the unequal glumes. Lower pale with 3 awns at the tip variously contorted.—Culms branching. Leaves narrow, often rolled. In sterile soils. Aug.—Oct.

§ Awns twisted and united below, jointed to
    the pale, very long .... No. 1
§ Awns distinct to the base and not jointed to
    the pale...(*a*)
  a Awns very unequal, the 2 lateral 4 times
    shorter (2'') and erect .... Nos. 2, 3
  a Awns unequal, the 2 lateral twice shorter
    (6'') and suberect .... No. 4

Order 155.—GRAMINEÆ.

a Awns about eq\[u\]al, spreading.—Lower gl. longer than the upper......Nos. 5—7
— Glumes equal, or the lower shorter....Nos. 8, 9

1 **A. tuberculòsa** N. Culm rigid, 8—20′, with *tubercles* in the axils of the numerous branches; pan. large and loose; glume linear, awned, 1′; triple awn (*d*) 2′, united half-way up, thence involved and spreading. ① Mountains, N. J., and W.

2 **A. dichótoma** Mx. (*a*) Culm 8—12′, *dichotomously* branched; gl. 3—4″; lateral awns erect, minute, the middle awn (*b*) as long as the pale (3″), twice bent to the form of a bayonet. ① Dry sandy fields: common.

3 **A. ramosíssima** Engelm. Culms *diffusely branched;* gl. 9—10″, awn-pointed; lateral awns 2′, middle awn 1′, spreading. ① Sands, Ill., Ky.

4 **A. grácilis** Ell. Very slender, ½—1½f; pan. virgate, 4—8′; glume and flower equal, (2½—5″); middle awn 9—10″, horizontal, the lateral erect. ① Sandy places.
  β. *cirgàta*. Taller (2—3f), pan. 1f; gl. and fl. shorter (2″). S. (*Chapman*).

5 **A. lanàta** Poir. Culms 2—4f, stout, branched from base; lvs. flat, with *woolly* sheaths; pan. 1—2f, woolly in its axils; upper glume, lower (purplish) pale and lateral awns each 4—5″, middle awn some longer. ♃ Sandy soils, S.

6 **A. spicifórmis** Ell. Culms 1—3f, rigid, simple; lvs. rolled, rigid, smooth; pan. *spike-form*, dense; flower 1′ long, awns as long, gl. much shorter. ♃ Wet sands, S.

7 **A. purpuráscens** Poir. (*c*) Culms slender, 2—3f; lvs. scarcely rolled; pan. 1f long, loosely spicate; glume and fl. 4—5″, *purplish;* awns 1′, spreading. ♃ Sandy.

8 **A. stricta** Mx. Culms 2—3f, strictly erect, with long rigid rolled lvs.; pan. loosely racemous, 1f; gl. 6—7‴, fl. 6″, lateral awns 7—9″, central 9—15″. ♃ Va., and S.

9 **A. oligántha** Mx. Culms 1—1½f, in tufts; raceme few-flowered; glume and fl. 9″, very slender, awns capillary, divaricate, 18—30″ long. Prairies, W. and S.

17. PASPALUM, L.

Spikelets plano-convex, in one-sided spikes. Glumes (apparently) 2, membranous, equal, ovate or orbicular, closely applied to the fertile flower. Grain coated with the smooth coriaceous pales. (But theoretically, the lower glume is obsolete, and its place supplied by the empty pale of an abortive flower. In Nos. 15—17 the lower glume appears, under a lens, as a mere rudiment.)—Spikes linear, the flowers in 2—4 rows.

§ Paspalum. Spikelets round or round-oval, obtuse. Spikes alternate...(*)
§ Digitaria. Spikelets ovate to lanceolate, acute. Spikes often digitate...(**)
  * Terminal spike mostly solitary, rarely 2, 1″ wide, long-stalked..........No. 1
  * Terminal spikes mostly 3 or 4, 2″ wide. Spikelets in 2 rows..........Nos. 2, 3
  * Terminal spikes mostly 4 or 5.—Spikelets close, in 3 or 4 rows.. ....Nos. 4—6
    — Spikelets in remote pairs ....No. 7
  ** Rachis leaf-like, broader than the spikelets. Spikes alternate .. ....Nos. 8, 9
  ** Rachis narrower than the spikelets. Spikes digitate...(*a*)
    a Glumes (gl. and pale) about equal, as long as the flower...(*b*)
    a Gls. (both conspicuous) one or both very short. Spks. 4—9. ...Nos. 16, 17
      b Spikes spreading, always two in Nos. 10, 11; two—six in..........No. 12
      b Spikes erect.—Rachis flat, spikelets by 1's or 2's, close......Nos. 13, 14
        — Rachis filiform, spikelets by 3's, loose ....No. 15

1 **P. setàceum** Mx. Culm slender, ascending, 1—2f, naked above; lvs. linear, flat, 2—3″ wide, soft, hairy; spikes very slender, 2—4′, 1 or 2 on the long peduncle, often

a sheatned axillary one below; spikelets small, ½″, *in pairs*, but seeming 2-rowed, very smooth ♃ Dry or wet, Mass. to Ill., and S. Aug.

2 **P. laeve** Mx. (*p*) Cu¹m erect, 1½—3f; lvs. broad-linear, hairy at base or smooth; spikes 3—5; spikelets (*a, b*) *single*, contiguous, in 2 rows on the narrow straightish rachis, round and smooth, 1½″. ♃ Grassy banks, Ct. to Ind., and S. Aug.
  β. *altissimum*. Strict, tall; sheaths flattened close on the spikes.

3 **P. angustifòlium** Le Cont. Culm. wiry, 2—3f; lvs. linear-filiform, compressed-carinate; spikes 2 or 3, 1—2″; rachis narrow, flexuous; spikelets round-oval, brown, 1″, in 2 rows. Whole plant glabrous. ♃ Wet places. Ga., Fla., La. [and S.
  β. *tenue*. Spikes 4 or 5, very slender, 3-rowed; lvs. and sheaths ciliate. N. J.,

4 **P. praecox** Walt. Culm erect, 3—4f; lvs. long, narrow, smooth; sheaths purple, smooth or hairy; spikes 3—6, bearded at base, dense; rachis straight and flat; spkl. orbicular, in 3 rows, often brown. ♃ Swamps, S. May, June.

5 **P. dasyphýllum** Ell. Culm rigid, erect, 2—3f; lvs. linear, and with the sheaths hairy all over; spikes 2—5, large, 2—4′; spkl. orbicular-oval, near 2″, in 2 or 3 rows under the very flexuous rachis. ♃ Dry fields, S. July—Oct.
  β. *Floridànum*. Lvs. long and narrow; spikelets in 3 rows. Damp, S.

6 **P. virgàtum** L. Culm 1½—3f; lvs. broad-linear, ciliate near the base; spikes 3—12, 2—4′; rachis broad, but narrower than the 3—4 rows of small (1″) roundish spikelets; glume 3-veined. ① Moist soils, S. July—Oct.
  β. *undulàtum*. Upper glume (pale) undulate-rugous at edge.
  γ. *latifòlium*. Lvs. very broad (6—9″); spikelets larger (1½″).

7 **P. racemulòsum** N. Culm erect, firm, 2—3f; lvs. long, linear, soft-hairy; spikes 3—4, raceme-like, 2—6′; spkl. oval, in remote pairs, 1½″, the glume 5-veined, tawny. ♃ Dry soils, S. (*P. interruptum* C-B.) Aug., Sept.

8 **P. fluitans** K. Culms floating or ascending, 12—20′; lvs. lance-linear, on open sheaths; spikes 20—50, 1—2½′, rachis 1″ wide, flat, pointed, out-running the minute white spikelets beneath them. ① River swamps, Ill. to Va., and S. Oct.

9 **P. Walterìànum** Schlt. Culm and lvs. as in *P. fluitans*. Spikes 3—5, 2—3′, partly sheathed; rachis not out-running the white (1″) spkl. Wet, N. J., and S. Jl.+

10 **P. Digitària** Poir. Assurgent, 1—2½f; lvs. broad-linear, flat, on long sheaths; spikes slender, 3—5′, a pair at top of the long ped. and some axillary sheathed below; spkl. lanceolate, rachis flattened vertically. ♃ Woods, Va., and S. Jl.—Sept.

11 **P. conjugàtum** Berg. Erect, 1—2f; lvs. short (2—4′); spikes a pair at top, (rarely axillary), very slender, 3′; spikelets minute, white, ovate. ① N. Orl. §

12 **P. glabrum** (Gaud.) Culms decumbent, spreading, 8—15′; lvs. short; spikes 2—4, spreading, 1—2′, slender; spkl. ovate, purple, ½″, 2-rowed; upper gl. equaling the fl., lower minute. ① Sandy fields, N. J., and S. § (*P. ambiguum*, DC.) Aug. +

3 **P. dístichum** L. Culms assurgent, 12—18′; lvs. broad-linear; spikes 2 or 3, erect, near the top, 1½—2½′; rachis linear, narrower than the 2 or 3 rows of whitish ovate 1½″ spikelets. ♃ Wet grounds, S. States. Plant smoothish. July, Aug.

14 **P. trístichum** Le C. Culm ascending, 1—2f; peduncles from the upper joint, 1—3, filiform, each bearing 3 filiform suberect spikes; spkl. whitish, lance-ovate, minute; rachis flexuous. Wet places. ♃ Ga., Fla., to La. Aug.

15 **P. filifórme** Swtz. Culm filiform, erect, 1—1½f; lvs. short; spikes 2—6, filiform, erect; rachis filiform; spkl. oblong, ½″, in 3's; lower glume obsolete, upper as long as the flower. Dry soils, ① Ms. to Ky., and S.

16 **P. serótinum** Figg. Decumbent, rooting, hairy-villous; lvs. short (1—2)′, lance-linear; branches each with 3—5 filiform digitate spikes; rachis straight; spkl. lance-ovate, striate, minute. ♃ Sandy fields, S. C. to La. Sept., Oct.

17 **P. sanguinàle** Lam. *Crab* or *Finger G.* (*d*) Erect, 1—2f, lvs. and sheaths oftener hairy; spikes 5—9, digitate, spreading, 4—6′; rachis flexuous; spkl. (*c*) oblong-lanceolate, 1½″, upper gl. (*c*) ½ as long as the flower, (*e*) lower one minute. ① Waste grounds Aug.—Oct. §

18. MÍLLIUM, L. Millet G. Spikelets awnless, consisting of 2 coriaceous pales enclosed in apparently 2 glumes, which are longer. (But theoretically the glumes are as in Paspalum.) Sta. 3. Grain coated by the pales. Panicle open.

M. effusum L. (*a*) Culm erect, 3—8f; lvs. flat, smooth; pan. diffuse, 6—9′ long; spkl. oblong, (*c*) scattered, acute, 1″. Woods, Can. to Ill. and Pa. Summer.

19. AMPHICÁRPUM, Kunth. Spikelets apparently 1-flwd., and perfect as in Millium, but of two kinds; the terminal deciduous and sterile, the radical under ground, and fertile. Gl. and pales sub-equal, lanceolate, acute. Panicle strict, erect. Radical fls. larger, solitary.

A. Púrshii K. (*f*) Culm 1f, erect; lvs. erect, hairy; sheaths hairy, the upper leafless; pan. on a long exserted ped.; ♂ spikelets 1½″ long, the ♀ radical, 2¼″, the grain terete, same length. Barrens, N. J., and S. Aug.

20. PÁNICUM, L. Panic G. Glumes 2, unequal, awnless, the lower much smaller. Fls. 2, dissimilar, the lower of 1 or 2 pales, neutral or ♂; the upper ♀ of 2 equal cartilaginous polished, concave, awnless pales coating the grain. Sta. 3. Stig. 2, plumous, purple. Spikelets in simple or compound panicles.

§ Spikelets acute, or acuminate, very numerous, racemed in large panicles...(*)
§ Spikelets obtuse, or barely acute, solitary, pedicillate, not numerous...(**)
  * Abortive fl. neutral, consisting of one pale...(*a*)
  * Abortive flower neutral, of 2 pales...(*b*)
  * Abortive flower ♂, of 2 pales. Culms erect, terete, with one panicle............Nos. 10, 11
    a Panicle ample, capillary, spikelets single on capillary pedicels.............Nos. 1, 2
    a Panicle not capillary, dense-flowered.....No. 3
      b Lower glume as long as the upper, 2″, both 3-veined.................................No. 4
      b Lower gl. very short, the upper 3-5-veined, 1″ or less.....Nos. 5, 6
      b Lower gl. very short, upper 7-9-veined, not tumid.........Nos. 7, 8
                —upper 11-veined, tumid at base, 2″........No. 9
  ** Abortive flower neutral, consisting of a single pale...   Nos. 12, 13
  ** Abortive flower of 2 pales, the upper small and scarious...(*c*)
    c Leaves narrow (1 5″ wide), obscurely veined...(*d*)
    c Leaves broad, 5—20″ wide, conspicuously veined...(*e*)
      d Spikelets silky-fringed. Lower glume obsolete. ♀ Fl. colored.....No. 14
      d Spikelets glabrous, or merely pubescent. Lower glume small...(*e*)
        e Spikelets less than 1″ long, round-oval. Glume 5-veined....Nos. 15, 16
        e Spikelets 1-1½″ long, oval. Glume 9-veined......   Nos. 17, 18

x Abortive fl. usually staminate. Spikelets obovate, 1½″..........Nos. 19, 20
x Abortive flower neutral, never with stamens...(y)
y Plant stout, soft-downy, except the smooth noder................No. 21
y Plant smoothish, or rough-hairy, branched or simple........Nos. 22, 23
Exotic, cultivated....No. 24

1 **P. capillàre** L. Culms thick at base, 1—2f; lvs. broad-linear, and with the sheaths bristly-hairy; panicle ample, pyramidal, capillary, loose; spkl. lance-ovate, acuminate, ⅛″, purple. ① Fields and waysides. Aug.

2 **P. autumnàle** Bosc. Culm slender, 10—20′; lvs. short, soon rolled, and with the long sheaths glabrous; pan. diffuse, bearded in the axils; ped. long (2—4′), capillary; spkl. lance-oblong; lower gl. minute. Ill. to Car.

3 **P. prolíferum** Lam. Glabrous, 2—3f; lvs. broad-linear, on tumid sheaths; pan. terminal and lateral, pyramidal, ped. sheathed; spkl. elliptic, 1″; lower gl. $^1/_4$ or $^1/_5$ as long as the upper; ♂ fl. pointed. Rich shady soils. Aug., Sept.

β. *geniculàtum.* Culm thick, geniculate below; pan. dense. Marshes.

4 **P. gymnocárpum** Ell. Culms 2—3f, stout, erect; lvs. lanceolate, 1′ wide; pan. large, expanding; spkl. lanceolate, 2″, in clusters of 3—5; glumes and neutral pales twice longer than the naked fertile fl. Banks, Ga., Fla., and W.

5 **P. hians** Ell. Slender, glabrous, decumbent at base, 2f; lvs. narrow; pan. of slender racemes; spkl. ⅛″, lower gl. ½—⅓ as long as the upper; both fls. coriaceous, divergent or *gaping* at apex. Damp barrens, S. Aug.—Oct.

6 **P. agrostoìdes** Muhl. (*a*) Culm 1½—3f, compressed; lvs. long, rough-edged; pan. term. and lateral, pyramidal, purplish, of dense racemes; spkl. (*b*) 1″, lance ovate; upper gl. 3-veined, ⅓ longer than the lower; neutral pales sub-equal. Jl. +

7 **P. anceps** Mx. Culm and lvs. as in No. 6. Pan. very large and open; spkl. 1½″, forked when ripe; upper gl. 5-veined, twice longer than the lower, shorter than the lower neutral pale, which is twice longer than the other pales. N. J., and S. Aug. +

8 **P. vilfifórme** Wood. Very glabrous; pan. at each joint, and term. of loose racemes; spkl. lance-ovate; up. gl. 9-veined, 1½″, lower neutral pale a little longer, the other 3 pales a little shorter, lower gl. ½ as long. Meadows, E. Tenn. Aug.

9 **P. gibbum** Ell. Culm 2—3f, assurgent; lvs. broad-linear, glabrous; pan. 5—6′, dense, spindle-form; spkl. tumid, near 2″; lower gl. very small, upper very large, 11-veined, *gibbous* at base; sterile fl. (♂, *Chapm.*) neutral. Wet. S. Jl.—Sept.

10 **P. amàrum** Ell. Culm terete, strict, 2—3f; lvs. rolled and rigid (*bitter* to taste), pan. 6—10′, contracted, its smooth branches appressed-erect; spkl. lance-ovate; glumes pointed, the lower 1″, upper nearly 2″; sterile fl. 1½″, anth. orange. Sands.

11 **P. virgàtum** L. Culm 3—5f, lvs. flat; pan. large, thin, at length diffuse, 10—20′ long; spkl. scattered, ovate, pointed, purplish; upper gl. 2″, sterile fl. 1½″, fertile fl. and lower gl. 1″, all divergent when ripe; anth. purple. N. Y., S., and W. Aug.

β. *obtùsum.* Panicle contracted; spikelets smaller, not pointed, obtusish. N. J.

12 **P. verrucòsum** Muhl. Slender, weak, decumbent below, 10—20′; lvs. lance-linear, short; pan. few-flowered; spikelets obovate, bluish, ½—⅔″, beset with fine warty (*verrucous*) points. ① Thickets and swamps, not rare. Aug.

13 **P. villòsum** Ell. Villous with soft white hairs throughout, 10—20′; lvs. flat, short; pan. small (2—3′ long), oblong, loose; spkl. oval, 1″, green; upper gl. and 2 fls. equal, lower glume ½ as long. Evergreen, damp. S. Apr., May.

14 **P. ciliatiflòrum** Wood. *Fringed G.* Erect, strict, 2—3f; lvs. narrow, rigid, flat, ciliate; pan. slender, strict, 3—4′; spkl. 1½″, oblong, silky-villous glume solitary, equaling the lower staminate pale, 5-veined. Barrens, S. Sept.

β. *rufum.* Lvs. glabrous, erect; sterile fl. neutral, hairs purple.

15 **P. dichótomum** L. Culm at first simple with one panicle, soon branched, slender, 8—20′; lvs. lance-linear, short, 1—4′ by 2—4″; terminal pan. oval, small (1—2′), stalked; spkl. few and small, ½″, round-oval; lower gl. ½—⅓ as long as the upper. Common in fields. June—Sept.

β. *nitidum.* Smooth, shining; lvs. narrow; ped. long; spkl. oval.
γ. *sphærocárpum.* Hairy; peduncle long; spkl. rounded, dark-purple.
δ. *barbulàtum.* Taller; nodes with a ring of deflexed hairs.
ε. *lanuginòsum.* Woolly; lvs. larger; spikelets green; pan. larger.
ζ. *spathàceum.* Hairy and leafy to the top; panicles sessile.

16 **P. depauperàtum** Muhl. Culm simple, strict, tufted, 6—12′; lvs. linear, erect, the upper elongated; pan. simple, sessile or becoming long-stalked; spkl. oval, ⅓—1″; lower gl. ½ as long as the upper 7-veined one. Hills and woods, common. June. Varies with lvs. hairy or smoothish, and

β. *involùtum*, with lvs. involute, ending in a long stiff point.

17 **P. paucifòrum** Ell. (*c*) Culm assurgent, 1—2f; lvs. lanceolate, 3—5′ by 5—7″ hirsute below as well as the sheaths, faintly 9-veined; pan. open; spkl. (*d, e*) few, large (1—1½″), oval; lower gl. ½ as long as the upper. (*x*, neutral fl.) Damp shades.

18 **P. pubéscens** Lam. Culm slender, branched, 2—3f; lvs. lance-linear, 3—6′ by 3—5″, 9-veined, retrorsely hirsute as well as the open sheaths; spkl. oval, 1¼″, pubescent, outer glume lanceolate, 1″, inner 9-veined. Dry fields. June.

19 **P. latifòlium** L. Erect, 1—2f; lvs. lanceolate, dilated and cordate-clasping at base, 3—5′ by 1′, smoothish, 11-13-veined; pan. exserted, 3′ long; spkl. obovate, 1½″; lower gl. ovate, ½″, upper gl. 9-veined; neutral pales sub-equal, usually with 3 stamens. In moist shady places: common. June, July.

20 **P. xanthophỳsum** Gr. Culm simple or branched below, 9—15′; lvs. lanceolate, 3—6′ by 5—7″, not dilated at the ciliate clasping base; pan. long-stalked, raceme-like; spkl. few, round-obovate, 1½″; lower gl. ovate, ½ as long as the upper 9-nerved one; sterile fl. often ♂. Dry. N. Eng. to Wis. June.

21 **P. víscidum** Ell. Hoary with a dense viscid pubescence, 2—4f, stout; joints with a smooth brown ring; lvs. lance-linear, 3—6′ by 6—16″; pan. 4—6′, loose; spkl. pale, oval, 1″; lower gl. and upper pale minute. Wet. N. J., and S. Aug.

22 **P. clandestìnum** L. Culm rigid, leafy, 2—3f; lvs. 3—6′ by 1′, dilated and cordate at base; sheaths scabrous or rough-hairy, enclosing the lateral and often the terminal dense panicle; spkl. elliptical, 1½″. Moist woods. July, Aug.

23 **P. microcárpon** Muhl. Erect, simple, glabrous; lvs. lanceolate, broad and clasping at base, veiny, 6—10″ wide; pan. long-stalked, diffuse; spkl. small (½), oval, numerous, purple; lower gl. minute. Pa., W., and S. July—Sept.

24 **P. miliàceum.** *Millet.* Lvs. lance-linear and sheaths hairy; pan. large, open, nodding; spkl. ovate, solitary; glumes pointed, sub-equal. Turkey.

21. PENICILLÀRIA SPICÀTA.
Erect, 4f, branching, with broad, flat leaves. Panicle cylindric-oblong, 1f in length, compact, consisting of innumerable simple branches, each with 2 or 1 spikelets at the end, and clothed with spreading hairs. Each spikelet bears at length a white ripened grain. ⊕ E. India.

22. OPLÍSMENUS, Beauv. COCK-SPUR G.
Spikelets in dense, spike-like, panicled racemes. Glumes and lower pale of the sterile fl. rough-pointed or awned. Otherwise as in Panicum.

1 **O. crus-gallí** L. (*a*) Culm terete, 3—4f; lvs. lance-linear, rough-edged, ligule none; pan. with its spike-form branches alternate or in pairs; rachis rough-hairy; glumes bristly, scarcely awned; awn of the pale (*b*) 6—18″ long, very rough. Sheaths generally smooth. Waste grounds: com. Aug., Sept. § (merely pointed.

β. *mùticus.* (*c*) Awns very short, or the hispid pale
γ. *híspidus.* Sheaths very bristly; awns very long. A very coarse variety.

2 O. Walteri (Ell). Culms slender, 2f; lvs. narrow and sheaths glabrous; spikes one-sided, ½—1' long, alternate; glumes hispid, pointed; the fls. somewhat pointed, the sterile with 3 stamens. Low grounds, Car. to Fla., and La. July.

3 O. hirtéllus R. & S. Decumbent, branched, ciliate; lvs. lanceolate, 1—2' by 4—6''; spikes erect, remote, one-sided, ½' long. few in the perfectly simple panicle; pale long-awned, glumes short-awned. Woods, South. Aug.—Oct.

23. SETÀRIA, Beauv. Bristly Foxtail. Fls. in cylindric spikes or spike-like panicles. Spikelets each subtended by a cluster of awn-like bristles (abortive pedicels) forming a bristly involucre. Otherwise as in Panicum. July, Aug.

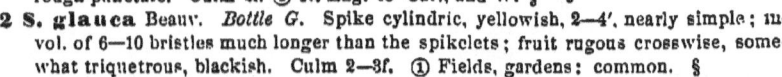

§ Bristles rough backward, in pairs, short..........No. 1
§ Bristles rough upward....(a)
    a 4—10 in each involucre.............Nos. 2—4
    a 1—3 in each involucre.............Nos. 5—7

1 S. verticillàta Beauv. Spicate pan. 2—3', composed of short divided branchlets seeming in many verticils; bristles little longer than the spikelets; fruit-pales rough-punctate. Culm 2f. ① N. Eng. to Car., and W. §

2 S. glauca Beauv. *Bottle G.* Spike cylindric, yellowish, 2—4', nearly simple; invol. of 6—10 bristles much longer than the spikelets; fruit rugous crosswise, somewhat triquetrous, blackish. Culm 2—3f. ① Fields, gardens; common. §

3 S. víridis Beauv. *Wild Timothy.* (a) Spike cylindric, 1—3', compound, green; invol. of 4—10 bristles much longer than the spikelets (b, c); fruit-pales striate lengthwise and dotted (under a lens). Culm 1—2f. ① Cultivated grounds, N. §

4 S. Germánica Beauv. *Millet. Bengal G.* Spike flattened, oblong-cylindric, compound, 3—5' by 9''; rachis bristly; invol. of 4—8 bristles, little longer than the spikelets, yellowish; ⚥ pales dull-rugous. Culm 3—4f. ① Fields. §

5 S. Itàlica K. Spicate pan. 6—18' long by 1—2' thick; invol. yellowish, of 2 or 3 bristles 8—10 times longer than the spikelets and half-concealing them; ⚥ pales smooth, polished, shining. Culm 4—6f. ① Swamps, S.

6 S. corrugàta Schul. Spicate pan. 3—6', cylindric, dense above; bristles 1 to each spikelet and thrice as long; ⚥ pales strongly corrugated. Fla., Ga.

7 S. compósita K. Spicate pan. loose, its lower clusters separated; bristles 1 or 2 under each spkl. and 5 times longer; ⚥ flower acute, smoothish. Fla.

24. CENCHRUS L. Burr G. Fls. racemed or spicate. Involucre a burr (a) beset with spines, becoming hard and pungent in fruit, and enclosing several (1—3) spikelets (b). Glumes and flowers as in Panicum, the sterile flower ♂. Culms branched. Aug.

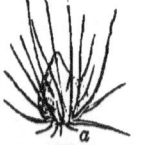

C. tribuloìdes L. Culms 1—2f, tufted, decumbent, spreading; lvs. as short as their open compressed sheaths; spikes several, 1—2' long; burrs adhering by their rough spines to everything passing. Sandy shores, N. J. to Ill., and N. (See *Addenda*.)

25. PHÁLARIS, L. Canary G. Spikelets 1-(theoretically 3)-flowered. Gl. 2, subequal, carinate, longer than the two shining pales of the ⚥ fl., all awnless. Neutral rudiments at base of the ⚥ fl. merely 2 single pales or hairy pedicels (b, c). Grain coated. Handsome flat-leaved grasses.

1 P. arundinàcea L. *Ribbon G.* A showy but not valuable grass, 2—5f; lvs. lance-linear; pan. contracted, dense, 3—6' long; glumes (a) 2¼'', pointed; rudiments

2, hairy, at the base of the ovate pales (*b*). ♃ Ditches
and swamps, Can. to Car., and Ky. July, August.

β. **picta.** *Striped G.* Lvs. endlessly variegated with
white and green. Cultivated.

2 P. Canariénsis L. *Canary G. Bird-seed.* Culm
terete, erect, 1—2f; lvs. lance-linear; pan. spicate, ovoid,
1—2′; gl. winged on the keel (*c*); rudiments smooth. ⓘ
Introduced into fields and gardens from Isle Fortunatus.

26. ANTHOXÁNTHUM, L. Sweet Vernal

G. Spikelets (*d*) 3-flowered, the central fl. ☿, the
two lateral neuter, each of 1 bearded pale. Gl. 2,
unequal. Pales 2, short, awnless. Sta. 2.

A. odorátum L. Slender, erect, 10—18′; lvs. short; panicle spicate, 1½—3′; neutral pales ciliate (*e*), one with a
bent awn from near the base, the other with a straight awn
from the back above. Fls. in May and June ill-scented,
but when cut as hay it is very fragrant. § (*x*, the ☿ fl.)

27. HIERÓCHLOA, Gmel. Seneca G. Spkl.

3-flwd. Gl. 2, scarious. Lateral fls. ♂
triandrous, central fl. ☿, with 2 (or 3) stamens. Inflor. paniculate. Sweet-scented.

1 H. boreális R. & S. (*f*) Very smooth;
simple, erect, 15—30′; root lvs. as long as the
culm, cauline lvs. lanceolate, short; pan. open,
few-flwd., 2—3′; spkl. (*g*) broad, subcordate, colored, awnless. ♃ Wet
meadows, Va., and North. May.

2 H. alpina R. & S. Smooth; culm erect,
6—8′, stout; lvs. lance-linear; pan. ovoid,
1—2′; spkl. purple, longer than their branchlets; lower fl. with an awn on the back as long as the
pales. ♃ High Mts., N. Eng., N. Y. June.

28. HOLCUS, L. Soft G. Spkl. 2-flwd., paniculate. Gl. herbaceous, boat-shaped, mucronate.
Fls. pedicellate, the lower ☿, awnless; the upper
♂ or neutral, awned on the back. July.

H. lanátus L. (*h*) Hoary-pubescent, 1½—2f; lvs. lance-linear; pan. oblong, dense, purplish-white; fls. (*i*)
shorter than the glumes (*k*); awn of the sterile fl.
curved, included. ♃ Wet meadows. A beautiful grass.

29. AIRA, L. Spkl. 2-flwd. without abortive
or sterile ones. Gl. 2, thin, shining, subequal.
One of the fls. pedicellate. Pales subequal, hairy
at base, the lower truncate at apex, and awned on
the back. Fls. in an open pan., silvery-purplish.

§ Glumes longer than the fls. Pale entire.........No. 1
§ Gl. about equalling the fls. Pale lacerated.. Nos 2, 3

1 A. atropurpùrea Wahl. In tufts, 1f, very slender; lvs. flat; pan. thin; awn stout, twice as long as the pale. ♃ High Mts., N. Eng. and N. Y. August.
2 A. flexuòsa L. (*l*) In large tufts, smooth, 1—2f; lvs. setaceous, mostly radical; pan. loose, with long flexuous spreading branches; awn geniculate, twice longer than the pale (*m*). ♃ Dry hills: common, June.
3 A. cæspitòsa L. (*n*) Tufted, glabrous, 18—30'; lvs. narrow-linear, flat; pan. oblong, finally diffuse; awn straight, as long as the pale, which is longer than the bluish glumes. (*o*, spikelet, *p*, fl.) ♃ Swamps, northward. May.

30. DANTHÒNIA, DC. Spkl. 2–7-flwd. Gl. 2, subequal, cuspidate, longer than the whole spikelet of fls. Pales hairy at base, lower one bidentate and awned at apex, upper obtuse, entire. Awn flattened and twisted at base. ♃ Fls. racemous.

1 D. spicàta R. & S. (*a*) Lvs. narrowly-linear, shorter than the internodes; culm 1—2f, slender; spkl. few (about 6), in a subsimple raceme; gl. 4—5''; fls. (*b*) about 7, pubescent. Lvs. mostly radical, in little tufts. Dry hills: com. June—Aug.
    β. **compréssa.** Lvs. longer than the internodes; spkl. about 4 in the simple raceme; gl. twice longer than the spikelet. Onondaga Co., N. Y. (*S. N. Cowles*). (D. compressa, Austin?) These characters are not constant.
2 D. serícea Nutt. Taller (2—2½f); lvs. and sheaths silky-hirsute; spkl. 9—17, evidently paniculate; gl. 8—9''; fls. about 7, densely clothed with silvery-silky hairs; awns brown at base (as in No. 1), very long. Rare N., common S. June.

31. AVÈNA, L. OAT. OAT G. Spkl. 2–5-flwd. Gl. 2, loose, thin, awnless, large. Pales 2, becoming coriaceous, the lower bifid, bearing (mostly) a bent or twisted awn on the back; upper pale coating the oblong grain. Fls. paniculate.

§ ARRHENÁTHERUM. Glumes unequal, 2-flowered, with a rudiment of a third; lower flower staminate and awned. Tall.....................No. 1
§ AIRÓPSIS. Gl. subequal, 2-flwd., both flowers ☿, no rudiment. Dwarf......................Nos. 2, 3
§ AVÈNA. Gl. equal, longer than the 2 perfect flowers, strongly striate...................Nos. 4, 5

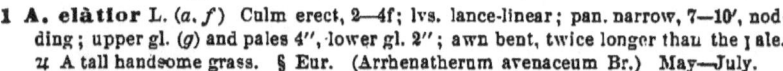

1 A. elàtior L. (*a*, *f*) Culm erect, 2—4f; lvs. lance-linear; pan. narrow, 7—10', nodding; upper gl. (*g*) and pales 4'', lower gl. 2''; awn bent, twice longer than the pale. ♃ A tall handsome grass. § Eur. (Arrhenatherum avenaceum Br.) May—July.
2 A. præcox Beauv. (*d*) Culms tufted, erect, 2—5'; lvs. setaceous; *pan. dense, oblong*, ¼—1'; gl. (*b*) equaling the fls. (*c*); awns bent, twice longer. ♃ N. Y. to Va. Jn.
3 A. caryophýlla L. Culms 5—10'; lvs. very narrow; *pan. loose, open;* glumes silvery-purple, scarce 1'', pales shorter, awns exserted. Dry fields, M. § Eur.
4 A. SATÌVA. *Common Oat.* Culm terete, erect, 2—4f; lvs. lance-linear; pan. loose, pyramidal; spkl. large, pendulous; both fls. ☿, 7'', the lower mostly awned; both pales coating the nutritious grain. Cultivated, common. June.
    β. **nìgra.** *Black Oats.* Pales dark brown, almost black, without awns.
    γ. **secúnda.** *Horse-mane Oat.* Panicle one-sided, nodding; awns short.
5 A. STÉRILIS. *Animated O.* Spkl. 5-flwd., 2 lower fls. each with hairy pales and a long bent awn which is so sensitive to moisture as to be kept in motion by the ordinary changes in the air. From Europe. Cult. as a curiosity. ⨁ 4f. July, August.

32. TRISÈTUM, L. Spkl. 2–5-flwd. Glumes 2, shorter than the fls. Lower pale with two bristles at the apex and a soft flexuous awn from above the middle of the back. Grain coated, furrowed. ♃ Fls. paniculate.

1 **T. purpuráscens** Torr. Spkl. (*p*) about 4-flwd., 6–8″, few (6–9) in the very simple purple panicle; fls. (*d*) separate, bearded at base; gl. (*g*) unequal; lvs. narrow-linear; culm erect, 2–3f. Mountain bogs, N. June.

2 **T. palústre** (Mx.) Spkl. (*a, b*) 3-flwd. 2¼″, the upper fl. abortive; middle fl. with a bent awn its own length; pan. narrow, 4–6′; lvs. very short (2–3′); culm slender, 2f. Plant smooth. Wet meadows. May—July. (*c*, pale.)

3 **T. molle** (Mx.) Spikelets 2-flwd., 3″; upper fl. with a bent awn its own length; g̃-lance-linear; panicle as in No. 2; lvs. broader and longer; plant 2f, minutely downy. Rocky hills, N. July.

33. BROMUS, L. Brome G. Spikelets 5–∞-flwd. Gl. unequally veined. Lower pale 5–9-veined, awned from below the mostly bifid tip. Upper pale ciliate on its 2 keels, adhering to the linear grain. Coarse grasses, with flat leaves, and large, nodding, panicled spikelets. June, July.

§ Glumes narrow, the lower 1-veined, upper 3-veined. Lower pale keeled...(*b*)
§ Glumes veiny, the lower 3–5, upper 5–7-veined. Lower pale convex...(*a*)
  a Awn much shorter than its pale. Panicle spreading...............Nos. 1, 2
  a Awn as long as its pale. Panicle erect, contracted in fruit......Nos. 3, 4
    b Lower pale compressed-carinate, awn very short..................No. 5
    b Lower pale rounded on the back, the awn conspicuous.........Nos. 6, 7

1 **B. Kálmii** Gr. *Wild Chess.* More or less hairy, 1½–3f; spkl. drooping, closely 7–12-flwd., densely silky; lower pale much the larger; pan. small. ♃ Dry.

2 **B. secalinus** L. *Cheat* or *Chess*. (*s*) Nearly glabrous, 2–1f; spkl. ovate, turgid, glabrous, 7–10-flwd., fls. (*a*) soon diverging, blunt, awned or not; panicle nearly simple. 4–8′ long, spikelets 8–10″ long, drooping. (ɪ) Fields. § Eur.

3 **B. racemòsus** L. *Erect Chess.* Spkl. ovate-oblong, glabrous; closely 8–12-flwd., awns straight, 4″; pan. simple; plant slender, some hairy. (ɪ) Fields. § Eur.

4 **B. mollis** L. *Downy Chess.* Plant downy, with spreading hairs; spkl. ovate, about 6-flwd., fls. closely imbricated; awns straight, 3–4″. (ɪ) (*d*) Fields: rare.

5 **B. unioloides** H. & K. *Rescue G.* Culm erect, 1½–3f, smoothish; pan. narrow, 6–10′, nodding; spkl. lance-oblong, compressed, 1′, 8–12-flwd. (ɪ) Cult. South.

6 **B. ciliàtus** L. Pan. compound, 5–8′, soon nodding; spkl. at first lance-fusiform (*b*), 7–11-flwd., the fls. soon separating; pale (*c*) compressed-carinate above, silky-haired at edge, twice longer than its straight awn; culm 2–4f; lvs. some hairy. ♃ Shady banks: common. July, August.
  β. *púrgans.* Plant finely and closely pubescent all over.

7 **B. stérilis** L. Pan. compound, soon 1-sided and nodding; ped. capillary; spkl linear-oblong, about 5-flwd., puberulent; fls. linear-subulate, scarcely as long as the awn. (ɪ) Banks, Pa., and N. Rare. §

8 **B. brizoïdes.** Culm 1f, erect; lvs. narrow, conduplicate, rigid; pan. erect, with a few large, hanging, ovate, awned spikelets; pale dilated, ear-shaped above. Cult.

Order 155.—GRAMINEÆ.

34. TRICÚSPIS, Beauv. Spkl. terete, or tumid, 3-9-flwd. Glumes unequal, awnless. Lower pale (*n, c*) conspicuously fringe-bearded on the 3 strong veins, tipped with 2 or 3 teeth, and 1 or 3 short awns or cusps; upper pale much shorter, 2-toothed (*n*). Fls. paniculate. Sheaths hairy at throat. Aug., Sept.

§ Windsòria. Culm erect, simple. Lower pale 3-cusped.................................Nos. 1, 2
§ Urálepis. Culm spreading, branched. Lower pale 1-cusped............................Nos. 3, 4

1 **T. seslerioìdes** (Mx). *False Red-top.* (*s, a, n, m*) Culm 3—5f; lvs. linear, involute when dry; pan. open, loose, 8—12′, the slender branches at length spreading; spkl. (*a*) oblong, 3″, 5- or 6-flwd., purple, shining. ♃ Beautiful.
  β. **flexuòsa.** Branches of the panicle flexuous; spkl. 3-5-flwd., 2″. Pa.
2 **T. ambígua** (Ell.) Culm 2—3f, wiry; lvs. narrow and rolled; pan. small (3—5′), few-flwd.; spkl. ovate, the 5—7 fls. divaricate. ♃ Pine-barrens, S.
3 **T. purpùrea** (Walt.) (*b*) Culm bearded at the nodes, 10—18′; lvs. subulate, short; panicles more or less sheathed; spkl. (*b*) 3-flwd., awn scarcely exceeding the eroded segments of its pale. ① Coast sands, Mass. to Fla. (*c*, lower pale.)
4 **T. cornùta** (Ell.) Culm 2f; lvs. and sheaths hairy; awn of the lower pale plumous, much longer than the lateral teeth, recurved. Dry sands, S.

35. ARÚNDO Donax. A gigantic ornamental grass from Italy, where it is cult. for vine-poles, fence-wood, fishing-rods, etc. Culm 10—15f high; lvs. broad, flat, smooth, and shining; pan. diffusely branched; gl. as long as the 3 fls.; rachis beset with long hairs; lower pale with a short awn in the cleft at apex. ♃
  β. versícolor. *Gardener's Garters.* Leaves striped with white.

36. GRAPHÉPHORUM, Desv. Spkl. of 2—5 remote fls. with sub equal glumes. Fls. bearded at base. Gls. and pales thin, lanceolate, awnless, convex, not keeled. ♃ Erect, glabrous. Lvs. flat. Panicle simple

  G. melicoìdes Beauv. Culm slender, 1—2f, with 2 or 3 short erect linear lvs.; pan. loose, 3—4′ long; spkl. 2-3-flwd., 3—4″ long. Upper Mich. (*C. E. and A. H. Smith*).
  β.? *triflorum* (Aira trif. Ell.) "Fls. somewhat woolly at base, not villous." Ga.

37. GYNÈRIUM argénteum. Pampas Grass. A magnificent reed from S. Am., becoming common. ② Leaves in a dense, radical cluster, recurved, narrow, channeled. Culms 10—18f, clustered, bearing dense, hairy panicles, which are 1½—2f, silvery white, with innumerable flowers and their long, silky hairs. Some of the panicles are fruitful (♀), others barren (♂).

38. DÁCTYLIS, L. Orchard G. Spkl. 3-5-flwd. compressed. Glumes unequal, shorter than the fls. Pales subequal, lance-acuminate, the lower (and glumes) carinate, awn-pointed. Lvs. channeled. Panicle composed of dense 1-sided clusters. June.

  D. glomeràta L. Culm 2—4f high; lvs. broad, glaucous; stipules lacerate; spkl. loose-flwd.; gl. very unequal. ♃ Shady fields. A good grass for hay or pasturage. §

39. KŒLÈRIA, Pers. Spkl. 2-7-flwd., compressed; gl. subequa' acute, scarcely shorter than the fls.; upper fl. pedicellate; lower pale

Order 155.—GRAMINEÆ.

(and gl.) carinate, often bristle-pointed. ♃ Culms tufted, erect, simple, with dense, narrow panicles.

K. cristàta Sm. Culm 20—30′, leafy below; lvs. flat, erect, pubescent, narrow, 2—3′ by 1—2″; pan. spike-like, 3—5′; spkl. (*a*) 2″, silvery, about 2-flwd., with an abortive pedicel. (*b*, a flower.) Mid., W., and N.

β. *grácilis.* Slender and delicate, with a simple pan. (K. nitida, N.)

40. DIARRHÈNA, Raf. Panicle simple, racemous. Glumes 2, very unequal, rigid, acuminate-mucronate, 2-5-flwd. (*d*) Pales (*e*) cartilaginous, lower cuspidate, 3″, upper much smaller, emarginate. Grain large, loose in its pericarp. Stam. 2. ♃ Culm rigidly erect, 15—30′. Lvs. mostly radical, broad-linear.

D. Americàna Beauv.—Woods and river-banks, O. to Ill. Aug. (Festuca, Mx.)

41. FESTÙCA, L. Fescue G. Spkl. 3-∞-flwd. Glumes unequal, mostly carinate. Pales firm, the lower rounded (not carinate) on the back, obscurely veined, awned from the tip, or awnless. Sta. 1—3. Grain mostly adhering to the upper pale. Spkl. panicled or racemed, the fls. remote, not webbed at base.

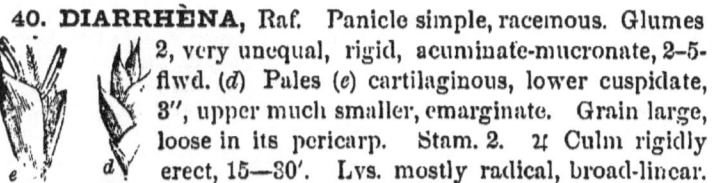

§ Flowers lanceolate to oblong, awnless. Culms tall, leaves flat.................................Nos. 5—7
§ Flowers subulate, awned at the tip. Leaves mostly involute...(*x*)
  x Awn much shorter than the flower. ♃....Nos. 3, 4
  x Awn as long as the fl. or much longer. ①..Nos. 1, 2

1 **F. Myùrus** L. Culm 5—12′; lvs. subulate, 2—3′; glumes minute, equal, 4-6-flwd.; awn 6″, twice longer than the pale; panicle slender. M., S. §

2 **F. tenélla** Willd. *Slender F.* (*a, b*) Culm wiry-filiform, often in tufts, 6—12′; lvs. linear-setaceous; pan. simple, narrow, 2—3′; spkl. 6-9-flwd., 4—6″ long; flowers puberulent, brown; awn about as long (2″). Sandy. June, July.

3 **F. ovìna.** L. *Sheep's F.* Culm erect, 6—10′; lvs. numerous below, very narrow, 2 -4′; pan. simple, narrow, 2—1′; spkl. ovate, 3-5-flwd.; fls. lance-oblong, 1¼″, the awn ¼—½ as long. ♃ Pastures and fields. A valuable grass. June, Europe.

β. *vivipara.* Spikelets transformed to leafy tufts. Mountains, N.

4 **F. duriúscula** L. *Hard F.* Culm erect, 12—18′; lvs. linear, flattish; pan. oblong, spreading, 3—5′; spkl. 5-8-flwd., teretish before flowering; fls. lance-subulate, 2¼″, the awn 1″ or less; pales equal. ♃ Valuable. Common. June, July.

β. *rubra.* Spikelets 7-9-flwd., fls. pubescent; the herbage reddish. N.

5 **F. praténsis** Huds. *Meadow F.* Culm erect, 2 -3′; lvs. lance-linear; pan. 4—6′ ,ong, narrow, with short branches; spkl. few (10—25) and large, teretish before flowering, 6—9″ long, 6-9-flwd.; pales 3″, barely pointed. ♃ A fine grass. June.

6 **F. elátior** L. Culm 2—4f, erect; lvs. lance-linear; pan. diffuse, nodding, compound, branches branched, and floriferous above, naked below; spkl. numerous, 3-5-flwd., 2—3″ long; fls. oblong, 1½″, acute; lower gl. 3-veined. ♃ Fields. §

7 **F. nùtans** Willd. *Nodding F.* (*c, d*) Culm slender, 2—4f, about 2-jointed; lvs. linear; pan. very open, with few long drooping branches floriferous at the end; spkl. 3″, lance-ovate (*c*) 4-6-flwd.; fls. (*d*) smooth, nearly veinless. ♃ Rocks.

β. *palústris.* Panicle less diffuse, spkl. 3-5-flwd. Between Nos. 6 and 7.

42. EATÒNIA, Raf. Spkl. mostly 2-flowered, numerous, panicled, silvery. Glumes unlike, the lower linear, 1-veined, the upper broadly obovate, rounded and 3-veined on the back. Pales obtuse, chartaceous, awnless. Grain oblong. ♃ Delicate grasses with simple culms.

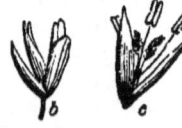

1 **E. obtusàta** (Mx.) Panicle narrow, dense, 3—5′ by ⅓—1′; branches short, appressed; spkl. (*a*, *b*) 1⅓″ long, 2-flwd., tumid; pales (*c*) scarious at tip, a little longer than the very obtuse upper glume. Dry. Penn. to Wis., and S. June, July. 2f.

2 **E. Pennsylvánica** (DC.) Panicle 5—10′, slender, open and loose; spkl. 1½″; upper gl. abruptly short-pointed, or obtuse; upper flower exserted half its length. Shady rocks and meadows. Elegant. Summer. 2f.

43. MÈLICA, L. MELIC G. Glumes unequal, obtuse, 2–5-flowered. Fls. exserted, the upper incomplete. Pales truncate, veiny as well as the glumes. Grain free. ♃ Lvs. flat; spkl. pedicellate, in a subsimple panicle.

M. mùtica Walt. Culm 3—4f; lvs. linear, flat; pan. few-flwd., inclined to one side; spkl. (*e*) 4—6″ long, with 2 fertile fls., and the third upper one contorted; pales (*f*) unequal, veined. Penn. to Wis., and S.

44. ERAGRÓSTIS, Beauv. Spkl. 2–∞ -flwd., membranous. Lower pale carinate, 3-veined, never webby at base, upper pale persistent on the flexuous rachis after the free grain and lower pale have fallen. Culm simple or branched. Leaves often rolled, bearded at the throat. Panicle with hairy axils.

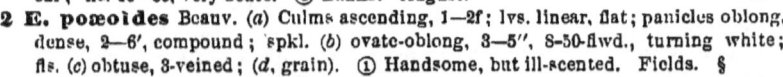

§ Culms branched, prostrate; spikelets sub-sessile ..... No. 1
§ Culms branched, ascending; panicles 1—3 .... Nos. 2—7
§ Culms simple, erect, shorter than its loose pan ... Nos. 8—11

1 **E. reptans** Nees. Culms creeping and rooting, 6—12′; lvs. subulate, 1—2′; panicles many, small, dense; spkl. lance-linear; fls. 10—30, very acute. ① Banks. August.

2 **E. pœoìdes** Beauv. (*a*) Culms ascending, 1—2f; lvs. linear, flat; panicles oblong, dense, 2—6′, compound; spkl. (*b*) ovate-oblong, 3—5″, 8-50-flwd., turning white; fls. (*c*) obtuse, 3-veined; (*d*, grain). ① Handsome, but ill-scented. Fields. §

3 **E. pilòsa** L. Culms in tufts, ascending, 4—12′; lvs. linear, flat, tender; panicles oblong, loose; spkl. linear, bluish, about as long (2—4″) as their pedicels; flowers 4—12. obtuse, with only the midvein apparent. ① Dry, sandy places. July. §

4 **E. Púrshii** Schr. Culms ascending, 6—12—20′; lvs. 1—3′, very narrow; panicles long and loose; ped. capillary; spkl. linear-oblong, 2—4″; fls. 5—12, acute or acutish, 3-veined. purplish. ① Dry fields, N. J., Penn., and S. Common. July, August.

5 **E. erythrógona** Nees. (E. Frankii Meyer.) Culms in tufts, much branched, ascending, 6—18′, *joints red*; pan. narrow, beardless, 2—4′; spkl. about 1″, their ped. much longer; gls. and pales very acute, obscurely 3-veined. ① Dry. Pa. to Ill., and S.

6 **E. ciliàris** (L.) Culms decumbent and ascending, 6—12′; pan. cylindrical, branches appressed. covered with the minute (⅓″) ovate spikelets; fls. 5—7, mucronate, upper pale ciliate-fringed. ① Waste grounds, South.

7 **E. confèrta** Trin. Culm stout, erect, 2—3f; lvs. broad-linear; pan. long (5—12′), narrow, branches erect, covered with innumerable small (1—1½″) spikelets; fls. 7—11. hyaline, obtuse, 3-veined, whitish. ① River banks, S. Aug., Sept.

Order 155.—GRAMINEÆ.

8 **E. tenuis** (Ell. Poa trichodes N.) Plant 1—3f high; pan. long (8—24), loose, capillary, bearded in the lower axils; spikelets 3(2-6)-flwd. (sometimes 7-9-flwd. *Gray*); pales and glumes lanceolate, hyaline, 3-veined, 1½″ long. ♃ Ill., and S.

9 **E. capillaris** (L.) Like E. tenuis, but the spikelets are minute (1—1½″), the fls. 2—4, acute, scabrous, with only the midvein apparent. ♃ Sandy fields. Aug.

10 **E. nitida** (Ell.) Plant 2—4f, glabrous and polished (except the bearded throat of the long, rolled lvs.); pan. 1½—3f long, narrow, branches some whorled; spkl. lance-linear, 3—4″, 5-12-flwd., on capillary divaricate pedicels; gl. and pales acute, 3-veined, often purplish, 1″ long. ♃ Marshes, Ill. (*J. Wolf*), and South.

11 **E. pectinacea** (Mx.) Gr. (E. hirsuta [Ell. etc.]). Culm 1—3f, rigid; sheaths some hairy; pan. very large, branches rigid, the lower deflexed in fruit; spkl. (*e, f*) oblong, purple, 2—3″; fls. 5—15, oval, acutish, strongly 3-veined. ♃ Sandy fields. July, Aug. (Poa spectabilis Ph.) A showy grass, sport of the winds when dry.

45. **POA**, L. Spear G. Meadow G. Spikelets 2—5(rarely –9)-flwd., compressed. Glumes subequal, pointless, shorter than the contiguous fls. Pales herbaceous, soft, awnless, the lower compressed-carinate, 5-veined, usually clothed at base with a cobweb-like wool. Grain free. Smooth grasses, with soft flat leaves, and panicled flowers.

§ Branches of the panicle in 2's, 3's, or often single..(*)
§ Branches of the panicle in about 5's, half-whorled..(**)
* Fls. not webbed, merely pubescent on the back...(*a*)
* Flowers webbed together at the base with gossamer-like wool...(*b*)
   a Annual or biennial. Panicle dense, spikelets subsessile..........No. 1
   a Perennial. Panicle loose, spikelets long-pedicelled..............Nos. 2, 3
     b Spikelets 2- or 3-flowered, on slender pedicels..............Nos. 4—7
     b Spikelets mostly 5-flowered, ovate, short-pedicelled..........Nos. 8, 9
  ** Spikelets 2-4-flowered, loosely pedicelled. Panicle large..........Nos. 10—12
  ** Spikelets 3-5-flowered, subsessile, panicles rather dense..........Nos. 13, 14

1 **P. ánnua** L. Low (3—8′), tender, spreading; culms flattened; lvs. 2—4′ by 1—2″; pan. 2—3′, dense; spikelets ovate-oblong, nearly sessile, loosely 5-7-flwd., 2—2½″; fls. lanceolate, acutish. ① ② Fields and lawns, forming a soft, dense turf. Com. Eur.

2 **P. flexuòsa** Muhl. Culms erect, 12—20′; lvs. linear, 2—5′; pan. very thin and open; branches filiform, often flexuous, long (2—3′), bearing the spikelets near the end; fls. 3—6, lance-linear, 2½″, 3-veined, remote. ♃ Woods, Va., Ky., and S.

3 **P. hexántha** Wood. Weakly erect, 1½—2f, leafy to the top; branches of the thin panicle filiform, suberect, straight, 2—4′; spkl. few, terminal, oblong, 3—4″; *fls. six* (5—7), oblong, 1½″, 5-veined, very obtuse. ♃ Meadows, Atlanta, Ga.

4 **P. brevifòlia** Muhl. Culm compressed, 1—2f, its lvs. generally short (½—2′), abruptly cuspidate, root lvs. long, pointed; pan. loose, branches filiform, spreading; spikelet ovate, purplish; fls. 3 or 4, 2½″, lanceolate, 5-veined, webbed. ♃ Pa. to Ill.

5 **P. débilis** Torr. (*d*) Culms terete, weak, 1½—2f; pan. loose, some spreading, branches capillary, in 2's and 3's; spkl. (*c*) few, ovate; fls. (*f*) 3(2-4), broadly oblong, very obtuse, 1½″, the glumes ovate, 1″; ligule oblong, acute. ♃ Woods, R. I., and W.

6 **P. dinántha** Wood. (*a*) Culm compressed, very slender, 1½—2f; lvs. long, 1″ wide; ligule short, truncate; pan. slender, branches in 1's and 2's, suberect; spkl. (*b*) ovate; fls. (*c*) 2(1—3), linear-oblong, acute, 1½″; gl. ½ as long. ♃ Fields, Ala. May.

7 **P. laxa** Henke. Culms tufted, 6—8′; lvs. erect, 1—3′, very narrow; pan. open, 1—2′ long; spkl. few, 2½″ long; glumes acuminate, as long as the (3) purplish fls. (1½″); lower pale villous on the keel. ♃ Mountains, N.

26

8 P. alpìna L. Culms erect, 6—12'; lvs. broad-linear, 1—2' by 2—3''; panicle equal, ovoid-oblong, loose, with rather large (3'') ovate spikelets; flowers about 5(4—9), ovate. ♃ Isle Royal, L. Superior (*Porter*), C. W., and North.

9 P. compréssa L. *Blue G.* Plant bluish green; culm *compressed*, decumbent at base, rigid. 12—18'; pan. contracted. 3' by 1', or less; spikelets glomerate, ovate oblong; fls. 3—7, 1'' long. ♃ Pastures, etc.: common. May, June.

10 P. sylvéstris Gr. Culm compressed, erect, 1—2f; lvs. linear, soft; pan. oblong pyramidal, thin; branches flexuous, the middle longest; spkl. oval, 1½''; fls. about 3, lance-oblong, 1''. obtuse. ♃ Woods, meadows, N. Y. to Va., and W.

11 P. cæsia Sm. (P. nemoralis Torr. P. alsodes Gr. P. Guadini K.) Culm compressed, 18—30. sheathed to near the top; pan. large (6—12' long), loose, roughish; spkl. lance-ovate, 2—2¼''; fls. 2 or 3, lance-linear, acute, as long as the very acute glumes (1½—1¾''); pales obscurely veined. ♃ Woods, N. H. to Penn., and Wis.

12 P. serótina Ehrh. *Foul Meadow. False Red-top.* Culms erect, weak, 2—3f; lvs. narrow, flat, long; ligules elongated, torn; pan. large, open, capillary; spkl. 2- or 3-flwd., 1½—2'' long, often tawny; gls. and fls. acute, narrow. ♃ Wet. N. July.

13 P. triviàlis L. *Rough Meadow G.* Culms roughish backward, 20—30'; lvs. rough-edged, the lower elongated; ligules long, pointed; pan. dense, lance-shaped, 3—5', spkl. subsessile, 2-3-flwd., fls. oblong, acute, strongly 5-veined. ♃ N. Jn., Jl.

14 P. praténsis L. *Spear G. June G.* Smooth; culm 1—2f, terete; ligules short, truncate; pan. open, egg-shaped, 3—10'; spkl. ovate, subsessile, 2''. about 4-flowered; fls. ovate, acute, close. ♃ Abundant and valuable. April, May.

46. BRYZOPÝRUM, Link.
Spikelets ∞-flowered, compressed, crowded in a spikelike panicle. Glumes unequal. Pales awnless, sub-coriaceous, not carinate, obsoletely many-veined. ♃ Leaves mostly rolled, smooth and rigid. Fls. diœcious.

B. spicàtum Hook. (*a*) Culm rigid, erect, 10—20', branched at base, beset with many bayonet-shaped lvs., 1—3', the highest exceeding the short, spikelike panicle (*a*); spkl. (*b, c*) 7-9-flwd. (*d*, pistillate flower, *e*, a stamen.) Salt marshes, Conn. to Car. July.

47. GLYCÈRIA, Br. MANNA G.
Spikelets ∞-flwd., teretish or turgid, rachis jointed. Glume subequal, pointless. Pales awnless, webless, herbaceous, the lower mostly 7-veined, rounded on the back, not carinate. Grain free. ♃ Smooth grasses in wet places, with creeping rhizomes and simple panicles. Sheaths mostly fistular (not split).

§ Salt marsh grasses. Lower pale 5-veined.
   Stigmas sessile, simply plumed..........Nos. 1, 2
§ In fresh swamps, etc. Lower pale 7-veined.
   Stigmas doubly plumous...(*a*)
  a Spikelets linear-lanceolate, in a very simple panicle.........................Nos. 3, 4
  a Spikelets linear-oblong, in compound, spreading panicles..................Nos. 5, 6
  a Spikelets ovate, short, turgid...(*b*)
    b In slender appressed panicles........Nos. 7, 8
    b In an open, recurved panicle......Nos. 9, 10

Order 155.—GRAMINEÆ.   403

1 G. marítima Wahl. Culm 1—1½f, terete; lvs. rolled; pan. erect, dense, the branches in pairs; spkl. terete, about 5-flwd., fls. obtuse. ♃ Mass. June.
2 G. distans Wahl. Culm 1—2f, terete, firm; lvs. flat; pan. spreading, the branches fascicled in 3's—5's; spkl. oblong, sessile, 3(3–6)-flowered. ♃ N. Y.
3 G. flùitans (L.) Culm flattened, 3—5f; lvs. broad-linear; ligule very large; pan. secund, virgate; spkl. linear, 8—10″; fls. 7—12, obtuse. Wet. June.
4 G. acutiflòra Torr. Culm flattened, 1—2f; lvs. narrow; pan. long, raceme-like; spkl. linear, 9—12″; fls. 4—6, distant, acute. ♃ Wet places, Penn., and N. June.
5 G. aquática (L.) (g) Stout, leafy, 3—5f; lvs. broad, soft; pan. diffuse, with spreading, flexuous branches in 3's—5's; spikelets (h) purple, 2—3″, wi.h 6—8 ovate, obtuse flowers (k). ♃ Wet places, Pa., and N. A handsome grass.
6 G. pállida Trin. Weak, ascending, 1—2½f; lvs. flat, with long ligules; pan. capillary, spreading; spkl. few, 3″; fls. 5—9; lower pale 5-toothed at apex, upper 2-toothed; the veins conspicuous. ♃ Swamps, Va., and N. June.
7 G. nervàta Trin. Culm 3—4f; lvs. broad-linear, ligules torn; pan. large, diffuse, branches in 2's and 3's, capillary, pendulous in fruit; fls. about 5, in the ovate-oblong spikelet, conspicuously veined. ♃ Wet, N. June.
8 G. elongàta Trin. Culm terete, erect. 3f; lvs. narrow, ligule very short; pan. raceme-like, nodding, 8—10′; branches solitary or in 2's, appressed; spkl. tumid, of about 2 obtuse, 5-veined fls. Meadows, N., M., and W. July.
9 G. obtùsa (Muhl.) Pan. dense, oblong, erect, 3—4′; spkl. ovate, acute, thick, of 5—7 ovate, obtuse fls.; lower pale obscurely 7-veined; culm 2—3f, lvs. often longer, dark green. ♃ Swamps, Penn., and N. Aug., Sept.
10 G. Canadénsis Trin. (m) Panicle large, 6—8′ long, branches flexuous, in half-whorls, spreading or recurved; spkl. (n) broad-ovate, 6-8-flwd.; upper pale (o) very obtuse, lower acute and longer. ♃ 3—4f. Shady, N. July.

48. BRIZA, L. Quaking G. Spikelets cordate, 6–9-flowered. Glumes 2, unequal, roundish. Pales ventricous, lower one cordate, embracing the shorter roundish upper one. Grain beaked. Paniculate, spkl. large, drooping on slender pedicels.

1 B. mèdia L. Pan. erect, spreading; spkl. soon cordate, of 5—9 flowers; gl. smaller than the greenish-purple veinless flowers. ♃ Meadows, coastward, N. Eng. to Penn. May. (b, c)
2 B. máxima. Pan. nodding at top; spikelets oblong-cordate, of 13—17 flowers. ⓘ Gardens. Cultivated for the curious spikes, which are light-brown, hyaline, ½′ in length. From Europe.
3 B. minor. Pan. erect, diffuse: spkl. triangular, 5-7-flwd.; glumes larger than the flowers. ⓘ From Europe. Small and pretty.

49. UNIÒLA, L. Union G. Spkl. compressed, and two-edged, 3-20-flwd. Lower fl. or fls. neutral, of 1 pale, similar to the 2 carinate gls. Pales awnless, the lower wing-keeled, upper doubly so. Sta. 1 or 3. Grain free. ♃ Smooth, erect, often branching.

§ Spikelets 6—16″ long, in large open panicles, drooping..........................Nos. 1, 2
§ Spikelets 2—6″, subsessile, in slender, spikelike panicles..........................Nos. 3, 4

1 U. latifòlia Mx. (a) Culm 2—4f; lvs. very broad, ½—1′ wide; spikelets oblong-ovate, 9—12″, flat, 9-13-flowered, drooping on slender pedicels; glumes (c) unequal, much smaller than the fls. (b) Sta. 1. ♃ Dry woods, M., W. Elegant. August.

2 **U. paniculàta** L. *Sea-side Oats.* Culm 4—8f; lvs. long, narrow, rolled, fringed at throat, spikelets ovate, short-pedicelled, 12—20-flwd.; lower pale obtuse, 9-veined; stamens 3. ♃ Sand-hills, coastward, Va. to Fla. July.

3 **U. nítida** Baldw. Culm wiry, 2—3f; lvs. narrow, flat; pan. simple; spkl. subsessile, broad, with about 7 long-pointed fls. Sta. 1. ♃ Ga. to La.

4 **U. grácilis** Mx. (*d*) Slender, 3—4f; lvs. broad-linear, flat; pan. long, simple, branches solitary, appressed; spkl. (*e*) 2′′, 3-4-flwd. Sea-coast, N. Y., and South.

50. **PHRÁGMITES**, Trin. REED. Fls. 3—6, the lowest sterile and monandrous; rachis beset with long silky hairs. Gl. acute, keeled, very unequal. Lower pale subulate, silky villous at base. Sta. 3. Grain free. ♃ Tall; lvs. broad and flat; panicle diffuse.

P. commùnis Trin. Culm erect, 6—12f, near 1′ thick; lvs. 1—2′ broad; pan. effuse, spkl. (*a*) 4-5-flwd., erect; fls. (*b*) colored, as long as the white hairs. Ponds. July.

51. **ARUNDINÀRIA**, Rich. CANE. Spkl. flattened, 5-12-flwd., fls. all ☿, triandrous, remote. Gl. (*a*) small. Lower pale lance-ovate, rounded, awn-pointed. *Stigmas* (*b*) 3. Grain (*c*) free. ♄ ♄ Tall, branching, leafy. Flowers in spikes or panicles.

A. macrospérma Mx. (*a*) Culm woody, from strong running root-stocks, 10—25f high, with fascicled branches; lvs. lanceolate, 1f and less; spkl. 1—2¼′ long, subsessile on leafless axillary or radical branches (from the rhizome) Swamps, Va. to Ky., and S., forming *the brakes*.

β. *tecta.* Culm 2—10f; lvs. lance-linear; spikes mostly radical.

52. **LEPTÙRUS**, Br. Spikelet 1 on each joint of the filiform rachis impressed into a cavity, 1- or 2-flwd. Gl. coriaceous, acute, subulate. Pales acute, subequal. Stam. 3. Grain linear, free. ① Culm branching, leaves very narrow. Spikes solitary or panicled.

L. paniculàtus N. (*c*) Culm ascending, 10—18′; lvs. near the base, filiform-subulate, short; rachis ⅔ of the culm, the slender spikes 2′, alternate, remote; spkl. 2′′, gls. lateral, shorter than the pales. Illinois to Louisiana.

53. **HÓRDEUM**, L. BARLEY. Spkl. 3 at each joint of the rachis, 2-flowered, the lateral imperfect or abortive. Gl. 2, subulate, awned, collateral, all 6 in front of the cluster. Lower pale long-awned, both adhering to grain.

1 **H. jubàtum** L. *Squirrel-tail G.* (*a*) Culm terete, 2f; lvs. broad-linear; spike 2—3′ long; spkl. (*b*) with the lateral fls. neuter, the 7 awns 6 times (2′) as long as the flowers. ② Marshes, N. Eng. to Mo., and N. June.

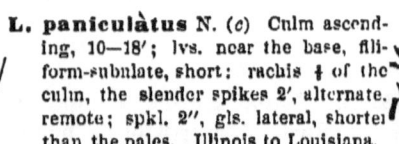

2 **H. pusillum** N. Culm ascending, 4—12′; lateral fls. awnless; central fl. ι with 3 subequal awns (7″); spike linear, 1—2′ long. ② Ohio, and W. May.
3 **H. vulgàre.** *Four-rowed B.* Culm 2—4f; lvs. broad, auricled at base; spike thick, 2—4′; fls. all fertile, fruit in 4 rows. ① Cultivated. May.
4 **H. distichum.** *Two-rowed B.* Culm and leaves as above. Lateral fls. abortive; fruit arranged in two rows. ① More common in cultivation. June.

54. ÉLYMUS, L. LYME G. WILD RYE. Spikelets 2—4 at each joint of the rachis, 2–6-flwd. Gl. 2, subulate, placed on the outer side of their spikelet, forming an involucre to the group, sometimes minute, or obsolete. Pales coriaceous, involving the grain, the lower acute or awned. (See Addenda.)

§ ELYMUS *proper.* Involucre present, consisting of the conspicuous glumes...(a)
¶ GYMNÓSTICHUM. Invol. glumes small or minute, or obsolete............................No. 6
a Spikelets 5-8-flowered, soft-pubescent, without awns..........................No. 5
a Spikelets 1-5-flowered, hard, rough, with conspicuous awns...(b)
b Spikelets glabrous, merely rough, 2- or 3-flowered.................Nos. 1, 2
b Spikelets hispid with hairs, 1-3- or 2-5-flowered..................Nos. 3, 4

1 **E. Virginicus** L. Culm erect, 3—4f, smooth; lvs. broad, flat, scabrous; spike 3—5′ long, thick, erect, often sheathed at base; gl. lance-linear, strongly veined, tipped (as well as the 2 or 3 fls.) with short (6—10″) awns. ♃ Banks. August.
β. *arcu͂tus.* (a) Glumes thickened and connate-arcuate at the base. S.
2 **E. Europœus** L. Culm erect, 3—5f; lvs. broad, flat, scabrous; spike suberect, 6—8′, exserted; spkl. in 3's, 2-flowered, scabrous, each with 4 long (1½—2′) straight awns; glumes linear, 5-veined. ♃ River banks, South.
3 **E. Canadénsis** L. (b) Spikes 4—8′ long, rather loose, nodding, hairy; spikelets (b) in 2's and 3's, 3-6-flwd.; awns of the flowers (c) usually curved, longer than (7—13″) those of the lance-linear glumes; culm 3—5f. ♃ Banks. August.
4 **E. striàtus** Willd. Spike 3—4′ long, dense, suberect; spikelets in pairs, 1-3-flwd., hispid-pubescent; awns subequal, 3 or 4 times longer than the flowers. ♃ Banks and rocky woods. Culm slender, 2—3f. August.
β. *villòsus.* Culm 3—4f, sheaths villous, and the glumes very hairy.
5 **E. mollis** Trin. Culm 2—4f, stout, soft-pubescent above, as well as the erect 5–8 spike; spikelets in pairs, about 7-flwd.; leaves and sheaths smooth. Shores, N-W.
6 **E. Hìstrix** L. *Hedgehog G.* Glabrous, tall (3—4f); spike erect, 4—6′; spikelets remote on the flexuous rachis, widely divergent, 2- or 3-flwd.; fls. subulate, ½′ long, their awns straight, 1′ or more; glumes commonly rudimentary. *Mr. J. Wolf* sends specimen from Illinois with awn-like glumes 4—8″ long. ♃ Woods. July.

55. LÓLIUM, L. DARNEL G. Spkl. ∞

flwd., sessile, remote, placed edgewise to the axis, the terminal one with 2 glumes, the lateral with but 1. Pales herbaceous, the lower awned or mucronate.

* **L. perenne** L. *Ray Darnel.* (a) Smooth, simple, 1—2f; spike 5—8′; spkl. 15—20, oblong, 5—6″, awnless. 7-13-flowered, flowers exceeding the glume. ♃ Fields. May, June ⚥

2 L. temuléntum L. *Poisonous D.* Smooth, 2f, simple; lvs. rough-edged; spkl.
5-7-flwd., remote on the scabrous rachis, shorter or not longer than their glume; fls.
twice shorter than their awn. ① Fields. Pa., and N. Grain *poison.* (b, c)

β. *Canadénse* (Mx.) Fls. awnless! or some of them short-awned; glume 1' long, much exceeding the flowers. Wayne Co., N. Y. *E. L. Hankenson.*

56. TRÍTICUM, L. WHEAT. Spikelets sessile in 2 rows on the teeth of the rachis, and sidewise to it, its upper fls. abortive. Gl. 2, equal, opposite, mucronate. Pales 2, the lower awned or mucronate. Spike simple, rarely branched.

§ AGROPYRUM. Glumes lanceolate, acute or awn-pointed . Nos. 1, 2
§ TRÍTICUM. Glumes ovate-oval, obtuse or truncate.. ...Nos. 3, 4

1 T. repens L. *Couch G. Quick G.* (a) Culms trailing at base, then erect, 1—2f, *from long creeping rhizomes* (Fig. 257, p. 78); spike (a) erect, 3—5'; spikelet remote, lance-oblong, 5-7-flowered; awns short or 0. ⚄ A vile weed, in gardens, etc. June, July. (b, a flower.)

β. *dasystáchyum.* Glaucous; spikelets hoary-pubescent. Lake shores, N-W.

2 T. violàceum Hornm. Erect, 2—3f; root fibrous; spike slender, dense, 2—4'; spkl. closely imbricated, 3-5-flwd.; awns 1—3" long, straight. Mts., Pa. (*Porter*), & N.

3 T. caninum L. *Dog's Couch G.* Ascending, 2—3f; rt. fibrous; sp. dense; spkl. 5-7-flwd.; awns (6") twice longer than the pale, some recurved. ⚄ Fields, Del. to Wis.

4 T. VULGÀRE. *Common Wheat.* Culm firm, 3—5f; leaves broad-linear; spike somewhat 4-sided; spkl. crowded, broad, 4-flwd.; gl. blunt, round-convex; flowers often awned; grain free. ① ② Varies as Summer Wheat, with awns, and sown in spring; and Winter Wheat, without awns, sown in autumn.

57. SECÀLE, L. RYE. Spikelets single on the teeth of the rachis 2–3-flwd., the 2 lower fls. fertile, sessile opposite, the upper one abortive. Gl. 2, opposite, subulate. Pales 2, herbaceous, the lower awned.

S. CEREÀLE. Culm firm, 4—6f high; lvs. glaucous; spike linear, flattened, 3—6', nodding; lower pale and its long straight awn ciliate-scabrous. ① ② Said to be native in the steppes of Caucasus. Cultivated from earliest times.

58. LEPTÓCHLOA, Beauv. Spkl. 3- ∞-flwd., subsessile, in one-sided, slender spikes. Gl. 2, keeled, awnless. Pales membranous, awnless or awned, the lower keeled, 3-veined. Lvs. flat and soft. Pan. composed of many long, slender spikes. Aug., Sept.

§ Spikelets 2-4-flowered. Lower pale simply acute....................................Nos. 1, 2
§ Spikelets 6-10-flowered. Lower pale mucronate and notched......................Nos. 3, 4

1 L. mucronàta K. Culm ascending, 2—3f; leaves broad-linear; pan. 1f or more; spikes filiform, 3—4', floriferous from base; spikelet of fls. minute, shorter than the mucronate glumes. ① Fields, Va. to Ill., & S.

2 L. filifórmis R. & S. (b) Tall, stout; pan. 1—2f; spikes filiform, straight, suberect, 5—8', very many; spk... of fls. (d) exceeding the acute glumes. ①? S-W.

3 L. fasciculàris (Lam.) (a) Tall, stout; pan. oblong, dense, 9—15'; spikes 2—3'; spkl. (c) lance-oblong, 2—3", *short-pedicelled;* lower pale strongly 3-veined, the veins excurrent into 2 teeth and a cusp between Marshes, N. Y. S and W.

4 L. Domingénsis Link. Culms simple, slender; lvs. linear-filiform; spikes few (6—12), distant; spikelets nearly as in No. 3.   S. Fla. *(Chapman).*   Oct.

59. GYMNOPOGON, Beauv.   Spikes setaceous, corymbously panicled. Spkl. remote, 1-flwd., with an awn-like rudiment. Gl. 2, keeled, lance-linear. Lower pale with a straight awn near the tip.   ♃ Low, reed-like.

1 G. racemòsum B. (*a*) Culm ascending, 1½—2f; lvs. lanceolate from a broad base, short; spikes erect but soon spreading, thread-form, 5—8′, floriferous from base; gl. (*b*) pungent; fertile flower and abortive rudiment (*c*), both long-awned. Sands, N. J., and S.

2 G. brevifòlium Trin. (*d*) Culm 8—16′; lvs. 1—2′; spikes bristle-form, 4—6′, flower-bearing only above the middle; fertile fl. awned (*e*), rudiment not. Md., and S.

60. MANISŬRUS, L. Lizard-tail G. Spikes terminal and lateral, their short stalks involved in sheaths. Spkl. in pairs, 1-flwd., the lower ☿, the upper neutral, consisting merely of 2 empty subequal glumes. ☿ Glumes coriaceous, the lower rounded, concave. Pales hyaline, thin. ①

M. granulàris Swtz. Culm 2—3f, branching; sheaths hairy; leaves flat; spikes ½—1′, colored; spkl. minute, the perfect globular, its gl. tessellated. Waysides. S. §

61. CŸNODON, Rich. Bermuda G. Sp. digitate, one-sided. Spkl. 1-flwd. (*c*), with a rudiment. Gl. 2 (*d*), persistent. Pales 2, membranous, the lower keeled. Rudiment an awn-like pedicel.

C. dáctylon Pers. (*a*) Diffusely creeping, sending up short branches; narrow lvs. and sheaths hairy; spikes (*b*) 4 or 5, 2—3′ long, spreading. ♃ Waste grounds. Evergreen. Pa., and S. §

62. CHLORIS, Swtz. (Eustachys, Desv.) Spikes digitate-fasciculate, rarely few. Spkl. sessile along one side of the rachis, 2-8-flwd., the lower 1 or 2 fls. ☿, the rest neutral or ♂. Gl. 2, persistent, acute or short-awned. Lower pale keeled, mucronate or awned below the tip. Culms flattened, often branched. Leaves obtuse.

1 C. petræa (Thunb.) Culms 1—2f; lvs. linear, 2—4′, flat, on carinate sheaths; spikes 3—6, straight, erect; spkl. 2-flwd., brown, ciliate, bearded at base. ♃ Brackish. S.
2 C. glauca (Chapm.) Glaucous, stout, 3—5f; leaves 18—24′ by ½′; spikes about 20; spkl. roundish, upper flower obovate; pales brown. ① Marshes, Fla. Aug.+
3 C. Floridàna (Chapm.) Slender, 2f; lvs. glaucous, 2—4′; spikes 1 or 2; spkl. 3-flwd., light brown, middle flower ♂, upper neutral, both smooth. Barrens, Fla., Jl. +
4 C. radiàta. From E. Ind. Cultivated for ornament. Culms leafy at base, scapo-like, bearing at top numerous long, slender, radiating spikes; spikelets 2-flowered, with 2 long awns, the fertile flower bearded at base, the sterile club-shaped.

63. ELEUSÌNE, Gaert. Crab G. Yard G. Spikes digitate, uni lateral. Spikelet 5-7-flwd., sessile. Gl. obtuse, the lower smaller Pales awnless, lower carinate, upper bicarinate. Grain ovate-triquetrous, free, loose in its pericarp. Lvs. flat.

E. Índica L. Culms clustered, ascending, 3—6—12'; leaves linear; spikes (a) 2—4, rarely 1, linear, straight, spreading, 2—4' by 2''; spkl. (b) closely imbricated on the under side of the rachis, smooth; fruit brown. ① Waysides: common M., S-W. August.

64. DACTYLOCTÈNIUM, Willd. EGYPTIAN G. Spikes several, digitate, unilateral. Spkl. 2 - ∞-flwd. Gl. compressed-carinate, the upper awned. Pales boat-shaped, acute-mucronate. Grain roundish, free.

D. Egypticum Willd. Culms creeping and ascending, 1—1½f; lvs. ciliate at base; spikes commonly 4 (cruciate), pointed; spkl. 3-flwd. ① Fields: com. Va. to Fla. §

65. SPARTÌNA, Schreb. MARSH G. CORD G. Spkl. flat, 1-flwd., closely imbricated in a double row on one side of the triquetrous rachis, forming dense spikes. Glumes keeled, coriaceous. Pales awnless. Style very long. ♃ Rigid marsh grasses.

* Upper glume decidedly awned. Lower pale rough-hispid on the keel ....No. 1
* Glumes merely pointed...(a)
 a Lower pale rough-hispid on the keel........Nos. 2, 3
 a Lower pale smooth. Spikes 1—12..........Nos. 4, 5

1 S. cynosuroìdes Willd. Culm 2—4f, slender but firm; lvs. long, narrow, involute-filiform above; spikes 5—30. in a raceme-like panicle, each 2—4' long; upper glume with its awn 8—10'', lower glume and subequal pales 4—5''. Brackish soils. August.

2 S. polystáchya Willd. Culm 4—8f, ½—1' in diameter; leaves broadly linear, flat; spikes 20—50, in a dense panicle, and 3—4'; upper pointed gl. 6'', lower gl. 2—3'', half as long as the equal pales. Salt marshes, chiefly southward. Aug., +(a,b.c)

3 S. grácilis Hook. Culm 1—2f; lvs. rolled, rigid, rush-like; spikes 15—30, very short (½'), closely imbricated into a spike-form panicle. Swamps, Fla. July, August.

4 S. júncea Willd. Culm 1—2f, slender; leaves rolled and rush-like or setaceous; spikes 1—6, subsessile, 1—1½' long; upper glume 4'', lower 1½'', pales 3½''; whole plant glabrous except the rough-keeled upper glume. Marshes along the coast.

5 S. alterniflòra Lois. *Soft Marsh G.* Culm 3—5f, juicy; leaves channeled, long; spikes 3—12, sessile, appressed, their rachis produced and pointed; upper gl. lin., obtuse, smooth as well as the entire plant; lower ½ as long. Salt marshes August.

66. BOUTELOÙA, Lagasca. MUSQUITE G. Spkl. sessile in two rows on one side of the rachis, forming dense spikes. Glumes keeled, the lower larger. Flowers several, the lowest ☿, the rest abortive. ☿ Lower pale 3-toothed, upper 2-toothed. Abortive flowers awned.

§ ATHEROPÓGON. Spikes numerous and short, forming an erect, virgate, one-sided raceme; spikelets 4—8......No. 1
§ CHONDRÒSIUM. Spikes 1 or few, dense; spkl. ∞ ...Nos. 2, 3

ORDER 155.—GRAMINEÆ. 409

1 **B. curtipéndula** (Mx.) (c) Culm ascending, 1—2f; leaves lance-linear ; spikes 20—40, near ½' long, deflexed ; spkl. (a) 2-flwd., abortive fl. 1-awned. ♃ M., W. Jl.
2 **B. hirsùta** Lag. Culms tufted, 1f; leaves at base lance-linear, flat; spikes 1—3; glumes (b) glandular-hispid, shorter than the 3 awns of the smooth (d) sterile flower. ① Sandy soils, Wis., and S.
3 **B. oligostáchya** (N.) Culm filiform, 6—12'; lvs. at base subulate-setaceous; gl. and lower pale downy, equaling the 3 awns of the villous ster. fl. ♃ Wis., and W.

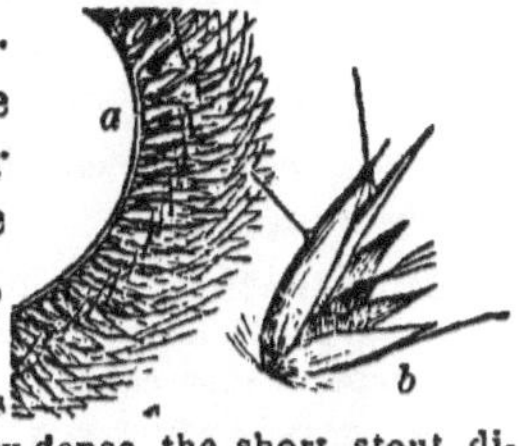

67. CTÈNIUM, Panner. TOOTH-ACHE G. Spkl. (b) 4–5-flwd., closely imbricated on one side of a flat rachis, middle fl. ☿, the upper and lower sterile. Upper gl. exterior, with an awned tubercle on the back. Lower ☿ pale awned near the apex, silky-fringed below. Spike solitary, recurved.

C. aromáticum (Ell.) Culm rigidly erect, 3—5f; leaves involute-setaceous above ; scorpoid spike (a) 4—6', very dense, the short, stout, divar. awns arranged in 3 rows. ♃ Sandy swamps, Va., and S. Curious. Herb. pung.

68. TRIPSACUM, L. SESAME G. Spikes ♂ above, ♀ below. Gl. coriaceous. ♂ Spkl. 2-flwd., inner fl. neuter. ♀ Spkl. 2-flwd., the lower abortive. Outer gl. covering the fls. in a cavity of the thick-jointed rachis, with an aperture each side at base.

T. dactyloides L. Culm solid with pith, 4—6f, stout ; lvs. broad and flat ; spikes (5—8') 2 or 3 together at top, and solitary in the sheaths, sometimes, in
β. *monostáchyon*, solitary at the top also. ♃ Banks and shores, Penn. to Ill.

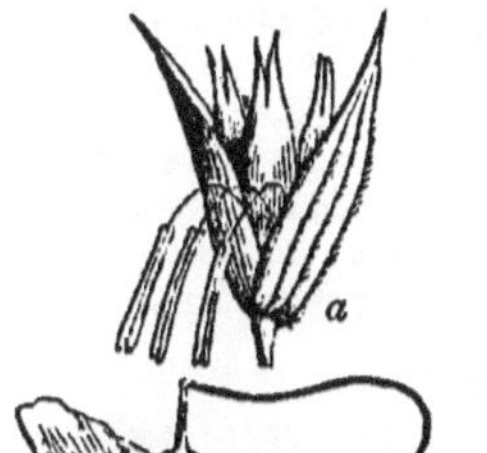

69. ZEA, L. INDIAN CORN. ♂ Fls. awnless. ♂ Fls. in a terminal panicle of racemes ; spkl. (a) 2-flwd. ♀ Fls. embedded in the thick axillary spadix (cob), which is enveloped in many bracts (husks) ; spikelets (b) 2-flowered, 1 fertile. Glumes roundish. Pistil thread-form (silk), very long, green. ① Culm solid.

Z. Mays L. Culm stout, erect, 5—15f, smooth, with many ample lin.-lanceolate lvs. Native of S. Am. Cultivated in many varieties. Grain always in even 8—24 rows in the *ear*, golden yellow, varying to br.-purple or pearl-wh.
β. JAPÓNICA. Leaves variegated with stripes of white and green. Gardens.

70. ROTTBŒLLIA, Br. RAT-TAIL G. Spkl. in pairs at each joint of a terete spike, one sessile in a cavity of the rachis, 2 flwd., the other pedicelled, abortive. Lower fl. of the sessile spkl. abortive. Gl. 2, subequal, the outer concave, coriaceous. Pales hyaline. ♃ Spikes pedunculate. Culm solid.

1 **R. cylindrica** (Mx.) Pedicellate spkl. a minute rudiment ; ☿ glume ovate acute, obscurely impressed-dotted in lines ; spikes *cylindric*, slender, single

culm terete, slender, 2—4f, with very narrow involute-setaceous leaves. Dry barrens, Fla. to La. July.+ (R. campéstris N.)

2 R. rugòsa (N.) Pedicellate spkl. neutral; ♂ gl. lanceolate, transversely *rugous*; spikes 2—3', terminal and axillary; culm compressed, 2—4f. Swamps, S. Sept.+

3 R. corrugàta Baldw. (*a*) Pedicellate spkl. (*d*) staminate; ♂ gl. (*c*) ovate, deeply reticulately pitted; spikes 3—6', colored; culm compressed, 2—4f. Lowlands. S.

71. STENOTÁPHRUM, Trin.

Spike flattened. Spkl. 2-flwd., in pairs at each joint, embedded, one pedicelled and sterile, the other sessile and constructed like Panicum (p. 391). ♃ Culm branched.

S. dimidiàtum (Thunb.) (*a*) Smooth, leafy, decumbent, 2—3f; leaves (*b*) lance-linear, flat; spikes single, lateral and terminal, 3' by 3'', joints *not* separating. Low lands, S. June.+

72. ERIÁNTHUS, Rich. PLUME G.

Spkl. all fertile, 2-flwd., in pairs at each joint of the slender rachis, one sessile, the other pedicelled, both involucrate at base with a tuft of hairs. Gl. subequal, exceeding the fls. Lower fl. neutral, of 1 hyaline pale, upper of 2, 1-awned. ♃ Stout, erect grasses, with flat leaves and tawny silky panicles.

§ Hairs of the invol. much longer than the spkl. . Nos. 1, 2
§ Hairs of the involucre short or none............ Nos. 3, 4

1 E. alopecuroìdes Ell. Culm (6—10f!) and broadlvs. silky-hirsute; panicle dense, oblong, 12—20'; hairs of the invol. twice longer than the (2¼'') spkl., thrice shorter than the straight awn which is terminal on its pale. Wet pine-barrens, N. J., W. and S. (a, b)

2 E. contórtus Ell. Culm (4—6f), and broad-linear leaves glabrous; panicle oblong, 6—10'; hairs of the invol. thrice longer than the (3'') spkl., twice shorter than the *contorted* awn issuing from the base of the 2-cleft pale. Wet grounds, S.

3 E. brevibárbis Mx. Culm and leaves as in the last; panicle dense, 8—14'; hairs shorter than the (4'') spkl.; awn some twisted, 8—10''; pale bifid. Low grounds, S. (c)

4 E. strictus Bald. Culm (4—7f) and long, narrow (3—5'') leaves glabrous; panicle *strict*, spike-form, 10—20', reddish brown; awn straight; invol. almost 0. Banks, S.

73. SACCHÀRUM, L. SUGAR-CANE.

Spkl. all fertile, awnless, in pairs, one sessile, the other pedicellate, 2-flwd., lower fl. neuter, of a single pale, upper fl. ☿ of 2 pales. Gl. 2, subequal. Pales 2, hyaline. Sta. 1—3. ♃ Gigantic tropical grasses with branching panicles. Spikelets cinctured at base with long silky hairs.

S. OFFICINÀRUM. Culm solid, short-jointed, erect, 8—20f; lvs. many, broad and flat; pan. 1—2f, of numerous racemes, richly clothed with the long, white, silky, involucrate hairs. Native of S. Asia. Cultivated far South.

74. ANDROPÒGON, L. BEARD G.

Spkl. in pairs at each joint of a slender rachis (*a*), one on a plumous-bearded pedicel (*d*) imperfect, the other (*e*) sessile, 2-flwd. Lower flower of 1 empty pale, upper flower ☿ of 2 hyaline pales, the lower tipped with an awn. Sta. 1—3. ♃ Culms erect, branched, coarse. Flowers spiked.

Order 155.—GRAMINEÆ.

§ Hairs copious silky, longer than the gl. Sta. 1 ....Nos. 1–3
§ Hairs shorter than the glumes. Sta. 3 (1 in No. 4)...(a)
  a Spikes digitate. 2—4 together at summits.................Nos. 4, 5
  a Spikes single, one at the top of each branch.. ....Nos. 6—8
  a Spikes clustered, paniculate; awns very long..................No. 9

1 **A. macroùrus** Mx. Culm erect, 2–3f, much branched; spkl. very delicate, in pairs, with a spathe, very many, forming a dense leafy, silky panicle; sterile spikelet only a pedicel; ☿ awn a straight bristle, 8″, hairs 4″. Wet grounds, N. Y., and S. Sept.+

2 **A. Virgínicus** L. Culm triangular, tall (3–-5f), the upper half loosely paniculate and nodding; spikes (like No. 1, light and feathery, 1′, two from each spathe) scattered; sterile spikelet a mere pedicel; awns 9″; spathe 2′. Dry soils. Sept., Oct.

3 **A. argénteus** Ell. Culm purplish, slender, 1–3f; branches 1 or 2 at each upper node, each with a pair of spikes 12—15″ long at top: fls. concealed by the silvery-white hairs; awn 7—8″. No spathe. Dry soils, Va., and S. Sept., Oct.

4 **A. tetrástychus** Ell. Culm erect, 2—3f; leaves and sheaths very hairy; branches 1 or 2 at each node, each with 4 (rarely 2) spikes at top; sterile spikelet an awnlike glume only; glume serrulate; awn 4 times its length. Low lands, S. Sept.

5 **A. furcàtus** Muhl. *Forked spike.* Culm erect, 4—7f; lvs. and sheaths glabrous; spikes purplish, digitate, in 2's—5's, 3—5′ long; spkl. appressed, the stalked one ♂; awn of the ☿ flower bent, 8—10″ long. Meadows and prairies: common. August.

6 **A. tener** (Nees). Culms 2—3f, slender, rigid; leaves narrow, rigid; spikes erect, 2, slender; spkl. appressed; pedicellate fl. neuter; ☿ awn bent, 4—6″. Dry barrens.

7 **A. ciliàtus** (Nutt.) Culms 3–-4f, with long linear lvs.; spikes 3—6′, on long pedicels; hairs close-pressed, white; spkl. awnless, the stalked one ♂, Damp, S.

8 **A. scopàrius** Mx. *Broom G.* (a) Culm 3f, erect, with erect, often fascicled branches; lvs. more or less hairy; spikes single on the filiform pedicels, loose, 6-12-flowered, hairs spreading nearly as long as the fls.; ☿ awns 6″ long, twisted; stalked flower (b) neuter, or (in β. *Halei*) (d) staminate. In dry fields, forming tufts.

9 **A. melanocárpus** (Muhl.) Culms 4—8f; lvs. glabrous; spikes numerous, clustered; spkl. many, large, each from a subulate spathe, the 2 lower spathes longest, glume-like ☿ awn 3—4′ long, twisted. Fields, Ga., Fl. Sept.+

75. SORGHUM, Pers. BROOM CORN. Spkl. in 2's and 3's, panicled, the middle spkl. complete, 2-flwd., lower fl. abortive. Lateral or lower spkl. sterile. Glumes coriaceous, pales membranous. Sta. 3. Otherwise like Andropogon. Culms simple.

1 **S. nutans** (L.) *Indian G. Wood G.* Culm 2—4f; pan. elongated, 10—20′, narrow, nodding; spkl. all tawny, the sterile reduced to mere pedicels in contact with the ☿, all bristly ciliate; awn contorted, longer than the flower. ♃ Dry: common.

2 **S. saccharàtum.** *Broom Corn.* Culm thick, solid, 6—10f; leaves broad, downy at base; panicle large, diffuse, with the slender branches whorled; ☿ glumes hairy, persistent. (T) E. Indies.

3 **S. vulgàre.** *Indian Millet.* Culm erect, 6—12f, round, solid; leaves broad, keeled pan. compact, erect, oval; glumes and pales caducous, fruit naked. (T) E. Ind.—The *Sugar Sorghum* is regarded as a variety of this species.

76. COIX LACRYMA. JOB'S TEARS. Culm 1—2f, solid, with erect, slender branches clustered in the upper sheaths; leaves lanceolate. Spikelets few in the short spikes, awnless, the lowest enclosed in an involucre which becomes ovoid, bony, polished, and bluish-white, likened to a falling tear. ① Garden. From E. Indies.

Subkingdom, CRYPTOGAMIA,

Or FLOWERLESS PLANTS. Vegetables destitute of true stamens and pistils, gradually descending to a mere cellular structure, with reproductive organs of 1 or 2 kinds, producing, instead of seeds, minute, dust-like bodies (spores) having neither integuments nor embryo.

PROVINCE, ACROGENS. Flowerless plants, having a regular stem or axis which grows by the extension of the apex only, without increasing in diameter, generally with leaves, and composed of cellular tissue and scalariform ducts. (Ferns, Mosses, Club-mosses, Horsetails, &c.)

ORDER CLVI. MARSILIACEÆ. PEPPERWORTS.

Herbs creeping or floating, with the *leaves* petiolate or sessile, circinate in vernation. *Fruit* (sporocarps) situated at the base of the leaves or leaf-stalks, containing the capsular *sporanges* of one kind with 2 kinds of spores, or of 2 kinds with the different spores separated.

* Leaves compound, on slender petioles, with 4 leaflets. Stems creeping............MARSILIA. 1
* Leaves simple, grass-like, radical. Stem a corm.................................ISOETES. 2
* Leaves minute, lobed, imbricated. Stem filiform, floating free.... ....AZOLLA. 3

1. MARSÍLIA, L. Sporocarps at the base of the leaf-stalks, of one kind, 2-celled, cells transversely many-celled, separating into two lobes at maturity. Sporangia inserted on each horizontal partition, of 2 kinds, some 1-spored, others ∞-spored. ♃ Stems creeping and rooting. *Leaves* petiolate, apparently radical, *of 4 whorled leaflets*, resembling clover.

1 **M. quadrifòlia** L. Lfts. round-cuneiform, as broad as long, glabrous; sporocarps oblong, smoothish, 1, 2, or 3 on each short peduncle, as large as a peppercorn. ♃ Petioles 3—5′ high. Margin of pond, Litchfield, Conn. (Prof. Eaton). Leaves floating.
2 **M. vestita** Hook & Grev. Lfts. cuneiform-obovate, longer than broad, glab.; sporocarps glob.-oval, 2½′′, hisp., 1 only on each short (3′′) peduncle, 2-*toothed* on back. S-W.
3 **M. uncinàta** Brann. Lfts. cuneiform-obovate, hispid, petioles 1—2′ high; sporocarps 2′′, subsessile at the base of the petioles, clothed with rust-colored wool. Iowa.

2. ISÒETES, L. QUILLWORT. Sporocarps oval, 1-celled, of 2 kinds, sessile in the axils of the radical lvs. and adhering to them. Spores in the *outer* sporangia larger, globular; in the *inner* minute, powdery. ♃ ≈ Leaves linear, grass-like, clustered on the short corm.

* Species growing under water, generally wholly submersed, in ponds, &c..Nos. 1—3
* Species growing in shallow water, or in damp grounds, emersed......... Nos. 4—7

1 **I. lacústris** L. Lvs. 2—6′, subulate, rigid, erect-spreading; sporocarps round-ovate, unspotted, the larger spores with crested ridges. Varies with the leaves setaceous-subulate and recurved. the sporocarps rarely a little spotted. N.

ORDER 157.—LYCOPODIACEÆ. 413

2 **I. echinóspora** Dur. Lvs. subulate, 3—10', red at base, 15—30 in number; sporocarps round-ovate, spotted, larger spores echinate with minute points. N. J., Pa , & N.
3 **I. fláccida** Shutt. Lvs. *flaccid*, 1—2f long, almost filiform, yellowish green; sporocarps oblong-ovate; spores not netted, minutely roughened. Ponds and lakes. Fla.
4 **I. ripària** Eng. Lvs. 10-30 in number, 4-8', lin. ; sporocarps oblong; spotted ; spores with a band of crested ridges, ash-colored ; leaves emersed. Del. R. (Porter), and N
5 **I. saccharàta** Eng. Leaves few (7—15), subulate-filiform, 2—3', recurved ; sporocarps ovate, spotless ; spores minutely tubercled. Wicomico R., Md. (Canby, Porter).
6 **I. melanópoda** J. Gay. Leaves very slender, 8—10', carinate on the back, brown at base ; sporocarps brown; spores smooth, smaller than in No. 5. Ill. (Prof. Porter).
7 **I. Engelmánni** Braun. Leaves 25—100, 10—20' long, filiform-linear, weak ; sporocarps oblong, spotless ; spores honeycombed all over. Shallow waters, E. and W.
  β. *grácilis*. Leaves about 10, very flaccid, 1f. N. E. to Ill. (J. Wolf).
  γ. *válida*. Lvs. very numerous, 2f, from a stock 6"—1' thick. Del. & Pa. (Porter).

3. **AZÓLLA**, Lam. Small floating plants, with filiform stems and minute imbricated leaves or fronds. Sporocarps of 2 kinds, sessile on the under side of the branches, the smaller sterile, filled with *antheridia*, the larger fertile, thin, containing *sporangia* on stalks, each with several spores.
A. Caroliniàna Willd. Lvs. ovate-oblong, obtuse, fleshy, ¼", reddish beneath ; sterile fruits 1 or 2 at the base of the fertile, and many times smaller. Still waters, N. & W.

ORDER CLVII. LYCOPODIACEÆ. CLUB MOSSES.

551. Lycopodium dendroideum. 552, A single spike. 553, A scale with its axillary sporange bursting. 554, Spores.

These are interesting evergreen creepers or runners, rarely erect, branching, abounding in ducts, with the *leaves* small, numerous, crowded, entire, lanceolate or subulate, 1-nerved. *Fruits* sessile, axillary or crowded into a spike, 2-valved, containing few rather large spores, or numerous minute ones appearing like powder.

1. **LYCOPÓDIUM**, L. CLUB MOSS. Spore-cases all of one kind, 1-celled, reniform, opening transversely, 2-valved ; spores numerous, minute, sulphur-yellow.—Leaves in 4, 8, or 16 ranks.

§ Fruit in pedunculated spikes (the fertile branches nearly leafless)...(c)
§ Fruit in sessile spikes (the branches leafy throughout)...(b)
§ Fruit scattered, axillary, forming no distinct spike........................Nos. 1, 2
  b Leaves of the spike bract-like, discolored..... ....Nos. 3, 4
  b Leaves of the spikes and stems all alike........................Nos. 5, 6
    c Spikes several (2—6) on each peduncle........................,Nos. 9, 10
    c Spike solitary on each peduncle. ....Nos. 7, 8

1 **L. Selàgo** L. *Fir Club Moss*. Erect, 2—6', fastigiately branched ; lvs. covering the branches, all alike, entire, acute and pungent, awnless. Tops of high mountains, N.

2 L. lucídulum Mx. *Shining C.* Ascending, forking, 8—16′; lvs. in 8 rows, linear-lanceolate, denticulate, shining, spreading or reflexed, pointed, large for the genus (3—4′′), the fruitful ones like the rest, as in No. 1. Damp woods.

3 L. inundàtum L. *Marsh C.* Stem creeping, often submersed, the simple solitary ped. 1—3′ (Conn., Mr. Bowles) or 4—7′ (Mass., Dr. Ricard); leaves soft and fine, curving upward; spike solitary, 1—1½′ long, leafy. Swamps, Can. to Car.

4 L. alopecuroides L. Sterile branches decumbent, shorter than the tall (7—20′) erect fertile ones; leaves crowded, subulate, awned; spikes leafy, 2—3′ long. Swamps in pine-barrens, N. J. to Fla. and La.

5 L. annótinum L. Creeping, branches twice forked, ascending 6—8′; leaves in 5 rows, lance-linear, spreading, denticulate; spikes solitary. Woods, N.

6 L. dendroìdeum Mx. *Tree C. Ground Pine.* Erect, about 8′, with its erect branches spirally arranged, forked and crowded; lvs. lance-linear, in 6 equal rows; spikes several but solitary, 1½′, yellow-brown. Woods. Very elegant.

β. *obscùrum.* Branches spreading; spikes 1 or 2, greenish brown.

7 L. Caroliniànum L. Stem and branches creeping and rooting; lvs. appearing 2-ranked, the lateral spreading while the others are appressed, lanceolate; peduncles simple, 2—4′, bearing each a single spike. Barrens, N. J., and S.

8 L. sabinæfòlium Willd. *Ground Fir.* Long, creeping; branches erect, short, with fastigiate branchlets; lvs. terete-subulate; ped. short. White Mts., and N.

9 L. complanàtum L. *Festoon Ground Pine.* Long, trailing; branches repeatedly forking, fan-shaped, spreading; leaves 4-ranked, the marginal connate, diverging, the others distinct, appressed; peduncles long, with 4—6 spikes. Woods.

10 L. clavàtum L. *Common C.* Extensively creeping, branches ascending; leaves scattered, incurved, bristly-acuminate; peduncles erect, remotely bracted, 3—5′, bearing a pair of straight spikes 2′ long. In shades; common.

2. **SELAGINÉLLA**, Spr. DWARF CLUB MOSS. Fruits of two kinds, viz., *antheridia*, which are 1-celled, opening at apex; and *oöphoridia*, larger, containing 1—4 (rarely 6) globous-angular grains.—A large genus. The species are cultivated in every greenhouse. Spikes quadrangular, bracts in 4 rows. (Lycopodium L.)

§ Leaves all alike and similarly imbricated all around. Native..............Nos. 1, 2
§ Leaves of 2 kinds, in 4 rows, those of the 2 lateral rows larger and spreading, of the 2 intermediate rows superficial, small, appressed...(*a*)
  a Slender rootlets produced along the stems.—*x* Leaves unequal-sided..Nos. 3—5
    —*x* Leaves equal-sided....Nos. 6—8
  No rootlets, &c.—*y* Stems erect, frond-like, simple, stalk-like below..Nos. 9—11
    —*y* Stems diffuse, branched from the base............Nos. 12, 13

1 S. rupéstre (L). Sts. ascending, 2—4′, divided into numerous tufted, mossy branches; leaves crowded, fine, blue-green, ciliate; spike indistinct, 6′′. Rocks.

2 S. selaginoìdes (L). Stem filiform, creeping, branches suberect, 3—6′, the fertile simple, 1-spiked; leaves lanceolate, yellow-green, ciliate. Woods, N.

3 S. apus Spr. Stem weak, loosely branched, with hair-like rootlets near the base; leaves ovate, slightly oblique, acutish, the smaller ones pointed. Damp. †

4 S. STOLONÍFERA. Sts. producing long threadform rootlets below, 3-4-pinnately branched; branchlets 2—4′′ broad; lvs. imbricated, ovate, entire, obtuse, the smaller ones with a filiform straight point. The older stems become zigzag. 6–10′. Com. (S. Mertensii.)

5 S. DENTICULÀTA (or Kraussiàna). Prostrate, delicate, remotely and somewhat 3-pinnately branched; leaves 1′′, oblong-ovate, *minutely denticulate*, acute, distant on the stem, crowded on the branchlets; smaller leaves with reflexed points. Very common.

β. VARIEGÀTA. Ends of the branchlets with their leaves white. Rootlets hair-like.

6 S. UNCINÀTA (or cæsia). Long-creeping, with hair-like rootlets. 2-3-pinnately branched, branchlets crowded, short, 2" wide; leaves crowded, oblong, entire, obtuse, the smaller ones with an *uncinate* (reflexed) slender point.

7 S. SERPENS. Stems prostrate, with hair-like rootlets, 2–3-pinnate; branchlets short and crowded, 1" wide; lvs. crowded, round-ovate, cordate, obtuse, entire, the smaller acute.

8 S. DELICATÍSSIMA. Sts. creeping, 5–8', rooting, filiform, loosely 2–3-pinnate, 1" wide; leaves ovate, obtuse, ciliate, not crowded, the middle ones scarcely smaller, acute.

9 S. CAULÉSCENS. Glabrous, suberect, 12–18', 3–4-pinnately branched, fern-like, and lanceolate in outline; branchlets close, 1½" wide; leaves close, ovate, entire, very acute, the points turned upward; smaller leaves mucronate; stem straw-colored.

10 S. WILLDENÒVII. Like the last as to stems and branches, but they are finely pubescent, and the leaves are less crowded, ovate, and obtuse. 6—12', ovate in outline.

11 S. ERÝTHROPUS. Stems red, with scattered, appressed leaves; frond wide-spread, somewhat palmate, with crowded branchlets and leaves, branchlets 1½" wide; leaves ovate-oblong, oblique, obtuse, ciliate, the smaller with long straight points.

12 S. CUSPIDÀTA. Stem or frond 3—6', densely and somewhat dichotomously branched; branchlets 1" wide; leaves closely imbricated, all nearly alike, elliptical, ciliate, bristle-pointed, with the point inclined upward.—A variety (perhaps the fertile stems) are lanceolate in outline, 2–3-pinnately branched.

13 S. LEPIDOPHÝLLA, *Resurrection Moss,* is a roundish ball when dry. In a cup of water it soon expands into a dense circle of dark-green, densely 2–3-pinnate fronds, with innumerable oval, obtuse, entire leaves. From Lower California.

3. PSILÒTUM, R. Br. Sporangia sessile, 3-celled, imperfectly 3-valved by terminal chinks, filled with farinaceous spores.—Stem fork-branched, with alternate, minute leaves, as if leafless.

P. triquetrum Swtz. Stem erect, 8–10', many times forked, and, with the branches, 3-angled; leaves remote, ½"; fruit 3-lobed, sessile along the branches. E. Fla

ORDER CLVIII. EQUISETACEÆ. HORSETAILS.

Plants leafless simple stems, or with whorled branches. *Stems* striate-sulcate, jointed, fistular between, and separable at the joints. *Sheaths* dentate, crowning each internode. Fructification a dense, oblong-cylindric, terminal, and cone-like spike, composed of 6-sided, peltate scales, arranged spirally, bearing beneath 4—7 spore-cases, which open laterally. *Spores* globular, each with 4 *elaters* attached, involving them spirally, or open when discharged. (See Figures.)

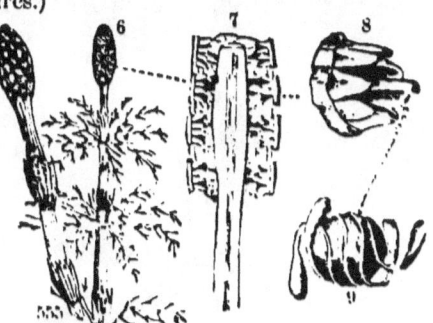

EQUISÈTUM, L. SCOURING RUSH. Character the same as that of the order.—The sheaths may be regarded as a whorl of united lvs. The cuticle abounds in silex.

555, Equisetum arvense. 556, E. sylvaticum. 557, Section of the spike, enlarged. 558, A peltate scale with 7 sporanges beneath (or one compound sporange), magnified. 559, A spore with its elaters highly magnified.

§ Species fruiting in Spring and decaying before the following Winter...(*a*)
§ Species fruiting in Summer and lasting through the following Winter ...(*b*)

ORDER 159.—FILICES.

a Fertile stems never branching, the sterile with simple, whorled branches..Nos. 1, 2
a Fertile stems at length, like the sterile, with compound, whorled branches..Nos. 3, 4
  b Stems with whorls of simple branches from the middle joints..........Nos. 5, 6
  b Stems mostly simple, large, 20–40-furrowed...........................Nos. 7–9
  b Stems always simple, very slender, 3–9-furrowed.....................Nos. 10, 11

1 **E. arvénse** L. Fertile stems erect, 6—8′, simple; sterile 12–14-furrowed, with simple, ascending, 4-angled branches; sheath cut into long dark-brown teeth; spike 6—12″, oblong. Can. to Va. and Ky. The sterile stems appear after the fertile.
  β. *serótinum*. Sterile plant also producing a late spike of fruit. Pa. (Porter).

2 **E. Telmateìa** Ehr. *Ivory H.* Sterile stem 2—5f, *white*, about 30-furrowed, its 30 branches 4-angled; fertile stems simple; sheaths with subulate teeth. L. Superior.

3 **E. sylváticum** L. Stems 12- or 13-furrowed, both kinds with *compound*, deflexed, angular branches, 9—16′. Woods and low grounds. North.

4 **E. praténse** Ehr. Stems 10–12-furrowed, both kinds soon producing *simple*, straight branches, in several whorls; branches 3-angled. N. W.

5 **E. limòsum** L. *Pipes.* Stems 2—3f, smooth, erect, 15–20-striate, mostly with a few irregular, simple, 5-sided branches near the middle; sheaths white above, with 15—20 teeth, tipped with black. Shores and swamps.

6 **E. palústre** L. Sts. 1—1½f, erect, with 6—8 prominent striæ; branches few, sheaths with as many pointed teeth as striæ. Marshes, N. Rare in the United States.

7 **E. lævigàtum** Braun. Stems 2—3f, erect, simple or some branched; sheaths long (6—7″), close, green, with 20—25 black teeth; branch sheaths 8-toothed. Miss. River.

8 **E. robústum** Braun. Sts. 2—4f, very stout, some branched above; sheaths short (3—4″), close, with 40 (in the branches 11) deciduous teeth, and a black band near the base, rarely with another above. River banks, W. States to California!

9 **E. hyemàle** L. *Scouring Rush.* Stems all simple, erect, 2f, very rough with silicious points; sheaths ashy-white, black at base and summit, short (2—3″), with about 20 subulate, awned, deciduous teeth. Conspicuous in wet shades.

10 **E. variegàtum** Schleicher. Simple (branched from base), slender, straight, 6—12′, 5–9-furrowed; sheaths very short, with brown bristle-tipped teeth. N. Rare.

11 **E. scorpioìdes** Mx. Sts. tufted, filiform, 4–8′, recurved, 3–4-furrowed; sheaths black, teeth 3 or 4, scarious and bristle-tipped. Woods, Penn., and N.

ORDER CLIX. FILICES. FERNS.

Stem a perennial, creeping, horizontal rhizome, or sometimes erect and tree-like. *Fronds* (fruit-bearing leaves) variously divided, rarely entire, with mostly forked veins and *circinate* vernation. *Fruit* occupying the back or margin of the fronds arising from the veins. *Sporangia* (spore-cases) of one kind, scattered, or clustered in *sori*, 1-celled, containing numerous minute spores.

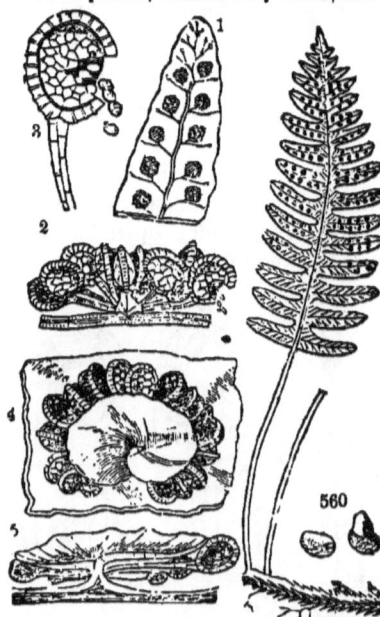

Fig. 560, Polypodium vulgare, frond pinnate. 561, A leaflet of the frond enlarged, showing the sori. 562, One of the sori enlarged, showing the sporangia. 563, One sporange further magnified, bursting and discharging its spores. 564, A sorus of Aspidium marginale covered with the indusium. 565, Side view of the same.

ORDER 159.—FILICES.

A large and interesting Order, distinguished for their elegant, plume-like foliage. They are usually a few inches to a few feet high, but some of the Tropical species, as the Cyatheæ, are 15 to 25 feet, vieing with the Palms in size and beauty.

☞ The *stipe* is the stalk of the frond, and the *rachis* its continuation through it. The *pinnæ* (or *pn.*) are the first divisions of a divided frond (often called leaflets). *Pinnulæ* (or *pnl.*) are the first divisions of the pinnæ when further divided. *Segments* (*seg.*) are the *final* divisions, and the partial divisions of the segments are *lobes*, &c. The *sori* (fruit-dots) are either *naked*, or covered with an *indusium* (see cut).

§ POLYPODIACEÆ. THE TRUE FERNS, with fronds mostly radical, circinate in bud. Sporangia in *sori*, pedicellate, with a vertical, elastic ring, opening transversely...(*f*)
§ CYATHEACEÆ. THE TREE FERNS, with fronds on an erect trunk. Sporangia as in § 1...(*e*)
§ HYMENOPHYLLACEÆ. PELLUCID FERNS ; sporangia in a cup and on a thread...(*d*)
§ SCHIZÆCEÆ. Very slender vines or fronds. Sporangia with a ring-crown at apex...(*c*)
§ OSMUNDIACEÆ. Fronds stout, radical. Sporangia with no ring, 2-valved...(*b*)
§ OPHIOGLOSSACEÆ. Frond single (in our species), on an erect stem. Sporangia with no ring...(*a*)

a Fruit in a spike. Frond entire, reticulate-veined......................................OPHIOGLOSSUM. 1
a Fruit in a panicle. Frond divided, fork-veined........................... ...BOTRYCHIUM. 2
b Fronds pinnate or bipinnate, with straight, forked veins........................OSMUNDA. 3
c Fronds palmately lobed. Stems climbing, 3–4f..LYGODIUM. 4
c Fronds linear-filiform, undivided, a few inches high......................................SCHIZÆA. 5
c Fronds 3-parted, middle division sterile, the lateral paniculate........................ANEIMIA. 6
d Fronds pellucid or opaque. Sporangia with a transverse ring................TRICHOMANES. 7
e Fruit-dots in little round cups. Trunk and leaves smooth........................ ....CYATHEA. 8
e Fruit-dots becoming entirely naked. Fronds prickly or hairy........................ALSOPHILA. 9
e Fruit-dots enclosed in the reflexed tip of the lobe, with two valves...............§ BALANTIUM. 23
f Sporangia scattered singly all over the surface (not in sori), naked...(*q*)
f Sporangia collected in dots (sori) growing from the veins...(*h*)
g Fronds simple or pinnate. Pinnæ on short petiolules..............................ACROSTICHUM. 10
g Fronds forked at the summit, entire below, the sterile different................PLATYCERIUM. 11
h Sori (fruit-dots) naked, having no covering of any kind...(*k*)
h Sori involved (at first) in the rolled segments of the panicled fertile frond...(*m*)
h Sori not involved, but invested with special coverings (called *indusia*)...(*n*)
k Fronds smooth or scaly, never powdery. Sori distinct, roundish..................POLYPODIUM. 12
k Fronds covered with powder on the back. Sori in many dorsal lines..........GYMNOGRAMMA. 13
k Fronds powdery or scaly on the back (bipinnate). Sori in a marginal line.......NOTHOLÆNA. 14
k Fronds linear, simple. Sori in a continuous line on the split margin. Fla......VITTARIA lineata
m Fertile frond bipinnate, segments berry-like. Veins reticulated.................ONOCLEA. 15
m Fertile frond pinnate, pinnæ moniliform. Veins forking..........................STRUTHIOPTERIS. 16
m Fertile fronds bipinnate, segments oblong, soon opening......................ALLOSORUS. 17
n Sori marginal, indusia only the reflexed altered margin of the frond...(*o*)
n Sori marginal, indusium double—a scale combined with the margin...(*p*)
n Sori dorsal, oblong or linear, indusium attached to the side of a vein...(*q*)
n Sori dorsal, round or roundish, indusium on the back or the tip of a vein ..(*r*)
o Fronds of 2 kinds, the fertile contracted. Sori continuous to apex......... . LOMARIA. 18
o Fronds all similar, smooth. Indusia continuous all around. Stipe green or brown..PTERIS. 19
o Fronds woolly, &c. Sori separate or continuous. Stipe brown, hairy......CHEILANTHES. 20
o Fronds smooth. Sori separate. Stipe black and polished..................ADIANTUM. 21
p Indusium a 2-lipped cup at the edge of the segments.......................DICKSONIA. 22
p Indusium an entire cup or goblet at the edge of the segments........ ....DAVALLIA. 23
q Sori parallel to the mid-vein, the indusia opening toward it...(*r*)
q Sori oblique to the mid-vein, borne laterally on the veinlets...(*s*)
r Sori linear, nearly continuous, in 2 rows, sunk in the frond........................WOODWARDIA. 24
r Sori oblong, remote, in two rows and superficial. Stipes black........................DOODIA. 25
r Sori linear, in 1 double row, the whole length of the segment........................BLECHNUM. 26
r Sori oblong, in 1 short double row. Frond finely cleft. ....ONYCHIUM. 27
s Indusia single, regularly arranged, in 2 rows............................ASPLENIUM. 28
s Indusia single, scattered irregularly. Frond simple or lobed............CAMPTOSORUS. 29
s Indusia double, regularly arranged. Frond simple........ ...........SCOLOPENDRIUM. 30

x Indusium :upform, fringed, fixed beneath all around the sorus.. ....WOODSIA. 31
x Indusium hoodform, fixed by the base and 2 sides............................CISTOPTERIS. 32
x Indusium reniform, opening only toward the margin of the segm. Fla...NEPHROLEPIS *exaltata.*
x Indusium round-reniform, fixed in the midst, open all around......................ASPIDIUM. 33

1. OPHIOGLÓSSUM, L. ADDER'S TONGUE.
Sporangia roundish, naked, opening transversely, arranged in two rows along the margins of the fertile, contracted, spike-like frond. Veins reticulated.

1 **O. vulgàtum** L. Root of thick fibres; stem simple, bearing 1 oblong-ovate, entire, smooth frond, 2—3', with no mid-vein, and a terminal spike, 1—2'. A curious little plant, in low grounds. Vernation straight, as in all this section,—not circinate.

2 **O. bulbòsum** L. Root a globular corm; frond ovate to reniform, on the stem close to the ground. Wet pine-barrens, N. J., and S. Often 2 stems from 1 corm.

2. BOTRÝCHIUM, Swartz. MOONWORT. GRAPE FERN.
Sporangia subglobous, 1-celled, 2-valved, distinct, coriaceous, smooth, adnate to the compound rachis of a racemous panicle. Valves opening transversely.

§ Frond ternately divided, segments palmately veined......................Nos. 1, 2
§ Frond pinnately divided, segments pinnately veined.....................Nos. 3—5

1 **B. lunarioìdes** Swtz. Scape 8—12', bearing a stalked frond near the base and a panicle of numerous little 2-ranked spikes at the top; frond in 3 bipinnatifid divisions; segment obliquely lanceolate, crenulate. Shady pastures and woods.
  β. **disséctum.** Frond more numerously dissected, almost tripinnatifid.

2 **B. simplex** Hitchcock. Frond simple, or 3-lobed or parted, segm. broad-wedge-obovate, small, incised or subentire, unequal; spike compound, interrupted, small. Dry hills, Vt., Mass. Whole plant 3—6'. Frond 6—12'', short-stalked, near the base.

3 **B. negléctum** Wood. Frond 1—2', simply pinnate, with oval or ovate incised pinnæ, short-stalked, on upper part of stem, which is 5—8' high. Pan. 1—2'. N. H., Vt., to Pa.—*Prof. Porter* regards both this and No. 2 as var. of B. matricariæfolium Braun

4 **B. lanceolatum** Angst. Frond bipinnatifid, closely sessile, triangular in outline with lanceolate, incised segments; panicle 2- or 3-pinnate. N. J., Pa., to L. Sup. (O. B. Wheeler). Certainly distinct from No. 3.

5 **B. Virgínicum** L. *Rattlesnake Fern.* Stem 1—2f, with the large (5—8') tripin., triangular frond sess. at or above the middle; ultimate segm. obtuse, 3-5-toothed; pan. decompound, 3—6', reddish br. A beautiful Fern, in damp woods, not uncom. Jn., Jl.

3. OSMÚNDA, L. FLOWERING FERN.
Sporangia globular, half 2-valved, roughened on the surface somewhat in lines, pedicellate and clustered on the lower surface of the frond or a portion of it, which is more or less contracted into the form of a panicle. Spores green. Tall, handsome Ferns. Veins forked, straight. June.

§ Frond bipinnate with distinct pinnæ, the upper part contracted and fertile....No. 1
§ Frond pinnate with pinnatifid pinnæ, partially or separately fertile........Nos. 2, 3

1 **O. regàlis** Mx. A large and beautiful Fern in meadows and swamps; fronds 3—4f, glabrous, bipinnate, fruiting above in an ample panicle; pinnæ with 6—9 pairs of distinct, oblong, serrulate, subsessile leaflets; fruit rust-colored.

2 **O. cinnamòme** L. Sterile fronds pinnate, in clumps 3—5f; pinnæ pinnatifid with ovate-oblong, obtuse, entire segments; fertile frond bipinnate, pinnæ all contracted, panicled, clothed with cinnamon-colored wool.

3 **O. Claytoniàna** L. Fronds ample, 2—3f, smooth, pinnate, the pinnæ lance-linear, pinnatifid, some of the intermediate ones fertile, contracted and raceme-like.

4. LYGÒDIUM, Swartz. CLIMBING FERN.
Sporangia sessile, arranged in 2-ranked spikelets issuing from the margin of the contracted frond, open-

ing on the inner side from the base to the summit. Indusium a scale-like veil covering each sporange. (Fig. 310.)

L. palmàtum Swtz. Smooth throughout; stem flexuous, thread-like or wire-like, climbing 3—5f; fronds palmately 5–7-lobed, 2 on each short stipe, lobes entire, obtuse; upper fronds contracted, fertile, each a cluster of spikelets. Abundant in a swamp in Windsor, Conn. (Dr. Wm. Wood); also rarely found in N. J., Ky., and S.

5. **SCHIZÆA**, Sm. Sporangia oval, crowned with a ring at top, sessile, opening laterally. Indusium continuous, formed of the inflexed margins of the lfts., which are contracted, spike-like, crowded at the top of the frond.
S. pusilla Ph. Fronds clustered, simple, linear-filiform, tortuous, 3—6′, the fertile bearing a few little spikelets at top in two rows. Barrens, Quaker Bridge, N. J. Aug.

6. **ANEÌMIA**, Swtz. Sporangia sessile, crowned with a ring, in 1-sided panicled spikes, in partially or wholly fertile fronds. Indusium none. Fronds erect.
1 **A. adiantifòlia** Sw. Fronds 6—12′, on a slender stipe, 3-parted, the middle division sterile, 2- or 3-pinnate, the lateral ones fertile panicles on long stalks. S. Fla. †
2 **A.** Mandioccàna. Fronds 12—15′, long-stiped, 3-parted like the other, but the sterile division simply pinnate with lance-oblong serrulate pinnæ. S. America.

7. **TRICHÓMANES**, L. Sporangia with a transverse complete ring, and arranged on the base of a thread-like receptacle, which is *in* and exserted *from* a cup at the edge of the pellucid frond.
1 **T. radicans** Sw. Fronds thin and delicate, 6′, lance-ovate, bipinnatifid, pinnæ triangular, obtuse, very oblique at base; receptacle exserted. South. Rare.
2 **T. élegans**. Sterile frond pinnate, fertile, long-linear, edged and fringed all around with the thread-like receptacles and their cups. From S. America.

8. **CYÀTHEA**, Sm. Sori globular, on the veins, wholly enclosed in an indusium, which soon opens and remains cupform. Sporangia subsessile on an elevated receptacle. ♄ With cylindrical trunks.
C. arbòrea. Trunk 10—20f, unarmed, simple, crowned with a spreading tuft of bipinnate fronds 6—8f long, gracefully arched; pinnulæ again pinnatifid or lobed, cups in 2 rows, smooth, round, entire. Grows near Panama! †

9. **ALSÓPHILA** áspera. Another Tree Fern, from W. Indies, cult. by Mr. Buchanan, at Astoria, N. Y., under the name of *Hematelia horrida*. Trunk 6—10f, bearing a splendid crown of fronds 4—5f long, arched and spreading, tripinnate. Piul. deeply lobed, lobes obtuse, each with a double row of fruit-dots, which at first are covered with jagged scales, but finally naked. Stipe and rachis prickly.—**A.** pruinàta, very elegant, with a trunk near 1f, clothed with light-brown woolly hairs, and a crown of light-green bipinnate fronds, 3f long, is growing with the other. •

10. **ACRÓSTICHUM**, L. Fronds simple or pinnate. Sporangia scattered (not in sori), occupying the under surface of the whole or a part of the frond. Veins netted.
A. aùreum L. A noble Fern, 3—6f high, coriaceous, evergreen, pinnate, with alternate, lance-oblong, entire pinnæ. Swamps, Fla., and in conservatories.

11. **PLATYCÈRIUM**, Desv. Stag-horn Fern. Fronds coriaceous, net-veined, forking at the summit. Sporangia in large patches on the under surface of the frond. From Africa, &c.

P. alcicórne. Sterile fronds roundish, lobed, spreading; fertile erect, 10—16', dark green above, pale beneath, fruiting on its 2—4 lanceolate segments. Curious.

12. POLYPÒDIUM, L. Polypody. Sori roundish, scattered on various parts of the under surface of the frond, with no indusium (cover or involucre).—Ferns of various habit.

> * Fronds simple and entire, pinni-veined, with cross veinulets..............Nos. 1, 2
> * Fronds pinnatifid or pinnate, with forking veinlets.......................Nos. 3—6
> * Fronds bipinnatifid, the veinlets forked (Phlegopteris)..................Nos. 7—9

1 **P. Phyllítidis** L. Fronds lance-linear, 1—2f, pointed, thin and papery, with the fruit-dots arranged in a double row between the veinlets. Fla., and W. Indies. †

2 **P. Língua.** Fronds lance-ovate, 6—12', obtuse, smooth above, rusty-downy beneath, and there covered with the innumerable sori, in rows. China.

3 **P. incànum** Ph. Fronds deeply pinnatifid, 3—6', thick, clothed with whitish scales beneath; pinnæ oblong-linear, the upper fruitful; sori distinct and separate: veins invisible. Grows on the mossy bark of trees, W. and S.

4 **P. vulgàre** L. Fronds deeply pinnatifid, smooth, 6—12', pinnæ linear-oblong, alternate, sori large, in 2 rows, distinct, yellow-brown. On shady rocks.

5 **P. Plumula** Willd. Fronds lance-linear, 1f × 1½'; pinnæ linear-oblong, very numerous, attached to the hairy rachis by a broad base. Fla., and cultivated.

6 **P. angustifòlia.** Fronds lin.-lanceolate, 18' × 2', bright green; pn. oblong, attached to the chaffy rachis by the mid-vein only, the base auriculed on the upper side.

7 **P. Phlegópteris** L. *Beech P.* Frond bipinnatifid, longer than wide (3—6'), the lower pinnæ curved, but scarcely larger than the middle ones; sori all marginal, about four on each segment; stipe hairy. Woods, Can. to Penn., and W.

8 **P. hexagonópterum** Mx. Frond bipinnatifid, broader than long, rachis peculiarly winged; lower panicle much enlarged, deflexed; sori partly marginal, many on each segment; stipe smooth. Woods. Rather common.

9 **P. Dryópteris** L. *Ternate P.* Frond ternate, the divisions stalked and bipinnate, light green, thin and delicate; sori marginal. Woods, Penn., and N.

   β. *calcàreum.* Divisions of the frond more rigid, erect. Northward.

13. GYMNOGRÁMMA, Desv. Fronds 2-3-pinnate, covered beneath with a white or yellow farinaceous powder. Sori arranged in rows along the veins. A beautiful genus, much cultivated. Tropical America.

> * *Golden Ferns,*—the fronds yellow-powdery beneath......................Nos. 1—3
> * *Silver Ferns,*—the fronds white-powdery beneath, 2-pinnate.................. No. 4

1 **G.** triangulàris. Stipes clustered, slender, 3—12', polished, ebony-brown; frond 5-angled, 1—3', pedately pinnate; pinnæ triangular-oblong, finally the fertile covered with the russet sori beneath. Common in California. Very fine.

2 **G.** sulphùrea. Stipe and rachis brown, at first powdery; frond 6—10', lanceolate, bipinnate; pinnæ lanceolate; segments cuneate, cut-lobed, crenate at the obtuse apex. From Jamaica (Rev. E. Wilson), and cultivated. Very delicate.

3 **G.** chrysophÝlla. Frond triangular-lanceolate, bipinnate; pinnæ lanceolate, nearly contiguous; pinnæ cut-crenate-lobed. Golden yellow beneath.

   β. Merténsii. Pinnæ rather remote, narrow lanceolate, long-pointed.

4 **G.** calomélanos. Frond 2—3f, lance-ovate, stipe and rachis brown, polished; segments entire or with a single tooth, cream-white beneath.

   β. Peruviàna has the lower segment hastate-lobed and very rich green.

14. NOTHOLÆNA, Br. Frond 1-2-pinnate, scattered, coriaceous, chaffy, or powdery beneath. Sori marginal, linear, continuous, naked. Sporangia short-stalked.

1 **N. nívea.** Very delicate, 6—12', bright green above, covered with a dense white powder beneath; frond bipinnate; pinnæ roundish, top one lobed; stipe black. Mex.
2 **N. Eckloniàna.** Rare and beautiful, clothed in white wool-like scales, bipinnate, pinnæ ovate, remote, pinnulæ pinnatifid, oblong, segments roundish. South Africa.

15. ONOCLÈA, L. Sensitive Fern.
Fronds scattered, net-veined, the sterile broad, the fertile contracted and panicled, its convolute segments berry-like, enclosing the sori, which are otherwise nearly naked.

O. sensíbilis L. Fronds 1—2f, common in low grounds, very *sensitive* to frost. The fertile dark-brown in color. Sterile fronds deeply pinnatifid, with few oblong entire or lobed pinnæ, the upper confluent. July.

β. *obtusíloba.* Fertile frond partially metamorphosed, the segments partly revolute on the fruit. Wendell, Mass. (Mrs. Piper), to N. Y. and Penn.

16. STRUTHIÓPTERIS, Willd. Ostrich Fern.
Fronds clustered, the sterile bipinnatifid, fork-veined, fertile much contracted, brown, with the pinnæ revolute into a necklace form, enclosing the sori, which are otherwise destitute of an indusium.

S. Germánica Willd. Sterile fronds in a circular clump, 3—5f; pinnæ numerous, long and crowded, with numerous oblong segments; fertile fronds much smaller, their crowded pinnæ 1—2' long, appearing later in the season.

17. ALLOSÒRUS, Bernh.
Fronds small, 2-3-pinnate, fork-veined; the fertile some contracted, margins of the leaflets reflexed and meeting over the confluent sori, but soon opening.

A. acrostichoïdes Spr. Fronds in tufts, bipinnate, 3—6', pale green with whitish stipes; seg. oblong, the sterile crenate, the fertile entire, petiolulate. 2—3'' long. Isle Royal, in L. Superior (Prof. Porter), W. to Washington Terr. (Rev. Mr. Gray).

18. LOMÀRIA, Willd.
Fronds clustered, of 2 forms, the fruitful contracted. Sori marginal, linear, continuous; indusium linear, scarious, the reflexed edge of the frond, opening toward the mid-vein.

1 **L. spicant.** Fronds pinnate, long, and narrow, the fertile nearly solitary in the midst of the numerous sterile ones, and twice as tall (2—3f) as they; stipe purple, polished. Europe, Oregon. Very elegant. (Blechnum boreale.)
2 **L. gibba.** Fronds oblong-lanceolate, pinnate, pinnæ linear-falcate, 1—3', their broad bases almost confluent.
3 **L. ciliatélla.** Fronds oval to oblong; pinnæ oblong, slightly lobed, truncate at apex, ciliolate-spinescent with the projecting veins.

19. PTÈRIS, L. Brake.
Sori borne on the ends of the veins forming a marginal line or band, covered with the membranous, reflected edge of the frond. Fronds once to thrice pinnate, or decompound.

§ PTERIS *proper.* Sori a mere line. Stipes greenish or pale...(*x*)
§ PELLÆA. Sori forming a broad band. Stipes purple or brown...(*y*)

 x Frond triangular, twice or thrice pinnate, lowest pinnæ long-stalked..........No. 1
 x Frond pedately pinnate, the pinnæ few and long ....Nos. 2, 3
 x Frond pedately bipinnatifid, the pinnæ numerous. ....No. 4
 x Frond simply pinnate, with numerous long pinnæ ....No. 5
 y Frond pedate and pinnatifid, as broad as long, 5-angled. † ....No. 6
 y Fronds pinnate, pinnæ few, the lower again divided. Native. ......Nos. 7, 8
 y Fronds simply pinnate, or completely tripinnate. Cultivated.. .....Nos. 9 -11

1 P. aquilina L. *Common Brake.* Frond 3-parted, branches bipinnate, segments oblong, obtuse, the terminal often elongated. Abundant everywhere. 2—6f.
   β. *caudàta.* The terminal segment linear-oblong. Common South.
2 P. Crética L. Pale-bright-green, 1—1½f, smooth; pinnæ lin.-lanceolate, the lower ones 2-parted and petiolulate, serrulate; fertile longer, linear. Fla. Cultivated.
   β. *albi-lineata.* Pinnæ white-banded in the midst along the mid-vein.
3 P. SERRULÀTA. Bright green, 1—1½f; pinnæ long-linear, decurrent on the rachis, except the lowest pair, which are 2- or 3-parted and short-stalked. China.
   β. CRISTÀTA. Each segment expanded at apex into a fan-shaped blade.
4 P. QUADRIAURÌTA. Frond ample, ovate, 1—3f! smooth; pinnæ distinct, pinnatifid, lobes contiguous, oblong, obtuse, with the forked veins conspicuous. Jamaica.
   β. ARGYRIA. Pinnæ whitened in the midst along the mid-vein.
5 P. longifòlia L. Tall, 2—3f r'g'd; pinnæ lance-linear petiolulate, obliquely truncate at base; stipe, rachis, and mid-veins chaffy-hairy. Fla., and cultivated.
6 P. PEDÀTA. Bright green, 4—6'. Frond 3-parted, as broad as long; lateral pinnæ 2-parted, all deeply lobed, sorl in a broad band all around. From the W. Indies.
7 P. grácilis Mx. Delicate, smooth and shining, 4—6'; fronds lanceolate, the sterile bipinnatifid, fertile bipinnate with narrow segments. Rocks, Vt., and W.
8 P. atropurpùrea L. *Rock Brake.* Coriaceous; rachis hairy; lower pn. ternate or pinnate; segments oppsite, oblong, margins conspicuously revolute, with edges often meeting behind, as in Allosorus, 3—6—12'. On lime rocks, N. and S.
   β. *Alabaménsis* (Buckley). Taller (10—20'), bipin. below, some pn. ½-auriculate. S.
9 P. ROTUNDIFÒLIA. Stipe, rachis, and chaffy hairs purple, 1—1½f; frond narrow, simply pinnate; segments small, round or oval, alternate. From New Zealand.
10 P. TRÉMULA. Bright green, 2—3f, tripinnate; pnl. or segments linear-oblong, obtuse, serrulate, the lower ones again pinnatifid. From N. S. Wales.
11 P. HASTÀTA. Frond bipinnate, 12—18'; pinnæ cordate-hastate; segments ovate, the terminal ones much larger, oblong or hastate, or 3-lobed. Varies much. From S. Afr.

20. CHEILÁNTHES, Swtz. LIP FERN.

Fronds small, mostly 2-3-pinnate, chaffy or hairy, mid-vein central. Sori on the ends of the veinlets, distinct, or some confluent, covered by an interrupted or continuous indusium from the edge of the frond. Stipes brown.—Hardly distinct from the preceding genus.

1 C. vestìta Sw. *Indusia separate,*—the reflexed, unchanged tips of the ovate segm.; fronds 5—12', bipinnate, lin.-oblong, hairy; pn. crenately lobed. Rocks, M. and S., rare.
2 C. tomentòsa Link. *Indusia continuous,*—the membranous margin of the small, obtuse segm.; fronds tripinnate, lance-oblong, rusty, 12—18'. N. C., and W.

21. ADIÁNTUM, L. MAIDEN-HAIR FERN.

Sori oblong or roundish, marginal. Indusia membranaceous, formed from the reflexed margins of distinct portions of the frond, and opening inwardly. Stipe ebony-black, polished. Ultimate segments often dimidiate, the mid-vein on the lower margin.—A large and beautiful genus, much cultivated.

* Fronds pedately divided, the divisions 1-3-pinnate; segments oblique....Nos. 1—4
* Fronds pinnately divided 2—4 times; segments subequilateral............Nos. 5—8
* Fronds simply pinnate, with very large opposite oblique segments..........No. 9

1 A. pedàtum L. Very smooth; branches of the regularly pedate frond pinnate; segments rhombic-oblong, 1', toothed on the upper side, obtuse at apex; sori oblong-lunulate. 8—14'. Damp, rocky woods. Our most elegant native Fern.
2 A. PUBÉSCENS. Stipe rough-pubescent; pn. 5—7, irregularly pedate, hispid beneath. 6—9' long; segments oblong, 6—8'', contiguous; sori round, crowded. N. Hol. 1f

3 A. trapeziofórme. Frond ample, decompound, glabrous, 2f; segments light green, large (12—18″ × 6—10″), *trapezoidal*, some of them fan-shaped; sori lunulate on 2 of the 4 margins; stipe jet-black. Superb! Jamaica (Rev. S. B. Wilson).

4 A. Sancta-Katrina, has large obliquely fan-shaped segments cut-lobed and toothed, with the veins uncommonly distinct. Cultivated in Bridgman's Garden, Astoria.

5 A. Capillus-Véneris L. Delicate, bright green, 6—18′, smooth, thrice pinnate at base; segments round-cuneate, lobed, or the sterile toothed; sori reniform, one on each lobe; stipe and branches *capillary*. Lime-rocks, S.: rare. Eur. Cultivated.

6 A. cuneàtum. Very delicate, 1f, 4 times pinnate at base, bright green; segm. very numerous, sharply cuneate, 2-4-cut-lobed, 4—6″; sori round-reniform. Brazil.

7 A. Æthiópicum, tinctum and **callópodes**, are greenhouse species or varieties, with roundish segments more or less oblique and lobed, 4—7″, with rounded sori, 6—12′.

8 A. alàtum, has the rachis narrowly winged, segm. sessile, obovate-long-wedge-shaped at base, coarsely toothed at apex. (Greenhouse of Bridgman & Wiegand.)

9 A. macrophýllum. Stipe jet-black, simple, bearing about 3 pairs of large, opposite, thick leaflets, and an odd one; leaflets triang.-hastate, oblique; sori linear. Jamaica.

22. DICKSÒNIA, L'Her. Dickson's Fern. Sori marginal, roundish, distinct, terminating a vein. Indusium double, the proper one cup-shaped, opening outward, the other formed of a reflected lobule of the margin, and opening inward.

1 D. pilosiúscula Willd. Frond bipinnate, lanceolate, 2—3f, with minute glandular hairs; pn. sessile, lanceolate; segm. finely pinnatifid, lobes toothed, each with a minute round sorus. Rocky pastures. Stipe yellowish.

2 D. (BALANTIUM) **antárctica.** A beautiful tree-fern from New Zealand, 3—20f, crowned with many long, heavy, dark-green, tripinnate fronds; pn. and pnl. sessile; segm. oval, 6-crenate; sori globular, with 2 distinct valves. (Buchanan's Conserv.)

23. DAVÁLLIA, Smith. Sori globous, marginal, on the end of a vein, in a goblet or pyxis, half of which is formed by the scarious indusium opening outward. Root-stock creeping above ground, chaffy.

1 D. tenuifòlia. Fronds delicate, 6—10′, tripinnate with few pinnæ, triangular-lanceolate; rachis narrowly winged; segments spatulate, toothed. China.

2 D. canariénsis. *Hare's-foot.* Fronds 3-parted, decompound, ultimate segments elliptical, decurrent, bearing 1 pyxis. 1—2f. Canaries.

3 D. disséota, is very different, irregularly pinnatifid, or almost entire.

24. WOODWÁRDIA, Sm. Sori straight, linear-oblong, on transverse veinulets, parallel to the mid-vein, in 2 rows. Indusium from the same veinulet, opening inwardly.

§ LORINSERIA. Fronds of 2 forms, net-veined throughout..................No. 1
§ ANCHISTEA. Fronds all similar, netted only close to mid-vein..........Nos 2, 3

1 W. angustifòlia Sm. Fertile fronds pinnate, with distant linear pinnæ covered with the fruit beneath; sterile lance-oblong in outline, deeply pinnatifid; segm. oblong, 2—3f. Resembles Onoclea. Mass. (Dr. S. Bowles), and S.

2 W. Virginiea Sm. Fronds glabrous, lanceolate, pinnate; pinnæ remote, pinnatifid, lance-linear; segments oblong, obtuse, 2—3f. Swamps, E. and S.

3 W. Japónica. Rachis chaffy; frond triangular, as broad as long; pinnæ lanceolate, pinnatifid, with ovate segments. Bright green. 1—2f.

25. DOÒDIA aspera. Fronds rough, lanceolate, pinnate, 1f, in clumps, the caudex a few inches above ground. Pinnæ oblong-linear, contiguous, with spinescent teeth. Sori in 1 or 2 rows each side.—**D. caudàta** has linear-lanceolate, pinnate fronds, with remote serrate segments, the terminal one elongated. Both from Australia.

ORDER 159.—FILICES.

26. BLECHNUM, L. Sori continuous on the cross veinulets, close to and parallel with the mid-vein. Indusia opening inward.

B. serrulàtum Mx. Fronds pinnate, lanceolate, erect; pinnæ sharply serrulate, those of the fertile fronds contracted. Florida.

27. ONÝCHIUM LUCIDUM (or JAPÓNICUM). Delicately beautiful, from A Ind., and of the easiest culture. Fronds 1—2f, alternately pinnate 3 or 4 times into innumerable linear-acute segments 2 or 3″ long. Few of the segments fertile with an oblong bivalved sorus on the mid-vein half its length.

28. ASPLÈNIUM, L. SPLEENWORT. Sori linear or oblong, straight (curved in No. 9), separate, regularly arranged, oblique to the mid-vein, each arising with its indusium from the forward side of a lateral vein and opening forward. Veins forked or pinnate.

* Fronds simple and entire, with regular linear fruit-dots.....................No. 1
* Fronds simply pinnate.—*a* Pinnæ roundish, nearly as broad as long.. ...Nos. 2, 3
  —*a* Pinnæ long,—much longer than wide..........Nos. 4, 5
* Fronds partly bipinnate, with few divisions. Ferns small, 2—8′ high.....Nos. 6, 7
* Fronds twice pinnate, with very many divisions. Large native Ferns....Nos. 8, 9
* Fronds twice or thrice pinnate. Exotic Ferns cult. in conservatories...Nos. 10—12

1 A. NIDUS. *Bird's-Nest.* Fronds thick and rigid, polished green, tongue-shaped, obtuse, 2—4f, clustered in a circle, forming as it were *a nest*. Oahu, &c. A noble Fern.

2 A. FLABELLIFÒLIUM. Fronds very delicate, long and narrow (12—16′); rachis pro longed some 5′ beyond the pn., and rooting at the end; pn. broad-cuneate, lobed and toothed, remote and alternate on the rachis. Australia. Suitable for baskets.

3 A. Trichómanes L. *Dwarf S.* Frond 3—6′, lance-linear, in tufts; pn. roundish, small. subsessile, bearing several sori each; stipe and rachis polished-black. Rocks.

4 A. ebéneum Wld. *Ebony S.* Fronds 8—14′, erect, lance-linear; pn. lance-oblong, 1′, some curved, serrate, auriculate on the upper side; stalk polished-brown. Dry.

5 A. angustifòlium Mx. Fronds 2—2½f, in tufts, the inner fertile; pn. lance-linear, alternate, short-stalked, 2—5′, of a thin texture; stalks green. Woods, E. and S.

6 A. Ruta-muràrìa L. *Wall-rue.* Very small and delicate, 2—3′, 2-pinnate at base, pinnate above; pn. petiolulate, cuneate, erose-dentate, few, 3—4″. Dry rocks.

7 A. montànum Willd. Glabrous, 2-pinnate; tufts 4—8′; pn. oblong-ovate, parted into a few (5 or 6) 2- or 3-toothed segm.; rachis green, winged. On cliffs, Penn., & S.

8 A. thelypteroìdes Mx. *Silvery S.* Fronds ample, ovate-acuminate, 1½—3f; stipe pale; pinnæ lance-linear, pointed, distinct, subsessile; segments oblong, obtuse, serrate, sessile on the *winged* rachis, with 2 rows of linear distinct sori. Shady banks.

9 A. Fìlix-fœmina Bernh. *Lady Fern.* Fronds ample, 1—2f, lance-oblong; pn. lanceolate-acuminate, rachis *not winged*; pnl. lance-linear, cut-pinnatifid; segments minute, sharply 2-toothed; sori oblong, curved, finally confluent. Moist woods.

10 A. GOVINGIÀNA. Slender and weak (in conservatories), 1f, lanceolate-acuminate; pn. lanceolate, long-pointed, stalked; rach. winged; seg. acute, sharp-serrat.; sori oblong.

11 A. BELÁNGERI. Fronds lance-linear, 1—2f×2—3′, pinnate with deeply pinnatifid pinnæ, segments linear, small, and very numerous, each with a sorus. From Java. Stipe stout, green. The upper base (or axillary) segments are 2-parted.

12 A. BULBÍFERUM. Frond lanceolate, bipinnate, 1—3f; pn. lanceolate from a broad base, deeply pinnatifid; seg. oblong, cut-lobed and toothed, bearing 1—6 bold sori,— 1 to a lobe. Often produces young plants from bulblets on the upper surface. N. Hol.

29. CAMPTOSÒRUS, Link. WALKING FERN. Frond lanceolate, entire, or pinnatifid, with the apex prolonged and inclined to root. Veins more or less netted. Sori oblong, irregularly scattered, with the indusia lateral on the veinlets. (Antigramma, C-B.)

1 C. rhizophyllus Lk. Frond 6—12′, subentire, at base stipitate, cordate, or truncate, or somewhat auriculate, the apex attenuated in a long thread-like acumination, arched, and rooting at the point. Rocky woods. Not common.

2 C. pinnatifidus (Nutt). Frond 4—8′, abrupt at base, pinnatifid, with a long attenuated apex inclined to root; sori large, at length confluent. Pa. to Tenn. Rare.

β *ebenoides*. Frond at base pinnate; stipe black and polished. Near Phila.

30. SCOLOPÉNDRIUM, Smith. HART'S-TONGUE.

Sori linear, transverse, scattered; indusium double (arising from 2 contiguous parallel veins), occupying both sides of the sorus, opening lengthwise along the middle.

S. officinàrum Willd. Frond simple, ligulate, acute, entire, cordate at base, 8—15′; stipe chaffy, 3—5′. Shady rocks, Chittenango, N. Y. (Sartwell).

31. WOÓDSIA, Brown. ROCK POLYPOD.

Sori roundish, scattered; indusium fixed beneath the sorus, early opening above it, with a multifid or fringed margin, including the pedicellate spore-cases, like a calyx. Small, tufted ferns, with pinnated fronds.

§ Indusium closed over the sorus at first, toothed when open..................No. 1
§ Indusium concealed under the sorus, fringed with cilliæ..................Nos. 2—4

1 W. obtùsa Torr. Fronds 6—12′, lance-oblong, smoothish, almost tripinnate; pn. distant, sessile; segments pinnatifid, lobes rounded, toothed, each bearing a round fruit-dot, which dots at length almost meet. Rocks and cliffs. Vt. to Car., and W.

2 W. ilvénsis Br. Frond 4—7′, lanceolate, *bipinnate*, the stipe, rachis, mid-veins and their bristly chaff rust-colored; pn. oblong-obtuse, sessile, with 13—17 obtuse, subentire segments. Dry or rocky woods, in tufts. Stipe as long as the frond.

3 W. glabélla Br. Frond glabrous, lance-linear, 2—5′, *pinnate*; pn. ovate, very obtuse, 2—4″, 3-7-lobed, the upper only crenate. Cliffs, N. Y., Vt., and N. No chaff.

4 W. Oregàna Eaton. Frond glabrous, lance-elliptic, 2—8′, *pinnate*; pn. pinnatifid, obtuse; segments ovate, obtuse, denticulate; indusia with very short ciliæ. L. Sup.

32. CISTÓPTERIS, Bernh. BLADDER FERN.

Sori roundish. Indusium hood-shaped, vaulted, fixed by the broad base (or by the base and sides), soon opening toward the forward end of the frond and thrown off. —Delicate Ferns, 2-3-pinnate.

1 C. fràgilis Bernh. Frond lance-oblong, 6—10′, on a slender stipe of the same length, with open divisions; pn. lance-ovate; segments pinnatifid below, only serrate above, oblong, with prominent veins and 4—10 sori. Shady rocks. Common.

2 C. bulbífera Bernh. Frond long-lanceolate, 12—18′, the stipe shorter; pn. triangular-ovate, the lowest pair longest; segments oblong, obtuse, pinnatifid below, toothed above, 1 sorus to each lobe. Bears some bulblets. Shades.

33. ASPÍDIUM, L. SHIELD FERN.

Sori orbicular, scattered, terminal or lateral on the pinnate veins. Indusium orbicular, peltate or reniform with a deep sinus, covering the sorus, opening all around.

§ ASPIDIUM. Indusium round, entire, centrally peltate. Pinnæ mostly auricled on the upper side at base.—*x* Fronds simply pinnate..........................Nos. 1—4
    —*x* Fronds bipinnate............................Nos. 5, 6
§ NEPHRODIUM. Indusium roundish, with a sinus on one side (subreniform).. (*a*)
  a Frond simply pinnate, with a few large pinnæ. Cultivated... ....No. 7
  a Frond once-and-a-half pinnate.—*y* Segments thin, quite entire...........Nos. 8—11
                        —*y* Segments thick, finely serrate..... .....Nos. 12, 13
  a Frond twice pinnate.—*z* Segments bluntly lobed, or crenate or entire... Nos. 14, 15
                —*z* Segments sharply serrate, or lobed or toothed..Nos. 16, 17

1 **A. acrostichoides** Swtz. Frond narrow-lanceolate, 15—18'; stipe chaffy; pn talcate-lanceolate, ciliate-serrulate, 1—2', auriculate on the upper side at base, the upper covered with fruit, smaller than the sterile. Rocky shades. Common.
  β. *incisum*. Segments incised and sharp-toothed, most of them fertile. N. Y., &c.
2 **A. Lonchitis** Sw. Frond linear-lanceolate, rigidly erect, 8—13': pn. triangular-ovate, auricled on the upper side at base, longest (1') in the middle, gradually lessened to apex and base, all densely fertile. Lake Superior, and N.
3 **A. munitum**. A splendid Fern from California, growing in clumps, 3—5f, smooth, rigid, evergreen, lance-linear; segm. oblong-falcate, spinulous-serrate; sori 2-rowed.
4 **A. falcàtum**. Frond thick, rich green, lanceolate, pinnate, 2—3f high, with ample, lance-acuminate pinnæ. A noble, hardy Fern from Japan.
5 **A. Floridànum** (Hook). Rigidly erect, lance-oblong, pinnate and barren below, bipinnate, fertile, and contracted above; lower pinnæ cut-pinnatifid; indusia large, *round, peltate*, as in No. 1. Ga., Fla., La. (A. Ludoviciàna C-B.)
6 **A. aculeàtum** Sw. β. *Braunii*. Fronds in tufts, dark green, 2—3f, pinnate, lanceolate, narrowed both ways; stipe short, shaggy with large scales; segm. ovate-falcate, auricled on the upper side, bristle-tipped. Mts., Vt. (Eaton), N. Y.
7 **A.** podophyllum (or Sieboldii). Fronds of two forms, thick, smooth, pinnate, with a few large oblong pinnæ, in the fertile contracted and covered with sori. China.
8 **A. Thelýpteris** Sw. '*Lady Fern*. Frond lance-ovate, 10—16'; pn. narrow, distant, deeply pinnatifid, the lowest pair as long as any; margins reflexed in fruit.
9 **A. Novaboracénse** Willd. *New York Fern*. Frond elliptic-lanceolate, 12—18'; pn. narrow, gradually shortening from the middle both ways; segm. oblong, obtuse, flat; sori close to the margin, at length confluent. Moist woods: com. Delicate.
10 **A. patens** Sw. Frond soft and thin, downy with rusty hairs, lance., 12—18': pn. linear-oblong, pinnatifid; segm. oblong, obtuse, entire; sori scattered. Dry, Fla.
11 **A.** molle, from S. Afr. and S. Am., is divided just like A. patens, and equally hairy, but is larger, finer, with straw-colored stipes, and the sori in regular marginal rows.
12 **A. cristàtum** Sw. Frond narrowly lanceolate, some 2f × 6'; pn. deeply pinnatifid, triangular-oblong or -ovate, acute; segm. toothed, bearing a single row of large sori each side of the mid-vein. A beautiful dark-green Fern, common in woods.
13 **A. Goldiànum** Hook. Frond oval or ovate, about 15 × 10', stipe same length; pn. broad (1½—2'), deeply pinnatifid; segm. subfalcate, crenate. Woods, E. and W.
14 **A. fragraus** Sw. Fronds linear-lanceolate, 6—12', tapering both ways, bipinnate; stipe short, chaffy; pn. ovate-oblong, 1—10''; segm. lin.-oblong, with a dozen roundish crenatures or lobes; sori confluent. Rocks, Northern Mich. and Wis.
15 **A. marginàle** Sw. Fern ovate to lance-ovate, thick, glabrous, 1—2f, bipinnate, stipe very chaffy at base; pn. lanceolate; segm. oblong-falcate, obtuse and entire at apex, the lower crenate-lobed; sori round, at or near the margin. Rocky woods.
16 **A. Filix-mas.** Fern lanceolate, 1—3f; stipe very chaffy; pn. triangular-lance.; segm. oblong, obtuse, serrate at apex; sori near the mid-vein. N. J. to Va. ? N. W.
17 **A. spinulòsum** Willd. Stipe elongated, soon smooth, the chaff deciduous; frond 1—2f, ovate, acuminate, nearly or quite tripinnate; pinnæ lanceolate, acuminate, the lower longest; pnl. oblong, acutish, segm. mucronate-serrate. Woods and pastures.
  β. *dilatàtum*. Stipe permanently chaffy; frond triangular-ovate; pnl. obtuse
  γ. *Boottii*. Stipe chaffy; frond oblong-lanceolate; pnl. rather acute.

LATIN INDEX:

INCLUDING ALSO A GLOSSARY OF THE GENERA.

Abelmoschus, 62. From the *Arabic*; a grain of musk.
Abies, 313. The ancient name.
Abronia, 279. *Greek*, delicate.
Abrotanum, 184. *Absinthium*, 184.
Abutilon, 61. Name of obscure origin.
Acacia, 99. *Gr.*, to sharpen; sc. the spines.
Acalypha, 296. *Gr.* word for the Nettle.
ACANTHACEÆ, 233.
Acanthus, 233. Classic for spine or thorn.
Acer, 74. The ancient name, sharp or strong.
Acerates, 273. *Gr.*, without horns.
Achæta, 178. *Gr.*, without chaff.
Achillea, 183. Named for Achilles.
Achimenes, 219. Meaning unknown.
Acnetla, 180. *Gr.*, a point; sense doubtful.
Aculda, 289. *Gr.*, negative of *stinging*.
Aconitum, 22. The ancient Greek name.
Acorus, 318. *Gr.*, a remedy for sore eyes.
ACROGENÆ, 412.
Acrostichum, 419. *Gr.*, a row at the top?
Actæa, 23. *Gr.*, resembling the Elder.
Actimeris, 178. Altered from the next.
Actinomeris, 178. *Gr.*, partly radiate.
Actinospermum, 182. *Gr.*, seed pappus radiate.
Adenocaulon, 160. *Gr.*, with stipitate glands.
Adiantum, 422, *Gr.*, not wetted by rain.
Adlumia, 33. Named for *John Adlum*.
Adonis, 19. Sacred to Adonis.
Æschynomene, 87. *Gr.*, modest, or sensitive.
Æsculus, 74. Name ancient and obscure.
Æthusa, 140. *Gr.*, to burn; poisonous.
Agapanthus, 345. *Gr.*, a lovely flower.
Agathæa, 160. *Gr.*, good, or excellent.
Agave, 333. *Gr.*, admirable.
Ageratum, 156. *Gr.*, fadeless; long in flower.
Agrimonia. 108. *Gr.*, prize of the field?
Agrostemma, 54. *Gr.*, crown of the field.
Agrostis, 384. *Gr.*, of the field. & 386.
Ailanthus, 72. *Chinese;* tree of Heaven.
Aira, 395. *Gr.*, a weapon; misapplied.
Albizzia, 82. For an Italian botanist.
Alchemilla, 108. *Arabic,* 'alkémelya.
Aletris, 335. *Gr.*, a miller's wife; sc. mealy.
Alisma, 323. *Celtic. alis*, water.
ALISMACEÆ, 322.
Allamanda, 271. To Dr. Allamand, of Leyden.
Allium, 343. *Celt.*, all, hot or burning.
Allosorus, 421. *Gr.*, changing sorus, or sort.
Alnus, 308. *Celt., al lan,* near the river.
Alonsoa, 222. To Zanoni Alonso.
Alopecurus, 387. *Gr.*, fox-tail. [Spain.
Aloysia, 236. To Maria Louisa, Queen of
Alpinia, 331. To P. Alpini, an Ital. botanist.
Alsine, 56. *Gr.*, in the grove Alsophila, 419.
Althæa, 60. *Gr.*, to cure; sc. medicinal.
Alyssum, 40. *Gr.*, allaying anger.

AMARANTACEÆ, 288.
Amarantus, 288. *Gr.*, unfading.
AMARYLLIDACEÆ, 332.
Amaryllis, 333. Dedicated to that nymph.
Amblygonum, 282. *Gr.*, around the joints; sc. ochreæ.
Ambrosia, 174. *Gr.*, food of the gods.
Amelanchier, 110. The French name.
Amianthium, 348. *Lat.*, flowers pure, or white.
Ammannia, 124, To John Ammann, a Russian.
Ammobium, 186. *Gr.*, living in sand.
Amorpha, 93. *Gr.*, formless or deformed.
Ampelopsis, 78. *Gr.*, resembling the Vine.
Amphianthus, 228. *Gr.*, flowers of two forms.
Amphicarpæa, 97. *Gr.*, fruit of two forms.
Amphicarpum, 391. *Gr.*, fruit of two forms.
Amsonia, 270. To Chas. Amson, of S. C.
Amygdalus, 102. The ancient name.
Amyris, 72. *Gr.*, myrrh; perfumed gum.
ANACARDIACEÆ, 72.
Anacharis, 324. *Gr.*, uncomely.
Anagallis, 213. *Gr.*, laughing, cheering.
Ananassa, 335. The name in Guiana is *anas*.
Anantherix, 273. *Gr.*, beardless.
Anchusa, 252. A name of obscure origin.
Andromeda, 201. Like Andromeda of old, bound by the waters' edge.
Andropogon, 410. *Gr.*, a man's beard.
Androsace, 211. *Gr.*, a man's buckler. [cence.
Aneimia, 419. *Gr.*, naked; sc. the inflores-
Anemone, 17. *Gr.*, wind; or Wind-flower.
Anethum, 136, 139. *Gr.*, burning, stimulating.
Angelica, 137. Name of excellence.
ANGIOSPERMÆ, 15.
ANONACEÆ, 26. [the bristles of the pappus.
Antennaria, 185. *Lat.*, antennæ; alluding to
Anthemis, 183. Flowering abundantly.
Anthoxanthum, 395. *Gr.*, yellow flower.
Antigramma, 424. *Gr.*, like writing.
Antirrhinum, 223. *Gr.*, like the nose.
Anychia, 57. Altered from Paronychia.
APETALÆ, 278.
Aphyllon, 217. *Gr.*, without leaves.
Apium, 140. *Celt., apon*, water.
Aplectrum, 328. *Gr.*, without a spur.
APOCYNACEÆ, 269.
Apocynum, 270. *Gr.*, repelling dogs.
Apogon, 190. *Gr.*, without beard; no pappus.
Apteria, 325. *Gr.*, without wings.
AQUIFOLIACEÆ, 207. 'eagles' talons.
Aquilegia, 22. *Lat.*, an eagle; petals like
Arabis, 37. Originally from Arabia.
ARACEÆ. 317.
Arachis, 87. *Gr.*, without branches.
Aralia, 142. Of unknown meaning.
ARALIACEÆ, 142.

428 LATIN INDEX.

Archangelica, 137. Name of excellence.
Archemora, 136. A fanciful name.
Arctostaphylos, 201. *Gr.*, Bear's Grape.
Arcyphyllum, 96. *Gr.*, arched leaf.
Arenaria, 55, (57). *Lat.*, a sand plant.
Arethusa, 331. Named for that nymph.
Argemone, 32. Remedy for sore eyes.
Arisæma, 318. Of unknown meaning.
Aristida, 388. *Lat.*, an ear of wheat.
Aristolochia, 278. *Gr.*, good in parturition.
ARISTOLOCHIACEÆ, 278.
Armeniaca, 102. Originally from Armenia.
Armeria, 215. Latin for the Sweet-William.
Armoracia, 41. Native of Armorica.
Arnica, 188. *Lat.*, lamb's skin.
Aronia. 112. [the staminate spikes awned.
Arrhenatherum, 396. *Gr.*, male—point—i. e.,
Artemisia, 184. To Artemis,=Diana.
ARTOCARPEÆ, 298.
Arundinaria, 404. Altered from the next.
Arundo, 396. *Lat.*, a reed. *Celt.*, *arn*, water.
Asarum, 278. Meaning unexplained.
ASCLEPIADACEÆ, 271.
Asclepias, 272. *Lat.*, Æsculapius.
Ascyrum, 48. *Gr.*, soft to the touch.
Asimina, 26. Of unknown meaning.
Asparagus, 347. *Gr.*, tearing; some are thorny.
ASPHODELEÆ, 341. [indusium.
Aspidium 425. *Gr.*, a little shield; sc. the
Asplenium, 424. *Gr.*, without the spleen.
Aster, 161. *Lat.*, a star.
ASTEROIDEÆ, 152.
Astilbe, 114. *Gr.*, not shining; opaque.
Astragalus, 94. *Gr.*, the vertebra.
Atragene, 16. *Gr.*, night-born.
Atriplex, 287. *Lat.*, black and straggling.
Atropa, 264. To Atropos, one of the Fates who
AURANTIACEÆ, 71. [cut the thread of life.
Avena, 396. *Celt.*, *a'an*, to eat?
Ayenia, 63. To the Duke of Ayen.
Azalia, 203. *Gr.*, arid; grows in dry places.
Azolla, 413. *Gr.*, killed by drought.
Baccharis, 171. Dedicated to Bacchus.
Baldwinia. 182. To Dr. Wm. Baldwin.
Ballota, 248. *Gr.*. to cast away; ill-scented.
BALSAMINEÆ, 67.
Baptisia, 84. *Gr.*, to dye,=to color.
Barbarea, 39. Dedicated to St. Barbara.
Bartonia, 268. To Dr. B. S. Barton, of Phila.
Batatas, 259. Indian name of Potato.
Batis, 303. The Indian name.
Batrachium, 19. *Gr.*, the frog; amphibious.
Begonia, 131. To Michael Begon, French, a
BEGONIACEÆ, 131. [promoter of Botany.
Bejaria, 204. To M. Bejar, a Spanish botanist.
Bellis, 165. *Lat.*, *bellus*, pretty.
Benzoin. 290. Fragrant like *benzoin*.
BERBERIDACEÆ, 27.
Berberis, 27. The ancient *Arabic* name.
Berchemia, 77. To M. Berchem, a French bot.
Berlandiera, 173. To M. Berlandier, French.
Beta, 280. *Celt.*, *bett*, signifying red.
Betonica, 249. *Celt.*, beutonic.
Betula, 308. From *betu*, its Celtic name.
BETULACEÆ, 307.
Bidens, 180. *Lat.*, two-toothed; sc. the seed.
Bigelovia, 169. To Dr. Jacob Bigelow, Boston.
Bignonia, 218. To Abbe Bignon, librarian to
BIGNONIACEÆ, 218. [Louis XIV.
Biotia, 161. *Gr.*, *bioō*. to live.
Blechnum, (421) 424. *Gr.*, *blechnon*. [calyx.
Blephilia, 245. *Gr.*, eyelash; sc. the fringed
Bletia, 328. To Louis Blet, a Spanish botanist.

Blitum, 286. *Gr.*, *bliton*,=insipid. [M D
Bocconia, 32. To Paolo Boccone, a Sicilian
Bœhmeria, 300. To G. R. Bœhmer, German.
Bœrhaavia, 279. To Boerhaave, of Holland.
Boltonia. 166. To J. B. Bolton, an English bot.
BORRAGINACEÆ, 250. [ing?
Borrago, 251. Altered from *cor ago*=nourish-
Borreria, 147. To J. W. Borrer, F. L. S.
Borrichia, 171. To Olof Borrich, Danish.
Botrychium. 418. *Gr.*. a cluster of grapes.
Boussingaultia, 285. To J. B. Boussingault a
Bouteloua, 403. [cel. German naturalist.
Bouvardia, 150. To Dr. Bouvard, of Paris.
Boykinia. 114. To Dr. Boykin, of Georgia
Brachychæta, 166. *Gr.*, short hair; sc. pappus
Brasenia, 29.
Brassica, 40. *Brassic* was the Celtic name.
Brickellia, 158. To Dr. Brickell, of Savannat.
Briza, 403. *Gr.*, to nod; sc. the spikelets.
Brizopyrum. 402. Briza and *pyros* (wheat).
BROMELIACEÆ. 335. [the Wild Oat.
Bromus, 397. *Gr.*, food; anciently applied tc
Broussonetia. 299. To P. N. V. Broussonet, Fr.
Browallia. 221. To J. Browallius, of Abo.
Brunella, 246. *German*, a throat-disease.
Brunfelsia, 221. To Otho Brunsfels, of Mentz
Brunnichia, 280. To F. Brunnich, Danish.
Bryonia. 130. *Gr.*, to grow (sc. rapidly).
Bryophyllum, 119. *Gr.*, growing from the leaf
Buchnera, 230. T. J. G. Buchner, German.
Buckleya, 291. To S. B. Buckley, Texas.
Bumelia, 210. Greek name of the Ash.
Bupleurum, 138. *Gr.*, ox-rib.
Burmannia. 325. To one Burmann, German
BURMANNIACEÆ, 325.
Bursera, 72. To Joachim Burser, Naples.
BURSERACEÆ, 72.
BUTTOMEÆ, 323.
Buxus. 298. *Gr.*, dense? sc. the wood.
CABOMBEÆ, 28. Cabomba. 29.
Cacalia. 186. *Gr.*, exceedingly pernicious.
CACTACEÆ, 132.
Cakile, 43. The Arabic name.
Caladium, 319. Altered from Calla.
Calamagrostis, 386. Calamus-Agrostis.
Calamintha, 243. *Gr.*, beautiful Mint.
Calampelis, 219. *Gr.*, pretty vine.
Calandrinia, 59. To J. L. Calandrini, Italian.
Calceolaria. 222. *Lat.*. a little clipper.
Calendula, 188. *Lat.*, *kalendæ*. the first of the
Calla, 318 (319). *Gr.*, beautiful. [month
Calliastrum, 161. *Gr.*, beautiful flower.
Callicarpa, 236. *Gr.*, beautiful fruit.
Callirrhoē, 60, 61. A Greek name.
Callistachys, 100. *Gr.*, beautiful spike.
Callistemon, 122. *Gr.*, beautiful stamens.
Callistephus, 165. *Gr.*, beautiful crown.
CALLITRICHACEÆ, 301.
Callitriche, 301. *Gr.*, beautiful hair.
Calluna, 200. *Gr.*, to sweep; sc. a broom.
Calochortus, 343. *Gr.*, beautiful grass.
Calonyction, 260. *Gr.*, "good-night."
Calophanes, 234. *Gr.*, appearing beautiful
Calopogon, 330. *Gr.*, beautiful beard.
Caltha, 21. Syncope for *calathos*, a goblet.
CALYCANTHACEÆ, 25.
Calycanthus, 25. *Gr.*, calyx flower.
Calycocarpum, 27. *Gr.*, calyx fruit.
Calypso, 326. Dedicated to that nymph.
Calyptranthes, 121. *Gr.*, calyptra flower.
Calystegia, 260. *Gr.*, calyx covered.
Camassia. 343 Indian, Quamass.
Camelina. 42. *Gr.*, dwarf Flax.

LATIN INDEX. 429

Camellia, 65. To Geo. J. Kamel, a Moravian
CAMELLIACEÆ, 64. [monk.
Campanula, 196. *Lat.*, a little bell.
CAMPANULACEÆ. 196.
Camptosorus, 424. *Gr.*, curved sorus.
CANELLACEÆ, 8.
Canna, 332. Celtic for cane or mat.
Cannabis, 301. The ancient name.
CAPPARIDACEÆ, 44.
Capparis, 44. Arabic for capers.
CAPRIFOLIACEÆ, 144.
Caprifolium, 145. *Lat.*, goat-leaf.
Capsella, 42. *Lat.*, a little capsule. [qualities.
Capsicum, 263. *Gr.*, to bite; sc. its pungent
Cardamine, 37. *Gr.*, heart-subduing.
Cardiospermum, 75. *Gr.*, heart-seed.
Carex, 368. *Lat.*, to want; upper spike want-
Cariceæ, 356. [ing seed.
Carphephorus, 156. *Gr.*, chaff-bearing.
Carpinus, 307. *Celtic*, head-wood; sc. good
Carthamus, 189. *Arab.*, to color. [for yokes.
Carum, 138. From Caria, in Asia Minor.
Carya, 304. *Gr.*, the walnut.
CARYOPHYLLACEÆ, 52.
Cassia, 83. *Heb.*, ketzioth; *Lat.*, cassia.
Cassiope, 201. *Gr.*, the mother of Andromeda.
Cassyta, 290. [Thessaly.
Castanea, 306. From Castanea, a province in
Castilleja, 232. To Don Castilleja, a Spanish
Catalpa, 218. The Indian name. [botanist.
Catananche, 192. *Gr.*, from necessity (must be admired).
Caulophyllum, 27. *Gr.*, stem-leaf.
Ceanothus, 77. *Gr.*, to prick; plant spiny.
Cedronella, 246. *Gr.*, fragrant like cedar.
Cedrus, 314. From the river Cedron, in Judæa.
CELASTRACEÆ, 75. [all winter.
Celastrus, 76. *Lat.*, winter; the fruit remains
Celosia, 288. *Gr.*, burnt; appearance of the fls.
Celtis, 299. Ancient name for the Lotus.
Cenchrus, 394. *Gr.*, oriental name of Millet.
Centaurea, 188. To the centaur Chiron.
Centradenia, 123. *Gr.*, spur-gland; sc. the ap-
Centrosema, 98. [pendages of the anthers.
Centunculus, 213. Ancient Latin name.
Cephalanthus, 150. *Gr.*, head-fl.; fls. in a head.
Cerastium, 54. *Gr.*, a horn; the shape of the capsules. [native region.
Cerasus, 102. From Cerasus, in Pontus, its
Ceratiola, 303. *Gr.*, a little horn; sc. the
CERATOPHYLLACEÆ, 302. [stigma.
Ceratophyllum, 302. *Gr.*, horn-leaf. [fruit.
Ceratoschœnus, 367. *Gr.*, horn-rush; sc. the
Cercis, 83. *Gr.*, a shuttle; sc. the legume.
Cereus, 133. *Lat.*, wax; the shoots are plastic.
Cestrum, 265. *Gr.* name for Betony.
Chærophyllum, 137. *Gr.*, rejoice, leaf; lvs. fra-
Chamælirium, 349. *Gr.*, dwarf lily. [grant.
Chamæmelum, 183. The Greek name.
Chamærops, 317. *Gr.*, dwarf stem. [ist.
Chaptalia, 194. To M. Chaptal, a French chem-
Chapmania, 87. To Dr. A. W. Chapman, the
CHARACEÆ. 14. [Southern botanist.
Cheilanthes, 422. *Gr.*, lip-flower; sc. the in-
Chelranthus, 38. *Gr.*, hand-flower. [dusium.
Chelidonium, 31. *Gr.*, a swallow; flowers with the arrival of that bird.
Chelone, 224. *Gr.*, tortoise; form of the
CHENOPODIACEÆ, 284. [flower.
Chenopodium, 287. Altered fr. Chenopodium.
Chenopodium, 285. *Gr.*, goose-foot; shape of the leaf. [ter-green.
Chimaphila, 206. *Gr.*, lover of winter; win-

Chiococca, 147. *Gr.*, winter berry.
Chiogenes, 199. *Gr.*, winter-born.
Chionanthus, 276. *Gr.*, snow (white) flower.
Chloris, 407. *Gr.*, green.
Chorozema, 100. *Gr.*, dance, drink; found near a spring in a thirsty land—N. Holland.
Chrysanthemum, 184. *Gr.*, golden flower.
Chrysobalanus, 101. *Gr.*, golden acorn, or fr.
Chrysogonum, 172. *Gr.*, golden joint; fls. in the axils.
Chrysopsis, 170. *Gr.*, golden appearance.
Chrysosplenium,113. *Gr.*, golden spleen (wort).
Chthamalia, 274. *Gr.*, on the ground; trailing
Cicer, 85. *Gr.*, strength; its nourishing quali-
CICHORACEÆ. 152. [ties.
Cichorium, 190. Greek name, adopted from the
Cicuta, 141. Name unexplained. [Egyptians.
Cimicifuga, 23. *Gr.*, bug-repelling.
Cineraria, 160. *Lat.*, ashes; clothed with ash colored down. See also 187.
Cinna, 385. An ancient name of a grass.
Circæa, 128. To the enchantress *Circe*.
Cirsium, 189. The old Greek name.
Cissus, 78. The Greek name for the Ivy.
CISTACEÆ, 47. [sinum inflated.
Cistopteris, 425. *Gr.*, bladder fern; sc. indu-
Citharexylum, 235. *Gr.*, harp-wood; fiddle-
Citrullus, 130. Derived from the next. [wood.
Citrus, 71. From Citron, in Judæa.
Cladastris, 84. *Gr.*, brittle branches?
Cladium, 367. *Gr.*, a branch or twig.
Clarkia, 126. To Captain Clark, the pioneer traveller in Oregon.
Claytonia, 59. To John Clayton, of Virginia.
Clematis, 16. *Gr.*, a tendril; the petioles act
Cleome, 41. *Gr.*, to shut; fls. closed. [as such.
Clethra, 204. The Greek name of the Alder.
Clianthus, 100. *Gr.*, the flower of glory.
Clintonia (195), 346. To Gov. De Witt Clinton,
Clitoria, 98. A fanciful name. [of N. Y.
Clusia, 8. To Charles de l'Ecluse, of Artois.
Cnicus, 189. *Gr.*, to prick.
Cnidoscolus, 296. *Gr.*, nettle-prickle.
Cobæa, 258. To B. Cobo, a Spanish botanist.
Cocculus, 27. *Lat.*, cochineal; berries red.
Coix, 411. A Greek name of a grass. [try.
Colchicum, 348. From Colchis, its native coun-
Coleus, 239. *Gr.*, a sheath; of the stamens.
Collinsia, 225. To Z. Collins, of Philadelphia.
Collinsonia, 241. To Peter Collinson, F. R. S.
Collomia, 257. *Gr.*, glue; referring to the seeds.
Colocasia, 319. [mens.
Colubrina, 76. *Gr.*, snake; the twisted sta-
Coluteu, 95. [character.
Comandra, 291. *Gr.*, hair stamens; see the
Comarum, 107. Greek name of the Arbutus.
COMBRETACEÆ, 12. [mous Dutch botanists.
Commelyna, 353. To J. and G. Commelyn, fa-
COMMELYNACEÆ, 353.
COMPOSITÆ, 152. [Bishop of London.
Comptonia, 309. To Henry Compton, Lord
CONIFERÆ, 312.
Conioselinum, 140. *i. e.*, Conium-Selinum.
Conium, 139. *Gr.*, dust; unexplained.
Conobea, 226. Name unexplained.
Conoclinium, 160. *Lat.*, conical receptacle
CONOIDEÆ, 311.
Conopholis, 217. *Gr.*, scale, cone.
Conostylis, 335. *Gr.*, cone, style.
Consolida, 22. *Lat.*, styles all in one?
Convallaria, 346. *Lat.*, a valley.
Convolvulus, 260. *Lat.*, to entwine, or involve
Conyza, 171. Unexplained.

Coptis, 21. *Gr.*, to cut ; sc. the cleft leaves.
Corallorhiza, 328. *Gr.*, coral-root.
Corchorus, 64. *Gr.*, to purge ; laxative.
Cordia, 250. To E. Cordius, a Germ. botanist.
Corema, 303. *Lat.*, a broom ; sc. the habit.
Coreopsis, 178. *Gr.*, bug-like ; sc. the seeds.
Coriandrum, 141. *Gr.*, bug ; from the odor.
Corispermum, 287. *Gr.*, bug-seed.
CORNACEÆ, 142. [of the wood.
Cornus, 143. *Lat.*, a horn ; from the hardness
Coronilla, 87. *Lat.*, a little crown.
Corydalis, 33. Greek name for Fumitory.
Corylus, 307. *Gr.*, a helmet ; the involucrate fr.
Corythium, 332. *Gr.*, a helmet ; sc. the flower.
Cosmanthus, 255. *Gr.*, elegant flower.
Cotula, 172. The old Latin name.
Cranichis, 330. Derivation uncertain.
Crantzia, 135. To Prof. Crantz, Eng.
Crassula, 119. *Lat.*, thick ; leaves fleshy.
CRASSULACEÆ, 117. [ness of the wood.
Cratægus, 110. *Gr.*, strength ; from the hard-
Crinum, 333. The Greek name of the Lily.
Crocus, 337. The name in *Chaldaic*.
Croomia, 339. To H. B. Croom, of Florida.
Crotalaria, 90. *Gr.*, a rattle ; sc. the sds. in pod.
Croton, 297. *Gr.*, a tick ; sc. the seeds.
Crotonopsis, 297. Croton-like.
CRUCIFERÆ, 34. [are in the sheaths.
Crypsis, 387. *Gr.*, concealed ; as the flowers
CRYPTOGAMIA, 412. [the calyx).
Cryptotænia, 138. *Gr.*, concealed border (of
Ctenium, 409. *Gr.*, a comb ; sc. the beard.
Cucumis, 131. *Lat.*, crooked ? (fruit).
Cucurbita, 130. *Lat.*, crookedness ; the fruit.
CUCURBITACEÆ, 129.
Cunila, 240.
Cuphea, 123. *Gr.*, curved ; sc. the capsule.
Cupressus, 315. *Gr.*, equal growth ; referring
CUPULIFERÆ, 304. [to the reg. branches.
Cuscuta, 260. Name from the Arabic.
Cyathea, 419. *Gr.*, little cup ; sc. indusium.
CYCADACEÆ, 311.
Cycas, 312. A name in Greek for a Palm.
Cyclamen, 212. *Gr.*, circular ; sc. the leaves.
Cycloloma, 285. *Gr.*, circle, border (of the cal.)
Cydonia, 112. From Cydon, in Crete.
Cynara, 188. *Gr.*, a dog ; involucre spiny.
Cynodon, 407 *Gr.*, dogtooth ; sc. the spikelets.
Cynoglossum, 251. *Gr.*, dogtongue ; sc. the lvs.
Cynthia, 191. A name of Diana.
CYPERACEÆ, 356.
Cyperus, 357. A name of Venus.
Cypripedium, 326. *Gr.*, Venus' slipper.
Cyrilla, 205. To Dom. Cyrillo, M. D., Naples.
Cyrtanthera, 235. *Gr.*, curved flower.
Cytisus, 100. First found in Isl. Cythrus.
Dactylis, 398. *Gr.*, a finger ; spikes digitate.
Dactyloctenium, 408. *Gr.*, finger comb ; the spikes digitate-pectinate.
Dahlia, 166. For A. Dahl, a Swedish botanist.
Dalea, 93. For Thos. Dale, an English botanist.
Dalibarda, 105. To Dalibard, a Fr. botanist.
Danthonia, 390. To M. Danthoine, a Fr. bot.
Daphne, 292. A *nymph* transformed by Apollo.
Dasystoma, 230. *Gr.*, hairy mouth ; sc. the cor.
Datura, 265. From the Arabic, *Totorah*.
Daucus, 139. The Greek name.
Davallia, 423. M. Davall, a Swiss botanist.
Decumaria, 116. Lat., *decem*, ten ; fis.10-parted.
Delphinium, 24. *Gr.*, a dolphin.
Dentaria, 37. *Lat.*, a tooth ; the root toothed.
Desmanthus, 82. *Gr.*, bundle (of) flowers.
Desmodium, 83. *Gr.*, a bond ; sc. the loment.

Deutzia, 116. For *Deutz*, a Dutch botanist.
DIALYPETALÆ, 15. [the pod
Diamorpha, 119. *Gr.*, peculiarly formed ; sc.
Dianthera, 234. *Gr.*, two anthers.
Dianthus, 52. *Gr.*, the flower of Jove.
Diapensia, 258. *Gr.*, flowers by 5's ; 5-cleft.
Diarrhena, 399. *Gr.*, two rough (keels in the
Dicentra, 33. *Gr.*, two spurs. [pales.
Dicerandra, 242. *Gr.*, anthers two-horned.
Dichondra, 260. *Gr.*, two grains (carpels).
Dichromena, 364. *Gr.*, two-colored. [amist.
Dicksonia, 423. To Jas. Dickson, cryptog-
Dicliptera, 234. *Gr.*, double-valved (capsule).
Dictamnus, 70. Greek name of the Ash.
Didiplis, 124. *Gr.*, twice double.
Dielytra, 33. *Gr.*, two wings.
Diervilla, 146. To M. Diervīlle, M.D., French.
Digitalis, 228. *Lat.*, finger of a glove.
Digitaria, 389. *Lat.*, a finger ; sc. the spikes.
Diodia, 149. *Gr.*, wayside (plants).
Dionæa, 51. A name of Venus.
Dioscorea, 338. To Pedacius Dioscorides, a
DIOSCOREACEÆ, 338. [Greek physician.
Diospyros, 209. *Gr.*, the pear of Jove.
Dipholis, 210. *Gr.*, two scales (bet. the petals).
Diphylleia, 28. *Gr.*, two-leaved.
Diplopappus, 164. *Gr.*, double pappus.
DIPSACEÆ, 151. [hold water.
Dipsacus, 151. *Gr.*, to thirst ; the leaf-axils
Dipteracanthus, 234. *Gr.*, 2-winged Acanthus.
Dirca, 292. *Gr.*, a fountain.
Discopleura, 141. *Gr.*, disk, ribs (united).
Dodecatheon, 211. *Gr.*, twelve deities (flowers).
Dodonæa, 74. To R. Dodonæus, M. D.
Dolichos 98. *Gr.*, long ; sc. the twining stems.
Doodia, 423. To S. Doody, botanist, London.
Downingia, 195. To J. Downing, florist, &c.
Draba, 41. *Gr.*, acrid or biting ; sc. the leaves.
Dracocephalum, 246. *Gr.*, dragon head.
Dracopsis, 176. *Gr.*, dragon-like.
Dracunculus, 184. *Gr.*, little dragon.
Drosera, 51. *Gr.*, dew (-drops on the leaves).
DROSERACEÆ, 50.
Dryas, 105. *Gr.*, Oak nymph ; sc. its leaves.
Dulichium, 356. First found on that island.
Duranta, 235. To Castor Durant, 1580.
Dysodia, 181. *Gr.*, ill-scented.
Eatonia, 400. To Prof. Amos Eaton, the well-
EBENACEÆ, 209. [known botanist.
Eccremocarpus, 218. *Gr.*, pendent fruit.
Echeveria, 119. To M. Echeveri, botanic artist.
Echinacea, 175. *Gr.*, hedgehog ; sc. the spines.
Echinocactus, 132. *Gr.*, hedgehog cactus.
Echinocystis, 129. *Gr.*, hedgehog bladder ; fr.
Echinodorus, 323. *Gr.*, hedgehog sac ; carpels.
Echinospermum, 251. Hedgehog seed.
Echites, 271. *Gr.*, a viper ; the smooth shoots.
Echium, 251. *Gr.*, a viper ; sc. the seeds.
Eclipta, 172. *Gr.*, deficient ; sc. no pappus.
Ehretia, 250. To D. G. Ehret, German artist.
ELÆAGNACEÆ, 292.
Elæagnus, 292. *Gr.*, the olive ; resemblance.
ELATINACEÆ, 51.
Elatine, 51. *Gr.*, the fir ; resemblance.
Eleocharis, 359. *Gr.*, marsh delight.
Elephantopus, 156. *Gr.*, elephant's foot.
Eleusine, 407. A name of Ceres.
Elliottia, 205. To Stephen Elliott, S. Car.
Ellisia, 254. To Joseph Ellis, F. R. S.
Elodea, 50. *Gr.*, a marsh. [in the sheath.
Elymus, 405. *Gr.*, enveloped ; sc. the spike
Elytraria, 235. *Gr.*, enveloped ; the fis.in bracts
EMPETRACEÆ, 302.

Empetrum, 303. *Gr.*, on a rock.
ENDOGENÆ, 316.
Enslenia, 273. To Aloysius Enslen.
Epidendrum, 331. *Gr.*, on a tree.
Epigœa, 200. *Gr.*, on the earth; trailing.
Epilobium, 124. *Gr.*, on the pod (sc. the fls.)
Epiphegus, 217. *Gr.*, on the Beech (roots).
Epiphyllum, 132. *Gr.*, on a leaf (sc. the fls.)
EQUISETACEÆ, 415.
Equisetum, 415. *Lat.*, horse-hair.
Eragrostis, 400. *Gr.*, lovely grass.
Erectites, 186. *Gr.*, to trouble.
Erianthus, 410. *Gr.*, wool-flower.
Erica, 200. *Lat.*, the old name.
ERICACEÆ, 197.
Erigenia, 140. *Gr.*, spring-born.
Erigeron, 165. *Gr.*, in spring (early) old.
Eriocaulon, 355. *Gr.*, woolly stem.
ERIOCAULONACEÆ, 355.
Eriogonum, 280. *Gr.*, woolly joint.
Eriophorum, 362. *Gr.*, wool-bearing.
Erithalis, 147. *Gr.*, to grow green.
Ernodea, 147. *Gr.*, branched; much branched.
Erodium, 68. *Gr.*, a heron's (bill).
Erophila, 41. *Gr.*, lover of Spring.
Eryngium, 135. *Gr.*, to belch; a remedy.
Erysimum, 39. *Gr.*, to draw (blisters).
Erythræa, 267. *Gr.*, red; sc. the flowers.
Erythrina, 97. Same as the last.
Erythronium, 341. Ditto.
Escallonia, 116. To Escallon, Spanish.
Eschscholtzia, 32. To Eschscholtz, German.
Eucalyptus, 121. *Gr.*, well covered; sc. the cal.
Eugenia, 122. To Prince Eugene, of Savoy.
Eulophus, 141. *Gr.*, handsome crest.
Euonymus, 76. *Gr.*, well named.
Eupatorium, 158. Named for Eupator.
Euphorbia, 293. To Euphorbus, of Mauritania.
EUPHORBIACEÆ, 293,
Euphrasia, 232. To the Muse Euphrosyne.
Eustachys, 407. *Gr.*, handsome spike.
Eustoma, 267. *Gr.*, handsome mouth.
Eutoca, 255. *Gr.*, fruitful.
Euxolus, 288. *Gr.*, well closed.
Evolvulus, 260. *Lat.*, to roll out, to trail.
Excœcaria, 296. *Lat.*, to blind; the poisonous
EXOGENÆ, 15. [juice destroys the sight.
Exostemma, 147. *Gr.*, stamens exserted?
Faba, 85. *Gr.*, to eat.
Fabiana, 265. To F. Fabiana, of Valencia.
Fagopyrum, 284. *Gr.*, beech-nut wheat.
Fagus, 307. The ancient name.
Fedia, 151. From *fedus*, a kid.
Fenzlia, 257. To Dr. Fenzl, a botanic author.
Festuca, 399. Celt.,*fest*, pasture.
FICOIDEÆ, 133.
Ficus, 299. The ancient Latin name.
Filago, 185. *Lat.*, thread-spinning; the plant
FILICES, 416. [is clothed in cotton.
Fimbristylis, 363. *Gr.*, fringed style.
Flœrkea, 68. To Flœrke, a German botanist.
FLORIDEÆ, 322.
Fœniculum, 139. *Lat.*, a kid; why?
Forestiera, 277. To M. Forestier, French.
Forsteronia, 270. To T. F. Forster, an Eng. bot.
Forsythia, 276. To Mr. Forsyth, horticulturist.
Fothergilla, 120. To J. Fothergill, M.D., Lond.
Fragaria, 106. *Lat.*, fragrant; sc. the fruit.
Franciscea, 221. To Francis, Emperor of Aust.
Franklinia, 65. [plants in the South.
Frasera, 268. To John Fraser, collector of
Fraxinus, 277. *Lat.*, a hedge; hedge plants.
Fritillaria, 342. *Lat.*, a chess-board.

Frœlichia, 290. To J. A. Frœlich, a Germ. bot
Fuchsia, 127. To Leonard Fuchs, German.
Fuirena, 359. To G. Fuiren, Danish.
Fumaria, 34. *Lat.*, smoke; sc. the smell.
FUMARIACEÆ, 33.
FUNGI, 14.
Funkia, 345. To Henry Funk, German.
Gaillardia, 181. To M. Gaillard, French.
Galactia, 97. *Gr.*, milk.
Galanthus, 334. *Gr.*, milk-flower.
Galax, 206. *Gr.*, milk; flowers milk-white?
Galeopsis, 248. *Gr.*, weasel-like; sc. the fl.
Galinsoga, 172. To M. Galinsoga, Madrid.
Galium, 148. *Gr.*, milk (to curdle).
GAMOPETALÆ, 144.
Gardoquia, 246. To Diego Gardoqui, Spanish.
Gaultheria, 201. To Dr. Gaulthier, Quebec.
Gaura, 126. *Gr.*, superb. [French chemist.
Gaylussacia, 198. To Gaylussac, the celebrated
Gazania, 181. *Lat.*, riches (richness).
Gelsemium, 269. Italian for Jessamine.
Genista, 90. Celt., *gen*, a bush.
Gentiana, 267. To Gentius, king of Illyria.
GENTIANACEÆ, 266.
GERANIACEÆ, 67.
Geranium, 68. *Gr.*, crane's (bill); sc. the fruit.
Gerardia, (230) 231. To John Gerard, English.
Gesneria, 219. To Conrad Gesner, German.
GESNERIACEÆ, 219. [of G. urbicum.
Geum, 105. *Gr.*, to give relish; sc. the roots
Gilia, 257. To P. S. Gill, Spanish.
Gillenia, 104. Named for A. Gille, German.
Ginkgo, 316. The name in Japanese.
Ginseng, 142. The name in Chinese.
Gladiolus, 338. *Lat.*, a little sword; sc. the lvs.
Glaucium, 31. *Gr.*, glaucous (in color).
Glaux, 212. Ditto.
Glechoma, 246. An old Greek name.
Gleditschia, 83. To Prof. G. Gleditsch, Berlin.
Glottidium, 93. *Gr.*, tongue; sc. the pods.
Gloxinia, 219. To P. B. Gloxin, of Colmar.
GLUMIFERÆ. 336,
Glyceria, 402. *Gr.*, sweet; sc. the herbage.
Gnaphalium, 185. *Gr.*, soft down.
Godetia, 125. To M. Godet, French.
Gomphrena, 289. *Gr.*, a club; sc. the flowers.
Gonolobus, 274. *Gr.*, angular pods.
GOODENIACEÆ, 10.
Goodyera, 330. To John Goodyer, English.
Gordonia, 65. To Alex. Gordon, London.
Gossipium, 63. *Arabic*, a softness.
GRAMINEÆ, 380.
GRAMINOIDEÆ, 336. [bearded at base
Graphephorum, 398. *Gr.*, pencil-bearing; de.
Gratiola, 227. *Lat.*, grace (medicinally).
GROSSULACEÆ (113).
Grossularia, 117. Name of doubtful meaning.
Guettarda, 147. To Etienne Guettard, French.
Guiacum, 67. The aboriginal name.
GUTTIFERÆ, 8.
Gymnadenia, 326. *Gr.*, naked gland.
Gymnocladus, 83. *Gr.*, naked branches.
Gymnogramma, 420. *Gr.*, naked writing (sori)
Gymnopogon, 407. *Gr.*, naked beard.
Gymnospermæ, 311. *Gr.*, naked seeds.
Gynandropsis, 44. *Gr.*, like gynandria.
Gynerium, 398. *Gr.*, style woolly.
Gypsophila, 53. *Gr.*, loving chalk (cliffs).
Habenaria, 326. *Lat.*, thong,=the long spur.
Habrothamnus, 265. *Gr.*, a gay branch.
HÆMODORACEÆ, 335.
Halenia, 268. A personal name.
Halesia, 209. To S. Hales, D.D., F.R.S.

HALORAGEÆ. 120.
HAMAMELACEÆ, 120.
Hamamelis, 120. *Gr.*, (flower) with the fruit.
Hamella, 147. To H. L. Duhamel. [berg.
Hardenbergia, 99. To the Countess of Harden-
Hedeoma, 241. The Greek name for Mint.
Hedera, 142. *Celt.*, a cord.
Hedychium, 331. *Gr.*, sweet snow (white fls.)
Hedysarum, 87. An old Greek name.
Helenium, 181. Dedicated to Helen.
Helianthella, 177. Diminutive of Helianthus.
Helianthemum, 47. *Gr.*, Sun-flower.
Helianthus, 176. Ditto.
Helichrysum, 186. *Gr.*, golden sun.
Heliophytum, 251. *Gr.*, Sun-plant.
Heliopsis, 175. *Gr.*, sun-like.
Heliotropium, 250. *Gr.* turning (with) the sun.
Helleborus, 21. *Gr.*, killing (poisonous) food.
Helonias, 349. *Gr.*, a marsh.
Helosciadium, 140. *Gr.*, marsh umbel.
Hematelia, 419.
Hemerocallis, 345. *Gr.*, beauty of a day.
Hemicarpha, 363. *Gr.*, half (of the) chaff.
Hepatica, 18. *Gr.*, of or resembling the liver.
HEPATICÆ, 14.
Heracleum, 136. Sacred to Hercules.
Herpestis, 226. *Gr.*, a creeper.
Hesperis. 39. *Gr.*, the evening. [anthers.
Heteranthera, 350. *Gr.*, other (two kinds of)
Heterotheca, 170. *Gr.*, other (2 kinds of) fruits.
Heuchera, 115. To Dr. H. Heucher, Wittembg.
Hibiscus, 62. From *ibis*, the stork.
Hieracium, 191. Gr., *hierax*, the hawk.
Hierochloa, 395. *Gr.*, holy Grass.
HIPPOCASTANEÆ, 73.
Hippomane, 293. *Gr.*, horse madness.
Hippophæ, 293. *Gr.*, horse destroyer.
Hippuris, 121. *Gr.*, mare's tail.
Holcus, 395. *Gr.*, to extract (thorns).
Holosteum, 54. *Gr.*, all bone (by antithesis).
Honkenya, 56. A personal name.
Hordeum, 404. *Gr.*, heavy (sc. bread).
Hottonia, 211. To Prof. P. Hotten, of Leyden.
Houstonia, 149. To Wm. Houston, M. D., Eng.
Hoya, 275. To Thos. Hoy, F. L. S.
Hudsonia, 48. To Wm. Hudson, F. R. S.
Humea, 194. To Lady Hume, of Wormleybury.
Humulus, 301. *Lat.*, on the ground,=trailing.
Hyacinthus, 344. A boy killed by Zephyrus.
Hydrangea, 116. *Gr.*, a water-vessel.
Hydranthelium, 228. *Gr.*, a little water-flower.
Hydrastis, 23. In or near water.
HYDROCHARIDACEÆ, 324.
Hydrocleis, 323. *Gr.*, enclosed in water.
Hydrocotyle, 135. *Gr.*, a water-vessel.
Hydrolea, 255. *Gr.*, water, oil; sc. an oily
HYDROPHYLLACEÆ, 253. [water-plant.
Hydrophyllum, 254. *Gr.*, water leaf.
Hygrophila, 234. *Gr.*, loving moisture.
Hymenopappus, 181. *Gr.*, membranous pap-
Hyoscyamus, 264. *Gr.*, hog-bean. [pus.
Hypelate, 74. Unexplained.
HYPERICACEÆ, 48.
Hypericum, 49. Not satisfactorily explained.
Hypobrychia, 124. [(the pod).
Hypoxis, 334. *Gr.*, sharp under; the base of
Hyptis, 239. *Gr.*, resupinate; sc. the cor. upper
Hyssopus, 241. The old Hebrew name. [lip.
Iberis, 42. From Iberia, now Spain.
Ilex, 207. The ancient name.
Illicium, 24. *Lat.*, alluring; sc. the perfume.
Ilysanthes, 227. *Gr.*, mud-flower. [touched.
Impatiens, 69. *Lat.*, impatient; not to be

Indigofera, 95. *Lat.*, indigo-bearing.
Inula, 171. A corruption of Hellenium.
Iodanthus, 36. *Gr.*, violet-flower.
Ipomæa, 259 (260). *Gr.*, like bindweed.
Ipomopsis, 257. *Gr.*, like Ipomæa.
Iresine, 289. Gr., *eiros*, wool.
IRIDACEÆ, 336.
Iris, 336. From its varied colors.
Isanthus, 239. *Gr.*, equal (regular) flower.
Isatis, 43. *Gr.*, to smooth (the skin); a cos
Isoëtes, 412. *Gr.*, equal (all the) year. [metic
Isopappus, 170. *Gr.*, equal pappus.
Isopyrum, 20. *Gr.*, equal wheat.
Itea, 115. Greek name of the Willow.
Iva, 174. Leaves resembling the Greek Iva.
Ixia, 337. *Lat.*, bird-lime; sc. sticky.
Jacquemontia, 258. To Victor Jacquemont.
Jasminum, 275. *Gr.*, violet smell; sc. fragrant
Jatropha, 296. *Gr.*, physician, food; sc. medi
cinal.
Jeffersonia, 28. To President Thos. Jefferson
JUGLANDACEÆ, 303. [walnut
Juglans, 304. *Gr.*, the nut of Jove; sc. the
JUNCACEÆ, 350.
JUNCAGINEÆ, 323. [of these rushes.
Juncus, 351. *Lat.*, to join; ropes were made
Juniperus, 314. *Celt.*, rough or rude.
Jussiæa, 125. To Antoine Jussieu, the elder.
Justicia. 235. To J. Justice, a Scotch botanist.
Kallstræmia, 67. A personal name.
Kalmia, 200. To Prof. Peter Kalm, of Abo.
Kennedya, 99. To Mr. Kennedy, of Ham-
mersworth.
Kerria, 104. To Mr. Kerr, botanist, Ceylon.
Kœleria, 398. To Prof. Kœler, of Mayence.
Kœlreuteria, 75. To J. G. Kœlreuter, German
Kosteletzkya, 62. A personal name. [botanist.
Krameria, 80. To J. G. and W. H. Kramer, Ger.
Krigia, 191. To Dr. David Kreig, German.
Kuhnia, 158. To Adam Kuhn, of Pennsylvania.
Kuhnistera, 93. From Kuhnia.
Kyllingia, 359. To P. Kylling, Danish, 1690.
LABIATÆ, 237. LABIATIFLORÆ, 153, 155
Laburnum, 91. The old Latin name.
Lachnocaulon, 355. *Gr.*, wool-stem.
Lachnanthes, 335. *Gr.*, wool-flower.
Lactuca, 193. Lat., *lac*,=milk; sc. milk-weed.
Lagenaria, 130. *Lat.*, a bottle; sc. the gourd.
Lagerstrœmia, 123. To Marcus Lagerstrœm,
Laguncularia. *Lat.*, a small bottle. [Ger.
Lamium, 248. *Gr.*, throat; sc. gaping-flowers.
Lampsana, 190. A personal name.
Lantana, 237. Old Latin name for Laburnum.
Lapithœa, 206.
Laportea, 300. To M. Laporte, French.
Lappa, 190. Old Latin name of Burdock.
Larix, 314. *Celt.*, fat or resinous; from *lar*.
Lathyrus, 85. *Gr.*, stimulating.
LAURACEÆ, 290. [made of lavender
Lavandula, 239. *Lat.*, to wash; from the use
Lavatera, 60. To the two Lavaters, of Zurich
Leavenworthia, 38. To Dr. Leavenworth, U.S.A
Lechea, 47. To G, Leche, Sweden, 1760.
Ledum, 204. An old Greek name. [nist.
Leersia, 383. To J. D. Leers, a German bota-
LEGUMINOSÆ, 80.
Leiophyllum, 204. *Gr.*, smooth leaf. [Florida.
Leitneria, 309. To Dr. Leitner, collector in
Lemna, 319. The Greek name of some water-
LEMNACEÆ, 319. [plant.
Lens, 100. The seeds are shaped like a *lens*.
LENTIBULACEÆ, 215.
Leonotis, 249. *Gr*, lion's ear; sc. the flowers

LATIN INDEX. 433

Leontodon, 191. *Gr.*, lion's-tooth; sc. the lvs.
Leonurus, 249. *Gr.*, lion's-tail; sc. the spike of flowers.
Lepachis, 176. From *lepis*, Gr. word for scale.
Lepidium, 42. *Gr.*, a little scale; sc. the sili-
Leptocaulis, 140. *Gr.*, slender stem. [cles.
Leptochloa, 406. *Gr.*, slender grass.
Leptopoda, 182. *Gr.*, slender foot or stem.
Leptosiphon, 257. *Gr.*, slender tube; sc. the flowers.
Lepturus, 404. *Gr.*, slender tail; sc. the spikes.
Lepuropetalon, 115. *Gr.*, husk petal. [ida.
Lespedeza, 89. To M. Lespedez, Gov. of Flor-
Leucanthemum. 183. *Gr.*, white flower.
Leucas, 238. *Gr.*, whiteness; sc. of the flowers.
Leucojum, 334. *Gr.*, white violet.
Liatris, 157. A name unexplained.
LICHENES, 14.
LIGULIFLORÆ, 152, 155.
Ligusticum, 140. Originally found in Liguria.
Ligustrum, 276. Lat., *ligare*, to tie; sc. its
LILIACEÆ, 341. [flexible branches.
Lilium, 342. Celt., *li*, whiteness.
Limnanthemum, 268. *Gr.*, marsh-flower.
Limnanthes, 68. Ditto.
Limnobium, 324. *Gr.*, marsh-life.
Limnocharis, 323. *Gr.*, marsh-joy.
Limosella, 228. *Gr.*, little mud (plant).
LINACEÆ, 66. [resembles.
Linaria, 222. From Linum, flax; which it
Lindera, 290. Name unexplained.
Linnæa, 144. To the great naturalist, Carl von
Linum, 66. Celt., *lin*, = a thread. [Linnæus.
Liparis, 329. *Gr.*, *liparos*, unctuous.
Lipocarpha, 363. *Gr.*, oil chaff; why ?
Lippia, 236. To Aug. Lippi, French traveller.
Liquidambar, 120. *Lat.*, liquid amber.
Liriodendron, 25. *Gr.*, lily-tree; sc. tulip-tree.
Listera, 329. To Dr. Martin Lister, English.
Lithospermum, 252. *Gr.*, stone-seed.
Loasa, 128. Name unexplained.
LOASACEÆ, 128. [to James I.
Lobelia, 194. To Matthew Lobel, physician
LOBELIACEÆ, 194. [nist.)
LOGANIACEÆ, 269. (Jas. Logan, Eng. bota-
Loiseleuria, 203. A mythological name.
Lolium, 405. The Celtic name is *loloa*. [sori.
Lomaria, 421. *Gr.*, the edge; position of the
Lonicera, 145. To Adam Lonicer, Germ., 1580.
Lophanthus, 245. *Gr.*, crest-flower.
Lophiola, 335. *Lat.*, diminutive; little crest.
Lophospermum, 223. *Gr.*, crest-seed.
LORANTHACEÆ, 291. *Lorinseria*, 371.
Ludwigia, 127. To Prof. C. D. Ludwig, Leipsic.
Lunaria, 40. *Lat.*, the moon; sc. the silicles.
Lupinus, 93. *Lat.*, a wolf; devours the soil ?
Luziola, 388. *Lat.*, *lux*, light; sparkling with
Luzula, 351. *Germ.*, the glow-worm. [dew.
Lychnis, 54. *Gr.*, a lamp (wick).
Lycium, 261. The old Greek name.
Lycopersicum, 362. *Gr.*, wolf-peach.
LYCOPODIACEÆ, 413.
Lycopodium, 413, (414). *Gr.*, wolf-foot.
Lycopsis, 251. *Gr.*, wolf-like; the flower is fancied to resemble a wolf's eye.
Lycopus, 240. *Gr.*, wolf-foot.
Lygodesmia, 193. *Gr.*, flexible band.
Lygodum, 418. *Gr.*, a flexible (vine).
Lysimachia, 212. *Gr.*, dissolution of strife;
LYTHRACEÆ, 123. [sc. loose-strife.
Lythrum, 123. *Gr.*, black blood; sc. purple.
Macbriden, 247. To Dr. Jas. McBride, of S. C.
Maclura, 299. To Wm. Maclure, Pennsylvania.

Macranthera, 230. *Gr.*, long anthers.
Macrotis, 23. *Gr.*, long ears; sc. racemes.
Madia, 173. The name in Chili.
Magnolia, 24. To Prof. Pierre Magnol, Mont-
MAGNOLIACEÆ, 24. [pelier, France.
Majanthemum, 346. *Lat.*, May-flower.
Malachodendron, 65. *Gr.*, Mallow-tree.
MALPIGHIACEÆ, 8.
Malus, 112. *Lat.*, the apple. [=soft
Malva, 60. Altered from the Greek *malache*
MALVACEÆ, 59.
Malvastrum, 61. From Malva.
Malvaviscus, 62. *Lat.*, glue mallow.
Mammilaria, 132. Lat., *mamma*, nipple; sc. the protuberances. [Ayres.
Mandevilla, 271. To H. B. Mandeville, Buenos
Manisurus, 407. *Gr.*, lizard's-tail. [1550.
Maranta, 331. To B. Maranti, M. D., Venice,
Marrubium, 249. *Hebrew*, bitter juice.
Marshallia, 182. To Humphrey Marshall, Phila.
Marsilia, 412. To Count F. Marsigli, Bologna.
MARSILIACEÆ, 412. [bridge, 1765.
Martynia, 219. To Prof. John Martyn, Cam-
Maruta, 183. Meaning unexplained.
Matricaria, 183. An anatomical word. [1750.
Matthiola, 38. To Dr. P. A. Matthioli, Italy,
Maurandia, 223. To Prof. Maurandi, Cartha-
Mayaca, 351. Name unexplained. [gena.
Maytenus, 76. The Chilian name.
Meconopsis, 32. *Gr.*, poppy-like.
Medeola, 340. From Medea, the sorceress.
Medicago, 92. An ancient name. [(branches).
Melaleuca, 122. *Gr.*, black (trunk), white
Melampyrum, 233. *Gr.*, black wheat.
MELANTHACEÆ, 347.
Melanthera, 174. *Gr.*, black anthers.
Melanthium, 348. *Gr.*, black flower.
MELASTOMACEÆ, 122. [Ash.
Melia, 65. The Greek name for the Manna
MELIACEÆ, 65.
Melica, 400. *Italian*, from *mel*, honey.
Melilotus, 92. *Lat.*, honey lotus.
Melissa, 243. *Lat.*, a bee; yields honey.
Melocactus, 133. *Gr.*, melon cactus.
Melothria, 130. The old Greek name.
MENISPERMACEÆ, 26.
Menispermum, 26. *Gr.*, moon-seed.
Mentha, 240. Minthe, daughter of Cocyton.
Mentzelia, 128. To C. Mentzel, of Brandenburg.
Menyanthes, 268 (269). *Gr.*, moon-flower.
Menziesia, 201. To Archibald Menzies, F.L.S.
Mercurialis, 297. Dedicated to Mercury.
Mertensia, 253. To Prof. F. C. Mertens, Bremen
Mesembryanthemum, 133. *Gr.*, mid-day flower
Metastelma, 274. *Gr.*, with a girdle.
Micranthemum, 227. *Gr.*, minute flower.
Microstylis, 329. *Gr.*, minute style.
Mikania, 160. To Prof. Joseph Mikan, Prague.
Milium, 391. *Lat.*, a thousand (seeds).
Mimosa, 82. *Gr.*, a mimic; sc. its motions.
Mimulus, 226. *Gr.*, an ape; sc. its flowers.
Mimuseops, 210. *Gr.*, ape-like.
Mirabilis, 279. *Lat.*, wonderful; sc. the fls.
Mitchella, 148. To Dr. John Mitchell, Va.
Mitella, 113. *Lat.*, a little mitre; sc. the fruit.
Mitreola, 269. Ditto.
Modiola, 61. *Lat.*, a little measure or cup.
Moenchia, 56. To the Germ. botanist, Moench
Moeringia, 55. To Dr. P. H. G. Mohrling, Germ..
Mollugo, 58. Name applied by Pliny. [1730.
Moluccella, 248. Natives of the Moluccas.
Monarda, 245. To Dr. N. Monardez, Seville.
Moneses, 206. From *monos*, = one; sc. 1 fld.

Monotropa, 206. *Gr.*, one, turning; flowers
Montelia, 289. [turned one way.
Morinda, 147. *i. e.*, Indian Mulberry.
Morus, 300. *Celt.*, black; sc. the fruit.
Muhlenbergia, 385. To Rev. Henry Muhlen-
MULISIACEÆ, 153. [berg, D. D.
Mulgedium, 193. Meaning unknown.
Musa, 331. To Antonius Musa.
MUSACEÆ, 331.
Muscari, 344. From *moschus*, musk.
MUSCI, 14.
Myginda, 76. To Francis von Mygind, Germ.
Mylocarium, 205. *Gr.*, mill-nut; form of the
fruit.
Myosotis, 252. *Gr.*, mouse-ear; sc. the lvs.
Myosurus, 20. *Gr.*, mouse-tail; sc. the torus.
Myrica, 309. *Gr.*, (On the banks of) flowing
MYRICACEÆ, 308. [(rivers).
Myriophyllum, 121. *Gr.*, a thousand leaves.
MYRSINACEÆ, 10. (*Gr.*, myrrh.)
MYRTACEÆ, 121.
Myrtus, 122. *Gr.*, perfume.
Nabalus, 192. The meaning unknown.
NAIADACEÆ, 320.
Najas, 320. *Gr.*, a water-nymph.
Napæa, 61. *Gr.*, dell-nymph. [on the nerves.
Narcissus, 332. From *narke*, stupor; its effect
Nardosmia, 160. *Gr.*, smell of nard, or spike-
Narthecium, 351. *Gr.*, a rod, or wand. [nard.
Nasturtium, 36. *Lat.*, twisted nose; on ac-
Naumbergia, 212. [count of its acridity.
Negundo, 74. Of unknown meaning.
Nelumbium, 29. Nelumbo is the *Cingalese*
Nomastylis, 337. *Gr.*, thread style. [name.
Nemesia, 222. An old name revived.
Nemopanthes, 208. *Gr.*, grove-flower.
Nemophila, 254. *Gr.*, loving the grove.
Nepeta, 215. From Nepet, a town in Tuscany.
Nephrodium, 425. *Gr.*, the kidney; sc. the sori.
Nephrolepis, 418. *Gr.*, kidney scale.
Neptunea, 82. Dedicated to Neptune.
Nerium, 271. *Gr.*, humid; sc. the habit.
Nesæa, 124. The name of a sea-nymph.
Neurophyllum, 136. *Gr.*, nerve-leaf.
Neviusia, 104. To Rev. R. Nevius.
Nicandra, 263. [duced tobacco into France.
Nicotiana, 265. To John Nicot, who intro-
Nierembergia, 264. To J. E. Nieremberg,
Nigella, 21. *Lat.*, black; the seeds. [Spanish.
Nolana, 262. *Lat.*, a little bell; sc. corolla.
Nolina, 343. To P. C. Nolin, American.
Notholæna, 420. *Gr.*, false cloak; the indusia.
Nuphar, 29. The Arabic name of Water-lily.
NYCTAGINACEÆ, 279.
NYMPHÆACEÆ, 28.
Nymphæa, 29. *Gr.*, a water-nymph.
Nyssa, 143. The name of a water-nymph.
Obione, 287. *Gr.*, a shield; the round leaves.
Obolaria, 263. *Gr.*, a piece of money.
Ocimum, 238. *Gr.*, to smell; strong-scented.
Œnothera, 125. *Gr.*, wine-hunting; incentive
OLACACEÆ, 10. [to wine-drinking.
Oldenlandia, 150. To H. B. Oldenland, Danish.
Olea, 276. The Greek name of the Olive. [1695.
OLEACEÆ, 275.
Omphalodes, 251. *Gr.*, navel-like.
ONAGRACEÆ, 124.
Oncidium, 328. *Gr.*, a tumor; sc. the form of
the depressed stem.
Onoclea, 421. *Gr.*, closed vessel; sc. the fruit.
Onopordon, 189. *Gr.*, an ass, to explode; its
supposed effects.
Onosmodium, 252. Compared to the Onosma.

Onychium, 424. *Gr.*, the finger nail; a fanci
ful name. [the frond.
Ophioglossum, 418. *Gr.*, serpent's tongue; sc.
Oplismenus, 393. *Gr.*, strong weapon; cock-
Opuntia, 132. From Opus, in Locris. [spur.
ORCHIDACEÆ, 325.
Orchis, 326. Name a physiological conceit.
Origanum, 242. *Gr.*, mountain joy.
Ornithogalum, 343. *Gr.*, bird milk.
OROBANCHACEÆ, 217. [sc. the Vetch.
Orobus, 100. *Gr.*, to excite (nourish) the ox;
Orontium, 318. Name adopted from the Greek.
Orthodanum, 96. *Gr.*, a true gift.
Oryza, 383. The Arabic name is *eruz.*=Rice.
Oryzopsis, 388. *Gr.*, Oryza-like.=Rice-like.
Osmanthus, 276. *Gr.*, fragrant flower.
Osmorhiza, 137. *Gr.*, fragrant root.
Osmunda, 418. *Osmunder* was a Celtic divinity.
Ostrya, 307. *Gr.*, a scale; sc. the scaly catkins.
Otophylla. 231. *Gr.*, ear-leaf.
OXALIDEÆ, 67. [taste.
Oxalis, 67. *Gr.*, acid; the plant has a sour
Oxybaphus, 279. *Gr.*, acid dye.
Oxycoccus, 199. *Gr.*, acid berry.
Oxydendrum, 203. *Gr.*, acid tree.
Oxyria, 280. *Gr.*, acid.
Pachysandra. 298. *Gr.*, thick stamens.
Pæonia, 23. To the physician Pæon. [ance.
Pæpalanthus, 355. *Gr.*, dust-flower; its appear-
Palafoxia. 181. To Palafox, a Spanish general
PALMACEÆ, 310.
Panax, 142. *Gr.*, all-healing; sc. the Ginseng.
Pancratium, 333. *Gr.*, all-potent.
Panicum, 391. *Lat.*, a panicle.
Papaver. 32. *Lat.*, pap, or thick milk; Poppy
PAPAVERACEÆ. 31. [seeds were used in pap
PAPILIONACEÆ. 80. [for children.
Pardanthus, 337. *Gr.*, leopard flower. [cality.
Parietaria, 301. *Gr.*, a wall; their frequent lo-
Parnassia. 115. Mt.Parnassus was feigned their
nativity. [dy for felon.
Paronychia. 57 (58). *Gr.*, near the nail; reme-
Parthenium, 173. *Gr.*, a virgin; sc. its medi-
cinal properties. [Millet.
Paspalum, 389. One of the Greek names for
Passiflora, 129. *Lat.*, passion-flower; the floral
organs resembling the *Cross and nails*.
PASSIFLORACEÆ, 129. [its form.
Pastinaca, 136. *Lat.*, a garden dibble; from
Paulownia, 223. To Paulownia, princess of Rus-
Pavia, 75. To Prof. Peter Paiv, Leyden. [sia.
Pedicularis, 232. *Lat.*, a louse; sc. Lousewort.
Pelargonium, 68. *Gr.*, a stork; sc. Stork-bill.
Pellæa, 421. *Gr.*, little cup. [character.
Peltandra, 318. *Gr.*, shield anther; from the
Penicillaria, 303. *Lat.*, a pencil; sc. the spikes.
Penthorum. 119. *Gr.*, five bounds; sc. 5 styles.
Pentstemon, 224. *Gr.*, five stamens.
Perilla, 240. A word unexplained.
Periploca. 274. *Gr.*, intertwining.
Persea, 290. Adopted from the Egyptian.
Persicaria. 282. *Lat.*, Peach-like.
PETALIFERÆ. 316. [mens.
Petalostemon, 93. *Gr.*, petals (joined to) sta-
Petiveria, 284. To Dr. J. Petiver, F. R. S.
Petunia, 264. Adopted from the Brazil'n *petun*.
Peucedanum, 136. *Gr.*, parched pine; sc. its
Phaca, 94. *Gr.*, to eat; food. [resinous smell.
Phacelia, 255. *Gr.*, a bundle; sc. the flowers.
PILÆNOGAMIA, 15.
Phalaris, 394. *Gr.*, brilliant; its shining seeds.
Pharbitis, 259. Meaning not known.
Phaseolus, 96. *Lat.*, a little boat; sc. the pods.

LATIN INDEX. 435

Phelipæa, 217. To L. & J. Phelipaux, French.
Philadelphus, 116. Adopted from Aristotle.
Phlegopteris, 368. *Gr.*, burning wing or fern.
Phleum, 387. Adopted from the Greek.
Phlomis, 243. *Gr.*, flame; used for lamp-wicks.
Phlox, 256. *Gr.*, flame; the appearance of the fls.
Phorodendron, 291. *Gr.*, thief of the tree; tree
Phragmites, 404. *Gr.*, a hedge; its use. [thief.
Phryma, 236. The meaning unknown.
Phygelius, 225. [on the leaf-like stems.
Phyllanthus, 297. *Gr.*, leaf-flower; the flowers
Phyllocactus, 133. *Gr.*, leaf Cactus. [leaves.
Phyllodendron, 319. *Gr.*, leaf-tree; immense
Phyllodoce, 201. A mythological name.
Physalis, 263. *Gr.*, a bladder; sc. the calyx.
Physostegia, 247. *Gr.*, bladder covering; calyx.
Phytolacca, 284. *Gr.*, plant lac; the crimson
PHYTOLACCACEÆ, 284. [fruit.
Pilea, 300. *Lat.*, a cap; one of the sepals.
Pimpinella, 139. Altered from *bipinnate*.
Pinckneya, 150. To Gen. Pinckney, of S. Car.
Pinguicula, 215. *Lat.*, fat; the greasy leaves.
Pinus, 312. The ancient Greek name.
Piriqueta, 129. Meaning unknown.
Pisonia, 279. To M. Piso, M. D., Amsterdam.
Pistia, 318. Meaning unexplained.
Pisum, 85. Celt., *pis*,=a pea.
PITTOSPORACEÆ, 9.
Planera, 299. To J. Planer, a German botanist.
PLANTAGINACEÆ, 213. [in footpaths.
Plantago, 213. *Lat.*, the sole of the foot; grows
PLATANACEÆ, 303.
Platanthera, 326. *Gr.*, broad anther.
Platanus, 303. *Gr.*, ample; the branches & lvs.
Platycerium, 419. *Gr.*, broad horn; the split
Platycodon, 197. *Gr.*, broad bell. [frond.
Pleea, 349. *Gr.*, the Pleiades; seven white fls.
Pluchea, 171. Meaning unexplained.
PLUMBAGINACEÆ, 214. [der of the eyes.
Plumbago, 215. A cure for *plumbago*, a disor-
Poa, 401. The general Greek word for grass.
Podocarpus, 316. *Gr.*, fruit-stalks (long).
Podophyllum, 28. *Gr.*, foot leaf; duck's-foot.
PODOSTEMIACEÆ, 302.
Podostemum, 302. *Gr.*, foot stem?
Podostigma, 273. *Gr.*, foot (stalked) stigma.
Pogonia, 330. *Gr.*, beard; flowers fringed.
Poinciana, 99. To M. de Poinci, gov. Antilles.
Polanisia, 41. *Gr.*, many unequal (stamens).
POLEMONIACEÆ, 256.
Polemonium, 257. *Gr.*, war; Pliny says that
two kings fought for its honors.
Polianthes, 334. *Gr.*, polished flower.
Polyanthes, 334. *Gr.*, many flowers.
Polycarpon, 57. *Gr.*, much fruit.
Polygala, 78. *Gr.*, much milk; effect on goats.
POLYGALACEÆ, 78.
POLYGONACEÆ, 280.
Polygonatum, 346. *Gr.*, many joints.
Polygonella, 282. From Polygonum.
Polygonum, 282. *Gr.*, many joints.
Polymnia, 172. The name of one of the Muses.
Polypodium, 420. *Gr.*, many feet (roots).
Polypogon, 386. *Gr.*, much beard.
Polypremum, 269. *Gr.*, many stems.
Polypteris, 181. *Gr.*, many wings.
Polytænia, 136. *Gr.*, many fillets (vittæ).
Pontederia, 350. To Prof. Julius Pontedera,
PONTEDERIACEÆ, 350. [of Padua.
Pouthieva, 330. To M. de Pouthieu, W. India.
Populus, 311. The *arbor populi* of the Romans.
Portulaca, 59. *Lat.*, to carry milk, or juice.
PORTULACACEÆ, 58.

Potamogeton, 221. *Gr.*, neighbor of the river.
Potentilla, 107. *Lat.*, powerful (in medicine).
Poterium, 108. *Lat.*, a cup; used in cool drinks.
Primula, 211. *Lat.*, the first; early flowering.
PRIMULACEÆ, 210.
Prinos, 208. The Greek name of the Holly.
Priva, 235. Derivation unknown. [dulous.
Prosartes, 347. *Gr.*, to suspend; sc. fls. pen-
Proserpinaca, 120. *Lat.*, to creep; sc. the roots.
Prunus, 101. The old Greek name.
Psilocarya, 364. *Gr.*, slender Carex.
Psilotum, 415. *Gr.*, naked (of leaves).
Psoralea, 92. *Gr.*, scurfy; from the appear-
Psycotria, 147. Gr., *psyche*, life? [ance.
Ptelea, 71. The Greek name for the Elm.
Pteris, 421. *Gr.*, a wing; the fronds.
Pterocaulon, 171. *Gr.*, winged stem.
Pterospora, 207. *Gr.*, winged seed.
Pulsatilla, 17. A coined name.
Punica, 123. *Lat.*, of or near Carthage.
Pycnanthemum, 241. *Gr.*, dense flowers.
Pyrethrum, 184. *Gr.*, fire; taste of the roots.
Pyrola, 205. From *Pyrus*, pear-tree; its lvs.
Pyrrhopappus, 193. *Gr.*, flame-colored pappus.
Pyrularia, 292. Meaning unexplained.
Pyrus, 112. *Peren* was the Celtic word for Pear.
Pyxidanthera, 258. *Gr.*, box anther. [*cyamos*.
Quamoclit, 258. Resembles the bean-vine,=
Quercus, 305. The orig. name, from the Celtic.
Randia. To J. Rand, a London botanist.
RANUNCULACEÆ, 15. [phibious.
Ranunculus, 19. *Lat.*, a little frog; sc. am-
Raphanus, 43. *Gr.*, quick to appear; rapid
Reseda, 45. *Lat.*, to calm, or soothe. [growth.
RESEDACEÆ, 44.
RHAMNACEÆ, 76.
Rhamnus, 77. The old name, from the Celtic.
Rheum, 281. First found on the banks of the
River Rha (Volga).
Rhexia, 122. *Lat.*, a rupture; an astringent.
Rhinanthus, 232. *Gr.*, snout-flower.
RHIZOPORACEÆ, 8.
Rhodanthe, 186. *Gr.*, rose-flower.
Rhododendron, 203. *Gr.*, rose-tree.
Rhodora, 204. *Gr.*, the rose; sc. the color.
Rhus, 72. From the Celtic *rhudd*, red.
Rhynchosia, 96. *Gr.*, a beak; flower beaked.
Rhynchospora, 365. *Gr.*, beak-seed.
Rhytiglossa, 231. *Gr.*, wrinkled tongue.
Ribes, 117. Adopted from the Arabic.
Richardia, 319. To L. C. Richard, French.
Ricinus, 297. *Lat.*, a tick; sc. the seeds.
Rivina, 281. To A. Q. Rivinus, of Saxony.
Robinia, 95. To Jean Robin, bot. to Henry IV.
Rochea, 119. To M. de la Roche, French.
Rosa, 108. *Celt.*, red; the prevailing color o'
ROSACEÆ, 101. [the flowers
Rosmarinus, 244. *Lat.*, dew of the sea.
Rottbœllia, 409. To C. F. Rottbœll, Danish.
Roubieva, 286. To G. J. Roubieu, French.
ROXBURGHIACEÆ, 339.
Rubia, 148. *Lat.*, red; the color of the roots.
RUBIACEÆ, 147.
Rubus, 104. *Celt.*, red; color of the fruit.
Rudbeckia, 175. To Prof. Olaf Rudbec, Upsal.
Ruellia, 233. To John Ruelle, bot to Francis I.
Rugelia, 188. To Mr. Rugel, collector in Flo.
Rumex, 281. *Lat.*, to suck; the lvs. allay thirst.
Ruppia, 321. To H. B. Ruppia, German.
Russelia, 225. To Alex. Russel, M.D., F. R. S.
Ruta, 70. *Gr.*, to flow; Eng., Rue.
RUTACEÆ, 70.
Sabal, 317. Word not explained.

Sabbatia, 266. To L. Sabbati, an Italian bot.
Saccharum, 410. The Arabic name is *soukar*.
Sageretia, 76. To M. Sageret, Fr. [Eng., *sugar*.
Sagina, 56 (55). *Lat.*, fatness; for pasturage.
Sagittaria, 323. *Lat.*, an arrow; shape of the
SALICACEÆ, 309. [leaves.
Salicornia, 287. *Lat.*, salt horn; the locality and shape. [Salisbury, Eng.
Salisburia, 316. To the distinguished R. A.
Salix, 309. *Celtic*, near the water. [style.
Salpiglossis, 221. *Gr.*, tube tongue; sc. the
Salsola, 288. *Lat.*, salt; grows in salt marshes.
Salvia, 244. *Lat.*, *salvo*, to save; salutary.
Sambucus, 146. *Lat.*, a musical instrument, made of elderwood.
Samolus, 213. *Celtic*, pig's food. [juice.
Sanguinaria, 31. *Lat.*, blood; filled with red
Sanguisorba, 108. *Lat.*, to absorb (stanch)
Sanicula, 135. *Lat.*, to heal. [blood.
SANTALACEÆ, 291.
SAPINDACEÆ, 73.
Sapindus, 75. Sapo Indicus; Indian soap.
Saponaria, 53. *Lat.*, soap; sc. Soapwort.
SAPOTACEÆ, 210. [corona.
Sarcostemma, 272. *Gr.*, fleshy crown; the
Sarracenia, 80. To Dr. Sarrasin, of Quebec.
SARRACENIACEÆ, 80.
Sassafras, 290. The aboriginal name.
Satureja, 242. The Arabic *Sattar*, a labiate
SAURURACEÆ, 301. [plant.
Saururus, 301. *Gr.*, lizard-tail.
Saxifraga, 113. *Lat.*, to break a stone: growing in the clefts of rocks.
SAXIFRAGACEÆ, 112.
Scabiosa, 152. *Lat.*, the itch: which it cures.
Scævola, 10. *Lat.*, the left hand; sc. the corolla.
Scandix, 137. The Greek name of an eatable plant. [a German botanist.
Schæfferia, 76. To Jos. Christian Schæffer,
Scheuchzeria, 324. To John and Jas. Scheuchzer, German. [flowers.
Schizæa, 419. *Lat.*, to cut: applied to the
Schizandra, 25. *Lat.*, to cleave (the stamens).
Schizanthus, 221. *Lat.*, cut flower.
Schizopetalon, 40. *Lat.*, cut petals.
Schizostylis, 337. *Lat.*, cut style.
Schœnocaulon, 348. *Gr.*, rush-stem.
Schœnolirion, 344. *Gr.*, Rush-lily.
Schollera, 350. To one Scholler, a Germ. bot.
Schrankia, 82. To F. de Paula Schrank, Germ.
Schwalbea, 232. To one Schwalb, Germ. bot.
Schweinitzia, 207. To Rev. Lewis de Schweinitz, North Carolina.
Scilla, 343. *Gr.*, to injure: bulb poisonous.
Scirpus, 361. Celt., *cirs*, rushes.
SCITAMINEÆ, 331.
Scleranthus, 58. *Gr.*, hard flower.
Scleria, 367. *Gr.*, hard; referring to the fruit.
Sclerolepis, 156. *Gr.*, hard scales.
Scolopendrium, 425. *Lat.*, a centipede; its appearance beneath.
Scrophularia, 224. Good in the scrofula.
SCROPHULARIACEÆ, 220. [sc. the calyx.
Scutellaria, 246. *Lat.*, a little cup, or vizor;
Scutia, 76. *Lat.*, a shield. [tian.
Sebastiania, 293 (296). Dedicated to St. Sebas-
Secale, 406. The ancient name of Rye.
Sedum, 118. *Lat.*, to sit; habit of the plants.
Selaginella, 414. Diminutive, from Selago, club-moss. [ley.
Selinum, 139. *Selinon* is the Greek for Pars-
Sempervivum, 119. *Lat.*, to live forever.
Senebiera, 43. To John de Senebier, Geneva.

Senecio, 187. *Lat.*, an old man; the receptacle
Sequoya, 315. The Indian name. [naked.
Sericocarpus, 160. *Lat.*, silken fruit.
Sesamum, 219. From the Egyptian, *Sempsen*.
Sesbania, 93. The Arabic name is *Sesban*.
Sesuvium, 133. Not explained.
Setaria, 394. *Lat.*, a bristle; sc. the involucre.
Seutera, 274. Not explained.
Seymeria, 230. To Henry Seymer, English.
Shepherdia, 293. To John Shepherd, Liver
Shortia, 206. To Dr. Short, Kentucky. [pool.
Sibbaldia, 107. To Prof. Robert Sibbald, Edin
Sicyos, 130. The Greek for Cucumber. [burgh
Sida, 61. Adopted from Theophrastus.
Sideroxylon, 210. *Gr.*, iron-wood. [tions.
Silene, 53. *Gr.*, saliva; from the viscid secre-
Silphium, 172. Adopted from the Greek.
Simaruba, 72. The name in the West Indies.
SIMARUBACEÆ, 71. [bage-plants.
Sinapis, 40. A general name in Greek for cab-
Siphonychia, 58. *Gr.*, tube, and Anychia.
Sisymbrium, 39 (37). The old Greek name.
Sisyrinchium, 337. *Gr.*, pig-snout; sc. the spathe.
Sium, 141 (140). From a Celtic word for water.
SMILACEÆ, 338.
Smilacina, 346. Derived from Smilax.
Smilax, 338. *Gr.*, a scraper; from its rough-
SOLANACEÆ, 261. [ness.
Solanum, 262. Etymology doubtful.
Solea, 45. To W. Sole, of England.
Solidago, 166. *Lat.*, to unite: good for wounds.
Soliva, 185. To Salvator Soliva, M. D., Spain.
Sonchus, 194. *Gr.*, hollow; its stems are hol-
Sophora, 100. Adopted from the Arabic. [low.
Sorbus, 112. Old name for Mountain Ash.
Sorghum, 411. The Italian name is Sorghi.
SPADICIFLORÆ, 316. [like leaves
Sparganium, 320. *Gr.*, a fillet; for the ribbon-
Spartina, 408. *Gr.*, a rope; the use of its lvs.
Spartium, 90. *Gr.*, a rope; use of its twigs.
Specularia, 196. *Lat.*, a mirror; suggested by the flowers.
Spergula, 57. *Lat.*, to scatter (its seeds).
Spergularia, 57. From Spergula.
Spermacoce, 149. *Gr.*, seed-points; the pod pointed with the calyx lobes.
Sphenogyne, 173. *Gr.*, wedge-shaped pistil.
Spigelia, 209. To Prof. Adrien Spigelius, Padua, 1620. [brow i.
Spilanthus, 180. *Gr.*, spot-flower; the d'sk
Spinacia, 287. *Lat.*, a spine or prickle.
Spiræa, 103. *Gr.*, to wind; sc. into wreaths.
Spiranthes, 329. *Gr.*, spiral fls.; spike twisted.
Spirodela, 319. *Gr.*, spiral bait; duck-meat.
Sporobolus, 384. *Gr.*, to cast the seeds; drop-
Sprekelia, 334. A personal name. [seed.
Stachys, 248. A spike (of flowers).
Stachytarpha, 235. *Gr.*, spikes dense. [dam
Stapelia, 275. To Dr. Boderus Stapel, Amster
Staphylea, 74. *Gr.*, a cluster (the scarlet fr.)
Statice, 215. *Gr.*, to stop; an astringent.
Stellaria, 55. *Lat.*, a star.
Stenanthium, 349. *Gr.*, narrow flower.
Stenotaphrum, 410.
Stephanotis, 275. *Gr.*, crown, ear; crown with ear-shaped segments.
Sterculia, 63. *Lat.*, *stercus*; from its bad odor.
STERCULIACEÆ, 63.
Stillingia, 296. To Dr. Benj. Stillingfleet, Eng.
Stipa, 388. *Lat.*, something silky or feathery.
Stipulicida, 57. *Lat.*, cut stipules.
Stokesia, 156. To Dr. Jonathan Stokes, Eng.

LATIN INDEX. 437

Strelitzia, 331. To the Queen of George III., of Mecklenburg-Strelitz.
Streptopus, 347. *Gr.*, twisted foot (-stalk).
Strumpfia, 147. A personal name.
Struthiopteris, 421. *Gr.*, ostrich-wing (fern).
Stuartia, 65. To John Stuart, Marquis of Bute.
Stylisma, 260. Refers to the two styles.
Stylosanthes, 87. *Gr.*, style, flower; style long.
STYRACACEÆ, 208.
Styrax, 209. The Arabic name is Assthiac.
Subularia, 42. *Subula* is the Latin for an awl.
Sullivantia, 114. To Wm. S. Sullivant, Ohio.
Swietenia, 66. To Gerard van Swieten, Hol-
SURIANACEÆ, 8. [land.
Symphoricarpus, 144. *Gr.*, to accumulate fruit.
Symphytum, 252. *Gr.*, to cause to unite; heal-
Symplocarpus, 318. *Gr.*, connected fruit. [ing.
Symplocos, 209. *Gr.*, connected (stamens).
Synandra, 247. *Gr.*, united anthers.
Syndesmon, 17. *Gr.*, with a bond.
Synthyris, 228. *Gr.*, door (valves) closed.
Syringa, 276. *Gr.*, a pipe; the slender shoots are filled only with pith. [god.
Tagetes, 188. Dedicated to Tages, a Tuscan
Talinum, 59. From *thalia*, a green branch?
Tamarix. 64. Found on the river Tamaris.
TAMARISCINEÆ, 63. [France.
Tanacetum, 183. Altered from Athanasia?
Taraxacum, 193. *Gr.*, a cathartic.
TAXACEÆ, 315.
Taxodium, 315. *Gr.*, like the Yew.
Taxus, 316. *Gr.*, the bow; used for making.
Tecoma, 218. The Mexican name. [flowers.
Telanthera, 289. *Gr.*, complete or perfect
Tephrosia, 94. *Gr.*, ash-colored (herbage).
Tetragonotheca, 175. *Gr.*, four-angled en-
Tetranthera, 291. *Gr.*, four anthers, [velope.
Teucrium, 239. To Teucer, founder of Troy.
Thalia, 332. To J. Thalius, M. D., Germ., 1585.
Thalictrum, 18. *Gr.*, to grow green.
Thaspium, 138. From the Isle of Thaspia or Thapsus.
Thea. 65. *Teha* is the Chinese for Tea.
THEOPHRASTACEÆ, 210.
Thermopsis, 85. *Gr.*, like a Lupine. [F. R. S.
Thunbergia, 233. To Charles P. Thunberg.
Thuya, 315. Gr. *thyou*, a sacrifice; the wood
Thuyopsis, 315. Like Thuya. [so used.
THYMELACEÆ, 232. [Thyme is reviving.
Thymus, 243. *Gr.*, courage; the smell of
Thysanella, 282. Gr. *thysanotus*, fringed.
Tiarella, 113. *Tiara*, a Persian diadem. [burg.
Tiedmannia, 136. To Prof. Tiedmann, Heidel-
Tigridia, 337. *Lat.*, like a tiger; fls. spotted.
Tilia, 64. Etymology unknown.
TILIACEÆ, 64.
Tillæa, 118. To M. A. Tilli, Italian.
Tillandsia, 335. To Prof. Elias Tillands, Abo.
Tipularia, 328. Lat., *Tipula*, the crane-fly.
Tofieldia, 349. Dedicated to a Mr. Tofield.
Torreya. 316. Dedicated to Dr. John Torrey.
Tournefortia, 250. To Joseph P. de Tournefort.
Tradescantia, 353. To J. Tradescant, gardener.
Tragia, 296. To Jerome Bock Tragus, German.
Tragopogon,191. *Gr.*,goat's beard; the pappus.
Trautvetteria, 19. To one Trautvetter, Germ.
Tribulus, 67. *Gr.*, 3-pointed; sc. each carpel.
Trichelostylis, 363. *Gr.*, triple style.
Trichomanes, 419. *Gr.*, soft hair; the stipes.
Trichostema, 239. *Gr.*, hair stamens.
Tricuspis, 398. *Gr.*, 3-cusped; the chaff.
Trientalis, 212. *Lat.*, *triens*, 3 inches (high).
Trifolium, 91 *Lat.*, three-leaf; lvs. 3-foliate.

Triglochin, 324. *Gr.*, three points; pod 3-angl.
Trigonella, 100. *Gr.*, 3-angled; so the corolla.
TRILLIACEÆ, 340.
Trillium, 340. Parts of the plant all in 3s.
Triosteum, 144. *Gr.*, three bones (bony seeds).
Tripsacum, 409. *Gr.*, to thresh.
Trisetum, 397. *Lat.*, three bristles (awns).
Triticum, 406. Lat. *trito*, to rub or grind.
Tritoma, 345. *Gr.*, thrice-cutting; lvs. 3-edged.
Trollius, 21. German, *trol*, something round.
Tropæolum, 69. *Gr.*, trophy; shield and hel-
Troximon. 193. *Gr.*, something eatable. [met.
TUBULIFLORÆ, 152, 153.
Tulipa, 311. The Persian name is *Thoulyban*.
Turnera, 129. To Wm. Turner, M. D., London,
TURNERACEÆ, 128. [1550.
Turritis, 36. *Lat.*, a tower; remarkably erect.
Tussilago. 160. Lat., *tussis*, a cough; cure for.
Typha, 320. *Gr.*, a marsh; the habitat.
TYPHACEÆ. 319.
ULMACEÆ, 298.
Ulmus. 298. The Saxon name was *ulm*.
UMBELLIFERÆ, 133.
Uniola, 403. *Lat.*, unity; many fls. in one?
Urtica, 300. *Lat.*, to burn (*uro*); stinging.
URTICACEÆ, 298.
Utricularia, 216. Lat., *utriculu*, a little bladder.
Uvularia, 347. Used for diseases of the *uvula*.
Vaccinium, 198. The ancient name.
Vachellia, 90. Not explained.
Valeriana, 150. To King Valerius.
VALERIANACEÆ. 150.
Valerianella, 151. Derived from Valeriana.
Vallesia, 270. To F. Vallesio, phys. to Philip II.
Vallisneria, 325. To Ant. Vallisner, Italy.
Vallota, 333. To Pierre Vallo, French. [root.
Veratrum, 348. *Lat.*, true black; the fls. or
Verbascum, 222. *Lat.*, beard; plant woolly.
Verbena, 235. From the Celtic *Ferfæn*.
VERBENACEÆ, 235.
Verbesina, 180. Same meaning as Verbena.
Vernonia, 155. To Wm. Vernon, collector in North America.
Veronica, 229. Not well explained.
Vesicaria, 42. *Lat.*, a blister; the inflated pods.
Viburnum, 146. *Lat.*, to tie; twigs pliant.
Vicia, 86. Lat., *vincio*, to bind; its tendrils.
Victoria. 30. To Queen Victoria, of England
Vigna, 96. To Dominic Vigni.
Vilfa, 384. Of unknown meaning.
Vinca, 270. Lat. *vinculum*, a band.
Vincetoxicum, 274. Meaning unexplained
Viola, 45. The old Latin name.
VIOLACEÆ, 45.
Visiania 276, To Prof. Visiani, Patavia.
VITACEÆ, 77. [ible.
Vitex, 237. Lat., *vieo*, to bind; branches flex-
Vitis, 77. Celtic, *gwyd*, = best of trees.
Vittaria, 417, Lat., *vitta*, a riband; its form.
Waldsteinia, 107. To Franz de Waldstein.
Waltheria, 63. To Prof. A. F. Walther, Leipsic.
Waren, 39. To Mr. Ware, its discoverer.
Whitlavia, 255. A personal name.
Wigela, 415. A personal name. [mala.
Wigandia. 256. To Bishop Wigand, of Pome-
Wistaria, 96. To Prof. Caspar Wistar, Phila.
Wolffia, 319. A personal name.
Woodsia, 425. To Joseph Woods, English.
Woodwardia, 423. To Thomas J. Woodward
Xanthium, 174. Said to dye the hair *yellow*.
Xanthosoma, 318. *Gr.*, yellow mouth.
Xanthoxylum. See Zanthoxylum.
Xeranthemum, 186. *Gr.*, dry flowers.

Xerophyllum, 349. *Gr.*, dry leaf. [monk.
Ximenia. 10. To F. Ximenes, a Spanish
Xylosteon, 145. *Gr.*, wood bone; hard wood.
XYRIDACEÆ, 354.
Xyris, 354. *Gr.*, acute; sc. the leaves.
Yucca, 345. The Peruvian name. [Italy.
Zannichellia, 521. To John J. Zannichelli,
Zanthorhiza, 21. *Gr.*, yellow root.
Zanthoxylum, 70. *Gr.*, yellow wood.
Zauschneria, 125. A personal name.

Zea, 409. *Gr.*, *zao*, to live; plants nutritive.
Zephyranthus, 333. *Gr.*, zephyr flower.
Zigadenus, 348. *Gr.*, joined glands (on the petals).
Zinnia, 175. To Prof. John G. Zinn, Gottingen.
Zizania, 383. A Greek name adopted.
Zizia, 138 (139). To J. B. Zizi, German.
Zornia, 86. To John Zorn, Bavaria.
Zostera, 321. *Gr.*, a riband · sc. the long lvs.
ZYGOPHYLLACEÆ, 66.

ENGLISH INDEX.

Abele Poplar. 311	Banana 331	Black Grass............ 352
Acacia, Rose.. 95	Baneberry 23	Black Haw 147
ACANTHADS 233	Banyan 299	Black Hoarhound 243
Aconite................. 23	Barley 404	Black Jack............. 305
ACROGENS 412	Basil..........(232, 243) 241	Blackroot 171
Adam and Eve 328	Bass-wood 64	Black Snakeroot........ 23
Adder's-tongue.... 418	Bastard Toad-Flax 291	Black Thorn 111
Agrimony.......... ... 108	Bath-flower 340	Bladder Campion 53
Alder.............(208) 308	Bay...... 24, 65, 124, 203, 290	Bladder Fern........... 425
Alexanders............. 138	Bayberry................ 309	Bladder-nut 74
All-seed................ 57	Bay-galls............... 290	Bladder-pod.... 12
Almond................. 103	Bayonet Rush........... 353	Bladder Senna 95
Aloe.................... 333	Bean ....(268) 96	Bladderwort 216
Alum-root.............. 115	BEAN CAPERS......... 66	Blazing Star.......157, 349
AMARANTHS.......... 288	Bear-berry 201	Bleeding-heart 33
Amaranth, Globe....... 289	Beard Grass......410... 388	Blessed Thistle 189
AMARYLLIDS 332	Beard-tongue 224	Blite..... 286
American Centaury 266	Bear's Grass........... 345	Blood-root 31
American Laurel........ 200	Bear's Thread.......... 345	BLOODWORTS 335
Angelica 137	Bed-straw.............. 148	Bluebell............... 196
Angelica-tree........... 142	Beech 307	Blueberry.............. 198
Angelico 140	Beechdrops........(207) 217	Blue-curls 246, 239
Anise.............(24) 139	Beet 285	Blue Dangles.......... 198
ANONADS.............. 26	Beetleweed 206	Blue-eyed Grass....... 337
Apple................... 112	Beggar-ticks.......180, 251	Blue Flag 336
Apple Haw 111	BEGONIADS 131	Blue Grass........... 402
Apple of Peru 263	BELLWORTS.......... 196	Blue-hearts 230
Apple of Sodom........ 263	Bellwort............... 347	Blue Palmetto 317
Apricot 103	Bent Grass............. 384	Bluets 149
ARALIADS............. 142	BERBERIDS 27	Bog Rush............ 367
Arbor-vitæ............. 315	Berberry 27	Boneset............(160) 158
AROIDS............... 317	Bergamot 245	Borrage 251
Arrow Grass........... 324	Betony................ 249	BORRAGEWORTS...... 250
Arrow-head............ 323	Bhotan Pine 312	Boston Iris.......... 336
Arrow-root............ 312	Big Laurel 24	Bottle-brush 122
Arrow-wood 146	Bilberry............... 198	Bouncing Bet........ 53
Artichoke ....(177) 188	Bindweed..........(232) 260	Boursault 110, 109
ASCLEPIADS 271	BINDWEEDS 258	Bowman's-root....... 104
Ash.............(70, 112) 277	Birch 308	Box Elder............ 74
Ash Maple 74	BIRCHWORTS........ 307	Boxwood 208
Aspen................. 311	Bird's-nest..........206, 424	Brake................ 421
Aster 160	Birthwort............. 278	Bramble 104
ASTERWORTS......... 152	BIRTHWORTS........ 278	Bridal Rose......... 105
Atamasco Lily 333	Bishop's-cap 113	Brier ....(82, 238) 109
Auricula 211	Bishopweed 141	Broccoli 40
Avens 105	Bitter Cress........... 37	Brome Grass......... 397
Awlwort............ 42	Bitter-nut 304	BROMELIADS 335
Bachelor's Button 188	Bittersweet 262	Brooklime 229
Balm........(72, 241, 311) 243	Bitter Vetch........... 100	Broom............90, 100
Balm-of-Gilead..... 72, 311	Black Alder........... 208	Broom Corn 411
Balsamine 70	Blackberry 104	Broom Grass........ 411
Baltimore Belle. 109	Blackberry Lily 337	Broom-rape 217

ENGLISH INDEX. 439

BROOMRAPES	217	Cedar-of-Lebanon	314	Crape Myrtle	123
Bryony	130	Celandine	31	Creeping Greenhead	150
Buck Bean	268	Celery	140	Cress	36, 37, 39, 43, 65
Buck-eye	74	Centaury	266	Crest-flower	335
Buckthorn	77	Century Plant	334	Crookneck Squash	130
BUCKTHORNS	76	Chaff-seed	232	CROWBERRIES	302
Buckwheat	284	Chamomile	183	Crowberry	303
Buckwheat-tree	205	Chaste-tree	237	Crowfoot	19
Buffalo-berry	293	Cheat	397	CROWFOOTS	15
Bugbane	23	Checkerberry	201	Crow Garlic	244
Bugleweed	240	CHENOPODS	284	Crownbeard	180
Bugloss	(251) 252	Cherry	(262, 263) 102	Crown Imperial	343
Bull Rush	361	Cherry Laurel	102	CRUCIFERS	34
Burdock	(173) 190	Chequered Lily	342	Cuckoo-flower	38
Burnet	108	Chervil	137	Cucumber	(130) 131
Burning-bush	76	Chess	397	Cucumber-root	340
Burr Grass	394	Chestnut	306	Cucumber-tree	24
Burr Marigold	180	Chick Pea	85, 86	CUCURBITS	129
Burr Reed	320	Chickweed	54, 55	Cudweed	185
Burr-seed	251	Chickweed Wintergreen	212	Culver's Physic	229
BURSERIDS	72	Chicory	190	Cup-plant	173
Bush Clover	89	China Aster	165	Cupseed	27
Bush Honeysuckle	146	Chinquapin	307	Currants	117
Bush Trefoil	88	Chokeberry	112	Cutflower	221
Buttercups	19	Chokecherry	102	Cut Grass	383
Butterfly-weed	273	Christmas Rose	21	CYCADS	311
Butternut	304	Cinnamon Fern	366	Cypress	(257) 315
Butterweed	187	Cinquefoil	107	Cypress Vine	258
BUTTERWORTS	215	Citron	(71) 130	Daffodil	333
Button-bush	150	Cives	344	Dahlia	166
Buttonwood	303	Cleavers	148	Dahoon	207
Cabbage	40	Climbing Boneset	160	Daisy	165
Cactus	132	Climbing Fern	418	Dandelion	(191) 193
Cajeput	122	Clotweed	174	DAPHNADS	292
Cale	40	Cloudberry	105	Darnel	405
Calaminth	243	Clover	(80, 92) 91	Daughter-of-Spring	140
Calamus	319	Club Moss	413	Day Lily	345
Calico-bush	200	CLUB MOSSES	413	Deadly Nightshade	264
California Poppy	32	Club Rush	361	Deerberry	198
CALYCANTHS	25	Cock's-comb	288	Deer-grass	122
CAMELLIAS	64	Cockspur Grass	393	Deer's-tongue	157
Campion	53, 54	Cocoa Plum	101	Dewberry	105
Canada Thistle	190	Coffee Bean	85	Dickson's Fern	423
Canary-bird	69	Coffee-tree	83	Dill	136
Canary Grass	394	Cohosh	27	Ditch Grass	321
Candleberry	309	Colic-root	335	Ditch Moss	324
Candytuft	42	Colocynth	131	Dittany	240
Cane	404	Colt's-foot	160	Dock	281
Canterbury Bells	196	Columbine	22	Dockmackie	146
CAPERS	66	Columbo	268	Dodder	260
Caper Spurge	295	Comfrey	252	Dogbane	270
CAPPARIDS	66	Cone-flower	175	DOGBANES	269
Caraway	138	CONIFERS	312	Dog Fennel	181, 183
Cardinal-flower	195	Coontie	312	Dogwood	73, 143
Cardoon	188	Coral-root	328	Doorweed	283
Carnation	52	Coriander	141	Dragonhead	246
Carolina Beech-drops	207	Corn Cockle	54	Dragon-root	318
Carpet Cress	43	Coruel	143	Drop-flower	192
Carpet-weed	58	CORNELS	142	Dropseed	388, 384
Carrion-flower	275, 339	Corn Flag	338	Dry Strawberry	107
Carrot	139	Cotton	63	Duckmeat	319
Cassena Tea	208	Cotton Grass	362	Dundee Rambler	169
Castor-oil Plant	297	Cotton Rose	185	Dutchman's Pipe	278
Catalpa	218	Cotton Thistle	189	Dwarf Clubmoss	414
Catchfly	53	Cottonwood	311	Dwarf Dandelion	191
Catchfly Grass	383	Couch Grass	406	Dwarf Pink	119
Cat-gut	94	Cowbane	136	Dyer's Broom	94
Catmint	245	Cowslip	(211) 21	Dyer's Cleavers	148
Catnep	246	Cow-wheat	233	Dyer's-weed	45
Cat-tail	387, 320	Crab Grass	390, 407	Ear-drop	33, 127
Cauliflower	40	Crab Tree	112	Earth-galls	182
Cayenne Pepper	263	Cranberry	(146) 199	EBONADS	201
Cedar	314, 315	Crane's bill	68	Eel-grass	325

ENGLISH INDEX.

Egg-plant	263	
Eglantine (110)	109	
Egyptian Calla	319	
Elder (74, 142, 174)	146	
Elecampane	171	
Elephant's-ear	131	
Elephant's-foot	156	
Elm	298	
Enchanter's Nightshade	128	
Endive	191	
ENDOGENS	316	
English Mint	183	
English Moss	118	
Eternal Flower	186	
Evening Primrose	125	
Everlasting	185	
Everlasting Pea	86	
EXOGENS	15	
Eyebright	232	
False Dogfennel	181	
False Flax	42	
False Goldenrod	166	
False Hellebore	348	
False Mermaid	68	
False Nettle	300	
False Pennyroyal	239	
False Pimpernel	213	
False Redtop	398, 402	
False Rice	383	
False Rocket	36	
False Rue-Anemone	20	
False Syringa	116	
False Violet	105	
False Wallflower	39	
Felwort	268	
Fennel	139	
Fennel-flower	21	
Fenugreek	100	
FERNS	416	
Fescue Grass	399	
Festoon Pine	414	
Fetter-bush	202	
Feverfew	183	
Feverwort	144	
Fig	299	
Figwort	224	
FIGWORTS	220	
Filbert	307	
Finger Grass	390	
Fireweed	186	
Fir	313	
Fir Balsam	314	
Flag	318, 336	
Flaming Pinxter	203	
FLAXWORTS, Flax	66	
Fleabane	165	
Fleur-de-lis	337	
Flixweed	39	
Floating-heart	268	
Florida Arrowroot	312	
Florin Grass	384	
Flower-de-luce	336	
Flowering Fern	418	
FLOWERING PLANTS	15	
FLOWERLESS PLANTS	412	
Flower-of-an-hour	63	
Fly-poison	348	
Fogfruit	236	
Fool's Parsley	140	
Forget-me-not	252	
Forked Spike	411	
Foul-meadow	402	
Four-o'clock	279	
Foxglove (230)	228	
Foxtail	394, 387	
Fraxinella	70	
French Mulberry	236	
Fringe Grass	392	
Fringe-tree	276	
FROGBITS	324	
Frost-plant	47	
FUMEWORTS	33	
Fumitory	34	
GALEWORTS	308	
Gale	309	
Galingale	357	
Gargetweed	284	
Garden Orache	287	
Garlic	343	
Gay-feather	157	
Gentian	267	
GENTIANWORTS	266	
GERANIA	67	
Geranium	68	
Germander	239	
GESNERWORTS	219	
Gilia	257	
Gill-over-the-ground	246	
Ginger, Wild	278	
GINGERWORTS	331	
Ginseng	142	
Glasswort	287	
Globe Amaranth	289	
Globe-flower (104)	21	
Glue Mallow	62	
Gnatbane	171	
Goat's-beard	104	
Goat's Rue	94	
Golden Alexanders	138	
Golden Bartonia	128	
Golden-chain	91	
Golden Club	318	
Golden Fern	420	
Goldenrod	166	
Goldthread	21	
Good-king-Henry	286	
Good-night	260	
Gooseberry	117	
Goosefoot	285	
GOOSEFOOTS	284	
Goosegrass 107,	148	
Gourd	130	
GRAMINOIDS	356	
Grape	77	
Grape Fern	418	
Grape Hyacinth	344	
GRASSES	380	
Grass of Parnassus	115	
Grass Pink	330	
Grass-poly	123	
Greek Valerian	257	
Green Brier	338	
Green Dragon	318	
Green-head	150	
Green Violet	45	
Gromwell	252	
Ground Cherry	263	
Ground Fir	414	
Ground Pine	414	
Ground Ivy	245	
Ground-nut 96,	142	
Groundsel	187	
Groundsel-tree	171	
Guava	122	
Gum-tree (120)	143	
GYMNOSPERMS	311	
Hair Grass	384, 386	
Hardhack	105	
Hare-bell	196	
Hare's-foot	91, 423	
Hart's-tongue	425	
Haw (146, 147)	111	
Hawthorn 110,	111	
Hawkweed	191	
Hazelnut	301	
Heart's-ease	47	
Heart-seed	73	
Heath, Heather	200	
HEATHWORTS	197	
Hedgehog	92	
Hedgehog Grass	405	
Hedge Hyssop 227,	245	
Hedge Mustard	39	
Hedge Nettle	248	
Hedge Bindweed	283	
Heliotrope	250	
Hellebore (348)	21	
Hemlock (139, 141)	313	
Hemp (289)	301	
Hemp Nettle	248	
Henbane	261	
Henbit	248	
Herb Robert	68	
Hercules' Club	142	
Herd's Grass	387	
Heron's-bill	68	
Hickory	304	
High Cranberry	146	
High-water Shrub	174	
HIPPURIDS	120	
Hoarhound (159, 248)	249	
Hobble-bush	146	
Hogweed	174	
HOLLYWORTS, Holly	207	
Hollyhock	60	
Honesty	40	
Honewort	138	
Honey Locust	83	
Honeysuckle	144-6	
HONEYSUCKLES	144	
Hoop-petticoat	333	
Hop	301	
Hop Hornbeam	307	
Hornbeam	307	
Horn Pondweed	321	
Horn Poppy	31	
HORNWORTS	302	
Horse Balm	241	
Horse Chestnut	74	
Horsemint 240,	245	
Horse Nettle	263	
Horse Radish	41	
Horse-tail	415	
Horse-weed	174	
Hound's-tongue	251	
Houseleek	119	
HOUSELEEKS	117	
Huckleberry	198	
Hyacinth	344	
Hydrangea	116	
HYDROPHYLLS	253	
Hyssop (227, 245)	241	
Immortal-flower	186	
Indian Corn	409	
Indian Cress	69	
Indian Cucumber-root	346	
INDIAN FIGS	132	
Indian Mallow	61	
Indian Millet	411	

ENGLISH INDEX. 441

Name	Page
Indian Physic	104
Indian Pipe	206
Indian Rice	383
Indian Shot	332
Indian Tobacco	195
Indian Turnip	318
India-rubber tree	299
India Wheat	284
Indigo Plant (84)	95
Inkberry	208
Innocence	149, 225
IRIDS	336
Ironweed	155
Ironwood	307
Ivy (73, 188)	142
Ivy, Poison	73
Jack-in-the-pulpit	318
Jacobœa	187
Jacobœa Lily	334
Japan Globe-flower	104
Japan Quince	112
Japan Rose	65
Japonica	65
Jersey Tea	77
Jerusalem Artichoke	177
Jerusalem Cherry	262
Jerusalem Sage	248
Jessamine (269, 392)	275
Jewelweed	69
Jimson-weed	265
Job's-tears	411
Jonquil	333
Judas-tree	83
July-flower	38
June Grass	402
Juniper	314
Kidney Bean	96
Knap-weed	188
Knawel	58
Knot Bindweed	283
Knotgrass	282
Labrador Tea	204
Lady-Fern	424, 426
Lady's Eardrop	127
Lady's Mantle	108
Lady's Slipper	326
Lady's Tresses	329
Lamb Lettuce	151
Larch	314
Larkspur	22
Laurel 24, 102,	200
LAURELS	290
Laurestine	147
Lavender	239
Lead Plant	93
Leadwort	215
LEADWORTS	214
Leaf-cup	172
Leather-flower	17
Leather-leaf	202
Leather-wood	292
Leek	344
LEGUMINOUS PLANTS	80
Lemon	71
Lentil	100
Lettuce (151, 192,)	193
Leverwood	307
Lignum-vitæ	67
Lilac	276
Lily (333-4, 337, 345,)	342
Lily-of-the-valley	346
LILYWORTS	341
Lime	71
Lime-tree (143)	64
Linden	64
LINDENBLOOMS	64
Lion's-ears	249
Lion's-foot	192
Lion's-heart	247
Lip Fern	422
Liquorice	148
Live-forever 118,	119
Liver-leaf, Liverwort	18
Lizard-tail 407,	301
LOASADS	128
LOBELIADS	194
Loblolly Bay	65
Locust (83)	95
Long Moss	335
Loosestrife	212
LOOSESTRIFES (127)	123
Lopseed	236
LORANTHS	291
Lousewort	262
Lovage	140
Love-lies-bleeding	289
Lucerne	92
Lungwort	253
Lupine	90
Lychnidea	256
Madder (283)	148
MADDERWORTS	147
Mad-dog Skull-cap	247
Madwort	40
MAGNOLIADS	24
Mahogany	66
Maidenhair	422
Maize	409
Mallow	60–62
MALLOWS	59
Mangel-wurzel	285
Manna Grass	402
Maple	74
MAPLEWORTS	73
Mare's-tail	121
Marigold (21)	188
Marjoram	242
Marsh Cress	36
Marsh Elder	174
Marsh Mallow	60
Marsh Marigold	21
Marsh Rosemary	215
Marvel-of-Peru	279
MARVELWORTS	279
MASTWORTS	304
Matrimony-vine	264
May Apple	28
May-flower	200
May-weed	183
Meadow-Rue	18
Meadow-sweet	104
Medick	92
MELANTHS	347
MELASTOMES	122
Melic Grass	400
Melilot	92
MENISPERMADS	26
Mercury	296
Mermaid	68
Mermaid-weed	120
MESEMBRYANTHS	133
Mexican Tea	286
Mexican-vine	285
Miami-mist	255
MIGNONETTES	44
Milkweed	272
MILKWORTS, Milkwort,	78
Milk Vetch	94
Millet 391, 393,	391
Millfoil (121)	183
Mint (183, 245)	240
Mistletoe	291
Mitrewort	113
Mockernut	304
Mock-Orange	116
Molucca Balm	248
Moneywort	212
Monkey-flower	226
Monk's-hood	22
Moon-seed	26
Moonwort	418
Morello	102
Morning-glory	259
Moss Campion	53
Moss Pink	257
Mother Carey	183
Motherwort	249
Moth Mullein	222
Mountain Ash	112
Mountain Fringe	33
Mountain Heath	201
Mountain Mint	245
Mountain Sorrel	280
Mourning Bride	152
Mouse-ear Chickweed	54
Mouse-tail	20
Moving-plant	89
Mud Purslane	51
Mudwort	228
Mugwort	184
Mulberry (105, 236, 299)	300
Mullein	222
Mullein Pink	54
Muscadine	78
Musk Melon	131
Musk-plant	226
Mustard (36, 39)	40
Myrtle (123, 204, 309)	122
MYRTLEBLOOMS	121
NAIADS	320
Nailwort	57
Narcissus	332
Nasturtion	69
Navelwort	251
Neapolitan	47
Neckweed	229
Nectarine	103
Nelumbo	29
Nettle (248, 263, 296)	300
Nettle-tree	299
NETTLEWORTS	298
New York Fern	426
Nightshade (128, 264)	262
NIGHTSHADES	261
Ninebark	103
Nipplewort	190
Noisette	110
Nonesuch	92
Nutmeg-flower	21
NYMPHADS	28
Oak	305
Oak, Poison	73
Oak-of-Jerusalem	286
Oat (347)	396
Ogeechee Lime	143
Okra	63
Oil-nut	292
Oil-seed	219
Oleander	271

Oleaster	292	
Olive	276	
OLIVEWORTS	275	
ONAGRADS	124	
Onion	343	
Opium Poppy	32	
Orache	287	
Orange	(116, 299) 71	
ORANGEWORTS	71	
ORCHIDS	325	
Orris-root	337	
Osage Orange	299	
Osier	(143) 309	
Ostrich Fern	421	
Oxeye	(171) 175	
Oxheart	102	
Pæony	23	
Painted-cup	232	
Palmetto	317	
PALMS	316	
Pampas Grass	398	
Panic Grass	391	
Pansy	45	
Paper Mulberry	299	
Pappoose-root	28	
Parsley	123	
Parsley-piert	108	
Parsnip	(141) 137	
Partridge-berry	148	
Pasque-flower	17	
Passion-flower	129	
PASSIONWORTS	129	
Pawpaw	26	
Pea	(83, 86) 85	
Peach	103	
Peanut	87	
Pear	112	
Pearlwort	56	
Pea-vine	97	
Pecan-nut	304	
Pellitory	301	
Pencil-flower	87	
Pennyroyal	(239) 241	
Pennywort	135, 268	
Pepper	263	
Pepper-and-Salt	140	
Pepper-bush	204	
Pepper-grass	42	
Peppermint	240	
PEPPERWORTS	412	
Periwinkle	270	
Persimmon	209	
Pettimorrel	142	
Pheasant's-eye	19, 52	
Phlox	(39) 256	
PHLOXWORTS	256	
Pickerel-weed	350	
Pie-plant	281	
Pigmy-weed	118	
Pignut	304	
Pigweed	(289) 285	
Pimpernel	213	
Pine	312	
Pineapple	335	
Pine-sap	206	
Pink	(54, 203, 257, 330) 52	
Pink-root	269	
PINKWORTS	52	
Pinweed	47	
Pinxter-bloom	203	
Pipes	416	
Pipewood	202	
Pipeworts	335	

Pipsessiwa	206	
Pitcher-plant	30	
Plane-tree	303	
Plantain	(323, 330) 213	
Plum	101	
Plume Grass	410	
Poet's Narcissus	333	
Poison Haw	146	
Poison Hemlock	139	
Poison Ivy	73	
Poison Oak	73	
Poke	284	
Pokeworts	284	
Polar-plant	173	
Polypody	(373) 420	
Pomegranate	123	
Pond Lily	29	
Pond Spice	291	
Pond-weed	321	
PONTEDERIADS	350	
Poor-man's-weather-glass	213	
Poplar	311	
Poppy	32	
POPPYWORTS	31	
Possum Haw	147	
Potato	(259) 262	
Poverty Grass	388	
Prairie Burdock	173	
Prairie Queen	109	
Prickly Ash	70	
Prickly Pear	132	
Prickly Poppy	32	
Pride-of-India	65	
Pride-of-Ohio	211	
Pride-of-the-Meadow	104	
Prim	276	
Primrose	(125) 211	
Primrose-peerless	333	
Primworts	210	
Prince's-feather	283 288	
Prince's Pine	206	
Privet	276	
Puccoon	252	
Pumpkin	130	
Purple Cone-flower	175	
Purple Jacobæa	187	
Purslane	(51, 127, 133) 59	
PURSLANES	58	
Patty-root	328	
Quake Grass	408	
Quamash	343	
Quassia	72	
QUASSIAWORTS	71	
Queen-of-the-Prairie	104	
Quillwort	412	
Quince	112	
Radish	43	
Ragged Lady	21	
Ram's-head	326	
Raspberry	105	
Rattle-pod	90	
Rattlesnake Fern	418	
Rattlesnake Plantain	330	
Red Bay	290	
Red-bud	83	
Red Osier	143	
Red Pepper	263	
Red-root	(77) 335	
Redtop	384	
Red-wood	315	
Reed	(385) 404	
Reed-mace	320	
Resurrection Moss	415	

Rheumatism-root	28	
Rhubarb	281	
RIBWORTS	213	
Rice	388	
Richweed	300	
Riverweed	302	
Robin's Plantain	165	
Rock Cress	37	
Rocket	(36, 43) 39	
Rock Polypod	425	
ROCK ROSES, Rock Rose	47	
Roman Wormwood	184	
Rose	(21, 105) 108	
Rose Acacia	95	
Rose Apple	122	
Rose Bay	124, 203	
Rose Campion	54	
Rosemary	(202, 215, 303) 244	
ROSEWORTS	101	
Rosin-weed	172	
Rue	(18, 94, 372) 70	
Rue Anemone	(20) 18	
RUEWORTS	70	
Rush	361, 413, 363	
Rushes	350	
Rutland Beauty	260	
Rye	406	
Saffron	189, 337	
Sage	248, 244	
Saltwort	212, 287, 288	
Samphire	287	
SANDALWORTS	291	
Sand-hill Rosemary	303	
Sand Myrtle	204	
Sand Orache	287	
Sand Reed	387	
Sand Spurry	57	
Sandwort	55	
Sanicle	135	
Sarsaparilla	142, 328	
Sassafras	290	
Satin-flower	41	
SAURURADS	301	
Savory	242	
Saxifrage	113	
SAXIFRAGES	112	
Scabish	152, 182	
Scarcity	285	
Scorpion Senna	87	
Scotch Broom	100	
Scouring Rush	415	
Scratch-grass	284	
Screw-stem	268	
Scuppernong	78	
Scurvy-grass	39	
Sea Aster	164	
Sea Oxeye	171	
Sea Purslane	132	
Sea Rocket	43	
Sea Wormwood	185	
Sea-wrack	321	
SEDGES, Sedge	356, 357	
Seed-box	127	
Self-heal	246	
Seneca Snakeroot	79	
Senna	(87, 95) 83	
Sensitive Brier	82	
Sensitive Fern	421	
Sensitive Pea	83	
Sensitive Plant	83, 82	
Serpent Cucumber	131	
Service-tree	110	
Sesame Grass	409	

ENGLISH INDEX. 443

Shaddock-tree	71	
Shad-flower	10	
Shagbark	304	
Shamrock	91	
Sheep-poison	200	
Sheep Sorrel	281	
Shell-flower	248	
Shepherd's-purse	42	
Shield Fern	425	
Shrub Trefoil	71	
Sickle-pod	87	
Sidesaddle-flower	30	
SILK COTTONS	63	
Silk-tree	82	
Silk-weed	272	
Silver-berry	292	
Silver Fern	420	
Silver-weed	107	
Single-seed Cucumber	130	
Skullcap	246	
Skunk Cabbage	318	
Sloe	147	
Slipper-flower	222	
Smartweed	283	
Smoke-tree	73	
Smooth Lungwort	253	
Snails	92	
Snake-head	224	
Snake-root	23, 79, 192, 278	
Snapdragon	223	
Snapdragon Catchfly	53	
Sneezewort	183	
Snowball	146	
Snowberry	144	
Snowdrop	334	
Snowdrop-tree	209	
Snow-flake	334	
Soapberry	75	
Soapwort	53	
SOAPWORTS	210	
Soft Grass	395	
Solomon's Seal	346	
Sorrel	(67) 281	
Sorrel-tree	203	
SORRELWORTS	280	
Southernwood	184	
Sow-Thistle	194	
Spanish-daggers	345	
Spanish-needles	180	
Spear Grass	401	
Spearmint	240	
Spearwort	19	
Speedwell	229	
Spice-wood	290, 291	
Spider-flower	41	
SPIDERWORTS	353	
Spinach, Spinage	287	
Spleenwort	(133) 424	
Sponge-tree	99	
Spoonwood	200	
Spring-beauty	59	
Spruce	313	
Spurge	293	
Spurge Nettle	296	
SPURGEWORTS	293	
Spurry	57	
Squash	130	
Squaw-root	217	
Squill	343	
Squirrel Corn	33	
Staff-tree	76	
STAFF-TREES	75	
Stagger-bush	202	
Standing Cypress	257	
Staghorn Fern	419	
St. Andrew's Cross	48	
Star Anise	24	
Star-of-Bethlehem	343	
Star-grass	334, 335	
Star Thistle	189	
STARWORTS	301	
St. John's-wort	49	
ST. JOHN'S-WORTS	48	
Stock	38	
Stone-crop	115, 119	
Stork's-bill	68	
St. Peter's-wort	48	
St. Peter's-wreath	103	
Strawberry	(107) 106	
Strawberry Blite	286	
Strawberry Tomato	263	
Succory (Chicory)	190	
Sugar-berry	299	
Sugar Cane	410	
Sugar-tree (Maple)	74	
SUMACS, Sumac	72	
Summer Savory	242	
SUNDEWS, Sundew	50	
Sunflower	176	
Supple Jack	77	
Swamp Laurel	200	
Swamp Pink	203	
Sweet Basil	238	
Sweet Brier	109	
Sweet Cicely	137	
Sweet Fern	309	
Sweet Flag	318	
Sweet Gale	309	
Sweet Gum	120	
Sweet Pea	86	
Sweet Pepperbush	204	
Sweet Potato	259	
Sweet Reed	395	
Sweet-scented Clover	92	
Sweet-scented Shrub	25	
Sweet Sultan	189	
Sweet Vernal Grass	395	
Sweet Viburnum	147	
Sweet William	54, 62	
Swine Cress	43	
Sycamore	74, 303	
SYCAMORES	303	
Tacmahac	311	
Tallow-tree	296	
Tamarac	314	
TAMARISKS	63	
Tansy	183	
Tansy Mustard	39	
Tares	86	
Tassel-flower	186	
Tassel-tree	62	
Tea	(77, 204, 286) 65	
TEAWORTS, Tea-Rose	64, 65	
TEASELWORTS, Teasel	151	
Thimbleberry	105	
Thistle	189	
Thorn	110	
Thorn Apple	265	
Thoroughwax	138	
Thoroughwort	152	
Threadfoot	302	
THREADFOOTS	302	
Three-birds	224, 330	
Three-seed Mercury	296	
Thrift	215	
Thyme	243	
Tick-seed	178
Tiger-flower	337
Timothy	(394) 387
Toad Flax	(231) 222
Tobacco	(195) 265
Tomato	(263) 262
Tongue-grass	43
Toothache Grass	409
Touch-me-not	69
Torch wood	72
Tower Mustard	36
Trailing Arbutus	206
Tree Fern	419, 423
Tree Hibiscus	65
Tree of Heaven	72
True Orchis	331
Trefoil	71, 88
TRILLIADS	340
Trumpet-flower	218
Trumpet-leaf	30
Trumpet Milkweed	193
Trumpet-tongue	221
Trumpet-weed	159
Tuberose	334
Tulip	34
Tulip-tree	25
Turk's-cap	342
Turmeric-root	23
Turnip	40
Turnip Beet	285
Turtle-head	224
Twayblade	329
Twin-flower	144
Twin-leaf	28
Twist-foot	347
TYPHADS	319
UMBELWORTS	133
Umbrella-leaf	28
Umbrella-tree	25
Unicorn-plant	219
Valerian	(257) 150
Vanilla-plant	157
Vegetable Marrow	130
Vegetable Oyster	191
Venus' Comb	137
Venus' Flytrap	51
Venus' Looking-glass	197
Vervain	235
Vetch	(94, 100) 86
Victoria Lily	30
VINES	77
VIOLETS, Violet	(105) 45
Viper's Bugloss	251
Virginia Creeper	78
Virginia Lass	109
Virginia snakeroot	278
Virginia Stonecrop	119
Virgin's-bower	16
Wake Robin	340
Walking Fern	424
Wall-flower	(39) 38
Wall Rue	424
Walnut	301
WALNUTS	302
Water-carpet	113
Water Cress	36
Water-feather	211
Water Hemlock	141
Water Hemp	289
Water Hoarhound	240
Water-leaf	254
Water Lily	29
Water Melon	130

444   ADDENDA.

Water Milfoil............ 121	Wild Elder............ ... 142	Woodbine.... 145
Water Nymph......... 320	Wild Foxglove......... 230	Wood Cress............ 36
Water Parsnip 141	Wild Ginger........... 278	Wood Nettle........... 300
Water Pepper 283	Wild Indigo............ 84	Wood Sorrel........... 67
WATER PEPPERS..... 51	Wild Liquorice......... 148	Woolmouth............ 230
Water Pimpernel 213	Wild Oats............ 347	Worm-seed............ 256
WATER PITCHERS..... 30	Wild Pink............ 54	Wormwood............ 184
WATER PLANTAINS... 322	Wild Potato........... 259	XYRIDS............... 354
Water Plantain......... 323	Wild Rosemary......... 202	Yam.................. 325
Water Purslane......... 127	Wild Sarsaparilla...... 142	Yarrow............... 183
Water Smartweed....... 283	Wild Sensitive-plant.... 83	Yellow-eyed Grass...... 354
Water Target........... 29	Wild Service.......... 110	Yellow Jessamine...... 269
Wax-plant............. 275	Willow 309	Yellow Phlox......... 39
Whahoo 299	Willow-herb.......... 124	Yellow Pond-lily....... 29
Wheat................ 406	WILLOW-WORTS...... 309	Yellow Poppy......... 32
Wheat-thief........... 253	Wind-flower........... 17	Yellow Rattle......... 232
Whistlewood 74	Winter-berry.......... 208	Yellow-root........... 21
White Bay............. 24	Winter Cress.......... 39	Yellow-seed........... 43
White Lettuce......... 192	Winter-green....201, 205	Yellow Sweet Sultan... 180
White-tipped Aster..... 160	Witch Grass.......... 350	Yellow-wood.......... 84
Whiteweed......165, 183	Witch Hazel.......... 120	Yew.................. 316
Whitewood........... 25	WITCH HAZELWORTS, 120	YEWS................ 315
Whitlow-grass......... 41	Woad................ 43	Yulan................ 25
Wild Basil............ 243	Woad-waxen.......... 90	Zigadene............. 343
Wild Bergamot........ 245	Wolfbane............. 22	Zizia................ 139
Wild Bugloss......... 251	Wolfberry............ 145	

ADDITIONAL INDEX.—Latin and English.

Agropyrum............ 406	Cunninghamia......... 315	*Pycreus*................ 357
Airopsis................ 396	*Decodon*.............. 124	Quick Grass........... 406
ALGÆ 14	*Deyeuxia* 386	Ragged Robin......... 54
Allspice............... 121	Dicotyledonous Plants.. 15	Rat-tail Grass......... 409
Alsophila.............. 419	Dog's Bent............ 384	Rescue Grass.......... 397
Althæa................ 68	*Echinocaulon* 282	Ribbon Grass.......... 394
Alstroemeria........... 334	Egyptian Grass........ 408	Saw Grass............. 367
Amethystea........... 239	*Euthamia* 107	Sea-side Oats......... 404
Ammophila 336	Feather Grass......... 388	Seneca Grass.......... 395
Anchistea............. 423	Gardener's Garters..... 398	Seven Sisters......... 109
Apios................. 96	Gardenia 445	*Sieversia*............. 106
Atheropogon.......... 408	German Ivy........... 188	Spiked Rush.......... 359
Avicennia 235	Giant-of-Battles....... 110	Squirrel-tail Grass..... 404
Balantium 423	Glumaceous Endogens.. 356	Star Chickweed....... 55
Beach Pea............ 86	*Gymnostichum*........ 405	Stellatæ.............. 147
Beach Plum........... 102	Gypsum Pink......... 53	Striped Grass......... 395
Belladonna........... 264	*Hypoporum*........... 367	*Suæda*............... 287
Bengal Grass 394	Ice Plant..... 133	Tarragon.............. 185
Bermuda Grass........ 407	Indian Grass.......... 411	*Tiniaria*.............. 282
Bird-seed............. 395	*Lorinseria*............ 423	Torenia............... 226
Black Oat-grass....... 388	Lyme Grass........... 405	*Tovaria*.............. 282
Blue-joint............. 387	*Madaria* 173	*Trichochloa*.......... 385
Bottle Grass.......... 394	Mahernia............. 445	*Trichophorum* 361
Brachyelytrum 385	*Mariscus* 357	*Trichodium*.......... 384
Brompton Stock....... 38	Marsh Fleabane 171	Tupelo 143
Burr-flower........... 254	Marsh Grass.......... 408	Union Grass.......... 403
Byrsonima............ 8	Meadow Beauty....... 122	*Uralepis*............. 398
Cape Jessamine 445	Meadow Grass........ 401	*Vaccaria*............. 53
Carduus 445	Monocotyledonous	*Villarsia*............. 269
Chætospora........... 368	Plants............. 316	Water Locust......... 83
Chondrosium......... 408	Monopetalous Exogens. 144	Weather Grass........ 388
Chrysastrum.... 166	Moss Plant........... 201	Whip Grass........... 367
Chrysoma............ 166	Mountain Rice........ 388	White Thorn.......... 111
Cinchoneæ........... 147	Muskit Grass......... 408	Wild Rye............. 405
Clementine 109	Nut Grass............ 358	*Windsoria*..... 398
Clot-grass............. 359	Nut Sedge............ 367	*Wintereæ*............ 24
Cloth-of-gold 110	Oat Grass............ 396	Wood Grass........... 411
Coccolobus........... 280	Orchard Grass........ 398	Wood Rush........... 351
CONVOLVULACEÆ... 258	Pavonia 62	Yard Grass........... 407
Cord-grass............ 408	*Poinsettia*............ 296	Youland-of-Aragon 110
Cow Parsnip.......... 136	Polypog Grass........ 186	

ADDENDA.

Page 42. After V. (Vesicària) Shórtii, add,

2 V. Lescùrii Gray. Pubescent; stems many, ascending 6—10'; lvs. oblong, clasping, with a sagittate base; flowers yellow, in lengthening terminal racemes; silicle roundish, hispid, twice longer than its style; seeds 1—4 in each cell. Meadows, Tenn. (*Mr. Hamlin.*)

Page 63. After S. (Stercùlia) platanifòlia, add,

2. MAHÉRNIA VERTICILLÀTA. A shrubby perennial from S. Africa, cultivated in conservatories. It has slender, vine-like branches, small pinnatifid leaves and stipules forming verticils. The flowers are small, yellow, bell-form, very sweet-scented, with 5 petals, stamens, and styles.

Page 68. After O. (Óxalis) versícolor, add,

7 O. cérnua. Leaflets 3, obcordate; scapes bearing umbels of many large, yellow, drooping flowers; styles very short. S. Afr.

Page 69. After T..(Tropæolum) perìgrinum, add,

5 T. (CHYMOCÁRPUS) PENTEPHŸLLUS. Climbing high; lvs. digitate, of 5 small lfts.; fls. curious, green and red, the spur 1' long; sepals valvate; petals 2, *small*; carpels 3 round berries. From Buenos Ayres.

Page 74. After A. (Acer) macrophýllum, add,

9 A. PLATANOÎDES. *Norway Maple.* Tree 40—50f; leaves bright green both sides, as broad as long, 5-lobed, lobes toothed and short-acuminate; corymbs nearly erect; fruit smooth, 2' long, wings very diverging.

Page 106. After G. (Geum) album, add,

β. *lùteum,* a variety with yellow flowers, rarely occurs in Pennsylvania.

Page 111. After C. (Cratægus) spatulàta, insert,

9a C. Pyracántha Pers. Shrub 10f, thorny; lvs. evergreen, lance-ovate or oblong, crenulate-serrate, smooth and shining. § Near Philadelphia, and southward.

Page 146. After D. (Diervílla) sessilifòlia, add,

4 D. JAPÓNICA, β. ROSEA. *Weigela.* Shrub from Japan, 4—6f, with straight branches; lvs. oblong-ovate, acuminate, large; flowers funnelform, rose-colored, 1' broad, covering the plant in Spring; ovaries and pods linear. Common in cultivation.

Page 150. After Bouvardia, add,

11. GARDÈNIA FLÓRIDA. *Cape Jessamine.* From China. Much cultivated South. Shrubby evergreen, 2—4f. Lvs. elliptical, acute both ways, very smooth. Flowers white, corolla 5-lobed or often many-lobed and double, salverform, 2' broad.

On page 175, after E. (Echinàcea) atrórubens, add,

4 E. Pórteri (Gray). Leaves lanceolate to lance-linear, remotely toothed, the highest entire; heads corymbed, 1' broad; scales about 9, lance-linear; rays 6—8, ovate lanceolate, *yellow;* chaff *spinescent.* Stone Mountain, Ga. (Prof. Porter). Has the habit of Rudbeckia, but its chaff is plainly that of Echinacea.

On page 190, before Lappa, may be inserted,

99a CÁRDUUS NUTANS, L. Bristles of the pappus *not plumous*, nearly naked. Stem 2f, slender; lvs. narrow, sinuate-spinescent, decurrent, 2—3'. Heads few, large, nodding, purple. Ach. linear-oblong, rugulous, 2", crowned with a many-bristled deciduous pappus three-quarters of an inch long. Harrisburg, Pa. (Prof. Porter). § Eur.

Page 208. After P. (Prinos) lævigàtus, add,

3a P. pubéscens Mx. Shrub 6—8f, with smooth, virgate branches; lvs. large, ovate, acuminate, serrulate, soft pubescent beneath; clusters umbellate, axillary, shorter than the petioles; berries dark red, 2—3" in diameter. Alleghanies, Pa.

Page 281. Next before R. (Rumex) crispus, insert,

1 R. patiéntia L. *Patience Dock*. Stem 3—5f, stout; leaves lance-oblong, 6'—2f; valves large (2—3"), broad-cordate, one of them bearing a small grain or *all* naked. Grows at New Baltimore, N. Y. (*Dr. Howe.*) § Eur.

Page 388. Next before S. (Stipa) avenàcea, insert,

1 S. Richardsònii Link. Culm 15—20', very erect and slender; lvs. shorter, filiform; pan. loose, 3—4'; glumes near 2", acutish; pales not bearded at the blunt base, the crooked awn about 6" in length. Mt. Marcy, N. Y. (*C. H. Peck.*)

Page 394. After C. (Cenchrus) tribuloìdes, add,

2 C. echinàtus L. Differs from No. 1, in the *globular, purplish*, downy involucres, beset above with rough, stiff bristles, and cleft into 8—10 segments inclosing 3—5 flowers; grain brown. South.

Page 44. After C. (Cleòme) pungens, add,

2 C. integrifòlia (Nutt.) Smooth, glaucous, 1—2f; lvs. 3-foliate, lfts. lance-oblong, entire, mucronate; rac. dense; calyx 5-toothed; pet. rose-color, subsessile, 4"; stam. 6, equal; pod much longer than its stipe. Banks of the Mississippi R., N Illinois. (*Mr. V. Friese.*) and Westward.

Page 340. After T. (Trìllium) cérnuum, add,

β. **atrórubens.** Petals brownish purple, ovate-lanceolate, acuminate. Hanover, Indiana. (*Mr. A. H. Young.*)

Page 291. After Phorodéndron, insert,

2. ARCEUTHÒBIUM, Bieb. Differs from Phorodéndron in having its anthers 1-celled, the ♀ perianth 2-toothed, the herbage yellowish and *leafless.*

A. Oxycèdri, β. abigenìum (Wood). Found growing on the branches of small starved spruce-trees (Abies nigra), in a marsh in Sandlake, N. Y. (*C. H. Peck*). Stems 3—9", jointed, each joint terminating in a truncated sheath. Fls. terminal and opposite; berry some 3-angled. The variety α grows on Pines and Cedars in Cal. and Oreg.! and is much larger.

Page 133. Under Sesuvium, insert,

2 S. pentándrum Ell. Lvs. spatulate-obovate; fls. sessile; *stamens* 5. ①? Sea coast, E. Hampton, L. I. (*J. S. Merriam*), Cape May (*C. F. Parker*), Cape Henlopen (*Dr. Leidy*), to Fla. Hitherto mistaken for S. Portulacastrum.

Page 164. After 45 A. (Aster) ericoides, insert,

β. **villòsus** (Mx.) Stem, branches, and often the leaves villous-hirsute.

Page 167. After 8 S. (Solidago) latifòlia, β. *pubens*, insert,

β. **ciliàta** (DC.) Upper racemes elongated and spreading. Ill. (*Mr. Wolf.*)

ADDENDA. 447

Page 168. After 30 S. (Solidago) Canadensis, insert,
β. *scabra.* Stem and leaves scabrous; leaves narrow, rigid, subentire.

Page 173. After 6 S. (Silphium) scaberrimum, insert,
β. *sessile.* Leaves nearly all sessile, lance-oblong to ovate. (S. Radula N.) Ill.

Page 180. After 5 B. (Bidens) connata, insert,
β. *petiolata.* Leaves more or less petiolate. (B. petiolata N.) Ill. (*Mr. Wolf.*)

Page 283. After 10 P. (Polygonum) Careyi, insert,
10a P. persicarioides K. Glabrous, 2—4f: stip. ciliate; lvs. lin.-lanceolate, subsessile, spotted, not acrid; spikes linear, erect, pale-purple; sta. 6—8; styles 3-cleft; ach. 3-angled, shining. Low ground. Ill. (*Mr. Wolf.*) New to our flora.

Page 346. At bottom insert,

22a MYRSIPHYLLUM ASPARAGOÏDES. A delicate vine, twining and climbing, from S. Africa. Cult. Branches very slender and smooth. Lvs. 1′ or more, ovate, pointed, thin, and polished. Ped. in pairs, with an empty bract-like one. Fls. similar to those of Asparagus, 6-parted, white. Filaments flattened. Popularly called *Smilax*.

Page 405. After E. (Elymus) Virginicus, β. *arcuatus*, add,
γ. *villosus.* Flowers villous-pubescent. (E. villosus Muhl.) Ill. (*Mr. Wolf.*)

ORDER LXXX. OLACACEÆ.

Trees or *shrubs* chiefly tropical, with alternate, ex-stipulate, petiolate, entire *leaves*, regular, hypogynous *flowers*, and drupe-like *fruit;* represented in our limits by the following genus only.

XIMENIA, Plum. Calyx small, 4-toothed. Petals 4, woolly within, barely united at the base. Stam. 8. Style 1, Ovary 4-celled, with several ovules, but forming a 1-seeded drupe. ⚥ ⚦ Thorny. Flowers axillary, single or in small corymbs.

X. Americana L. Leaves smooth, coriaceous, oval or oblong, obtuse; peduncles several-flowered, shorter than the leaves; petals oblanceolate, thick, spreading above, 4—5″ long.—Fla. from Picolata (*Mr. Fry*) and S. Fls. yellow, fragrant. Drupe as large as a plum, yellow, well-flavored. Thorns ½ an inch.

Page 76, under Celastreceae, insert,
3. PACHYSTIMA, Raf. Petals and stam. 4, inserted on the throat of the 4-lobed calyx. Style very short, expanded at base into the disk which covers the ovary and lines the calyx tube. Caps. oval, 2-celled, seeds 2—4, inclosed in a white dissected aril.—Low shrubs, with opposite, crowded, short-petioled, evergreen leaves, and minute axillary flowers.

P. myrsinites Raf. β Canbyi (Gray). Stems and branches creeping, ascending, bark blackish; lvs. oblong and linear-oblong, obtuse, with a few minute teeth; caps. obtuse.—Mountain bogs, Wytheville, Va. (*H. Shriver.*) Stems 8—15′. Lvs. 6—9″, margins revolute.

Page 234, after R. (Ruellia) strepens L., insert,
β. *micrantha* (Eng. and Gr.). Flowers crowded in the axils, with corolla reduced to a slender tube with an obsolete lip-shaped border, or quite apetalous, fertilized in the bud.—In ponds, Mount Carmel, Ill. (*Dr. Schneck.*)

Page 253, under Lithospermum, insert,

8 L. lutéscens Coleman. Minutely strigous; lvs. lanceolate, pointed, roughish above, about 5-veined; sepals subulate, shorter than the conspicuous yellow corolla.—Grand Rapids, Mich. (*N. Coleman.*) Allied to L. latifolium.

9 L. tuberosum Rugel. Hispid-bristly, erect, branching; lvs. obovate-oblong. dotted above with white glands, the upper lance-oblong; calyx lobes linear, as long as the yellowish corolla, twice as long as the polished nutlet.—Fla. to La. (*Dr. Joor.*)

Page 256, under Hydrophyllaceae, insert,

8. NAMA, L. Calyx 5-parted. Cor. tubular-funnelform, 5-cleft, Stam. 5 equal, included, styles 2 distinct. Caps. oblong. Seeds ∞, pitted.—Hairy diffuse herbs. Lvs. alternate, entire. Fls. cyanic.

N. Jamaicénsis L, Pubescent, prostrate, branched; stems angular; lvs. obovate, obtuse; fls. 1—3 in the axils; calyx lobes linear, as long (5″) as the corolla; caps. 2-, then 4-valved and the placentæ free.—Ditches, etc., Baton Rouge, La. (*Dr. Joor.*)

Page 263, under Solanum, insert,

14 S. verbascifòlium L. Shrubby, hoary-tomentous; lvs. large, ovate-oblong, entire; cymes dense-flowered, on a long stout forking peduncle; flowers in bud obovoid, cor. lobes obtuse; anthers lin.-oblong; ovary woolly.—Picolata, Fla. (*Mr. Fry*) and southward.

Page 140, after 3 A. (Apium) nodiflorum, read,

3a A. angustifolium Wood. Weakly erect 8—20′; lvs. pinnate, elongated; lfts. toothed, cut, or pinnatifid, oblong in outline; ped. as long as the rays; invol. and involucels 5—7-bracted; fr. round-oval, ribs and vittæ obscured by the thick pericarp. —Wet places, Peoria, Ill., (*Dr. Stewart*) and W. Used as celery. (Sium, L. Berula, Kotch.)

Page 178, after Silphium, insert,

41a. ACANTHOSPÉRMUM, Schrank. Heads radiate, rays (small) ♀ fertile, disk ☿ sterile. Invol. herbaceous, inner scales closely investing the ray cypselæ. Recep. chaffy. Cyp. few, oblong, without pappus, each enclosed in the hardened prickly scale.—① Diffusely branching. Lvs. opposite, toothed or incised. Fls. yellow.

A. xanthoides DC. Stems creeping, rooting at base; scabrous-pubescent; lvs. ovate or obovate, the lower petiolate; heads stalked; rays about 5; cyp. 5, spreading, 6″ long, the sack muricate.—Atlanta, Ga. (*T. B. Goulding.*) § S. Am. Jl. Aug.

Page 237, after Vitex, may be inserted,

7. CLERODÉNDRUM, L. Corolla salverform, limb some unequal, 5-cleft. Drupe baccate, of 4 (or fewer) 1-celled, 1-seeded drupes.—Shrubs or trees. Lvs. simple, entire, opposite or ternate. Cymes axillary, or terminal, trichotomous.

C. Siphonánthus R. Br. Glabrous, virgate, erect 4—8f.; lvs. whorled in 3s and 4s, long-lanceolate, pointed at both ends; cymes once or twice trichotomous; cor. white, tube 4′ long, limb 1′ broad; stam. long-exserted.—Macon, Ga., naturalized in fields, waysides. (*Dr. J. Mercer Green.*)

Page 358, after 17 C. (Cyperus) divergens, read,

17a C. Wolfii Wood. Glabrous, slender, erect 2—3f.; lvs. at base, narrowly linear, 3f, of the invol. 2f; rays about 5, very unequal, each bearing a dense globular head; spikes many, 4—5-flowered, oblong, *scales imbricated, obtuse*, 9—11-veined; rachis broadly winged.—Anna, Ill. (*J. Wolf.*)

www.ingramcontent.com/pod-product-compliance
Lightning Source LLC
Chambersburg PA
CBHW032005300426
44117CB00008B/910